T0226759

INSECT PHARMACOLOGY

CHANNELS, RECEPTORS, TOXINS AND ENZYMES

INSECT PHARMACOLOGY

CHANNELS, RECEPTORS, TOXINS AND ENZYMES

EDITED BY

LAWRENCE I. GILBERT
SARJEET S. GILL

AMSTERDAM • BOSTON • HEIDELBERG • LONDON • NEW YORK • OXFORD
PARIS • SAN DIEGO • SAN FRANCISCO • SINGAPORE • SYDNEY • TOKYO
Academic Press is an imprint of Elsevier

ACADEMIC
PRESS

Academic Press, 32 Jamestown Road, London, NW1 7BU, UK
30 Corporate Drive, Suite 400, Burlington, MA 01803, USA
525 B Street, Suite 1800, San Diego, CA 92101-4495, USA

© 2010 Elsevier B.V. All rights reserved

The chapters first appeared in *Comprehensive Molecular Insect Science*, edited by
Lawrence I. Gilbert, Kostas Iatrou, and Sarjeet S. Gill (Elsevier, B.V. 2005).

The following article is a US government work in the public domain and is not subject to copyright:
A5 Addendum: Baculoviruses: Biology, Biochemistry, and Molecular Biology
Published by Elsevier B.V., 2010.

*All rights reserved. No part of this publication may be reproduced or transmitted in any form or by any
means, electronic or mechanical, including photocopy, recording, or any information storage and
retrieval system, without permission in writing from the publishers.*

*Permissions may be sought directly from Elsevier's Rights Department in Oxford, UK:
phone (+44) 1865 843830, fax (+44) 1865 853333, e-mail permissions@elsevier.com.*

Requests may also be completed on-line via the homepage (http://www.elsevier.com/locate/permissions).

Library of Congress Cataloging-in-Publication Data
Insect pharmacology : channels, receptors, toxins and enzymes / editors-in-chief: Lawrence I. Gilbert,
Sarjeet S. Gill. – 1st ed.
p. cm.
Includes bibliographical references and index.
ISBN 978-0-12-381447-0 (alk. paper)
1. Insects–Physiology. 2. Insecticides–Physiological effect. 3. Pharmacology. 4. Insect biochemistry.
I. Gilbert, Lawrence I. (Lawrence Irwin), 1929- II. Gill, Sarjeet S.
QL495.I497 2010
632'.9517–dc22
2010010549

A catalogue record for this book is available from the British Library

ISBN 978-0-12-381447-0

Cover Images: (Top) Important pest insect targeted by neonicotinoid insecticides: Colorado
potato beetle from Jeschke and Nauen, *Insect Control*, 2010; (Bottom Left) A diagram summarizing
G protein-coupled receptor (GPCR) signalling, from Addendum A9; (Bottom Center) A molecular model of
the GABAA receptor a1 subunit, from Chapter 2; (Bottom Right) Structural model of the housefly sodium
channel showing the putative binding site for pyrethroids and DDT (Reproduced with permission, from
O'Reilly, A.O., Khambay, B.P.S., Williamson, M.S., Field, L.M., Wallace, B.A., Davies, T.G.E., 2006.
Modelling insecticide-binding sites in the voltage-gated sodium channel. *Biochem. J. 396*, 255–263.
*ª*The Biochemical Society).

Printed and bound by CPI Group (UK) Ltd, Croydon, CR0 4YY
Transferred to digital print 2013

CONTENTS

PREFACE

When Elsevier published the seven-volume series *Comprehensive Molecular Insect Science* in 2005, the original series was targeted mainly at libraries and larger institutions. While this gave access to researchers and students of those institutions, it has left open the opportunity for an individual volume on one of the most popular areas in entomology: pharmacology. Such a volume is of considerable value to an additional audience in the insect research community — individuals who either had not had access to the larger work or desired a more focused treatment of these topics.

As two of the three editors of the *Comprehensive* series, we felt that it was time to update the field of insect pharmacology by providing a volume for professional researchers and students that was updated by adding addenda to the original chapters. The new summaries for each chapter provide an overview of developments in the related article since its original publication. As editors of the original chapters, we expended a great deal of effort in finding the best available authors for each of those chapters. In most instances, authors who contributed to *Comprehensive Molecular Insect Science* also provided updates to our new volume, *Insect Pharmacology*.

The chapters included in *Insect Pharmacology* incorporate major targets for many of the insecticides currently in use. Most insecticides in use affect a small set of molecular targets. Chapters in this volume include "Sodium Channels," which are targets for pyrethroids, indoxacarb, and peptide toxins; "GABA Receptors," a target for fipronil; and "Esterases," some of which are targeted by organophosphates and carbamates. Other chapters cover "G-protein Coupled Receptors" that modulate peptide function; "Amino Acid Transporters" that play a role regulating the transport of neurotransmitters and amino acids critical in insect function; and "Glutathione Transferases," some of which are involved in insecticide detoxification and resistance. Chapters on "Avermectins," "Spider Toxins," and "Baculoviruses" address compounds that have been used successfully or show potential for insect control.

Chapters in the companion volume, *Insect Control*, would also be of interest to readers. That volume includes chapters on "Neonicotinoid Insecticides," which discusses insect nicotinic receptors, and "Insect Growth Regulators and Development Disrupting Insecticides," which includes recent progress on target sites involved in insect growth and regulation. Chapters on "*Bacillus thuringiensis*" and "*Bacillus sphaericus*" provide comprehensive discussion of the molecular targets for toxins produced by these bacteria.

Several years of effort was expended by both of us and our colleagues in choosing topics for the seven-volume series, in the selection of authors, and in the editing of the original manuscripts and galley proofs. Each and every chapter in those volumes was important, and even essential, to make it a "Comprehensive" series. Nevertheless, we feel strongly that having this volume with the updated material and many references on these important aspects of insect pharmacology will be of great help to professional insect biologists, to graduate students conducting research for advanced degrees, and even to undergraduate research students contemplating an advanced degree in insect science.

– LAWRENCE I. GILBERT,
Department of Biology,
University of North Carolina,
Chapel Hill

– SARJEET S. GILL,
Cell Biology and Neuroscience,
University of California,
Riverside

CONTRIBUTORS

M E Adams
University of California, Riverside, CA, USA

B C Bonning
Iowa State University, Ames, IA, USA

D Boudko
Rosalind Franklin University of Medicine and Science, North Chicago, IL, USA

S D Buckingham
University of Oxford, Oxford, UK

P M Campbell
CSIRO Entomology, Canberra, Australia

C Claudianos
Visual Neuroscience, Queensland Brain Institute, University of Queensland, QLD, Australia

K Dong
Michigan State University, East Lansing, MI

B C Donly
Agriculture and Agri-Food Canada, London, ON, Canada

S Gill
University of California, Riverside, CA

R L Harrison
Invasive Insect Biocontrol and Behavior Laboratory, USDA Agricultural Research Service, Plant Sciences Institute, MD, USA

W R Harvey
University of Florida, St. Augustine, FL, USA

J Hemingway
Liverpool School of Tropical Medicine, Liverpool, UK

V Herzig
Institute for Molecular Bioscience,
The University of Queensland, Queensland, Australia

R K Jansson
Centocor, Malvern, PA, USA

G F King
University of Connecticut Health Center, Farmington, CT, USA

F Maggio
University of Connecticut Health Center, Farmington, CT, USA

R D Newcomb
HortResearch, Auckland, New Zealand

J G Oakeshott
CSIRO Entomology, Black Mountain, Canberra, Australia

Y Park
Kansas State University, Manhattan, KS, USA

H Ranson
Liverpool School of Tropical Medicine, Liverpool, UK

D Rugg
Fort Dodge Animal Health, Princeton, NJ, USA

R J Russell
CSIRO Entomology, Black Mountain, Canberra, Australia

D B Sattelle
University of Oxford, Oxford, UK

D M Soderlund
Cornell University, Geneva, NY, USA

B L Sollod
University of Connecticut Health Center,
Farmington, CT, USA

B R Stevens
University of Florida, Gainesville, FL, USA

H W Tedford
University of Connecticut Health Center,
Farmington, CT, USA

1 Sodium Channels

D M Soderlund, Cornell University,
Geneva, NY, USA

© 2005, Elsevier BV. All Rights Reserved.

1.1. Introduction

1.1.1. Function and Structure of Voltage-Sensitive Sodium Channels

Voltage-sensitive sodium channels mediate the transient increase in sodium ion permeability that underlies the rising phase of the electrical action potential in most types of excitable cells (Hille, 2001). In vertebrates, sodium channels are found in neurons and cardiac and skeletal muscle cells as well as glial and neuroendocrine cells (Goldin, 2002). In contrast, voltage-sensitive sodium channels in insects appear to be limited in distribution to neurons (Littleton and Ganetzky, 2000).

The principal structural element of voltage-sensitive sodium channels is a large (~260 kDa) α subunit that forms the ion pore and confers the functional and pharmacological properties of the channel (Catterall, 2000; Yu and Catterall, 2003). Sodium channel α subunits are pseudotetrameric proteins that contain four internally homologous domains, each of which contains six hydrophobic transmembrane helices and additional hydrophobic segments that contribute to the formation of the ion pore (**Figure 1**). The four internally homologous domains form a radially symmetrical assembly with the ion pore at the center (Guy and Seetharamulu, 1986; Yellen, 1998; Lipkind and Fozzard, 2000).

Voltage-sensitive sodium channel α subunits are part of a larger family of cation channels that also includes cyclic nucleotide-gated channels and voltage-sensitive potassium and calcium channels (Strong *et al.*, 1993). Whereas voltage-sensitive sodium and calcium channels share the pseudotetrameric four-domain structure shown in **Figure 1**, voltage-sensitive potassium channels and cyclic nucleotide-gated channels are composed of four separate subunits, each of which corresponds to a single domain of sodium and calcium channels. The phylogenetic distribution of voltage-sensitive cation channels and amino acid sequence comparisons among and between channel classes suggest that the earliest channels were tetramers of four separate subunits, like voltage-sensitive potassium channels (Strong *et al.*, 1993; Goldin, 2002). These data also suggest that four-domain calcium channels arose from such channels via two rounds of intragenic duplication. The four-domain voltage-sensitive sodium channels, which appear later in evolution than calcium channels, are likely to have arisen from duplication and divergence of voltage-sensitive calcium channels.

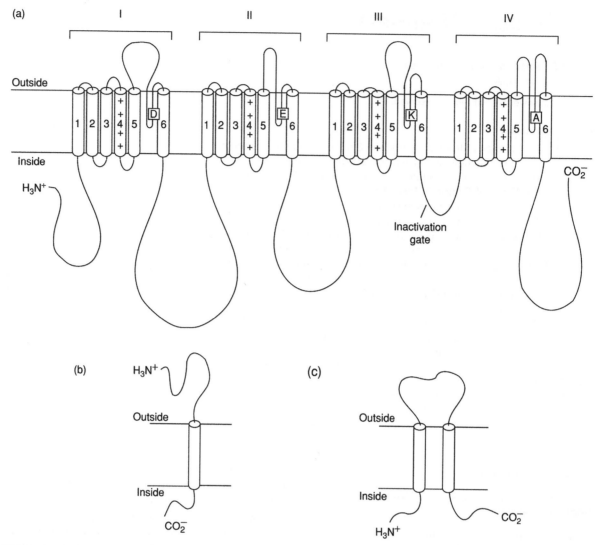

Figure 1 Diagrammatic representations of the structure of vertebrate and insect sodium channel subunits. (a) Sodium channel α subunit, showing the four internally homologous domains (I–IV), each containing six hydrophobic transmembrane helices. Also illustrated are other structural elements relevant to sodium channel function: the multiple positively charged residues (+) in the four S4 helices that constitute the voltage sensor, the four amino acids constituting the selectivity filter (the "DEKA motif") located in the pore-forming regions of the four internally homologous domains, and the inactivation gate peptide between the third and fourth domains. (b) Vertebrate sodium channel auxiliary β subunit. (c) Insect sodium channel auxiliary tipE subunit.

Further sodium channel gene duplication and divergence in the course of mammalian evolution has resulted in at least 10 sodium channel α subunit genes that differ in amino acid sequence, developmental and anatomical distribution, and biophysical and pharmacological properties (Goldin *et al.*, 2000; Goldin, 2002). In mammalian brain and skeletal muscle, sodium channel α subunits are associated with one or more smaller auxiliary β subunits that modulate the expression and functional properties of sodium channel α subunits (Catterall, 2000; Goldin, 2001).

1.1.2. Sodium Channels as Targets for Neurotoxicants

Voltage-sensitive sodium channels are the site of action of a wide structural variety of naturally occurring neurotoxins that contribute to the chemical ecology of predation and defense (Cestèle and Catterall, 2000; Blumenthal and Seibert, 2003; Wang and Wang, 2003). These sites, together with binding sites for synthetic neurotoxicants and drugs, identify at least 10 distinct binding domains associated with the voltage-sensitive sodium channel.

Five principal neurotoxin recognition sites, designated as sites 1–5 (Catterall, 1988, 1992), have been identified in both functional assays and in radioligand binding experiments (see **Table 1**). These sites are thought to represent physically non-overlapping domains of the sodium channel protein that interact allosterically as the channel protein changes conformation in response to the binding of one or more neurotoxins. Site 1 binds the water-soluble toxins tetrodotoxin (TTX) and saxitoxin (STX), which interact at or near the extracellular opening of the ion pore and block ion transport through the channel. Site 2 binds a structurally heterogeneous group of lipophilic toxins that alter both the opening (activation) and closing (inactivation) of sodium channels. Site 2 neurotoxins shift the voltage dependency of sodium channel activation toward more negative membrane potentials, thereby increasing the probability that channels will open at normal membrane resting potentials. These toxins also slow or completely block sodium channel

inactivation, thereby prolonging the open state of the channel. Site 3 binds a group of polypeptide toxins, called α-toxins, isolated from scorpion venoms or sea anemone nematocysts. These toxins selectively prolong sodium channel inactivation without affecting the rate or voltage dependency of channel opening and allosterically enhance the action of compounds acting at site 2. Site 4 binds a second group of polypeptide scorpion toxins, called β-toxins, which selectively enhance sodium channel activation and do not interact allosterically with site 2 compounds. Site 5 binds the brevetoxins and ciguatoxin, lipophilic compounds isolated from marine dinoflagellates, which shift the voltage dependency of sodium channel activation and prolong inactivation in a manner similar to site 2 neurotoxins. However, these compounds bind at a site distinct from site 2 and interact allosterically with compounds acting at site 2.

Subsequent research has identified two additional binding domains that are labeled with specific radioligands and shown to be distinct from sites 1 to 5. Site 6 (**Table 1**) binds δ-conotoxins and *Conus striatus* toxin (Gonoi et al., 1987; Fainzilber et al., 1994). Site 7 (**Table 1**) (also designated site 6 in some classifications) binds dichlorodiphenyltrichloroethane (DDT) and pyrethroids (Bloomquist and Soderlund, 1988; Lombet et al., 1988; Trainer et al., 1997). At least three other binding domains (arbitrarily designated as sites 8–10 in **Table 1**) have been postulated to account for the actions on sodium channels of μO-conotoxins (Terlau et al., 1996), *Goniopora* coral toxin (Gonoi et al., 1986), and local anesthetics, class I antiarrhythmics, and class I anticonvulsants (Catterall, 1987).

Sodium channels are widely recognized as the principal target site for diphenylethane (e.g., DDT) and pyrethroid insecticides (Sattelle and Yamamoto, 1988; Bloomquist, 1993; Soderlund, 1995; Narahashi, 1996). In addition, other toxicant-binding domains on the sodium channel are important target sites for the continued discovery and development of novel insect control agents. Two additional classes of synthetic insecticides alter sodium channel function by binding to sites that are distinct from the pyrethroid recognition site. Synthetic N-alkylamide insecticides, which are analogs of insecticidal natural products, produce excitatory effects by binding to site 2 (Ottea et al., 1989, 1990), whereas pyrazolines (also called dihydropyrazoles) and structurally related insecticides suppress normal nerve activity by the voltage-dependent blockade of sodium channels in a manner similar to anticonvulsants (Salgado, 1990, 1992; Deecher et al., 1991;

Table 1 Identified and inferred binding domains on the voltage-sensitive sodium channel

Site[a]	Active neurotoxins[b]	Physiological effect
Binding domains identified with specific radioligands		
1	Tetrodotoxin	Inhibit ion transport
	Saxitoxin	
	μ-Conotoxin	
2	Veratridine	Persistent activation
	Batrachotoxin	
	Aconitine	
	Grayanotoxins	
	Pumiliotoxin-B	
	N-alkylamides	
3	α Scorpion toxins	Prolong inactivation
	Sea anemone toxins	
4	β Scorpion toxins	Enhance activation
5	Brevetoxins	Persistent activation
	Ciguatoxin	
6	δ-Conotoxins	Prolong inactivation
	Conus striatus toxin	
7	DDT and pyrethroids	Prolong inactivation
Binding domains inferred but not characterized with radioligands		
8	μO-conotoxins	Inhibit ion transport
9	Gonioporatoxin	Prolong inactivation
10	Local anesthetics	Inhibit ion transport
	Anticonvulsants	
	Antiarrhythmics	
	Pyrazolines	

[a]Sites 1–5 after Catterall (1988); sites 6 and 7 after Zlotkin (1999); sites 8–10 are assigned arbitrarily to distinguish them from sites 1–7.
[b]Insecticides are capitalized.

Deecher and Soderlund, 1991). The recent development of indoxacarb, the first commercial insecticide
derived from pyrazoline-type compounds (Harder
et al., 1996; Wing et al., 1998), underscores the
continued relevance of the sodium channel as a target
for insecticide development. Sodium channels are also
the target of insect-selective polypeptide toxins from
scorpion venoms (Zlotkin et al., 1995) which have
been genetically engineered into insect baculoviruses;
two potential biopesticide products based on this
technology have proceeded to the stage of large-
scale field tests (Cory, 1999).

1.1.3. Progress in Insect Sodium Channel Research Since 1985

The report in 1984 of the first DNA sequence for a
voltage-sensitive ion channel, the voltage-sensitive
sodium channel of the eel (*Electrophorus electricus*)
electric organ (Noda et al., 1984), marked the
beginning of a period of explosive growth in the
application of molecular techniques to the study of
ion channels that continues to the present. Since that
time, studies of mammalian ion channels have identified the existence of gene families for both the α
and β subunits of the voltage-sensitive sodium channel and have elucidated the structural domains of
sodium channel α subunits that confer many of the
biophysical and pharmacological properties of
channel isoforms (Catterall, 2000; Goldin, 2001).
Biomedical applications resulting from this research
include the identification of sodium channel gene
mutations that cause heritable diseases and a
renewed interest in voltage-sensitive sodium channels as targets for therapeutic agents (Cannon,
2000; Clare et al., 2000; Baker and Wood, 2001;
Goldin, 2001).

The corresponding expansion of information on
voltage-sensitive sodium channels in insects during
this period has occurred in two stages. The first of
these involved the integration of molecular biology with classical genetic approaches in *Drosophila
melanogaster* to further illuminate the nature of
gene mutations that were implicated genetically as
determinants of sodium channel function and to
determine the molecular structures of the genes
defined by these mutations. The second phase, in
which the molecular and functional characterization of sodium channels was extended to include
numerous other insect taxa, was driven principally
by research to define the role of sodium channel
gene mutations in insecticide resistance. This chapter reviews the literature on insect voltage-sensitive
sodium channels since 1985 from both of these
perspectives.

1.2. Sodium Channel Genes in Insects

1.2.1. Sodium Channel Genes in *Drosophila melanogaster*

1.2.1.1. *dsc1* The first putative insect sodium
channel α subunit gene to be identified, *dsc1*, was
isolated from *D. melanogaster* DNA libraries by
low-stringency hybridization to a probe from the *E.
electricus* sodium channel (Okamoto et al., 1987;
Salkoff et al., 1987; Ramaswami and Tanouye,
1989). Sequence analysis of partial genomic DNA
and cDNA clones identified sequences encoding the
four internally homologous domains (Salkoff et al.,
1987), but sequences of the 5′ and 3′ termini and the
segments lying between homology domains I and II
and homology domains II and III, all of which are
regions of low sequence conservation in vertebrate
genes, were not defined until the recent report of a
full-length *dsc1* coding sequence (Kulkarni et al.,
2002). The *dsc1* locus was initially mapped to
cytogenetic region 60D-E on chromosome 2R
(Okamoto et al., 1987; Salkoff et al., 1987;
Ramaswami and Tanouye, 1989) and subsequently
localized to cytogenetic region 60E5 by genome
sequencing (Adams et al., 2000). Expression of the
dsc1 gene is under both anatomical and developmental regulation: expression in embryos and larvae
is restricted to very few cells, some of which may be
nonneuronal, whereas *dsc1*-derived transcripts are
found at much higher levels in pupae and adults,
where they are widely expressed in the central nervous system and retina (Tseng-Crank et al., 1991;
Hong and Ganetzky, 1994). A recent genome-wide
screen identified *dsc1* transcripts as targets for RNA
editing at a site in the "inactivation gate" sequence
(**Figure 1**) between homology domains III and IV
(Hoopengardner et al., 2003).

Alignment of the dsc1 protein sequence with those
of vertebrate sodium channels suggests that dsc1
may exhibit atypical ion selectivity. Two conserved
amino acid residues, a lysine in the pore-forming
region of homology domain III and an alanine in
the corresponding location in homology domain IV,
are crucial for the sodium ion selectivity typical of
voltage-sensitive sodium channels. When either of
these is mutated to a glutamate residue the resulting
channel becomes permeable to calcium and other
monovalent and divalent cations that normally do
not permeate sodium channels (Heinemann et al.,
1992). The dsc1 protein sequence contains a glutamate residue rather than the conserved lysine residue
in the pore-forming region of homology domain III
and therefore may form a channel that is permeable
to calcium as well as sodium. This observation is
intriguing in light of recent studies identifying a

novel sodium current in dorsal unpaired median neurons of the cockroach *Periplaneta americana* that is biophysically distinct from the more typical sodium channels present in the same neurons and is permeable to calcium as well as sodium (Grolleau and Lapied, 2000; Defaix and Lapied, 2001).

1.2.1.2. *para* The *para^{ts}* (*paralytic–temperature-sensitive*) locus at cytogenetic location 14D on the X chromosome of *D. melanogaster* was first identified as the site of a temperature-sensitive paralytic mutation (Suzuki *et al.*, 1971), which was subsequently found to cause temperature-sensitive impairment of action potential conduction in nerves (Siddiqi and Benzer, 1976; Wu and Ganetzky, 1980). Sequencing of genomic DNA and cDNA clones derived from the *para* locus showed that *para* encodes a protein having all of the structural hallmarks of voltage-sensitive sodium channel α subunits and exhibiting approximately 50% overall amino acid

sequence identity to vertebrate sodium channel α subunits (Loughney *et al.*, 1989). The *para* gene product is quite divergent from the *dsc1* gene product in that the four conserved internal homology domains of the two *D. melanogaster* genes are only as similar to each other (∼50% amino acid sequence identity) as they are to the corresponding regions of vertebrate sodium channel proteins.

Sequence analyses of partial cDNAs derived from the *para* loci of *D. melanogaster* and the related species *D. virilis* identified alternative mRNA splicing at eight sites involving seven optional exons (designated *a*, *b*, *e*, *f*, *h*, *i*, and *j*) and one pair of mutually exclusive exons (designated *c/d*), resulting in 256 possible unique structural variants from the *para* gene (Thackeray and Ganetzky, 1994, 1995; O'Dowd *et al.*, 1995; Warmke *et al.*, 1997). The locations of alternatively spliced exons in relation to other structural landmarks in the para protein are illustrated in **Figure 2**. Alternative splicing at

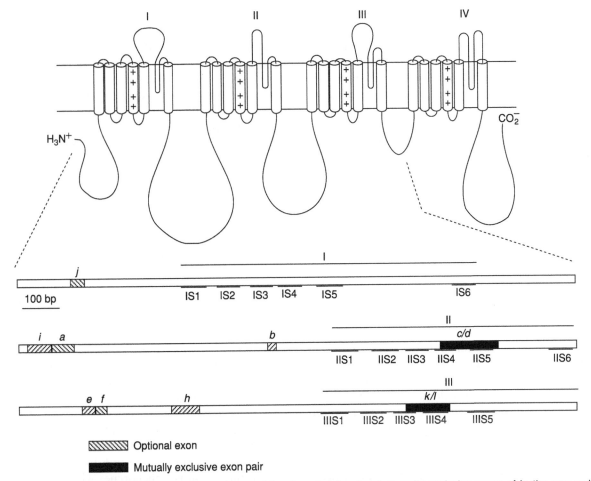

Figure 2 Diagram showing the approximate size and location of optional and mutually exclusive exons of in the *para* and *Vssc1* sodium channel genes in relation to other sodium channel structural landmarks. (Modified with permission from Lee, S.H, Ingles, P.J., Knipple, D.C., Soderlund, D.M., **2002**. Developmental regulation of alternative exon usage in the housefly *Vssc1* sodium channel gene. *Invert. Neurosci. 4*, 125–133.)

these sites produces a heterogeneous family of sodium channel α subunit transcripts. Additional transcript heterogeneity in these species is conferred by posttranscriptional RNA editing at three sites (Hanrahan *et al.*, 2000). Transcripts from the *para* locus are expressed strongly in the nervous system throughout development, but analysis of transcript pools from different developmental stages and anatomical regions revealed significant developmental and anatomical regulation of both alternative splicing and RNA editing (Tseng-Crank *et al.*, 1991; Hong and Ganetzky, 1994; Thackeray and Ganetzky, 1994, 1995; Hanrahan *et al.*, 2000).

1.2.1.3. *tipE*   The *tipE* (*temperature-induced paralysis, locus E*) gene, at cytogenetic region 64B2, is the site of a temperature-sensitive paralytic mutation having a phenotype similar to that of *para* mutants (Kulkarni and Padhye, 1982; Jackson *et al.*, 1986; Feng *et al.*, 1995b). Sequence analysis of cDNAs derived from the *tipE* locus identified an open reading frame encoding a 452-amino acid protein that is clearly not a member of the voltage-sensitive ion channel family and exhibits no significant sequence similarity to any other proteins in sequence databases (Feng *et al.*, 1995a). The tipE protein contains two hydrophobic segments having properties consistent with the formation of transmembrane helices that flank a loop that contains five potential N-glycosylation sites. Transcripts from the original *tipE* mutant strain contain a premature stop codon between the two hydrophobic domains and therefore express a truncated tipE protein. Expression of the *tipE* gene in *D. melanogaster* is restricted to the nervous system and exhibits strong developmental regulation, so that transcript levels are very low in embryonic and larval stages but increase markedly in the late pupal stage. Genetic and functional criteria (see Section 1.3.1) suggest that the tipE protein may be a sodium channel auxiliary subunit, functionally analogous but structurally unrelated to vertebrate sodium channel β subunits.

1.2.1.4. Genetic and functional characterization of *D. melanogaster* mutants   Gene dosage and interaction studies with *para^ts* mutants provide evidence for a crucial physiological role for the para protein. Homozygous *para^ts* mutants that exhibit temperature-sensitive paralysis in a wild-type background are unconditionally lethal in the presence of either *nap^ts* (*no action potential, temperature-sensitive*) or *tipE*, both of which also impair nerve function at nonpermissive temperatures (Ganetzky, 1984, 1986). Similarly, decreased dosage of *para^+* is unconditionally lethal in a *nap^ts* background,

but increased dosage of *para^+* causes leg-shaking behavior, a sign of neuronal excitation, and suppresses the temperature sensitivity of *nap^ts* (Stern *et al.*, 1990).

The *smellblind* locus was originally identified as the site of mutations causing olfactory defects and was mapped close to *para* on the X chromosome (Lilly and Carlson, 1989). Two viable *smellblind* alleles exhibit temperature-dependent lethality to embryos at high temperatures and to adults at low temperatures, whereas four other alleles are recessive lethals (Lilly and Carlson, 1989; Lilly *et al.*, 1994b). None of the six *smellblind* alleles complements the lethality of two unconditionally lethal *para* alleles, and two of the recessive lethal alleles of *smellblind* contain rearrangements within the primary *para* transcript (Lilly *et al.*, 1994a). Taken together, these results show that *smellblind* is a novel class of *para* mutation. Interestingly, *smellblind* alleles do not exhibit the rapid, temperature-sensitive paralysis of *para^ts* mutants, and the *para^ts* mutants examined so far do not exhibit olfactory defects (Lilly *et al.*, 1994a).

Electrophysiological and pharmacological studies provide additional evidence for the functional importance of the *para* gene product. The temperature-sensitive failure of nerve conduction associated with *para^ts* alleles is presynaptic to the cervical giant fibers, which remain electrically excitable at nonpermissive temperatures (Nelson and Wyman, 1989; Elkins and Ganetzky, 1990). Cultured neurons from embryos homozygous for different *para^ts* alleles exhibit an allele-dependent reduction in sodium current density; in the case of *para^ST76* the expression of sodium currents was reduced by 98% (O'Dowd *et al.*, 1989). One of these alleles, *para^ts2*, also caused a depolarizing shift in the voltage dependence of activation of sodium channels in cultured embryonic neurons. Cultured larval neurons from *para^ts* individuals exhibit temperature-dependent resistance to veratridine, which binds to site 2 on the sodium channel and causes persistent channel activation (Suzuki and Wu, 1984). Also, several *para^ts* strains are resistant to DDT and pyrethroid insecticides (Pittendrigh *et al.*, 1997).

The *tipE* mutation causes a temperature-dependent paralytic phenotype similar to that caused by *para^ts* alleles (Jackson *et al.*, 1986). Three additional mutant *tipE* alleles, generated by γ-ray mutagenesis and characterized as deletions or translocations disrupting the *tipE* locus, confer the temperature-sensitive phenotype when heterozygous with the original ethylmethanesulfonate-induced *tipE* allele (Feng *et al.*, 1995b). This finding suggests that the original *tipE* mutation, identified as a premature stop codon yielding a truncated gene product (Feng *et al.*, 1995a), is a loss-of-function mutation. In addition to conferring the conditional paralytic phenotype, the *tipE* mutation causes temperature-sensitive impairment of conduction in adult, but not larval, nerves (Ganetzky, 1986; Elkins and Ganetzky, 1990) and reductions of approximately 50% in the sodium current density of cultured embryonic neurons and in density of [³H]saxitoxin binding sites in adult head preparations (Jackson *et al.*, 1986). Studies using cultured embryonic neurons show that *tipE*⁺ is required for the sustained repetitive firing of sodium-dependent action potentials (Hodges *et al.*, 2002). Sodium currents in neurons from animals carrying the *tipE* mutation recover more slowly, and the minimum interpulse interval needed to produce a second full-amplitude action potential is longer, than those in neurons from wild-type animals. Expression of the wild-type *tipE*⁺ transgene rescues all of the functional deficits conferred by the *tipE* mutation in these assays.

Interactions between *tipE* and *para*^ts^ alleles provide further insight into the functional role of the tipE protein. As noted above, *tipE* interacts with various *para*^ts^ alleles to cause unconditional lethality at temperatures that are permissive for single mutants (Ganetzky, 1986). It is interesting that some alleles of *para* exhibit partial viability when combined with *tipE*, whereas other allelic combinations are unconditionally lethal. These observations have been interpreted as evidence that the tipE and para proteins may interact physically (Ganetzky, 1986).

In contrast to the large body of information available about the functional significance of *para* and *tipE*, there is very little information on the role of *dsc1*. Embryonic neurons homozygous for a deficiency that includes the *dsc1* locus exhibit normal sensitivity to veratridine (Sakai *et al.*, 1989), thereby indicating the presence of functional sodium channels in these cells. Moreover, these neurons express voltage-sensitive sodium currents that are indistinguishable from those in wild-type cells (Germeraad *et al.*, 1992). These studies provide further evidence that *para* rather than *dsc1* encodes the sodium channels found in embryonic neurons.

Until recently, there were no mutations associated with the *dsc1* locus that might illuminate the function of the dsc1 protein. However, a series of *smell-impaired* (*smi*) mutants, created by single P element insertion mutagenesis and isolated on the basis of the loss of an olfactory avoidance behavioral phenotype, included one mutation (*smi60E*) mapping close to the *dsc1* locus (Anholt *et al.*, 1996). Subsequent studies showed that this mutant strain contains a P element transposon within the second intron of the *dsc1* gene (Kulkarni *et al.*, 2002). The reduction of olfactory avoidance behavior in this strain is correlated with a reduction in the abundance of *dsc1* transcripts, and excision of the P element restores wild-type behavior and *dsc1* transcript levels. These results provide the first clear indication of a physiologically defined role for the dsc1 protein.

1.2.2. Sodium Channel Genes in Other Insect Species

1.2.2.1. Orthologs of *para*   The identification of the *para* gene product as a physiologically important voltage-sensitive sodium channel in *D. melanogaster* (Loughney *et al.*, 1989) and the concurrent development of the polymerase chain reaction (PCR) as a means of gene identification (Saiki *et al.*, 1988) provided the conceptual and technical framework for the isolation of *para*-orthologous DNA sequences from other arthropod species (Doyle and Knipple, 1991; Knipple *et al.*, 1991). The likely importance of voltage-sensitive sodium channels as sites of mutations conferring resistance to DDT and to pyrethroid insecticides provided further impetus for this effort (Soderlund and Knipple, 2003). As a result, full-length or partial genomic DNA or cDNA sequences now are known for *para*-orthologous sodium channel genes from at least 15 additional arthropod species.

Two concurrent and independent efforts led to the characterization of the full-length coding sequence of *Vssc1* (also called *Msc*), the *para* ortholog of the housefly (*Musca domestica*) (Ingles *et al.*, 1996; Williamson *et al.*, 1996). The inferred sequence of the Vssc1 protein is 90% identical to that of the most similar splice variant of the *para* gene product (Loughney *et al.*, 1989; Thackeray and Ganetzky, 1994). The initial sequence analyses of *Vssc1* cDNA clones obtained from adult fly heads did not identify alternatively spliced transcripts (Ingles *et al.*, 1996; Williamson *et al.*, 1996). A subsequent study investigated the issue of alternative splicing at the *Vssc1* locus in greater detail by examining the heterogeneity of multiple cDNA clones derived from *Vssc1* transcripts in three different developmental stages and two different adult body regions (Lee *et al.*, 2002). This investigation identified multiple alternative exons in the *Vssc1* gene, including a new pair of mutually exclusive exons (designated *k* and *l*) (**Figure 2**) encompassing part of transmembrane domain IIS3 and all of transmembrane domain IIS4. This splice site was not previously identified in the analysis of *para* transcripts in *Drosophila* species, which contain a segment homologous to exon *l* (Thackeray and Ganetzky, 1994, 1995;

O'Dowd *et al.*, 1995; Warmke *et al.*, 1997), but a search of the *para* genomic sequence (Adams *et al.*, 2000) identified a segment upstream of the exon *l*-like sequence with substantial sequence similarity to exon *k* of *Vssc1* (Lee *et al.*, 2002). The amino acid sequences of exons *k* and *l* are highly divergent, differing at 16 of 41 amino acid residues, but retain conserved features of the voltage sensor structure in domain IIIS4. Other novel aspects of alternative splicing of *Vssc1* transcripts in *M. domestica* included the apparent constitutive expression of optional exons *h* and *i* and the apparent inactivation of exon *c* by a stop codon, so that all transcripts encoding full-length channel sequences contained only exon *d* at this site. Alternative splicing at all of the sites identified either in *Vssc1* or in the *para* genes of *D. melanogaster* or *D. virilis* could theoretically generate up to 512 structurally unique sodium channel splice variants. However, only a small subset of these variants (five or fewer) comprised more than 90% of the transcripts in each *M. domestica* cDNA pool examined.

The *para* ortholog of *Blattella germanica* (designated *para*^CSMA^) encodes an inferred protein sequence that is 76–78% identical to the two dipteran sequences (Dong, 1997; Tan *et al.*, 2002a). Analysis of multiple *para*^CSMA^ sequences identified novel aspects of alternative exon usage at some of the sites of alternative splicing found in *D. melanogaster* and *M. domestica*. For example, *para*^CSMA^ transcripts contain two examples of segments corresponding to optional exon *j*, suggesting that the *B. germanica* gene may incorporate multiple variants of this exon. Similarly, the *B. germanica* transcripts contain three different exons (designated G1, G2, and G3) at the exon *k/l* splice site in homology domain III (Tan *et al.*, 2002a). Exons G1 and G2 correspond to exons *l* and *k* of *para* and *Vssc1*, whereas exon G3 exhibits no obvious sequence similarity to either exon G1 or G2 and contains an in-frame stop codon. Exon G3 corresponds exactly to an alternative exon in vertebrate Na$_v$1.6 sodium channels that contains a premature stop codon at the same position (Plummer *et al.*, 1997).

The structure of *hscp*, the ortholog of *para* in *Heliothis virescens*, was determined by sequence analyses of genomic DNA and PCR-amplified cDNA segments (Park *et al.*, 1999). The reported *hscp* coding sequence, which is incomplete at the 3′ end, gives a predicted protein sequence that is 80% identical to the corresponding coding region of *para*. Fifteen of the 19 introns identified in the *hscp* gene were conserved in location with respect to the corresponding introns in *para*, and alternatively spliced transcripts containing optional exon *j* and mutually exclusive exons *c* and *d* were detected. This study also

identified several transcripts lacking sequences from the linker between homology domains I and II that contained premature stop codons. Unlike the truncation of the *M. domestica* and *B. germanica* transcripts by premature stop codons located in alternatively spliced exons, these *H. virescens* transcripts appear to be the result of splicing errors that resulted in missing exons and frameshifts.

The ortholog of *para* in the head louse, *Pediculus capitis*, encodes a voltage-sensitive sodium channel α subunit protein that is 73% identical to *para* and the *B. germanica* and *M. domestica* orthologs (Lee *et al.*, 2003). Full-length sequences from two strains of head louse and one strain of body louse (*P. humanus*) were 99.6% identical at the nucleotide level, providing further evidence that the head louse and body louse are conspecific. Analysis of multiple transcripts provided evidence for alternative splicing of a segment homologous to exon *j* of *para*. All of the transcripts contained segments corresponding to *para* optional exons *a*, *b*, and *h* and mutually exclusive exons *d* and *l* but did not contain segments corresponding to optional exons *i*, *e*, and *f*.

The ortholog of *para* in the varroa mite, *Varroa destructor* (designated *VmNa*) encodes a voltage-sensitive sodium channel α subunit that is only 51% identical to the *para* gene product of *D. melanogaster* (Wang *et al.*, 2003). Sequence analysis of multiple *VmNa* transcripts identified several novel aspects of alternative exon usage in this gene. Some *VmNa* transcripts contained an optional exon (designated exon 3) corresponding to the pair of mutually exclusive introns *k* and *l* in *Drosophila* species and *M. domestica* (**Figure 2**). Some *VmNa* transcripts also contained option exons at three sites not previously identified as sites of alternative splicing: exon 1, a segment of homology domain II containing part of the IIS2 and all of the IIS3 transmembrane helices; exon 2, a short inserted sequence in the domains II–III intracellular linker; and a retained intron in the C terminus containing an alternative stop codon. Channels lacking exon 1 or 3 would also lack critical determinants of channel organization and function and therefore may be inactive, whereas the alternative splicing of exon 2 and the alternative C terminus could be involved in the regulation of channel expression or functional properties.

Partial sequences of varying length also exist for the *para* orthologs from 10 additional arthropod species: *Anopheles gambiae* (Martinez-Torres *et al.*, 1998; Ranson *et al.*, 2000); *Bemisia tabaci* (Morin *et al.*, 2002); *Boophilus microplus* (He *et al.*, 1999); *Culex pipiens* (Martinez-Torres *et al.*, 1999a); *Frankliniella occidentalis* (Forcioli *et al.*, 2002); *Helicoverpa armigera* (Head *et al.*, 1998); *Hematobia irritans* (Guerrero *et al.*, 1997); *Leptinotarsa decemlineata* (Lee *et al.*, 1999b); *Myzus persicae* (Martinez-Torres *et al.*, 1999b); and *Plutella xylostella* (Schuler *et al.*, 1998).

1.2.2.2. Orthologs of *dsc1* The search for the *para* ortholog of *B. germanica* also yielded a partial cDNA of the ortholog of *dsc1* in this species (Dong, 1997). Subsequent work described the complete coding sequence of this gene (designated *BSC1*) and identified three regions of alternative exon usage (Liu *et al.*, 2001). Reverse transcription (RT)-PCR analyses documented the expression of *BSC1* transcripts not only in the nerve cord but also in muscle, gut, fat body, and ovary, a much broader pattern of expression than was found by *in situ* hybridization of *dsc1*-derived probes to *D. melanogaster* tissues (Hong and Ganetzky, 1994). RT-PCR assays also showed that the expression of *BSC1* splice variants was both tissue-specific and developmentally regulated (Liu *et al.*, 2001). So far, the only other ortholog of *dsc1* described in the literature is a partial genomic DNA sequence from *H. virescens* (Park *et al.*, 1999).

1.2.2.3. Orthologs of *tipE* The *tipE* gene is of interest because of its clear involvement in modifying sodium channel expression and function and the suggestion that it may encode a novel type of sodium channel auxiliary subunit. The ortholog of *tipE* in *Musca domestica*, designated *Vsscβ*, encodes a predicted protein that exhibits 72% overall amino acid sequence identity to the tipE protein and 97% identity within the hydrophobic regions identified as probable transmembrane domains (Lee *et al.*, 2000a). These results suggest that *tipE* and *Vsscβ* are substantially more divergent in sequence than are the two sodium channel α subunit genes from these species.

1.3. Functional Expression of Cloned Insect Sodium Channels

1.3.1. Expression of Functional Sodium Channels in *Xenopus* Oocytes

Unfertilized oocytes of the African clawed frog, *Xenopus laevis*, are a powerful tool for confirming the functional roles of neurotransmitter receptors and ion channels and for correlating channel structure with functional and pharmacological properties (Lester, 1988). Oocytes injected with cRNA (synthetic mRNA prepared from cloned cDNA) express sodium channels in the cell membrane that can be detected by conventional electrophysiological

assays, such as two-electrode voltage clamp analysis of the currents carried by the expressed proteins in response to changes in membrane potential or the application of native or exogenous ligands (Goldin, 1992; Stühmer, 1992). This system, coupled with site-directed mutagenesis of cloned cDNAs, has been widely employed in structure–function studies of vertebrate sodium channels (Catterall, 2000; Goldin, 2001; Yu and Catterall, 2003).

Initial efforts to express functional sodium channel from synthetic *para* cRNA injected into oocytes produced voltage-gated sodium currents of very low amplitude, but coexpression of *para* and *tipE* cRNAs stimulated sodium current expression approximately 30-fold (Feng *et al.*, 1995a). Subsequent studies (Warmke *et al.*, 1997) showed that injection of large amounts of *para* cRNA was required to obtain more robust currents in the absence of *tipE* cRNA. These authors also showed that para/tipE channels inactivated more rapidly than those expressed from *para* alone. Thus, coexpression of *para* and *tipE* reconstitutes the functional properties of the native *D. melanogaster* voltage-sensitive sodium channel and provides evidence that the tipE protein may function as a sodium channel auxiliary subunit.

Efforts to express functional sodium channels in oocytes from *Vssc1* cRNA produced results similar to those with *para*, requiring large amounts of injected cRNA to yield low levels of sodium current expression, but coexpression with *tipE* cRNA resulted in the robust expression of voltage-gated sodium currents (Smith *et al.*, 1997; Lee *et al.*, 2000a). These Vssc1/tipE channels, like para/tipE channels, inactivated more rapidly than channels expressed from either *para* or *Vssc1* cRNA alone. Coexpression of Vssc1 with its conspecific auxiliary subunit, Vsscβ, was more effective than coexpression with tipE in enhancing the level of sodium current expression and accelerating the inactivation kinetics of Vssc1 sodium channels (Lee *et al.*, 2000a). Coexpression with *tipE* cRNA was also found to be necessary for the expression of functional sodium channels from *para*^{CSMA} cRNA in oocytes (Tan *et al.*, 2002b). However, this study did not document the expression of functional channels in oocytes from *para*^{CSMA} cRNA alone.

1.3.2. Pharmacological Properties of Expressed Insect Sodium Channels

Insect sodium channels expressed in oocytes retain sensitivity to insecticides and other natural and synthetic toxicants that alter sodium channel function. Insect sodium channels expressed in oocytes also are sensitive to blockade by nanomolar concentrations of tetrodotoxin (Warmke *et al.*, 1997; Smith *et al.*,

1998; Tan *et al.*, 2002a). These channels are also susceptible to modification by the alkaloid batrachotoxin (Lee and Soderlund, 2001) and polypeptide toxins from sea anemone (Warmke *et al.*, 1997) and scorpion (Shichor *et al.*, 2002) venoms.

The effects of pyrethroids on expressed insect sodium channels have been extensively investigated in the context of the functional characterization of sodium channel mutations associated with pyrethroid resistance (see Section 1.4.5). When assayed under voltage clamp conditions, type I pyrethroids (i.e., cismethrin and permethrin) appear to bind predominantly to resting or inactivated channels, shifting the voltage dependence of activation to more negative potentials and causing a slowly activating sodium current. These compounds also produce characteristic sodium tail currents following a depolarizing pulse that decay with first-order time constants (Smith *et al.*, 1997, 1998; Warmke *et al.*, 1997; Zhao *et al.*, 2000). In contrast to these results, type II pyrethroids (i.e., [1R,cis,αS]-cypermethrin and deltamethrin) exhibit profound use-dependent modification of sodium currents, which implies that these compounds bind preferentially to activated sodium channel states (Smith *et al.*, 1998; Vais *et al.*, 2000; Tan *et al.*, 2002b). The tail currents caused by these compounds are more persistent than those caused by cismethrin or permethrin. In the case of deltamethrin, tail current decay is biphasic, rather than monophasic (Vais *et al.*, 2000; Tan *et al.*, 2002b); this finding has been interpreted as evidence for two binding sites on insect sodium channels with different affinities for this ligand (Vais *et al.*, 2000).

1.3.3. Functional Characterization of Sodium Channel Splice Variants

The conservation of alternative exon structure and the developmental and anatomical regulation of alternative exon usage in the *para* sodium channel gene of *D. melanogaster* and its orthologs in other insect species imply that alternative splicing may generate a family of sodium channel proteins with differing functional properties, as has been found for other ion channels and receptors (Harris-Warwick, 2000). Most of the optional exons identified in *para* orthologs to date (see **Figure 2**) are located in intracellular domains of the sodium channel protein. These exons may therefore be involved in the regulation of sodium channel expression or function as the result of interactions with protein kinases or G proteins (Cukierman, 1996; Cantrell and Catterall, 2001). This interpretation is consistent with the existence in exons *a* and *i* of consensus protein kinase A phosphorylation sites (O'Dowd *et al.*, 1995; Ingles *et al.*, 1996). Similarly, exons

2 and 4 of the *VmNa* sodium channel are located in intracellular domains, and exon 2 contains contains a consensus protein kinase C phosphorylation site (Wang *et al.*, 2003). The sole exception so far to the intracellular location of optional exons is exon 1 of the *VmNa* sodium channel sequence, which is located in the transmembrane region of domain II (Wang *et al.*, 2003).

So far, there is little direct evidence bearing on the functional significance of the alternative splicing of optional exons. In embryonic *D. melanogaster* neurons, functional sodium channels were detected only in those cells having *para* transcripts containing exon *a* (O'Dowd *et al.*, 1995). This study also documented enhanced sodium current expression in cells expressing channels that contain both exons *a* and *i*. Whereas these findings imply a critical role for exon *a* alone and the combination of exons *a* and *i* together in sodium channel regulation, the direct comparison in functional expression assays using *X. laevis* oocytes of variants of *para* that differ only by the presence or absence of exon *a* did not find any effects of exon *a* on sodium current expression or properties in this system (Warmke *et al.*, 1997).

In contrast to the optional exons, the mutually exclusive exons in *para* orthologs occur within the transmembrane regions of homology domains II and III (**Figure 2**). There is no information on the functional role of the alternative splicing of exons *c* and *d*. In *D. melanogaster*, these exons differ by only two of 55 amino acid residues (Loughney *et al.*, 1989). In *M. domestica*, all functional channels apparently contain only exon *d* because exon *c* contains an in-frame stop codon (Lee *et al.*, 2002). These observations suggest that alternative splicing at the *c/d* site may play a role in posttranscriptional regulation rather than in the generation of functionally distinct channel variants.

The most significant functional effects of alternative exon usage have been documented for splice variants at the exon *k/l* site. Unlike exons *c* and *d*, exons *k* and *l* (corresponding to exons G2 and G1 in *B. germanica*) differ substantially in amino acid sequence (Lee *et al.*, 2002; Tan *et al.*, 2002a). Expression of *para*^CSMA^ variants containing either exon G1 or G2 in oocytes documents differences in the voltage dependence of both activation and inactivation of these channels (Tan *et al.*, 2002a). Unexpectedly, this study also found substantial differences between these variants in their sensitivity to the pyrethroid insecticide deltamethrin. These results provide the first experimental evidence for functional differences between splice variants of insect sodium channels. Alternative splicing at the *k/l* site in

B. germanica and *V. destructor* also appears to be involved in the expression of inactive channel variants, in that exon G3 in the *para^{CSMA}* sequence encodes a truncated channel (Tan *et al.*, 2002a) and the absence of exon 3 in the *VmNa* sequence encodes a channel lacking one of the four voltage sensor regions (Wang *et al.*, 2003).

1.4. Sodium Channels and Knockdown Resistance to Pyrethroids

1.4.1. Knockdown Resistance

The knockdown resistance (*kdr*) trait, which confers resistance to the rapid knockdown action and lethal effects of DDT and pyrethrins, was first documented in houseflies in 1951 (Busvine, 1951) and isolated genetically in 1954 (Milani, 1954). The *kdr* trait confers resistance to both the rapid paralytic and lethal actions of all known pyrethroids, as well as the pyrethrins and DDT, but does not diminish the efficacy of other insecticide classes (Oppenoorth, 1985). Electrophysiological assays employing a variety of nerve preparations from larval and adult *kdr* insects (Bloomquist, 1988) provide direct evidence for reduced neuronal sensitivity as the basis for the *kdr* trait. A second resistance trait in the housefly (designated *super-kdr*) that confers much greater resistance to DDT and pyrethroids than that found in *kdr* strains has also been isolated genetically and mapped to chromosome 3 (Sawicki, 1978; Farnham *et al.*, 1987). The *kdr* and *super-kdr* traits are widely presumed to represent allelic variants at a single resistance locus on the basis of their similar spectra of resistance and their common localization to chromosome 3 in the housefly.

Resistance mechanisms similar to *kdr* have been inferred in a number of agricultural pests and disease vectors on the basis of cross-resistance patterns and the absence of synergism by compounds known to inhibit the esterase and monooxygenase activities involved in pyrethroid metabolism (Soderlund and Bloomquist, 1990; Bloomquist, 1993; Soderlund, 1997; Soderlund and Knipple, 1999). Confirming electrophysiological evidence for reduced neuronal sensitivity to pyrethroids also exists for at least six species: *H. virescens*, *Spodoptera littoralis*, *Culex quinquefasciatus*, *A. stephensi*, *B. germanica*, and *P. xylostella* (Bloomquist, 1988, 1993; Schuler *et al.*, 1998).

Pyrethroids are known to exert their insecticidal effects by altering the function of voltage-sensitive sodium channels in nerve membranes (see Section 1.1.2). Therefore, studies of *kdr*-like resistance have focused on mechanisms that might affect the

regulation, pharmacology, or function of the sodium channel.

1.4.2. Altered Sodium Channel Regulation as a Mechanism of Knockdown Resistance

The possible role of reduced insecticide receptor density in knockdown resistance was initially implicated on the basis of resistance-associated reductions in the density of binding sites for [3H]STX (Rossignol, 1988), a radioligand that specifically labels site 1 of the sodium channel (see **Table 1**). However, further investigation of [3H]STX binding in susceptible, *kdr*, and *super-kdr* housefly strains (Grubs *et al.*, 1988; Sattelle *et al.*, 1988; Pauron *et al.*, 1989) revealed that a reduction in sodium channel density is not an obligatory component of the *kdr* or *super-kdr* phenotypes of the housefly. Moreover, comparisons of *Vssc1* transcript and protein levels in susceptible and *kdr* housefly strains did not document any differences between strains (Castella *et al.*, 1997).

Although these results rule out sodium channel downregulation as the mechanism underlying the *kdr* and *super-kdr* traits in the housefly, reduced sodium channel density could, in principle, produce a *kdr*-like phenotype. The relationship between sodium channel density and *kdr*-like resistance has been evaluated directly using the *napts* strain of *D. melanogaster*, in which the density of sodium channels (measured as binding sites for [3H]STX in head membrane preparations) is approximately half that of wild-type flies (Jackson *et al.*, 1984). Flies homozygous for *nap^{ts}* exhibit modest (~threefold) resistance to the lethal effects of DDT and pyrethroids that is also evident in delayed onset of paralysis and reduced physiological sensitivity of the adult central nervous system (Kasbekar and Hall, 1988; Bloomquist *et al.*, 1989). These results suggest that if such a mechanism were present in other insects that exhibit knockdown resistance, the magnitude of resistance observed would require a profound and readily detectable reduction in sodium channel density (Grubs *et al.*, 1988). Because reductions in sodium channel density more severe than that observed in the *nap^{ts}* strain of *D. melanogaster* would be anticipated to compromise viability, it is unlikely that reduced target density can account for *kdr*-like traits that confer significant levels of resistance.

1.4.3. Genetic Linkage between Knockdown Resistance and Sodium Channel Genes

Two studies employed restriction fragment length polymorphisms (RFLPs) in the *Vssc1* gene coupled with discriminating dose bioassays with DDT to demonstrate tight genetic linkage (within ~1 map unit) of the *kdr* and *super-kdr* resistance trait and the *Vssc1* gene of *M. domestica* (Williamson *et al.*, 1993; Knipple *et al.*, 1994). In addition to providing strong genetic evidence for mutations at a sodium channel structural gene as the cause of knockdown resistance in the housefly, these two studies also provided the first experimental evidence for the widely presumed allelism of the *kdr* and *super-kdr* traits in this species.

Conceptually similar approaches were employed to investigate linkage between knockdown resistance traits and *para*-orthologous sodium channel sequences in other species. In the case of *B. germanica*, knockdown resistance was tightly linked (within ~0.2 map units) to an RFLP located with the *para*-orthologous sodium channel gene (Dong and Scott, 1994). The use of RFLP markers to assess the linkage between knockdown resistance and sodium channel gene sequences in *H. virescens* (Taylor *et al.*, 1993) was complicated by the use of a strain with multiple resistance mechanisms. Nevertheless, results of these assays suggested that one component of resistance was linked to an RFLP in the *para*-orthologous sodium channel gene of this species. In *L. decemlineata* (Lee *et al.*, 1999b) and *B. tabaci* (Morin *et al.*, 2002), sequencing of DNA from individual insects of susceptible and resistant phenotypes has identified sequence variants within sodium channel coding regions that are genetically linked with resistant phenotypes. Finally, some mutant alleles of *para* that exhibit the temperature-sensitive paralytic phenotype also exhibit resistance to pyrethroids at permissive temperatures (Hall and Kasbekar, 1989; Pittendrigh *et al.*, 1997).

1.4.4. Identification of Resistance-Associated Mutations

The selective alteration of sodium channel pharmacology in knockdown resistant insects and the genetic linkage of knockdown resistance traits and sodium channel gene sequences provided a strong impetus for the identification and functional characterization resistance-associated mutations in insect sodium channel genes. Results of these efforts are the subject of two recent comprehensive reviews (Soderlund and Knipple, 1999, 2003).

Comparison of partial and complete sequences from 15 housefly strains representing multiple examples of susceptible, *kdr*, and *super-kdr* phenotypes consistently identified two point mutations that were associated with resistant phenotypes (**Figure 3**): mutation of leucine to phenylalanine at

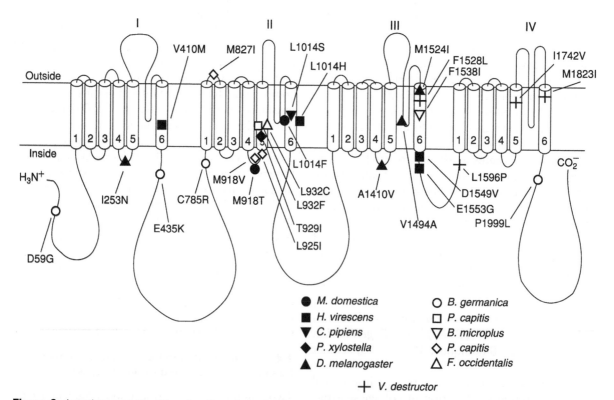

Figure 3 Locations and identities of sodium channel point mutations associated with knockdown resistance to pyrethroids. Symbols indicate the species in which each mutation was first identified (see also **Table 2**). (Modified with permission from Soderlund, D.M., Knipple, D.C. **2003**. The molecular biology of knockdown resistance to pyrethroid insecticides. *Insect Biochem. Mol. Biol. 33*, 563–577.)

amino acid residue 1014 (designated L1014F) in all *kdr* and *super-kdr* strains, and the additional mutation of methionine to threonine at residue 918 (designated M918T) only in *super-kdr* strains (Ingles *et al.*, 1996; Miyazaki *et al.*, 1996; Williamson *et al.*, 1996). Mutations in *para*-orthologous sodium channel gene sequences corresponding to the L1014F mutation in the housefly have been also identified to date in eight additional pest species (**Table 2**): *B. germanica* (Miyazaki *et al.*, 1996; Dong, 1997), *F. occidentalis* (Forcioli *et al.*, 2002), *H. irritans* (Guerrero *et al.*, 1997), *A. gambiae* (Martinez-Torres *et al.*, 1998), *P. xylostella* (Schuler *et al.*, 1998), *L. decemlineata* (Lee *et al.*, 1999b), *M. persicae* (Martinez-Torres *et al.*, 1999b), and *C. pipiens* (Martinez-Torres *et al.*, 1999a). (For clarity and consistency and to facilitate comparisons between species, including those for which full-length sodium channel sequences are not available, all resistance-associated mutations in this report are numbered according to their positions in the amino acid sequence of the most abundant splice variant of the housefly Vssc1 sodium channel α subunit (GenBank

Accession no. U38813).) In contrast to the many examples of mutations in other species corresponding to the L1014F mutation in the housefly, a mutation corresponding to the second-site M918T mutation that is associated with the *super-kdr* trait of the housefly has been found to date only in highly resistant populations of *H. irritans* (Guerrero *et al.*, 1997).

The search for sodium channel gene mutations associated with knockdown resistance has also identified numerous novel mutations (**Figure 3; Table 2**). Studies with pyrethroid-resistant *H. virescens* populations identified a second mutation at sequence position 1014, L1014H (Park and Taylor, 1997), as well as three new mutations: V410M (Park *et al.*, 1997), D1549V, and E1553G (Head *et al.*, 1998). The latter two mutations were found together in sodium channel sequences from pyrethroid-resistant strains of both *H. virescens* and *H. armigera* (Head *et al.*, 1998). Studies with knockdown-resistant populations of *C. pipiens* and *A. gambiae* identified a third variant at position 1014, L1014S (Martinez-Torres *et al.*, 1999a;

Table 2 Sodium channel amino acid sequence polymorphisms associated with knockdown resistance in arthropod species

Species	Mutations identified[a]	Reference
Anopheles gambiae	L1014F	Martinez-Torres *et al.* (1998)
	L1014S	Ranson *et al.* (2000)
Bemisia tabaci	M918V; L925I	Morin *et al.* (2002)
Blattella germanica	L1014F[b]	Miyazaki *et al.* (1996), Dong (1997)
	L1014F+E435K+C785R[b]	Liu *et al.* (2000)
	L1014F+D59G+E435K+C785R+P1999L[b]	Liu *et al.* (2000)
Boophilus microplus	F1538I	He *et al.* (1999)
Culex pipiens	L1014F; L1014S	Martinez-Torres *et al.* (1999)
Drosophila melanogaster	I253N; A1410V; A1494V; M1524I	Pittendrigh *et al.* (1997)
Frankliniella occidentalis	L1014F+T929C	Forcioli *et al.* (2002)
Helicoverpa armigera	D1549V+E1533G	Head *et al.* (1998)
Heliothis virescens	L1014H	Park and Taylor (1997)
	V410M	Park *et al.* (1997)
	D1549V+E1533G	Head *et al.* (1998)
Hematobia irritans	L1014F+M918T	Guerrero *et al.* (1997)
Leptinotarsa decemlineata	L1014F	Lee *et al.* (1999b)
Musca domestica	L1014F; L1014F+M918T	Ingles *et al.* (1996), Miyazaki *et al.* (1996), Williamson *et al.* (1996)
Myzus persicae	L1014F	Martinez-Torres *et al.* (1999b)
Pediculus capitis	M827I+T929I+L932F[c]	Lee *et al.* (2000b), (2003)
Plutella xylostella	L1014F+T929I	Schuler *et al.* (1998)
Varroa destructor	L1596P+M1823I	Wang *et al.* (2002)
	F1528L+L1596P+I1742V+M1823I	

[a]Positions numbered according to the amino acid sequence of the most abundant splice variant of the housefly Vssc1 sodium channel protein (Ingles *et al.*, 1996; Williamson *et al.*, 1996).
[b]These mutations correspond to the D58G, E434K, C764R, L993F, and P1880L mutations in the full-length *Blattella germanica para*[CSMA] sodium channel (Dong, 1997; Liu *et al.*, 2000).
[c]These mutations correspond to the M815I, T917I, and L920F mutations in the full-length *Pediculus capitis* sodium channel (Lee *et al.*, 2003).
Modified from Soderlund, D.M., Knipple, D.C. **2003**. The molecular biology of knockdown resistance to pyrethroid insecticides. *Insect Biochem. Mol. Biol. 33*, 563–577.

Ranson *et al.*, 2000). A novel mutation at Met918 (M918V) was recently identified in some pyrethroid-resistant *B. tabaci* populations (Morin *et al.*, 2002). Characterization of sodium channel sequences from pyrethroid-resistant *P. xylostella* identified a novel putative second-site mutation (T929I) associated with the more commonly observed L1014F mutation in strains with high pyrethroid resistance (Schuler *et al.*, 1998). The T929I mutation is also observed in combination with two other novel mutations, M827I and L932F, in pyrethroid-resistant *P. capitis* (Lee *et al.*, 2000b, 2003). A second mutation at position 929 (T929C) was found in combination with the L1014F mutation in highly resistant populations of *F. occidentalis* (Forcioli *et al.*, 2002). Recent studies with *B. tabaci* have also identified another novel mutation in this region of the sodium channel protein (L925I) that is tightly linked to pyrethroid resistance (Morin *et al.*, 2002). Four novel putative second-site mutations (D59G, E435K, C785R, and P1999L) are found together with the L1014F mutation in one or more strains of *B. germanica* that exhibit high levels of pyrethroid resistance (Liu *et al.*, 2000). Finally, a screen for pyrethroid resistance in strains of *D. melanogaster* having temperature-sensitive paralytic phenotypes that map to the *para* sodium channel identified four novel resistance-associated mutations: I253N, A1410V, A1494V, and M1524I (Pittendrigh *et al.*, 1997).

The identification of mutations associated with pyrethroid resistance in *para*-orthologous sodium channel gene sequences also extends to noninsect arthropod species. Sodium channel sequences from populations of *B. microplus* that exhibit very high levels of pyrethroid resistance contain the F1538I mutation (He *et al.*, 1999). Also, the *para*-orthologous sodium channel sequences obtained from populations of the mite *V. destructor* that are resistant to the pyrethroid fluvalinate contain four novel mutations (F1528L, P1596L, I1742V, and V1823I) (Wang *et al.*, 2002), which include the first resistance-associated mutations identified in homology domain IV (**Figure 3**).

Among the 26 unique sodium channel amino acid sequence polymorphisms associated so far with pyrethroid resistance, those occurring at five sites have been found as single mutations in resistant populations: Val410 (V410M in *H. virescens*), Met918 (M918V in *Bemisia tabaci*); Leu925 (L925I in *B. tabaci*), Leu1014 (L1014F in several species, L1014H in *H. virescens*, and L1014S in *C. pipiens* and *A. gambiae*); and Phe1538 (F1538I in *B. microplus*). Mutations at six sites (M918T in *M. domestica* and *H. irritans*; T929I in *P. xylostella*

and T929C in *F. occidentalis*; D59G, E435K, C785R, and P1999L in *B. germanica*) have been found in combination with the L1014F mutation in highly resistant strains and therefore have been hypothesized to function as second-site mutations that produce additive or synergistic enhancement the resistance caused by the L1014F mutation. The functional status of the remaining resistance-associated mutations is more ambiguous.

1.4.5. Functional Analysis of Resistance-Associated Mutations

The *X. laevis* oocyte expression system has been employed to characterize the effects of several resistance-associated mutations on the pyrethroid sensitivity and functional properties of expressed sodium channels. Most of these efforts have focussed on the mutations identified in sodium channels of *kdr* and *super-kdr* houseflies, but a more limited group of studies have also examined additional putative primary resistance mutations and certain putative secondary mutations. All of these studies involve the insertion of candidate mutations into wild-type sodium channel cDNAs, expression of wild-type and specifically mutated channels in oocytes, and direct comparison of the pharmacological properties of the expressed channels.

1.4.5.1. Functional analysis of the kdr and super-kdr mutations The effects of the L1014F mutation, the sodium channel gene mutation most commonly associated with knockdown resistance, on pyrethroid sensitivity have been examined in several sodium channel sequence contexts and with different pyrethroids as probes. Housefly Vssc1 sodium channels containing the L1014F substitution, coexpressed in oocytes with the *D. melanogaster* tipE protein, were approximately tenfold less sensitive to modification by cismethrin, a type I pyrethroid, than wild-type Vssc1/tipE channels (Smith *et al.*, 1997). The L1014F mutation also accelerated the rate of decay of cismethrin-induced sodium tail currents. Similar experiments with wild type and specifically mutated *D. melanogaster* para sodium channels coexpressed with the tipE protein and deltamethrin as the test pyrethroid confirmed and extended these findings (Vais *et al.*, 2000). Comparison of the actions of deltamethrin on wild-type para/tipE channels and channels containing the L1014F mutation reduced the sensitivity of expressed channels to deltamethrin approximately 17-fold and accelerated the rate of deltamethrin-induced tail current decay. More detailed analysis showed that mutated channels exhibited lower affinity for deltamethrin as well as a reduced availability of open channel states due to enhanced closed-state inactivation. The effects of the L1014F mutation on deltamethrin sensitivity were also examined using *B. germanica* para^CSMA sodium channels coexpressed with tipE (Tan *et al.*, 2002b). In this study, the L1014F mutation conferred approximately sixfold resistance to deltamethrin but, unlike the results obtained with Vssc1/tipE and para/tipE channels, the reduction in pyrethroid sensitivity of mutated para^CSMA/tipE channels was not accompanied by an acceleration of tail current decay.

The discovery that the rat $Na_v1.8$ (also called SNS or PN3) sodium channel isoform was highly sensitive to both type I and type II pyrethroids (Smith and Soderlund, 2001) provided the opportunity to examine the impact of the L1014F mutation in a sodium channel sequence environment that is otherwise substantially divergent (i.e., only ∼40% identical at the amino acid sequence level) from para-orthologous channels of insects. The L1014F mutation, introduced at the cognate conserved leucine residue of the rat $Na_v1.8$ channel, reduced the sensitivity of expressed channels to cismethrin more than tenfold and increased the rate of decay of the cismethrin-induced tail current. These effects were qualitatively identical to the effects of the L1014F mutation on the cismethrin sensitivity of Vssc1/tipE channels (Smith *et al.*, 1997).

The effects of the M918T mutation in the presence of the L1014F mutation, the combination of mutations associated with the *super-kdr* resistance trait of the housefly, have also been examined in both insect and mammalian sodium channel sequence contexts with both type I and type II pyrethroids as probes. Incorporation of the M918T and L1014F mutations into the Vssc1 protein gave rise to Vssc1/tipE channels that were completely insensitive to both cismethrin and [1*R*,*cis*,α*S*]-cypermethrin at the highest concentrations that could be achieved given the low aqueous solubility of these compounds (Lee *et al.*, 1999c). Similarly, rat $Na_v1.8$ sodium channels containing the M918T/L1014F double mutation were completely insensitive to modification by the highest attainable concentration of cismethrin (Soderlund and Lee, 2001). In the para/tipE sequence context the M918T/L1014F double mutation decreased the sensitivity of expressed channels to deltamethrin approximately 100-fold and also produced tail currents with monophasic, rather than biphasic, decay kinetics (Vais *et al.*, 2000). The latter effect was interpreted as evidence that the double mutation reduced the number of deltamethrin binding sites per channel from two to one.

The effects of the M918T single mutation on pyrethroid sensitivity has also been evaluated in multiple sequence contexts. Insertion of the M918T mutation into para/tipE channels gave channels that were twofold more resistant to deltamethrin than the doubly mutated (M918T/L1014F) channels; the M918T channels also exhibited monophasic tail current decay kinetics (Vais *et al.*, 2001). Insertion of the M918T mutation in Vssc1/tipE sodium channels significantly impaired sodium current expression in oocytes and produced channels that were not detectably modified by cismethrin (Lee *et al.*, 1999c). In rat $Na_v1.8$ channels, the M918T mutation caused a degree of resistance to cismethrin that was equivalent to that caused by the single L1014F mutation (Soderlund and Lee, 2001). These results suggest that the M918T mutation does not enhance resistance caused by the L1014F mutation in an additive or synergistic manner but rather provides a high level of resistance that supercedes the effect of the L1014F mutation alone.

In light of the profound reduction in pyrethroid sensitivity caused by the M918T single mutation in these assays, it is surprising that it has not been identified as a single mutation in pyrethroid-resistant insect populations. A likely explanation for this situation has emerged from studies of alternative exon usage in *para* and its orthologs (see Sections 1.2.1.2 and 1.2.2.1). In the *para* sequence of *D. melanogaster*, the Met918 residue occurs in mutually exclusive exons *c* and *d*, each of which is incorporated into a subpopulation of sodium channel sequence variants (Thackeray and Ganetzky, 1994). Exons homologous to exons *c* and *d* of *D. melanogaster* are also found among transcripts from the *para*-orthologous sodium channel gene of *H. virescens* (Park *et al.*, 1999) and *C. pipiens* (Martinez-Torres *et al.*, 1999a). In these situations, two independent point mutations (one each in exons *c* and *d*) would be required to insure that all channel variants contained the M918T substitution. In contrast, exon *c* in the housefly *Vssc1* gene contains a stop codon, so that all full-length *Vssc1* transcripts are derived solely from exon *d* (Lee *et al.*, 2002). Thus, a single point mutation in exon *d* of the housefly is sufficient to insure that all functional sodium channel splice variants in this species contain the M918T substitution. This interpretation suggests that the M918T mutation and other mutations at this site (such as the M918V mutation in *B. tabaci*) can only be selected in species in which alternative splicing at the exon *c/d* locus produces functional channels from only one of these alternative exons. Although consistent with available data on alternative splicing at the *Vssc1* locus in the housefly, this analysis does not explain the absence of the single M918T mutation in resistant strains of this species. It therefore remains possible that a functional deficit associated with the M918T mutation is somehow complemented or rescued by the presence of the L1014F mutation (Lee *et al.*, 1999c).

1.4.5.2. Functional analysis of other putative primary resistance mutations

The effects of the V410M mutation, a single mutation associated with knockdown resistance in some *H. virescens* populations, has been examined in para/tipE, Vssc1/tipE, and para^{CSMA}/tipE sodium channels expressed in oocytes. Insertion of the V410M mutation into para/tipE channels resulted in a >10-fold reduction in permethrin sensitivity coupled with a ~45-fold acceleration in the rate of sodium tail current decay (Zhao *et al.*, 2000). Similar results were obtained in Vssc1/tipE channels containing the V410M mutation, which decreased channel sensitivity to cismethrin by ~20-fold and accelerated the rate of tail current decay ~10-fold (Lee and Soderlund, 2001). In para^{CSMA}/tipE channels, the V410M mutation caused ~17-fold resistance to deltamethrin (Liu *et al.*, 2002). In contrast to the effects of this mutation on the kinetics of channel modification by permethrin and cismethrin, resistance to deltamethrin in para^{CSMA}/tipE channels containing the V410M mutation was not accompanied by a significant acceleration in tail current decay kinetics. Although Met410 occurs in a region of the sodium channel known to be involved in the binding and action of batrachotoxin, the V410M mutation did not affect the potency of batrachotoxin as a modifier of Vssc1/tipE sodium channels in oocytes (Lee and Soderlund, 2001).

The V410M mutation is the only resistance mutation that has been examined both in the oocyte expression system and in its native cellular context. Assays with cultured adult central neurons from *H. virescens* homozygous for the V410M mutation documented a ~21-fold reduction in permethrin sensitivity of sodium channels in the resistant strain (Lee *et al.*, 1999a). Although these authors did not report the time constants for the decay of permethrin-induced sodium tail current, inspection of their data suggests that reduced permethrin sensitivity in cultured neurons was apparently not correlated with a significant acceleration of tail current decay as was observed in some of the studies of the V410M mutation in oocytes (Zhao *et al.*, 2000; Lee and Soderlund, 2001). In cultured neurons, channels containing the V410M mutation were approximately 2.6-fold more sensitive to the polypeptide scorpion toxin LqhαIT (Lee *et al.*, 1999a).

The characterization of mutations at Leu1014 was expanded to include the L1014H mutation, another single mutation found in some knockdown-resistant *H. virescens* populations (Zhao *et al.*, 2000). Incorporation of the L1014H mutation into para/tipE sodium channels conferred >tenfold resistance to permethrin, a type I pyrethroid. Permethrin-induced sodium tail currents recorded from channels containing the L1014H mutation decayed 57-fold more rapidly than those carried by wild-type para/tipE channels.

The effects of the F1538I mutation, identified in pyrethroid-resistant populations of *B. microplus*, have been assessed only in rat Na$_v$1.4 sodium channels that were also mutated to enhance baseline pyrethroid sensitivity (Wang *et al.*, 2001). Deltamethrin caused use-dependent modification of these rat Na$_v$1.4 sodium channels. However, introduction of the F1538I mutation into these channels caused a loss of use-dependent modification by deltamethrin as well as a reduction in sensitivity to this compound.

1.4.5.3. Functional analysis of putative secondary mutations

The effects of the T929I mutation, a putative second-site resistance mutation in *P. xylostella* and *P. capitis*, has been examined in para/tipE sodium channels expressed in oocytes (Vais *et al.*, 2001). As a single mutation, the T929I substitution reduced the sensitivity of para/tipE channels to deltamethrin approximately tenfold. Deltamethrin-induced tail currents carried by channels containing the T929I mutation decayed rapidly with first-order kinetics. Channels containing both the T929I and L1014F mutations, the combination found in highly resistant populations of *Plutella xylostella* were more than 10 000-fold resistant to deltamethrin when compared to wild-type channels.

The effects of the E435K and C785R mutations, putative second-site mutations identified in several highly resistant *B. germanica* populations, on pyrethroid sensitivity have been examined in para^{CSMA}/tipE channels singly, together, and in combination with two primary resistance mutations. Channels containing the C785R mutation alone were identical to wild-type para^{CSMA}/tipE channels in their sensitivity to deltamethrin, but channels containing E435K single mutation and the E435K/C785R double mutation were more sensitive to deltamethrin than wild-type channels (Tan *et al.*, 2002b). When either the E435K or C785R mutations were combined with the L1014F resistance mutation, the resulting double mutants were approximately 20-fold less sensitive to deltamethrin than channels containing the single L1014F mutation and approximately 100-fold less sensitive than wild-type para^{CSMA}/tipE channels. Channels containing all three resistance mutations, a situation found in highly resistant *B. germanica* populations, were more than 500-fold less sensitive to deltamethrin than wild-type channels.

In a companion study, the E435K and C785R mutations were also evaluated as modifiers of deltamethrin resistance conferred by the V410M mutation (Liu *et al.*, 2002). Insertion of either the E435K or C785R mutation into para^{CSMA}/tipE channels containing the V410M mutation did not significantly affect the level of resistance conferred by the V410M mutation, but channels containing all three resistance mutations were approximately sixfold less sensitive to deltamethrin than channels containing only the V410M mutation and 100-fold less sensitive than wild-type para^{CSMA}/tipE channels.

1.5. Conclusions

1.5.1. Unique Features of Insect Sodium Channels

Although sodium channel function and structure is strongly conserved between evolutionarily divergent animal species, the molecular characterization of insect sodium channels has identified several distinctive features. First, in contrast to the family of sodium channel α subunit genes in mammalian species, insects appear to have only a single gene (*para* in *D. melanogaster* and its orthologs) that encodes voltage-gated, sodium-selective channels that are involved in action potential generation. However, insects are capable of generating a remarkable diversity of sodium channel variants from this locus by alternative exon usage and RNA editing. The number of sodium channel variants generated by these posttranscriptional modifications found in any given species appears to be only a fraction of the theoretical maximum, and the patterns of exon usage do not appear to be strongly conserved between species. So far there is scant information on the functional significance of alternative exon usage.

Insects also lack the sodium channel β subunit family of vertebrates and instead appear to use a novel family of glycoproteins, encoded by the *tipE* gene of *D. melanogaster* and its orthologs, as sodium channel auxiliary subunits. Genetic and cytochemical localization studies with *D. melanogaster* imply that the tipE protein is restricted in expression to the adult nervous system, suggesting that it may function as a modulator of sodium channel function and expression rather than as an obligatory auxiliary subunit. So far only two examples of *tipE*-like genes have been described, but their broader existence in insects is inferred by the heterologous

coassembly of the *D. melanogaster* tipE protein with the *para*-orthologous sodium channel gene of *B. germanica*. It will be of interest to determine the breadth of distribution of *tipE*-like genes in the Insecta and in other animal taxa.

Finally, insect genomes also contain a second sodium channel α subunit-like gene, *dsc1* of *D. melanogaster* and its orthologs, that is equally divergent from the *para*-like sodium channel genes of insects and the vertebrate sodium channel α subunit gene family. The dsc1 protein is clearly not the sodium channel principally responsible for nerve action potential generation or the primary target for sodium channel-directed insecticides, and it may lack the stringent selectivity for sodium ions that is a hallmark of typical voltage-sensitive sodium channels. The functional significance of dsc1 and homologous proteins has remained obscure, but recent studies point to an as-yet undefined role in chemosensory pathways.

1.5.2. Sodium Channels and Pyrethroid Resistance

Molecular and genetic studies over the past decade have provided convincing evidence that point mutations in insect voltage-sensitive sodium channel genes are the primary cause of knockdown resistance to pyrethroids. Genetic linkage experiments and targeted DNA sequencing studies have consistently identified resistance-associated mutations in sodium channel genes that are orthologous to the *para* gene of *D. melanogaster*. Heterologous expression assays have documented the ability of many of these mutations, introduced into wild-type insect sodium channels either alone or in the combinations found in highly resistant populations, to reduce the sensitivity of expressed channels to pyrethroids exemplifying the type I and type II structural classes. Finally, the magnitude of resistance conferred by these mutations in heterologous expression assays *in vitro* has consistently been found to be in good agreement with the magnitude of resistance observed in resistant insect populations carrying the same mutations. These results not only identify the molecular mechanism of the *kdr* trait but also provide important confirmation that sodium channels encoded by the *Vssc1* gene are the principal target site for the toxic actions of DDT analogs and pyrethroids. Although the actions of DDT and pyrethroids on sodium channels have been characterized in great depth and detail over the past three decades, numerous other sites of action for some or all of these insecticides have also been proposed (Soderlund and Bloomquist, 1989; Bloomquist, 1993). The toxicological impact of the knockdown resistance mutations in *para*-orthologous sodium channels implies that actions of DDT and pyrethroids at this target alone are sufficient to account for the toxic effects of these compounds in whole insects.

The search for sodium channel gene mutations in knockdown-resistant strains of various arthropod species has revealed the existence of resistance-associated polymorphisms at an unanticipated diversity of sites. Although not all of these polymorphisms have been shown to cause pyrethroid resistance, the multiplicity of sodium channel sequence polymorphisms associated with knockdown resistance contrasts with the situation for cyclodiene resistance in insects, which involves a mutation at a single amino acid residue in the subunit of the γ-aminobutyric acid receptor–chloride ionophore encoded by the *Rdl* gene in all cases that have been examined (ffrench-Constant, 1994). The diversity of sequence polymorphisms that are potentially involved in knockdown resistance poses a significant challenge for the use of this information in pyrethroid resistance monitoring and management.

The surprisingly large number of amino acid residues implicated as sites of resistance-causing mutations also poses challenges for the use of these mutations to map the pyrethroid binding site on the sodium channel. Not all of these residues are likely to interact physically with the pyrethroid molecule; mutations at some sites may indirectly alter the architecture of the pyrethroid site by altering or restricting the conformational flexibility of the sodium channel protein. As shown in **Figure 3**, most putative resistance mutations are located in the S5 and S6 helices and closely associated regions. Current models of sodium channel structure (Guy and Seetharamulu, 1986; Yellen, 1998; Lipkind and Fozzard, 2000) place the S5 and S6 domains in close proximity to each other adjacent to the inner pore of the channel. In this context, the clustering of a large number of resistance-associated mutations in the S5–S6 region of homology domain II implies an important role for this region of the sodium channel in determining, either directly or indirectly, the binding of pyrethroids. Recently, the crystal structure of a simple bacterial potassium channel was employed to generate a three-dimensional model of the S5–pore–S6 regions of the rat $Na_v1.4$ voltage-sensitive sodium channel (Lipkind and Fozzard, 2000). The use of such a model to identify the spatial relationships among the resistance-associated mutations in insect sodium channels may clarify the functional role of these residues in pyrethroid binding and action.

1.5.3. Future Exploitation of the Sodium Channel as an Insecticide Target

Despite the long and widespread use of DDT and then pyrethroids in the control of pests and disease vectors and the existence of knockdown resistance traits in populations of many pest and vector species, the pyrethroids remain an important and effective class of insecticides. Moreover, three key factors contribute to the continued value of the sodium channel as a target for future insecticides. First, toxicologically relevant sodium channels in insects are the products of a single gene and therefore exhibit conserved pharmacology in all neuronal tissues and insect life stages. Second, the value of sodium channel disruption as a mode of insecticidal action is amply demonstrated by the efficacy of sodium channel-directed natural toxins and insecticides. Third, toxin binding sites on the sodium channel other than the pyrethroid site (see **Table 1**) appear to be unaffected by mutations that confer pyrethroid resistance. In this context, each sodium channel binding domain can be envisioned as a separate potential target for insecticide discovery and development.

These factors in favor of the continued exploitation of sodium channels for insect control are counterbalanced by the evolutionary conservation of sodium channel structure, function, and pharmacology across animal taxa. Intrinsic selectivity of agents for insect sodium channels is uncommon, and the development of novel insecticides directed toward this target is complicated by the potential for toxicity to nontarget species. The example of pyrethroids is instructive in this regard: the notable safety of pyrethroids for humans is based principally on differential metabolism rather than differential target sensitivity, and the use of pyrethroids has been limited in some contexts by undesirable toxicity to aquatic vertebrate and invertebrate species. The development of indoxacarb, which exploits differential metabolic bioactivation as a mechanism of selective toxicity, illustrates the potential for the discovery of novel sodium channel-directed insecticides that exhibit acceptable safety and selectivity.

References

Adams, M., Celniker, S., Holt, R., *et al.*, 2000. The genome sequence of *Drosophila melanogaster*. *Science* 287, 2185–2195.

Anholt, R.R.H., Lyman, R.F., Mackay, T.F.C., **1996**. Effects of single *P*-element insertions on olfactory behavior in *Drosophila melanogaster*. *Genetics* 143, 293–301.

Baker, M.D., Wood, J.N., 2001. Involvement of Na$^+$ channels in pain pathways. *Trends Pharmacol. Sci. 22*, 27–31.

Bloomquist, J.R., 1988. Neurophysiological assays for the characterization and monitoring of pyrethroid resistance. In: Lunt, G.G. (Ed.), Neurotox '88: The Molecular Basis of Drug and Pesticide Action. Elsevier, Amsterdam, pp. 543–551.

Bloomquist, J.R., 1993. Neuroreceptor mechanisms in pyrethroid mode of action and resistance. In: Roe, M., Kuhr, R.J. (Eds.), Reviews in Pesticide Toxicology. Toxicology Communications, Raleigh, NC, pp. 181–226.

Bloomquist, J.R., Soderlund, D.M., 1988. Pyrethroid insecticides and DDT modify alkaloid-dependent sodium channel activation and its enhancement by sea anemone toxin. *Mol. Pharmacol. 33*, 543–550.

Bloomquist, J.R., Soderlund, D.M., Knipple, D.C., 1989. Knockdown resistance to dichlorodiphenyltrichloroethane and pyrethroid insecticides in the *nap^{ts}* mutant of *Drosophila melanogaster* is correlated with reduced neuronal sensitivity. *Arch. Insect Biochem. Physiol. 10*, 293–302.

Blumenthal, K.M., Seibert, A.L., 2003. Voltage-gated sodium channel toxins: poisons, probes, and future promise. *Cell Biochem. Biophys. 38*, 215–237.

Busvine, J.R., 1951. Mechanism of resistance to insecticide in houseflies. *Nature 168*, 193–195.

Cannon, S.C., 2000. Spectrum of sodium channel disturbances in the nondystrophic myotonias and periodic paralyses. *Kidney Int. 57*, 772–779.

Cantrell, A.R., Catterall, W.A., 2001. Neuromodulation of Na$^+$ channels: an unexpected form of cellular plasticity. *Nature Rev. 2*, 397–407.

Castella, C., Castells-Brooke, N., Bergé, J.-B., Pauron, D., 1997. Expression and distribution of voltage-sensitive sodium channels in pyrethroid-susceptible and pyrethroid-resistant *Musca domestica*. *Invert. Neurosci. 3*, 41–47.

Catterall, W.A., 1987. Common modes of drug action on sodium channels: local anesthetics, antiarrhythmics and anticonvulsants. *Trends Pharmacol. Sci. 8*, 57–65.

Catterall, W.A., 1988. Structure and function of voltage-sensitive ion channels. *Science 242*, 50–61.

Catterall, W.A., 1992. Cellular and molecular biology of voltage-gated sodium channels. *Physiol. Rev. 72*, S15–S48.

Catterall, W.A., 2000. From ionic currents to molecular mechanisms: structure and function of voltage-gated sodium channels. *Neuron 26*, 13–25.

Cestèle, S., Catterall, W.A., 2000. Molecular mechanisms of neurotoxin action on voltage-gated sodium channels. *Biochimie 82*, 883–892.

Clare, J.J., Tate, S.N., Nobbs, M., Romanos, M.A., 2000. Voltage-gated sodium channels as therapeutic targets. *Drug Discov. Today 11*, 506–520.

Cory, J.S., 1999. Use and risk assessment of genetically modified baculoviruses. In: Beadle, D.J. (Ed.), Progress in Neuropharmacology and Neurotoxicology of Pesticides and Drugs. Royal Society of Chemistry, Cambridge, pp. 117–123.

Cukierman, S., 1996. Regulation of voltage-dependent sodium channels. *J. Membrane Biol. 151*, 203–214.

Deecher, D.C., Payne, G.T., Soderlund, D.M., 1991. Inhibition of [^{3}H]batrachotoxinin A 20-α-benzoate binding to mouse brain sodium channels by the dihydropyrazole insecticide RH 3421. *Pestic. Biochem. Physiol. 41*, 265–273.

Deecher, D.C., Soderlund, D.M., 1991. RH 3421, an insecticidal dihydropyrazole, inhibits sodium channel-dependent sodium uptake into mouse brain preparations. *Pestic. Biochem. Physiol. 39*, 130–137.

Defaix, A., Lapied, B., 2001. Characterization of a novel low-voltage-activated, tetrodotoxin-sensitive sodium channel permeable to calcium in insect pacemaker neurosecretory cells. *J. Physiol. 533P*, 54P–55P.

Dong, K., 1997. A single amino acid change in the *para* sodium channel protein is associated with knockdown-resistance (*kdr*) to pyrethroid insecticides in the German cockroach. *Insect Biochem. Mol. Biol. 27*, 93–100.

Dong, K., Scott, J.G., 1994. Linkage of *kdr*-type resistance and the *para*-homologous sodium channel gene in German cockroaches (*Blattella germanica*). *Insect Biochem. Mol. Biol. 24*, 647–654.

Doyle, K.E., Knipple, D.C., 1991. PCR-based phylogenetic walking: isolation of *para*-homologous sodium channel gene sequences from seven insect species and an arachnid. *Insect Biochem. 21*, 689–696.

Elkins, T., Ganetzky, B., 1990. Conduction in the giant nerve fiber pathway in temperature-sensitive paralytic mutants of *Drosophila*. *J. Neurogenet. 6*, 207–219.

Fainzilber, M., Kofman, O., Zlotkin, E., Gordon, D., 1994. A new neurotoxin receptor site on sodium channels is identified by a conotoxin that affects sodium channel inactivation in molluscs and acts as an antagonist in rat brain. *J. Biol. Chem. 269*, 2574–2580.

Farnham, A.W., Murray, A.W.A., Sawicki, R.M., Denholm, I., White, J.C., 1987. Characterization of the structure–activity relationship of *kdr* and two variants of *super-kdr* to pyrethroids in the housefly (*Musca domestica* L.). *Pestic. Sci. 19*, 209–220.

Feng, G., Deak, P., Chopra, M., Hall, L.M., 1995a. Cloning and functional analysis of TipE, a novel membrane protein that enhances *Drosophila para* sodium channel function. *Cell 82*, 1001–1011.

Feng, G., Dèak, P., Kasbekar, D.P., Gil, D.W., Hall, L.M., 1995b. Cytogenetic and molecular localization of *tipE*: a gene affecting sodium channels in *Drosophila melanogaster*. *Genetics 139*, 1679–1688.

ffrench-Constant, R.H., 1994. The molecular and population genetics of cyclodiene insecticide resistance. *Insect Biochem. Mol. Biol. 24*, 335–345.

Forcioli, D., Frey, D., Frey, J.E., 2002. High nucleotide diversity in the *para*-like voltage-sensitive sodium channel gene sequence in the western flower thrips (Thysanoptera: Thripidae). *J. Econ. Entomol. 95*, 838–848.

Ganetzky, B., 1984. Genetic studies of membrane excitability in *Drosophila*: lethal interaction between two temperature-sensitive paralytic mutations. *Genetics* 108, 897–911.

Ganetzky, B., 1986. Neurogenetic analysis of *Drosophila* mutations affecting sodium channels: synergistic effects on viability and nerve conduction in double mutants involving *tip-E. J. Neurogenet.* 3, 19–31.

Germeraad, S.E., O'Dowd, D.K., Aldrich, R.W., 1992. Functional assay of a putative *Drosophila* sodium channel gene in homozygous deficiency neurons. *J. Neurogenet.* 8, 1–16.

Goldin, A.L., 1992. Maintenance of *Xenopus laevis* and oocyte injection. *Meth. Enzymol.* 207, 266–297.

Goldin, A.L., 2001. Resurgence of sodium channel research. *Annu. Rev. Physiol.* 63, 871–894.

Goldin, A.L., 2002. Evolution of voltage-gated Na$^+$ channels. *J. Exp. Biol.* 205, 575–584.

Goldin, A.L., Barchi, R.L., Caldwell, J.H., Hofmann, F., Howe, J.R., *et al.*, 2000. Nomenclature of voltage-gated sodium channels. *Neuron 28*, 365–368.

Gonoi, T., Ashida, K., Feller, D., Schmidt, J., Fujiwara, M., *et al.*, 1986. Mechanism of action of a polypeptide neurotoxin from the coral *Goniopora* in mouse neuroblastoma cells. *Mol. Pharmacol.* 29, 347–354.

Gonoi, T., Ohizumi, Y., Kobayashi, J., Nakamura, H., Catterall, W.A., 1987. Action of a polypeptide toxin from the marine snail *Conus striatus* on voltage-sensitive sodium channels. *Mol. Pharmacol.* 32, 691–698.

Grolleau, F., Lapied, B., 2000. Dorsal unpaired median neurones in the insect central nervous system: towards a better understanding of the ionic mechanisms underlying spontaneous electrical activity. *J. Exp. Biol.* 203, 1633–1648.

Grubs, R.E., Adams, P.M., Soderlund, D.M., (1988). Binding of [^{3}H]saxitoxin to head membrane preparations from susceptible and knockdown-resistant houseflies. *Pestic. Biochem. Physiol.* 32, 217–223.

Guerrero, F.D., Jamroz, R.C., Kammlah, D., Kunz, S.E., 1997. Toxicological and molecular characterization of pyrethroid-resistant horn flies, *Hematobia irritans*: identification of *kdr* and *super-kdr* point mutations. *Insect Biochem. Mol. Biol.* 27, 745–755.

Guy, H.R., Seetharamulu, P., 1986. Molecular model of the action potential sodium channel. *Proc. Natl Acad. Sci. USA 83*, 508–512.

Hall, L.M., Kasbekar, D.P., 1989. *Drosophila* sodium channel mutations affect pyrethroid sensitivity. In: Narahashi, T., Chambers, J.E. (Eds.), Insecticide Action: From Molecule to Organism. Plenum, New York, pp. 99–114.

Hanrahan, C.J., Palladino, M.J., Ganetzky, B., Reenan, R.A., 2000. RNA editing of the *Drosophila para* Na$^+$ channel transcript: evolutionary conservation and developmental regulation. *Genetics 155*, 1149–1160.

Harder, H.H., Riley, S.L., McCann, S.F., Irving, S.N., 1996. DPX-MP062: a novel broad-spectrum, environmentally soft, insect control compound. *Brighton Crop Protection Conference: Pests and Diseases*, 449–454.

Harris-Warwick, R.M., 2000. Ion channels and receptors: molecular targets for behavioral evolution. *J. Comp. Physiol. A 186*, 605–616.

He, H., Chen, A.C., Davey, R.B., Ivie, G.W., George, J.E., 1999. Identification of a point mutation in the *para*-type sodium channel gene from a pyrethroid-resistant cattle tick. *Biochem. Biophys. Res. Commun. 261*, 558–561.

Head, D.J., McCaffery, A.R., Callaghan, A., 1998. Novel mutations in the *para*-homologous sodium channel gene associated with phenotypic expression of nerve insensitivity resistance to pyrethroids in heliothine lepidoptera. *Insect Mol. Biol. 7*, 191–196.

Heinemann, S.H., Terlau, H., Stühmer, W., Imoto, K., Numa, S., 1992. Calcium channel characteristics conferred on the sodium channel by single mutations. *Nature 356*, 441–443.

Hille, B., 2001. Ion Channels of Excitable Membranes. Sinauer, Sunderland, MA.

Hodges, D.D., Lee, D., Preston, C.F., Boswell, K., Hall, L.M., *et al.*, 2002. *tipE* regulates Na$^+$-dependent repetitive firing in *Drosophila* neurons. *Mol. Cell. Neurosci.* 19, 402–416.

Hong, C.-S., Ganetzky, B., 1994. Spatial and temporal expression patterns of two sodium channel genes in *Drosophila. J. Neurosci.* 14, 5160–5169.

Hoopengardner, B., Bhalla, T., Staber, C., Reenan, R., 2003. Nervous system targets of RNA editing identified by comparative genomics. *Science 301*, 832–836.

Ingles, P.J., Adams, P.M., Knipple, D.C., Soderlund, D.M., 1996. Characterization of voltage-sensitive sodium channel gene coding sequences from insecticide-susceptible and knockdown-resistant housefly strains. *Insect Biochem. Mol. Biol. 26*, 319–326.

Jackson, F.R., Wilson, S.D., Hall, L.M., 1986. The *tip-E* mutation of *Drosophila* decreases saxitoxin binding and interacts with other mutations affecting nerve membrane excitability. *J. Neurogenet.* 3, 1–17.

Jackson, F.R., Wilson, S.D., Strichartz, G.R., Hall, L.M., 1984. Two types of mutants affecting voltage-sensitive sodium channels in *Drosophila melanogaster. Nature* 308, 189–191.

Kasbekar, D.P., Hall, L.M., 1988. A *Drosophila* mutation that reduces sodium channel number confers resistance to pyrethroid insecticides. *Pestic. Biochem. Physiol.* 32, 135–145.

Knipple, D.C., Doyle, K.E., Marsella-Herrick, P.A., Soderlund, D.M., 1994. Tight genetic linkage between the *kdr* insecticide resistance trait and a voltage-sensitive sodium channel gene in the housefly. *Proc. Natl Acad. Sci. USA 91*, 2483–2487.

Knipple, D.C., Payne, L.L., Soderlund, D.M., 1991. PCR-generated conspecific sodium channel gene probe for the housefly. *Arch. Insect Biochem. Physiol.* 16, 45–53.

Kulkarni, N.H., Yamamoto, A.H., Robinson, K.O., Mackay, T.F.C., Anholt, R.R.H., 2002. The DSC1 channel, encoded by the *smi60E* locus, contributes to odor-guided behavior in *Drosophila melanogaster*. *Genetics 161*, 1507–1516.

Kulkarni, S.J., Padhye, A., **1982**. Temperature-sensitive paralytic mutations on the second and third chromosomes of *Drosophila melanogaster*. *Genet. Res. 40*, 191–199.

Lee, D., Park, Y., Brown, T.M., Adams, M.E., **1999a**. Altered properties of neuronal sodium channels associated with genetic resistance to pyrethroids. *Mol. Pharmacol. 55*, 584–593.

Lee, S.H., Dunn, J.B., Clark, J.M., Soderlund, D.M., **1999b**. Molecular analysis of *kdr*-like resistance in a permethrin-resistant strain of Colorado potato beetle. *Pestic. Biochem. Physiol. 63*, 63–75.

Lee, S.H., Gao, J.-R., Yoon, K.S., Mumcuoglu, K.Y., Taplin, D., *et al.*, **2003**. Sodium channel mutations associated with knockdown resistance in the human head louse, *Pediculus capitis* (De Geer). *Pestic. Biochem. Physiol. 75*, 79–91.

Lee, S.H., Ingles, P.J., Knipple, D.C., Soderlund, D.M., **2002**. Developmental regulation of alternative exon usage in the housefly *Vssc1* sodium channel gene. *Invert. Neurosci. 4*, 125–133.

Lee, S.H., Smith, T.J., Ingles, P.J., Soderlund, D.M., **2000a**. Cloning and functional characterization of a putative sodium channel auxiliary subunit gene from the housefly (*Musca domestica*). *Insect Biochem. Mol. Biol. 30*, 479–487.

Lee, S.H., Smith, T.J., Knipple, D.C., Soderlund, D.M., **1999c**. Mutations in the housefly *Vssc1* sodium channel gene associated with *super-kdr* resistance abolish the pyrethroid sensitivity of Vssc1/tipE sodium channels expressed in *Xenopus* oocytes. *Insect Biochem. Mol. Biol. 29*, 185–194.

Lee, S.H., Soderlund, D.M., **2001**. The V410M mutation associated with pyrethroid resistance in *Heliothis virescens* reduces the pyrethroid sensitivity of housefly sodium channels expressed in *Xenopus* oocytes. *Insect Biochem. Mol. Biol. 31*, 19–29.

Lee, S.H., Yoon, K.-S., Williamson, M.S., Goodson, S.J., Takano-Lee, M., *et al.*, **2000b**. Molecular analysis of *kdr*-like resistance in permethrin-resistant strains of head lice, *Pediculus capitis*. *Pestic. Biochem. Physiol. 66*, 130–143.

Lester, H.A., **1988**. Heterologous expression of excitability proteins: route to more specific drugs? *Science 241*, 1057–1063.

Lilly, M., Carlson, J., **1989**. *smellblind*: a gene required for *Drosophila* olfaction. *Genetics 124*, 293–302.

Lilly, M., Kreber, R., Ganetzky, B., Carlson, J.R., **1994a**. Evidence that the *Drosophila* olfactory mutant *smellblind* defines a novel class of sodium channel mutation. *Genetics 136*, 1087–1096.

Lilly, M., Riesgo-Escovar, J., Carlson, J., **1994b**. Developmental analysis of *smellblind* mutants: evidence for the role of sodium channels in *Drosophila* development. *Devel. Biol. 162*, 1–8.

Lipkind, G.M., Fozzard, H.A., **2000**. KcsA crystal structure as framework for a molecular model of the Na^+ channel pore. *Biochemistry 39*, 8161–8170.

Littleton, J.T., Ganetzky, B., **2000**. Ion channels and synaptic organization: analysis of the *Drosophila* genome. *Neuron 26*, 35–43.

Liu, Z., Chung, I., Dong, K., **2001**. Alternative splicing of the *BSC1* gene generates tissue-specific isoforms in the German cockroach. *Insect Biochem. Mol. Biol. 31*, 703–713.

Liu, Z., Tan, J., Valles, S.M., Dong, K., **2002**. Synergistic interaction between two cockroach sodium channel mutation and a tobacco budworm sodium channel mutation in reducing channel sensitivity to a pyrethroid insecticide. *Insect Biochem. Mol. Biol. 32*, 397–404.

Liu, Z., Valles, S.M., Dong, K., **2000**. Novel point mutations in the German cockroach *para* sodium channel gene are associated with knockdown resistance (*kdr*) to pyrethroid insecticides. *Insect Biochem. Mol. Biol. 30*, 991–997.

Lombet, A., Mourre, C., Lazdunski, M., **1988**. Interactions of insecticides of the pyrethroid family with specific binding sites on the voltage-dependent sodium channel from mammalian brain. *Brain Res. 459*, 44–53.

Loughney, K., Kreber, R., Ganetzky, B., **1989**. Molecular analysis of the *para* locus, a sodium channel gene in *Drosophila*. *Cell 58*, 1143–1154.

Martinez-Torres, D., Chandre, F., Williamson, M.S., Darriet, F., Bergé, J.B., *et al.*, **1998**. Molecular characterization of pyrethroid knockdown resistance (*kdr*) in the major malaria vector *Anopheles gambiae s.s. Insect Mol. Biol. 7*, 179–184.

Martinez-Torres, D., Chevillon, C., Brun-Barale, A., Bergé, J.B., Pasteur, N., *et al.*, **1999a**. Voltage-dependent Na^+ channels in pyrethroid-resistant *Culex pipiens* L mosquitoes. *Pestic. Sci. 55*, 1012–1020.

Martinez-Torres, D., Foster, S.P., Field, L.M., Devonshire, A.L., Williamson, M.S., **1999b**. A sodium channel point mutation is associated with resistance to DDT and pyrethroid insecticides in the peach-potato aphid, *Myzus persicae* (Sulzer) (Hemiptera: Aphididae). *Insect Mol. Biol. 8*, 339–346.

Milani, R., **1954**. Comportamento mendeliano della resistenza alla azione abbattante del DDT: correlazione tran abbattimento e mortalita in *Musca domestica* L. *Riv. Parassitol. 15*, 513–542.

Miyazaki, M., Ohyama, K., Dunlap, D.Y., Matsumura, F., **1996**. Cloning and sequencing of the *para*-type sodium channel gene from susceptible and *kdr*-resistant German cockroaches (*Blattella germanica*) and housefly (*Musca domestica*). *Mol. Gen. Genet. 252*, 61–68.

Morin, S., Williamson, M.S., Goodson, S.J., Brown, J.K., Tabashnik, B.E., *et al.*, **2002**. Mutations in the *Bemisia tabaci para* sodium channel gene associated with resistance to a pyrethroid plus organophosphate mixture. *Insect Biochem. Mol. Biol. 32*, 1781–1791.

Narahashi, T., **1996**. Neuronal ion channels as the target sites of insecticides. *Pharmacol. Toxicol. 78*, 1–14.

Nelson, J.C., Wyman, R.J., **1989**. Examination of paralysis in *Drosophila* temperature-sensitive paralytic mutations affecting sodium channels: a proposed mechanism of paralysis. *J. Neurobiol. 21*, 453–469.

Noda, M., Shimizu, S., Tanabe, T., Takai, T., Kayano, T., *et al.*, **1984**. Primary structure of *Electrophorus electricus* sodium channel deduced from cDNA sequence. *Nature 312*, 121–127.

O'Dowd, D.K., Gee, J.R., Smith, M.A., **1995**. Sodium current density correlates with expression of specific alternatively spliced sodium channel mRNAs in single neurons. *J. Neurosci. 15*, 4005–4012.

O'Dowd, D.K., Germeraad, S.E., Aldrich, R.W., **1989**. Alteration in the expression and gating of *Drosophila* sodium channels by mutations in the *para* gene. *Neuron 2*, 1301–1311.

Okamoto, H., Sakai, K., Goto, S., Takasu-Ishakawa, E., Hotta, Y., **1987**. Isolation of *Drosophila* genomic clones homologous to the eel sodium channel gene. *Proc. Japan Acad. Ser. B 63*, 284–288.

Oppenoorth, F.J., **1985**. Biochemistry and genetics of insecticide resistance. In: Kerkut, G.A., Gilbert, L.I. (Eds.), Comprehensive Insect Physiology Biochemistry and Pharmacology, vol. 12. Pergamon, Oxford, pp. 731–773.

Ottea, J.A., Payne, G.T., Soderlund, D.M., **1989**. Activation of sodium channels and inhibition of [^{3}H]batrachotoxinin A-20-α-benzoate binding by an *N*-alkylamide neurotoxin. *Mol. Pharmacol. 36*, 280–284.

Ottea, J.A., Payne, G.T., Soderlund, D.M., **1990**. Action of insecticidal *N*-alkylamides at site 2 of the voltage-dependent sodium channel. *J. Agric. Food Chem. 38*, 1724–1728.

Park, Y., Taylor, M.F.J., **1997**. A novel mutation L1029H in sodium channel gene *hscp* associated with pyrethroid resistance for *Heliothis virescens* (Lepidoptera: Noctuidae). *Insect Biochem. Mol. Biol. 27*, 9–13.

Park, Y., Taylor, M.F.J., Feyereisen, R., **1997**. A valine421 to methionine mutation in IS6 of the *hscp* voltage-gated sodium channel associated with pyrethroid resistance in *Heliothis virescens* F. *Biochem. Biophys. Res. Commun. 239*, 688–691.

Park, Y., Taylor, M.F.J., Feyereisen, R., **1999**. Voltage-gated sodium channel genes *hscp* and *hDSC1* of *Heliothis virescens* F. genomic organization. *Insect Mol. Biol. 8*, 161–170.

Pauron, D., Barhanin, J., Amichot, M., Pralavorio, M., Berge, J.-B., *et al.*, **1989**. Pyrethroid receptor in the insect Na$^+$ channel: alteration of its properties in pyrethroid-resistant flies. *Biochemistry 28*, 1673–1677.

Pittendrigh, B., Reenan, R., ffrench-Constant, R.H., Ganetzky, B., **1997**. Point mutations in the *Drosophila* sodium channel gene *para* associated with resistance to DDT and pyrethroid insecticides. *Mol. Gen. Genet. 256*, 602–610.

Plummer, N.W., McBurney, M.W., Meisler, M.H., **1997**. Alternative splicing of the sodium channel *SCN8A* predicts a truncated protein in fetal brain and nonneuronal cells. *J. Biol. Chem. 272*, 24008–24015.

Ramaswami, M., Tanouye, M.A., **1989**. Two sodium-channel genes in *Drosophila*: implications for channel-diversity. *Proc. Natl Acad. Sci. USA 86*, 2079–2082.

Ranson, H., Jensen, B., Vulule, J.M., Wang, X., Hemingway, J., *et al.*, **2000**. Identification of a point mutation in the votage-gated sodium channel gene of Kenyan *Anopheles gambiae* associated with resistance to DDT and pyrethroids. *Insect Mol. Biol. 9*, 491–497.

Rossignol, D.P., **1988**. Reduction in number of nerve membrane sodium channels in pyrethroid resistant houseflies. *Pestic. Biochem. Physiol. 32*, 146–152.

Saiki, R.K., Gelfand, D.H., Stoffel, S., Scharf, S.J., Higuchi, R., *et al.*, **1988**. Primer-directed enzymatic amplification of DNA with a thermostable DNA polymerase. *Science 239*, 487–491.

Sakai, K., Okamoto, H., Hotta, Y., **1989**. Pharmacological characterization of sodium channels in the primary culture of individual *Drosophila* embryos: neurons of a mutant deficient in a putative sodium channel gene. *Cell Diff. Devel. 26*, 107–118.

Salgado, V.L., **1990**. Mode of action of insecticidal dihydropyrazoles: selective block of impulse generation in sensory nerves. *Pestic. Sci. 28*, 389–411.

Salgado, V.L., **1992**. Slow voltage-dependent block of sodium channels in crayfish nerve by dihydropyrazole insecticides. *Mol. Pharmacol. 41*, 120–126.

Salkoff, L., Butler, A., Wei, A., Scavarda, N., Giffen, N., *et al.*, **1987**. Genomic organization and deduced amino acid sequence of a putative sodium channel gene in *Drosophila*. *Science 237*, 744–749.

Sattelle, D.B., Leech, C.A., Lummis, S.C.R., Harrison, B.J., Robinson, H.P.C., *et al.*, **1988**. Ion channel properties of insects susceptible and resistant to insecticides. In: Lunt, G.G. (Ed.), Neurotox '88: The Molecular Basis of Drug and Pesticide Action. Elsevier, Amsterdam, pp. 563–582.

Sattelle, D.B., Yamamoto, D., **1988**. Molecular targets of pyrethroid insecticides. *Adv. Insect Physiol. 20*, 147–213.

Sawicki, R.M., **1978**. Unusual response of DDT-resistant houseflies to carbinol analogues of DDT. *Nature 275*, 443–444.

Schuler, T.H., Martinez-Torres, D., Thompson, A.J., Denholm, I., Devonshire, A.L., *et al.*, **1998**. Toxicological, electrophysiological, and molecular characterisation of knockdown resistance to pyrethroid insecticides in the diamondback moth, *Plutella xylostella* (L.). *Pestic. Biochem. Physiol. 59*, 169–192.

Shichor, I., Zlotkin, E., Ilan, N., Chikashvili, D., Stühmer, W., *et al.*, **2002**. Domain 2 of *Drosophila para* voltage-gated sodium channel confers insect properties to a rat brain channel. *J. Neurosci. 22*, 4364–4371.

Siddiqi, O., Benzer, S., **1976**. Neurophysiological defects in temperature-sensitive paralytic mutants of *Drosophila melanogaster*. *Proc. Natl Acad. Sci. USA 73*, 3253–3257.

Smith, T.J., Ingles, P.J., Soderlund, D.M., **1998**. Actions of the pyrethroid insecticides cismethrin and cypermethrin on housefly *Vssc1* sodium channels expressed in *Xenopus* oocytes. *Arch. Insect Biochem. Physiol.* 38, 126–136.

Smith, T.J., Lee, S.H., Ingles, P.J., Knipple, D.C., Soderlund, D.M., **1997**. The L1014F point mutation in the housefly *Vssc1* sodium channel confers knockdown resistance to pyrethroids. *Insect Biochem. Mol. Biol.* 27, 807–812.

Smith, T.J., Soderlund, D.M., **2001**. Potent actions of the pyrethroid insecticides cismethrin and cypermethrin on rat tetrodotoxin-resistant peripheral nerve (SNS/PN3) sodium channels expressed in *Xenopus* oocytes. *Pestic. Biochem. Physiol.* 70, 52–61.

Soderlund, D.M., **1995**. Mode of action of pyrethrins and pyrethroids. In: Casida, J.E., Quistad, G.B. (Eds.), Pyrethrum Flowers: Production, Chemistry, Toxicology, and Uses. Oxford University Press, New York, pp. 217–233.

Soderlund, D.M., **1997**. Molecular mechanisms of insecticide resistance. In: Sjut, V. (Ed.), Molecular Mechanisms of Resistance to Agrochemicals. Springer, Berlin, pp. 21–56.

Soderlund, D.M., Bloomquist, J.R., **1989**. Neurotoxic actions of pyrethroid insecticides. *Annu. Rev. Entomol.* 34, 77–96.

Soderlund, D.M., Bloomquist, J.R., **1990**. Molecular mechanisms of insecticide resistance. In: Roush, R.T., Tabashnik, B.E. (Eds.), Pesticide Resistance in Arthropods. Chapman and Hall, New York, pp. 58–96.

Soderlund, D.M., Knipple, D.C., **1999**. Knockdown resistance to DDT and pyrethroids in the housefly (Diptera: Muscidae): from genetic trait to molecular mechanism. *Ann. Entomol. Soc. America* 92, 909–915.

Soderlund, D.M., Knipple, D.C., **2003**. The molecular biology of knockdown resistance to pyrethroid insecticides. *Insect Biochem. Mol. Biol.* 33, 563–577.

Soderlund, D.M., Lee, S.H., **2001**. Point mutations in homology domain II modify the sensitivity of rat $Na_v1.8$ sodium channels to the pyrethroid cismethrin. *Neurotoxicology* 22, 755–765.

Stern, M., Kreber, R., Ganetzky, B., **1990**. Dosage effects of a *Drosophila* sodium channel gene on behavior and axonal excitability. *Genetics* 124, 133–143.

Strong, M., Chandy, K.G., Gutman, G.A., **1993**. Molecular origin of voltage-sensitive ion channel genes: on the origins of electrical excitability. *Mol. Biol. Evol.* 10, 221–242.

Stühmer, W., **1992**. Electrophysiological recording from *Xenopus* oocytes. *Methods Enzymol.* 207, 319–339.

Suzuki, D.T., Grigliatti, T., Williamson, R., **1971**. Temperature-sensitive mutations in *Drosophila melanogaster*, VII. A mutation (*para^ts*) causing reversible adult paralysis. *Proc. Natl Acad. Sci. USA* 68, 890–893.

Suzuki, N., Wu, C.-F., **1984**. Altered sensitivity to sodium channel-specific neurotoxins in cultured neurons from temperature-sensitive paralytic mutants of *Drosophila*. *J. Neurogenet.* 1, 225–238.

Tan, J., Liu, Z., Nomura, Y., Goldin, A.L., Dong, K., **2002a**. Alternative splicing of an insect sodium channel gene generates pharmacologically distinct sodium channels. *J. Neurosci.* 22, 5300–5309.

Tan, J., Liu, Z., Tsai, T.-D., Valles, S.M., Goldin, A.L., *et al.*, **2002b**. Novel sodium channel gene mutations in *Blattella germanica* reduce the sensitivity of expressed channels to deltamethrin. *Insect Biochem. Mol. Biol.* 32, 445–454.

Taylor, M.F.J., Heckel, D.G., Brown, T.M., Kreitman, M.E., Black, B., **1993**. Linkage of pyrethroid insecticide resistance to a sodium channel locus in the tobacco budworm. *Insect Biochem. Mol. Biol.* 23, 763–775.

Terlau, H., Stocker, M., Shon, K.-J., McIntosh, J.M., Olivera, B.M., **1996**. μO-conotoxin MrVIA inhibits mammalian sodium channels, but not through site I. *J. Neurophysiol.* 76, 1423–1429.

Thackeray, J.R., Ganetzky, B., **1994**. Developmentally regulated alternative splicing generates a complex array of *Drosophila para* sodium channel isoforms. *J. Neurosci.* 14, 2569–2578.

Thackeray, J.R., Ganetzky, B., **1995**. Conserved alternative splicing patterns and splicing signals in the *Drosophila* sodium channel gene *para*. *Genetics* 141, 203–214.

Trainer, V.L., McPhee, J.C., Boutelet-Bochan, H., Baker, C., Scheuer, T., *et al.*, **1997**. High affinity binding of pyrethroids to the α subunit of brain sodium channels. *Mol. Pharmacol.* 51, 651–657.

Tseng-Crank, J., Pollock, J.A., Hayashi, I., Tanouye, M.A., **1991**. Expression of ion channel genes in *Drosophila*. *J. Neurogenet.* 7, 229–239.

Vais, H., Williamson, M.S., Devonshire, A.L., Usherwood, P.N.R., **2001**. The molecular interactions of pyrethroid insecticides with insect and mammalian sodium channels. *Pest. Mgt. Sci.* 57, 877–888.

Vais, J., Williamson, M.S., Goodson, S.J., Devonshire, A.L., Warmke, J.W., *et al.*, **2000**. Activation of *Drosophila* sodium channels promotes modification by deltamethrin: reductions in affinity caused by knock-down resistance mutations. *J. Gen. Physiol.* 115, 305–318.

Wang, R., Huang, Z.Y., Dong, K., **2003**. Molecular characterization of an arachnic sodium channel gene from the varroa mite (*Varroa destructor*). *Insect Biochem. Mol. Biol.* 33, 733–739.

Wang, R., Liu, Z., Dong, K., Elzen, P.J., Pettis, J., *et al.*, **2002**. Association of novel mutations in a sodium channel gene with fluvalinate resistance in the mite, *Varroa destructor*. *J. Apicult. Res.* 40, 17–25.

Wang, S.-Y., Barile, M., Wang, G.K., **2001**. A phenylalanine residue at segment D3-S6 in $Na_v1.4$ voltage-gated Na^+ channels is critical for pyrethroid action. *Mol. Pharmacol.* 60, 620–628.

Wang, S.-Y., Wang, G.K., **2003**. Voltage-gated sodium channels as primary targets of diverse lipid-soluble neurotoxins. *Cell. Signal.* 15, 151–159.

Warmke, J.W., Reenan, R.A.G., Wang, P., Qian, S., Arena, J.P., *et al.*, **1997**. Functional expression of *Drosophila para* sodium channels: modulation by the membrane protein *tipE* and toxin pharmacology. *J. Gen. Physiol. 110*, 119–133.

Williamson, M.S., Denholm, I., Bell, C.A., Devonshire, A.L., **1993**. Knockdown resistance (*kdr*) to DDT and pyrethroid insecticides maps to a sodium channel gene locus in the housefly (*Musca domestica*). *Mol. Gen. Genet. 240*, 17–22.

Williamson, M.S., Martinez-Torres, D., Hick, C.A., Devonshire, A.L., **1996**. Identification of mutations in the housefly *para*-type sodium channel gene associated with knockdown resistance (*kdr*) to pyrethroid insecticides. *Mol. Gen. Genet. 252*, 51–60.

Wing, K.D., Schnee, M.E., Sacher, M., Connair, M., **1998**. A novel oxadiazine insecticide is bioactivated in lepidopteran larvae. *Arch. Insect Biochem. Physiol. 37*, 91–103.

Wu, C.-F., Ganetzky, B., **1980**. Genetic alteration of nerve membrane excitability in temperature-sensitive paralytic mutants of *Drosophila melanogaster*. *Nature 286*, 814–816.

Yellen, G., **1998**. The moving parts of voltage-gated ion channels. *Q. Rev. Biophys. 31*, 239–295.

Yu, F.H., Catterall, W.A., **2003**. Overview of the voltage-gated sodium channel family. *Genome Biol. 4*, 207.1–207.7.

Zhao, Y., Park, Y., Adams, M.E., **2000**. Functional and evolutionary consequences of pyrethroid resistance mutations in S6 transmembrane segments of a voltage-gated sodium channel. *Biochem. Biophys. Res. Commun. 278*, 516–521.

Zlotkin, E., **1999**. The insect voltage-gated sodium channel as target of insecticides. *Annu. Rev. Entomol. 44*, 429–455.

Zlotkin, E., Moskowitz, H., Hermann, R., Pelhate, M., Gordon, D., **1995**. Insect sodium channel as the target for insect-selective neurotoxins from scorpion venom. In: Clark, J.M. (Ed.), Molecular Action of Insecticides on Ion Channels. American Chemical Society, Washington, DC, pp. 57–85.

A1 Addendum: Recent Progress in Insect Sodium Channel Research

K Dong, Michigan State University, East Lansing, MI

© 2010, Elsevier BV. All Rights Reserved.

In Chapter 1, David Soderlund provided a comprehensive review of insect sodium channels (Soderlund, 2005). The intent of this addendum is to provide a brief update to the chapter.

A1.1. How Many Sodium Channel Genes Are in a Given Insect Species?

It is well established that *para* in *D. melanogaster* and *para*-like genes in other insects encode voltage-gated sodium channels. In addition to *para*, the genome of *D. melanogaster* contains another sodium channel-like gene, *DSC1* (Littleton and Ganetzky, 2000). However, functional characterization of *BSC1* (a DSC1 ortholog from the German cockroach) in *Xenopus* oocytes showed that this gene does not encode a sodium channel; rather, it encodes a novel voltage-gated calcium-selective cation channel (Zhou *et al.*, 2004). A more recent characterization of the DSC1 channel in oocytes reveals the functional characteristics of the DSC1 channel similar to that of the BSC1 channel (Zhang and Dong, unpublished data). Therefore, only one insect sodium channel gene is present in *D. melanogaster*, and the role of DSC1 and DSC1-like channels in insect neurophysiology largely remains to be explored.

In *Periplaneta americana*, in addition to the *para* ortholog *PaNav1*, a second *para*-like gene, *Periplaneta americana four P-domain channel* (*PaFPC*), was recently isolated by Moignot *et al.* (2009). PaNav1 and PaFPC share 57% amino acid sequence identity. Both the inner (DEKA motif) and outer (EEMD motif) rings that determine Na⁺ selectivity are present in the PaFPC protein (Moignot *et al.*, 2009). However, the MFM motif in the third linker connecting domains III and IV, which is critical to fast inactivation of sodium channels, is not conserved in PaFPC. No *PaFPC* ortholog has been found in any of the currently sequenced insect genomes (Moignot *et al.*, 2009), and the functionality

of the PaFPC channel remains to be determined. Thus, currently it seems that most insect species have only a single sodium channel gene.

A1.2. More Auxiliary Subunits of Insect Sodium Channels?

TipE enhances the expression of insect sodium channels heterologously expressed in *Xenopus* oocytes and is thought to be an insect sodium channel auxiliary subunit functionally analogous to vertebrate sodium channel β subunits (see Section 1.2.1.3 in Soderlund, 2005). Interestingly, four *TipE*-homologous (*TEH*) genes have been identified in *D. melanogaster* (Derst *et al.*, 2006), and TEH orthologs have been found in the sequenced genomes of other insects. Like TipE, TEH1–4 proteins contain two transmembrane segments connected by a large extracellular linker and cytosolic N- and C-termini. Similarly, TEH1–3 significantly enhances the expression of sodium currents in oocytes, and TEH1 modulates steady-state inactivation and recovery from fast inactivation. The presence of multiple putative sodium channel auxiliary subunits offers exciting possibilities as to their role in neuronal plasticity (as seen with mammalian sodium channel auxiliary subunits).

A1.3. Generation of Sodium Channel Diversity by Alternative Splicing and RNA Editing

As discussed in Soderlund (2005), extensive alternative splicing has been reported in insect sodium channel gene transcripts. Recent cDNA sequence analysis of sodium channel genes from *Anopheles gambiae* (Davies *et al.*, 2007), *Plutella xylostella* (Sonoda *et al.*, 2008), and *Bombyx mori* (Shao *et al.*, 2008) reveal that most splicing sites appear to be conserved, and suggest the potential universality of alternative splicing in insect sodium channel genes. Additionally, recent evidence has shed light on the functional consequences of alternative splicing. Electrophysiological characterization of a large number of full-length *Drosophila DmNa$_v$* (i.e., *para*) and cockroach *BgNav* cDNAs in *Xenopus* oocytes revealed that these splicing variants exhibit an extraordinary variety of gating properties. Specific alternative splicing events alter the amplitude of sodium current (Tan *et al.*, 2002; Song *et al.*, 2004), change voltage-dependence of activation and inactivation, and even generate persistent current (Olson *et al.*, 2008; Lin *et al.*, 2009). Splicing variants also exhibit differential sensitivities to pyrethroid insecticides (Tan *et al.*, 2002; Du *et al.*, 2009a). These findings, therefore, provide strong evidence that alternative splicing may be a major mechanism by which insects achieve the functional diversity of sodium currents observed *in vivo* (see Dong, 2007 and reference therein).

Recent evidence also indicates that sodium current diversity is also achieved through RNA editing. To date, several U-to-C editing events have been identified in *DmNa$_v$* and *BgNav* transcripts. As a result, these editing events produce sodium channel variants with distinct gating properties (such as persistent currents) (Liu *et al.*, 2004; Song *et al.*, 2004). An A-to-I editing event in *DmNa$_v$* generates a unique low-voltage-activated sodium channel (Olson *et al.*, 2008). Thus, unlike mammals, which express multiple sodium channel genes as a major means of generating sodium channel functional diversity in various tissues and cell types, insects rely on alternative splicing and RNA editing of apparently one sodium channel gene to achieve a similar goal.

A1.4. Sodium Channel Mutations and Pyrethroid Resistance in Insects

As summarized in Figure 3 and Table 2 in Soderlund (2005), multiple common and distinct point mutations are found in pyrethroid-resistant populations of pest species. A few new mutations can be added to this list. For example, in *Aedes aegypti*, an isoleucine to methionine or valine mutation in IIS6 upstream of the first *kdr* mutation; and a valine to isoleucine or glycine mutation in IIS6 downstream of the first *kdr* mutation were detected in various pyrethroid-resistant populations (Brengues *et al.*, 2003; Saavedra-Rodriguez *et al.*, 2007).

Perhaps the most significant recent work in the study of sodium channel-mediated pyrethroid resistance involves computer modeling of the pyrethroid-binding site. O'Reilly *et al.* (2006) used the X-ray structure of the K$_v$1.2 potassium channel as a template to predict the open conformation of the house fly sodium channel. This interesting model predicts that the pyrethroid receptor site is located in a hydrophobic cavity delimited by the IIS4-S5 linker and IIS5 and IIIS6 helices. Recent systematic site-directed mutagenesis of these regions provided further experimental support for this model (Usherwood *et al.*, 2007; Du *et al.*, 2009b). Indeed, several key *kdr* mutations, M918T, L925I, L932F, and F1538I in Figure 3 and Table 2 (Soderlund, 2005), are situated in these regions, further supporting the findings from pharmacological studies that these mutations likely reduce pyrethroid binding to the receptor site on the sodium channel (Vais *et al.*, 2000, 2001, 2003; Tan *et al.*, 2005). But how *kdr* mutations that are not found near this predicted receptor site alter the interaction of pyrethroids with sodium channels remains elusive.

References

Brengues, C., Hawkes, N.J., Chandre, L., McCarroll, S., Guillet, P., Manguin, S., Morgan, J.C., Hemingway, J., **2003**. Pyrethroid and DDT cross-resistance in *Aedes aegypti* is correlated with novel mutations in the voltage-gated sodium channel gene. *Med. Vet. Ent. 17*, 87–94.

Davies, T.G, Field, L.M., Usherwood, P.N., Williamson, M.S., **2007**. A comparative study of voltage-gated sodium channels in the Insecta: implications for pyrethroid resistance in Anopheline and other Neopteran species. *Insect Mol. Biol. 16*, 361–375.

Derst, C., Walther, C., Veh, R.W., Wicher, D., Heinemann, S.H., **2006**. Four novel sequences in *Drosophila melanogaster* homologous to the auxiliary Para sodium channel. *Biochem. Biophys. Res. Commun. 339*, 938–948.

Dong, K., **2007**. Insect sodium channels and insecticide resistance. *Invert. Neurosci. 7*, 17–30.

Du, Y., Nomura, Y., Luo, N., Liu, Z., Lee, J.E., Khambay, B., Dong, K., **2009**. Molecular determinants on the insect sodium channel for the specific action of type II pyrethroid insecticides. *Toxicol. Appl. Pharmacol. 234*, 266–272.

Du, Y., Lee, J.E., Nomura, Y., Zhang, T., Zhorov, B.S., Dong, K., **2009**. Identification of a cluster of residues in transmembrane segment 6 of domain III of the cockroach sodium channel essential for the action of pyrethroid insecticides. *Biochem. J. 419*, 377–385.

Lin, W.H., Wright, D.E., Muraro, N.I., Baines, R.A., **2009**. Alternative splicing in the voltage gated sodium channel DmNav regulates activation, inactivation, and persistent current. *J Neurophysiol. 102*, 1994–2006.

Littleton, J.T., Ganetzky, B., **2000**. Ion channels and synaptic organization: analysis of the *Drosophila* genome. *Neuron 26*, 35–43.

Liu, Z., Song, W., Dong, K., **2004**. Persistent tetrodotoxin-sensitive sodium current resulting from U-to-C RNA editing of an insect sodium channel. *Proc. Natl. Acad. Sci. U.S.A. 101*, 11862–11867.

Moignot, B., Lemaire, C., Quinchard, S., Lapied, B., Legros, C., 2009. The discovery of a novel sodium channel in the cockroach *Periplaneta americana*: evidence for an early duplication of the para-like gene. *Insect. Biochem. Mol. Biol. 39*, 814–823.

Olson, R., Liu, Z., Nomura, Y., Song, W., Dong, K., 2008. Molecular and functional characterization of voltage-gated sodium channel variants from *Drosophila melanogaster*. *Insect Biochem. Mol. Biol. 38*, 604–610.

O'Reilly, A.Q., Khambay, B.P.S., Williamson, M.S., Field, L.M., Wallace, B.A., Davies, T.G.E., 2006. Modeling insecticide binding sites at the voltage-gated sodium channel. *Biochem. J. 396*, 255–263.

Saavedra-Rodriguez, K., Urdaneta-Marquez, L., Rajatileka, S., Moulton, M., Florest, A.E., Fernandez-Salas, I., Bisset, J., Rodriguez, M., Mccall, P.J., Donnelly, M.J., Ranson, H., Hemingway, J., Black, W.C., IV, 2007. A mutation in the voltage-gated sodium channel gene associated with pyrethroid resistance in Latin American *Aedes aegypti*. *Insect Mol. Biol. 16*, 785–798.

Shao, Y.M., Dong, K., Zhang, C.X., 2008. Molecular characterization of a sodium channel gene from the silkworm *Bombyx mori*. *Insect Biochem. Mol. Biol. 39*, 145–151.

Sonoda, S., Tsukahara, Y., Ashfaq, M., Tsumuki, H., 2008. Genomic organization of the *para*-sodium channel α-subunit genes from the pyrethroid-resistant and susceptible strains of the diamondback moth. *Arch. Insect Biochem. Physiol. 69*, 1–12.

Soderlund, D., 2005. Sodium channels. In: Gilbert, L.I, Iatrou, K., Gill, S.S (Eds.), Comprehensive Insect Science. Pharmacology, Vol. 5. Elsevier, Amsterdam, pp. 1–24.

Song, W., Liu, Z., Tan, J., Nomura, Y., Dong, K., 2004. RNA editing generates tissue-specific sodium channels with distinct gating properties. *J. Biol. Chem. 279*, 2554–2561.

Tan, J., Liu, Z., Nomura, Y., Goldin, A.L., Dong, K., 2002. Alternative splicing of an insect sodium channel gene generates pharmacologically distinct sodium channels. *J. Neurosci. 22*, 5300–5309.

Tan, J., Liu, Z., Wang, R., Huang, Z.Y., Chen, A.C., Gurevitz, M., Dong, K., 2005. Identification of amino acid residues in the insect sodium channel critical for pyrethroid binding. *Mol. Pharmacol. 67*, 513–522.

Usherwood, P.N., Davies, T.G., Mellor, I.R., O'Reilly, A.O., Peng, F., Vais, H., Khambay, B.P., Field, L.M., Williamson, M.S., 2007. Mutations in DIIS5 and the DIIS4-S5 linker of *Drosophila melanogaster* sodium channel define binding domains for pyrethroids and DDT. *FEBS Lett. 581*, 5485–5492.

Vais, H., Williamson, M.S., Goodson, S.J., Devonshire, A.L., Warmke, J.W., Usherwood, P.N.R., Cohen, C., 2000. Activation of *Drosophila* sodium channels promotes modification by deltamethrin: reductions in affinity caused by knock-down resistance mutations. *J. Gen. Physiol. 115*, 305–318.

Vais, H., Williamson, M.S., Devonshire, A.L., Usherwood, P.N., 2001. The molecular interactions of pyrethroid insecticides with insect and mammalian sodium channels. *Pest Manag. Sci. 57*, 877–888.

Vais, H., Atkinson, S., Pluteanu, F., Goodson, S.J., Devonshire, A.L., Williamson, M.S., Usherwood, P.N., 2003. Mutations of the para sodium channel of *Drosophila melanogaster* identify putative binding sites for pyrethroids. *Mol. Pharmacol. 64*, 914–922.

Zhou, W., Chung, I., Liu, Z., Goldin, A., Dong, K., 2004. A voltage-gated calcium-selected channel encoded by a sodium channel-like gene. *Neuron 42*, 101–112.

2 GABA Receptors of Insects

S D Buckingham and D B Sattelle, University of
Oxford, Oxford, UK

© 2005, Elsevier BV. All Rights Reserved.

2.1. Introduction

Receptors for the neurotransmitter γ-aminobutyric acid (GABA) are abundant in the nervous systems of insects (Sattelle, 1990) where, as in vertebrates, they play a major role in inhibition. In 1985 when the first set of volumes of this series entitled *Comprehensive Insect Physiology, Biochemistry, and Pharmacology* was published, little was known about these proteins, despite the already established use of insect preparations to address fundamental questions in neurobiology, and despite the fact that GABA receptors (GABARs) were already considered to be possible insecticide targets. Since then, our understanding of insect GABARs has been transformed.

Since 1985, studies on GABARs have benefited from the application of molecular biology (molecular cloning, functional expression, and determination of their gene structure), electrophysiology (single-channel and whole-cell patch clamp recordings), integrative physiology (demonstrating a key role for such receptors in particular insect behaviors), genome analysis (yielding a complete set of ligand gated anion channels for the fruit fly, *Drosophila melanogaster* and the mosquito, *Anopheles gambiae*, and genetics (providing a detailed understanding of mutations that underly insecticide resistance). Radioligand binding studies have generated a detailed description of the pharmacology of insect GABAR binding sites. Electrophysiological studies, some using identified neurons, have contributed important information on

GABARs, both functional native and expressed recombinant receptors. Patch clamp methods, in particular, have provided key insights, enabling: (1) the analysis of single channel properties; and (2) description of the effects of chemicals on receptor kinetics. A number of studies have examined the distribution of GABARs and other molecular components of insect GABAergic synapses using antibodies to GABA, enzymes involved in GABA synthesis, and, more recently, antibodies to cloned GABAR subunits. Thus, arange of complementary experimental approaches have led to new discoveries on the roles of GABARs in sensory receptive field tuning, learning, and even a possible role in the insect circadian clock.

A major advance in the study of insect GABARs resulted from the cloning of three candidate *Drosophila* ionotropic GABAR subunits, one of which resistant to DieLdrin(RDL) readily forms a functional, recombinant, homomeric GABA gated chloride channel. This was followed by the discovery of RDL orthologs in several insect orders. Thus, for the first time, it was possible to study in detail the physiology and pharmacology of an insect GABAR of known molecular composition. Consequently, in the years since the publication of the first edition of *Comprehensive Insect Physiology, Biochemistry and Molecular Biology*, there has been considerable growth in our understanding of insect GABARs (Hosie *et al.*,

1997) and their vertebrate counterparts (Whiting, 2003).

Insect GABAR subtypes were not well established in 1985, although even then data were beginning to emerge, to question earlier notions that benzodiazepine binding sites were not present in insects (Nielsen *et al.*, 1978). Today, we know that insect GABARs rival the complexity of their vertebrate counterparts, and possess a number of sites for modulation by not only benzodiazepines, but also several other allosteric modulators. Again, in 1985, there was no evidence for the existence of an insect metabotropic receptor – today, three candidate insect metabotropic GABARs have been cloned and their characterization is now in progress.

This chapter concentrates on new information that has accumulated since the previous edition of this series. After a brief review of findings up to 1985, we discuss the more recent findings on the physiology and pharmacology of insect ionotropic GABARs, examining each major chemical binding site in turn. Evidence for distinct subtypes of native insect GABARs is also discussed. Attention is then drawn to findings derived from the cloning and functional expression of recombinant insect GABARs, and their contributions to our understanding of (1) native receptors and (2) insecticide modes of action. Next, we consider the roles of ionotropic GABARs in insect behavior. Finally, new findings on metabotropic GABARs are described.

By 1985, GABA was already known to be an inhibitory neurotransmitter at nerve–muscle junctions and in the central nervous system (CNS) of arthropods (Kuffler and Edwards, 1958; Usherwood and Grundfest, 1965; Otsuka *et al.*, 1966), but very little is still known of insect GABARs. Most information was based on electrophysiology performed on native nerve and muscle receptors. However, GABA had been shown to induce hyperpolarizing responses in (mostly unidentified) central neurons of several insect species, including postsynaptic membranes of interneurons (Callec and Boistel, 1971; Callec, 1974; Hue *et al.*, 1979) and unidentified cell bodies (Kerkut *et al.*, 1969; Pitman and Kerkut, 1970; Walker *et al.*, 1971; Roberts *et al.*, 1981) in the terminal abdominal ganglion of the cockroach, *Periplaneta americana*. Dissociated cells of locust thoracic ganglia (Giles and Usherwood, 1985), as well as identified dorsal unpaired median neurons (DUMETi) that innervate the *Extensor tibiae* muscle of the embryonic grasshopper *Schistocerca nitens* (Goodman and Spitzer, 1980), were also found to respond to GABA with hyperpolarizations. Peripheral GABARs located on somatic muscle fibers had been described in some

considerable detail (Cull-Candy and Miledi, 1981; Cull-Candy, 1982). These early studies established that GABA mediates hyperpolarizing, largely inhibitory responses in insect neurons, but other than showing a sensitivity to picrotoxinin, little was known of the details of the pharmacology of these responses.

2.2. Ionotropic GABARs of Insects

2.2.1. Lessons from Vertebrate GABARs

Our understanding of both insect and vertebrate GABARs is based on binding studies, electrophysiology of *in situ* receptors and electrophysiology of expressed recombinant receptors, each approach offering its own particular advantages. Electrophysiological responses of neurons in insect or vertebrate nervous systems as well as cultured neurons provide a detailed account of the pharmacology, ion permeability, and kinetic properties of native GABARs. It is particularly convenient where identified neurons, or classes of neurons, are studied, as the analysis is restricted to a circumscribed (in some cases, perhaps homogeneous) class of receptors. Electrophysiological studies using heterologously expressed recombinant receptors offer the further advantage that the subunit composition of the receptors is known. Binding studies complement these functional approaches. However, when applied to native tissues the technique often pools data from multiple GABARs (as well as any other binding sites for the radioligand) from all cell types in the tissue under investigation. As functional and binding experiments yield different types of information, care must be exercised when comparing data obtained using these different techniques. For example, electrophysiological and binding studies typically address different states of the same receptor, which can also contribute to some of the apparent discrepancies.

Nineteen ionotropic GABAR subunit molecules have been cloned from vertebrates (Whiting, 2003). They are members of the dicysteine-loop (cys-loop) family of ionotropic neurotransmitter receptors that also includes the nicotinic acetylcholine receptors (nAChRs) (Karlin, 2002), serotonin type 3 (5-HT_3) receptors ($5HT_3Rs$) (Reeves and Lummis, 2002), and strychnine sensitive glycine receptors (GlyRs) (Betz *et al.*, 1999) of vertebrates as well as glutamate gated chloride channels (GluCls), a 5-HT gated chloride channel and a histamine gated chloride channel of invertebrates (review: Raymond and Sattelle, 2002). Each receptor molecule consists of five subunits arranged symmetrically around an integral, hydrophobic ion pore (**Figure 1**). Each polypeptide

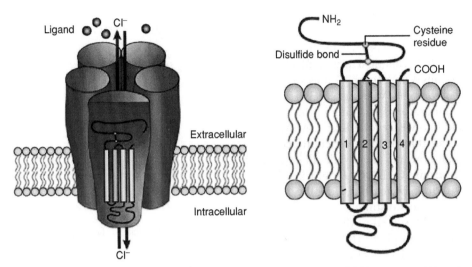

Figure 1 A schematic representation of the three-dimensional structure of ionotropic GABA receptors. Although more data are available for the structure of vertebrate ionotropic GABA receptors, it is probable that insect GABA receptors have a similar structure. (From Raymond, V., Sattelle, D.B., **2002**. Novel animal-health drug targets from ligand-gated chloride channels. *Nat. Rev. Drug Discov. 6*, 427–436.)

subunit possesses a long N-terminal extracellular domain, which contains residues that contribute to the neurotransmitter (GABA) binding site and four transmembrane regions (M1–M4) (**Figure 1**). The second transmembrane domain (M2) contains most of the residues that line the chloride channel (Karlin and Akabas, 1995; Smith and Olsen, 1995). Vertebrate ionotropic GABARs may be divided into two pharmacological categories: GABA$_A$ receptors, which are all antagonized by the alkaloid, bicuculline, and regulated by a wide range of allosteric modulators such as benzodiazepines, barbiturates, and pregnane steroids; and GABA$_C$ receptors, which are insensitive to bicuculline as well as the majority of modulators of GABA$_A$ receptors (Bormann, 2000; Zhang *et al.*, 2001). GABA$_C$ receptors are less liable to desensitize than GABA$_A$ receptors (Johnston, 1996). This physiological and pharmacological characterization is mirrored by corresponding structural differences between the two receptor subtypes, in that they are composed of distinct subunits. GABA$_A$ receptors are composed of subunits from several different classes (α_{1-6} β_{1-3} γ_{1-3}, δ, ε, π, and θ) (Hedblom and Kirkness, 1997; Thompson *et al.*, 1998; Neelands *et al.*, 1999; Bonnert *et al.*, 1999; Sinkkonen *et al.*, 2000; Whiting, 2003), whereas GABA$_C$ receptors are believed to be composed of the three known isoforms of the ρ subunit, ρ_{1-3} (Sieghart, 1995; Djamgoz, 1995; McKernan and Whiting, 1996). GABA$_A$ receptors are complex allosteric proteins and contain, in addition to the agonist binding site, distinct sites for the actions of barbiturates, steroids, loreclezole, furosemide, picrotoxin (PTX), zinc, lanthanum, volatile

anaesthetics, benzodiazepines, and the anaesthetic, propofol. Although it should be noted that there is not perfect consensus on this division of vertebrate GABARs into A and C subtypes, this has been the view of the majority of workers to date and has received recent support (Bormann, 2000). This view will therefore be adopted for the purposes of this review.

Functional expression studies, in which various combinations of subunits are heterologously expressed in *Xenopus laevis* oocytes or cell lines, have shown that the pharmacology of such recombinant vertebrate receptors is strongly influenced by subunit composition. For example, the specific subtype of β subunit present is a critical determinant of agonist pharmacology, whereas the α and γ subunits are important in conferring sensitivity to benzodiazepines. When the α_1 subtype is present in the heterologously expressed receptor it exhibits a pharmacology that resembles benzodiazepine type 1 pharmacology defined on the basis of *in situ* native receptor studies. Also, the inclusion of either the α_2 or α_3 subunit confers to the receptor the classical benzodiazepine type 2 pharmacology (Pritchett *et al.*, 1989). The γ_2 subunit is essential for benzodiazepine activity (Pritchett *et al.*, 1989), and the specific subtype of γ subunit present in the receptor plays a major role in determining the receptor's responsiveness to allosteric modulators, such as benzodiazepines. Taken together, experiments using recombinant expression have suggested that native vertebrate GABA$_A$ receptors are composed of at least α, β, and γ subunits, a conclusion supported by several immunoprecipitation studies

(Mohler *et al.*, 1992; Khan *et al.*, 1994; Araujo *et al.*, 1996; Jechlinger *et al.*, 1998; Nusser *et al.*, 1999). About 35% of GABA$_A$ receptors in the brain contain $\alpha_1\beta_2\gamma_2$ subunits, 20% contain $\alpha_3\beta_3\gamma_2$, and 15% contain $\alpha_2\beta_3\gamma_2$ subunits. Other combinations are present, but they are less common (Whiting, 2003).

With fewer laboratories working in the field the growth in our knowledge of insect GABARs has been slower than that of their vertebrate counterparts. It is clear, however, that while all vertebrate and insect ionotropic GABARs are blocked by the plant derived toxin, picrotoxinin, a number of pharmacological differences distinguish insect GABARs from vertebrate GABA$_A$ and GABA$_C$ receptors. For example, insect ionotropic GABARs do not readily fit the vertebrate classification. There are two principal differences. First, the majority of insect GABARs, unlike GABA$_A$ receptors, are insensitive to bicuculline, although bicuculline sensitive insect GABARs do exist. Secondly, most insect GABARs are pharmacologically distinct from GABA$_C$ receptors in that they are subject to allosteric modulation by benzodiazepines and barbiturates (Sattelle, 1990). There are other minor differences in aspects of their detailed pharmacology, each of which will be addressed in the following sections.

2.2.2. Pharmacology of *In Situ* Insect GABARs

2.2.2.1. Agonist site   Early binding studies on cockroach (*P. americana*) nerve cord preparations using [³H]GABA and [³H]muscimol indicated that insect GABARs, unlike vertebrate GABA$_A$ receptors, were insensitive to 3-aminopropanesulfonic acid (3-APS). Also, muscimol is of the same, or higher, potency as GABA on insect GABARs (Lummis and Sattelle, 1985, 1986). The most potent displacers of [³H]GABA binding to cockroach nerve cord preparations were found to be GABA itself and muscimol, with IC$_{50}$s of 130 and 710 nM respectively. In contrast, the other vertebrate agonists tested (thiomuscimol, 4,5,6,7-tetrahydroisoxazolo[5,4-c] pyridin-3-ol (THIP), 3-APS, and isoguvacine) had IC$_{50}$s greater than 30 mM (Lummis and Sattelle, 1986). The dissociation rate for GABA binding was estimated to be around 0.116 min^{-1} and the estimated association rate was approximately 316 mM^{-1} min^{-1}. The molecular structures of some GABAR agonists are illustrated in **Figure 2**.

Later, electrophysiological studies were performed on GABARs in an identified *Periplaneta* motor neuron (the fast coxal depressor, D_f) (Sattelle *et al.*, 1988), and on cloned *Drosophila* GABARs containing only RDL subunits heterologously expressed in oocytes or insect cell lines (Millar *et al.*, 1994; Hosie and

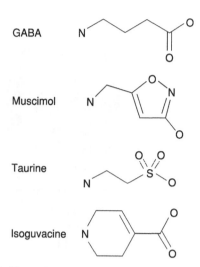

Figure 2   The molecular structures of some common agonists of vertebrate and insect ionotropic GABA receptors.

Sattelle, 1996a, 1996b; Buckingham *et al.*, 1996). The only major difference between the [³H]GABA displacement profile and the functional studies is that isoguvacine is a potent agonist on native, functional insect GABARs (Sattelle *et al.*, 1988), providing a further distinction from vertebrate GABARs. The reason for this apparent discrepancy between physiology and binding data is unclear, but may be attributable to the fact that the two methods probe different receptor states, and that the two approaches may probe different pools of receptor subtypes. The sensitivity of insect GABARs to ZAPA ((Z)-3-[(aminoiminomethyl)thio]prop-2-enoic acid) (Taylor *et al.*, 1993) also distinguishes them from vertebrate GABA$_A$ receptors, which are largely insensitive to this compound (Woodward *et al.*, 1993; Qian and Dowling, 1993).

Reports from several laboratories show a lack of action of established vertebrate GABA$_B$ agonists or antagonists on insect GABARs (Lees *et al.*, 1987; Sattelle *et al.*, 1988; Murphy and Wann, 1988; Bermudez *et al.*, 1991; Wolff and Wingate, 1998). However, this does not preclude the existence of GABA$_B$ receptors with unusual pharmacology. For example, the GABA$_B$R agonist baclofen and the GABA$_B$ antagonist saclofen were inactive on GABARs of D_f (Sattelle *et al.*, 1988). However, more recent experiments have identified GABA$_B$-like responses, which differ significantly in their pharmacology from vertebrate GABA$_B$ receptors in being insensitive to baclofen (Bai and Sattelle, 1995). The presence of these insect GABA$_B$ receptors was only detected when responses mediated through ionotropic GABARs were blocked by PTX.

Most insect ionotropic GABARs are relatively insensitive to the GABA analog CACA (*cis*-aminocrotonic acid), which is an agonist of vertebrate GABA$_C$, but not GABA$_A$, receptors. For example, visual interneurons in the fly, *Calliphora erythrocephala*, are CACA insensitive (Brotz and Borst, 1996). However, it should be noted that CACA activates GABARs that mediate inhibition by identified filiform hair receptors of an identified projection interneuron, a circuit that mediates wind elicited responses (Gauglitz and Pfluger, 2001).

Taurine (2-aminoethanesulfonic acid), an abundant amino acid in both vertebrates and insects, may act as a coagonist with GABA in certain insects. It is abundant in the nervous system of the locust, *Schistocerca gregaria* (Whitton *et al.*, 1987), where its distribution changes following periods of intense muscular activity. When ionophoretically applied to isolated locust neurons in culture, taurine evokes membrane conductance increases very similar to those evoked by GABA (Whitton *et al.*, 1994). Indeed, responses to both GABA and taurine are both blocked by PTX and enhanced by flunitrazepam, consistent with their acting through the same receptor. It was therefore suggested that taurine may act at the same site as GABA, serving to modulate the GABA response. In the migratory locust, *Locusta migratoria*, taurine is coexpressed with GABA in local and intersegmental inhibitory interneurons but not in efferent neurosecretory or paracrine cells, suggesting a role in preventing hyperexcitation during stressful conditions (Stevenson, 1999).

2.2.2.2. Competitive antagonists A consistent feature of most, but not all, insect GABARs is that they are insensitive to the plant derived alkaloid, bicuculline (**Figure 3**), a defining blocker of vertebrate GABA$_A$ receptors (Curtis *et al.*, 1971; Watson and Girdlestone, 1995). Extrasynaptic GABARs on the motor neuron, D_f (Sattelle *et al.*, 1988; Buckingham *et al.*, 1994), and synaptic GABARs on the giant interneuron 2 of *P. americana* (Buckingham *et al.*, 1994) are insensitive to bicuculline, as are many locust neuron ionotropic GABARs (Benson, 1988; Lees *et al.*, 1987). Bicuculline does not block either the cloned *Drosophila* GABAR, RDL (Millar *et al.*, 1994), or the cloned *Heliothis virescens* RDL-like GABAR (Wolff and Wingate, 1998). GABA mediated inhibitory postsynaptic potentials (IPSPs) recorded in an identified locust (*S. gregaria*) interneuron are also bicuculline insensitive (Watson and Burrows, 1987). Similarly, bicuculline insensitivity is a feature of GABARs that mediate inhibition by identified filiform hair receptors of an identified projection interneuron, a circuit that

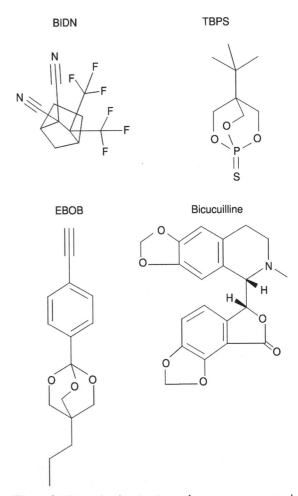

Figure 3 The molecular structures of some common convulsant antagonists of vertebrate and insect ionotropic GABA receptors.

mediates wind elicited responses (Gauglitz and Pfluger, 2001). Responses to muscimol in motion sensitive visual interneurons of the fly, *C. erythrocephala*, are also insensitive to bicuculline (Brotz and Borst, 1996). Furthermore, insensitivity to bicuculline is also seen for certain spider GABARs (Panek *et al.*, 2002).

However, there is a body of evidence that bicuculline does block ionotropic GABARs of some adult *Manduca sexta* abdominal ganglion neurons (Roberts *et al.*, 1981; Waldrop *et al.*, 1987; Waldrop 1994) as well as electrically or odor evoked IPSPs and GABA responses in projection neurons of antennal lobes of this species (Christensen *et al.*, 1998). Recently, Sattelle *et al.* (2003) have also shown that bicuculline insensitive GABARs are present in the larval nervous system of *M. sexta*. Further, the observation that bicuculline is effective in recovering behaviors disrupted by the systemic administration of GABA transporter blockers (Leal and Neckameyer, 2002) suggests that, despite the lack of bicuculline action on RDL, bicuculline may nonetheless block some *D. melanogaster* GABARs. In addition, bicuculline also blocks certain nicotinic acetylcholine receptors (nAChRs) of insects (Benson, 1988; Buckingham *et al.*, 1994) and vertebrates (Rothlin *et al.*, 1999).

One attempt to resolve the issue of bicuculline sensitivity tested the possibility that such divergent reports may reflect pharmacological differences between extrasynaptic and synaptic GABARs (Buckingham *et al.*, 1994). The cockroach, *P. americana*, offers a convenient model for such an approach because of the number of indentified neurons with ready access to both their cell bodies and their synapses. Motor neuron D_f has a population of well-characterized GABARs on its cell body membrane, and giant interneuron 2 (GI 2) not only presents an accessible cell body, but also receives synaptic inputs from nerve X. IPSPs measured in GI 2 using the oil gap recording method in response to electrical stimulation of nerve X were blocked by both bicuculline (100 μM) and PTX (0.01 μM). However, bicuculline injected directly into the neuropil in the region of these synapses did not block IPSPs. The apparent block of synaptic GABARs by bicuculline was found to be due to inhibition of nAChRs upstream in the synaptic pathway. While the study failed to find differences in the pharmacology of synaptic and extrasynaptic GABARs, it did show that three separate populations of GABARs exist: synaptic receptors in the neuropilar branches of GI 2, extrasynaptic GABARs on the cell body of GI 2, and extrasynaptic GABARs on D_f, all of which are bicuculline insensitive. The finding that bicuculline acts on nAChRs does not, however, explain the block of IPSPs by bicuculline in *M. sexta* interneurons (Waldrop *et al.*, 1987), since ACh responses in this preparation are insensitive to direct bicuculline application (Waldrop and Hildebrand, 1988).

Thus, there is some crossover between the pharmacophore for ligand binding sites of insect ionotropic GABARs and nAChRs. Whereas most insect GABARs are insensitive to bicuculline, many insect ACh receptors, in contrast, are blocked by bicuculline (Buckingham *et al.*, 1994; Waldrop, 1994). At the same time, imidacloprid, an insecticide acting as a partial agonist of nAChRs, can also partially block GABARs of honeybee Kenyon cells, albeit at relatively high concentrations (Deglise *et al.*, 2002).

In addition to their abundant expression in the CNS, and in contrast to vertebrates, insect GABARs are also expressed on muscle fibers. In many respects, these muscle insect GABARs resemble those of the nervous system in that they are sensitive to muscimol and to the GABA channel blocker, picrotoxinin

(Scott and Duce, 1987; Murphy and Wann, 1988). However, until 1997, no detailed comparison had been reported for muscle and neuronal GABARs in the same animal for any insect species. One study, therefore, set out directly to compare GABARs on fibers of the coxal levator muscle (182c,d) of the cockroach, *P. americana* (Schnee *et al.*, 1997), with the well-characterized GABARs on the motor neuron, D_f, of the same species (Sattelle *et al.*, 1988). Both were sensitive to PTX and insensitive to bicuculline, but there were significant pharmacological differences between them. Unlike the neuronal receptors, the muscle GABARs were insensitive to isoguvacine. The EC_{50} of muscimol relative to GABA was also much lower for muscle than for D_f. The muscle GABARs were also sensitive to certain cyclodienes: 12-ketoendrin, endrin, and heptachlor epoxide all reduced GABA responses with IC_{50}s in the 1–10 nM range. Similarly, micromolar concentrations of the cyclodiene, heptachlor, produced a complete block of muscle receptors, whereas concentrations as high as 100 μM resulted in about 50% block of GABARs on cockroach motor neuron D_f (Lummis *et al.*, 1990). Micromolar *tert*-butyl-bicyclo[2.2.2]phosbrothionate (TBPS) and tert-butyl-bicylo orthobenzoate (TBOB) were almost without effect on muscle receptors. Hence, in the cockroach there are some clear differences between neuronal and muscle GABARs. The observation that muscle GABARs are significantly more sensitive to cyclodienes than the GABARs of motor neuron D_f is interesting in view of the fact that mutants of RDL are resistant to cyclodienes, yet RDL is expressed exclusively in the nervous system. This discrepancy remains to be addressed.

2.2.2.3. Noncompetitive antagonists and convulsants

Several convulsants (**Figure 3**) are noncompetitive antagonists of insect GABARs and this site is the target of many insecticides, including the cyclodienes (such as endrin, dieldrin), hexachlorocyclohexanes (such as lindane), and bicycloorthobenzoates (such as 1-(4-ethynylphenyl)4-*n*-propyl-2,6, 7-trioxabicyclo[2.2.2]octane (EBOB)) (Matsumura and Ghiasuddin, 1983; Lummis and Sattelle, 1986; Wafford *et al.*, 1989; Lummis *et al.*, 1990; Deng *et al.*, 1993; Anthony *et al.*, 1994). The earliest evidence that cyclodienes act at the convulsant site came from binding studies using [^{3}H]dihydropicrotoxinin, which indicated that the cyclodiene and PTX binding sites were similar (Matsumura and Ghiasuddin, 1983). Further, houseflies resistant to dieldrin show cross-resistance to many polychlorocycloalkanes and phenylpyrazoles, and to some bicycloorthobenzoates, but not to avermectins

(Deng *et al.*, 1991; Deng and Casida 1992; Cole *et al.*, 1993; Casida, 1993).

Initial studies using labeled probes for the convulsant site were hampered by the limitations of [^{3}H]dihydropicrotoxinin as a probe. These were largely overcome as improved radioligands became available. Specific, saturable binding of [^{35}S]TBPS to cockroach nerve cord membranes yielded a K_d of around 18 nM and B_{max} of around 177 fmol mg^{-1} protein (Lummis and Sattelle, 1986). This binding was inhibited by GABA and dihydroavermectin B$_{1a}$ at micromolar concentrations, suggesting the presence of an allosteric linkage between these sites in insect GABARs.

Radioligand binding and toxicity approaches used by John Casida's laboratory have been particularly fruitful not only in identifying the sites of action of many insecticidally relevant compounds, but also in providing a pharmacological dissection of the convulsant site. The [^{3}H]EBOB binding site of housefly heads was found to be identical to, or to overlap with, a wide array of structurally distinct insecticides (Deng *et al.*, 1991) including cyclodienes, bicycloorthobenzoates, and PTX, but not avermectins. This was pursued further in a comprehensive study of the convulsant site examining the displacement of labeled EBOB and TBPS by 29 different compounds (Deng *et al.*, 1993). Binding of [^{35}S]TBPS to housefly head membranes had a K_d of 145 nM and B_{max} of 2.4 pmol mg^{-1} protein, and both [^{35}S]TBPS and [^{3}H]EBOB were displaced by a number of polychlorocycloalkanes, bicycloorthobenzoates, and phenylpyrazoles. The binding parameters resembled those of vertebrate tissues with respect to pH dependence, anion specificity and kinetics of association, and dissociation. Comparing the modes of competition of each compound class (i.e., competitive versus noncompetitive) and their relevance to toxicity, this study concluded that four subsites can be defined pharmacologically: site A (the EBOB site) overlaps with site C (the phenylpyrazole site) and with site B (the TBPS site), but sites B and C do not overlap. The avermectin site (site D) is distinct from all of the other three sites, but is allosterically linked to sites A and C.

Similar studies added weight to the conclusion that insect GABARs differ significantly from their vertebrate counterparts at the noncompetitive binding sites. For instance, they are significantly less sensitive to the polychlorocycloalkane, cage convulsants (such as TBPS and TBOB), and PTX, compounds that by contrast are potent inhibitors of vertebrate GABARs (Rauh *et al.*, 1990). A single exception to this rule is furnished by the polychlorocyclohexane insecticide, heptachlor, and its epoxide

metabolite, which more potently displaced labeled TBPS from *Musca domestica* membranes than from rat (Lummis *et al.*, 1990). However, when muscle GABARs of the cockroach, *P. americana*, were compared to those on an identified motor neuron of the same species (Schnee *et al.*, 1997), muscle GABARs were found to be more sensitive to lindane and to heptachlor and its epoxide metabolite.

One fertile approach to understanding the convulsant site has been to compare the efficacies of a systematically varying set of PTX analogs in either blocking the receptor or in displacing radiolabeled probes. Such studies on vertebrate tissues led to the proposal (Ozoe and Matsumura, 1986) that the key elements in determining picrotoxane activity are the presence of a bulky lipophilic group and at least two electronegative centers. Several studies have emphasized the importance of two groups in the PTX molecule (the hydroxyl and the isopropenyl groups) in conferring high lethality (Jarboe *et al.*, 1968), on displacing [^{3}H]-dihydropicrotoxin (Ticku *et al.*, 1978), and as well as affecting electrophysiological responses of *in situ* (Kudo *et al.*, 1984) and recombinant GABARs (Anthony *et al.*, 1993). These conclusions were drawn in particular from the observation that fluoropicrotoxinin, which substitutes a fluorine at the hydroxyl position, is almost equally active as the parent molecule, whereas PTX acetate is much reduced in potency, and that picrotin, which adds a hydroxyl group to the isopropenyl, is also comparatively ineffective. Such findings suggest that for vertebrates, the hydroxyl group may act as a hydrogen bond acceptor, that substitutions at the hydroxyl group are liable to steric hindrance, and that the isopropenyl group must be lipophilic or electron-rich (Anthony *et al.*, 1993) for maximal activity. To determine whether the same is true for insect GABARs, one study (Anthony *et al.*, 1994) examined the actions of a series of PTX analogs on the well-defined GABARs on the identified motor neuron, D$_f$, of the cockroach, *P. americana*. The rank order of potency was the same as for expressed chick optic lobe GABARs, suggesting that the convulsant sites of insect and vertebrate GABA$_A$ receptors share certain common key features. Similar findings were obtained for RDL, a cloned insect GABAR (see Section 2.2.4.3.2).

The reduced sensitivity of insect GABARs to TBPS slowed the development of our understanding of insect convulsant site probes. Recently, however, two compounds have been deployed with selectivity for insect over vertebrate GABARs, BIDN, and fipronil (**Figures 3** and **4**).

BIDN ([^{3}H]3,3-bis-trifluoromethyl-bicyclo[2,2,1] heptane-2,2-dicarbonitrile) was developed using a combined biochemical and pharmacophore modeling approach (Rauh *et al.*, 1997a, 1997b). A large library of compounds was systematically screened for their ability to displace [^{35}S]TBPS from rat brain membranes. Several bicyclic dinitriles were identified and further screened for their selectiviy for insect GABARs over rat GABARs. This approach identified the common structural features of known convulsant site ligands, which helped to determine the key pharmacophore features (**Figure 5**) This was done by manually aligning each of a series of polycyclic dinitriles over a representation of the PTX molecule using the Sybyl software package (SYBYL Molecular Modeling Software, Tripos Associates, Inc., St. Louis, MO). Key pharmacophore elements identified were the presence of a polarizable moiety separated by a fixed distance from two hydrogen bond-accepting (i.e., electronegative) elements (**Figure 5**), confirming and extending the model proposed earlier by Ozoe and Matsumura (1986). A number of new potential ligands were developed from this pharmacophore model. Of these, the most useful was BIDN (Rauh *et al.*, 1997a, 1997b).

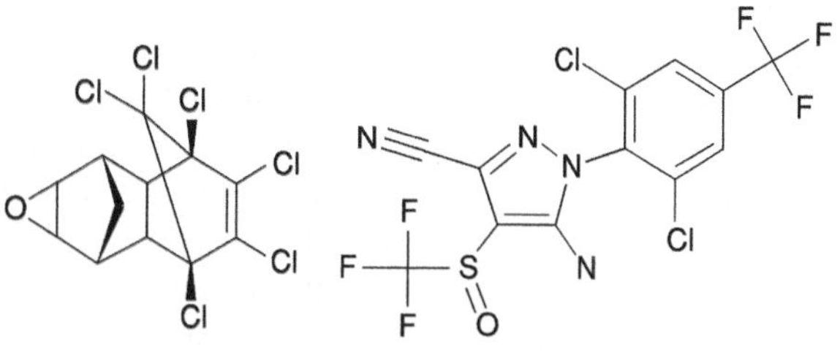

Binding of [^{3}H]BIDN to southern corn rootworm (*Diabrotica undecimpunctata*) membranes was saturable and specific with high affinity (26 nM) and a Hill coefficient of unity suggesting that BIDN binds to a single class of receptor sites (Rauh *et al.*, 1997b). Such high-affinity, specific binding was confirmed in the same study for five other insect species, including *P. americana* and *M. domestica*, whereas binding of BIDN to rat membranes was of lower affinity (around 250 nM). The same study examined the actions of BIDN using electrophysiological techniques, which showed that BIDN blocks GABARs of the identified cockroach motor neuron D_f as well as those present on the coxal levator muscles of the same species. BIDN also blocks GABA evoked depolarizations of thoracic neurons in the southern corn rootworm, (Rauh *et al.*, 1997b). The estimated IC$_{50}$ for BIDN action against central (on the cell body of motor neuron, D_f) and peripheral (on coxal levator muscle fibers) GABARs of *P. americana* was 0.9 µM and 2.5 µM, respectively. Although showing a degree of selectivity for insect over vertebrate receptors, the vertebrate toxicity of BIDN was a factor precluding its further development as a commercial pesticide.

Another insect GABAR ligand is the potent, widely used insecticide active on insect GABARs, the phenyl pyrazole, fipronil. Fipronil is a potent antagonist of cloned insect GABARs (Hosie *et al.*, 1997; Grolleau and Sattelle, 2000) and has also been used as the basis for the development of a photo-affinity probe for the convulsant site (Sirisoma *et al.*, 2001). There is abundant evidence that BIDN and fipronil act at the convulsant site. Fipronil competes for EBOB binding to housefly head membranes (Ratra and Casida, 2001). BIDN (Hosie *et al.*, 1995b), picrotoxinin (Shirai *et al.*, 1995), and fipronil (Hosie

Figure 4   The molecular structures of two potent antagonists of insect ionotropic GABA receptors, dieldrin and fipronil, both of which have been used as insecticides.

Dieldrin

Fipronil

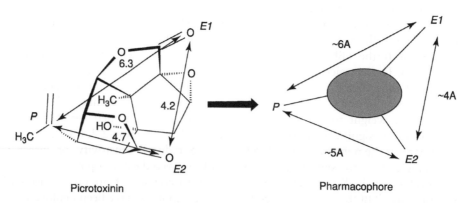

Picrotoxinin

Pharmacophore

Figure 5   The ideal pharmacophore for the picrotoxin molecule. This was derived from a number of structure–function studies that have revealed the structural features that confer potency to picrotoxin analogs. This was used to predict that a series of bicyclic dinitriles would have potency at the picrotoxinin site. (From Rauh, J.J., Holyoke, C.W., Kleier, D.A., Presnail, J.K., Benner, E.A., *et al.*, **1997b**. Polycyclic dinitriles: a novel class of potent GABAergic insecticides provides a new radioligand, [^{3}H]BIDN. *Invert. Neurosci.* 3, 261–268.)

et al., 1995b) all block RDL (see below) in a voltage independent, partly competitive and partly noncompetitive manner. However, although dieldrin displaces [^{3}H]BIDN binding competitively, PTX does not (Sattelle et al., 1995), suggesting that the PTX and the dieldrin site are not identical. This confirms an earlier report (Deng et al., 1993) that identified no less than four convulsant antagonist binding sites in the housefly GABAR that can be differentiated by [^{3}H]EBOB and [^{35}S]TBPS. It is therefore likely that BIDN, fipronil, EBOB, and PTX act at distinct but overlapping sites. Together, these compounds therefore provide a powerful tool-kit with which to further dissect the actions of insecticides on GABARs. Single channel properties of RDL expressed in a Drosophila cell line suggested that BIDN and fipronil both reduce the mean open time of GABAR chloride channels, but do so by acting at distinct sites (Grolleau and Sattelle, 2000) – see Section 2.2.4.3.2.

There are no doubt a number of naturally occurring ligands with GABAergic actions that await discovery. An interesting class of compounds acting at this site is the α and β thujones, the active ingredients of absinthe and present in certain herbal medicines. These compounds act at the convulsant site to block GABARs noncompetitively, and have insecticidal activity (Hold et al., 2000, 2001). It will be of interest to study possible actions of other plant derived compounds on insect GABARs.

2.2.2.4. Allosteric modulators

2.2.2.4.1. Benzodiazepines
In 1985, there was still some controversy as to whether insect nervous tissues possess receptors for benzodiazepines. One report suggested a lack of benzodiazepine receptors in invertebrates (Neilsen et al., 1978) but this was followed by a number of reports (Abalis et al., 1983; Lummis and Sattelle, 1986; Robinson et al., 1985) of sensitivity of insect GABARs to benzodiazepines. All these reports indicated that the pharmacology of the insect benzodiazepine site differs from most of its vertebrate counterparts. Specific [^{3}H]flunitrazepam binding to cockroach (P. americana) nerve cord membranes yielded an estimated K_d of 383 nM and B_{max} of 5.5 pmol mg^{-1} protein (Lummis and Sattelle, 1986). The most potent displacer of this binding was Ro5-4864 (IC$_{50}$ = 1 μM), which was only slightly more potent than flunitrazepam (IC$_{50}$ = 1.6 μM). Clonezapam was much less effective (IC$_{50}$ = 25 μM). The pharmacological profile of the [^{3}H]flunitrazepam binding site in P. americana was Ro5-4864 (IC$_{50}$ = 0.6 μM) > diazepam (IC$_{50}$ = 1.0 μM) > flunitrazepam (IC$_{50}$ = 1.6 μM) ≫ clonazepam (IC$_{50}$ = 25 μM) ≫ Ro15-1788 (IC$_{50}$ = 100 μM). Allosteric linkage between the agonist and benzodiazepine

sites was confirmed by the fact that 0.1 mM GABA enhanced flunitrazepam binding by some 150%. Furthermore, flunitrazepam enhanced responses to GABA in locust neuronal somata (Lees et al., 1987; Whitton et al., 1994) and in the motor neuron D_f of the cockroach at micromolar concentrations (Sattelle et al., 1988). The mode of action of benzodiazepines seems to be, as in vertebrates, an increase in the frequency of opening of the GABA gated channel (Shimahara et al., 1987).

The potent displacement of [^{3}H]flunitrazepam binding in cockroach by Ro5-4864 and its potentiation of GABA responses suggests that cockroach CNS benzodiazepine receptors resemble vertebrate peripheral benzodiazepine binding sites rather than central ones (Lummis and Sattelle, 1986). These peripheral binding sites in vertebrates are located principally in the kidney and liver, where they are associated with calcium channels. Nevertheless, the observation that flunitrazepam and GABA binding are allosterically linked, along with physiological evidence discussed below, confirm that the [^{3}H]flunitrazepam binding site in insect CNS is indeed part of a GABAR.

Flunitrazepam and diazepam have no effect upon GABARs of coxal levator muscle of cockroach, even at concentrations of up to 100 μM (Schnee et al., 1997) or upon the cloned H. virescens GABAR (Wolff and Wingate, 1998). Flunitrazepam does, however, evoke responses to GABA in the cockroach D_f motor neuron (Sattelle et al., 1988) and other neuronal preparations (review: Sattelle, 1990). That there are differing accounts of the actions of benzodiazepines upon insect GABARs is not surprising, given the great diversity in benzodiazepine pharmacology of vertebrate GABARs and the acute sensitivity of its pharmacology on subunit composition. A sensitivity to Ro5-4864 is probably the most significant distinguishing feature since the site of action of this compound in vertebrates is not on GABARs. Further, the lack of evidence for actions of benzodiazepines on insect muscle GABARs may represent an additional distinction between muscle and neuronal GABAR subtypes in insects, although more benzodiazepines would have to be tested to confirm this hypothesis.

2.2.2.4.2. Barbiturates
Although there have been relatively few studies, there is evidence that insect GABARs possess a site for the action of barbiturates. Enhancement of responses to GABA recorded in locust neurons by barbiturates has been reported (Lees et al., 1987), and phenobarbital enhances responses mediated by the heterologously expressed H. virescens GABAR (Wolff and Wingate, 1998). In contrast, sodium pentobarbital at 100 μM failed to modify

responses to GABA recorded in cockroach coxal levator muscle (Schnee *et al.*, 1997).

2.2.2.4.3. Other allosteric modulators

In addition to the benzodiazepine and barbiturate allosteric modulator sites, there are reports of modulatory actions of other compounds. Pregnane steroids displace [^{35}S]TBPS binding from *M. domestica* membranes, but with low affinity, while the insect steroids ecdysone and 20-hydroxyecdysone are almost without effect (Rauh *et al.*, 1990). Other allosteric modulators of insect GABARs include thymol (Gonzalez-Coloma *et al.*, 2002). Priestley *et al.* (2003) have shown that thymol acts at a novel allosteric site. Propofol, widely used as an anaesthetic, enhances GABA responses mediated through RDL by up to about nine-fold with an EC$_{50}$ of about 8 μM (Pistis *et al.*, 1996a). The action of propofol on vertebrate GABA$_A$ receptors is critically dependent upon the presence of a specific amino acid in the second transmembrane domain of the receptor (Pistis *et al.*, 1999).

2.2.3. Diversity: Existence of Subtypes

One of the most convincing demonstrations of the existence of multiple GABAR subtypes is provided where different tissues within the same animal possess receptors of distinct pharmacology. Bicuculline sensitive and insensitive GABARs are present in the nervous system of the moth, *M. sexta* (Waldrop, 1994; Waldrop *et al.*, 1987a,b; Sattelle *et al.*, 2003). In *P. americana*, muscle receptors differ from nervous system receptors in their insensitivity to isoguvacine and higher sensitivity to channel blockers, along with differences in their sensitivity to barbiturates and benzodiazepines, although the latter is less likely to be useful in defining the muscle subtype given the wide variation in benzodiazepine pharmacology among different GABARs. Taken together, however, the differences in pharmacology between insect muscle and neuronal receptors is sufficient to justify considering them as distinct receptor subtypes.

An even more convincing demonstration of the existence of GABAR subtypes is furnished where two types of GABARs can be demonstrated on the same cells. Here, work on identified cells is particularly compelling. DUM (dorsal unpaired median) neurons of the cockroach have been shown to possess at least two subtypes of ionotropic GABAR, one of which is regulated by intracellular calcium ions through a calmodulin pathway, while the other is sensitive to CACA but seemingly not regulated by intracellular calcium (Alix *et al.*, 2002).

Clearly, not only are insect GABARs to be distinguished pharmacologically from the major vertebrate subtypes, but there is also considerable evidence to suggest that insect GABARs are not of one pharmacologically uniform type. However, there is not enough data available yet to group all the known insect GABARs into a clearly delimited classification scheme. We would argue, however, that a classification into bicuculline sensitive versus insensitive and a subset of bicuculline insensitive muscle receptors is justified on the basis of currently available evidence.

2.2.4. Molecular Biology

2.2.4.1. Cloned receptors

Ionotropic GABAR subunits have been cloned from eight species of insect. These fall into three distinct types: RDL, which has been found in a number of insect species, and lignad-gated chloride channel 3 (LCCH3) and GABAA and glycine receptor-like subunit of *Drosophila* (GRD), which are only known to date in *Drosophila*. All three are structurally similar to vertebrate GABA$_A$ receptor subtypes, and more generally to the cys-loop family of neurotransmitter receptors: they have four transmembrane domains of which the second possesses an amphipathic helix and a dicysteine motif in the extracellular domain (see Hosie *et al.*, 1997). GRD resembles both ionotropic GABA and glycine receptors of vertebrates (Harvey *et al.*, 1994). It has 33–44% identity to vertebrate GABAR subunits, but is distinguished by a large insertion between the dicysteine loop and the first transmembrane spanning region. Such a feature is not seen in any other ligand gated chloride channel. Its highest sequence identity is to vertebrate GABA$_A$ α subunits (40–44% identity), but it has almost the same degree of sequence identity to vertebrate glycine receptor α subunits (40–41% identity). GRD does not form receptors in *Xenopus* oocytes, but its presumed M2 region contains seven of the eight amino acid motif (TTVLTMTT) that is a signature of ligand gated chloride channels (Harvey *et al.*, 1994). Furthermore, *Xenopus* oocytes, injected with mRNA encoding the GRD subunit, do not respond to a wide range of amino acids and glutamate receptor agonists, neither does GRD form functioning receptors when coinjected with bovine GABA$_A$ α1 or β1 subunit or rat glycine β subunits (Harvey *et al.*, 1994). In light of the failure of GRD to support GABA responses, it is probably significant that GRD lacks any equivalent to the ligand binding domain II (TGSY) of GABA$_A$ receptors. There is, therefore, reason to suspect that GRD may not even be a GABAR subunit. LCCH3 has high (approximately 47%) identity to vertebrate GABA$_A$ β subunits (Henderson *et al.*, 1993, 1994).

2.2.4.2. RDL: a model insect ionotropic GABAR

The *rdl* (resistance to dieldrin) gene was originally isolated from a naturally occurring dieldrin resistant strain of *D. melanogaster* (ffrench-Constant *et al.*, 1991, 1993a). The *rdl* locus maps to position 66F of chromosome III. It has nine exons, alternative splicing of two of which gives rise to four gene products (**Figure 6**), of which the most extensively studied are RDLac and RDLdb. The *rdl* gene is not limited to *Drosophila*, nor indeed even to the Diptera. RDL-like GABAR subunits with 85–99% amino acid identity have also been cloned from *Myzus persicae* (Anthony *et al.*, 1998), *H. virescens* (Wolff and Wingate, 1998), *Aedes aegypti* (Thompson *et al.*, 1993a), *Drosophila simulans* (ffrench-Constant *et al.*, 1993a), *M. domestica* (Thompson *et al.*, 1993a), *Blatella germanica*

(Thompson *et al.*, 1993b; Kaku and Matsumura, 1994), *Tribolium castaneum* (Miyazaki *et al.*, 1995), and *Hypothenemus hampei* (Andreev *et al.*, 1994), indicating that it is widespread throughout insect orders. The RDL polypeptide has between 30% and 38% identity with vertebrate GABAR subunits (about the same identity among different vertebrate subunit types).

RDL GABAR subunits encoded by the *rdl* gene are distributed throughout the adult (Harrison *et al.*, 1996) and embryonic (Aronstein and ffrench-Constant, 1995) nervous system of *Drosophila*, particularly in the optic lobes, ellipsoid body, fan-shaped body, ventrolateral cerebrum, and glomeruli of antennal lobes. RDL is not expressed in muscle. Furthermore, in *Drosophila*, RDL expression is largely confined to the neuropil, with no discernible expression on neuronal cell body membranes or muscle (Harrison *et al.*, 1996), although cell bodies were clearly stained in *P. americana* (Sattelle *et al.*,

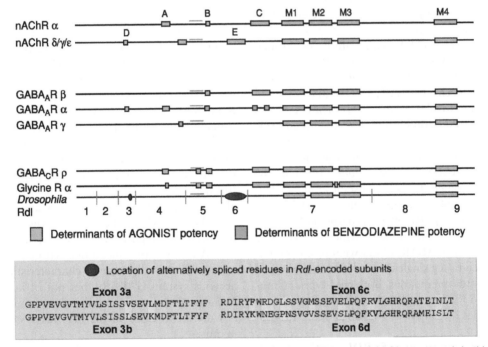

Figure 6 The locations of the determinants of agonist potency of cys-loop receptors are highly conserved. In this schematic alignment of cys-loop receptor subunits, the location of the cysteine-loop is indicated by the line above the subunit. The location of known determinants of agonist potency in nACh receptors, termed loops A–E, are marked yellow, while those of benzodiazepine potency are marked green. The exon boundaries of Rdl-encoded polypeptides are marked by vertical red lines. As a result of the alternative splicing of two exons, the *Drosophila* Rdl gene encodes four polypeptides, each of which exhibits characteristic features of GABA$_A$ receptor subunits. The alternatively spliced exons (3 and 6) encode regions of the extracellular N-terminal domain. There are two variant forms of each exon; those of exon 3 are termed "a" and "b" and differ by two residues, while the alternate forms of exon 6, termed "c" and "d", differ at 10 residues. Thus, depending on the splice variants present in a given polypeptide, the different Rdl encoded subunits may be referred to as RDL$_{ac}$, RDL$_{bd}$, etc. mRNAs encoding all four splice variants have been identified in embryonic *D. melanogaster*. The positions of these variant residues are marked in purple. The two alternate residues encoded by exon 3 lie close to a known determinant of the agonist potency of GABA receptors, while the 10 variant residues encoded by splice variants of exon 6 span a region that is poorly conserved in vertebrate GABA receptor subunits, but which corresponds to determinants of agonist potency in nAChRs. (Reproduced with permission from Hosie, A.M., Aronstein, K., Sattelle, D.B., ffrench-Constant, R.H., **1997**. Molecular biology of insect neuronal GABA receptors. *Trends Neurosci. 20*, 578–583.)

2000) and *Acheta domesticus* (Strambi *et al.*, 1998). RDL is very strongly expressed (Harrison *et al.*, 1996), where the pattern of staining is compartmentalized, suggesting that some, but not all, Kenyon cells (the principal interneurons of the mushroom bodies) receive GABAergic inputs (Brotz *et al.*, 1997). The significance of this pattern of expression is discussed in the following section. The distribution of RDL staining corresponds closely to staining for glutamic acid decarboxylase (GAD; the enzyme involved in the synthesis of GABA) (Buchner *et al.*, 1988), GABA, and synaptotagmin (DiAntonio *et al.*, 1993), suggesting that they contribute to the formation of synaptic receptors. Similar patterns of distribution in the mushroom bodies have been observed in the case of the cockroach *P. americana* (Sattelle *et al.*, 2000) and the cricket *A. domesticus* (Strambi *et al.*, 1998), where again RDL distribution closely matched GABA staining, although details of the pattern of staining differed somewhat across different species.

When expressed in *Xenopus* oocytes, RDL forms receptors (presumably homomeric, as no endogenous GABAR subunits have been reported in these cells) that respond to GABA with thresholds around 1–10 μM (ffrench-Constant *et al.*, 1993a, 1993b; Hosie *et al.*, 1995a, 1995b) as well as in baculovirus transfected sf9 cells (Lee *et al.*, 1994) and a *Drosophila* S2 cell line (Millar *et al.*, 1994). These currents reverse close to the equilibrium potential for chloride ions. Responses are dose-dependent and there is little evidence of rapid (<20 s) desensitization. The responses to GABA are very robust, often in the microampere range (ffrench-Constant *et al.*, 1993a, 1993b). The stable, inducible expression of RDL in the S2-RDL *Drosophila* cell line is particularly convenient: it allows long-term maintenance of a cell line expressing insect GABARs of known molecular composition, facilitating single channel kinetic analysis of ligand–receptor interactions. Both wild-type and mutant (RDL A302S) receptors have been expressed in these cell lines (Lee *et al.*, 1994; Millar *et al.*, 1994).

2.2.4.2.1. *Comparison of RDL with in situ GABARs of insects*

Currently there is little information on *Drosophila* native GABARs with which to compare the pharmacology or physiology of recombinant RDL receptors. However, it has been shown that GABA responses of *Drosophila* motor neurons reverse around the resting potential and are blocked by PTX (Rohrbough and Broadie, 2002). Neurons cultured from embryonic *Drosophila* form synapses with one another, which include GABAergic synapses (Lee and O'Dowd, 1999). More recently, GABAergic synapses between embryonic *Drosophila* neurons have been shown to be PTX-sensitive, and this sensitivity is reduced in neurons derived from the dieldrin insensitive mutant strain (Lee *et al.*, 2003), suggesting that these synaptic connections involve RDL. Also, *Drosophila* CNS midline precursors in culture show responses to GABA (Schmidt *et al.*, 2000). Similarly, GABA gated currents of cultured *Drosophila* neurons are also blocked by PTX but are insensitive to bicuculline (Zhang *et al.*, 1994). Thus, what little is known of *Drosophila* GABARs *in situ* suggests that they resemble typical bicuculline insensitive insect GABARs.

2.2.4.3. Pharmacology of heterologously expressed RDL

2.2.4.3.1. *The agonist site*

RDL of *D. melanogaster*, and its equivalent from the mosquito, *Aedes aegypti*, form functional homomers when expressed in *Xenopus* oocytes (ffrench-Constant *et al.*, 1993a, 1993b; Hosie *et al.*, 1996; Shotkoski *et al.*, 1994) as well as when stably expressed in a *Drosophila* cell line (Millar *et al.*, 1994). *Drosophila melanogaster* RDL expressed in *Xenopus* oocytes has a rank order of agonist potency of muscimol = GABA = TACA > CACA ≫ glycine (Hosie *et al.*, 1995a, 1995b). Similarly, the rank order of potency of agonists acting on RDL stably expressed in an insect cell line is GABA = TACA > CACA ≫ glycine = taurine (Buckingham *et al.*, 1996). The lack of activity of glycine is of interest, since the amino acid sequence of RDL more closely resembles that of glycine receptors than it does of many GABAR subunits (ffrench-Constant and Rocheleau, 1992). RDL is also insensitive to 3-APS yet is activated by isoguvacine (Hosie and Sattelle, 1996a). The high potency of muscimol (**Figure 2**), the sensitivity of RDL to CACA and isoguvacine, along with its insensitivity to 3-APS, are all features characteristic of insect nervous system GABARs, but not of insect muscle (see above). This supports the observation that RDL is widely expressed in the nervous system but not in muscle.

An interesting set of findings resulting from a chance observation of atypical agonist induced currents in RDL led to the discovery of the strong pH sensitivity of RDL. The GABA agonists THIP (4,5,6,7-tetrahydroisoxazolo[5,4-c]pyridin-3-ol) and ZAPA (Z-3-[(aminoiminomethyl)thio]prop-2-enoic acid) appeared to lead to a desensitizing response, followed by a pronounced "off" response upon agonist washout (Matsuda *et al.*, 1996). Further investigation showed this to be the result of acidification of the test saline by the agonists. When this was corrected, more typical responses were observed.

What is particularly significant about this, however, is that the apparent off response was due to a pH dependency of the efficacy of these two compounds–preacification of the saline greatly reduced the amplitude of the responses to these agonists. The actions of other agonists tested on RDL, including GABA, were pH independent. Other studies had shown that protonation of histidine residues underlies pH dependent actions of GABA on recombinant mammalian receptors (Krishek *et al.*, 1996), as well as the pH dependency of the action of zinc ions (Wang *et al.*, 1995). This points to further differences between the agonist binding sites of RDL and vertebrate GABA$_A$ receptors. From the pKa of GABA and THIP at the pH of the uncorrected test solutions (about 4.5), about 76% of the hydroxyls of GABA would be dissociated and about 50% of the hydroxyls of THIP would be dissociated. Thus, some of the differences in pH sensitivity of these drugs may be due to the importance of the dissociated state of the acidic group in ligand binding. Alternatively, these differences may also be attributable to pH dependent states of the histidine residues that form part of the receptor. There are seven histidine residues in the extracellar N-terminal of RDL, and future studies of the effects of point mutations involving each of these residues may help uncover new insights into the agonist actions on RDL.

Another key feature in which RDL resembles *in situ* insect GABARs is its insensitivity to bicuculline. This insensitivity to bicuculline is seen whether RDL is expressed transiently in *Xenopus* oocytes (Hosie and Sattelle, 1996a) or stably expressed in a *Drosophila* cell line (Millar *et al.*, 1994). Bicuculline insensitivity is a feature also possessed by the *rdl* gene cloned from *H. virescens* (Wolff and Wingate, 1998). Since, as argued above, most insect GABARs are bicuculline insensitive, RDL thus resembles the majority of insect GABARs, including those expressed on cultured *Drosophila* neurons (Zhang *et al.*, 1994).

In some respects, RDL is more similar to GABA$_C$ receptors and the recombinant receptors formed by expression of the vertebrate ρ subunit in their insensitivity to bicuculline and the ability to form functional homo-oligomers. However, it is clear that the agonist site of RDL is dissimilar to that of GABA$_C$ receptors in its preferred pharmacophore. At the sequence level, RDL bears no more closer resemblance to GABA$_C$ receptors than it does to GABA$_A$ receptors, and in fact more closely resembles glycine receptors (ffrench-Constant and Rocheleau, 1992). Thus, while RDL homomers are to be distinguished from GABA$_A$ receptors by their bicuculline insensitivity and the low potency of 3-aminosulfonic acid, they are also to be distinguished from GABA$_C$

receptors by the high efficacy of muscimol and isoguvacine.

2.2.4.3.2. The convulsant antagonist site The RDL channel expressed in *Xenopus* oocytes or the *Drosophila* S2 cell line is effectively blocked by 1–10 μM concentrations of PTX, TBPS, EBOB, dieldrin, BIDN, and fipronil (ffrench-Constant *et al.*, 1993a, 1993b; Millar *et al.*, 1994; Shirai *et al.*, 1995; Buckingham *et al.*, 1996). For structures of these molecules see **Figures 3–5**. The IC$_{50}$ for PTX block of RDL expressed in *Drosophila* cells was estimated to be 2.5 μM, whereas 10 μM TBPS reduced GABA evoked currents by less than 50%. This higher efficacy of PTX compared to TBPS is also typical of *in situ* insect GABARs, whereas the reverse is true for vertebrate GABARs. Curiously, the RDL homolog from *H. virescens* is insensitive to dieldrin (Wolff and Wingate, 1998). The block of RDL mediated GABA responses by PTX has both competitive and noncompetitive features, in that it reduces the maximal response to GABA and shifts the EC$_{50}$ of the GABA response (Shirai *et al.*, 1995). A similar action of PTX was reported for a crustacean GABAR (Smart and Constanti, 1986). As is the case for insect GABARs *in situ*, PTX is a more potent blocker of RDL than TBPS (Buckingham *et al.*, 1996), whereas the reverse is true for vertebrate GABA$_A$ receptors.

The convenience of the availability of a robust, functional homomeric GABAR was exploited by Shirai and coworkers (Shirai *et al.*, 1995) to determine the key structural elements of the PTX molecule that determine its potency. Using the same approach as that on expressed vertebrate GABA$_A$ receptors (Anthony *et al.*, 1993) and *in situ* insect receptors (Anthony *et al.*, 1994), the study agreed with earlier ones in that picrotin and PTX acetate were found to be of less efficacy in blocking the GABA response, whereas fluoropicrotoxin was of similar potency to PTX. Thus, the PTX recognition site of RDL resembles that of *in situ* insect GABARs. The potential that RDL offers over *in situ* receptors, namely that the subunit composition is known, could be exploited further in elucidating the mechanisms of action of PTX and related compounds.

These structure–function studies did not uncover any significant difference between vertebrate and insect GABARs at the convulsant site, yet such differences presumably exist, as is inferred from the selectivity of certain insecticides known to act at the convulsant site. Such differences did, however, emerge in a subsequent study by Sattelle and colleagues (Hosie *et al.*, 1996), which used a related series of PTX-like compounds, the picrodendrins, to probe the convulsant antagonist site. Picrodendrins and related tutin analogs are terpenoid toxins present in extracts of a plant of the Euphorbiaceae, *Picrodendron baccatum*. These plants are known as "mata becerro," or calf-killer, in the Dominican Republic where they have been used to kill bed bugs and lice (Ozoe *et al.*, 1994). The picrodendrins are structurally similar to PTX, and share its polycyclic lactone picrotoxane skeleton, acting at a similar site in the channel of vertebrate GABA$_A$ receptors (Ozoe *et al.*, 1994). The series of terpenoids studied differed mainly in their substituents at the C9 position. In most cases, these substituents were a spiro-γ-butyrolactone ring, providing one of at least two electronegative groups believed to be necessary for picrotoxane activity (Ozoe and Matsumura, 1986). Alteration of these substituents produced differing effects on their potency on RDL when compared to their effectiveness in displacing [^{35}S]TBPS from rat membranes (Ozoe *et al.*, 1994). First of all, most of the terpenoids were significantly less effective at blocking RDL than at displacing TBPS from rat membranes. More importantly, however, there were considerable differences in their rank order of potency on RDL compared to rat GABARs. Molecules that were most potent against RDL shared an olefin binding that brought the side chains closer to the plane of the spiro-γ-butyrolactone ring than the other compounds. This study succeeded in showing the importance of the position of the electronegative centers in the PTX-like family, which had not emerged from studies restricted to the C4 substituents and the isopropenyl moiety.

The bicyclic dinitrile BIDN, developed in the face of a need for a new, specific ligand for the insect convulsant site, is a potent blocker of recombinant RDL GABA receptors. At 1 μM, BIDN reversibly reduces the amplitude of RDL mediated GABA responses in *Xenopus* oocytes, with only a slight dependence on membrane potential (Hosie *et al.*, 1995b). This action was dose dependent, with an IC$_{50}$ that depended upon the concentration of GABA used. When tested against 20 μM GABA the IC$_{50}$ for BIDN action was estimated to be 1.0 μM, whereas when tested against 100 μM GABA it was estimated at 20 μM. This is suggestive of allosteric interactions between the BIDN and GABA ligand binding sites. Like PTX, the action of BIDN displayed both competitive and noncompetitive elements, in that it reduced the maximum GABA induced currents and shifted the dose–response curve to the right, pointing to allosteric actions of BIDN. BIDN also reversibly blocks RDL mediated responses in the S2-RDL and S2-RDL(A302S) cell lines (Buckingham *et al.*, 1996). In both the

Xenopus oocyte and S2 cell line expression systems, it was noticed that GABA responses in the presence of BIDN lacked the desensitization that was so clearly present in control responses (Buckingham *et al.*, 1996). One plausible interpretation of this might simply be that, like PTX, BIDN exerts its effects through an enhancement of desensitization, but favoring a different state than PTX. Evidence that BIDN and fipronil both operate by reducing open-state probability, but through different mechanisms, also comes from single channel studies (Grolleau and Sattelle, 2000).

RDL expressed in *Xenopus* oocytes and in a *Drosophila* cell line is sensitive to block by the phenylpyrazole insecticide, fipronil (Millar *et al.*, 1994; Hosie *et al.*, 1995a). In *Xenopus* oocytes, RDL mediated responses to 50 μM GABA are reduced by up to 80% by 100 μM fipronil. Recovery from fipronil block was very slow, and still incomplete after 10 min washout. This differs from the actions of both PTX and BIDN, which reverse completely and rapidly. Like the block by BIDN (Hosie *et al.*, 1995b), fipronil block was voltage independent (Hosie *et al.*, 1995a). A further similarity to the actions of both BIDN and PTX is that the action of fipronil consists of both competitive and noncompetitive elements, and is also dependent on the concentration of agonist. Thus, fipronil, BIDN, and PTX share many features (voltage independence, agonist dependence, and mixed mode of action) suggesting that they operate through similar mechanisms. However, it must be noted that whereas PTX and the cyclodienes are competitive inhibitors of [^{3}H]EBOB binding, fipronil is a noncompetitive displacer of [^{3}H]EBOB binding, at least in *M. domestica* (Deng *et al.*, 1991, 1993). The possible significance of this in understanding the mechanisms of insecticide resistance is discussed below.

Differences in the action of fipronil and BIDN on RDL were examined in a single channel patch-clamp study (Grolleau and Sattelle, 2000). The ability of RDL to form functional homomers simplifies the interpretation of experiments on the effects of the A302S mutation on single channel properties of the channel. When applied to outside-out patches from S2-RDL cells, GABA (50 μM) evoked inward currents that were completely blocked by 100 μM PTX. As had been reported for outside-out patches of *Xenopus* oocytes expressing RDL, the current–voltage relationship showed slight inward rectification at positive potentials. The conductance of the single channel currents was around 36 pS, but interestingly a 71 pS conductance was observed when the application of GABA was longer than 80 ms. When either 1 μM fipronil or 1 μM BIDN was present in the external saline, all the GABA gated channels were completely blocked. At lower concentrations, near their respective IC_{50} values, the duration of channel openings was shortened. Interestingly, the blocking action of BIDN resulted in the appearance of a novel channel conductance (17 pS). Examination of the effects of coapplication of these two convulsants (applying 300 nM BIDN with fipronil concentrations ranging from 100 to 1000 nM) revealed an additional BIDN induced dose dependent reduction in the maximum open probability. Thus, while both BIDN and fipronil shorten the duration of wild-type RDLac GABAR channel openings, they appear to do so by acting at distinct sites.

The findings reported to date on the actions of convulsants on RDL agree very closely with their actions on *in situ* insect GABARs. The agonist dependence and mixed competitive/noncompetitive action are consistent with a model of PTX action in which the PTX molecule binds preferentially to an agonist-bound conformational state of the receptor and hence increases the probability of a transition into a desensitized state. Such a conclusion was advanced to explain similar properties of PTX action at the crayfish neuromuscular junctional GABARs (Smart and Constanti, 1986) and the finding that the rate of development of a steady-state blockade of GABARs on rat sympathetic neurons was enhanced when GABA is applied in brief, rapidly repeated doses (Newland and Cull-Candy, 1992).

RDL has also been used to predict cross-resistance to insecticides acting at the convulsant site. Dieldrin resistant RDL mutant GABARs are not only comparatively insensitive to PTX, but are also insensitive to BIDN (Hosie *et al.*, 1995b). Similar results were obtained for fipronil (Hosie *et al.*, 1995a) although other workers reported different results (Wolff and Wingate, 1998). Furthermore, a novel tricyclic dinitrile, KN244, blocks RDL and the sensitivity of dieldrin resistant RDL to it is reduced 100-fold (Matsuda *et al.*, 1999).

2.2.4.3.3. Allosteric modulators

Responses mediated by RDL are sensitive to potentiation by benzodiazepines, but their benzodiazepine pharmacology differs from *in situ* insect receptors. Insect GABA responses are potentiated by Ro 5-4864 and flunitrazepam. Responses to GABA mediated by RDL expressed in *Xenopus* oocytes (Hosie and Sattelle, 1996b) were potentiated by micromolar concentrations of Ro 5-4864. In contrast, the same receptor expressed in the *Drosophila* S2 cell line appeared to be insensitive to Ro 5-4864 (Millar

et al., 1994), although enhancement was obtained when the concentration of the benzodiazepine was increased to 100 μM (Buckingham *et al.*, 1996b). This illustrates that caution must be exercised in interpreting negative data until a wide range of ligands is tested. However, flunitrazepam, which is active at GABARs on cockroach motor neuron, D_f, at micromolar concentrations, was found to be ineffective on RDL in either expression system even at concentrations of 100 μM (Millar *et al.*, 1994; Hosie and Sattelle, 1996b). This represents a departure from most *in situ* insect receptors studied to date. Significantly, the identified motor neuron D_f of the cockroach stains positively to the RDL antibody (Sattelle, personal communication), suggesting that in this cell at least, RDL may coassemble with another subunit. Flunitrazepam is also reported to have no effect on a GABAR cloned from *H. virescens* (Wolff and Wingate, 1998). Nonetheless, the comparatively low sensitivity of RDL to benzodiazepines, coupled with its sensitivity to the "peripheral" benzodiazepine, Ro 5-4864, are both characteristic of insect receptors compared to that of vertebrates.

That RDL should differ from *in situ* receptors principally at the benzodiazepine site is not surprising, given the wide variation in benzodiazepine pharmacology of GABARs generally, and may indicate that RDL coassembles with another subunit subtype *in situ*. Since only two other GABARs have been cloned from *Drosophila* to date, these are the obvious candidates for the other subunit. However, it is unlikely that LCCH3 is the additional subunit, since, as described above, receptors formed by coexpression of RDL with LCCH3 are quite unlike insect receptors. In any case, the lack of overlap of expression of these two subunits at any stage of development makes it highly unlikely that they coassemble *in vivo*. The question of whether RDL forms a homomer *in situ* is discussed in a separate section below. Unfortunately, coexpression of RDL with GRD has yet to be described. While expression studies on the agonist pharmacology of three of the four RDL splice variants has been described in some detail (Chen *et al.*, 1994; Hosie and Sattelle, 1996a), benzodiazepine pharmacology of these variants, or combinations of them, has not yet been assayed.

Like *in situ* insect receptors, RDL is potentiated by barbiturates. In both the *Xenopus* oocyte and the *Drosophila* S2 cell line expression systems, RDL mediated GABA responses were enhanced by either sodium phenobarbitone or sodium pentobarbitone by up to 50% of control values (Hosie and Sattelle, 1996b; Buckingham *et al.*, 1996). Similar results were reported for the RDLbd splice variant, whose responses were enhanced up to fivefold (Belelli *et al.*,

1996a). Similarly, the RDL equivalent cloned from *H. virescens* is also subject to barbiturate (phenobarbital) enhancement (Wolff and Wingate, 1998). In contrast to vertebrate GABARs, which are also sensitive to barbiturates, elevation of the barbiturate to high (1 mM) concentrations did not evoke *Drosophila* RDL mediated currents in the absence of GABA (Belelli *et al.*, 1996a; Buckingham *et al.*, 1996). It is therefore likely that the two components of barbiturate action on vertebrate receptors – the enhancement of the GABA response and the direct opening of the channel in the absence of GABA – are mediated by two separate sites, of which only one may be present on insect GABARs.

RDL is comparatively insensitive to steroids. The steroid, 5α-pregnan-3α-ol-20-one, enhances human $\alpha_3\beta_1\gamma_{2L}$ receptors, but at high concentrations had only a minimal effect on RDLbd (Belelli *et al.*, 1996) and RDLac (Millar *et al.*, 1994; Hosie and Sattelle, 1996b). Insect GABARs *in situ* are also highly insensitive to steroids (Rauh *et al.*, 1990).

The anticonvulsant, loreclezole, is a potentiator of vertebrate recombinant GABARs containing the β_2 or β_3 subunit (Wafford *et al.*, 1995), but not of those containing a β_1 subunit (Wingrove *et al.*, 1994). This is attributed to the presence of a serine in the β_2 and β_3 subunits, which is an asparagine at the equivalent position in the β_1. Curiously, although a mutation of this serine to a methionine abolishes loreclezole sensitivity (Stevenson *et al.*, 1995), RDL (which has methionine at this position) is sensitive to loreclezole (Belelli *et al.*, 1995; Hosie and Sattelle, 1996b).

RDL is also sensitive to a number of other allosteric modulators. Isoflurane enhances GABA responses mediated by expressed RDL and blocks at higher concentrations. This block is not seen for the cyclodiene resistant A302S mutant, suggesting that the block is due to an interaction at the PTX site (Edwards and Lees, 1997).

These data amply illustrate the usefulness of the cloned RDL receptor in advancing our understanding of insect GABARs. More than this, however, RDL, because of its ability to form functional homomers, has also been of use in structure/function studies on allosteric modulator sites of vertebrate GABARs. For instance, RDL is insensitive to the anesthetic etomidate, as are vertebrate β_1 subunits. Vertebrate β_2 and β_3 subunits, however, do confer sensitivity to etomidate and this is attributable to the presence of an asparagine residue at equivalent positions in a transmembrane-spanning domain, which is a methionine at the equivalent position in RDL. Mutating M314 to an asparagine in RDL rendered the expressed receptors sensitive to

enhancement by etomidate (McGurk *et al.*, 1998). As in vertebrate recombinant receptors, this sensitivity is stereospecific, whereas the native sensitivity to pregnane steroids or pentobarbitone remained unchanged. Conversely, the N289M mutation introduced into vertebrate β_2 abolished the etomidate sensitivity. Such complementarity of effects of mutation encourages confidence in interpreting experiments using an invertebrate receptor as a model of vertebrate receptors, especially in view of the comparative simplicity of interpreting data obtained using homomers.

2.2.4.3.4.   Other (RDL) ligands   RDL, like GABARs of the motor neuron, D_f, of the cockroach (Bai, 1994), is insensitive to zinc (Hosie and Sattelle, 1996b), which blocks many vertebrate GABA$_A$ receptors (Hosie *et al.*, 2003), as well as ρ receptors.

2.2.4.4.   Conclusions   There can be little doubt that the availability of RDL has accelerated the pace of research into the molecular pharmacology of insect GABARs. The chief advantage that it confers, in addition to the convenience of performing experiments on oocytes, is that hitherto studies aimed at uncovering the mechanisms of drug–receptor interaction for insect GABARs have been hindered by the fact that in no case is the subunit composition of any *in situ* GABAR known. It is of great significance that a number of different laboratories have confirmed independently that whatever the expression system chosen (*Xenopus* oocytes, *Drosophila* S2 cells, *Spodoptera* Sf9 cells), the pharmacology of the expressed RDL receptor does not differ significantly. Indeed one paper (Buckingham *et al.*, 1996) examined this issue directly. This encourages confidence that RDL is not coexpressing with GABAR subunits native to the respective expression systems, and that we can be reasonably certain that the subunit composition of the expressed RDL receptor is a homomeric pentamer. To appreciate the significance of such an advantage, we have only to look at the advances in understanding of nicotinic receptors that emerged from research using another homomer forming receptor, the vertebrate nicotinic α_7 receptor.

Since RDL homomers so closely mimic *in situ* GABARs, and since RDL antibody staining corresponds closely to GABA immunoreactivity, it is likely that RDL contributes to most of the GABARs in the insect brain. In view of the conclusion drawn above that there is pharmacological diversity of subtypes of GABARs in insects, the question of the source of this diversity remains to be addressed. The occurrence of alternative splicing and RNA editing may add to functional diversity of the small *Drosophila* GABAR gene family. Evidence for this possibility is discussed in the following section.

2.2.4.5. Alternative splicing in GABA receptor subunits

RDL is expressed in the form of four splice variants, arising from alternative splicing of two of its nine exons (exons 3 and 6). These four distinct polypeptides show characteristic structural features of a GABAR subunit. The regions in which the splice variants differ align closely with regions known to be determinants of agonist potency in vertebrate ionotropic receptor subunits. This is highly unusual. Although there are many examples of alternatively spliced GABAR subunit genes in vertebrates, the splicing usually occurs in their large intracellular loops. All four of these splice variants are expressed in embryonic *Drosophila* (ffrench-Constant and Rocheleau, 1992), yet their functional roles remain to be determined.

Since the pharmacological diversity observed among vertebrate GABA$_A$ receptors arises at least in part from the assorted assembly from different subunit isoforms, the alternate splicing of the *rdl* gene may serve a physiological role by increasing receptor diversity. Since RDL appears to be a component of the majority of insect GABARs, such a splicing mechanism may represent another route to subunit diversity in a small GABAR gene family. Evidence that alternative splicing confers pharmacological diversity comes from expression studies that have described pharmacological differences between these variants at the agonist site. For example, the three splice variants tested to date in functional expression studies (RDLac, RDLad, RDLbd) all differ in their EC$_{50}$ values for GABA (Hosie and Sattelle, 1996a). The region in which these variants differ include loop F of the GABA binding site.

2.2.4.6. Does RDL form a homomer *in situ*?

Despite our limited knowledge of the pharmacology of native GABARs of *Drosophila*, it is clear that RDL homomers mimic closely the pharmacology of insect GABARs. The general features that distinguish most insect GABARs from those of vertebrates – insensitivity to bicuculline, comparative insensitivity to benzodiazepines, sensitivity to the peripheral benzodiazepine Ro 5-4864, insensitivity to TBPS, and sensitivity to isoguvacine – are all seen in RDL homomers. The only challenge to this, the insensitivity of RDL to flunitrazepam, is perhaps not particularly serious in view of the diversity of benzodiazepine pharmacology amongst GABA$_A$ receptors studied to date and the acute dependence of benzodiazepine pharmacology upon subunit composition.

Where this difference in benzodiazepine pharmacology gains significance, though, lies in the observation that a *Drosophila* anti-RDL antibody stains motor neuron D$_f$ of the cockroach (Sattelle, personal observation), which is known to possess GABARs sensitive to micromolar concentrations of flunitrazepam (Sattelle *et al.*, 1988). Since RDL, which is flunitrazepam insensitive, is expressed on D$_f$, which is flunitrazepam sensitive, this difference hints at the existence of another subunit subtype in these cells.

Single channel recordings also provide reason to suspect that RDL may not always exist as a homomer *in situ*. Single, GABA gated channels recorded from *Xenopus* oocytes injected with *rdl* cRNA were inwardly rectifying, with inward and outward currents of around 21 pS and 12 pS, respectively (Zhang *et al.*, 1994). GABA gated single channels recorded from *Drosophila* larval neurons in culture, although also inwardly rectifying, had inward and outward conductances of around 28 and 29 pS, respectively (Zhang *et al.*, 1994), a value significantly different from expressed RDL homomers. Mean open times of RDL homomers in *Xenopus* oocytes were significantly shorter than single GABA gated channels in cultured *Drosophila* neurons: around 62 ms compared to 118 ms, respectively. A simple explanation for these findings would be that RDL coexpresses with another subunit *in situ*.

However, single channels recorded from a *Drosophila* S2 cell line stably expressing RDL were estimated to be 36 pS (Grolleau and Sattelle, 2000), a value also significantly different from values for RDL recorded in oocytes. Both these studies used the same concentration of GABA (50 μM) and both used brief applications of GABA to outside-out patches. So it is likely that this difference is due to the different expression environments or perhaps the different ionic compositions of the recording media. Given that no other differences in pharmacology have been detected to date between the two expression systems, and that both studies identified a slight inward rectification, the latter is the more likely explanation. It is interesting to note that Grolleau and Sattelle (2000) reported the emergence of a larger conductance when the duration of the GABA pulse was prolonged to over 80 ms, suggesting that the conductance properties of this channel may be more complex than hitherto suspected.

Does RDL coexpress with the other two insect GABAR subunits known to date, LCCH3 or GRD? LCCH3 (Henderson *et al.*, 1993) closely resembles vertebrate GABA$_A$ receptor β subunits. Neither LCCH3 nor GRD form functional homooligomeric receptors in *Xenopus* oocytes (Harvey *et al.*, 1994; Zhang *et al.*, 1994). When coexpressed with RDL in baculovirus transfected *Spodoptera frugiperda* Sf21 cells, however, LCCH3 forms a receptor that exhibits a biphasic response (Zhang *et al.*, 1994), which only becomes clear during prolonged applications of GABA (>1 s). The first component was a rapidly activating, rapidly desensitizing current with kinetics indistinguishable from those of RDL expressed alone. The second component, however, was of much slower kinetics of activation and desensitization, and was attributed to the RDL/LCCH3 combination. The coapplication of 500 µM PTX completely abolished the fast component and blocked about half of the slow current. In contrast, bicuculline (100 µM) completely abolished the slow component leaving the fast component unaffected. One simple explanation of these findings is that the fast component is attributable to RDL homomers that, in accord with other reports, is PTX sensitive and bicuculline insensitive, and that the slow response is mediated by RDL/LCCH3 multimers. This provides evidence that LCCH3 can indeed form a receptor when coexpressed with RDL, but the resulting receptor differs significantly from the pharmacology of all known insect GABARs with respect to picrotoxinin. Their sensitivity to bicuculline further distinguishes them from GABARs of cultured *Drosophila* neurons (Zhang *et al.*, 1994). Furthermore, expression of LCCH3 and RDL do not overlap in the adult–LCCH3 is expressed in cell bodies (Aronstein *et al.*, 1996) whereas RDL expression is restricted to neuropil and is not observed in *Drosophila* cell bodies (Aronstein *et al.*, 1995), although it is observed in cell bodies in other species (Harrison *et al.*, 1996). We therefore conclude that it is unlikely that RDL coexpresses with LCCH3 *in vivo*. It would be most interesting to establish whether perhaps LCCH3 contributes to the formation of bicuculline sensitive GABARs of the kind that are known to exist in *M. sexta* (Waldrop *et al.*, 1987).

One intriguing recent finding suggests that the additional subunit, from which we have excluded LCCH3 as a candidate, may not even be a GABAR subunit. Immunoprecipitation studies using antibodies to a *D. melanogaster* glutamate gated chloride channel (DmGluCl) showed that: (1) DmGluCl alpha antibodies precipitated all of the ivermectin and nodulisporic acid receptors solubilized by detergent from *Drosophila* head membranes; and (2) they also immunoprecipitated all solubilized nodulisporic receptors, but only approximately 70% of the ivermectin receptors (Ludmerer *et al.*, 2002). These data suggest that both DmGluCl α and RDL are components of nodulisporic acid and ivermectin receptors.

Finally, it is of interest to note the recent finding by O'Dowd and colleagues that RDL is involved at GABAergic synapses formed in cultures of embryonic *Drosophila* neurons (Lee *et al.*, 2003). Furthermore, the authors conclude that other subunit(s) are also involved.

2.2.5. Intracellular Modulation of GABARs

Ionotropic receptors are potentially susceptible to modulation by intracellular factors. This may alter their density, kinetics, or pharmacology, and may be an important component in long-term changes in synaptic function, such as may be involved in learning. One type of GABAR present on cockroach (*P. americana*) DUM neurons is regulated by changes in intracellular calcium through a calcium dependent protein kinase pathway (Alix *et al.*, 2002). Similarly, maintenance of GABA responses in isolated neurons from *P. americana* abdominal ganglia requires phosphorylation as well as, possibly, a nucleotide recognition site unrelated to PKA-dependent phosphorylation (Watson and Salgado, 2001). However, although RDL contains possible sites for intracellular phosphorylation (ffrench-Constant *et al.*, 1991), no rundown of RDL mediated responses is seen in *Drosophila* S2 cells in whole-cell patch clamp mode, even when no ATP is included in the pipette (Buckingham, personal observation). This does not, however, exclude the possibility that RDL is subject to modulation by intracellular signaling pathways. The latter is likely to provide a fertile area of study in the future, especially in view of the observations that suggest that RDL may be closely implicated in learning.

2.2.6. GABARs as Targets for Insecticides

Insect GABARs are the major targets for several chemically distinct classes of insecticidally active molecules: trioxabicyclooctanes, dithianes, silatranes, lindane, toxaphene, bi- and tricyclic dinitriles, chlorinated cyclodienes, and picrotoxinin (Deng *et al.*, 1991; Rauh *et al.*, 1997). These sites overlap, since [^{3}H]EBOB is displaced by all these compounds. The importance of the convulsant site for the insecticidal action is illustrated by the close linear correlation between the effectiveness of their displacing [^{3}H]EBOB binding from house fly head membranes and their LD$_{50}$ (Deng *et al.*, 1991).

2.2.6.1. Picrotoxin   Picrotoxin was used as an early insecticide in the preservation of lard (Bentley and Trimen, 1875). Picrotoxin contains picrotin and picrotoxinin. The radiolabeled form of picrotoxinin ([^{3}H]DHPTX) was used as an early probe of insect GABARs, but was later superseded by [^{35}S]TBPs.

2.2.6.2. Polychlorocyclohexanes Chemicals of this group, notably lindane, benzene hexachloride, and α-endosulfan, were used intensively as insecticides in the postwar years, until recent environmental concerns led to their disuse, with the exception of α-endosulfan. Binding (Lawrence and Casida, 1984) and electrophysiological (Wafford *et al.*, 1989) studies confirmed that they act at ionotropic GABARs.

2.2.6.3. Bicyclophosphorus esters and bicycloorthobenzoates The radioligand probe [^{35}S]TBPS overtook [^{3}H]DHPTX as the most useful probe of vertebrate GABA gated chloride channels. However, insect GABARs are comparatively insensitive to TBPS (Lummis and Sattelle, 1986) compared to vertebrate GABA$_A$ receptors, and the bicyclophosphorus esters have comparatively low insecticidal activity (Casida *et al.*, 1988). The next development arose from the discovery that the addition of a substitute benzene ring to the bicyclic ring greatly increased insecticidal potency (Palmer and Casida, 1985), as well as affinity for the convulsant site (Lummis and Sattelle, 1986). Selection for the optimal substituents gave rise to the bicycloorthobenzoates, of which the most frequently used are EBOB (1-(4′-ethyphenyl)-4-[2,3]propyl-2,6,7-trioxabicyclo [2.2.2]octane) (**Figure 3**) and TBOB. EBOB has proven particularly useful in characterizing the convulsant site of houseflies (Cole and Casida, 1992; Deng *et al.*, 1993) and *Drosophila* (Cole *et al.*, 1995).

2.2.6.4. Phenylpyrazoles The phenylpyrazole insecticide fipronil (**Figure 4**) is an effective pest control agent even at low doses (Tingle *et al.*, 2003). It has many characteristics that make it suitable: it degrades slowly with a half-life of 36 h to 7 months and it leaches out slowly into groundwater. Unfortunately, its toxicity to certain vertebrates may limit the use of fipronil insecticides.

Fipronil competes for the [^{3}H]EBOB binding site, along with α-endosulfan and lindane (Ratra and Casida, 2001), indicating that it acts at the convulsant site. RDL from *Drosophila* and *Heliothis* are both blocked micromolar concentrations of fipronil (Hosie *et al.*, 1995a; Wolff and Wingate, 1998), and fipronil reduces GABA gated currents in RDL by shortening the duration of openings (Grolleau and Sattelle, 2000). In addition to GABA gated chloride channels, fipronil also blocks glutamate gated chloride channels on DUM neurons of cockroach (Raymond *et al.*, 2000), but not histamine gated chloride channels of *Drosophila* eye (Zheng *et al.*, 2002). GLC-3, a homomer-forming recombinant glutamate gated chloride channel from *C. elegans*, is blocked by fipronil.

That fipronil and dieldrin both act at similar sites on insect GABARs allowed the prediction that cross-resistance between these insecticides will occur (Hosie *et al.*, 1995a). This was confirmed in a recent study that showed cross-resistance to dieldrin in a fipronil resistant mosquito line, and cross-resistance to fipronil in a dieldrin resistant mosquito line (Kolaczinski and Curtis, 2004).

2.2.6.5. Avermectins Avermectins (**Figure 7**) have been used as insecticides (e.g., abamectin) as well as endectocides (e.g., ivermectin) (review: Bloomquist, 1996 and (see **Chapter**). In addition to their actions on inhibitory glutamate receptors, avermectins also displace binding of EBOB to *Drosophila* membranes at 20 nM. This is also true of the dieldrin resistant mutant form of RDL (Cole *et al.*, 1995). Ivermectin binds to neuronal membranes of *D. melanogaster* and the locust, with nanomolar K_ds (Rohrer *et al.*, 1995). Although glutamate gated chloride channels are probably the principal site of action of avermectins, their insecticidal actions may also be due, at least in part, to actions on GABARs (Bloomquist, 1996). Their insecticidal actions can have a deleterious ecological effect where they are used as endectocides on cattle, as the presence of avermectins excreted in the dung is detrimental to dung dwelling insects (McCracken, 1993; Strong, 1993), slowing normal biodegradation.

2.2.7. Mechanisms of Insecticide Resistance

Resistance to insecticides is common in field populations of many insect species (Georghiou, 1986). Early binding studies provided evidence that resistance to cyclodienes is due largely to changes in the target site. The dissociation constant and saturation level of a cyclodiene resistant strain (Lpp) of the cockroach, *B. germanica*, differed from those of the wild-type (Matsumura *et al.*, 1987). These differences corresponded to an approximately tenfold reduction in affinity. A number of studies on geographically separate populations of *D. melanogaster* provided clear evidence of cross-resistance between cyclodiene insecticides and PTX (Kadous *et al.*, 1983; Matsumura and Ghiasuddin, 1983; Wafford *et al.*, 1989; Deng *et al.*, 1991; ffrench-Constant and Roush, 1991), again implicating changes in the convulsant site as a mechanism of resistance to the action of insecticides.

In the *rdl* (resistant to dieldrin) *Drosophila* strain, insecticide resistance arises from the substitution of a single amino acid (A302S), which contributes to

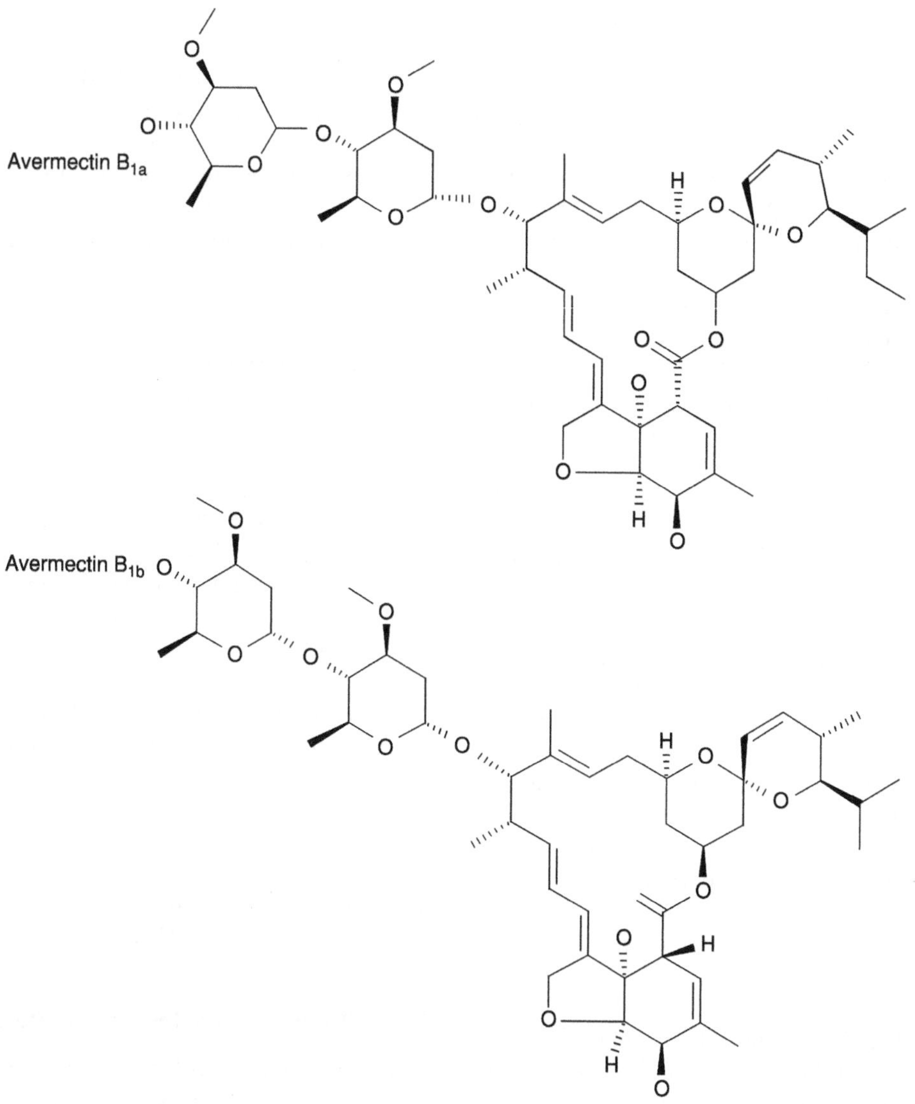

Figure 7 The molecular structures of avermectin B$_{1a}$ and avermectin B$_{1b}$.

the M2 pore forming region of the GABAR channel. It is close to a site that determines charge selectivity in nAChRs (Imoto *et al.*, 1988; Leonard *et al.*, 1988). A similar mutation (A285S) confers dieldrin insensitivity to a cloned *H. virescens* RDL equivalent (Wolff and Wingate, 1998). Similarly, cyclodiene resistance in the aphid, *Nasonovia ribisnigri*, a main pest of salad crops, is also due to an A302S mutation (Rufingier *et al.*, 1999). In the peach aphid, *M. persicae*, cyclodiene resistance is due to a mutation of A302, but in this case three mutations 5 A302Gly, Ser (TCG), and Ser (AGT), can generate resistance (Anthony *et al.*, 1998). Study of wild *T. castaneum* suggests multiple origins of cyclodiene resistance (Andreev *et al.*, 1999), whereas the mutation probably has a single origin in the case of *D. melanogaster* (ffrench-Constant *et al.*, 2000).

Indeed, the central importance of the RDL A302S mutation is reflected in the observation that a mutation of this amino acid is implicated in cyclodiene resistance in at least 60 strains of *D. melanogaster* and *D. simulans* collected worldwide (ffrench-Constant *et al.*, 1993a).

The A302S mutant receptor has a greatly reduced sensitivity to PTX. When expressed in *Xenopus* oocytes, wild-type RDL supports GABA responses that are completely blocked by 10 μM PTX, whereas RDL-A302S mediated responses are only about 50% blocked by the same concentration (ffrench-Constant *et al.*, 1993a, 1993b). The same study showed that sensitivity to dieldrin is also reduced. Dieldrin at 10 μM reduced the amplitude of wild-type responses by some 75%, whereas the mutant was almost completely insensitive to the concentrations tested. Similar results are obtained regardless of the expression system. GABA responses mediated by RDL stably expressed in a *Drosophila* cell line (S2) were blocked by PTX with an IC_{50} of approximately $2\,\mu M$, whereas in the case of an S2 cell line stably expressing the A302S mutant, $100\,\mu M$ PTX produced a maximum block of only about 30% (Buckingham *et al.*, 1996). The A302S mutation is not known to have any effect on agonist pharmacology.

The A302S substitution confers resistance to a variety of naturally occurring (e.g., picrotoxinin, picrodendrin-O) and man-made (e.g., fipronil) insecticides, which act allosterically as noncompetitive antagonists of insect GABARs. The A302S substitution also renders RDL homomers and native receptors resistant to the antagonists TBPS, lindane, and BIDN (Belelli *et al.*, 1995; Buckingham *et al.*, 1996; Hosie *et al.*, 1995b). However, these compounds interact noncompetitively in radioligand binding studies, suggesting that they have distinct, if overlapping, binding sites. The A285S mutant RDL of *H. virescens* is also resistant to fipronil and picrotoxinin to similar degrees as the *Drosophila* RDL mutant (Wolff and Wingate, 1998).

Studies on vertebrate $GABA_A$ receptors have suggested that PTX blocks the channel via an allosteric mechanism in which it binds preferentially to activated channels and stabilizes them in the agonist-bound, closed conformation (Smart and Constanti, 1986; Newland and Cull-Candy, 1992). It appears that one effect of the RDL A302S mutation is to stabilize the open conformation and reduce the rate of desensitization (Zhang *et al.*, 1994). In addition, the A302S mutation lowers the binding affinity for PTX (Cole *et al.*, 1995; Lee *et al.*, 1994). Since the sites for all the convulsant antagonists overlap, we can assume that similar mechanisms apply to other cyclodienes. In this context, it is interesting to observe that although the substitution of a serine for an alanine does not introduce any change in net charge, the added hydroxyl group is more bulky and may restrict access of the PTX molecule by steric interference (ffrench-Constant *et al.*, 1993a, 1993b). It has also been observed (Hosie *et al.*, 1995a) that the A302S mutation both alters the rate of desensitization and reduces the potency of PTX as well as that of fipronil. PTX and dieldrin both displace EBOB binding competitively in *M. domestica*, whereas fipronil's displacement of EBOB is noncompetitive. It has been suggested (Zhang *et al.*, 1994) that the A302S mutation must either alter both the fipronil and PTX binding sites, or perhaps instead, alters the allosteric linking mechanism by which convulsants reduce GABA responses, or both.

2.2.8. Molecular Basis of Selectivity of Insecticides for Insect GABARs

Many currently used insecticides have been developed for their selective toxicity to insects compared to vertebrates. Much of this is attributable to differences between the sites of action within the channel pore. For example, fipronil is more toxic to insects than to vertebrates because it has a higher affinity for insect GABARs (Hainzl *et al.*, 1998). However, binding of EBOB, fipronil, lindane, and endosulfan to housefly head membranes is comparable to the binding to human recombinant β_3 GABARs, attributable to sequence similarities at key points on their receptors. Studies of recombinant human $GABA_ARs$ show subunit dependency of insecticide affinity (Ratra and Casida, 2001). Similarly, the potency of lindane, fipronil, EBOB, and α-endosulfan on human recombinant $GABA_A$ receptors is dependent on the specific type of α or γ subunit present (Ratra *et al.*, 2001). A detailed study on the actions of picrodendrin and tutin antagonists on RDL expressed in oocytes (Hosie *et al.*, 1996) suggested that insect and vertebrate GABARs differ in their ideal convulsant site pharmacophore. The same conclusion is confirmed by similar studies using systematic variations of a class of molecules. For instance, acyclic esters compete with EBOB and changes to these molecules result in increases in potency against housefly head membranes, whereas their potency against rat GABARs is unchanged or reduced (Hamano *et al.*, 2000). A study using a series of 28 picrotoxane terpenoids concluded that rat GABARs require a spiro-γ as butyrolactone moiety at the 13 position and certain substituents at the 4-position, whereas these are not critical for interactions with *Musca* receptors. The electronegativity of the 16-carbon atom and the presence or absence of the 4- and 8-hydroxyl groups are important determinants of nor-diterpenes for *Musca* GABARs, again suggestive of differences in rat binding sites compared to *M. domestica* (Ozoe *et al.*, 1998).

Differences in receptor sites are not the only mechanisms of insecticide selectivity. Fipronil is an interesting case, in that metabolic detoxification contributes positively to insect selectivity. Although selective toxicity of fipronil for insects is at least partly due to its higher affinity for the convulsant site, it is also due to differences in the rate at which it is broken down to its more active sulfone metabolite (Hainzl *et al.*, 1998).

2.2.9. Single Channel Properties of Insect GABARs

It has proved difficult to obtain patch-clamp recordings of single GABA gated chloride channels of insects, and this has been attributed to a heterogenous distribution of GABARs over the cell membrane (Pichon and Beadle, 1988). Noise analysis of cultured embryonic cockroach (*P. americana*) brain neurons gave an estimate of single channel conductance of 18.6 pS and mean open time of 11.8 ms (Shimahara *et al.*, 1987). Single channel recordings of GABA gated currents in cultured adult *P. americana* neurons indicated two open conductance states, with single channel conductances of around 17 and 11 pS and two mean open times of around 0.28 and 1.43 ms, as well as three closed states with mean durations of 0.25, 2.22, and 43.7 ms (Malecot and Sattelle, 1990). An RDL splice variant, RDL_{ac}, expressed in a *Drosophila* cell line has single channel conductance of 36 pS. Mean open time distribution was best fitted to two time constants, suggesting that there are two open states (Grolleau and Sattelle, 2000).

2.2.10. Distribution

GABARs are present throughout the insect brain (Sattelle, 1990), but this distribution is not uniform. GABAR localization at the tissue level reflects not only behavioral roles, such as in learning and receptive field sharpening, but possibly also its role in development. Fine details of distribution also provide clues to their possible functional roles. For example, GABARs are distributed over the terminals of primary auditory afferents in certain crickets, and are located near output synapses, suggesting that they play a role in presynaptic inhibition (Hardt and Watson, 1999).

A number of studies have indicated particularly dense GABAR expression in brain structures associated with learning, in accord with the observations described below, which point to a central role of GABAergic inhibitory pathways in the consolidation of memory. It has been mentioned above that antibodies to RDL, a *D. melanogaster* GABAR subunit, shows strong immunoreactivity in cockroach, *P. americana* (Harrison *et al.*, 1996) and cricket *A. domesticus* (Strambi *et al.*, 1998) mushroom bodies. Similar results were obtained in the cricket *Gryllus bimaculatus* (Schurmann *et al.*, 2000), and the fly *C. erythrocephala* (Brotz *et al.*, 1997). The three types of GABA immunoreactive neurons in the mushroom bodies of the cockroach (Yamazaki *et al.*, 1998) are described below. We can therefore conclude that high levels of expression of RDL-like GABARs is a common feature of insect mushroom bodies.

In addition to high level GABAR expression, there is evidence that GABAergic intrinsic neurons in the mushroom bodies form distinct compartments. In the fly *C. erythrocephala*, a RDL antibody stains concentrically around an unstained core in the pedunculus and α and β lobes of the mushroom body, whereas the γ-lobes showed a compartmentalized RDL staining. That only some of the neurons, presumably Kenyon cells, should be stained suggests that only some Kenyon cells have GABARs and are therefore non homogeneous (Brotz *et al.*, 1997). Similar compartmentalization is seen in the arborizations of GABAergic input neurons in *G. bimaculatus* and *D. melanogaster* (Schurmann, 2000). The functional significance of this compartmentalization is as yet unknown, but may be related to the finding that each Kenyon cell specifically connects corresponding layers of the calyces and lobes (Grunewald, 1999).

In addition to intrinsic neurons of the mushroom body, certain input and output neurons are also GABAergic (Nishino and Mizunami, 1998), although most of them are cholinergic, at least in *D. melanogaster* and *G. bimaculatus* (Schurmann, 2000). The calycal giants (the input neurons to the Kenyon cells of the mushroom bodies) are GABA immunoreactive (Nishino and Mizunami, 1998). In *D. melanogaster* and *G. bimaculatus*, GABA immunoreactive presynaptic fibers of extrinsic neurons intermingle with nonimmunoreactive fibers in the mushroom bodies (Schurmann, 2000).

There may be close association between GABA expression and expression of nitric oxide (NO) in the insect nervous system. The synthesis of NO is under the control of NADPH-diaphorase. In the locust, almost all GABAergic neurons in the brain were also found to be NADPH-diaphorase positive (Seidel and Bicker, 1997), suggesting that GABA and NO could be cotransmitters.

Distribution of GABAergic neurons is also likely to be determined by developmental constraints. GABA immunoreactive neurons are linearly determined in *M. sexta* (Witten and Truman, 1991). In this species, and probably in eight other insect orders, GABA immunoreactive neurons are restricted to six lineages (Witten and Truman, 1998). Similarly, the approximately 360 GABAergic neurons in the larval midbrain of *T. molitor* form ten distinct clusters. The number of GABAergic somata, but not the number of clusters, increases during development (Wegerhoff, 1999). As NADPH-diaphorase activity in grasshoppers is detected as early as embryonic stage 50%, and since all NADPH-diaphorase positive local interneurons of adult antennal lobe also express GABA immunoreactivity, this is the earliest reported appearance of GABA immunoreactivity in the embryonic antennal lobe (Seidel and Bicker, 2002).

Antibodies to GABA and GABAR associated proteins also stain other parts of the insect CNS that perform roles in sensory and motor processing. In the locust *S. gregaria*, GABA immunoreactivity and GAD immunoreactivity are both observed in about 100 bilateral pairs of tangential neurons that connect the lateral accessory lobes to distinct layers of the central body in the central complex, a structure that plays a role in motor control and visual orientation (Homberg *et al.*, 1999). The brain of the moth *M. sexta* contains some 20 000 GABAergic neurons, most of which are optic lobe neurons (Homberg *et al.*, 1987), although there is also abundant staining in the mushroom bodies (Homberg *et al.*, 1987) and the antennal lobes (Hoskins *et al.*, 1986). In locusts GABA has also been shown to be colocalized with locustatachykinin immunoreactive local interneurons, a distinct subset of antennal lobe neurons (Ignell, 2001). In *D. melanogaster*, the RDL antibody stains the calyces of mushroom bodies, glomeruli of antennal lobes, lower central body and corpora cardiaca, as well as medulla and lobula regions of the optic lobe: regions associated with olfactory, visual, and mechanosensory processing (Harrison *et al.*, 1996). In the same species, immunoreactivity to GABA is reported in the larval antennal lobe and the tritocerebral-subesophageal ganglion. Whereas choline acetyltransferase (ChAT) is expressed only in subsets of olfactory and gustatory afferents, GABA is expressed in most, if not all, local interneurons (Python and Stocker, 2002). Immunostaining for an insect GABA transporter in *M. sexta* closely matches that for GABA, with staining in parts of the optic and antennal lobes, mushroom body, lateral protocerebrum, and central complex (Umesh and Gill, 2002).

GABARs are also present on identified neurons of certain insects, notably the fast coxal depressor motor neuron (D_f) (Sattelle *et al.*, 1988), the GI2 escape neuron of the cockroach *P. americana* (Hue, 1991; Buckingham *et al.*, 1994), and fg1 (a neuron in the frontal ganglion of *M. sexta*), as well as unidentified dorsal unpaired median neurons of *P. americana* (Sattelle *et al.*, 2003).

2.2.11. GABARs and Behavior

By 1985 evidence had already accumulated pointing to the roles of GABARs in fast inhibitory responses on giant interneurons that process sensory input in the cockroach *P. americana*, and in auditory processing in the cricket. Recent work has expanded considerably on these findings, as well as added to the list of roles played by GABARs in generating insect behavior.

Demonstrations of the involvement of GABARs in behavior usually consist of descriptions of GABARs or GABAergic neurons in circuits or structures known to be involved in specific behaviors, or reports of effects of GABAergic drugs exerting specific actions on discrete aspects of behavior. Two areas that have received considerable attention over the past 10 years have been insect learning, where GABARs have been shown to be particularly abundant in the antennal lobe and mushroom bodies, and the tuning of sensory encoding, in which GABAergic inhibitory pathways have been shown to sharpen receptive fields or enhance sensory discrimination.

2.2.11.1. Insect learning There have been a number of convincing demonstrations that structures in the insect brain known to be involved in the acquisition, consolidation, or retrieval of memory are rich in ionotropic GABARs. RDL antibody stains structures in the brain of the cricket *A. domesticus* that are associated with learning (Strambi *et al.*, 1998). Particularly dense staining was observed in the mushroom bodies (especially the upper part of the peduncle and the two arms of the posterior calyx) and antennal lobe. Subsequent electrophysiological studies have further provided direct evidence that picrotoxinin sensitive GABARs are present on Kenyon cells of *A. domesticus* (Cayre *et al.*, 1999).

Although it is difficult to determine from immunohistochemical data the exact role that GABAergic neurons play in learning, there are indications that one such role is in providing inhibitory feedback. In the honeybee, *A. mellifera*, 50% of the approximately 110 GABA immunoreactive neurons in the mushroom body appear to be feedback neurons (Grunewald, 1999). Here they connect specific layers of the output regions (α and β lobes and pedunculi) with the input regions (the calyces). In the same species, GABA immunoreactive processes were found to synapse onto fine fibers (70%) and large, non-GABAergic boutons (probably from intrinsic and extrinsic mushroom body neurons, respectively) (Ganeshina and Menzel, 2001). These synapses are components of microcircuits, probably involving feed-forward and feedback loops. Similar microcircuits have been identified in the calyx neuropil of the *D. melanogaster* mushroom body (Yasuyama *et al.*, 2002). Here also, extrinsic GABA immunoreactive neurons appear to form a network of fine fibers codistributed with mainly Kenyon cell dendrites, onto which they make divergent synaptic inputs as well as onto boutons of projection neurons. Each microcircuit takes the form of a glomerulus consisting of a large cholinergic bouton at its core, surrounded by tiny, vesicle-free Kenyon cell dendrites and input from GABA fibers. The pattern of staining suggests that Kenyon cells receive excitatory input from cholinergic cells along with postsynaptic inhibition from GABA cells.

GABAergic pathways appear also to provide substantial input to, and output from, the mushroom bodies. In the cockroach, *P. americana*, Yamazaki and colleagues (Yamazaki *et al.*, 1998) identified three types of GABA immunoreactive neurons in cockroach mushroom body. The first type consisted of a number of neurons (7–9) that ascended from the circumesophageal connective and projected into the calyces, suggesting that they were input neurons. A second type consisted of a larger number of cells (around 40), with dendrites ramifying in the junction between the pedunculus and the lobes, which are thought possibly to be inhibitory input neurons. Finally, a small number of large cells arborize around the alpha-lobe and project into large areas of calyces.

Mushroom bodies of the cockroach are also known to receive inputs from four giant interneurons, the calycal giants. These, too, are immunoreactive to GABA and input directly onto Kenyon cells. They are spontaneously rhythmically active, and this activity is inhibited by olfactory, visual, tactile, and air current stimuli (Nishino and Mizunami, 1998), and so may play a role in sensitizing Kenyon cells to olfactory stimuli through disinhibition.

Thus, inhibitory pathways are important in insect learning, but lack of functional data has so far precluded a detailed analysis of the different roles of GABARs. However, a number of exciting studies suggest that GABAergic circuits might contribute to the establishment of olfactory memory through the coordination of spike timing. For example, GABA affects the synchronous oscillation of intracellular calcium in Kenyon cells in *Drosophila*, an activity thought to be involved in memory consolidation (Rosay *et al.*, 2001). The possible role of GABARs in this activity is discussed in the following section.

2.2.11.2. Stimulus encoding and tuning
2.2.11.2.1. Odor representation in the antennal lobes
Neurons in the antennal lobes of locusts and honeybees respond to certain odor stimuli (**Figure 8**) with synchronized, oscillatory activity (MacLeod and Laurent, 1996; review: Kauer, 1998). There is ample evidence that GABARs play a key role in the formation of this synchronization. For instance, the injection of GABA antagonists into the first olfactory relay neuropil of locust abolishes synchronization of odor specific neural assemblies in the antennal lobe (MacLeod and Laurent, 1996). Significantly, the temporal response patterns of individual neurons

remain unaffected, even though such patterns include some hyperpolarization. Thus, this action of fast, ionotropic GABARs is specific to the formation of synchronization. Similar findings have been reported for the honeybee (review: Kauer, 1998).

The use of optical imaging techniques, such as calcium imaging, and the use of fluorescent, voltage sensitive dyes, allows the activity of large numbers of cells to be monitored simultaneously. In the honeybee, *A. mellifera*, spatiotemporal odor response patterns observed in olfactory glomeruli of the antennal lobes in response to odor stimuli can be mapped to identified glomeruli. By examining the effects of GABA or PTX injection in the antennal lobe, it can be shown that there exist two separate inhibitory networks that render such response patterns more confined, i.e., they enhance their spatial contrast (Sachse and Galizia, 2002). One of these networks is GABAergic and modulates overall antennal lobe activity, and the other is PTX insensitive and glomerulus specific. Confirmation that GABAergic connections underlie such synchronized oscillations is provided by computer simulations. For example, simulations show that GABA synapses forming inhibitory connections between local interneurons and projection interneurons of the antennal lobe allow the formation of dynamic assemblies of neurons, which oscillate together and synchronize transiently (Bazhenov *et al.*, 2001a). Thus, GABA-ergic synapses coordinate the activity of a large number of neurons to produce rhythmic, coordinated firing.

2.2.11.2.2. Fine-tuning of odor representation
In many insect sensory systems, lateral inhibition serves the role of sharpening stimuli and enhancing the contrast of stimulus representation. In the antennal lobe, odor identity and duration of stimulation are represented both spatially, in the pattern of activity in antennal lobe glomeruli, and temporally, in the pattern of activity in groups of neurons. Spatial encoding is seen in response to electrical stimulation of the antennal nerve in the silkworm moth, *B. mori*, which is followed by postsynaptic depolarizations in the antennal lobe. These depolarizations can be monitored using optical recordings with voltage sensitive dyes (Ai *et al.*, 1998). GABA mediated IPSPs occur in the antennal lobe within 3 ms of these postsynaptic depolarizations, and these IPSPs, like the depolarizations, are not uniform over the antennal lobe, suggesting that the pattern of excitation is specific. GABAergic IPSPs also occur in the macroglomerular complex and some ordinary glomeruli in the antennal lobe. Recent evidence using computer simulations suggests that in the locust

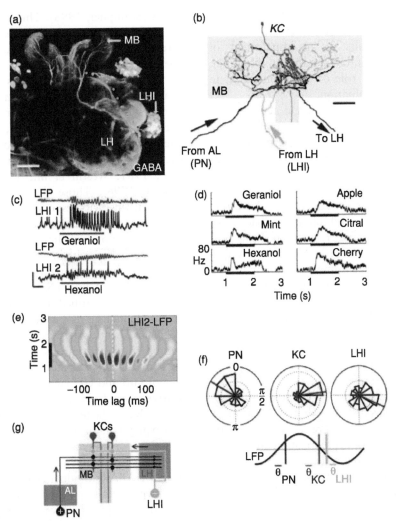

Figure 8 Feed-forward inhibition of KCs by LHIs. (a) Immunolabeling by antibody to GABA (37). Cluster of about 60 reactive somata (LHI) and tract of LHI axons running to the MB (stipples) are shown. The terminals of one of these axons in the MB are shown in (b). Scale bar = 100 μ. (b) PN axon (black) projects to the MB calyx (orange) and to the lateral horn (LH) (41). LHI axon (green) projects to the calyx (this study). PN and LHI axons terminate on KC dendrites (red). Neurons were stained by iontophoresis of cobalt hexamine (KC, PN) or neurobiotin (LHI) in separate preparations and were drawn with a camera lucida. Note varicosities in LHI and PN axon collaterals. Asterisk, KC axon. Scalebar = 50 μ. (c) Representative odor evoked responses of two LHIs and simultaneously recorded LFPs (5 to 40 Hz bandpass). Note membrane potential oscillations, locked to the LFP. Identity and delivery (1 s long) of stimulus indicated by black bar. LHI, 20 mV; LFP, 400 μV; 200 ms. (d) Instantaneous firing rate of LHI1 (in (C)) in response to various odors. Lower edge of profile shows mean instantaneous rate averaged across trials; profile thickness, SD. All LHIs responded to all odors tested, with response profiles that varied little across different odors. (e) Sliding cross-correlation between LFP and LHI2 traces (spikes clipped). Red, maxima; blue, minima. Strong locking is present throughout the response (odor delivery, vertical bar). Lower edge of correlation stripes just precedes stimulus onset due to width of the correlation window (200 ms). (f) Phase relationships between PN, KC, and, LHI action potentials, and LFP. (Upper) Polar plots. LFP cycle maxima defined as 0 rad, minima as π rad (PNs: 3 cell-odor pairs, 388 spikes; LHIs: 17 cell-odor pairs, 2632 spikes; KCs: 18 cells, 862 spikes). Mean phases are shown in red. Gridlines are scaled in intervals of 0.10 (probability per bin). (Lower) Schematic diagram showing LFP and mean firing phases. (g) Circuit diagram. (Reprinted with permission from Perez-Orive, J., Mazor, O., Turner, G. C., Cassenaer, S., Wilson, R.I., **2002**. Oscillations and sparsening of odor representations in the mushroom body. *Science 297*(5580), 359–365.)

these characteristic spatial patterns of activity seen in response to odor stimuli arise from competition among neurons with GABAergic outputs. GABAergic interconnections between local interneurons create competition between them that results in coordinated, phase-locked activity in separate groups of neurons. Thus, GABARs contribute to spatial and temporal patterns of activity that encode the stimulus (Bazhenov *et al.*, 2001b).

In addition to spatial representation of the odor stimulus, temporal synchronization plays a role in odor discrimination, as well as in memory.

In honeybees (Stopfer *et al.*, 1997; Hosler *et al.*, 2000), block of odor evoked oscillatory synchronization by PTX results in an impairment of discrimination of similar, but not that of dissimilar, odors. This points to a role of synchronization in sharpening odor discrimination.

The circuitry by which GABAergic neurons perform this is unknown. However, in the antennal lobe of the cockroach, *P. americana*, GABAergic neurons form connections that are suggestive of both feedforward and feedback inhibition (Distler *et al.*, 1998). In cockroach antennal lobe glomeruli, a major processing center in olfaction, GABAergic local interneurons mediate inhibitory feedback connections onto both antennal receptor neurons and uniglomerular projection neurons. Each GABAergic neuron innervates a large number of antennal lobe glomeruli, thus potentially passing information to neighboring glomeruli (Distler and Boeckh, 1997, 1998). It is presumably these connections between local interneurons and excitatory projection neurons that create the coordinated, distributed oscillations in response to odor stimulation, which, in locusts at least, encode odor identity.

In addition to the identity of an odor stimulus, antennal lobe neurons also encode the duration of the odor. Hence, continuous application of an odor evokes a complex wave of depolarization and hyperpolarization, whereas a pulsatile stimulus evokes a simple train of spikes (Christensen *et al.*, 1998). Here again, GABARs play a key role. GABA evoked IPSPs in *M. sexta* projection neurons of the antennal lobe resemble not only electrically evoked IPSPs, but also odor evoked IPSPs. For example, both GABA and odor evoked IPSPs are blocked by bicuculline. At the same time, administration of bicuculline also reversibly alters the temporal pattern of odor evoked activity in the projection neurons, providing circumstantial evidence that GABARs help shape the temporal pattern of activity (Christensen *et al.*, 1998). This may also indicate a particular role for the bicuculline sensitive subtype of GABAR.

2.2.11.2.3. Involvement of GABA inputs in fine-tuning of sensory pathways may be a general phenomenon
In a manner reminiscent of the sharpening of the spatial patterns of odor responses described above, GABAergic inputs appear to be responsible for fine tuning of sensory inputs in other sensory systems. An identified auditory neuron, AN1, of the bushcricket, *Ancistrura nigrovittata*, is finely tuned to the fundamental frequency of the male song. GABA input to this neuron greatly sharpens its selectivity for the fundamental frequency, severely attenuating subthreshold inputs in response to lower, and more so to higher, frequency sounds (Stumpner, 1998). This effect is highly specific in that other inputs through other neurotransmitters are responsible for side and frequency dependent inhibitions.

A similarly striking example is seen in the escape behavior of the cockroach, *P. americana*. Escape responses to wind stimuli are mediated by giant interneurons (Levi and Camhi, 2000), some of which respond selectively to wind from a particular direction, while others are omnidirectional (Buno *et al.*, 1981; Westin *et al.*, 1988; Okumah and Kondoh, 1996). Those that show a pronounced directional sensitivity respond to wind in a linear manner whereas those that are omnidirectional in their sensitivity respond nonlinearly. The synaptic input onto these cells results in a depolarization followed by a delayed hyperpolarization. This hyperpolarization was blocked by PTX , resulting in an altered response of an identified neuron (Neuron 101) from linear to nonlinear (Okumah and Kondoh, 1996). Hence, GABAergic synapses contribute to the transfer function of these neurons. Furthermore, Hill and Blagburn (1998) demonstrated that the receptive fields of GI6 and GI7 (which means the directional sensitivity) are sharpened by GABA mediated inputs.

GABARs on the terminals of afferent fibers probably also play a major role in the modulation of sensory input through presynaptic inhibition. In the locust, presynaptic inhibition of tegula afferents mediated through GABARs modulates the amplitude of postsynaptic potentials in the hind-wing motor neurons in this monosynaptic pathway in a phase dependent manner (Buschges and Wolf, 1999). GABARs also mediate inhibition by identified filiform hair receptors of an identified projection interneurone in a circuit that mediates wind elicited responses (Gauglitz and Pfluger, 2001). In addition, GABA ergic neurons modulate transmitter release at synapses between the fore-wing stretch receptor and wing depressor motor neurons of *L. migratoria* (Judge and Leitch, 1999). Consistent with this role is the observation that GABARs are distributed over the terminals of primary auditory afferents in certain crickets, and are located near output synapses, again suggesting that they play a role in presynaptic inhibition (Hardt and Watson, 1999).

Clearly, then, GABARs of insects probably play major roles in sharpening receptive fields of various sensory systems, as well as acting at "higher" sensory integration centers in aiding stimulus discrimination and the consolidation of memory.

2.2.11.3. Other behaviors involving GABARs
While attention has already been drawn to the role for GABARs (presumably including the cloned *Drosophila* GABAR subunit, RDL) in memory

consolidation, the pattern of staining of cockroach brains with an RDL antibody also suggests a wider involvement with olfactory, visual, and mechano-sensory processing (Sattelle *et al.*, 2000). In the locust, *S. gregaria*, GABA and GAD immunoreactivity is observed in about 100 bilateral pairs of tangential neurons that connect the lateral accessory lobes to distinct layers of the central body in the central complex, a structure that plays a role in motor control and visual orientation (Homberg *et al.*, 1999).

GABA is not only present in the insect nervous system, but also occurs in the periphery. GABARs have been demonstrated on skeletal muscle of cockroach (Schnee *et al.*, 1997), and those present on the flexor tibiae muscle of locust have received considerable attention (Cull-Candy and Miledi, 1981; Cull-Candy, 1982). GABA also plays a role in visceral functions. For instance, GABA decreases the heart rate of pupal *Drosophila*, although curiously it has no effect upon Hear rate of larval or adult flies (Zornik *et al.*, 1999). GABAergic neurons inhibit the prothoracic gland of *P. americana* up to the 17th day of the molt cycle, after which cessation of their input elicits competence of the prothoracic gland through disinhibition (Richter and Bohm, 1997).

Recent evidence also points to a role of GABA-ergic neurons in higher levels of behavior. GABA-immunoreactive neurons connect the noduli of accessory medulla to the medulla and to the lamina via processes in the distal tract of the cockroach, *Leucophaea maderae*. Interestingly, the accessory medulla of *D. melanogaster* and of *L. maderae* is thought to be the location of the circadian pacemaker. Injection of GABA into the vicinity of the accessory medulla resets the circadian motor activity of the cockroach in a stable, phase-dependent manner, suggesting that, along with Mas-allatotropin, GABA plays a role in circuits relaying photic information from circadian photoreceptors to the central pacemaker (Petri *et al.*, 2002).

Roles of GABARs in behavior can also be gauged indirectly through alterations in GABA transporter function. Disrupting GABA transporter function *in vivo*, thereby prolonging the effects of GABA at the synapse, in adult female *D. melanogaster* reduced locomotor activity, disrupted geotaxis, and induced convulsions with secondary loss of locomotor activity, without damaging feeding activity or female sexual receptivity (Leal and Neckameyer, 2002).

2.3. Metabotropic GABARs of Insects

Very little is known of insect $GABA_B$ (metabotropic) receptors, largely because baclofen is ineffective in many invertebrate preparations (Sattelle *et al.*, 1988;

Benson, 1989). However, $GABA_B$-like responses were observed in giant interneurons of the terminal abdominal ganglion of the cockroach (Hue, 1991). Similarly, the motor neuron, D_f, of the same species was thought not to possess $GABA_B$ receptors, until it was shown that responses to $GABA_B$ receptor agonists could be observed when the ionotropic GABARs are blocked by PTX (Bai and Sattelle, 1995). These responses, which appeared to be mediated by a potassium conductance, differed in their pharmacology from vertebrate $GABA_B$ receptors in their insensitivity to baclofen (a property shared by $GABA_B$ responses in the cockroach GI2 and reduced sensitivity to the highly potent vertebrate $GABA_B$ agonist, SK&F97541. It was also shown recently that cultured Kenyon cells of *A. domesticus* possess $GABA_B$ receptors, in addition to PTX sensitive receptors (Cayre *et al.*, 1999). This was soon followed by the cloning, sequencing, and functional expression of $GABA_B$-like receptors from *D. melanogaster*.

2.3.1. A Cloned Insect Metabotropic ($GABA_B$) Receptor

Three second-messenger linked, $GABA_B$-like receptors have been cloned from *Drosophila* to date (Mezler *et al.*, 2001). Two of them have high sequence similarity to mammalian $GABA_BR1$ and $GABA_BR2$, whereas the third has no known mammalian counterpart. All three are expressed in the embryonic nervous system, and two are expressed in similar regions and may coexpress *in vivo* to form the functional receptor. R1 and R2 coexpressed in oocytes produce receptors with pharmacology dissimilar to vertebrates in that they are insensitive to baclofen. R3, however, did not produce functional receptors in any combination. These *Drosophila* $GABA_B$ receptors have similar intron positions and exon size to human $GABA_BRs$ (Martin *et al.*, 2001). More detailed physiological, pharmacological, and immunohistochemical experiments are needed in order to establish functional roles for these receptors.

2.4. Conclusions

With the cloning of ionotropic insect GABARs, along with detailed immunohistochemical studies and improved optical recording techniques, we have seen remarkable advances in our understanding of insect GABARs over the past 15 years. In addition, the recent cloning of metabotropic GABARs opens up a vast new area of research aimed at establishing the functional role of these receptors in the insect nervous system.

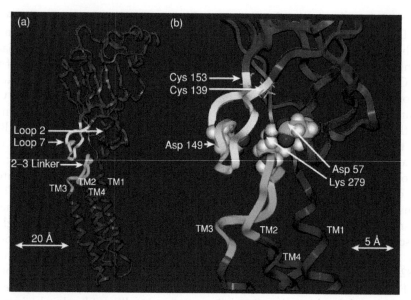

Figure 9   A molecular model of the GABA$_A$ receptor α1 subunit: (a) a model of a whole GABA$_A$ receptor α1 subunit viewed from the center of the ion channel; (b) detailed view of potential contacts between the extracellular and transmembrane domains. (From Kash, T.L., Jenkins, A., Kelley, J.C., Trudell, J.R., Harrison, N.L., **2003**. Coupling of agonist binding to channel gating in the GABAA receptor. *Nature 421*, 272–274.)

Our survey of the more recent findings in the area of insect GABARs leads us to suggest that these receptors are more complex than once was thought, that they form a class distinct from any vertebrate ionotropic GABAR, and that several distinct subtypes exist. Also, a single subunit, RDL, has both alternative splicing and RNA editing. It appears that GABARs play important roles, through inhibition, in stimulus discrimination, receptive field sharpening and memory formation.

Future investigations into this exciting class of neurotransmitter receptors will undoubtedly benefit from the developments in genomics and proteomics to facilitate the study of coexpressed genes and interacting gene products. Other areas that require further attention include the role played by GABARs in insect development, and the specific roles of GABAR subunit variants. The highly tractable development patterns of insects are likely to provide key insights into these strategic questions.

The recently developed technique of double-stranded RNA interference (RNAi) offers some exciting opportunities for future research into insect GABARs. For instance, this approach could be used to assess the roles of GRD and LCCH3, as well as probing further the role of RDL in synaptic transmission, insect learning, sensory tuning, and the circadian clock.

The recent crystal structure of the acetylcholine binding protein (Brjck *et al.*, 2001) now permits homology modeling of the N-terminal domain of a variety of cys-loop neurotransmitter receptors, including insect GABARs. In the future, it will be of interest to attempt molecular modeling of those parts of the RDL extracellular N-terminal domain (see **Figure 9**) that contribute to the agonist binding sites. This would permit elucidation of the mechanisms by which ligands dock to the four alternative splice variants of RDL to be compared, and introduces the possibility of undertaking parallel site-directed and *in silico* mutagenesis studies. Such a model may be of considerable use in the rational design of a future generation of insecticides incorporating both an improved level of safety and a reduced level of impact on nontarget species.

References

Abalis, I.M., Eldefrawi, F.E., Eldefrawi, A.T., **1983**. Biochemical identification of putative GABA/benzodiazepine receptors in house fly thorax. *Pestic. Biochem. Physiol. 20*, 39–48.

Ai, H., Okada, K., Hill, E.S., Kanzaki, R., **1998**. Spatiotemporal activities in the antennal lobe analyzed by an optical recording method in the male silkworm moth *Bombyx mori*. *Neurosci. Lett. 24*, 135–138.

Alix, P., Grolleau, F., Hue, B., **2002**. Ca^{2+}/calmodulin-dependent protein kinase regulates GABA-activated Cl$^-$ current in cockroach dorsal unpaired median neurons. *J. Neurophysiol. 87*, 2972–2982.

Andreev, D., Rocheleau, J., Phillips, T.W., Beeman, R., ffrench-Constant, R.H., **1994**. A PCR diagnostic for cyclodiene insecticide resistance in the red flour beetle, *Tribolium castaneum*. *Pestic. Sci. 41*, 345–349.

Andreev, D., Kreitman, M., Phillips, T.W., Beeman, R.W., ffrench-Constant, R.H., **1999**. Multiple origins of cyclodiene insecticide resistance in *Tribolium casta-neum* (Coleoptera: Tenebrionidae). *J. Mol. Evol. 48*, 615–624.

Anthony, N.M., Holyoke, C.W., Sattelle, D.B., **1993**. Actions of picrotoxin analogues on a chick optic lobe GABA$_A$ receptor expressed in *Xenopus* oocytes. *Mol. Pharmacol. 3*, 63–67.

Anthony, N.M., Holyoke, C.W., Sattelle, D.B., **1994**. Blocking actions of picrotoxin analogues on insect (*Periplaneta americana*) GABARs. *Neurosci. Lett. 171*, 67–69.

Anthony, N., Unruh, T., Ganser, D., ffrench-Constant, R., **1998**. Duplication of the Rdl GABAR subunit gene in an insecticide-resistant aphid, *Myzus persicae. Mol. Gen. Genet. 260*, 165–175.

Araujo, F., Tan, S., Ruano, D., Schoemaker, H., Benavides, J., et al., **1996**. Molecular and pharmacological characterization of native cortical gamma-amino-butyric acid$_A$ receptors containing both alpha1 and alpha3 subunits. *J. Biol. Chem. 271*, 27902–27911.

Aronstein, K., ffrench-Constant, R., **1995**. Immunocytochemistry of a novel GABAR subunit Rdl in *Drosophila melanogaster. Invert. Neurosci. 1*, 25–31.

Aronstein, K., Auld, V., ffrench-Constant, R., **1996**. Distribution of two GABAR-like subunits in the *Drosophila* CNS. *Invert. Neurosci. 2*, 115–120.

Bai, D., **1994**. Neurotransmitter Receptor Subtypes on Identifiable Insect Neurons. Ph.D. thesis. University of Cambridge.

Bai, D., Sattelle, D.B., **1995**. A GABA$_B$ receptor on an identified insect motor neurone. *J. Exp. Biol. 198*, 889–894.

Bazhenov, M., Stopfer, M., Rabinovich, M., Huerta, R., Abarbanel, H.D., et al., **2001a**. Model of transient oscillatory synchronization in the locust antennal lobe. *Neuron 30*, 553–567.

Bazhenov, M., Stopfer, H., Rabinovich, M., Abarbanel, H.D., Sejnowski, T.J., et al., **2001b**. Model of cellular and network mechanisms for odor-evoked temporal patterning in the locust antennal lobe. *Neuron 30*, 569–581.

Belelli, D., Hope, A.G., Callachan, H., Hill-Venning, C., Peters, L.A., et al., **1995**. A mutation in the putative M2 domain of a *Drosophila* GABAR subunit differentially affects antagonist potency. *Br. J. Pharmacol. 116*, 442.

Belelli, D., Callachan, H., Hill-Venning, C., Peters, J.A., Lambert, J.J., **1996**. Interaction of positive allosteric modulators with human and *Drosophila* recombinant GABARs expressed in *Xenopus laevis* oocytes. *Br. J. Pharmacol. 118*, 563–576.

Benson, J.A., **1988**. Bicuculline blocks the response to acetylcholine and nicotine but not to muscarine or GABA in isolated insect neuronal somata. *Brain Res. 458*, 45–71.

Benson, J.A., **1989**. A novel GABAR in the heart of a primitive arthropod, *Limulus polyphemus. J. Exp. Biol. 147*, 421–438.

Bentley, R., Trimen, H., **1875**. *Anamirta paniculata*. In: Medicinal Plants, Being Descriptions With Original Figures of the Principal Plants Employed in Medicine

and an Account of their Properties and Uses, Part 2, Item 14. J and A Churchill, London, p. 5.

Bermudez, I., Hawkins, C.A., Taylor, A.M., Beadle, D.J., **1991**. Actions of insecticides on the insect GABAR complex. *J. Recept. Res. 11*, 221–232.

Betz, H., Kuhse, J., Schmieden, V., Laube, B., Kirsch, J., et al., **1999**. Structure and functions of inhibitory and excitatory glycine receptors. *Ann. New York Acad. Sci. 868*, 667–676.

Bloomquist, J.R., **1996**. Ion channels as targets for insecticides. *Annu. Rev. Entomol. 41*, 163–190.

Bonnert, T.P., McKernan, R.M., Farrar, S., le Bourdelles, B., Heavens, R.P., et al., **1999**. Theta, a novel gamma-aminobutyric acid type A receptor subunit. *Proc. Natl Acad. Sci. USA 96*, 9891–9896.

Bormann, J., **2000**. The 'ABC' of GABARs. *Trends Pharmacol. Sci. 21*, 16–19.

Brejc, K., van Dijk, W.J., Klaassen, R.V., Schuurmans, M., van Der Oost, J., et al., **2001**. Crystal structure of an ACh-binding protein reveals the ligand-binding domain of nicotinic receptors. *Nature 411*, 269–276.

Brotz, T.M., Borst, A., **1996**. Cholinergic and GABAergic receptors on fly tangential cells and their role in visual motion detection. *J. Neurophysiol. 76*, 1786–1799.

Brotz, T.M., Bochenek, B., Aronstein, K., ffrench-Constant, R.H., Borst, A., **1997**. Gamma-aminobutyric acid receptor distribution in the mushroom bodies of a fly (*Calliphora erythrocephala*): a functional subdivision of Kenyon cells? *J. Comp. Neurol. 23*, 42–48.

Buchner, E., Bader, R., Buchner, S., Cox, J., Emson, P.C., et al., **1988**. Cell specific immunoprobes for the brain of normal and mutant *Drosophila melanogaster*. I. Wild-type visual system. *Cell Tissue Res. 253*, 357–370.

Buckingham, S.D., Hue, B., Sattelle, D.B., **1994**. Actions of bicuculline on cell body and neuropilar membranes of identified insect neurones. *J. Exp. Biol. 186*, 235–244.

Buckingham, S.D., Matsuda, K., Hosie, A.M., Baylis, H.A., Squire, M.D., et al., **1996**. Wild-type and insecticide-resistant homo-oligomeric GABARs of *Drosophila melanogaster* stably expressed in a *Drosophila* cell line. *Neuropharmacology 35*, 1393–1401.

Buno, W. Jr, Crispino, L., Monti-Bloch, L., Mateos, A., **1981**. Dynamic analysis of cockroach giant inter-neuron activity evoked by forced displacement of cercal thread-hair sensilla. *J. Neurobiol. 12*, 561–578.

Buschges, A., Wolf, H., **1999**. Phase-dependent presynaptic modulation of mechanosensory signals in the locust flight system. *J. Neurophysiol. 81*, 959–962.

Callec, J.J., **1974**. Synaptic transmission in the central nervous system of insects. In: Treherne, J.E. (Ed.), Insect Neurobiology. Amsterdam, Elsevier/North Holland, pp. 119–185.

Callec, J.J., Boistel, J., **1971**. Role possible du GABA comme mediateur inhibiteur du systeme de fibres geantes chez la blatte (*Periplaneta americana*). *J. Physiol. Paris 63*, 119A.

Casida, J.E., **1993**. Insecticide action at the GABA-gated chloride channel: Recognition, progress, and prospects. *Arch. Insect Biochem. Physiol. 22*, 13–23.

Casida, J.E., Nicholson, R.A., Palmer, C.J., **1988**. Trioxabicyclooctanes as probes for the convulsant site of the GABA-gated chloride channel in mammals and arthropods. In: Lunt, G.G. (Ed.), Neurotox 88: Molecular Basis of Drug and Insecticide Action. Elsevier, New York, pp. 125–144.

Cayre, M., Buckingham, S.D., Yagodin, S., Sattelle, D.B., **1999**. Cultured insect mushroom body neurons express functional receptors for acetylcholine, GABA, glutamate, octopamine, and dopamine. *J. Neurophysiol.* 81, 1–14.

Chen, R., Belelli, D., Lambert, J.J., Peters, J.A., Reyes, A., *et al.*, **1994**. Cloning and functional expression of a *Drosophila* gamma-aminobutyric acid receptor. *Proc. Natl Acad. Sci. USA* 91, 6069–6073.

Christensen, T.A., Waldrop, B.R., Hildebrand, J.G., **1998**. Multitasking in the olfactory system: context-dependent responses to odors reveal dual GABA-regulated coding mechanisms in single olfactory projection neurons. *J. Neurosci.* 18, 5999–6008.

Cole, L.M., Casida, J.E., **1992**. GABA-gated chloride-channel binding site for 4′-ethynyl-4-normal-[2,3-h-3(2)]propylbicycloorthobenzoate ([^{3}H]EBOB) in vertebrate brain and insect head. *Pestic. Biochem. Physiol.* 44, 1–8.

Cole, L.M., Nicholson, R.A., Casida, J.E., **1993**. Action of phenylpyrazole insecticides at the GABA-gated chloride channel. *Pestic. Biochem. Physiol.* 46, 47–54.

Cole, L.M., Roush, R.T., Casida, J.E., **1995**. *Drosophila* GABA-gated chloride channel: Modified [^{3}H]EBOB binding site associated with Ala→ Ser or Gly mutants of Rdl subunit. *Life Sci.* 56, 757–767.

Cull-Candy, S.G., **1982**. Properties of postsynaptic channels activated by glutamate and GABA in locust muscle fibers. In: Evered, D., O'Connor, M., Whelan, J. (Eds.), Neuropharmacology of Insects. Ciba Foundation Symposium, Vol. 88. Pitman, London, pp. 70–82.

Cull-Candy, S.G., Miledi, R., **1981**. Junctional and extrajunctional membrane channels activated by GABA in locust muscle fibers. *Proc. R. Soc. Lond. B* 211, 527–535.

Curtis, D.R., Duggan, A.W., Felix, D., Johnston, G.A., **1971**. Bicuculline, an antagonist of GABA and synaptic inhibition in the spinal cord of the cat. *Brain Res.* 32(1), 69–96.

Deglise, P., Grunewald, B., Gauthier, M., **2002**. The insecticide imidacloprid is a partial agonist of the nicotinic receptor of honeybee Kenyon cells. *Neurosci. Lett.* 324, 86.

Deng, Y., Casida, J.E., **1992**. House fly head GABA-gated chloride channel: toxicologically relevant binding site for avermectins coupled to site for ethynylbicycloorthobenzoate. *Pestic. Biochem. Physiol.* 43, 116–122.

Deng, Y., Palmer, C.H., Casida, J.E., **1991**. House fly brain γ-aminobutyric acid-gated chloride channel: target for multiple classes of insecticides. *Pestic. Biochem. Physiol.* 41, 60–65.

Deng, Y., Palmer, C.J., Casida, J.E., **1993**. House fly head GABA-gated chloride channel: four putative insecticide binding sites differentiated by [^{3}H]EBOB and [^{35}S]TBPS. *Pestic. Biochem. Physiol.* 47, 98–112.

DiAntonio, A., Burgess, R.W., Chin, A.C., Deitcher, D.L., Scheller, R.H., *et al.*, **1993**. Identification and characterization of *Drosophila* genes for synaptic vesicle proteins. *J. Neurosci.* 13, 4924–4935.

Distler, P.G., Boeckh, J., **1997**. Synaptic connections between identified neuron types in the antennal lobe glomeruli of the cockroach, *Periplaneta americana*: II. Local multiglomerular interneurons. *J. Comp. Neurol.* 383, 529–540.

Distler, P.G., Boeckh, J., **1998**. An improved model of the synaptic organization of insect olfactory glomeruli. *Ann. New York Acad. Sci.* 855, 508–510.

Distler, P.G., Gruber, C., Boeckh, J., **1988**. Synaptic connections between GABA-immunoreactive neurons and uniglomerular projection neurons within the antennal lobe of the cockroach, *Periplaneta americana*. *Synapse* 29, 1–13.

Djamgoz, M.B.A., **1995**. Diversity of GABARs in the vertebrate outer retina. *Trends Neurosci.* 18, 118–120.

Edwards, M.D., Lees, G., **1997**. Modulation of a recombinant invertebrate gamma-aminobutyric acid receptor-chloride channel complex by isoflurane: effects of a point mutation in the M2 domain. *Br. J. Pharmacol.* 122, 726–732.

ffrench-Constant, R.H., Rocheleau, T.A., **1992**. *Drosophila* cyclodiene resistance gene shows conserved genomic organization with vertebrate γ-aminobutyric acidA receptors. *J. Neurochem.* 59, 1562–1565.

ffrench-Constant, R.H., Rocheleau, T.A., Steichen, J.C., Chalmers, A.E., **1993b**. A point mutation in a *Drosophila* GABAR confers insecticide resistance. *Nature* 363, 449–451.

ffrench-Constant, R.H., Steichen, J.C., Rocheleau, T.A., Aronstein, K., Roush, R.T., **1993a**. A single-amino acid substitution in a γ-aminobutyric acid subtype A receptor locus is associated with cyclodiene insecticide resistance in *Drosophila* populations. *Proc. Natl Acad. Sci. USA* 90, 1957–1961.

ffrench-Constant, R.H., Roush, R.T., **1991**. Gene mapping and cross resistance in cyclodiene insecticide resistant *Drosophila melanogaster* (Mg). *Genet. Res.* 57, 17–21.

ffrench-Constant, R.H., Mortlock, D.P., Shaffer, C.D., MacIntyre, R.J., Roush, R.T., **1991**. Molecular cloning and transformation of cyclodiene resistance in *Drosophila*: An invertebrate γ-aminobutyric acid subtype A receptor locus. *Proc. Natl Acad. Sci. USA* 88, 7209–7213.

ffrench-Constant, R.H., Anthony, N., Aronstein, K., Rocheleau, T., Stilwell, G., **2000**. Cyclodiene insecticide resistance: from molecular to population genetics. *Annu. Rev. Entomol.* 45, 449–466.

Ganeshina, O., Menzel, R., **2001**. GABA-immunoreactive neurons in the mushroom bodies of the honeybee: an electron microscopic study. *J. Comp. Neurol.* 437, 335–349.

Gauglitz, S., Pfluger, H.J., 2001. Cholinergic transmission via central synapses in the locust nervous system. *J. Comp. Physiol. A 187*, 825–836.

Georghiou, G.P., 1986. The magnitude of the resistance problem. In: Pesticide Resistance: Strategies and Tactics for Management. National Academy Press, Washington, DC, pp. 14–43.

Giles, D.P., Usherwood, P.N.R., 1985. The effect of putative amino acid neurotransmitters on somata isolated from neurons of the locust central nervous system. *Comp. Biochem. Physiol. C 80*, 231–236.

Gonzalez-Coloma, A., Valencia, F., Martin, N., Hoffmann, J.J., Hutter, L., *et al.*, 2002. Silphinene sesquiterpenes as model insect antifeedants. *J. Chem. Ecol. 28*, 117–129.

Goodman, C.S., Spitzer, N.C., 1980. Embryonic development of neurotransmitter receptors in grasshoppers. In: Sattelle, D.B., Hall, L.M., Hildebrand, J.G. (Eds.), Receptors for Neurotransmitters, Hormones and Pheromones in Insects. Amsterdam, Elsevier/North-Holland, pp. 195–209.

Grolleau, F., Sattelle, D.B., 2000. Single channel analysis of the blocking actions of BIDN and fipronil on a *Drosophila melanogaster* GABAR (RDL) stably expressed in a *Drosophila* cell line. *Br. J. Pharmacol. 130*, 1833–1842.

Grunewald, B., 1999. Morphology of feedback neurons in the mushroom body of the honeybee, *Apis mellifera*. *J. Comp. Neurol. 404*, 114–126.

Hainzl, D., Cole, L.M., Casida, J.E., 1998. Mechanisms for selective toxicity of fipronil insecticide and its sulfone metabolite and desulfinyl photoproduct. *Chem. Res. Toxicol. 11*, 1529–1535.

Hamano, H., Nagata, K., Fukada, T.N., Shimotahira, H., Ju, X.L., *et al.*, 2000. 5-[4-(3,3-Dimethylbutoxycarbonyl)phenyl]-4-pentynoic acid and its derivatives inhibit ionotropic gamma-aminobutyric acid receptors by binding to the 4′-ethynyl-4-n-propylbicycloorthobenzoate site. *Bioorg. Med. Chem. 8*, 665–674.

Hardt, M., Watson, A.H., 1999. Distribution of input and output synapses on the central branches of bushcricket and cricket auditory afferent neurones: immunocytochemical evidence for GABA and glutamate in different populations of presynaptic boutons. *J. Comp. Neurol. 403*, 281–294.

Harrison, J.B., Chen, H.H., Sattelle, E., Barker, P.J., Huskisson, N.S., *et al.*, 1996. Immunocytochemical mapping of a C-terminus anti-peptide antibody to the GABAR subunit, RDL in the nervous system in *Drosophila melanogaster*. *Cell Tissue Res. 284*, 269–278.

Harvey, R.J., Schmitt, B., Hermans-Borgmeyer, I., Gundelfinger, E.D., Betz, H., *et al.*, 1994. Sequence of a *Drosophila* ligand-gated ion-channel polypeptide with an unusual amino-terminal extracellular domain. *J. Neurochem. 62*, 2480–2483.

Hedblom, E., Kirkness, E.F., 1997. A novel class of GABA$_A$ receptor subunit in tissues of the reproductive system. *J. Biol. Chem. 272*, 15346–15350.

Henderson, J.E., Soderlund, D.M., Knipple, D.C., 1993. Characterization of a putative γ-aminobutyric acid (GABA) receptor β subunit gene from *Drosophila melanogaster*. *Biochem. Biophys. Res. Commun. 193*, 474–482.

Henderson, J.E., Knipple, D.C., Soderlund, D.M., 1994. PCR-based homology probing reveals a family of GABAR-like genes in *Drosophila melanogaster*. *Insect Biochem. Mol. Biol. 24*, 363–371.

Hill, E.S., Blagburn, J.M., 1998. Indirect synaptic inputs from filiform hair sensory neurons contribute to the receptive fields of giant interneurons in the first-instar cockroach. *J. Comp. Physiol. A 183*, 467–476.

Hold, K.M., Sirisoma, N.S., Ikeda, T., Narahashi, T., Casida, J.E., 2000. Alpha-thujone (the active component of absinthe): gamma-aminobutyric acid type A receptor modulation and metabolic detoxification. *Proc. Natl Acad. Sci. USA 97*, 4417–4418.

Hold, K.M., Sirisoma, N.S., Casida, J.E., 2001. Detoxification of alpha- and beta-Thujones (the active ingredients of absinthe): site specificity and species differences in cytochrome P450 oxidation *in vitro* and *in vivo*. *Chem. Res. Toxicol. 14*, 589–595.

Homberg, U., Kingan, T.G., Hildebrand, J.G., 1987. Immunocytochemistry of GABA in the brain and suboesophageal ganglion of *Manduca sexta*. *Cell Tissue Res. 248*, 1–24.

Homberg, U., Vitzthum, H., Muller, M., Binkle, U., 1999. Immunocytochemistry of GABA in the central complex of the locust *Schistocerca gregaria*: identification of immunoreactive neurons and colocalization with neuropeptides. *J. Comp. Neurol. 409*, 495–507.

Hosie, A.M., Aronstein, K., Sattelle, D.B., ffrench-Constant, R.H., 1997. Molecular biology of insect neuronal GABA receptors. *Trends Neurosci. 20*, 578–583.

Hosie, A.M., Baylis, H.A., Buckingham, S.D., Sattelle, D.B., 1995a. Actions of the insecticide, fipronil, on dieldrin-sensitive and -resistant GABARs of *Drosophila melanogaster*. *Br. J. Pharmacol. 115*, 909–912.

Hosie, A.M., Dunne, E.L., Harvey, R.J., Smart, T.G., 2003. Zinc-mediated inhibition of GABA(A) receptors: discrete binding sites underlie subtype specificity. *Nat. Neurosci. 6*, 362–369.

Hosie, A.M., Sattelle, D.B., 1996a. Agonist pharmacology of two *Drosophila* GABAR splice variants. *Br. J. Pharmacol. 119*, 1577–1585.

Hosie, A.M., Sattelle, D.B., 1996b. Allosteric modulation of an expressed GABA-gated chloride channel of *Drosophila melanogaster*. *Br. J. Pharmacol. 117*, 1229–1237.

Hosie, A.M., Shirai, Y., Buckingham, S.D., Rauh, J.J., Roush, R.T., *et al.*, 1995b. Blocking actions of BIDN, a bicyclic dinitrile convulsant compound, on wild-type and dieldrin-resistant GABAR homo-oliomers of *Drosophila melanogaster* expressed in *Xenopus* ootyes. *Brain Res. 693*, 257–260.

Hosie, A.M., Ozoe, Y., Koike, K., Ohmoto, T., Nikaido, T., *et al.*, 1996. Actions of picrodendrin antagonists on dieldrin-sensitive and - resistant *Drosophila* GABARs. *Br. J. Pharmacol.* 119, 1569–1576.

Hoskins, S.G., Homberg, U., Kingan, T.G., Christensen, T.A., Hildebrand, J.G., **1986**. Immunocytochemistry of GABA in the antennal lobes of the sphinx moth, *Manduca sexta*. *Cell Tissue Res.* 244, 243–252.

Hosler, J.S., Buxton, K.L., Smith, B.H., **2000**. Impairment of olfactory discrimination by blockade of GABA and nitric oxide activity in the honey bee antennal lobes. *Behav. Neurosci.* 114, 514–525.

Hue, B., **1991**. Functional assay for GABAR subtypes of a cockroach giant interneuron. *Arch. Insect Biochem. Physiol.* 18, 147–157.

Hue, B., Pelhate, M., Chanelet, J., **1979**. Pre- and postsynaptic effects of taurine and GABA in the cockroach central nervous system. *Can. J. Sci. Neurol.* 6, 243–250.

Ignell, R., **2001**. Monoamines and neuropeptides in antennal lobe interneurons of the desert locust, *Schistocerca gregaria*: an immunocytochemical study. *Cell Tissue Res.* 306, 143–156.

Imoto, K., Busch, C., Sakmann, B., Mishina, M., Konno, T., *et al.*, **1988**. Rings of negatively charged amino acids determine the acetylcholine receptor channel conductance. *Nature* 335, 645–648.

Jarboe, C.H., Porter, L.A., Buckler, R.T., **1968**. Structural aspects of picrotoxinin action. *J. Med. Chem.* 11, 729–731.

Jechlinger, M., Pelz, R., Tretter, V., Klausberger, T., Sieghart, W., **1998**. Subunit composition and quantitative importance of hetero-oligomeric receptors: GABA$_A$ receptors containing alpha6 subunits. *J. Neurosci.* 18, 2449–2457.

Johnston, G.A., **1996**. GABAc receptors: relatively simple transmitter-gated ion channels? *Trends Pharmacol. Sci.* 17, 319–323.

Judge, S., Leitch, B., **1999**. Modulation of transmitter release from the locust forewing stretch receptor neuron by GABAergic interneurons activated via muscarinic receptors. *J. Neurobiol.* 40, 420–431.

Kadous, A.A., Ghiasuddin, S.M., Mantusura, F., Scott, J.G., Tanaka, K., **1983**. Difference in the picrotoxin receptor between the cyclodiene-resistant and susceptible strains of the German cockroach. *Pestic. Biochem. Physiol.* 19, 157–166.

Kaku, K., Matsumura, F., **1994**. Identification of the site of mutation within the M2 region of the GABAR of the cyclodiene-resistant German cockroach. *Comp. Biochem. Physiol. Biochem. Mol. Biol.* 108, 367–376.

Karlin, A., Akabas, M.H., **1995**. Toward a structural basis for the function of nicotinic acetylcholine receptors and their cousins. *Neuron* 15, 1231–1244.

Karlin, A., **2002**. Emerging structure of the nicotinic acetylcholine receptors. *Nat. Rev. Neurosci.* 3, 102–114.

Kash, T.L., Jenkins, A., Kelley, J.C., Trudell, J.R., Harrison, N.L., **2003**. Coupling of agonist binding to channel gating in the GABAA receptor. *Nature* 421, 272–274.

Kauer, J.S., **1998**. Olfactory processing, a time and place for everything. *Curr. Biol.* 8, R282–R283.

Kerkut, G.A., Walker, R.J., Pitman, R., **1969**. Ionophoretic application of ACh and GABA onto insect central neurones. *Comp. Biochem. Physiol.* 31, 611–633.

Khan, Z.U., Gutierrez, A., De Blas, A.L., **1994**. The subunit composition of a GABA$_A$/benzodiazepine receptor from rat cerebellum. *J. Neurochem.* 63, 371–374.

Kolaczinski, J.H., Curtis, C.F., **2004**. Chronic illness as a result of low-level exposure to synthetic pyrethroid insecticides, a review of the debate. *Food Chem. Toxicol.* 42, 697–706.

Krishek, B.J., Amato, A., Connolly, C.N., Moss, S.J., Smart, J.G., **1996**. Proton sensitivity of the GABA$_A$ receptor is associated with the receptor subunit composition. *J. Physiol* 492, 431–443.

Kudo, Y., Niwa, H., Tanaka, A., Yamada, K., **1984**. Actions of picrotoxin and related compounds on the frog spinal cord: the role of a hydroxyl group at the 6-position in antagonizing the actions of amino acids and presynaptic inhibition. *Br. J. Pharmacol.* 81, 373–380.

Kuffler, E.W., Edwards, C., **1958**. Mechanisms of gamma aminobutyric acid (GABA) action and its relation to synaptic inhibition. *J. Neurophysiol.* 32, 589–610.

Lawrence, J.J., Casida, J.E., **1984**. Interactions of lindane, toxaphene and cyclodienes with brain-specific *t*-butyl-bicyclophosphorothionate receptor. *Life Sci.* 35, 171–178.

Leal, S.M., Neckameyer, W.S., **2002**. Pharmacological evidence for GABAergic regulation of specific behaviors in *Drosophila melanogaster*. *J. Neurobiol.* 50, 245–261.

Lee, H.J., Rocheleau, T., Zhang, H.G., Jackson, M.B., ffrench-Constant, R.H., **1994**. Expression of a *Drosophila* GABAR in a baculovirus insect cell system–functional expression of insecticide susceptible and resistant GABARs from the cyclodiene resistance gene Rdl. *FEBS Lett.* 335, 315–318.

Lee, D., O'Dowd, D.K., **1999**. Fast excitatory synaptic transmission mediated by nicotinic acetylcholine receptors in *Drosophila* neurons. *J. Neurosci.* 19, 5311–5321.

Lee, D., Su, H., O'Dowd, D., **2003**. GABA receptors containing Rdl subunits mediate fast synaptic transmission in *Drosophila* neurones. *J. Neurosci.* 23, 4625–4634.

Lees, G., Beadle, D.J., Neumann, R., Benson, J.A., **1987**. Responses to GABA by isolated insect neuronal somata-pharmacology and modulation by a benzodiazepine and a barbiturate. *Brain Res.* 401, 267–278.

Leonard, R.J., Labarca, G.G., Charnet, P., Davidson, N., Lester, H.A., **1988**. Evidence that the M2 membrane-spanning region lines the ion channel pore of the nicotinic receptor. *Science* 242, 1578–1581.

Levi, R., Camhi, J.M., **2000**. Population vector coding by the giant interneurons of the cockroach. *J. Neurosci.* 20, 3822–3829.

Ludmerer, S.W., Warren, V.A., Williams, B.S., Zheng, Y., Hunt, D.C., *et al.*, **2002**. Ivermectin and nodulisporic acid receptors in *Drosophila melanogaster* contain both gamma-aminobutyric acid-gated Rdl and glutamate-gated GluCl alpha chloride channel subunits. *Biochemistry* 41, 6548–6560.

Lummis, S.C.R., Sattelle, D.B., **1985**. Insect CNS GABA receptors. *Neurosci. Lett. 60*, 13–18.

Lummis, S.C.R., Sattelle, D.B., **1986**. Binding sites for [^{3}H]GABA, [^{3}H]flunitrazepam and [^{35}S]TBPS in insect CNS. *Neurochem. Int. 9*, 287–293.

Lummis, S.C.R., Buckingham, S.D., Rauh, J.J., Sattelle, D.B., **1990**. Blocking actions of heptachlor at an insect central nervous system GABAR. *Proc. R. Soc. B. 240*, 97–106.

MacLeod, K., Laurent, G., **1996**. Distinct mechanisms for synchronization and temporal patterning of odor-encoding. *Science 274*, 976–979.

Malecot, C.O., Sattelle, D.B., **1990**. Single-channel recordings from insect neuronal GABA-activated chloride channels. *J. Exp. Biol. 151*, 495–501.

Martin, S.C., Russek, S.J., Farb, D.H., **2001**. Human GABA(B)R genomic structure: evidence for splice variants in GABA(B)R1 but not GABA(B)R2. *Gene 278*, 63–79.

Matsuda, K., Hosie, A.M., Buckingham, S.D., Squire, M.D., Baylis, H.A., *et al.*, **1996**. pH-dependent actions of THIP and ZAPA on an ionotropic *Drosophila melanogaster* GABAR. *Brain Res. 739*, 335–338.

Matsuda, K., Hosie, A.M., Holyoke, C.W., Rauh, J.J., Sattelle, D.B., **1999**. Cross-resistance with dieldrin of a novel tricyclic dinitrile GABAR antagonist. *Br. J. Pharmacol. 127*, 1305–1307.

Matsumura, F., Ghiasuddin, S.M., **1983**. Evidence for similarities between cyclodiene-type insecticides and picrotoxin in their action mechanisms. *J. Environ. Sci. Hlth B. 18*, 1–14.

Matsumura, F., Tanaka, K., Ozoe, Y., **1987**. GABA-related systems as targets for insecticides. In: Hollingworth, R.M., Green, M.B. (Eds.), Sites of Action of Neurotoxic Pesticides, ACS Symposium Series 356. American Chemical Society, Washington DC, pp. 44–70.

McCracken, D.I., **1993**. The potential for avermectins to affect wildlife. *Vet. Parasitol. 48*, 273–280.

McGurk, K.A., Pistis, M., Belelli, D., Hope, A.G., Lambert, J.J., **1998**. The effect of a transmembrane amino acid on etomidate sensitivity of an invertebrate GABAR. *Br. J. Pharmacol. 124*, 13–20.

McKernan, R., Whiting, P., **1996**. Which GABA$_A$-receptor subtypes really occur in the brain? *Trends Neurosci. 19*, 139–143.

Mezler, M., Muller, T., Raming, K., **2001**. Cloning and functional expression of GABA(B) receptors from *Drosophila*. *Eur. J. Neurosci. 13*, 477–486.

Millar, N.S., Buckingham, S.D., Sattelle, D.B., **1994**. Stable expression of a functional homo-oligomeric *Drosophila* GABAR in a *Drosophila* cell line. *Proc. R. Soc. Lond. B. Biol. Sci. 258*, 307–314.

Miyazaki, M., Matsumura, F., Beeman, R.W., **1995**. DNA sequence and site of mutation of the GABAR of cyclodiene-resistant red flour beetle, *Tribolium castaneum*. *Comp. Biochem. Physiol. Biochem. Mol. Biol. 111*, 399–406.

Mohler, H., Benke, D., Mertens, S., Fritschy, J.M., **1992**. GABA$_A$-receptor subtypes differing in alpha-subunit composition display unique pharmacological properties. *Adv. Biochem. Psychopharmacol. 47*, 41–53.

Murphy, V.F., Wann, K.T., **1988**. The action of GABAR agonists and antagonists on muscle membrane conductance in *Schistocerca gregaria*. *Br. J. Pharmacol. 95*, 713–722.

Neelands, T.R., Fisher, J.L., Bianchi, M., Macdonald, R.L., **1999**. Spontaneous and gamma-aminobutyric acid (GABA)-activated GABA(A) receptor channels formed by epsilon subunit-containing isoforms. *Mol. Pharmacol. 55*, 168–178.

Newland, C.F., Cull-Candy, S.G., **1992**. On the mechanism of action of picrotoxin on GABAR channels in dissociated symphathetic neruones of the rat. *J. Physiol. 447*, 191–213.

Nielsen, M., Braestrup, C., Squires, R.F., **1978**. Evidence for a late evolutionary appearance of brain-specific benzodiazepine receptors: an investigation of 18 vertebrate and 5 invertebrate species. *Brain Res. 141*, 342–346.

Nishino, H., Mizunami, M., **1998**. Giant input neurons of the mushroom body: intracellular recording and staining in the cockroach. *Neurosci. Lett. 246*, 57–60.

Nusser, Z., Ahmad, Z., Tretter, V., Fuchs, K., Wisden, W., *et al.*, **1999**. Alterations in the expression of GABA$_A$ receptor subunits in cerebellar granule cells after the disruption of the alpha6 subunit gene. *Eur. J. Neurosci. 11*, 1685–1697.

Okuma, J., Kondoh, Y., **1996**. Neural circuitry underlying linear representation of wind information in a nonspiking local interneuron of the cockroach. *J. Comp. Physiol. A 179*, 725–740.

Otsuka, M., Iversen, L.L., Hall, Z.W., Kravitz, E.A., **1966**. Release of gamma-aminobutyric acid from inhibitory nerves of lobster. *Proc. Natl Acad. Sci. USA 56*, 1110–1115.

Ozoe, Y., Akamatsu, M., Akamatsu, M., Higata, T., Ikeda, I., *et al.*, **1998**. Picrodendrin and related terpenoid antagonists reveal structural differences between ionotropic GABARs of mammals and insects. *Bioorg. Med. Chem. 6*, 481–492.

Ozoe, Y., Hasegawa, H., Mochida, K., Koike, K., Suzuki, Y., *et al.*, **1994**. Picrodendrins, a new group of picrotoxane terpenoids: structure-activity profile of action at the GABA$_A$ receptor-coupled picrotoxinin, binding site in rat brain. *Biosci. Biotech. Biochem. 58*, 1506–1507.

Ozoe, Y., Matsumura, F., **1986**. Structural requirements for bridged bicyclic compounds acting on picrotoxin receptor. *J. Agric. Food Chem. 34*, 126–134.

Palmer, C.H., Casida, J.E., **1985**. 1,4-Disubstituted 2,6,7-trioxabicyclo[2.2.2]octanes: a new class of insecticides. *J. Agric. Food. Chem. 33*, 976.

Panek, I., French, A.S., Seyfarth, E.A., Sekizawa, S., Torkkeli, P.H., **2002**. Peripheral GABAergic inhibition of spider mechanosensory afferents. *Eur. J. Neurosci. 16*, 96–104.

Perez-Orive, J., Mazor, O., Turner, G.C., Cassenaer, S., Wilson, R.I., *et al.*, **2002**. Oscillations and sparsening of odor representations in the mushroom body. *Science 297(5580)*, 359–365.

Petri, B., Homberg, U., Loesel, R., Stengl, M., 2002. Evidence for a role of GABA and Mas-allatotropin in photic entrainment of the circadian clock of the cockroach *Leucophaea maderae. J. Exp. Biol.* 205, 1459–1469.

Pichon, Y., Beadle, D., 1988. Acetylcholine, GABA and glutamate receptor channels in cultured insect neurones. In: Lunt, G.G. (Ed.), Neurotox '88: Molecular Basis of Drug and Pesticide Action. Elsevier Science Publishers, BV, Amsterdam, pp. 325–337.

Pistis, M., Belelli, D., McGurk, K., Peters, J.A., Lambert, J.J., 1999. Complementary regulation of anaesthetic activation of human (α6-β3-γ2L) and *Drosophila* (RDL) GABA receptors by a single amino acid residue. *J. Physiol.* 515(Pt 1), 3–18.

Pitman, R.M, Kerkut, G.A., 1970. Comparison of the actions of iontophoretically-applied acetylcholine and gamma aminobutyric acid with the EPSP and IPSP in cockroach central neurons. *Comp. Gen. Pharmac.* 1, 221–230.

Priestley, C.M., Williamson, E.M., Wafford, K.A., Sattelle, D.B., 2003. Thymol, a constituent of thyme essential oil, is a positive allosteric modulator of human GABA(A) receptors and a homo-oligomeric GABA receptor from *Drosophila melanogaster. Br. J. Pharmacol.* 140(8), 1363–1372.

Pritchett, D.B., Luddens, H., Seeburg, P.H., 1989. Type I and type II GABA$_A$-benzodiazepine receptors produced in transfected cells. *Science* 245, 1389–1392.

Python, F., Stocker, R.F., 2002. Immunoreactivity against choline acetyltransferase, gamma-aminobutyric acid, histamine, octopamine, and serotonin in the larval chemosensory system of *Drosophila melanogaster. J. Comp. Neurol.* 390, 455–469.

Qian, H., Dowling, J.E., 1993. Novel GABA responses from rod-driven retinal horizontal cells. *Nature 361,* 162–164.

Ratra, G.S., Casida, J.E., 2001. GABAR subunit composition relative to insecticide potency and selectivity. *Toxicol. Lett.* 122, 215–222.

Ratra, G.S., Kamita, S.G., Casida, J.E., 2001. Role of human GABA(A) receptor beta3 subunit in insecticide toxicity. *Toxicol. Appl. Pharmacol.* 172, 233–240.

Rauh, J.J., Lummis, S.C.R., Sattelle, D.B., 1990. Pharmacological and biochemical properties of insect GABARs. *Trends Pharmacol. Sci.* 11, 325–329.

Rauh, J.J., Benner, E., Schnee, M.E., Cordova, D., Holyoke, C.W., et al., 1997a. Effects of [³H]-BIDN, a novel bicyclic dinitrile radioligand for GABA-gated chloride channels of insects and vertebrates. *Br. J. Pharmacol.* 121, 1496–1505.

Rauh, J.J., Holyoke, C.W., Kleier, D.A., Presnail, J.K., Benner, E.A., et al., 1997b. Polycyclic dinitriles: a novel class of potent GABAergic insecticides provides a new radioligand, [³H]BIDN. *Invert. Neurosci.* 3, 261–268.

Raymond, V., Sattelle, D.B., Lapied, B., 2000. Co-existence in DUM neurones of two GluCl channels that differ in their picrotoxin sensitivity. *Neuroreport* 230, 45–48.

Raymond, V., Sattelle, D.B., 2002. Novel animal-health drug targets from ligand-gated chloride channels. *Nat. Rev. Drug Discov.* 6, 427–436.

Reeves, D.C., Lummis, S.C., 2002. The molecular basis of the structure and function of the 5-HT$_3$ receptor: a model ligand-gated ion channel. *Mol. Membr. Biol.* 19, 11–26.

Richter, K., Bohm, G.A., 1997. The molting gland of the cockroach *Periplaneta americana*: secretory activity and its regulation. *Gen. Pharmacol.* 29, 17–21.

Roberts, C.J., Krogsgaard-Larsen, P., Walker, R.J., 1981. Studies of the action of GABA, muscimol and related compounds on *Periplaneta* and *Limulus* neurons. *Comp. Biochem. Physiol.* 69C, 7–11.

Robinson, T.N., MacAllan, D., Lunt, G.G., Battersby, M., 1985. γ-aminobutyric acid receptor complex of insect CNS: characterization of a benzodiazepine binding site. *J. Neurochem.* 47, 1955–1962.

Rohrbough, J., Broadie, K., 2002. Electrophysiological analysis of synaptic transmission in central neurons of *Drosophila* larvae. *J. Neurophysiol.* 88, 847–860.

Rohrer, S.P., Birzin, E.T., Costa, S.D., Arena, J.P., Hayes, E.C., et al., 1995. Identification of neuron-specific ivermectin binding sites in *Drosophila melanogaster* and *Schistocerca americana. Insect Biochem. Mol. Biol.* 25, 11–17.

Rosay, P., Armstrong, J.D., Wang, Z., Kaiser, K., 2001. Synchronized neural activity in the *Drosophila* memory centers and its modulation by amnesiac. *Neuron 30,* 759–770.

Rothlin, C.V., Katz, E., Verbitsky, M., Elgoyhen, A.B., 1999. The alpha9 nicotinic acetylcholine receptor shares pharmacological properties with type A gamma-aminobutyric acid, glycine, and type 3 serotonin receptors. *Mol. Pharmacol.* 55, 248–254.

Rufingier, C., Pasteur, N., Lagnel, J., Martin, C., Navajas, M., 1999. Mechanisms of insecticide resistance in the aphid *Nasonovia ribisnigri* (Mosley) (Homoptera: Aphididae) from France. *Insect. Biochem. Mol. Biol.* 29, 385–391.

Sachse, S., Galizia, C.G., 2002. Role of inhibition for temporal and spatial odor representation in olfactory output neurons, a calcium imaging study. *J. Neurophysiol.* 87, 1106–1117.

Sattelle, D.B., 1990. GABA receptors of insects. *Adv. Insect Physiol.* 22, 1–113.

Sattelle, D.B., Bai, D., Chen, H.H., Skeer, J.M., Buckingham, S.D., 2003. Bicuculline-insensitive GABA-gated Cl⁻ channels in the larval nervous system of the moth *Manduca sexta. Invert. Neurosci.* 5(1), 37–43.

Sattelle, D.B., Pinnock, R.D., Wafford, K.A., David, J.A., 1988. GABARs on the cell-body membrane of an identified insect motor neurone. *Proc. R. Soc. Lond. B 232,* 443–456.

Sattelle, D.B., Harrison, J.B., Chen, H.H., Bai, D., Takeda, M., 2000. Immunocytochemical localization of putative gamma-aminobutyric acid receptor subunits in the head ganglia of *Periplaneta americana* using an anti-RDL C-terminal antibody. *Neurosci. Lett. 289,* 197–200.

Sattelle, D.B., Holyoke, C.W., Schnee, M.E., Cordova, D., Bennet, E.A., *et al.,* **1995.** A bicyclic dinitrile radioligand for studying GABA-gated chloride channels. *J. Physiol. 483,* 193P.

Schmidt, H., Luer, K., Hevers, W., Technau, G.M., **2000.** Ionic currents of *Drosophila* embryonic neurons derived from selectively cultured CNS midline precursors. *J. Neurobiol. 44,* 392–413.

Schnee, M., Rauh, J., Buckingham, S.D., Sattelle, D.B., **1997.** Pharmacology of skeletal muscle GABA-gated chloride channels in the cockroach *Periplaneta americana. J. Exp. Biol. 200,* 2947–2955.

Schurmann, F.W., **2000.** Acetylcholine, GABA, glutamate and NO as putative transmitters indicated by immunocytochemistry in the olfactory mushroom body system of the insect brain. *Acta. Biol. Hung. 51,* 355–362.

Schurmann, F.W., Ottersen, O.P., Honegger, H.W., **2000.** Glutamate-like immunoreactivity marks compartments of the mushroom bodies in the brain of the cricket. *J. Comp. Neurol. 418,* 227–239.

Scott, R.H., Duce, I.R., **1987.** Pharmacology of GABA receptors on skeletal muscle fibres of the locust (*Schistocerca gregaria*). *Comp. Biochem. Physiol. C 86(2),* 305–311.

Seidel, C., Bicker, G., **1997.** Colocalisation of taurine-with transmitter-immunoreactivities in the nervous system of the migratory locust. *Brain Res. 769,* 273–280.

Seidel, C., Bicker, G., **2002.** Developmental expression of nitric oxide/cyclic GMP signaling pathways in the brain of the embryonic grasshopper. *Brain. Res. Devel. Brain Res. 138,* 71–79.

Shimahara, T., Pichon, Y., Lees, G., Beadle, C.A., Beadle, D.J., **1987.** Gamma-aminobutyric acid receptors on cultured cockroach brain neurones. *J. Exp. Biol. 131,* 231–244.

Shirai, Y., Hosie, A.M., Buckingham, S.D., Holyoke, C.W.Jr, Baylis, H.A., *et al.,* **1995.** Actions of picrotoxin analogues on an expressed, homo-oliomeric GABAR of *Drosophila melanogaster. Neurosci. Lett. 188,* 1–4.

Sieghart, W., **1995.** Structure and pharmacology of gamma-aminobutyric acidA receptor subtypes. *Pharmacol. Rev. 47,* 181–234.

Sinkkonen, S.T., Hanna, M.C., Kirkness, E.F., Korpi, E.R., **2000.** GABA(A) receptor protein epsilon and theta subunits display unusual structural variation between species and are enriched in the rat locus ceruleus. *J. Neurosci. 20,* 3588–3595.

Sirisoma, N.S., Ratra, G.S., Tomizawa, M., Casida, J.E., **2001.** Fipronil-based photoaffinity probe for *Drosophila* and human beta 3 GABARs. *Bioorg. Med. Chem. Lett. 11,* 2979–2981.

Smart, T.G., Constanti, A., **1986.** Studies on the mechanism of action of picrotoxin and other convulsants at the crustacean muscle GABAR. *Proc. R. Soc. Lond. Ser. B 227,* 191–216.

Smith, G.B., Olsen, R.W., **1995.** Functional domains of GABA$_A$ receptors. *Trends Neurosci. 16,* 162–168.

Stevenson, P.A., **1999.** Colocalisation of taurine with transmitter-immunoreactivities in the nervous system of the migratory locust. *J. Comp. Neurol. 404,* 86–96.

Stevenson, A., Wingrove, P.B., Whiting, P.J., Wafford, K.A., **1995.** Beta-carboline gamma-aminobytyric acid (A) receptor inverse agonists modulate gamma-aminobutyric acid via the loreclezole binding site as well as the benzodiazepine site. *Mol. Pharmacol. 49,* 965–969.

Stopfer, M., Bhagavan, S., Smith, B.H., Laurent, G., **1997.** Impaired odour discrimination on desynchronization of odour-encoding neural assemblies. *Nature 390,* 70–74.

Strambi, C., Cayre, M., Sattelle, D.B., Augier, R., Charpin, P., *et al.,* **1998.** Immunocytochemical mapping of an RDL-like GABAR subunit and of GABA in brain structures related to learning and memory in the cricket *Acheta domesticus. Learn. Mem. 5,* 78–89.

Strong, L., **1993.** Overview: the impact of avermectins on pastureland ecology. *Vet. Parasitol. 48,* 3–17.

Stumpner, A., **1998.** Picrotoxin eliminates frequency selectivity of an auditory interneuron in a bushcricket. *J. Neurophysiol. 79,* 2408–2415.

Taylor, C.P., Vartanian, M.G., Yuen, P.W., Bigge, C., Suman-Chauhan, N., *et al.,* **1993.** Potent and stereospecific anticonvulsant activity of 3-isobutyl GABA relates to *in vitro* binding at a novel site labeled by tritiated gabapentin. *Epilepsy Res. 14(1),* 11–15.

Thompson, S.A., Bonnert, T.P., Whiting, P.J., Wafford, K.A., **1998.** Functional characteristics of recombinant human GABA(A) receptors containing the epsilon-subunit. *Toxicol. Lett. 100–101,* 233–238.

Thompson, M., Shotkoski, F., ffrench-Constant, R., **1993a.** Cloning and sequencing of the cyclodiene insecticide resistance gene from the yellow fever mosquito *Aedes aegypti.* Conservation of the gene and resistance associated mutation with *Drosophila. FEBS Lett. 325,* 187–190.

Thompson, M., Steichen, J.C., ffrench-Constant, R.H., **1993b.** Conservation of cyclodiene insecticide resistance-associated mutations in insects. *Insect Mol. Biol. 2,* 149–154.

Ticku, K., Ban, M., Olsen, R.W., **1978.** Binding of [^{3}H]a-dihydropicrotoxinin, a γ-aminobytyric acid synaptic antagonist to rat brain membranes. *Mol. Pharmacol. 14,* 391–402.

Tingle, C.C., Rother, J.A., Dewhurst, C.F., Lauer, S., King, W.J., **2003.** Fipronil: environmental fate, ecotoxicology, and human health concerns. *Rev. Environ. Contam. Toxicol. 176,* 1–66.

Umesh, A., Gill, S.S., **2002.** Immunocytochemical localization of a *Manduca sexta* gamma-aminobutyric acid transporter. *J. Comp. Neurol. 448,* 388–398.

Usherwood, P.N.R., Grundfest, H., **1965.** Peripheral inhibition in skeletal muscle of insects. *J. Neurophysiol. 28,* 497–518.

Wafford, K.A., Sattelle, D.B., Gant, D.B., Eldefrawi, A.T., Eldefrawi, M.E., **1989.** Noncompetitive inhibition of

GABARs in insect and vertebrate CNS by endrin and lindane. *Pest. Biochem. Physiol. 33*, 213.

Wafford, K.A., Bain, C.J., Quirk, K., McKernan, R.M., Wingrove, P.B., et al., **1995**. A novel allosteric modulatory site on the GABA$_A$ receptor γ subunit. *Neuron 12*, 775–782.

Waldrop, B., **1994**. Physiological and pharmacological properties of responses of GABA and ACh by abdominal motor neurons in *Manduca sexta. J. Comp. Physiol. A 175*, 754–785.

Waldrop, B., Christensen, T.A., Hildebrand, J.G., **1987**. GABA-mediated synaptic inhibition of projection neurons in the antennal lobes of the sphinx moth, *Manduca sexta. J. Comp. Physiol. A 161*, 23–32.

Waldrop, B., Hildebrand, J.G., **1988**. Physiology and pharmacology of acetylcholinergic responses of interneurones in the antennal lobes of the sphinx moth, *Manduca sexta. J. Comp. Physiol. A 164*, 433–441.

Walker, R.J., Crossman, A.R., Woodruff, G.N., Kerkut, G.A., **1971**. The effects of bicuculline on the gamma-aminobutyric acid (GABA) receptors of neurons of *Periplaneta americana* and *Helix aspersa. Brain Res. 33*, 75–82.

Wang, T.-L., Hackam, A., Guggino, W.B., Cutting, G.R., **1995**. A single histidine residue is essential for zinc inhibition of GABA rho receptors. *J. Neurosci. 15*, 7684–7691.

Watson, A.H.D., Burrows, M., **1987**. Immunocytochemical and pharmacological evidence for GABAergic spiking local interneurones in the locust. *J. Neurosci. 7*, 1741–1751.

Watson, S., Girdlestone, D., **1995**. TiPS on nomenclature. *Trends Pharmacol. Sci. 16*(1), 15–16.

Watson, G.B., Salgado, V.L., **2001**. Maintenance of GABAR function of small-diameter cockroach neurons by adenine nucleotides. *Insect Biochem. Mol. Biol. 31*, 207–212.

Wegerhoff, R., **1999**. GABA and serotonin immunoreactivity during postembryonic brain development in the beetle *Tenebrio molitor. Microsc. Res. Tech. 45*, 154–164.

Westin, J., Ritzmann, R.E., Goddard, D.J., **1988**. Wind-activated thoracic interneurons of the cockroach: I. Responses to controlled wind stimulation. *J. Neurobiol. 19*, 573–588.

Whiting, P.J., **2003**. GABA-A receptor subtypes in the brain: a paradigm for CNS drug discovery? *Drug Discov. Today 8*, 445–450.

Whitton, P.S., Strang, R.H.C, Nicholson, R.A., **1987**. The distribution of taurine in the tissues of some species of insects. *Insect Biochem. 17*, 573–577.

Whitton, P.S., Nicholson, R.A., Strang, R.H.C., **1994**. Electrophysiological responses of isolated locust (*Schistocerca gregaria*) somata to taurine and GABA application. *J. Insect Physiol. 40*, 195–199.

Wingrove, P.B., Wafford, K.A., Bain, C., Whiting, P.J., **1994**. The modulatory action of loreclezole at the γ-aminobutyric acid type A receptor is determined by a single amino acid in the β2 and β3 subunit. *Proc. Natl Acad. Sci. USA 91*, 4569–4573.

Witten, J.L., Truman, J.W., **1991**. The regulation of transmitter expression in postembryonic lineages in the moth *Manduca sexta*. II. Role of cell lineage and birth order. *J. Neurosci. 11*, 1990–1997.

Witten, J.L., Truman, J.W., **1998**. Distribution of GABA-like immunoreactive neurons in insects suggests lineage homology. *J. Comp. Neurol. 398*, 515–528.

Wolff, M.A., Wingate, V.P., **1998**. Characterization and comparative pharmacological studies of a functional gamma-aminobutyric acid (GABA) receptor cloned from the tobacco budworm, *Heliothis virescens* (Noctuidae: Lepidoptera). *Invert. Neurosci. 3*, 305–315.

Woodward, R.M., Polenzani, L., Miledi, R., **1993**. Characterization of bicuculline/baclofen-insensitive (ρ-like) γ-aminobutyric acid receptors expressed in *Xenopus* oocytes. II. Pharmacology of γ-aminobutyric acid$_A$ and γ-aminobutyric acid$_B$ receptor agonists and antagoinsts. *Mol. Pharmacol. 43*, 609–625.

Yamazaki, Y., Nishikawa, M., Mizunami, M., **1998**. Three classes of GABA-like immunoreactive neurons in the mushroom body of the cockroach. *Brain Res. 788*, 80–86.

Yasuyama, K., Meinertzhagen, I.A., Schurmann, F.W., **2002**. Synaptic organization of the mushroom body calyx in *Drosophila melanogaster. J. Comp. Neurol. 183*, 467–476.

Zhang, H.-G., ffrench-Constant, R.H., Jackson, M.B., **1994**. A unique amino acid of the *Drosophila* GABAR with influence on drug sensitivity by two mechanisms. *J. Physiol. 479*, 65–75.

Zhang, D., Pan, Z.H., Awobuluyi, M., Lipton, S.A., **2001**. Structure and function of GABA(C) receptors: a comparison of native versus recombinant receptors. *Trends Pharmacol. Sci. 22*, 121–132.

Zheng, Y., Hirschberg, B., Yuan, J., Wang, A.P., Hunt, D.C., et al., **2002**. Identification of two novel *Drosophila melanogaster* histamine-gated chloride channel subunits expressed in the eye. *J. Biol. Chem. 277*, 2000–2005.

Zornik, E., Paisley, K., Nichols, R., **1999**. Neural transmitters and a peptide modulate *Drosophila* heart rate. *Peptides 20*, 45–51.

A2 Addendum: Recent Progress in Insect GABA Receptors

S Gill, University of California, Riverside, CA

© 2010, Elsevier BV. All Rights Reserved.

Buckingham and Sattelle (2005) gave an extensive review of insect GABA receptors. This addendum provides a brief update on the previous comprehensive review.

A2.1. GABA$_A$ Receptor Diversity by Alternative Splicing and RNA Editing

Jones and Sattelle (2007) in a recent review of the cys-loop ligand-gated ion channel gene superfamily surmise that the honeybee, *Apis mellifera*, and the flour beetle, *Tribolium castaneum*, each have three GABA-gated chloride channels – the *Drosophila* equivalents of RDL, GRD, and LCCH3. The latter two form heteromeric channels, while RDL functions as a homomeric channel when expressed in *Xenopus* oocytes (Gisselmann *et al.*, 2004; Jones and Sattelle, 2007). Alternative splicing, as observed in exons 3 and 6 of *rdl*, increases GABA receptor diversity (Jones and Sattelle, 2007), and all the four alternative splicing variants are expressed (Buckingham and Sattelle, 2005; McGonigle and Lummis, 2009). Two spliced variants, RDL$_{ac}$ and RDL$_{bc}$, show greater sensitivity to GABA than variants RDL$_{ad}$ and RDL$_{bd}$; the latter showing the lowest GABA sensitivity (Jones *et al.*, 2009; McGonigle and Lummis, 2009). However, both RDL$_{ad}$ and RDL$_{bd}$ are more highly expressed in the adult nervous system than the other two variants.

Analysis of *Nasonia vitripennis, Anopheles gambiae, Aedes aegypti*, and *Culex pipiens* genomes demonstrates that the *rdl, grd*, and *lcch3* genes are similarly present in these insects with alternative splicing observed or predicted in *rdl* orthologs. Consequently, at least in dipteran species, alternative splicing results in a larger number of potential GABA receptor transcripts.

Additional RDL receptor diversity was observed by RNA editing, similar to that detected in a number of other targets of neurotransmission in *D. melanogaster* whereby pre-mRNA editing by adenosine deaminases acting on RNA (ADAR) results in amino acid substitutions. Four of these editing changes were observed in three *rdl* exons −4, 7, and 8 (Hoopengardner *et al.*, 2003). One of these amino acid changes, R122G, resulting from an A-to-I editing, showed decreased RDL sensitivity to GABA in fipronil susceptible wild-type flies and in a fipronil resistant line, A301G/T350M *rdl* mutant (Es-Salah *et al.*, 2008). Such editing resulted in decreased sensitivity, in both wild-type and resistant lines, to fipronil (Es-Salah *et al.*, 2008), an insecticide in wide use in agriculture and in animal health. A more extensive analysis of RNA editing, recently performed on alternatively spliced variants of *rdl* (Jones *et al.*, 2009), showed that splicing and editing may be linked. Greater editing was observed in the RDL$_{bd}$ splice variant than in the RDL$_{ad}$, the more abundant transcript in adult CNS (Jones *et al.*, 2009).

Thus, a combination of alternative splicing and RNA editing leads to the expression of a larger diversity of GABA channels in insects. These diverse channels respond differentially to insecticides (Es-Salah *et al.*, 2008) and behavioral inputs. A similar diversity was also observed in sodium channels (see Addendum A1 by Dong in this volume).

A2.2. GABA$_B$ Receptors

In *D. melanogaster*, three genes encode metabotropic GABA receptors (Family C) (Park and Adams, 2005). The three mosquito genomes sequenced to date, *An. gambiae, Ae. aegypti*, and *C. pipiens*, have a similar complementation of genes. However, to date there is little functional characterization of metabotropic GABA receptors in insects.

A2.3. GABA$_A$ Receptor Mutations and Resistance in Insects

Fipronil and the chlorinated insecticides lindane and α-endosulfan inhibit the human GABA β3 homomeric receptor just as they do the *D. melanogaster* RDL receptor (Casida and Tomizawa, 2008; Chen *et al.*, 2006). Co-assembly of the human β3 subunit with other subunits shows decreased sensitivity to fipronil and other non-competitive antagonists (Casida and Tomizawa, 2008; Chen *et al.*, 2006). Mutations of amino acid residues in transmembrane 2 (TM2) that line the pore of the GABA β3 homomeric receptor, results in reduced binding (Casida and Tomizawa, 2008; Chen *et al.*, 2006). In insect RDL, identical TM2 amino acids are involved.

Mutation of one of these TM2 residues, A302S, led to the characterization of the *D. melanogaster rdl* strain (*resistance to dieldrin*) (ffrench-Constant *et al.*, 1993). Similar mutations have been isolated in other insect RDLs. Alanine-to-glycine and alanine-to-serine substitutions were detected in the mosquito malarial vectors *Anopheles gambiae and An. arabiensis*, respectively (Du *et al.*, 2005). However, an alanine-to-glycine mutation showed weak association with dieldrin resistance in an *An. gambiae* population from Ghana, suggesting that other resistance mechanisms or other active site mutations may be involved (Brooke *et al.*, 2006). An A-to-S substitution at this site was also partially associated with fipronil resistance in the diamond back moth in China (Li *et al.*, 2006) and target-site insensitivity may be involved in resistance in a plant hopper (Tang *et al.*, 2009). A mutation at his site, A310G, was also observed in a laboratory-selected strain of *D. simulans* that showed high levels of resistance to fipronil, but an additional mutation in TM3, T350M, was also detected (Le Goff *et al.*, 2005). While A310G showed significant decreased sensitivity to GABA and fipronil, the double mutant showed the greatest insensitivity to GABA and fipronil, suggesting that mutations at other sites may also contribute to fipronil resistance (Le Goff *et al.*, 2005). Interestingly, a homologous A-to-S mutation in TM2 of cat flea *rdl* showed no change in susceptibility to fipronil (Brunet *et al.*, 2009), further indicating a role for additional critical amino acid residues in RDL function.

Interestingly, fipronil also blocks glutamate-activated chloride channels in insects but not those in mammals (Narahashi *et al.*, 2007). In contrast, as noted earlier, vertebrate GABA channels, in particular β3 homomeric channels, are sensitive to fipronil. Thus, it is possible that insect selectivity of fipronil resides not only on its action of GABA-gated chloride channels but also on its action on glutamate-gated chloride channels (Narahashi *et al.*, 2007).

A2.4. Role of GABA$_A$ Receptors in Behavior and Sleep

As noted in the original review (Buckingham and Sattelle, 2005), RDL plays an important role in insect behavior and it is highly expressed in mushroom bodies. Analysis of its expression levels and its silencing showed that the RDL GABA receptor suppresses olfactory learning, via a conditioned stimulus, which enhances memory acquisition but not its stability (Liu *et al.*, 2009; Liu *et al.*, 2007). Moreover, it is likely that A-to-I editing, potentially of *rdl*, plays a role in this coordinated behavior since loss of ADAR activity affects such activity (Jepson and Reenan, 2009).

Recent evidence also implicates a key role for this gene in sleep in the fruit fly (Agosto *et al.*, 2008; Chung *et al.*, 2009; Parisky *et al.*, 2008). These studies show that RDL expression in pigment-dispersing factor (PDF) neurons reduces sleep. More interestingly, use of the A302S RDL mutant shows that the regulation of sleep onset is uncoupled from that of sleep duration (Agosto *et al.*, 2008).

References

Agosto, J., Choi, J.C., Parisky, K.M., Stilwell, G., Rosbash, M., Griffith, L.C., **2008**. Modulation of GABA$_A$ receptor desensitization uncouples sleep onset and maintenance in Drosophila. *Nat. Neurosci.* 11, 354–359.

Brooke, B.D., Hunt, R.H., Matambo, T.S., Koekemoer, L.L., Van Wyk, P., Coetzee, M., **2006**. Dieldrin resistance in the malaria vector *Anopheles gambiae* in Ghana. *Med. Vet. Entomol.* 20, 294–299.

Brunet, S., Le Meter, C., Murray, M., Soll, M., Audonnet, J.C., **2009**. Rdl gene polymorphism and sequence analysis and relation to in vivo fipronil susceptibility in strains of the cat flea. *J. Econ. Entomol. 102*, 366–372.

Buckingham, S.D., Sattelle, D.B., **2005**. GABA receptors in insects. In: Gilbert, I.K., Gill, S.S. (Eds.), Comprehensive Molecular Insect Science, vol. 5. Elsevier, Amsterdam, pp. 107–142.

Casida, J.E., Tomizawa, M., **2008**. Insecticides interactions with GABA-aminobutyric acid and nicotinic acid receptors: predictive aspects of structural models. *J. Pestic. Sci. 33*, 4–8.

Chen, L., Durkin, K.A., Casida, J.E., 2006. Structural model for gamma-aminobutyric acid receptor noncompetitive antagonist binding: widely diverse structures fit the same site. *Proc Natl Acad Sci USA 103*, 5185–5190.

Chung, B.Y., Kilman, V.L., Keath, J.R., Pitman, J.L., Allada, R., 2009. The GABA(A) receptor RDL acts in peptidergic PDF neurons to promote sleep in Drosophila. *Curr. Biol. 19*, 386–390.

Du, W., Awolola, T.S., Howell, P., Koekemoer, L.L., Brooke, B.D., Benedict, M.Q., Coetzee, M., Zheng, L., 2005. Independent mutations in the Rdl locus confer dieldrin resistance to *Anopheles gambiae* and *An. arabiensis*. *Insect Mol. Biol. 14*, 179–183.

Es-Salah, Z., Lapied, B., Le Goff, G., Hamon, A., 2008. RNA editing regulates insect gamma-aminobutyric acid receptor function and insecticide sensitivity. *Neuroreport 19*, 939–943.

ffrench-Constant, R.H., Rocheleau, T.A., Steichen, J.C., Chalmers, A.E., 1993. A point mutation in a Drosophila GABA receptor confers insecticide resistance. *Nature 363*, 449–451.

Gisselmann, G., Plonka, J., Pusch, H., Hatt, H., 2004. *Drosophila melanogaster* GRD and LCCH3 subunits form heteromultimeric GABA-gated cation channels. *Br. J. Pharmacol. 142*, 409–413.

Hoopengardner, B., Bhalla, T., Staber, C., Reenan, R., 2003. Nervous system targets of RNA editing identified by comparative genomics. *Science 301*, 832–836.

Jepson, J.E., Reenan, R.A., 2009. Adenosine-to-inosine genetic recoding is required in the adult stage nervous system for coordinated behavior in Drosophila. *J. Biol. Chem. 284*, 31391–31400.

Jones, A.K., Buckingham, S.D., Papadaki, M., Yokota, M., Sattelle, B.M., Matsuda, K., Sattelle, D.B., 2009. Splice-variant- and stage-specific RNA editing of the Drosophila GABA receptor modulates agonist potency. *J. Neurosci. 29*, 4287–4292.

Jones, A.K., Sattelle, D.B., 2007. The cys-loop ligand-gated ion channel gene superfamily of the red flour beetle, *Tribolium castaneum*. *BMC Genomics 8*(327).

Le Goff, G., Hamon, A., Berge, J.B., Amichot, M., 2005. Resistance to fipronil in *Drosophila simulans*: influence of two point mutations in the RDL GABA receptor subunit. *J. Neurochem. 92*, 1295–1305.

Li, A., Yang, Y., Wu, S., Li, C., Wu, Y., 2006. Investigation of resistance mechanisms to fipronil in diamondback moth (Lepidoptera: Plutellidae). *J. Econ. Entomol. 99*, 914–919.

Liu, X., Buchanan, M.E., Han, K.A., Davis, R.L., 2009. The GABA$_A$ receptor RDL suppresses the conditioned stimulus pathway for olfactory learning. *J. Neurosci. 29*, 1573–1579.

Liu, X., Krause, W.C., Davis, R.L., 2007. GABAA receptor RDL inhibits Drosophila olfactory associative learning. *Neuron 56*, 1090–1102.

McGonigle, I., Lummis, S.C., 2009. RDL receptors. *Biochem. Soc. Trans. 37*, 1404–1406.

Narahashi, T., Zhao, X., Ikeda, T., Nagata, K., Yeh, J.Z., 2007. Differential actions of insecticides on target sites: basis for selective toxicity. *Hum. Exp. Toxicol. 26*, 361–366.

Parisky, K.M., Agosto, J., Pulver, S.R., Shang, Y., Kuklin, E., et al., 2008. PDF cells are a GABA-responsive wake-promoting component of the Drosophila sleep circuit. *Neuron 60*, 672–682.

Park, Y., Adams, M.E., 2005. Insect G protein-coupled receptors: recent discoveries and implications. In: Gilbert, I.K., Gill, S.S. (Eds.), Comprehensive Molecular Insect Science, vol. 5. Elsevier, Amsterdam, pp. 143–171.

Tang, J., Li, J., Shao, Y., Yang, B., Liu, Z., 2010. Fipronil resistance in the whitebacked planthopper (*Sogatella furcifera*): possible resistance mechanisms and cross-resistance. *Pest Manag. Sci. 66*, 121–125.

3   The Insecticidal Macrocyclic Lactones

D Rugg, Fort Dodge Animal Health, Princeton, NJ, USA
S D Buckingham and D B Sattelle, University of Oxford, Oxford, UK
R K Jansson, Centocor, Malvern, PA, USA

© 2005, Elsevier BV. All Rights Reserved.

3.1. Discovery

The naturally occurring avermectins and milbemycins are fermentation products of actinomycetes in the genus *Streptomyces*. They are 16-membered, macrocyclic lactones, which have structural similarities to antibacterial macrolides and antifungal polyenes but lack their antifungal or antibacterial activities and do not inhibit protein or chitin synthesis (Fisher and Mrozik, 1989). Avermectins are produced by the soil microorganism, *Streptomyces avermitilis* MA-4680 (NRRL 8165), which was first isolated at the Merck Research Laboratories in 1976 from a soil sample collected near a golf course in Japan by researchers at the Kitasato Institute (Campbell *et al.*, 1984). Milbemycins were first described from a culture of *S. hygroscopicus* and are structurally similar to the avermectins but lack the disaccharide substituent at C13 (Takiguchi *et al.*, 1980). Mishima *et al.* (1975) first reported the acaricidal activity of milbemycins. The anthelmintic activity of these milbemycins was later elucidated following testing of individual members of the milbemycin family (Mrozik *et al.*, 1983). Milbemycins were also isolated from cultures of *S. cyanogriseus* in 1984 by workers at American Cyanamid and from *S. thermoarchaensis* by Glaxo researchers (Rock *et al.*, 2002).

Four homologous pairs of closely related compounds are produced through the fermentation of *S. avermitilis* MA-4680 (NRRL 8165), each pair comprising a major and minor component, usually produced in the approximate ratio of 8:2 (Lasota and Dybas, 1991). A mixture of one of these pairs, avermectin B_1 (>80% B_{1a} and <20% B_{1b}) commonly called abamectin, was found to be active against nematodes (Burg *et al.*, 1979; Egerton *et al.*, 1979), insects (Ostlind *et al.*, 1979; James *et al.*, 1980; Wright, 1984), and mites (Putter *et al.*, 1981).

Ivermectin, the semisynthetic 22,23-dihydro derivative of avermectin B_1, was the first commercially produced avermectin, introduced in 1981 for use against a range of nematode and arthropod parasites (Jackson, 1989). Like abamectin, ivermectin is a mixture of two homologs known as ivermectin B_{1a} (>80%) and B_{1b} (<20%) (Campbell et al., 1983). The natural avermectin abamectin was introduced as an antiparasitic drug and an agricultural insecticide and miticide in 1985 (Campbell, 1989). The natural milbemycins milbenock and milbemectin have been used as an agricultural miticide (Tsukamoto et al., 1995) and in canine heartworm therapy (Jung et al., 2003), respectively. An oxime derivative of milbemycin has been used widely for filarid and nematode control in dogs. Moxidectin, a synthetically modified milbemycin derived from the fermentation product nemadectin (Carter et al., 1988), is used for insect and helminth control in animal health applications. In 1993, doramectin was introduced, also for use in animal health. This compound was derived from an induced mutation of S. avermitilis (Goudie et al., 1993) and is structurally closer to abamectin than to ivermectin (Shoop et al., 1995). A subsequent semisynthetic derivative, selamectin (the monosaccharide oxime), is used to control insect, acarid and helminth parasites of cats and dogs (Bishop et al., 2000).

A targeted analoging program around abamectin resulted in a five-step semisynthetic avermectin, emamectin, in 1984 (Cvetovich et al., 1994). MK-243 (the hydrochloride salt of emamectin) was discovered after screening several hundred avermectin derivatives in an in vivo screen using tobacco budworm, Heliothis virescens and southern armyworm, Spodoptera eridania (Dybas and Babu, 1988; Dybas et al., 1989; Mrozik et al., 1989; Mrozik, 1994). This compound was subsequently selected for further development for control of lepidopterans in crop protection. Synthesized aglycon derivatives of ivermectin have been reported to be potent flea insecticides (Cvetovich et al., 1997), but have not been developed commercially.

Numerous papers have been published on the avermectins and milbemycins (Kornis, 1995). Several authors have reviewed the discovery, structure–activity relationships, environmental fate, spectrum of activity, and applications for their anthelminthic, insecticidal, and acaricidal activity (Campbell et al., 1984; Fisher and Mrozik, 1984, 1989, 1992; Davies and Green, 1986; Strong and Brown, 1987; Dybas, 1989; Lasota and Dybas, 1991; Mrozik, 1994; Kornis, 1995; Shoop et al., 1995; Fisher, 1997; Jansson and Dybas, 1998; Vercruysse and Rew, 2002a). In this chapter the chemistry, mode of action, biological activity, and applications of macrocyclic lactones, with special emphasis on their use as insecticides are reviewed.

3.2. Chemistry

3.2.1. Chemical Structure

The chemical structure of avermectin B_{1a} is shown in Figure 1. The milbemycins share a similar basic 16-carbon lactone but lack the disaccharide moiety of the avermectins (e.g., moxidectin) (Figure 2). These compounds have strong ultraviolet absorption at 245 nm, a feature which is of value in analytical detection. Avermectins and milbemycins are highly lipophilic, being soluble in most organic solvents but almost insoluble in water (Fisher and Mrozik, 1989). Milbemycins tend to be more lipophilic than avermectins (Hennessy and Alvinerie, 2002). Ultraviolet light rapidly decomposes

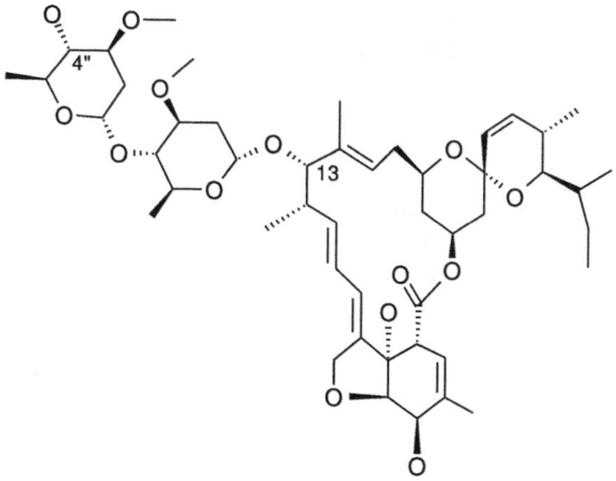

Figure 1 Chemical structure of avermectin B_{1a} (major component of abamectin).

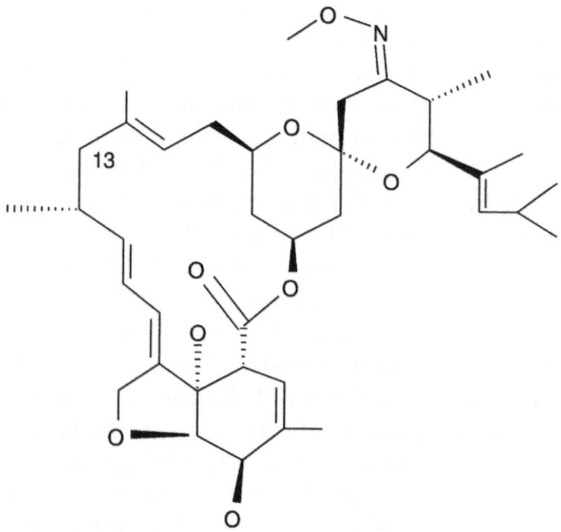

Figure 2 Chemical structure of moxidectin.

avermectins, leaving a large number of less toxic products (Halley *et al.*, 1989a) that lack any ultraviolet absorption (Mrozik *et al.*, 1988) as a residue. The half-life of ivermectin, exposed to sunlight as a thin film on glass, is about 3 h (Halley *et al.*, 1989b).

Ivermectin B_{1a} differs from abamectin in the loss of the 22,23 double bond, so that 22 and 23 are dihydrocarbons. The B_{1b} components differ from their respective B_{1a} structures by the replacement of a CH_2CH_3 group with a CH_3 (Fisher and Mrozik, 1989). Ivermectin is also commonly referred to as 22,23-dihydroavermectin B_1. Emamectin, an epimethyl amino substitution at the 4″ position of the sugar (**Figure 3**) was derived from abamectin using a five-step synthesis (Cvetovich *et al.*, 1994).

Doramectin was generated via a biosynthetic program. Mutated strains of *S. avermitilis* were used to generate novel avermectins (Goudie *et al.*, 1993). Doramectin (**Figure 4**) was selected for development as an endectocide (a compound that controls both internal and external parasites) from a number of compounds that differed from avermectins in

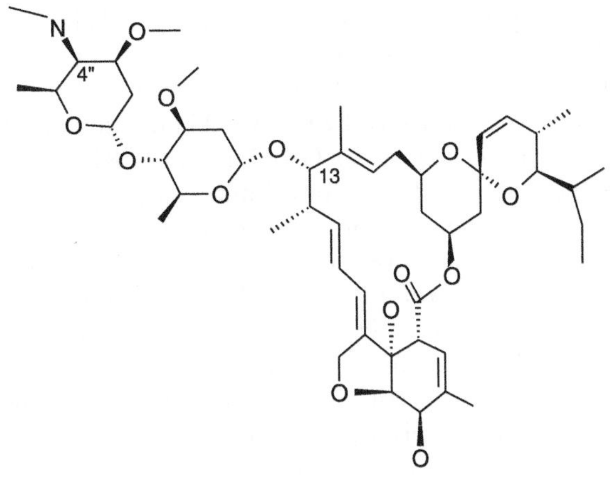

Figure 3   Chemical structure of emamectin.

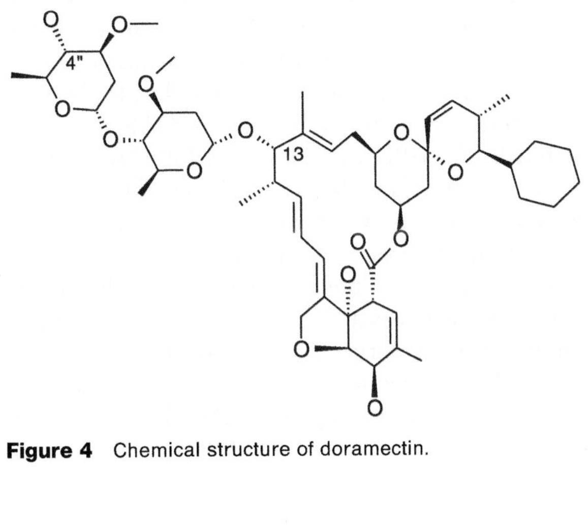

Figure 4   Chemical structure of doramectin.

substituents at the C25 position. Selamectin was chosen from a targeted chemistry research program aimed at identifying an avermectin that retained anthelminthic activity but with enhanced insecticidal activity, especially against fleas (Banks *et al.*, 2000).

Targeted synthesis programs with milbemycins have also led to commercial products. Moxidectin (**Figure 2**) was synthesized from the natural milbemycin Fα (nemadectin) via a multistep process (Carter *et al.*, 1988).

3.2.2. Structure–Activity Relationships

Structure–activity relationships of the avermectins and milbemycins are complicated due to the broad spectrum of activity against arthropods and helminths. However, reviews of the relationship between the activity of these compounds and chemical structure have shown some consistent patterns (Shoop *et al.*, 1995; Fisher, 1997). The major structural difference between avermectins and milbemycins is the lack of the disaccharide moiety at position C13 of the macrolide ring. Within avermectins, removal of the distal sugar results in a slight decrease in activity against nematodes. Removal of both sugars results in an avermectin aglycone with a hydroxy group at the C13 position with little useful activity against arthropods or helminths (Chabala *et al.*, 1980). When the oxygen is removed from the C13 position, the resultant deoxyavermectins recover potency and are structurally equivalent to milbemycins (Mrozik *et al.*, 1983; Shoop *et al.*, 1995). Generally, avermectins and milbemycins with lipophilic substituents at C13 are highly active, while polar substituents result in a loss of activity (Mrozik *et al.*, 1989). Similar structure–activity patterns have been seen with insects and mites.

For insecticidal activity, the most dramatic development was the synthesis of 4″-epi-amino avermectins (Fisher, 1997). In these compounds, the replacement of the hydroxy group on the distal sugar with an epimethylamino group resulted in a marked increase in potency against a variety of lepidopterans and reduced activity against mites.

3.3. Mode of Action

3.3.1. Biochemical and Molecular Action

Avermectins are potent endectocides. Ivermectin, the first and most commercially important member of this class, is the world's most successful animal health drug of all time (Raymond and Sattelle, 2002). While the molecular mode of action of avermectins initially proved difficult to define

(Jackson, 1989), more recent evidence indicates that they act primarily on glutamate-gated chloride channels of helminths and insects, with additional effects (especially at high concentrations) on γ-aminobutyric acid (GABA) receptors.

Early evidence for GABA receptors as a site of action came from studies performed mostly on insects (Sattelle, 1990). Dihydroavermectin B_{1a}-sensitive GABA-binding sites were identified in the cockroach *Periplaneta americana* using [^{3}H]GABA binding (Lummis and Sattelle, 1985). Furthermore, avermectins were shown to act as a partial agonist at a GABA-binding site in honeybees (*Apis mellifera*) defined by [^{3}H]muscimol binding (Abalis and Eldefrawi, 1986). Deng and Casida (1992) examined the interactions of radiolabeled avermectin with nerve membranes from housefly (*Musca domestica*) heads and hypothesized that avermectins may either have multiple binding sites at each chloride channel, or may bind to GABA-gated channels at low concentrations and at other chloride channels at higher concentrations.

Evidence from vertebrate studies also pointed to GABA receptors as a site of action. Administered to rats, ivermectin acts as a GABA-ergic anxiolytic drug (Spinosa *et al.*, 2002). Another study compared the potency of 25 avermectin analogs using a mouse seizure model with their direct action on recombinant $GABA_A$ receptors (Dawson *et al.*, 2000). Interestingly, while their binding affinities to brain membrane and to recombinant $GABA_A$ receptors stably expressed in cell lines were found to be closely similar, suggesting $GABA_A$ receptors as the target, the rank order of affinity of these analogs at the [^{3}H]ivermectin binding site differed from that of their anticonvulsant activities, suggesting that their activities differed between subtypes of $GABA_A$ receptor. When the binding of the analogs to cell lines expressing known combinations of $GABA_A$ receptor subunits were measured, only modest differences in affinity (up to fivefold) were observed, compared to differences in anticonvulsant activity in the seizure model of up to 40-fold. Furthermore, their efficacy in potentiating GABA-evoked currents in *Xenopus laevis* oocytes expressing $GABA_A$ receptors correlated well with their anticonvulsant activity. The conclusion drawn, as reported earlier (Arena *et al.*, 1993), was that the GABA receptor β subunit is a key determinant of avermectin potency. The study also suggested the presence of two distinct avermectin binding sites. Arena *et al.* (1993) had observed that there are twice as many avermectin binding sites as benzodiazepine sites correlating well with there being two β subunits in many $GABA_A$ receptors (Sigel and Buhr, 1997).

Certain evidence also suggested that avermectins act at sites other than $GABA_A$ receptors. For example, [^{3}H]22,23-dihydroavermectin-B_1 binding sites that were GABA-insensitive were identified in membrane preparations from *Caenorhabditis elegans* (Schaeffer and Haines, 1989). Tanaka and Matsumura (1985) found that avermectin stimulated chloride ion uptake in cockroach (*P. americana*) leg muscle; they suggested that this was not due to activity at the GABA binding site but rather a direct action on the chloride ion channel. Tanaka (1987) determined that abamectin and milbemycin act directly on a neuronal chloride channel of *P. americana*, and that this action could not be accounted for in terms of action on the GABA, picrotoxinin, or benzodiazepine sites on GABA-gated chloride channels. Nicholson *et al.* (1988) also deployed *P. americana* in studies of acetylcholine release from a central nervous system synaptosome preparation. The authors concluded that avermectins acted on a non-GABA-mediated chloride channel, with the resultant depolarization of the nerve terminal leading to increased acetylcholine release. Avermectin B_{1a} and moxidectin inhibited binding of 4-*n*-propyl-4′-ethnylbicycloorthobenzoate (EBOB) a noncompetitive blocker of the GABA-gated chloride channel in *Drosophila melanogaster* and *D. simulans* (Cole *et al.*, 1995). Their efficacy in displacing EBOB was unaffected by the A302S mutation in the cloned *D. melanogaster* GABA receptor, RDL, that reduces sensitivity to convulsants, suggesting that either the main site of action of avermectin B_{1a} is not RDL, or that avermectin binding does not affect the RDL convulsant site. The dieldrin-resistant Maryland strain of *D. melanogaster*, which has the RDL A302S mutation, is susceptible to avermectin MK-243, abamectin, and abamectin 8,9-oxide (Bloomquist, 1994), again suggesting that these avermectins do not act at the convulsant site of the GABA receptor encoded by the *rdl* gene.

Evidence that glutamate-gated chloride channels are the primary site of the anthelminthic activity of avermectins came from a number of studies on nematodes. These studies showed that the interaction of avermectins with glutamate-gated chloride channels is distinct from any effect upon GABA-sensitive channels (Schaeffer and Haines, 1989; Arena *et al.*, 1992, 1995). They also showed that the effects of ivermectin on its interaction with glutamate were concentration dependent. At low concentrations, ivermectin potentiated the response of the channel to glutamate, whereas at high concentrations, ivermectin opened the chloride channel directly (Arena *et al.*, 1992). Thus, in nematodes, avermectins are believed to potentiate or directly open the glutamate-gated chloride channel (Arena *et al.*, 1995).

Studies on cloned, expressed glutamate-gated chloride channels from nematodes and insects provide strong evidence that avermectins act upon these receptors. Cully *et al.* (1994) cloned two complementary DNAs encoding avermectin-sensitive glutamate-gated chloride channel subunits from the nematode *C. elegans*. These two genes (*GluClα1* and *GluClβ*) are newly discovered members of the cys-loop family of the ligand-gated ion channel superfamily. The α1 subunit gene is alternatively spliced to produce two subunits (Laughton *et al.*, 1997b). Cully *et al.* (1994) demonstrated that the potency of various avermectin compounds *in vivo* was correlated with their ability to potentiate glutamate-gated currents recorded when *Glu-Clα1* and *GluClβ* are coexpressed in *Xenopus* oocytes, suggesting that these two subunits are components of the target of avermectins in nematodes. The α subunit possesses a binding site for glutamate but is not coupled to the opening of the channel (Etter *et al.*, 1996). The *GluClα1* gene is expressed in the pharyngeal muscle cells pm4 and pm5, and in some neurons (Dent *et al.*, 1997) and *GluClβ* is also expressed in pm4 (Laughton *et al.*, 1997a). When a transposon element is used to introduce a deletion in *GluClα1*, to eliminate its function, sensitivity of the animals to ivermectin remained but mRNA extracted from them and expressed in *Xenopus* oocytes resulted in the expression of glutamate receptors with reduced sensitivity to both ivermectin phosphate and to glutamate (Vassilatis *et al.*, 1997). Furthermore, in the mutant, binding of [3H]ivermectin showed reduced binding affinity, indicating a loss of ivermectin binding sites. The same study also reported the cloning and functional expression of another glutamate-gated chloride channel, *GluClα2*, which when expressed in *Xenopus* oocytes responds reversibly to glutamate and irreversibly to avermectin.

Arguably, the strongest evidence that glutamate-gated chloride channels are targets for avermectins derives from genetics. Simultaneous mutation of genes encoding glutamate-gated chloride channel α-type subunits (*GluClα1, GluClα2*, and *GluClα3*) confers high-level resistance to ivermectin. Interestingly, only low-level (or no) resistance is conferred by mutation of any two of these three glutamate receptor genes (Dent *et al.*, 2000). It may prove to be a sound strategy to search for drug targets encoded by a multigene family because the development of drug resistance could be slower if several target genes have to be mutated simultaneously to confer resistance.

Similar results have been obtained using glutamate-gated chloride channels from other nematodes.

Of the four glutamate-gated chloride channel subunits known in *Haemonchus contortus* (Delany *et al.*, 1998; Jagannathan *et al.*, 1999), two bind [3H]ivermectin with high affinity when expressed in monkey kidney, COS-7, cells, which is not displaced by GABA, picrotoxin, or fipronil (convulsants that block GABA-gated chloride channels) or glutamate; and for the one channel examined, there was no difference in sequence between that from a wild-type and that from an ivermectin-resistant strain (Cheeseman *et al.*, 2001). The recently described tissue distribution of the four *H. contortus* avermectin-sensitive glutamate receptor subunits showed that these subunits are differentially expressed in the motor nervous system, as well as in other neurons (Portillo *et al.*, 2003).

Horoszok *et al.* (2001) reported the cloning of GLC-3, a glutamate-gated chloride channel from *C. elegans* that forms homooligomeric receptors when expressed in *Xenopus* oocytes. Ivermectin potently and irreversibly activated this receptor (EC$_{50}$ the concentration that evokes a half-maximal response – 0.4 μM). The availability of a new homomeric avermectin-sensitive receptor from a helminth offers exciting possibilities for structure–function research into the site of action of these compounds.

Rohrer *et al.* (1995) identified the binding sites of ivermectin in two arthropods, *D. melanogaster* and *Schistocerca americana*. One subunit, DrosGluClα, was recently cloned from *D. melanogaster*. This subunit is sensitive to both glutamate and avermectins (Cully *et al.*, 1996). The same authors also found transcripts related to DrosGluClα in other insects, including cat flea, *Ctenocephalides felis*, fall armyworm, *Spodoptera frugiperda*, and cotton bollworm, *Helicoverpa zea*, but not in the two-spotted spider mite, *Tetranychus urticae*. It remains to be determined whether or not additional GluCl subunits are present in insects.

An exciting recent discovery may explain, in part, why avermectins appear to act at both GABA and glutamate receptors. Immunoprecipitation studies using antibodies to a *D. melanogaster* glutamate-gated chloride channel (*Dm*GluCl) showed that *Dm*GluClα antibodies precipitated all of the ivermectin and nodulisporic acid receptors solubilized by detergent from *Drosophila* head membranes, and that *Dm*RDL antibodies also immunoprecipitated all solubilized nodulisporic acid receptors and approximately 70% of the ivermectin receptors (Ludmerer *et al.*, 2002). The data led the authors to suggest that both *Dm*GluClα and RDL are components of nodulisporic acid and ivermectin receptors.

It has been suggested that avermectins may bind to a common site on all ligand-gated chloride channels similar to the benzodiazepine site on mammalian GABA-receptors (Huang and Casida, 1997), and act in a tissue-specific manner either as positive or negative allosteric modulators. This would help to explain some of the additional effects that avermectins exert on chloride channels, such as their inhibition of strychnine binding to the mammalian glycine receptor (Graham *et al.*, 1982); activation of multiple-gated chloride channels (glutamate, GABA, and acetylcholine) in crayfish (Zufall *et al.*, 1989); and both potentiation and blocking of GABA-gated chloride channels in arthropods (Duce and Scott, 1985; Bermudez *et al.*, 1991).

There is evidence for actions of avermectins at sites other than glutamate- or GABA-gated chloride channels. Avermectin C and avermectin A_1 increased calcium-dependent chloride currents of *Chara corallina* cells at low concentrations but blocked them at high concentrations (Driniaev *et al.*, 2001). Avermectins A_2, B_1 (abamectin), B_2 and 22,23-dihydroavermectin B_1 (ivermectin) in the concentration range studied did not affect the chloride currents of *C. corallina* cells. There is also evidence for nicotinic acetylcholine receptor actions of avermectins. Ivermectin potentiates responses to acetylcholine mediated by heterologously expressed $\alpha7$ nicotinic receptor subunits (Krause *et al.*, 1998) with high potency (30 μM ivermectin potentiated responses to 30 μM acetylcholine (ACh) nearly threefold) through a presumed allosteric mechanism. However, responses to ACh mediated by ACR-16 (formerly known as Ce21), an $\alpha7$-like homomer-forming nicotinic acetylcholine receptor subunit from *C. elegans* (Ballivet *et al.*, 1996), with 47% identity with chicken $\alpha7$ at the amino acid level, were not enhanced by the same concentration of ivermectin, but rather were slightly attenuated (Raymond *et al.*, 2000).

In summary, avermectins affect invertebrates either by directly activating receptors for the neurotransmitters glutamate and GABA or by potentiating their actions, resulting in an influx of chloride ions into nerve cells and muscle. It appears that the anthelminthic properties of the avermectins are due predominantly to potentiation and/or direct opening of glutamate-gated chloride channels, whereas in insects, it is likely that avermectins bind to multiple sites, including both glutamate- and GABA-gated chloride channels as well as other insect chloride channels. However, not all insect chloride channels are sensitive to avermectins: one of the two chloride channels spontaneously active in cultured *D. melanogaster* larval neurons – the large conductance (35pS) "slow" channel – is insensitive

to dihydroavermectin B_{1a} (Yamamoto and Suzuki, 1988). Nevertheless, there is now a wealth of evidence that the chloride ion flux resulting from the opening by avermectins of chloride channels in invertebrate nerve and muscle, particularly those gated by glutamate and GABA, results in disruption of activity and loss of function in these excitable cells and accounts for the potent endectocidal actions of these molecules.

3.3.2. Physiological Activity

3.3.2.1. **Insects** While avermectins are effective through both topical and oral routes in lepidopteran larvae, toxicity is generally greater through ingestion than by residual contact (Wright *et al.*, 1985; Anderson *et al.*, 1986; Bull, 1986; Corbitt *et al.*, 1989). McKellar and Benchaoui (1996) surmised that the oral route made the major contribution to uptake of macrocyclic lactones in ectoparasitic arthropods. The higher *in vivo* activity of these compounds against sucking lice than against biting lice, and the potent effects on blood-feeding mites (Benz *et al.*, 1989), supports this hypothesis. Similarly, abamectin, although having relatively poor contact activity, is potent when administered orally against cockroaches (Rust, 1986), lice (Rugg and Thompson, 1993), ants (Glancey *et al.*, 1982; Baker *et al.*, 1985), termites (Su *et al.*, 1987), and vespid wasps (Chang, 1988).

In comparison with other neurotoxic agents, avermectins are considered to be relatively slow acting (Wright, 1986; Strong and Brown, 1987; Abro *et al.*, 1988). Affected organisms usually die slowly, often over a period of days, and there is no quick knockdown effect like that achieved with the use of some pyrethroids and organophoshates (Jackson, 1989). Death usually follows paralysis and immobility. Agee (1985) described the symptoms of abamectin poisoning in *H. zea* larvae; the insects showed a gradual loss of locomotor coordination followed by lethargy and complete loss of locomotor activity. This slow action could possibly be due to either progressive paralysis of the organism's feeding, resulting in slow accumulation of the toxin, or starvation (Jackson, 1989). However, it is apparent that there is a direct neurotoxic effect, as adults of *H. zea* died more rapidly from abamectin poisoning than from starvation (Agee, 1985) and poisoned German cockroaches (*Blattella germanica*) ate less than controls, but the starved controls did not die (Cochran, 1985). The blood-sucking bug *Rhodnius prolixus* can survive for months after a blood meal, but dies within 9 days of ingestion of ivermectin (De Azambuja *et al.*, 1985).

3.3.2.2. Acarids and helminths Symptoms of poisoning in acarids and helminths are similar to those of insects. The affected organisms cease feeding, suffer paralysis, and subsequently die (McKellar and Benchaoui, 1996). For most species, the oral route is considered the major means of uptake, although for some gastrointestinal and filarial nematodes, transcuticular absorption may be as significant as ingestion (Court *et al.*, 1988).

3.3.2.3. Vertebrates Of the vertebrates, only fish are uniformly susceptible to macrocyclic lactone toxicity (Wislocki *et al.*, 1989). Mammals are relatively insensitive to these compounds (Jackson, 1989). The signs of macrocyclic lactone poisoning in mammals are consistent with central nervous system toxicity. Symptoms include salivation, ataxia, muscle fasciculation, lingual paralysis, recumbency, apparent blindness, tremors, and depression. Lethal doses produce ataxia, convulsions, and coma (Campbell and Benz, 1984; Lankas and Gordon, 1989). These signs suggest poisoning of the central nervous system and are consistent with stimulation of GABA-gated chloride channels which are likely to be the active sites of macrocyclic lactones in mammals (Pong *et al.*, 1980; Pong and Wang, 1982; Drexler and Sieghart, 1984a, 1984b, 1984c).

3.3.3. Sublethal Toxicity

Sublethal effects of macrocyclic lactones on insects include inhibition of feeding, developmental abnormalities, and disruption of reproduction (Strong and Brown, 1987). Cessation of feeding appears to be a symptom of toxicity as overt repellency is rarely observed with avermectins (Schuster and Everett, 1983; Cochran, 1985; Orton *et al.*, 1992; Allingham *et al.*, 1994). In most instances, reduction in the prolonged feeding activity of insects is due to the onset of poisoning (Strong and Brown, 1987).

Sublethal effects on development are varied. Cockroach nymphs that had been fed sublethal doses of abamectin were able to complete development (Cochran, 1985), but juvenile *R. prolixus* failed to molt (De Azambuja *et al.*, 1985). Sublethal doses of abamectin fed to neonate tufted apple moth, *Platynota idaeusalis*, increased the developmental time from neonate to adult eclosion and greatly affected fecundity (Biddinger and Hull, 1999). In other lepidopterans, sublethal doses of abamectin may interfere with pupation and metamorphosis without marked disruption of larval development (Reed *et al.*, 1985; Bull, 1986) and similar effects have been observed for ivermectin in dipterans (Meyer *et al.*, 1980; Spradbery *et al.*, 1985; Strong, 1986). Rugg (1995) found that abamectin, ivermectin, and moxidectin

had some residual postfeeding effects resulting in death in the pupal stage of a dung-feeding fly, *Tricharea brevicornis*. Sublethal residues of avermectins have been shown to markedly delay the development of dung beetle larvae (Lumaret *et al.*, 1993; Kruger and Scholtz, 1997) and produce morphological developmental abnormalities in dung-feeding flies (Clarke and Ridsill-Smith, 1990; Strong and James, 1993) and dung beetles (Sommer *et al.*, 1993).

Sublethal doses of these insecticides also produce various effects on reproduction. Female *R. prolixus* suffered a permanent reduction in egg production following ingestion of ivermectin (De Azambuga *et al.*, 1985). In some Coleoptera, exposure to abamectin resulted in fewer eggs laid and lowered levels of hatching (Wright, 1984; Reed and Reed, 1986). Following abamectin ingestion, female *B. germanica* had fewer oocytes and produced fewer oothecae (Cochran, 1985). Also, female serpentine leafminer (*Liriomyza trifolii*) exposed to abamectin-treated leaves laid fewer eggs, possibly due to malfunction of the ovipositor after contact to abamectin (Schuster and Everett, 1983). They noted that ingestion of abamectin might have occurred and this could have an effect on toxicity to *Liriomyza*. Adult Australian blowfly *Lucilia cuprina* females fed sheep dung containing ivermectin residues show reduced fecundity, survival, and delayed reproductive development (Cook, 1991, 1993; Mahon and Wardhaugh, 1991). Orton *et al.* (1992) investigated the activity of avermectins as oviposition suppressants in *L. cuprina*. While the compounds tested were effective in suppressing oviposition, this was related to general toxic effects and was only apparent at levels that produced significant mortality. Cook (1993) found that sublethal doses of ivermectin affected mating behavior in male *L. cuprina*. The number of mating attempts was reduced and the duration of mating was longer in treated males than in nontreated males. When tsetse flies (*Glossina morsitans*) were fed on blood containing ivermectin, they aborted eggs and larvae. However, these flies produced normal larvae after refeeding on normal blood (Langley and Roe, 1984). Lepidopteran larvae treated with sublethal doses of abamectin that survived to adults displayed reduced mating, egg production, and hatching (Reed *et al.*, 1985; Robertson *et al.*, 1985; Wright, 1986). Sublethal doses of abamectin have been shown to irreversibly sterilize queens of the ant *Solenopsis invicta* (Glancey *et al.*, 1982).

3.3.4. Metabolism

Oxidative metabolism is believed to be the major route for detoxification and excretion of macrocyclic lactones in insects. Three main oxidative metabolites have been identified in mammals (Chiu and Lu, 1989; Zeng *et al.*, 1996). These metabolites are more polar than the parent compounds and greater hydrophilicity can enhance excretion of xenobiotics in insects. Studies with resistant and susceptible Colorado potato beetle (*Leptinotarsa decemlineata*) have implicated oxidative metabolism as a major mechanism in resistance to avermectins (Argentine, 1991; Argentine *et al.*, 1992; Yoon *et al.*, 2002).

3.4. Biological Activity

3.4.1. Spectrum and Potency

3.4.1.1. Insects The macrocyclic lactones are among the most potent insecticidal, acaricidal, and anthelminthic agents known. Avermectins have been shown to be toxic to most insects although potency and speed of action can vary widely (Strong and Brown, 1987). These compounds are used commercially to control pests in the orders Isoptera, Blattodea, Phthiraptera, Thysanoptera, Coleoptera, Siphonaptera, Diptera, Lepidoptera, and Hymenoptera. Abamectin, the compound most widely used for insect control, has been shown to be a potent stomach toxin in cockroaches (Cochran, 1985; Rust, 1986; Koehler *et al.*, 1991), fire ants (Glancey *et al.*, 1982), Argentine ants (Baker *et al.*, 1985), vespid wasps (Chang, 1988), termites (Su *et al.*, 1987), various blood-feeding Diptera (Miller *et al.*, 1986; Allingham *et al.*, 1994), and sheep-biting lice, *Bovicola ovis* (Rugg and Thompson, 1993). Among agricultural pests, the compound is highly toxic to tobacco hornworm (*Manduca sexta*), diamondback moth (*Plutella xylostella*), tomato pinworm (*Keiferia lycopersicella*), tobacco budworm (*H. virescens*), Colorado potato beetle (*L. decemlineata*), and the serpentine leafminer (*L. trifolii*) (lethal concentration (LC$_{90}$) values range between 0.009 and 0.19 μg ml^{-1}). However, it is less potent against certain Homoptera and most Lepidoptera (LC$_{90}$ values range between 1 and >25 μg ml^{-1}) (Dybas, 1989; Lasota and Dybas, 1991; Jansson and Dybas, 1998). Although abamectin is toxic to certain aphids (e.g., LC$_{90}$ values against black bean aphid, *Aphis fabae* and cotton aphid, *Aphis gossypii* range from 0.4 to 1.5 μg ml^{-1}) (Putter *et al.*, 1981; Dybas and Green, 1984), it was not as effective at controlling aphids in translaminar assays (Wright *et al.*, 1985). The reduced efficacy at controlling aphids is probably due to its prime activity as a stomach poison and its reduced concentrations in phloem tissue, where aphids actively feed. Poor residual control of aphids

in the field has been confirmed in several studies (Dybas, 1989 and references therein).

Emamectin benzoate is a recently introduced avermectin that is highly potent to a broad spectrum of lepidopterous pests (Jansson and Dybas, 1998). LC$_{90}$ values for emamectin benzoate against a variety of lepidopterous pests range between 0.002 and 0.89 μg ml^{-1} (Dybas, 1989; Cox *et al.*, 1995b; Jansson and Dybas, 1998). The hydrochloride salt of emamectin was up to 1500-fold more potent against armyworm species, e.g., beet armyworm (*Spodoptera exigua*), than abamectin (Trumble *et al.*, 1987; Dybas *et al.*, 1989; Mrozik *et al.*, 1989). Emamectin hydrochloride was also 1720-, 884-, and 268-fold more potent to *S. eridania* than methomyl, thiodicarb, and fenvalerate, respectively, and 105- and 43-fold more toxic to cotton bollworm (*H. zea*) and tobacco budworm (*H. virescens*) larvae than abamectin (Dybas and Babu, 1988). Recent studies showed that emamectin benzoate was 875- to 2975-fold and 250- to 1300-fold more potent than tebufenozide to *H. virescens* and *S. exigua*, respectively. Emamectin benzoate was also 12.5- to 20-fold and 250- to 500-fold more potent than λ-cyhalothrin and 175- to 400-fold and 2033- to 8600-fold more potent than fenvalerate to these two Lepidoptera, respectively (Jansson *et al.*, 1998). Emamectin benzoate was 1.2- to 4.8-orders of magnitude more potent to lepidopterous pests of cole crops (e.g., *S. exigua*, *P. xylostella*, and cabbage looper, *Trichoplusia ni*), than other new insecticides, including chlorfenapyr, fipronil, and tebufenozide (Jansson and Dybas, 1998; Argentine *et al.*, 2002b).

Emamectin benzoate is markedly less toxic to most nonlepidopteran arthropods. Emamectin is 8- to 15-fold less toxic to the serpentine leafminer, *L. trifolii* and the two-spotted spider mite, *T. urticae*, respectively, than abamectin (Dybas *et al.*, 1989; Cox *et al.*, 1995a). Emamectin benzoate and abamectin are comparable in their potency against Mexican bean beetle, *Epilachna varivestis*, and Colorado potato beetle, *L. decemlineata* (Dybas, 1989). Emamectin benzoate is markedly less toxic to black bean aphid, *A. fabae*, than abamectin (Jansson and Dybas, 1998).

The macrocyclic lactones are widely used in animal health applications. The use of these compounds for antiparasitic therapy has been recently reviewed (Raymond and Sattelle, 2002; Vercruysse and Rew, 2002a). James *et al.* (1980) demonstrated the initial insecticidal activity of crude extracts and partially purified avermectins against Australian sheep blowfly, *L. cuprina*. Hughes and Levot (1990) examined the potencies of abamectin, ivermectin, and 4-deoxy-4-epi(methyl amino) avermectin B$_1$ in *in vitro* assays with *L. cuprina*. All three avermectins were equally potent against newly hatched larvae, and an order of magnitude more potent than diazinon when it was first introduced for blowfly control. Similarly, ivermectin is about two orders of magnitude more potent than deltamethrin against pyrethroid-susceptible larvae of *L. cuprina* (Rugg *et al.*, 1995a). Ostlind *et al.* (1997) determined that ivermectin and emamectin were more potent than a number of common insecticides in a bioassay using larvae of *Lucilia sericata*. However, these workers found that abamectin was about 100-fold less toxic than ivermectin.

Avermectins have similar potency against most myiasis producing dipterans. Abamectin or ivermectin concentrations of 5–20 ng ml^{-1} in rearing medium were toxic to early stage larvae of *Cochliomia macellaria* (Chamberlain, 1982) and *Chrysomya bezziana* (Spradbery *et al.*, 1985). Similarly, avermectins are highly potent against parasitic larvae in the families Cuterebridae (Ostlind *et al.*, 1979; Roncalli, 1984), Gasterophilidae (Klei and Torbert, 1980; Craig and Kunde, 1981) and Oestridae (Preston, 1982; Yazwinski *et al.*, 1983).

Avermectins are also potent against hematophagous dipterans. Miller *et al.* (1986) demonstrated that the therapeutic anthelminthic (subcutaneous) dose of ivermectin would be effective against the blood-feeding horn fly, *Haematobia irritans*, for more than 2 weeks. Standfast *et al.* (1984) found that this same dose in cattle produced 99% mortality in engorged ceratopogonid midges 48 h after feeding for up to 10 days after treatment. *In vitro* studies have demonstrated that the subcutaneous dose of ivermectin in cattle kills 98% of buffalo flies fed for 7 days on blood taken 1 day after treatment (Kerlin and East, 1992). *In vitro* blood-feeding assays demonstrate that ivermectin is highly toxic to horn fly (88 h LC$_{50}$ was 3–7 ng ml^{-1}) (Miller *et al.*, 1986) and buffalo fly (*H. irritans exigua*) (48 h LC$_{50}$ was 30–60 ng ml^{-1}) (Allingham *et al.*, 1994). Adult tsetse fly (Langley and Roe, 1984), stable fly (*Stomoxys calcitrans*) (Miller *et al.*, 1986), and mosquitoes (Pampiglione *et al.*, 1985) tend to be less susceptible and generally survive feeding on animals treated at therapeutic dose rates.

Fleas are generally considered to be relatively insensitive to avermectins or milbemycins systemically administered at, or slightly above, doses that provide ectoparasite control in production animals (Strong and Brown, 1987; Zakson-Aiken *et al.*, 2001). However, Zakson-Aiken *et al.* (2001) examined a milbemycin and 19 avermectin aglycones, monosaccharides, and disaccharides for toxicity to fleas in an artificial membrane feeding system and found that 19 of these compounds produced between 52% and 97% mortality in fleas fed a dose of 20 µg ml^{-1}. The potencies of these compounds, which included abamectin, selamectin, ivermectin, and milbemycin D, were remarkably similar, yet the most potent compound was inactive when tested at 1 mg kg^{-1} subcutaneously in the dog (Zakson-Aiken *et al.*, 2001). By contrast, selamectin provides effective flea control for up to 4 weeks after topical application at 6 mg kg^{-1} on dogs (Banks *et al.*, 2000; Bishop *et al.*, 2000).

Sucking lice on cattle are effectively controlled by therapeutic doses of avermectins, but biting lice are rarely affected by oral or injectable doses (Benz *et al.*, 1989). However, topical administration to cattle does give effective control of biting lice (Benz *et al.*, 1989). *In vitro* studies with sheep biting lice indicate that efficacy is not due to direct contact activity, but rather to the differential distribution following application resulting in the availability of a toxic oral dose. Rugg and Thompson (1993) examined the toxicity of abamectin, ivermectin, and cypermethrin to the sheep-biting, louse *Damalinia ovis*. These workers found that avermectins were an order of magnitude less potent than cypermethrin when assessed in a contact bioassay. However, in an assay incorporating ingestion and contact action both macrocyclic lactones were an order of magnitude more potent than cypermethrin.

Macrocyclic lactones are also highly toxic to many dung-feeding insects, but toxicity may vary widely between life stages. Dung-feeding Diptera and Coleoptera are highly susceptible; however, adult dung beetles are relatively insensitive. Rugg (1995) found that adult *Onthophagus gazella* were four orders of magnitude less susceptible to ivermectin or moxidectin than larvae in *in vitro* bioassays. This relative insensitivity in adult dung beetles with no significant effects on adult mortality seen in mature beetles fed dung from cattle treated with ivermectin has been reported previously (Ridsdill-Smith, 1988; Wardhaugh and Rodriguez-Mendez, 1988; Fincher, 1992). However, newly emerged and immature adult beetles may be affected negatively by ivermectin residues (Wardhaugh and Rodriguez-Mendez, 1988; Houlding *et al.*, 1991). Adult dung beetles are sexually immature at emergence and have an initial period of intense feeding. It is during this period that they would be most susceptible to residues in feces. Nevertheless, the relative insensitivity of adult dung beetles is attributable to ingestion being the main route of toxicity, to the feeding behavior of the beetles, and the characteristics of the compounds. Adult dung beetles ingest only the liquid and colloidal constituents of dung, as their mouthparts are unsuitable for solids (Bornemissza, 1970). Avermectins and similar compounds bind tightly to soil and organics, and are virtually insoluble in water (Halley *et al.*, 1989b) and could be expected to partition differentially to the solid components of dung. Thus, adults would be expected to ingest relatively low concentrations of these compounds when feeding on the liquid portion of dung.

Milbemycins generally tend to have lower toxicities to insects than the avermectins, but similar or higher potencies against some nematodes and acarids (McKellar and Benchaoui, 1996). *In vitro* bioassays have shown that moxidectin is 50- to 100-fold less toxic than ivermectin and abamectin to the sheep blowfly, *L. cuprina*, and about sixfold less potent than these two avermectins against sheep-biting lice (Rugg, 1995). This difference is clearly demonstrated in the effects of these compounds on dung feeding insects. An *in vitro* comparison of moxidectin and abamectin (Doherty *et al.*, 1994) found that abamectin was about 50-fold more toxic to larvae of a dung beetle (*O. gazella*) and buffalo fly than moxidectin when added to cattle feces. Rugg (1995) observed a similar result with both ivermectin and abamectin, which were at least an order of magnitude more potent than moxidectin to adults and larvae of a dung beetle and larvae of a dung-feeding fly (*T. brevicornis*). While ivermectin residues are generally considered to be toxic to some dung beetle larvae and fly larvae for up to 4 weeks after the treatment of cattle, there are differing reports for moxidectin. Strong and Wall (1994) found that moxidectin residues had no toxic effects on the larvae of beetles and cyclorrhaphous flies in cattle dung, and similarly, Fincher and Wang (1992) concluded that dung from moxidectin treated animals did not affect two species of dung beetles. However, Webb *et al.* (1991) reported that feces from cattle treated with a low, controlled-release dose of moxidectin were toxic to larvae of the face fly, *Musca autumnalis*, and concluded that the therapeutic injectable dose of moxidectin would be an effective control of *M. autumnalis* larvae in dung. Miller *et al.* (1994) reported significant mortality in horn fly larvae in dung for up to 28 days after oral or injectable treatment of cattle with moxidectin. Rank ordering of the negative effects of various avermectins and moxidectin on insect development in the dung of treated cattle was doramectin >ivermectin ≥ eprinomectin ≫ moxidectin (Floate *et al.*, 2001, 2002).

While moxidectin is generally less potent against insects than the avermectins in *in vitro* evaluations, it is commonly used at the same or similar therapeutic dose rates in production animals. Interestingly, the spectrum of insects controlled by these products is remarkably similar (Shoop *et al.*, 1995; McKellar and Benchaoui, 1996). The only macrocyclic lactones that have been specifically developed for primary insect control on animals are avermectins: ivermectin for blowfly and lice control on sheep (Eagleson *et al.*, 1993a, 1993b; Rugg *et al.*, 1993, 1995a, 1995b; Thompson *et al.*, 1994), and selamectin for flea control on cats and dogs (Benchaoui *et al.*, 2000; McTier *et al.*, 2000a, 2000b).

3.4.1.2. Acarids, copepods, and helminths

Macrocyclic lactones are highly potent acaricides. Abamectin and natural milbemycins are widely used as agricultural miticides. Abamectin is highly effective against a broad spectrum of phytophagous mites and is one of the most potent acaricides with LC_{90} values ranging between 0.02 and 0.24 $\mu g\,ml^{-1}$ for mites in the families Tetranychidae, Eriophyidae, and Tarsonemidae (Lasota and Dybas, 1991). Eriophyid mites, such as the citrus rust mite, *Phyllocoptruta oleivora*, and various tetranychid mites, such as *T. urticae*, are among the most susceptible (Dybas, 1989). Other phytophagous mites, such as citrus red mite, *Panonychus citri* ($LC_{90} =$ 0.24 $\mu g\,ml^{-1}$) are slightly more tolerant (Dybas, 1989). Similarly, ticks and parasitic mites are variably susceptible to the macrocyclic lactones. Much of the variation in potency against ticks relates to the different life cycle of ticks, the relatively slow action of these compounds against ticks, and the interactions with bioavailability of the compound in the host. These compounds tend to be much more effective against single host ticks; and are often considered relatively poor compounds against multihost ticks (McKellar and Benchaoui, 1996). In multihost ticks, adults are the infestive stage and are the least sensitive to macrocyclic lactones. Parasitic mites show similar variation: the deep burrowing sarcoptic mange mites that feed directly on blood and body fluids are highly sensitive, while chorioptic mange mites that live and feed on the superficial dermal surface are relatively insensitive (Benz *et al.*, 1989).

Macrocyclic lactones are also toxic to a number of parasitic copepods and ivermectin, emamectin benzoate, and doramectin have been used for control of parasitic sea lice in farmed salmon (Burka *et al.*, 1997; Davies and Rodger, 2000; Roth, 2000).

The macrocyclic lactones are highly potent nematicides, but have no useful activity against trematodes or cestodes. These compounds are especially toxic to the microfilarial stages of many filaroid nematodes (McKellar and Benchaoui, 1996). This extreme potency against microfilaria is best demonstrated in the use of these compounds for heartworm control in dogs. Oral doses of 3–$12\,\mu g\,kg^{-1}$ of moxidectin or ivermectin once a month provide effective protection against heartworm infection in dogs (Guerrero et al., 2002). The oral use rate of these compounds for general nematode control in ruminants is about $200\,\mu g\,kg^{-1}$ (McKellar and Benchaoui, 1996).

3.4.2. Bioavailability

3.4.2.1. Crop protection uses
Macrocyclic lactones are very susceptible to rapid photodegradation. The half-life of abamectin as a thin film under artificial light or simulated sunlight was 4–6 h and the rate of degradation of abamectin on petri dishes and on leaves was markedly greater in light than in dark environments (MacConnell et al., 1989). The half-life of emamectin benzoate has been estimated to be 0.66 days on celery (Feely et al., 1992) and is expected to be even shorter on cole crops. Photoirradiation of avermectins leads to the production a large number of decomposition products (Mrozik et al., 1988), and a number of these photodegrades have also been identified for abamectin (Crouch et al., 1991) and emamectin benzoate (Feely et al., 1992). Many of these photodegrades show impressive insecticidal activity against larval lepidopterans (Argentine et al., 2002a).

Despite the rapid degradation of avermectin insecticides in sunlight, significant amounts of these compounds are taken up rapidly via translaminar movement into sprayed foliage (Wright et al., 1985; Dybas, 1989 and references therein). When abamectin was applied to plants held in the dark, the prolonged stability resulted in greater penetrability into leaves and more effective mite control (MacConnell et al., 1989). Wright et al. (1985) demonstrated that this translaminar movement of abamectin produced good control of mites, but aphids were not controlled. They surmised that abamectin residues were probably abundant in the parenchyma tissues of leaves where mites feed, whereas abamectin probably did not distribute into the phloem tissue where aphids feed. The excellent residual efficacy of abamectin against several dipteran and lepidopteran leafminers is probably also due to translaminar movement and the presence of abamectin reservoirs in parenchyma tissue (Jansson and Dybas, 1998). This lack of true systemic activity (distribution into the phloem) in plants following spray application is demonstrated by the lack of residual efficacy against aphids. Spray applications of abamectin that had good contact efficacy provided little if any residual control of the aphids Myzus persicae (Johnson, 1985) and A. fabae

(Green and Dybas, 1984). Similarly, the lack of downward transport of abamectin sprayed on foliage results in poor control of root knot nematodes (Stretton et al., 1987; Cayrol et al., 1993). By contrast, root dip, soil application or stem injections of avermectins have shown the potential to provide effective control of these parasites (Sasser et al., 1982; Jansson and Rabatin, 1997, 1998).

3.4.2.2. Animal health uses
In animal health, the high potency and lipophilic nature of the macrocyclic lactones allows extensive systemic delivery through a variety of routes. These compounds are used in oral, injectable, and topical formulations. The route of delivery, however, influences the bioavailability and distribution of the compounds in the host animal and thus the efficacy against different target parasites. Similarly, bioavailability can vary dramatically between species and be markedly affected by feed quantity (Ali and Hennessy, 1996) or composition (Taylor et al., 1992) when orally administered. Efficacy against parasites may be dependent on transport in the plasma and distribution to vascular or nonvascular sites that constitute the parasite habitat. The pharmacokinetics of the macrocyclic lactones were reviewed by Hennessy and Alvinerie (2002).

Following most means of administration, macrocyclic lactones are eliminated fairly rapidly from the treated animals; elimination half-lives in various species range from about 3 to 10 days (see Hennessy and Alvinerie, 2002). Regardless of the means of administration or the species treated, the main route of elimination of the compounds is via feces with the compounds passing into the gastrointestinal tract in the bile (Hennessy and Alvinerie, 2002). A number of controlled release devices (Shoop and Soll, 2002) or sustained release formulations (Rock et al., 2002; Shoop and Soll, 2002) using macrocyclic lactones have been used to provide long-term protection against parasites.

The most dramatic differences between application methods are seen with certain insect ectoparasites. While a number of blood-sucking parasites, such as sucking lice of cattle, are controlled by therapeutic doses of macrocyclic lactones administered orally or by injection, chewing lice are generally not susceptible. Topical applications (generally at about a 2.5-fold higher dose rate) are usually highly effective against chewing lice (Strong and Brown, 1987; Vercruysse and Rew, 2002b). Similar results have also been observed in lice on sheep (Coop et al., 2002). While the efficacy and spectrum of these topical formulations against internal parasites is similar to the oral or injectable treatments, this greater efficacy against chewing lice is thought to be a result of a high concentration of the active ingredient on the surface of the skin where these parasites feed (Titchener *et al.*, 1994). Similarly, most of the macrocyclic lactone pour-on formulations for cattle have demonstrated a persistent efficacy against the blood-feeding horn fly and buffalo fly that is not seen with oral or injectable formulations (Vercruysse and Rew, 2002b).

This apparent greater efficacy of pour-on formulations against external parasites is presumably the result of deposition of the active compound on, and in, the skin. The high lipophilicity of the macrocyclic lactones is likely to result in the formation of depots of active ingredient in skin lipids and oil and fat secretions. This characteristic has been exploited in topical formulations of ivermectin that provide long-term residual control of blowfly and lice on sheep (Eagleson *et al.*, 1993a, 1993b; Thompson *et al.*, 1994), and selamectin which controls fleas on dogs and cats for at least 1 month (Benchaoui *et al.*, 2000). Even greatly exaggerated oral or injectable doses of macrocyclic lactones are considered unlikely to provide residual control of fleas (Zakson-Aiken *et al.*, 2001). In sheep, topically applied ivermectin is thought to bind to skin lipids and secretions and may be passively distributed around the sheep's body in this medium following application to a discrete site (Rugg and Thompson, 1997). Selamectin is thought to selectively partition into sebaceous glands of dogs and cats from where it is slowly released to the skin over an extended period (Hennessy and Alvinerie, 2002).

3.4.3. Safety and Selectivity

3.4.3.1. Crop protection uses   Abamectin is generally considered to be nonphytotoxic under normal use even in sensitive ornamental plants (Wislocki *et al.*, 1989), although mild foliar spotting has been noted in some ferns, daisies, and carnations (Dybas, 1989). Phytotoxicity was not affected by the addition of spray oils (Green *et al.*, 1985), although, the addition of spray oils to abamectin has produced epidermal damage in pears that is not seen with the use of abamectin alone (Hilton *et al.*, 1992).

Integrated pest management (IPM) involves the use of all available means (chemical, biological, cultural, physical, etc.) to achieve effective control of pests or pest damage. Central to the design of IPM programs are selective insecticides that reduce pest abundance and yet are innocuous to beneficial predators or parasites by allowing their survival and range expansion. The suitability of pesticides for use in IPM programs is based on the differential toxicity of a compound to pest and beneficial arthropod

populations. Differential toxicity can occur through pharmacokinetic or metabolic differences between pests and beneficial organisms (physiological selectivity) or by a compound's unique qualities that result in toxic exposure of phytophagous pests with concomitant reduced exposure of beneficial organisms (ecological selectivity). The macrocyclic lactone insecticides possess both of these qualities, thereby making them highly selective and compatible with IPM.

The ecological selectivity of the avermectins and milbemycins results from a number of qualities of these compounds following application in the field. As noted earlier, avermectin insecticides are readily taken up by plant foliage. Foliar uptake via translaminar movement results in a reservoir of the toxicant inside the leaf (MacConnell *et al.*, 1989). Any remaining compound on the outside of foliage is rapidly degraded by sunlight, thereby resulting in minimal residues of the compound on the plant surface soon after application. These qualities of the avermectin insecticides that result in the selective availability of the compounds to phytophagous pests and reduced exposure to nonphytophagous organisms are the main reasons for their compatibility with beneficial arthropods and IPM programs.

A number of researchers have shown that applications of abamectin did not disrupt the *Liriomyza* leafminer–parasitoid complex on a variety of vegetable crops (see Dybas, 1989). Trumble and Alvarado Rodriguez (1993) demonstrated that because of the reduced toxicity of abamectin to natural enemies, it was an important component of a multiple-pest IPM program on fresh market tomato that was based on intensive sampling, parasitoid releases, mating disruption, microbial pesticides, and abamectin. Abamectin has also been shown to be less detrimental to certain life stages of two leafminer parasitoids, *Diglyphus intermedius* and *Neochrysocharis punctiventris*, compared with other commonly used insecticides (Schuster, 1994).

The differential toxicity of abamectin to a variety of pest and predatory mite species is also well documented (see Lasota and Dybas, 1991). Hoy and Cave (1985) showed that although abamectin was toxic to the predator *Metaseiulus occidentalis*, it was more toxic to the two-spotted spider mite, *T. urticae*. Similar results have been found with other predatory mite species (Grafton-Cardwell and Hoy, 1983; Zhang and Sanderson, 1990). Hoy and Cave (1985) also showed that 2–4 day field-aged residues of abamectin were safe for *M. occidentalis*. Several other studies demonstrated that abamectin was less toxic to naturally occurring and introduced natural enemies of pest mites, and for this reason, it selectively kills target pests and conserves their natural enemies (Dybas, 1989; Lasota and Dybas, 1991).

Zchorifein *et al.* (1994) found that abamectin was compatible with *Encarsia formosa*, an important component of IPM programs for greenhouse whitefly, *Trialeurodes vaporariorum*. No mortality was observed when adult parasitoids were exposed to 24 h residues of abamectin. They found that a management program based on the combined treatment of abamectin with releases of *E. formosa* maintained lower densities of the whitefly on poinsettia throughout the season and required fewer applications of abamectin compared with a management program based on applications of abamectin alone. Similarly, Brunner *et al.* (2001) found that while abamectin was highly toxic when applied topically to two parasitoids of leafrollers, 1-day-old residues were not toxic.

Like abamectin, emamectin benzoate is less toxic to beneficial arthropods such as honeybees, parasitoids, and predators. Kok *et al.* (1996) reported that emamectin hydrochloride had minimal adverse effects on two hymenopterous parasitoids (*Pteromalus puparum* and *Cotesia orobenae*) of Lepidoptera on broccoli. Contact activity of emamectin benzoate residues on alfalfa against two species of bees declined rapidly and was negligible 24 h after treatment (Chukwudebe *et al.*, 1997). They determined that bee mortality was directly related to transient dislodgeable residues and thus these beneficial insects could be expected to survive and colonize treated crops within relatively short intervals after applications of emamectin benzoate. Sechser *et al.* (2003) found that emamectin benzoate was relatively safe to a number of insect predators of sucking pests on cotton.

The selectivity of abamectin has been widely investigated in mites. Trumble and Morse (1993) showed that abamectin was compatible with *Phytoseiulus persimilis* in field-grown strawberry. They found that the highest net economic return was achieved when several releases of the predatory mite *P. persimilis* were integrated with suitably timed applications of abamectin. In a laboratory leaf-disc bioassay, adult females of the predatory mite *Amblyseius womersleyi* dipped in abamectin solution showed low mortality (16.6%), while all *T. urticae* females died within 24 h after dipping (Park *et al.*, 1996). In apple, abamectin was highly compatible with the phytoseiid mites *M. occidentalis* and *Amblyseius fallacis*. Abamectin was effective at reducing and maintaining numbers of European red mites (*Panonychus ulmi*) at or below the economic threshold for at least 3 weeks after application. Although predatory mite numbers declined initially, they quickly resurged resulting in favorable predator to prey ratios that were comparable or superior to those found in nontreated plots, and aided in mite suppression for the remainder of the growing season (Jansson and Dybas, 1998).

In glasshouse roses, Sanderson and Zhang (1995) found that full canopy applications of abamectin at 3–5 day spray intervals effectively controlled *T. urticae* populations, but were detrimental to the predatory mite *P. persimilis*. However, by applying abamectin to the upper canopy of plants only, excellent mite control in the marketable portion of the plant was achieved and predatory mite populations in the lower canopy were conserved. This demonstrated the potential for integrating abamectin with natural predators for control of a target pest, even under intense abamectin pressure.

3.4.3.2. Animal health uses The macrocyclic lactones are generally considered to be relatively safe in mammals. Glutamate-gated chloride channels have not been reported in mammals. Clinical signs are associated with neurotoxicity of the central nervous system as this is the site of GABA-gated chloride channels in mammals. The poor distribution of these compounds through the blood–brain barrier and/or their rapid removal by the transmembrane protein, P-glycoprotein, are thought be responsible for their lack of toxicity to mammals (Shoop and Soll, 2002; Abu-Qare *et al.*, 2003). Increased susceptibility to avermectins has been observed in Murray Grey cattle (Seaman *et al.*, 1987) and in some collie-breed dogs (Paul *et al.*, 1987). It is thought that a deficiency of P-glycoprotein in these animals results in increased accumulation of avermectins in the central nervous system and the resultant neurotoxicity (Shoop and Soll, 2002; Roulet *et al.*, 2003).

In animal health, nontarget effects of macrocyclic lactones are generally minimized due to their main use as systemic parasiticides. The compounds are applied discretely to the target host and not released into the environment. The major nontarget effects in animal health uses are on the insect fauna of dung of treated animals. Regardless of the route of administration, the major means of elimination of these compounds from animals is via the feces and the major component in feces is usually the parent compound (Campbell *et al.*, 1983; Chiu and Lu, 1989). Feed-through compounds for the control of pestiferous Diptera in dung have been in use, especially in cattle, for many years, yet little attention has been paid to the effects on nontarget dung fauna (Strong, 1992). Similarly, most of the compounds used as external parasiticides for cattle are toxic to dung insects, but organophosphorus and pyrethroid compounds are more toxic to dung-inhabiting insects than avermectins (Bianchin et al., 1992). However, there has been a great deal of attention given to possible environmental consequences of systemic parasiticides, especially avermectins and related compounds, with entire journal volumes (e.g., *Veterinary Parasitology* 48, 1993) and dedicated regulatory or scientific workshops (National Registration Authority, 1998; Alexander and Wardhaugh, 2001) devoted to this subject.

This rise in interest is probably related to the widespread use of these compounds. Their broad spectrum, encompassing both nematodes and arthropods, has resulted in their extensive use, and especially where gastrointestinal worms are the prime target, most arthropods could be considered nontarget organisms. Certainly, in countries where the introduction of dung beetles to control dung build-up and dung breeding pests have occurred, any compounds with adverse effects on these beneficial agents will gain some attention. Also, the differing toxicities of related compounds marketed by different companies may be perceived as providing a commercial advantage. Hence, authors have reported the effects of ivermectin in controlling dung-breeding pestiferous Diptera (e.g., Meyer et al., 1980; Miller et al., 1981; Marley et al., 1993), or its toxicity to dung beetles (e.g., Ridsdill-Smith, 1988; Sommer and Overgaard Nielsen, 1992; Sommer et al., 1992, 1993), and that dung degradation may be retarded (e.g., Wall and Strong, 1987; Sommer et al., 1992) or not affected (e.g., Jacobs et al., 1988; McKeand et al., 1988; Wratten et al., 1993). Other reports have appeared directly comparing the effects of avermectins and moxidectin on dung fauna (Fincher and Wang, 1992; Doherty et al., 1994; Floate et al., 2001, 2002).

There is little doubt that avermectins and related compounds have the potential to affect dung-inhabiting insects. Moxidectin has little if any impact on dung insects, but also is less efficacious against pestiferous fly larvae. The selection of these compounds for certain animal health uses should be evaluated on the basis of their efficacy against pest target and on concomitant environmental concerns. Thus, for example, ivermectin use in a controlled-release device for sheep provides added protection against breech strike (Rugg et al., 1998a), and residues in dung likely reduce populations of the adult blowfly population (Cook, 1991, 1993), while exerting little impact on the rate of degradation of sheep dung where insects play a relatively minor role

(Cook, 1993; King, 1993). Where there are concerns over possible adverse effects of treatment for gastro-intestinal worm infestation on cattle dung fauna, moxidectin would be the compound of choice (Strong and Wall, 1994). If control of arthropod ectoparasites or their dung-feeding stages is of importance (Marley et al., 1993), an avermectin may be preferred because of its greater toxicity to insects. Even so, a study examining the effects of commercial ivermectin use in cattle in South Africa (Scholtz and Krüger, 1995) found that dung insect populations recovered rapidly following an initial depression. They concluded that the long-term effects on dung fauna might not be as severe as previous studies have claimed. However, simulation of the impact of eprinomectin on dung beetle populations using computer modeling suggests that, in the absence of immigration, a single treatment of eprinomectin could reduce beetle activity by 25–35% in the next generation (Wardhaugh et al., 2001). In studies directly comparing the effects of macrocyclic lactones in cattle dung, avermectins result in lethal and sublethal effects on dung fauna in feces excreted for a number of weeks after treatment, while moxidectin residues are relatively innocuous (Fincher and Wang, 1992; Doherty et al., 1994; Floate et al., 2001, 2002; Wardhaugh et al., 2001).

3.5. Uses

3.5.1. Crop Protection

Abamectin is registered worldwide for control of mites and certain insect pests on a variety of ornamental and horticultural crops and on cotton. Abamectin is effective for controlling agromyzid leafminers, *Liriomyza* spp., Colorado potato beetle, *L. decemlineata*, diamond back moth, *P. xylostella*, tomato pinworm, *K. lycopersicella*, citrus leafminer, *Phyllocnistis citrella*, and pear psylla, *Cacopsylla pyricola*. Abamectin has also been incorporated into bait to control red imported fire ants and cockroaches. These and other urban pest applications were reviewed by Lasota and Dybas (1991).

Recent studies have shown that avermectins have potential as a component in Africanized honeybee abatement programs (Danka et al., 1994). Another novel use is the injection of abamectin into trees which controls certain phytophagous arthropods, such as elm leaf beetle (*Pyrrhalta luteola*), for up to 83 days after application (Harrell and Pierce, 1994). The main commercial agricultural use of macrocyclic lactones is as acaricides. These compounds have demonstrated mite control and residual activity superior to most acaricides on a variety of ornamental crops, food crops, and cotton

(Putter *et al.*, 1981; Dybas and Green, 1984; Dybas, 1989). Macrocyclic lactones are also highly effective for the control of plant parasitic nematodes following injection application (Jansson and Rabbatin, 1997, 1998; Takai *et al.*, 2003) or soil incorporation (Blackburn *et al.*, 1996). Emamectin benzoate, milbemectin, and nemadectin are used commercially for the control of pine wilt nematode in Japan.

Emamectin benzoate is used for broad-spectrum control of lepidopteran pests on certain vegetable crops. This compound is very effective against numerous lepidopteran pests of a variety of crops. It is expected that the compound will be used to control lepidopterous pests on a wide variety of vegetable, tree, and row crops and also for control of thrips on eggplant and tea in Japan (Jansson and Rabatin, 1997).

3.5.2. Animal and Human Health

The macrocyclic lactones have had a dramatic impact on animal health. Their potency and broad spectrum has resulted in this chemistry dominating the parasiticide market. Macrocyclic lactones have been commercialized for the control of nematodes and arthropod parasites for most common food-producing and companion animals and are widely used "off-label" for parasite control in many other species. Avermectins (e.g., emamectin) are also used for control of copepod parasites of farmed salmon (Davies and Rodger, 2000). In human health, ivermectin has been used since 1987 in a compassionate program in Africa and Central and South America to control *Onchocerca volvulus* which is the causative agent of river blindness in man (Shoop and Soll, 2002). The program was expanded recently to include the reduction of spread of elephantiasis caused by lymphatic filarid nematodes. Recently, moxidectin has been demonstrated to be safe and well tolerated in humans (Cotreau *et al.*, 2003). Moxidectin is more efficacious than ivermectin in *Onchocerca* models and is currently undergoing clinical trials prior to inclusion in filarid control programs (Molyneux *et al.*, 2003). Ivermectin has been approved for treatment of intestinal strongiloidosis scabies in humans, and is also useful for the treatment of louse infestations (Elgart and Meinking, 2003). These authors consider that there may be a number of additional uses for the macrocyclic lactones as antiparasitic medications for parasites of skin in humans.

3.6. Resistance

3.6.1. Insects

Resistance to the avermectins in insects has been reviewed by Clark *et al.* (1995). In general, a variety of biochemical and pharmacokinetic mechanisms may contribute to avermectin resistance in arthropods, although biochemical mechanisms tend to be more important among most arthropod systems studied to date. However, there are many conflicting reports on the importance of different resistance mechanisms. Similarly, cross resistance between avermectin insecticides/acaricides and other classes of chemistry is not well understood, and there are probably equal numbers of reports demonstrating cross-resistance as there are those refuting it.

Cross-resistance between abamectin and pyrethroids has been reported for field strains of houseflies (Scott, 1989; Geden *et al.*, 1990). Abro *et al.* (1988) proposed that there was a low level of cross resistance to abamectin in a field strain of diamondback moth that was highly resistant to DDT, malathion, and cypermethrin when assessed in a topical assay. However, the same population was equally susceptible to abamectin as the baseline colony in an ingestion bioassay. Roush and Wright (1986) found no evidence of cross-resistance to avermectins in houseflies resistant to diazinon, dieldrin, DDT, or permethrin, and Parella (1983) reported no cross-resistance to avermectins in a pyrethroid-resistant agromyzid fly. Campanhola and Plapp (1989) determined that there was no cross-resistance to abamectin in pyrethroid-resistant tobacco budworm, where both target site resistance and metabolic resistance to pyrethroids were present. Similarly, Cochran (1990) concluded that pyrethroid-resistant field populations of German cockroach were not cross-resistant to abamectin. Argentine and Clark (1990) examined a susceptible laboratory strain and a multiple-resistant field strain of Colorado potato beetle, and found no difference in their dose responses to abamectin. No cross-resistance to abamectin was observed in a laboratory-selected pyrethroid-resistant strain of citrus thrip, although cross-resistance was detected to DDT and some organophosphates and carbamates (Immaraju and Morse, 1990). Rugg and Thompson (1993) reported that pyrethroid-resistant sheep biting lice (*B. ovis*) were not cross-resistant to ivermectin.

Beeman and Stuart (1990) examined a number of pesticides against a field-collected strain of red flour beetle (*Tribolium castaneum*) resistant to lindane that had been further selected with dieldrin in the laboratory. These workers found cross-resistance to other cyclodienes, but not to abamectin nor to organophosphates. Resistance to cyclodienes, accounted for by channel mutants in the Rdl GABA-gated chloride channel subunit (review: ffrench-Constant, 1994), usually extends to all insecticides that block chloride ion channels, and this resistance is not conferred to avermectins as these activate chloride channels and therefore act at a separate site (Bloomquist, 1993). Rohrer *et al.* (1995) showed that fipronil, which interacts with GABA-gated chloride channels (as an antagonist) did not affect ivermectin binding either in *Drosophila* head membranes or in locust neuronal membranes.

Ismail and Wright (1991) examined the toxicities of a number of insect growth regulators, and abamectin, to a susceptible strain of diamondback moth and to strains selected for resistance to the growth regulators chlorfluazuron and teflubenzuron. They found that cross-resistance to other growth regulators was variable and did not detect cross-resistance to abamectin. Recently, Zhao *et al.* (2002) showed that the resistance to spinosad in a field population of diamondback moth was not cross-resistant to emamectin benzoate.

Scott (1989) proposed that cross-resistance to abamectin in pyrethroid-resistant houseflies was polygenic and due to decreased cuticular penetration and enhanced metabolism mediated by mixed-function oxidases. Subsequently, Scott *et al.* (1991) reported laboratory selection of high-level abamectin resistance in these field-collected houseflies; after seven selections resistance ratios of >60 000-fold (determined by topical application) and 36-fold (determined by residual exposure) were reached. In their study the synergists piperonyl butoxide, diethyl maleate, and *S,S,S*-tributylphosphorotrithioate did not affect the toxicity of abamectin, indicating that resistance was not due to enhanced metabolism mediated by mixed function oxidases, glutathione transferases, or hydrolases. Konno and Scott (1991) found that resistance in this selected strain resulted from decreased cuticular penetration (>1700-fold) and altered abamectin binding (35-fold). This strain was also cross-resistant to two abamectin analogs and resistance was highly recessive.

Laboratory selection of abamectin against two strains of Colorado potato beetle resulted in resistance levels of 21-fold for a mutagen-induced resistance and 38-fold by direct selection. Both resistance factors were incompletely dominant and polyfactorial (Argentine and Clark, 1990). Oxidative metabolism and possibly carboxylesterase activity were responsible, in part, for resistance to abamectin (Argentine, 1991; Argentine *et al.*, 1992). Lamine (1994) found that abamectin-resistant larvae of *L. decemlineata* were seven- to tenfold less susceptible to emamectin benzoate than the baseline colony in a topical bioassay. No cross-resistance between abamectin and emamectin benzoate was evident in adult beetles. Recent studies showed

that *L. decemlineata* populations which were 15- to 23-fold less sensitive to abamectin in a contact, topical bioassay (Clark *et al.*, 1995) were classified as susceptible to abamectin in a diet-based ingestion assay (Dively and Jansson, 1996). Recently, oxidative metabolism was confirmed as a principal resistance mechanism in laboratory-selected, abamectin-resistant Colorado potato beetle (Yoon *et al.*, 2002; Gouamene-Lamine *et al.*, 2003).

Hughes and Levot (1990) reported that a laboratory-selected strain of *L. cuprina* resistant to organophosphates and carbamates exhibited a low level cross-resistance (two- to threefold) to abamectin and ivermectin when compared to susceptible flies. However, emamectin benzoate was equally toxic to both strains. This strain also exhibited a low level of cross-resistance to pyrethroids (Sales *et al.*, 1989). A laboratory-selected pyrethroid-resistant strain of *L. cuprina* that showed cross-resistance between an organophosphate and a carbamate insecticide (Sales *et al.*, 1989) was subsequently selected with deltamethrin, diflubenzuron, and butacarb resulting in levels of resistance in excess of 1000-fold to all three compounds (Kotze and Sales, 1994). These workers also found that metabolic inhibitors significantly synergized the three selected strains, indicating the involvement of both monooxygenases (mixed-function oxidases) and esterases in these resistances. Rugg *et al.* (1995a) showed that the same pyrethroid- and carbamate-resistant laboratory strain of *L. cuprina* was not cross-resistant to ivermectin. Organophosphate-resistant field strains of *L. cuprina* have also been shown to have no cross-resistance to avermectins (Hughes and Levot, 1990; Rugg *et al.*, 1998b).

Rugg *et al.* (1998b) selected larvae of *L. cuprina* with ivermectin for over 60 generations and achieved an eightfold increase in tolerance compared to the parental strain. This low-level resistance reverted rapidly in the absence of selection pressure, and although enzymatic metabolism was implicated, no specific resistance mechanisms were determined. Kotze (1995) found that the induction of monooxygenase and glutathione transferase enzymes with phenobarbital in insecticide-susceptible *L. cuprina* larvae resulted in increased tolerance to butacarb, diazinon, and diflubenzuron. In similar experiments, Rugg *et al.* (1998a) found no increase in tolerance to ivermectin in induced flies that had reduced susceptibility to diazinon.

Resistant field populations of *P. xylostella* from Malaysia were not cross-resistant to *Bacillus thuringiensis* (Iqbal *et al.*, 1996). Lasota *et al.* (1996) showed that there was no cross-resistance between abamectin, emamectin benzoate, permethrin, and methomyl in *P. xylostella* using an ingestion bio-assay. Following laboratory selection of a field population of *P. xylostella* from China, testing of F_1 progeny from reciprocal crosses between abamectin-resistant and abamectin-susceptible strains indicated that resistance was autosomal and incompletely recessive and backcross studies suggested that the resistance was probably controlled by more than one gene (Liang *et al.*, 2003). Further, there was little cross-resistance between abamectin and four pyrethroids (deltamethrin, β-cypermethrin, fenvalerate, and bifenthrin) and no cross-resistance between abamectin and the acylureas chlorfluazuron or flufenoxuron.

Morse and Brawner (1986) reported LC_{50} values ranging from 7 to 21 mg l^{-1} in strains of citrus thrips that had not been exposed to abamectin. Immaraju and Morse (1990) retested one of these strains that had been maintained in the laboratory and also selected for resistance to fluvalinate, and determined LC_{50} values of 0.9 and 0.4 mg ml^{-1}, respectively. Responses to abamectin in nonexposed populations varied by over 50-fold suggesting that thrips may have a large inherent variation in susceptibility to abamectin. Immaraju *et al.* (1992) reported the occurrence of two abamectin-resistant populations of western flower thrips in Californian glasshouses. Resistance ratios (LC_{90}) of 18- and 798-fold were attributed to intensive use of abamectin during the previous year (five or ten applications, respectively).

In summary, resistance to avermectins has been demonstrated in a number of insects. In all instances, although not clearly determined, resistance is believed to be polygenic, relatively slow to develop, and able to revert rapidly in the absence of selection pressure. Metabolic mechanisms seem to be the most important route of resistance in the arthropod systems studied to date, and in all cases, a number of biochemical and sometimes physiological mechanisms are implicated. Cross-resistance between avermectins and other insecticide classes is similarly ill-defined; however, it is apparent that there is little significant cross-resistance to avermectins except possibly in one strain of housefly (Scott, 1989). The variations in responses of field strains resistant to other insecticide classes to avermectins may in some cases be due to enhanced metabolism resulting from previous exposure to other insecticides, although there is evidence to both support and refute this supposition. A number of species have also demonstrated inherent population differences in response to avermectins without the influence of any predisposing factors. A wide variation among field-collected strains has been seen in several insects which does not constitute cross-resistance,

nor does it indicate that this variation predisposes insects to an increased likelihood for resistance development to avermectins.

Cross-resistance between the avermectins and milbemycins (specifically abamectin, ivermectin, doramectin, and moxidectin) has been demonstrated in a number of nematode species. Because abamectin is the only avermectin compound used widely in agriculture, there have been a few studies conducted to examine cross-resistance within the class. Sensitivity to emamectin benzoate in select abamectin-resistant strains has been assessed in several insects (e.g., *L. trifolii*, *L. decemlineata*, *P. xylostella*) and only minimal, if any, cross-resistance has been detected (Jansson and Dybas, 1998; Jansson and Rugg, unpublished data). The reasons for the lack of cross-resistance to these closely related compounds are unclear. It is possible that the changes in the derivatization of emamectin benzoate may result in the loss or protection of the chemical bonds that are subject to enzymatic cleavage in abamectin resistance. Additionally, because macrocyclic lactones have multiple sites of action and may affect a number of different receptors to activate or potentiate chloride ion influx, subtle differences in chemistry between the different avermectins and milbemycins may alter the relative activities of compounds at the different sites. This could affect the patterns and levels of cross-resistance within the class and also be a factor in the subtle differences seen in the spectra and potencies of these compounds.

3.6.2. Acarids

Laboratory selection of up to 15 generations of two species of tetranychid mites with abamectin resulted in no significant increase in resistance compared with a nonselected baseline colony (Hoy and Conley, 1987). Conversely, 20 rounds of selection against a predatory phytoseiid mite resulted in a gradual and modest shift in susceptibility to abamectin (Hoy and Ouyang, 1989). A recent survey of field populations of two-spotted spider mite in California (Campos *et al.*, 1995) found that mites varied widely in their susceptibility to abamectin, with up to a 658-fold difference in susceptibility in a residual exposure assay. Levels of resistance were correlated with the amount and time of abamectin usage, and laboratory selection of a single population over 38 generations resulted in more than a 100-fold increase in resistance. Despite the higher level of resistance in a residual assay, selected mites were still considered susceptible to abamectin in a direct contact bioassay. Further, despite the wide variation in responses among field populations of mites, no field failures of the product were reported.

In *T. urticae*, resistance to abamectin has been attributed to increased excretion and decreased absorption (Clark *et al.*, 1995), monooxygenases (Campos *et al.*, 1996), esterases (Campos *et al.*, 1996, 1997; Jansson *et al.*, unpublished data), glutathione-*S*-transferases (Clark *et al.*, 1995), and decreased penetration (Clark *et al.*, 1995). Levels of resistance in two spotted spider mite populations collected from the field were shown to drop markedly within 2–6 weeks after collection when maintained in the absence of selection pressure (Campos *et al.*, 1995; Jansson *et al.*, unpublished data). Bergh *et al.* (1999) reported wide variation in susceptibility to abamectin in citrus rust mite (*P. oleivora*) field populations with no clear relationship between susceptibility and product efficacy. While some populations from commercial groves with a history of exposure to abamectin were less susceptible to abamectin, mortality was only slightly lower than that of a pristine population. Also bioassay results indicated that the responses of these field populations did not change whether they were assessed before or after abamectin spray applications.

3.6.3. Helminths

Ivermectin resistance was initially detected in nematodes of sheep and goats, and was generally a result of intensive use over a number of years (Shoop, 1993). Ivermectin-resistant nematodes have been shown to have cross-resistance to other avermectins and milbemycins (Shoop *et al.*, 1993). However, moxidectin was significantly more toxic to many parasitic nematodes (Shoop, 1993; Shoop *et al.*, 1993; Kieran, 1994) and this difference in potency has resulted in moxidectin providing economic control of ivermectin-resistant parasites in field use situations. Conder *et al.* (1993) showed that the modes of action of ivermectin and moxidectin were qualitatively similar by examining changes in membrane conductance in crab muscle fibers. These workers also demonstrated cross-resistance to moxidectin in an ivermectin-resistant strain of *H. contortus* and confirmed that moxidectin was about fourfold more potent than ivermectin against this parasite in a jird model. Recently, Ranjan *et al.* (2002) reported the results of concurrent selection of *H. contortus* for 22 generations with ivermectin and moxidectin. These workers found that worms that had developed resistance through selection with either of these macrocyclic lactones were cross-resistant to the other compound. However, the rate of development of resistance differed between the two compounds and occurred more slowly with moxidectin, which was also markedly more potent than ivermectin. See Pritchard (2002) for the most recent, detailed review of the status of resistance to macrocyclic lactones in nematodes. Briefly, resistance is still a major concern for parasites of sheep and has also been detected in some cattle parasites. Resistance has been found to be associated with enhanced P-glycoprotein expression resulting in increased efflux from the target site, but there are thought to be a number of possible resistance mechanisms including various amino acid-gated anion subunit genes, genes associated with glutamate receptors and transporters, and genes involved in amphid structure and function. The multiplicity of possible sites for macrocyclic lactone binding and activity and their variable sensitivity to different compounds may be responsible for the differential potencies seen within and between avermectins and milbemycins and suggests that products of several genes may be involved in the mechanism of action of macrocyclic lactones. Similarly, these subtle differences in the ways in which ivermectin and moxidectin act at the molecular level could result in the differences seen between these compounds in potency, rates of development of resistance, and levels of cross-resistance.

3.7. Summary

The macrocyclic lactones are potent insecticides, nematicides, and acaricides. The class consists of two closely related groups of compounds, the avermectins and milbemycins, which are natural fermentation products or synthetic derivatives of natural products derived from actinomycetes. The broad spectrum of activity against a variety of insects, acarids, and nematodes has led to the widespread use of these compounds in animal and human health, crop protection, and urban pest control. Ivermectin has been the most successful antiparasitic drug introduced in animal health and abamectin has been the premier agricultural miticide in recent years.

Macrocyclic lactones affect invertebrates by either directly activating receptors for the neurotransmitters glutamate and GABA, or by potentiating their actions, resulting in an influx of chloride ions into nerve cells and muscle. It is currently believed that the anthelminthic properties of the avermectins are due predominantly to potentiation and/or direct opening of glutamate-gated chloride channels, whereas in insects, it is likely that avermectins bind to multiple sites, including glutamate and GABA-gated chloride channels as well as other insect chloride channels (Sattelle, 1990;

Bloomquist, 1993). The chloride ion flux resulting from the opening by these compounds of chloride channels in invertebrate nerve and muscle, particularly those gated by glutamate and GABA, results in disruption of activity and loss of function in these excitable cells and accounts for the potent actions of these molecules (review: Raymond and Sattelle, 2002). Toxicity is generally greater through ingestion than by residual contact and compared with other neurotoxic agents, macrocyclic lactones are considered to be relatively slow acting. Intoxicated organisms usually die slowly, often over a period of days, and there is no quick knockdown effect. Death usually follows paralysis and immobility. The comparatively poor contact activity of these compounds plus their short environmental persistence generally results in good compatibility with beneficial and nontarget organisms. These compounds have low relative toxicities to vertebrates and are not phytotoxic.

Resistance to avermectins has been demonstrated in a number of arthropods. In most instances, although not clearly determined, resistance is apparently polygenic, develops relatively slowly, and tends to revert rapidly in the absence of selection pressure. Metabolic resistance mechanisms seem to be the most important in the arthropod systems studied to date, and in all these cases a number of biochemical and physiological mechanisms are implicated. Cross-resistance between macrocyclic lactones and other insecticide classes is similarly ill-defined; however, it is apparent that there is little significant cross-resistance. The variations in responses of field strains resistant to other insecticide classes to avermectins may in some cases be due to enhanced metabolism resulting from previous exposure to other insecticides. A number of species have demonstrated wide inherent variation in response to avermectins without the influence of any predisposing factors. This natural variation in susceptibility does not constitute cross-resistance, nor is there evidence to indicate that this variation results in predisposition to develop increased resistance to avermectins.

Despite the widespread, intensive use of abamectin against a number of pests with a high propensity to develop resistance, development of resistance has been relatively slow and rare. The rapid reversion of resistance in the absence of selection pressure and the lack of any confirmed cases of cross-resistance make them amenable for use in resistance management programs. However, in order to prolong the longevity of these compounds in the market place, it will be necessary to minimize selection pressure in practice, especially in high use situations, through proper rotation, good pest management practices, and proper resistance management strategies.

References

Abalis, I.M., Eldefrawi, A.T., **1986**. [^{3}H]muscimol binding to a putative GABA receptor in honey bee brain and its interaction with avermectin B$_{1a}$. *Pestic. Biochem. Physiol. 25*, 279–287.

Abro, G.H., Dybas, R.A., Green, A.St.J., Wright, D.J., **1988**. Toxicity of avermectin B1 against a susceptible laboratory strain and an insecticide-resistant strain of *Plutella xylostella* (Lepidoptera: Plutellidae). *J. Econ. Entomol. 81*, 1575–1580.

Abu-Qare, A.W., Elmasry, E., Abou-Donia, M.B., **2003**. A role for P-glycoprotein in environmental toxicology. *J. Toxicol. Environ. Health 6*, 279–288.

Agee, H.R., **1985**. Neurobiology of the bollworm moth: effects of abamectin on neuronal activity and on sensory systems. *J. Agric. Entomol. 2*, 325–336.

Alexander, M., Wardhaugh, K., **2001**. Workshop on the effects of parasiticides on dung beetles. *CSIRO Entomol. Tech. Rep. 89*, 1–37.

Ali, D.N., Hennessy, D.R., **1996**. The effect of level of feed intake on the pharmacokinetic distribution, excretion and biotransformation of ivermectin in sheep. *J. Vet. Pharmacol. Therap. 19*, 89–94.

Allingham, P.G., Kemp, D.H., Thompson, D.R., Rugg, D., **1994**. Effect of ivermectin on three populations and a laboratory strain of *Haematobia irritans exigua* (Diptera: Muscidae). *J. Econ. Entomol. 87*, 573–576.

Anderson, T.E., Babu, J.R., Dybas, R.A., Mehta, H., **1986**. Avermectin B$_1$: ingestion and contact toxicity against *Spodoptera eridania* and *H. virescens* (Lepidoptera: Noctuidae). *J. Econ. Entomol. 79*, 197–201.

Arena, J.P., Liu, K.K., Paress, P.S., Frazier, E.G., Cully, D.F., *et al.*, **1995**. The mechanism of action of avermectins in *Caenorhabditis elegans*: correlation between activation of glutamate-sensitive chloride current, membrane binding, and biological activity. *J. Parasitol. 81*, 286–294.

Arena, J.P., Liu, K.K., Paress, P.S., Schaeffer, J.M., Cully, D.F., **1992**. Expression of a glutamate-activated chloride current in *Xenopus* oocytes injected with *Caenorhabditis elegans* RNA: evidence for modulation by avermectin. *Brain Res. Mol. Brain Res. 15*, 339–348.

Arena, J.P., Whiting, P.J., Liu, K.K., McGurk, J.F., Paress, P.S., *et al.*, **1993**. Avermectins potentiate GABA-sensitive currents in *Xenopus* oocytes expressing cloned GABA$_A$ receptors. *Biophys. J. 64*, 325.

Argentine, J., **1991**. Two abamectin-resistant strains of Colorado potato beetle. *Resist. Pest Mgt 3*, 30–31.

Argentine, J.A., Clark, J.M., **1990**. Selection for abamectin resistance in Colorado potato beetle (Coleoptera: Chrysomelidae). *Pestic. Sci. 28*, 17–24.

Argentine, J.A., Clark, J.M., Lin, H., **1992**. Genetics and biochemical mechanisms of abamectin resistance in two isogenic strains of Colorado potato beetle. *Pestic. Biochem. Physiol. 44*, 191–207.

Argentine, J.A., Jansson, R.K., Starner, V.R., Halliday, W.R., 2002a. Toxicities of emamectin benzoate homologues and photodegradates to Lepidoptera. *J. Econ. Entomol.* 95, 1185–1189.

Argentine, J.A., Jansson, R.K., Halliday, W.R., Rugg, D., Jany, C.S., 2002b. Potency, spectrum and residual activity of four new insecticides under glasshouse conditions. *Fla. Entomol.* 85, 552–562.

Baker, T.C., van Vorhis Key, S.E., Gaston, L.K., 1985. Bait-preference tests for the Argentine ant (Hymenoptera: Formicidae). *J. Econ. Entomol.* 78, 1083–1088.

Ballivet, M., Alliod, C., Bertrand, S., Bertrand, D., 1996. Nicotinic acetylcholine receptors in the nematode *Caenorhabditis elegans*. *J. Mol. Biol.* 258, 261–269.

Banks, B.J., Bishop, B.F., Evans, N.A., Gibson, S.P., Goudie, A.C., et al., 2000. Avermectins and flea control: structure activity relationships and the selection of selamectin for development as an endectocide for companion animals. *Bioorg. Med. Chem.* 8, 2017–2025.

Beeman, R.W., Stuart, J.J., 1990. A gene for lindane + cyclodiene resistance in the red flour beetle (Coleoptera: Tenebrionidae). *J. Econ. Entomol.* 83, 1745–1751.

Benchaoui, H.A., Clemence, R.G., Clements, P.J.M., Jones, R.I., Watson, P., et al., 2000. Efficacy and safety of selamectin against fleas on dogs and cats presented as veterinary patients in Europe. *Vet. Parasitol.* 91, 223–232.

Benz, G.W., Roncalli, R.A., Gross, S.J., 1989. Use of ivermectin in cattle, sheep, goats and swine. In: Campbell, W.C. (Ed.), Ivermectin and Abamectin. Springer, New York, pp. 215–229.

Bergh, J.C., Rugg, D., Jansson, R.K., McCoy, C.W., Robertson, J.L., 1999. Monitoring the susceptibility of citrus rust mite (Acari: Eriophyidae) populations to abamectin. *J. Econ. Entomol.* 92, 781–787.

Bermudez, I., Hawkins, C.A., Taylor, A.M., Beadle, D.J., 1991. Actions of insecticides on the insect GABA receptor complex. *J. Recept. Res.* 11, 221–232.

Bianchin, I., Honer, M.R., Gomes, A., Koller, W.W., 1992. Effect of some antitick compounds/insecticides on *Onthophagus gazella*. *EMBRAPA Comunicado Technico* 45, 1–7.

Biddinger, D.J., Hull, L.A., 1999. Sublethal effects of selected insecticides on growth and reproduction of a laboratory susceptible strain of tufted apple bud moth (Lepidoptera: Tortricidae). *J. Econ. Entomol.* 92, 314–324.

Bishop, B.F., Bruce, C.I., Evans, N.A., Goudie, A.C., Gration, K.A.F., et al., 2000. Selamectin: a novel broad-spectrum endectocide for dogs and cats. *Vet. Parasitol.* 91, 163–176.

Blackburn, K., Alm, S.R., Yeh, T.S., 1996. Avermectin B$_1$, isazofos, and fenamiphos for control of *Hopolaimus galeatus* and *Tylenchorhynchus dubius* infesting *Poa annua*. *J. Nematol.* 28(Suppl.), 687–694.

Bloomquist, J.R., 1993. Toxicology, mode of action and target site-mediated resistance to insecticides acting on chloride channels. *Comp. Biochem. Physiol.* 106, 301–314.

Bloomquist, J.R., 1994. Cyclodiene resistance at the insect GABA receptor/chloride channel complex confers broad cross resistance to convulsants and experimental phenylpyrazole insecticides. *Arch. Insect Biochem. Physiol.* 26, 69–79.

Bornemissza, G.F., 1970. Insectary studies on the control of dung breeding flies by the activity of the dung beetle, *Onthophagus gazella* F. (Coleoptera: Scarabaeinae). *J. Austral. Entomol. Soc.* 9, 31–41.

Brunner, J.F., Dunley, J.E., Doerr, M.D., Beers, E.H., 2001. Effect of pesticides on *Colpoclypeus florus* (Hymenoptera: Eulophidae) and *Trichogramma platneri* (Hymenoptera: Trichogrammatidae), parasitoids of leafrollers in Washington. *J. Econ. Entomol.* 94, 1075–1084.

Bull, D.L., 1986. Toxicity and pharmacodynamics of avermectin in the tobacco budworm, corn earworm, and fall armyworm (Noctuidae: Lepidoptera). *J. Agric. Food Biochem.* 34, 74–78.

Burg, R.W., Miller, B.M., Baker, E.E., Birnbaum, J., Currie, S.A., et al., 1979. Avermectins, new family of potent antihelminthic agents: producing organism and fermentation. *Antimicrob. Agents Chemother.* 15, 361–367.

Burka, J.F., Hammell, K.L., Horsberg, T.E., Johnsons, G.R., Rainnie, D.J., et al., 1997. Drugs in salmonid aquaculture: a review. *J. Vet. Pharmacol. Therap.* 20, 333–349.

Campanhola, C., Plapp, F.W., Jr., 1989. Toxicity and synergism of insecticides against susceptible and pyrethroid-resistant neonate larvae and adults of the tobacco budworm (Lepidoptera: Noctuidae). *J. Econ. Entomol.* 82, 1527–1533.

Campbell, W.C. (Ed.) 1989. Ivermectin and Abamectin. Springer-Verlag, New York.

Campbell, W.C., Benz, G.W., 1984. Ivermectin: a review of efficacy and safety. *J. Vet. Pharmacol. Therap.* 7, 1–16.

Campbell, W.C., Burg, R.W., Fisher, M.H., Dybas, R.A., 1984. The discovery of ivermectin and other avermectins. In: Magee, P.S., Kohn, G.K., Menn, J.J. (Eds.), Pesticide Synthesis through Rational Approaches. American Chemical Society, Washington, DC, pp. 5–20.

Campbell, W.C., Fisher, M.H., Stapley, E.O., Albers-Schonberg, T., Jacob, T.A., 1983. Ivermectin: a potent new antiparasitic agent. *Science* 221, 923–928.

Campos, F.C., Dybas, R.A., Krupa, D.A., 1995. Susceptibility of twospotted spider mite (Acari: Tetranychidae) populations in California to abamectin. *J. Econ. Entomol.* 88, 225–231.

Campos, F.C., Krupa, D.A., Dybas, R.A., 1996. Susceptibility of populations of twospotted spider mites, *Tetranychus urticae* Koch (Acari: Tetranychidae) from Florida, Holland, and the Canary Islands to abamectin and characterization of abamectin resistance. *J. Econ. Entomol.* 89, 594–601.

Campos, F.C., Krupa, D.A., Jansson, R.K., 1997. The evaluation of a petri plate assay for the assessment of abamectin susceptibility in *Tetranychus urticae* Koch (Acari: Tetranychidae). *J. Econ. Entomol.* 90, 742–746.

Carter, G.T., Nietsche, J.A., Hertz, M.R., **1988**. LL-F28249 antibiotic complex: a new family of antiparasitic macrocyclic lactones: isolation, characterization and structures of LL-F28249α,β,γ,λ. *J. Antibiotics 41*, 519–529.

Cayrol, J.C., Dijan, C., Frankowski, J.P., **1993**. Efficacy of abamectin B$_1$ for control of *Meloidogyne arenaria*. *Fund. Appl. Nematol. 16*, 239–246.

Chabala, J.C., Mrozik, H., Tolman, R.L., Eskola, P., Lusi, A., *et al.*, 1980. Ivermectin, a new broad-spectrum antiparasitic agent. *J. Med. Chem. 23*, 1134–1136.

Chamberlain, W.F., **1982**. Evaluation of toxicants against the secondary screwworm. *Insectic. Acaric. Tests 7*, 255.

Chang, V., **1988**. Toxic baiting of the western yellowjacket (Hymenoptera: Vespidae) in Hawaii. *J. Econ. Entomol. 81*, 228–235.

Cheeseman, C.L., Delany, N.S., Woods, D.J., Wolstenholme, A.J., **2001**. High-affinity ivermectin binding to recombinant subunits of the *Haemonchus contortus* glutamate-gated chloride channel. *Mol. Biochem. Parasitol. 114*, 161–168.

Chiu, S.-H.L., Lu, A.Y.H., **1989**. Metabolism and tissue residues. In: Campbell, W.C. (Ed.), Ivermectin and Abamectin. Springer, New York, pp. 131–143.

Chukwudebe, A.C., Cox, D.L., Palmer, S.J., Morneweck, L.A., Payne, L.D., *et al.*, 1997. Toxicity of emamectin benzoate foliar dislodgeable residues to two beneficial insects. *J. Agric. Food Chem. 45*, 3689–3693.

Clark, J.M., Scott, J.G., Campos, F., Bloomquist, J.R., **1995**. Resistance to avermectins: extent, mechanisms, and management implications. *Annu. Rev. Entomol. 40*, 1–30.

Clarke, G.M., Ridsdill-Smith, T.J., **1990**. The effect of avermectin B$_1$ on developmental stability in the bush fly, *Musca vetustissima*, as measured by fluctuating asymmetry. *Entomol. Exp. Appl. 54*, 265–269.

Cochran, D.G., **1985**. Mortality and reproductive effects of avermectin B$_1$ fed to German cockroaches. *Entomol. Exp. Appl. 37*, 83–88.

Cochran, D.G., **1990**. Efficacy of abamectin fed to German cockroaches (Dictyoptera: Blattellidae) resistant to pyrethroids. *J. Econ. Entomol. 83*, 1243–1245.

Cole, L.M., Roush, R.T., Casida, J.E., **1995**. *Drosophila* GABA-gated chloride channel: modified [³H]EBOB binding site associated with Ala→Ser or Gly mutants of Rdl subunit. *Life Sci. 56*, 757–765.

Conder, G.A., Thompson, D.P., Johnson, S.S., **1993**. Demonstration of co-resistance of *Haemonchus contortus* to ivermectin and moxidectin. *Vet. Rec. 132*, 651–652.

Cook, D.F., **1991**. Ovarian development in females of the Australian sheep blowfly *Lucilia cuprina* (Diptera: Calliphoridae) fed on sheep faeces and the effects of ivermectin residues. *Bull. Entomol. Res. 81*, 249–256.

Cook, D.F., **1993**. Effect of avermectin residues in sheep dung on mating of the Australian sheep blowfly *Lucilia cuprina*. *Vet. Parasitol. 48*, 205–214.

Coop, R.L., Barger, I.A., Jackson, F., **2002**. The use of macrocyclic lactones to control parasites of sheep and goats. In: Vercruysse, J., Rew, R.S. (Eds.), Macrocyclic Lactones in Antiparasitic Therapy. CAB International, Wallingford, pp. 303–321.

Corbitt, T.S., Green, A.S.J., Wright, D.J., **1989**. Relative potency of avermectin B$_1$ against larval stages of *Spodoptera littoralis* and *Heliothis virescens*. *Crop Protect. 8*, 127–132.

Cotreau, M.M., Warren, S., Ryan, J.L., Fleckenstein, L., Vanapalli, S.R., *et al.*, 2003. The antiparasitic moxidectin: safety, tolerability, and pharmacokinetics in humans. *J. Clin. Pharmacol. 43*, 1108–1115.

Court, J.P., Murgatroyd, R.C., Livingstone, D., Rohr, E., **1988**. Physiochemical characteristics of non-electrolytes and their uptake by *Brugia pahangi* and *Dipetalonema viteae*. *Mol. Biochem. Parasitol. 38*, 183–227.

Cox, D.L., Remick, D., Lasota, J.A., Dybas, R.A., **1995a**. Toxicity of avermectins to *Liriomyza trifolii* (Diptera: Agromyzidae) larvae and adults. *J. Econ. Entomol. 88*, 1415–1419.

Cox, D.L., Knight, A.L., Biddinger, D.G., Lasota, J.A., Pikounis, B., *et al.*, 1995b. Toxicity and field efficacy of avermectins against codling moth (Lepidoptera: Tortricidae) on apples. *J. Econ. Entomol. 88*, 708–715.

Craig, T.M., Kunde, J.M., **1981**. Controlled evaluation of ivermectin in Shetland ponies. *Am. J. Vet. Res. 42*, 1422–1429.

Crouch, L.S., Feely, W.F., Arison, B.H., Vanden Heuvel, W.J., Colwell, L.F., *et al.*, 1991. Photodegradation of avermectin B$_{1a}$ thin films on glass. *J. Agric. Food Chem. 39*, 1310–1319.

Cully, D.F., Vassilatis, D.K., Liu, K.K., Paress, P.S., Van der Ploeg, L.H., *et al.*, 1994. Cloning of an avermectin-sensitive glutamate-gated chloride channel from *Caenorhabditis elegans*. *Nature 371*, 707–711.

Cully, D.F., Paress, P.S., Liu, K.K., Schaeffer, J.M., Arena, J.P., **1996**. Identification of a *Drosophila melanogaster* glutamate-gated chloride channel sensitive to the antiparasitic agent avermectin. *J. Biol. Chem. 271*, 20187–20191.

Cvetovich, R.J., Kelly, D.H., DiMichele, L.M., Shuman, R.F., Grabowski, E.J.J., **1994**. Synthesis of 4″-epi-amino-4″-deoxyavermectins B$_1$. *J. Org. Chem. 59*, 7704–7708.

Cvetovich, R.J., Senanayake, C.H., Amato, J.S., DiMichelle, L.M., Bill, T.J., *et al.*, 1997. Practical synthesis of 13-O-[(2-methoxyethoxy) methyl]-22,23-dihydroavermectin B$_1$ aglycon [dimedectin isopropanol, MK-324] and 13-epi-O-(2-methoxymethyl)-22,23-dihydroavermectin B$_1$ aglycon [L-694,554], flea active ivermectin analogues. *J. Org. Chem. 62*, 3989–3993.

Danka, R.G., Loper, G.M., Villa, J.D., Williams, J.L., Sugden, E.A., *et al.*, 1994. Abating feral Africanized honey bees (*Apis mellifera* L.) to enhance mating control of European queens. *Apidologie 25*, 520–529.

Davies, G.H., Green, R.H., **1986**. Avermectins and milbemycins. *Nat. Prod. Rep. England 3*, 87–121.

Davies, I.M., Rodger, G.K., **2000**. A review of the use of ivermectin as a treatment for sea lice [*Lepeophtheirus* *salmonis* (Kroyer) and *Caligus elongatus* (Nordmann)] infestation in farmed Atlantic salmon (*Salmo salar* L.). *Aquaculture Res. 31*, 869–883.

Dawson, G.R., Wafford, K.A., Smith, A., Marshall, G.R., Bayley, P.J., *et al.*, 2000. Anticonvulsant and adverse effects of avermectin analogs in mice are mediated through the gamma-aminobutyric acid(A) receptor. *J. Pharmacol. Exp. Ther. 295*, 1051–1060.

De Azambuja, P., Gomes, J.E.P.L., Lopes, F., Garcia, E.S., 1985. Efficacy of ivermectin against the bloodsucking insect, *Rhodnius prolixus* (Hemiptera, Triatominae). *Mem. Inst. Oswaldo Cruz. 80*, 439–442.

Delany, N.S., Laughton, D.L., Wolstenholme, A.J., 1998. Cloning and localisation of an avermectin receptor-related subunit from *Haemonchus contortus*. *Mol. Biochem. Parasitol. 97*, 177–187.

Deng, Y., Casida, J.E., 1992. House fly head GABA-gated chloride channel: toxicologically relevant binding site for avermectins coupled to a site for ethynylbicycloorthobenzoate. *Pestic. Biochem. Physiol. 43*, 116–122.

Dent, J.A., Davis, M.W., Avery, L., 1997. Avr-15 encodes a chloride channel subunit that mediates inhibitory glutamatergic neurotransmission and ivermectin sensitivity in *Caenorhabditis elegans*. *EMBO J. 16*, 5867–5879.

Dent, J.A., Smith, M.M., Vassilatis, D.K., Avery, L., 2000. The genetics of ivermectin resistance in *Caenorhabditis elegans*. *Proc. Natl Acad. Sci. USA 97*, 2674–2697.

Dively, G.P., Jansson, R.K., 1996. Baseline monitoring of susceptibility to abamectin in the Colorado potato beetle (*Leptinotarsa decemlineata* (Say)). *Pestic. Resist. Newslett. 8*, 8–12.

Doherty, W.M., Stewart, N.P., Cobb, R.M., Kieran, P.J., 1994. *In vitro* comparison of the larvicidal activity of moxidectin and abamectin against *Onthophagus gazella* (F.) (Coleoptera: Scarabaeidae) and *Haematobia irritans exigua* De Meijere (Diptera: Muscidae). *J. Austral. Entomol. Soc. 33*, 71–74.

Drexler, G., Sieghart, W., 1984a. Properties of high affinity binding site for tritium-labelled avermectin B$_{1a}$. *Eur. J. Pharmacol. 99*, 269–277.

Drexler, G., Sieghart, W., 1984b. Evidence for association of a high affinity avermectin binding site with benzodiazepine receptor. *Eur. J. Pharmacol. 101*, 201–207.

Drexler, G., Sieghart, W., 1984c. Sulphur-35-labelled *tert*-butyl-bicyclophosphorothionate and avermectin bind to different sites associated with the gamma-aminobutyric acid-benzodiazepine receptor complex. *Neurosci. Lett. 50*, 273–277.

Driniaev, V.A., Mosin, V.A., Kruglyak, E.B., Sterlina, T.S., Kataev, A.A., *et al.*, 2001. Effect of avermectins on Ca^{2+}-dependent Cl$^-$ currents in plasmalemma of *Chara corallina* cells. *J. Membr. Biol. 182*, 71–79.

Duce, I.R., Scott, R.H., 1985. Actions of dihydroavermectin B$_{1a}$ on insect muscle. *Br. J. Pharmacol. 85*, 395–401.

Dybas, R.A., 1989. Abamectin use in crop protection. In: Campbell, W.C. (Ed.), Ivermectin and Abamectin. Springer, New York, pp. 287–310.

Dybas, R.A., Babu, J.R., 1988. 4″-Eoxy-4″-methylamino-4″-epiavermectin B$_1$ hydrochloride (MK-243): a novel avermectin insecticide for crop protection. In: Proc. Br. Crop Protect. Conf., Pests and Diseases, vol. 1. pp. 57–64.

Dybas, R.A., Green, A.S.J., 1984. Avermectins: their chemistry and pesticidal activity. *Proc. Br. Crop Protect. Conf., Pests and Diseases 9B-3*, 947–954.

Dybas, R.A., Hilton, N.J., Babu, J.R., Preiser, F.A., Dolce, G.J., 1989. Novel second-generation avermectin insecticides and miticides for crop protection. In: Demain, A.L., Somkuti, G.A., Hunter-Cevera, J.C., Rossmoore, H.W. (Eds.), Novel Microbial Products for Medicine and Agriculture. Society for Industrial Microbiology, Fairfax, VA, pp. 203–212.

Eagleson, J.S., Thompson, D.R., Scott, P.G., Cramer, L.G., Barrick, R.A., 1993a. Field trials to confirm the efficacy of ivermectin jetting fluid for control of blowfly strike in sheep. *Vet. Parasitol. 51*, 107–112.

Eagleson, J.S., Thompson, D.R., Scott, P.G., Cramer, L.G., 1993b. The efficacy of ivermectin jetting fluid for control of the sheep biting louse (*Damalinia ovis*). *Austral. J. Exp. Agric. 33*, 843–845.

Egerton, J.R., Ostlind, D.A., Blair, L.S., Eary, C.H., Suhayda, D., *et al.*, 1979. Avermectins, new family of potent antihelminthic agents: efficacy of the B$_{1a}$ component. *Antimicrob. Agents Chemother. 15*, 372–378.

Elgart, G.W., Meinking, T.L., 2003. Ivermectin. *Dermatol. Clin. 21*, 277.

Etter, A., Cully, D.F., Schaeffer, J.M., Liu, K.K., Arena, J.P., 1996. An amino acid substitution in the pore region of a glutamate gated chloride channel enables the coupling of ligand binding to channel gating. *J. Biol. Chem. 271*, 16035–16039.

Feely, W.F., Crouch, L.S., Arison, B.H., Vanden Heuvel, W.J.A., Colwell, L.F., *et al.*, 1992. Photodegradation of 4″-(epimethylamino)-4″-deoxyavermectin B$_{1a}$ thin films on glass. *J. Agric. Food Chem. 40*, 691–696.

ffrench-Constant, R.H., 1994. The molecular and population genetics of cyclodiene insecticide resistance. *Insect Biochem. Mol. Biol. 24*, 335–345.

Fincher, G.T., 1992. Injectable ivermectin for cattle: effects on some dung-inhabiting insects. *Environ. Entomol. 21*, 871–876.

Fincher, G.T., Wang, G.T., 1992. Injectable moxidectin for cattle: effects in two species of dung burying beetles. *Southwest. Entomol. 17*, 303–306.

Fisher, M.H., 1997. Structure–activity relationships of the avermectins and milbemycins. In: Hedin, P.A., Hollingworth, R.M., Masler, E.P., Miamoto, J., Thompson, D.G. (Eds.), Phytochemicals for Pest Control. American Chemical Society, Washington, DC, pp. 220–238.

Fisher, M.H., Mrozik, H., 1984. The avermectin family of macrolide-like antibiotics. In: Omura, S. (Ed.), Macrolide Antibiotics. Academic Press, New York, pp. 553–606.

Fisher, M.H., Mrozik, H., 1989. Chemistry. In: Campbell, W.C. (Ed.), Ivermectin and Abamectin. Springer, New York, pp. 1–23.

Fisher, M.H., Mrozik, H., **1992**. The chemistry and pharmacology of avermectins. *Annu. Rev. Pharmacol. Toxicol. 32*, 537–553.

Floate, K.D., Spooner, R.W., Colwell, D.D., **2001**. Larvicidal activity of endectocides against pest flies in the dung of treated cattle. *Med. Vet. Entomol. 16*, 1–4.

Floate, K.D., Colwell, D.D., Fox, A.S., **2002**. Reductions of non-pest insects in dung of cattle treated with endectocides: a comparison of four products. *Bull. Entomol. Res. 92*, 471–481.

Geden, C.J., Steinkraus, D.C., Long, S.J., Rutz, D.A., Shoop, W.L., **1990**. Susceptibility of insecticide-susceptible and wild house flies (Diptera: Muscidae) to abamectin on whitewashed and unpainted wood. *J. Econ. Entomol. 83*, 1935–1939.

Glancey, B.M., Lofgren, C.S., Williams, D.F., **1982**. Avermectin B_{1a}: effects on the ovaries of red imported fire ant queens (Hymenoptera: Formicide). *J. Med. Entomol. 19*, 743–747.

Gouamene-Lamine, C.N., Yoon, K.S., Clark, J.M., **2003**. Differential susceptibility to abamectin and two bioactive avermectin analogs in abamectin-resistant and -susceptible strains of Colorado potato beetle, *Leptinotarsa decemlineata* (Say) (Coleoptera: Chrysomelidae). *Pestic. Biochem. Physiol. 76*, 15–23.

Goudie, A.C., Evans, N.A., Gration, K.A.F., Bishop, B.F., Gibson, S.P., et al., **1993**. Doramectin: a potent novel endectocide. *Vet. Parasitol. 49*, 5–15.

Grafton-Cardwell, E.E., Hoy, M.A., **1983**. Comparative toxicity of avermectin B1 to the predator *Metaseiulus occidentalis* (Nesbitt) (Acarina: Phytoseiidae) and the spider mites *Tetranychus urticae* Koch and *Panonychus ulmi* (Koch) (Acari: Tetranychidae). *J. Econ. Entomol. 76*, 1216–1220.

Graham, D., Pfeiffer, F., Betz, H., **1982**. Avermectin B_{1a} inhibits the binding of strychnine to the glycine receptor of rat spinal cord. *Neurosci. Lett. 29*, 173–176.

Green, A.St.J., Dybas, R.A., **1984**. The control of *Liriomyza* leafminers and other insect species with avermectin B_1. *Proc. Brit. Crop Protect. Conf., Pests and Diseases 3*, 1135–1141.

Green, A.St. J., Heijne, B., Schreurs, J., Dybas, R.A., **1985**. Serpentine leafminer (*Liriomyza trifolii* (Burgess) control with abamectin (MK-936) in Dutch ornamentals, a review of the processes involved in the evolution of the use directions, and a summary of the results of phytotoxicity evaluations. *Med. Fac. Landbouww. Rijksuniv. Gent 50/2b*, 623–632.

Guerrero, J., McCall, J.W., Genchi, C., **2002**. The use of macrocyclic lactones in the control and prevention of heartworm and other parasites in dogs and cats. In: Vercruysse, J., Rew, R.S. (Eds.), Macrocyclic Lactones in Antiparasitic Therapy. CAB International, Wallingford, pp. 353–369.

Halley, B.A., Jacob, T.A., Lu, A.Y.H., **1989a**. The environmental impact of the use of ivermectin: environmental effects and fate. *Chemosphere 18*, 1543–1563.

Halley, B.A., Nessel, R.J., Lu, A.Y.H., **1989b**. Environmental aspects of ivermectin usage. In: Campbell, W.C. (Ed.), Ivermectin and Abamectin. Springer, New York, pp. 162–172.

Harrell, M.O., Pierce, P.A., **1994**. Effects of trunk-injected abamectin on the elm leaf beetle. *J. Arboric. 20*, 1–3.

Hennessy, D.R., Alvinerie, M.R., **2002**. Pharmacokinetics of the macrocyclic lactones: conventional wisdom and new paradigms. In: Vercruysse, J., Rew, R.S. (Eds.), Macrocyclic Lactones in Antiparasitic Therapy. CAB International, Wallingford, pp. 97–123.

Hilton, R.J., Riedl, H., Westigard, P.H., **1992**. Phytotoxicity response of pear to application of abamectin–oil combinations. *Hortscience 27*, 1280–1282.

Horoszok, L., Raymond, V., Sattelle, D.B., Wolstenholme, A.J., **2001**. GLC-3: a novel fipronil and BIDN-sensitive, but picrotoxinin-insensitive, L-glutamate-gated chloride channel subunit from *Caenorhabditis elegans*. *Br. J. Pharmacol. 132*, 1247–1254.

Houlding, B., Ridsdill-Smith, T.J., Bailey, W.J., **1991**. Injectable ivermectin causes a delay in Scarabaeinae dung beetle egg-laying in cattle dung. *Austral. Vet. J. 68*, 185–186.

Hoy, M.A., Cave, F.E., **1985**. Laboratory evaluation of avermectin as a selective acaricide for use with *Metaseiulus occidentalis* (Nesbitt) (Acarina: Phytoseiidae). *Exp. Appl. Acarol. 1*, 139–152.

Hoy, M.A., Conley, J., **1987**. Selection for abamectin resistance in *Tetranychus urticae* and *T. pacificus* (Acari: Tetranychidae). *J. Econ. Entomol. 80*, 221–225.

Hoy, M.A., Ouyang, Y.-L., **1989**. Selection of the western predatory mite, *Metaseiulus occidentalis* (Acari: Phytoseiidae), for resistance to abamectin. *J. Econ. Entomol. 82*, 35–40.

Huang, J., Casida, J.E., **1997**. Avermectin B_{1a} binds to high- and low-affinity sites with dual effects on the gamma-aminobutyric acid-gated chloride channel of cultured cerebellar granule neurons. *J. Pharmacol. Exp. Therap. 281*, 261–266.

Hughes, P.B., Levot, G.W., **1990**. Toxicity of three avermectins to insecticide susceptible and resistant larvae of *Lucilia cuprina* (Wiedemann) (Diptera: Calliphoridae). *J. Austral. Entomol. Soc. 29*, 109–111.

Immaraju, J.A., Morse, J.G., **1990**. Selection for pyrethroid resistance, reversion, and cross-resistance with citrus thrips (Thysanoptera: Thripidae). *J. Econ. Entomol. 83*, 698–704.

Immaraju, J.A., Paine, T.D., Bethke, J.A., Robb, K.L., Newman, J.P., **1992**. Western flower thrips (Thysanoptera: Thripidae) resistance to insecticides in coastal California greenhouses. *J. Econ. Entomol. 85*, 9–14.

Iqbal, M., Verkerk, R.H.J., Furlong, M.J., Ong, P.C., Rahman, S.A., et al., **1996**. Evidence for resistance to *Bacillus thuringiensis* (Bt) subsp. *kurstaki* HD-1, Bt subsp. *aizawai* and abamectin in field populations of *Plutella xylostella* from Malaysia. *Pestic. Sci. 48*, 89–97.

Ismail, F., Wright, D.J., **1991**. Cross-resistance between acylurea insect growth regulators in a strain of *Plutella xylostella* L. (Lepidoptera: Yponomeutidae) from Malaysia. *Pestic. Sci. 33*, 359–370.

Jackson, H.C., **1989**. Ivermectin as a systemic insecticide. *Parasitol. Today 5*, 146–156.

Jacobs, D.E., Pilkington, J.G., Fisher, M.A., Fox, M.T., **1988**. Ivermectin therapy and degradation of cattle faeces. *Vet. Rec. 123*, 400.

Jagannathan, S., Laughton, D.L., Critten, C.L., Skinner, T.M., Horoszok, L., *et al.*, **1999**. Ligand-gated chloride channel subunits encoded by the *Haemonchus contortus* and *Ascaris suum* orthologues of the *Caenorhabditis elegans* gbr-2 (avr-14) gene. *Mol. Biochem. Parasitol. 103*, 129–140.

James, P.S., Picton, J., Riek, R.F., **1980**. Insecticidal activity of the avermectins. *Vet. Rec. 106*, 59.

Jansson, R.K., Dybas, R.A., **1998**. Avermectins: biochemical mode of action, biological activity, and agricultural importance. In: Ishaaya, I., Degheele, D. (Eds.), Insecticides with Novel Modes of Action: Mechanism and Application. Springer, New York, pp. 152–170.

Jansson, R.K., Rabatin, S., **1997**. Curative and residual efficacy of injection applications of avermectins for control of plant parasitic nematodes on banana. *J. Nematol. 29* (Suppl.), 695–702.

Jansson, R.K., Rabatin, S., **1998**. Potential of foliar, dip, and injection applications of avermectins for control of plant parasitic nematodes. *J. Nematol. 30*, 65–75.

Jansson, R.K., Halliday, W.R., Argentine, J.A., **1998**. Comparison of miniature and high volume bioassays for screening insecticides. *J. Econ. Entomol. 90*, 1500–1507.

Jansson, R.K., Brown, R., Cartwright, B., Cox, D., Dunbar, D.M., *et al.*, **1996**. Emamectin benzoate: a novel avermectin derivative for control of lepidopterous pests. In: Proc. 3rd Intern. Workshop Mgt Diamondback Moth and Other Crucifer Pests, Kuala Lumpur, Malaysia, pp. 171–177.

Johnson, A.W., **1985**. Abamectin for tobacco insect control. *Tobacco Sci. 29*, 135–138.

Jung, M., Saito, A., Buescher, G., Maurer, M., Graf, J.-F., **2003**. Chemistry, pharmacology and safety: milbemycin oxime. In: Vercruysse, J., Rew, R.S. (Eds.), Macrocyclic Lactones in Antiparasitic Therapy. CAB International, Wallinford, pp. 51–74.

Kerlin, R.L., East, I.J., **1992**. The survival and fecundity of buffalo flies after treatment of cattle with three anthelminthics. *Austral. Vet. J. 69*, 283–285.

Kieran, P.J., **1994**. Moxidectin against ivermectin-resistant nematodes-a global view. *Austral. Vet. J. 71*, 18–20.

King, K.L., **1993**. Methods for assessing the impact of avermectins on the decomposer community of sheep pastures. *Vet. Parasitol. 48*, 87–97.

Klei, T.R., Torbert, B.J., **1980**. Efficacy of ivermectin (22,23-dihydroavermectin B$_1$) against gastro-intestinal parasites in ponies. *Am. J. Vet. Res. 41*, 1747–1750.

Koehler, P.G., Atkinson, T.H., Patterson, R.S., **1991**. Toxicity of abamectin to Cockroaches (Dictyoptera: Blattellidae, Blattidae). *J. Econ. Entomol. 84*, 1758–1762.

Kok, L.T., Lasota, J.A., McAvoy, T.J., Dybas, R.A., **1996**. Residual foliar toxicity of 4″-epimethylamino-4″-deoxyavermectin B1 hydrochloride (MK-243) and selected commercial insecticides to adult hymenopterous parasites, *Pteromalus puparum* (Hymenoptera: Pteromalidae) and *Cotesia orobenae* (Hymenoptera: Braconidae). *J. Econ. Entomol. 89*, 63–67.

Konno, Y., Scott, J.G., **1991**. Biochemistry and genetics of abamectin resistance in the house fly. *Pestic. Biochem. Physiol. 41*, 21–28.

Kornis, G.I., **1995**. Avermectins and milbemycins. In: Godfrey, C.R.A. (Ed.), Agrochemicals from Natural Products. Dekker, New York, pp. 215–255.

Kotze, A.C., **1995**. Induced insecticide tolerance in *Lucilia cuprina* (Diptera: Calliphoridae) larvae following dietary phenobarbital treatment. *J. Austral. Entomol. Soc. 34*, 205–209.

Kotze, A.C., Sales, N., **1994**. Cross-resistance spectra and effects of synergists in insecticide-resistant strains of *Lucilia cuprina* (Wiedemann) (Diptera: Calliphoridae). *Bull. Entomol. Res. 84*, 355–360.

Krause, R.M., Buisson, B., Bertrand, S., Corringer, P.-J., Galzi, J.-L., *et al.*, **1998**. Ivermectin: a positive allosteric effector of the α7 neuronal nicotinic acetylcholine receptor. *Molec. Pharmacol. 53*, 283–294.

Kruger, K., Scholtz, C.H., **1997**. Lethal and sublethal effects of ivermectin on the dung-breeding beetles *Euoniticellus intermedius* (Reiche) and *Onitis alexis* Klug (Coleoptera, Scarabaeidae). *Agric. Ecosys. Environ. 61*, 123–131.

Lamine, C.N.G., **1994**. Biochemical factors of resistance and management of Colorado potato beetle, *Leptinotarsa decemlineata* (Say) (Coleoptera: Chrysomelidae). MS thesis, University of Massachusetts, Amherst.

Langley, P.A., Roe, J.M., **1984**. Ivermectin as a possible control agent for tsetse fly, *Glossina morsitans*. *Entomol. Exp. Appl. 36*, 137–143.

Lankas, G.R., Gordon, L.R., **1989**. Toxicology. In: Campbell, W.C. (Ed.), Ivermectin and Abamectin. Springer, New York, pp. 89–112.

Lasota, J.A., Dybas, R.A., **1991**. Avermectins, a novel class of compounds: implications for use in arthropod pest control. *Annu. Rev. Entomol. 36*, 91–117.

Lasota, J.A., Shelton, A.M., Bolognese, J.A., Dybas, R.A., **1996**. Baseline toxicity of avermectins to diamondback moth (Lepidoptera: Plutellidae) populations: implications for susceptibility monitoring. *J. Econ. Entomol. 89*, 33–38.

Laughton, D.L., Lunt, G.G., Wolstenholme, A.J., **1997a**. Reporter gene constructs suggest the *Caenorhabditis elegans* avermectin receptor β-subunit is expressed solely in the pharynx. *J. Exp. Biol. 200*, 1509–1514.

Laughton, D.L., Lunt, G.G., Wolstenholme, A.J., **1997b**. Alternative splicing of a *Caenorhabditis elegans* gene produces two novel inhibitory amino acid receptor subunits with identical ligand binding domains but different ion channels. *Gene 201*, 119–125.

Liang, P., Gao, X.W., Zheng, B.Z., **2003**. Genetic basis of resistance and studies on cross-resistance in a population of diamondback moth, *Plutella xylostella* (Lepidoptera : Plutellidae). *Pest Mgt Sci. 59*, 1232–1236.

Ludmerer, S.W., Warren, V.A., Williams, B.S., Zheng, Y., Hunt, D.C., et al., 2002. Ivermectin and nodulisporic acid receptors in *Drosophila melanogaster* contain both gamma-aminobutyric acid-gated Rdl and glutamate-gated GluCl alpha chloride channel subunits. *Biochemistry 41*, 6548–6560.

Lumaret, J.P., Galante, E., Lumberas, C., Mena, J., Bertrand, M., et al., 1993. Field effects of ivermectin residues on dung beetles. *J. Appl. Ecol. 30*, 428–436.

Lummis, S.C.R., Sattelle, D.B., 1985. Binding sites for 4-aminobutyric acid and benzodiazepines in the central nervous system of insects. *Pestic. Sci. 16*, 695–697.

MacConnell, J.G., Demchak, R.J., Preiser, F.A., Dybas, R.A., 1989. Relative stability, toxicity, and penetrability of abamectin and its 8,9-oxide. *J. Agric. Food. Chem. 37*, 1498–1501.

Mahon, R.J., Wardhaugh, K.G., 1991. Impact of dung from ivermectin-treated sheep on oogenesis and survival of adult *Lucilia cuprina. Austral. Vet. J. 68*, 173–177.

Marley, S.E., Hall, R.D., Corwin, R.M., 1993. Ivermectin cattle pour-on: duration of a single late spring treatment against horn flies, *Haematobia irritans* (L.) (Diptera: Muscidae) in Missouri, USA. *Vet. Parasitol. 51*, 167–172.

McKeand, J., Bairden, K., Ibarra-Silva, A.M., 1988. The degradation of bovine pats containing ivermectin. *Vet. Rec. 122*, 587–588.

McKellar, Q.A., Benchaoui, H.A., 1996. Avermectins and milbemycins. *J. Vet. Pharmacol. Therap. 19*, 331–351.

McTier, T.L., Jernigan, A.D., Rowan, T.G., Holbert, M.S., Smothers, C.D., et al., 2000a. Dose selection of selamectin for efficacy against adult fleas (*Ctenocephalides felis felis*) on dogs and cats. *Vet. Parasitol. 91*, 177–185.

McTier, T.L., Jones, R.L., Holbert, M.S., Murphy, M.G., Watson, P., et al., 2000b. Efficacy of selamectin against adult flea infestations (*Ctenocephalides felis felis* and *Ctenocephalides canis*) on dogs and cats. *Vet. Parasitol. 91*, 187–199.

Meyer, J.A., Simco, J.S., Lancaster, J.L., 1980. Control of face fly larval development with ivermectin, MK-933. *Southwest. Entomol. 5*, 207–209.

Miller, J.A., Kunz, S.E., Oehler, D.D., Miller, R.W., 1981. Larvicidal activity of Merck MK-933, an avermectin, against the horn fly, stable fly, face fly and house fly. *J. Econ. Entomol. 74*, 608–611.

Miller, J.A., Oehler, D.D., Scholl, P. J., 1994. Moxidectin: pharmacokinetics and activity against horn flies (Diptera: Muscidae) and trichostrongyle nematode egg production. *Vet. Parasitol. 53*, 133–143.

Miller, J.A., Oehler, D.D., Siebernaler, A.J., Kunz, S.E., 1986. Effect of ivermectin on survival and fecundity of horn flies and stable flies (Diptera: Muscidae). *J. Econ. Entomol. 79*, 1564–1569.

Mishima, H., Kurabayashi, M., Tamura, C., Sato, S., Kuwano, H., et al., 1975. Structures of milbemycin beta1, beta2, and beta3. *Tetrahedron Lett. 16*, 711–714.

Molyneux, D.H., Bradley, M., Hoerauf, A., Kyelem, D., Taylor, M.J., 2003. Mass drug treatment for lymphatic filariasis and onchocerciasis. *Trends Parasitol 19*, 516–522.

Morse, J.G., Brawner, O.L., 1986. Toxicity of pesticides to *Scirtothrips citri* (Thysanoptera: Thripidae) and implications to resistance management. *J. Econ. Entomol. 79*, 565–570.

Mrozik, H., 1994. Advances in research and development of avermectins. In: Hedin, P.A., Menn, J.J., Hollingsworth, R.M. (Eds.), Natural and Engineered Pest Management Agents. American Chemical Society, Washington, DC, pp. 54–73.

Mrozik, H., Chabala, J.C., Eskola, P., Matzuk, A., Waksmunski, F., et al., 1983. Synthesis of milbemycins from avermectins. *Tetrahedron Lett. 24*, 5333–5336.

Mrozik, H., Eskola, P., Linn, B.O., Lusi, A., Shih, T.L., et al., 1989. Discovery of novel avermectins with unprecedented insecticidal activity. *Experentia 45*, 315–316.

Mrozik, H., Eskola, P., Reynolds, G.F., Arison, B.H., Smith, G.M., et al., 1988. Photoisomers of avermectins. *J. Organ. Chem. 53*, 1820–1823.

National Registration Authority, 1998. *NRA special review of macrocyclic lactones. NRA Special Review Series 98.3* 1–79.

Nicholson, R., Robinson, P.S., Palmer, P.J., Casida, J.E., 1988. Ivermectin-stimulated release of neurotransmitter in the insect central nervous system: modulation by external chloride and inhibition by a novel trioxabicyclooctane and two polychlorocycloalkane insecticides. In: Neurotoxicology '88, Abstr. 96.

Orton, C.J., Watts, J.E., Rugg, D., 1992. Comparative effectiveness of avermectins and deltamethrin in suppressing oviposition in *Lucilia cuprina* (Diptera: Calliphoridae). *J. Econ. Entomol. 85*, 28–32.

Ostlind, D.A., Cifelli, S., Lang, R., 1979. Insecticidal activity of the antiparasitic avermectins. *Vet. Rec. 105*, 168.

Ostlind, D.A., Felchetto, T., Misura, A., Ondeyka, J., Smith, S., et al., 1997. Discovery of a novel indole diterpene insecticide using first instars of *Lucilia sericata. Med. Vet. Entomol. 11*, 407–408.

Pampiglione, S., Majori, G., Petrangeli, G., Romi, R., 1985. Avermectins, MK-933 and MK936, for mosquito control. *Trans. Roy. Soc. Trop. Med. Hyg. 79*, 797–799.

Parella, M.P., 1983. Evaluation of selected insecticides for control of permethrin-resistant *Liriomyza trifolii* (Diptera: Agromyzidae) on chrysanthemums. *J. Econ. Entomol. 76*, 1460–1464.

Park, C.G., Yoo, J.K., Lee, J.O., 1996. Toxicity of some pesticides to two-spotted spider mite (Acari: Tetranychidae) and its predator *Amblyseius womersleyi* (Acari: Phytoseiidae). *Korean J. Appl. Entomol. 35*, 232–237.

Paul, A.J., Tranquilli, W.J., Seward, R.L., Todd, K.S., DiPietro, J.A., 1987. Clinical observations in collies given ivermectin orally. *Am. J. Vet. Res. 48*, 684–685.

Pong, S.-S., Wang, C.C., 1982. Avermectin B_{1a} modulation of gamma-aminobutyric acid receptors in rat brain membranes. J. Neurochem. 38, 375–379.

Pong, S.-S., Wang, C.C., Fritz, L.C., 1980. Studies on the mechanism of action of avermectin B_{1a}: stimulation of release of gamma-aminobutyric acid from brain synaptosomes. J. Neurochem. 34, 351–358.

Portillo, V., Jagannathan, S., Wolstenholme, A.J., 2003. Distribution of glutamate-gated chloride channel subunits in the parasitic nematode Haemonchus contortus. J. Comp. Neurol. 462, 213–222.

Preston, J.M., 1982. The avermectins: new molecules for use in warble fly control. In: Boulard, C., Thornberry, H. (Eds.), Warble Fly Control in Europe. A.A. Balkema, Rotterdam, pp. 17–20.

Pritchard, R.K., 2002. Resistance against macrocyclic lactones. In: Vercruysse, J., Rew, R.S. (Eds.), Macrocyclic Lactones in Antiparasitic Therapy. CAB International, Wallingford, pp. 163–182.

Putter, I., MacConnell, J.G., Preiser, F.A., Haidri, A.A., Ristich, S.S., et al., 1981. Avermectins: novel insecticides, acaracides and nematicides from a soil microorganism. Experentia 37, 963–964.

Ranjan, S., Wang, G.T., Hirschlein, C., Simkins, K.L., 2002. Selection for resistance to macrocyclic lactones by Haemonchus contortus in sheep. Vet. Parasitol. 103, 109–117.

Raymond, V., Mongan, N.P., Sattelle, D.B., 2000. Anthelmintic actions of homomer-forming nicotinic acetylcholine receptor subunits: chicken α7 and ACR-16 from the nematode Caenorhabditis elegans. Neuroscience 101, 785–791.

Raymond, V., Sattelle, D.B., 2002. Novel animal-health drug targets for ligand-gated chloride channels. Nature Rev. Drug Disc. 1, 427–436.

Reed, D.K., Reed, G.L., 1986. Activity of avermectin B_1 against the striped cucumber beetle (Acalymma vittatum) (Coleoptera: Chrysomelidae). J. Econ. Entomol. 79, 943–947.

Reed, D.K., Tromley, N.J., Reed, G.L., 1985. Activity of avermectin B_1 against codling moth (Lepidoptera: Olethreutidae). J. Econ. Entomol. 78, 1067–1071.

Ridsdill-Smith, T.J., 1988. Survival and reproduction of Musca vetustissima Walker (Diptera: Muscidae) and a scarabaeinae dung beetle in dung of cattle treated with avermectin B1. J. Austral. Entomol. Soc. 27, 175–178.

Robertson, J.L., Richmond, C.E., Preisler, H.K., 1985. Lethal and sublethal effects of avermectin B_1 on the western spruce budworm (Lepidoptera: Tortricidae). J. Econ. Entomol. 78, 1129–1132.

Rock, D.W., DeLay, R.L., Gliddon, M.J., 2002. Chemistry, pharmacology and safety: moxidectin. In: Vercruysse, J., Rew, R.S. (Eds.), Macrocyclic Lactones in Antiparasitic Therapy. CAB International, Wallingford, pp. 75–96.

Rohrer, S.P., Birzin, E.T., Costa, S.D., Arena, J.P., Hayes, E.C., et al., 1995. Identification of neuron-specific ivermectin binding sites in Drosophila melanogaster and Schistocerca americana. Insect Biochem. Mol. Biol. 25, 11–17.

Roncalli, R.A., 1984. Efficacy of ivermectin against Oestrus ovis in sheep. Vet. Med. Small Anim. Clinic. 79, 1095–1097.

Roth, M., 2000. The availability and use of chemotherapeutic sea lice control products. Contrib. Zool. 69, http://dpc.uba.uva.nl/ctz/vol69/nr01/a12.

Roulet, A., Puel, O., Gesta, S., Lepage, J.F., Drag, M., et al., 2003. MDR1-deficient genotype in collie dogs hypersensitive to the P-glycoprotein substrate ivermectin. Eur. J. Pharmacol. 460, 85–91.

Roush, R.T., Wright, J.E., 1986. Abamectin toxicity to houseflies (Diptera: Muscidae) resistant to synthetic organic insecticides. J. Econ. Entomol. 79, 562–564.

Rugg, D., 1995. Toxicity and tolerance to ivermectin in some insects associated with livestock and the interaction of ivermectin with the skin and fleece of sheep. PhD thesis, University of Sydney.

Rugg, D., Thompson, D.R., 1993. A laboratory assay for assessing the susceptibility of Damalinia ovis (Schrank) (Phthiraptera: Trichodectidae) to avermectins. J. Austral. Entomol. Soc. 32, 1–3.

Rugg, D., Thompson, D.R., 1997. The interactions of ivermectin with the fleece and skin of sheep. In: Proc. 4th Int. Congr. Sheep Vet., Armidale, NSW, pp. 393–397.

Rugg, D., Kotze, A.C., Thompson, D.R., Rose, H.A., 1998b. Susceptibility of laboratory-selected and field strains of Lucilia cuprina (Diptera: Calliphoridae) to ivermectin. J. Econ. Entomol. 91, 601–607.

Rugg, D., Sales, N., Kotze, A.C., Levot, G.W., 1995a. Susceptibility to ivermectin of pyrethroid-resistant Lucilia cuprina (Wiedemann) (Diptera: Calliphoridae). J. Austral. Entomol. Soc. 34, 69–70.

Rugg, D., Thompson, D.R., Boyle, R., Eagleson, J.S., 1995b. Field efficacy of an ivermectin jetting fluid for control of the sheep body louse Bovicola (Damalinia) ovis in New Zealand. N. Z. Vet. J. 43, 48–49.

Rugg, D., Thompson, D.R., Scott, P.G., Cramer, L.G., Barrick, R.A., 1993. Efficacy of ivermectin jetting fluid against strike by some primary and secondary blowflies of sheep. Austral. Vet. J. 70, 180–182.

Rugg, D., Thompson, D.R., Gogolewski, R.P., Allerton, G.R., Barrick, R.A., et al., 1998a. Efficacy of ivermectin in a controlled-release capsule for the control of breech strike in sheep. Austral. Vet. J. 76, 350–354.

Rust, M.K., 1986. Managing household pests. In: Bennett, G.W., Owens, J.M. (Eds.), Advances in Urban Pest Management. Van Nostrand Reinhold, New York, pp. 335–386.

Sales, N., Levot, G.W., Hughes, P.B., 1989. Monitoring and selection of resistance to pyrethroids in the Australian sheep blowfly, Lucilia cuprina. Med. Vet. Entomol. 3, 287–291.

Sanderson, J.P., Zhang, Z.Q., 1995. Dispersion, sampling, and potential for integrated control of two-spotted spidermite (Acari: Tetranychidae) on greenhouse roses. J. Econ. Entomol. 88, 343–351.

Sasser, J.N., Kirkpatrick, J.L., Dybas, R.A., 1982. Efficacy of avermectins for root knot control in tobacco. *Plant Dis.* 66, 961–965.

Sattelle, D.B., 1990. GABA receptors of insects. *Adv. Insect Physiol.* 22, 1–113.

Schaeffer, J.M., Haines, H.W., 1989. Avermectin binding in *Caenorhabditis elegans*: A two-state model for the avermectin binding site. *Biochem. Pharmacol.* 38, 2329–2338.

Scholtz, C.H., Krüger, K., 1995. Effects of ivermectin residues in cattle dung on dung insect communities under extensive farming conditions in South Africa. In: Stork, N.E., Harrington, R. (Eds.), Insects in a Changing Environment. Academic Press, New York, pp. 465–471.

Schuster, D.J., 1994. Life-stage specific toxicity of insecticides to parasitoids of *Liriomyza trifolii* (Burgess) (Diptera, Agromyzidae). *Int. J. Pest Mgt* 40, 191–194.

Schuster, D.J., Everett, P.H., 1983. Response of *Liriomyza trifolii* (Diptera: Agromyzidae) to insecticides on tomato. *J. Econ. Entomol.* 76, 1170–1174.

Scott, J.G., 1989. Cross-resistance to the biological insecticide abamectin in pyrethroid resistant strains of house flies. *Pestic. Biochem. Physiol.* 34, 27–31.

Scott, J.G., Roush, R.T., Liu, N., 1991. Selection of high-level abamectin resistance from field-collected house flies, *Musca domestica. Experentia* 47, 288–291.

Seaman, J.T., Eagleson, J.S., Corrigan, M.J., Webb, R.F., 1987. Avermectin B₁ toxicity in a herd of Murray Grey cattle. *Austral. Vet. J.* 64, 271–278.

Sechser, B., Ayoub, S., Monuir, N., 2003. Selectivity of emamectin benzoate to predators of sucking pests on cotton. *J. Plant Dis. Protect.* 110, 184–194.

Shoop, W.L., 1993. Ivermectin resistance. *Parasitol. Today* 9, 154–159.

Shoop, W.L., Haines, H.W., Michael, B.F., Eary, C.H., 1993. Mutual resistance to avermectins and milbemycins: oral activity of ivermectin and moxidectin against ivermectin-resistant and susceptible nematodes. *Vet. Rec.* 133, 445–447.

Shoop, W.L., Mrozik, H., Fisher, M.H., 1995. Structure and activity of avermectins and milbemycins in animal health. *Vet. Parasitol.* 59, 139–156.

Shoop, W., Soll, M., 2002. Ivermectin, abamectin and eprinomectin. In: Vercruysse, J., Rew, R.S. (Eds.), Macrocyclic Lactones in Antiparasitic Therapy. CAB International, Wallingford, pp. 1–29.

Sigel, E., Buhr, A., 1997. The benzodiazepine binding site of GABA$_A$ receptors. *Trends Pharmacol. Sci.* 18, 425–429.

Sommer, C., Overgaard Nielsen, B., 1992. Larvae of the dung beetle *Onthophagus gazella* Fabricus (Coleoptera: Scarabaeidae) exposed to lethal and sublethal ivermectin concentrations. *J. Appl. Entomol.* 114, 502–509.

Sommer, C., Gronvold, J., Holter, P., Nansen, P., 1993. Effects of ivermectin on two afrotropical dung beetles, *Onthophagus gazella* and *Diastellopalpus quinquedens* (Coleoptera: Scarabaeidae). *Vet. Parasitol.* 48, 171–179.

Sommer, C., Steffanson, B., Overgaard Nielsen, B., Gronvold, J., Vagn Jensen, K.M., et al., 1992. Ivermectin excreted in cattle dung after subcutaneous injection or pour-on treatment: concentrations and impact on dung fauna. *Bull. Entomol. Res.* 82, 257–264.

Spinosa, H., de, S., Stilck, S.R., Bernardi, M.M., 2002. Possible anxiolytic effects of ivermectin in rats. *Vet. Res. Commun.* 26, 309–321.

Spradbery, J.P., Tozer, R.S., Drewett, N., Lindsay, M.J., 1985. The efficacy of ivermectin against larvae of the screw-worm fly (*Chrysomya bezziana*). *Austral. Vet. J.* 62, 311–314.

Standfast, H.A., Muller, M.J., Wilson, D.D., 1984. Mortality of *Culicoides brevitarsis* (Diptera: Ceratopogonidae) fed on cattle treated with ivermectin. *J. Econ. Entomol.* 77, 419–421.

Stretton, A.O.W., Campbell, W.C., Babu, J.R., 1987. Biological activity and mode of action of avermectins. In: Veech, A., Dickson, D.W. (Eds.), Vistas on Nematology. Society for Nematologists, Marceline, MO, pp. 136–146.

Strong, L., 1992. The use and abuse of feed-through compounds in cattle treatments. *Bull. Entomol. Res.* 82, 1–4.

Strong, L., 1986. Inhibition of pupariation and adult development in *Calliphora vomitoria* treated with ivermectin. *Entomol. Exp. Appl.* 41, 157–164.

Strong, L., Brown, T.A., 1987. Avermectins in insect control and biology: a review. *Bull. Entomol. Res.* 77, 357–389.

Strong, L., James, S., 1993. Some effects of ivermectin on the yellow dung fly, *Scatophaga stercoraria. Vet. Parasitol.* 48, 181–191.

Strong, L., Wall, R., 1994. Effects of ivermectin and moxidectin on the insects of cattle dung. *Bull. Entomol. Res.* 84, 403–409.

Su, N.-Y., Tamashiro, M., Haverty, M.I., 1987. Characterization of slow-acting insecticides for the remedial control of the Formosan termite (Isoptera: Rhinotermitidae). *J. Econ. Entomol.* 80, 1–4.

Takiguchi, Y., Mishna, H., Okuda, M., Tenao, M., 1980. Milbemycins, a new family of macrolide antibiotics: fermentation, isolation and physico-chemical properties. *J. Antibiotics* 33, 1120–1127.

Takai, K., Suzuki, T., Kawazu, K., 2003. Development and preventative effect against pine wilt disease of a novel liquid formulation of emamectin benzoate. *Pest Mgt Sci.* 59, 365–370.

Tanaka, K., 1987. Mode of action of insecticidal compounds acting at an inhibitory synapse. *J. Pestic. Sci.* 12, 549–560.

Tanaka, K., Matsumura, F., 1985. Action of avermectin B$_{1a}$ on the leg muscles and the nervous system of the American cockroach. *Pestic. Biochem. Physiol.* 29, 124–135.

Taylor, S.M., Mallon, T.R., Blanchflower, W.J., Kennedy, D.G., Green, W.P., 1992. Effects of diet on plasma concentrations of oral anthelminthics for cattle and sheep. *Vet. Rec.* 130, 264–268.

Thompson, D.R., Rugg, D., Scott, P.G., Cramer, L.G., Barrick, R.A., 1994. Rainfall and breed effects on the efficacy of ivermectin jetting fluid for the prevention of fly strike and treatment of infestations of lice in long-woolled sheep. *Austral. Vet. J. 71*, 161–164.

Titchener, N.R., Parry, J.M., Grimshaw, W.T.R., **1994**. Efficacy of formulations of abamectin, ivermectin and moxidectin against sucking and biting lice of cattle. *Vet. Rec. 134*, 452–453.

Tsukamoto, T., Sato, K., Kinoto, T., Yanai, T., Nishida, A., et al., **1995**. Syntheses of 8,9-epoxy- and 5-O-acyl-8,9-epoxymilbemycin A4 and their activity against *Tetranychus urticae. Biosci. Biotech. Biochem. 59*, 226–230.

Trumble, J.T., Alvarado Rodriguez, B., **1993**. Development and economic evaluation of an IPM program for fresh market tomato production in Mexico. *Agric. Ecosyst. Environ. 43*, 267–284.

Trumble, J.T., Morse, J.P., **1993**. Economics of integrating the predaceous mite *Phytoseiulus persimilis* (Acari: Phytoseiidae) with pesticides in strawberries. *J. Econ. Entomol. 86*, 879–885.

Trumble, J.T., Moar, W.J., Babu, J.R., Dybas, R.A., **1987**. Laboratory bioassays of the acute and antifeedant effects of avermectin B1 and a related analogue on *Spodoptera exigua* (Hubner). *J. Agric. Entomol. 4*, 21–28.

Vassilatis, D.K., Arena, J.P., Plasterk, R.H., Wilkinson, H.A., Schaeffer, J.M., et al., **1997**. Genetic and biochemical evidence for a novel avermectin-sensitive chloride channel in *Caenorhabditis elegans*: isolation and characterization. *J. Biol. Chem. 272*, 33167–33174.

Vercruysse, J., Rew, R.S. (Eds.) **2002a**. Macrocyclic Lactones in Antiparasitic Therapy. CAB International, Wallingford.

Vercruysse, J., Rew, R.S., **2002b**. General efficacy of the macrocyclic lactones to control parasites of cattle. In: Vercruysse, J., Rew, R.S. (Eds.), Macrocyclic Lactones in Antiparasitic Therapy. CAB International, Wallingford, pp. 185–222.

Wall, R., Strong, L., **1987**. Environmental consequences of treating cattle with the antiparasitic drug ivermectin. *Nature 327*, 418–421.

Wardhaugh, K.G., Rodriguez-Mendez, H., **1988**. The effects of the antiparasitic drug, ivermectin, on the development and survival of the dung breeding fly, *Orthelia cornicina* (F.) and the scarabaeine dung beetles, *Copris hispanus* L., *Bubas bubalus* (Oliver) and *Onitis belial* F. *J. Appl. Entomol. 106*, 381–389.

Wardhaugh, K.G., Longstaff, B.C., Morton, R., **2001**. A comparison of the development and survival of the dung beetle, *Onthophagus taurus* (Schreb.) when fed on the faeces of cattle treated with pour-on formulations of eprinomectin or moxidectin. *Vet. Parasitol. 99*, 155–168.

Webb, J.D., Burg, J.G., Knapp, F.W., **1991**. Moxidectin evaluation against *Solenoptes capillatus* (Anoplura: Linognathidae), *Bovicola bovis* (Mallophaga: Trichodectidae), and *Musca autumnalis* (Diptera: Muscidae) on cattle. *J. Econ. Entomol. 84*, 1266–1269.

Wislocki, P.G., Grosso, L.S., Dybas, R.A., **1989**. Environmental aspects of use in crop protection. In: Campbell,

W.C. (Ed.), Ivermectin and Abamectin. Springer, New York, pp. 182–214.

Wratten, S.D., Mead-Briggs, M., Gettinby, G., Ericsson, G., Baggot, D.G., **1993**. An evaluation of the potential effects of ivermectin on the decomposition of cattle dung pats. *Vet. Rec. 133*, 365–371.

Wright, J.E., **1984**. Biological activity of AVMB$_1$ against the boll weevil (Coleoptera: Curculionidae). *J. Econ. Entomol. 77*, 1029–1032.

Wright, D.E., **1986**. Biological activity and mode of action of avermectins. In: Ford, M.D., Lunt, G.G., Deay, R.C., Usherwood, P.N.R. (Eds.), Neuropharmacology and Pesticide Action. Ellis Horwood, Chichester, pp. 174–202.

Wright, J.E., Jenkins, J.N., Villavaso, E.J., **1985**. Evaluation of avermectin B$_1$ (MK-936) against *Heliothis* spp. in the laboratory and in field plots and against the boll weevil in field plots. *Southwest. Entomol. 7* (Suppl.), 11–16.

Yamamoto, D., Suzuki, N., **1988**. Patch-clamp analysis of single chloride channels in primary neurone cultures of *Drosophila*. In: Lunt, G.G. (Ed.), Neurotoxicology '88. pp. 375–381.

Yazwinski, T.A., Greenway, T., Preston, B.L., Pote, L.M., Featherstone, H., et al., **1983**. Antiparasitic activity of ivermectin in naturally parasitized sheep. *Am. J. Vet. Res. 44*, 2186–2187.

Yoon, K.S., Nelson, J.O., Clark, J.M., **2002**. Selective induction of abamectin metabolism by dexamethasone, 3-methylcholanthrene, and phenobarbital in Colorado potato Beetle, *Leptinotarsa decemlineata* (Say). *Pestic. Biochem. Physiol. 73*, 74–86.

Zakson-Aiken, M., Gregory, L.M., Meinke, P.T., Shoop, W.L., **2001**. Systemic activity of the avermectins against the cat flea (Siphonaptera: Pulicidae). *J. Med. Entomol. 38*, 576–580.

Zchorifein, E., Roush, R.T., Sanderson, J.P., **1994**. Potential for integration of biological and chemical control of greenhouse whitefly (Homoptera: Aleyrodidae) using *Encarsia formosa* (Hymenoptera: Aphelinidae) and abamectin. *Environ. Entomol. 23*, 1277–1282.

Zeng, Z., Andrew, N.W., Woda, J.M., Halley, B.A., Crouch, L.S., et al., **1996**. Role of cytochrome P450 isoforms in the metabolism of abamectin and ivermectin in rats. *J. Agric. Food Chem. 44*, 3374–3378.

Zhang, Z., Sanderson, J.P., **1990**. Relative toxicity of abamectin to the predatory mite *Phytoseiulus persimilis* (Acari: Phytoseiidae) and the two-spotted spider mite (Acari: Tetranychidae). *J. Econ. Entomol. 83*, 1783–1790.

Zhao, J.Z., Li, Y.X., Collins, M.L., Gusukuma-Minuto, L., Mau, R.F.L., et al., **2002**. Monitoring and characterization of diamondback moth (Lepidoptera : Plutellidae) resistance to spinosad. *J. Econ. Entomol. 95*, 430–436.

Zufall, F., Franke, C., Hatt, H., **1989**. The insecticide avermectin B$_{1a}$ activates a chloride channel in crayfish muscle membrane. *J. Exp. Biol. 142*, 191–205.

A3 Addendum: The Insecticidal Macrocyclic Lactones

D Rugg, Global Development and Operations Pfizer
Animal Health, Kalamazoo, MI, USA

© 2010, Elsevier BV. All Rights Reserved.

Macrocyclic lactones continue to be widely used for insect, acarid, and nematode control in crop protection, animal and human health, and urban pest control since the first commercialization of ivermectin as an animal health parasiticide in 1981. These compounds are still considered to be the most important veterinary anthelmintics available today (McCavera et al., 2009). Insecticidal use of macrocyclic lactones has expanded through novel applications or formulations. For example, abamectin has been incorporated into a plastic ear tag for long-term control of horn fly and a number of tick species on cattle, and emamectin benzoate, applied as a single trunk injection in pine trees, is effective against coneworm, Dioryctria spp. for over 3 years (Grosman et al., 2002) and effectively prevents attack by Ips spp. bark beetles and cerambycid larvae for at least 5 months (Grosman and Upton, 2006). Also, the introduction of generic macrocyclic lactones has reduced cost and encouraged greater use of this chemistry.

Despite their widespread use (and abuse), the introduction of generics and increased use as prices dropped, resistance to macrocyclic lactones has developed relatively slowly. Resistance has been demonstrated in a number of arthropods and nematodes and most often develops under intensive selection pressure, for example, under high density, high value cropping system such as greenhouse ornamentals. In mites, resistance to abamectin has arisen in regions with intensive and repeated exposure. Sato et al. (2005) evaluated selection, cross-resistance, and stability of abamectin resistance in a field strain of Tetranychus urticae from strawberry in Brazil that had been exposed to intensive abamectin exposure for over 10 years. These workers found that the field strain was moderately resistant to abamectin (~25-fold) and that in the absence of selection pressure, the mites rapidly reverted to susceptibility. After five laboratory selections, resistance to abamectin increased about 13-fold making the LC_{50} over 340 times higher than susceptible mites. These mites were cross-resistant to the macrocyclic lactone, milbemectin, but not to other acaricide classes. In Cyprus, T. urticae populations resistant to abamectin (and most other acaricides) were found on greenhouse roses, while populations on cucumbers (greenhouse and outdoor) were susceptible (Stavrinides and Hadjistylli, 2009). Cross-resistance to milbemectin, a macrocyclic lactone acaricide not yet registered in Cyprus, was confirmed in these populations. These workers surmised that resistance in the greenhouse situation was likely due to intensive use in a perennial crop; mites had been exposed to weekly or biweekly acaricide applications for a number of years. The presence of susceptible mites on perennial cucumber crops reflected a lower selection pressure and also illustrated that judicial use and management of existing acaricides could potentially minimize the development of resistance. He et al. (2009) investigated the genetics of abamectin resistance in a laboratory-selected population of Tetranychus cinnabarinus that had an 8.7-fold resistance after selection for 42 generations. These workers confirmed previous findings in other insect and mite species that resistance was polygenic, incompletely recessive and backcross data indicated that maternal or cytoplasmic effect had a role in resistance inheritance in this mite.

Zhao et al. (2006) reported multiyear monitoring for diamondback moth resistance to emamectin benzoate in the US and Mexico. In the initial survey in 2001, tolerance ratios for nine field populations ranged from 2.1 to 60.5-fold of that of a susceptible, laboratory population. Subsequent monitoring over 3 years showed that tolerance ratios remained stable or decreased with time. It is likely that the lack of resistance development was due to judicious use with alternative chemistries available and indicates the benefits of effective management of available insecticides. Avermectins are highly lipophilic, partition into membranes and affect membrane fluidity. Recently, the effects of avermectins on the fluidity of mitochondrial membranes were investigated in resistant and susceptible diamond-back moths (Hu *et al.*, 2008). These workers generated a population with 1078-fold resistance to abamectin through laboratory selection that exhibited high-level cross-resistance to emamectin benzoate and ivermectin. They found differences in membrane fluidity and membrane response to abamectin that indicated that resistance may have been due to reduced abamectin membrane disrobing. However, practical implications of this mechanism in field resistance to avermectins in diamondback moth are unknown. In another recent investigation, Qian *et al.* (2008) showed that a strain of diamondback moth selected for resistance to the ecdysone receptor agonist, tebufenozide (99-fold) was cross-resistant to abamectin (29-fold). When selected with abamectin, resistance to this compound increased but resistance to tebufenozide decreased. Piperonyl butoxide synergized the activity of abamectin and monooxygenase activity was enhanced in the resistant strains, indicating that increased monooxygenase activity was the mechanism conferring cross-resistance to abamectin. This laboratory-selected resistant strain had a relatively low fitness of 0.3-fold of the parental strain (Cao and Han, 2006) and reverted to susceptibility rapidly in the absence of selection (Qian *et al.*, 2008). Emamectin benzoate has been used in an integrated resistance management program for *Helicoverpa* spp. control on cotton in Australia for a number of years; resistance monitoring from 2002 to 2008 shows that resistance frequencies to this compound in the highly labile *Helicoverpa armigera* have remained very low (Rossiter *et al.*, 2008).

Emamectin benzoate has been used as an in-feed systemic for the control of copepod sea lice in farmed salmon since 2000; the ease of application and efficacy of the product has resulted in its extensive use. A recent study of trends in efficacy from 2002 to 2006 in Scotland provided evidence that efficacy varied across regions and tended to reduce over time (Lees *et al.*, 2008). The authors surmised that lack of efficacy may have been due to a number of cultural factors such as fish appetite and feeding rates, sub-therapeutic dosing, etc. or resistance, and highlighted an ongoing need for resistance monitoring and management programs.

In nematodes, the variety and differential expression of glutamate-gated chloride channels may be responsible for many of the biological effects of macrocyclic lactones and the diverse responses among species (Wolstenholme and Rogers, 2005). It is probable that multiplicity of glutamate-gated chloride channel subtypes and their different functionalities are mirrored by a variety of resistance mechanisms. Changes in the sequence and expression of ligand-gated chloride channels can cause resistance to macrocyclic lactones in model organisms, such as *Caenorhabditis elegans* and *Drosophila melanogaster*, but mutations in multiple glutamate-gated chloride channel subunit genes are required for high-level macrocyclic lactone resistance in *C. elegans* (McCavera *et al.*, 2007). Polymorphisms in several subunits have been reported from resistant isolates of the parasitic nematode, *Haemonchus contortus*. McCavera *et al.* (2009) confirmed marked differences in the glutamate-gated chloride channels of different nematode species, supporting the differential effects of various macrocyclic lactones on different species and variations in their patterns of resistance. Other possible mechanisms for macrocyclic lactone resistance in nematodes include over expression of efflux mechanisms such as P-glycoprotein and other ABC transporters and selection of the non-ligand-gated chloride channel, β-tubulin (Prichard and Roulet, 2007).

References

Cao, G., Han, Z., **2006**. Tebufenozide resistance selected in *Plutella xylostella* and its cross-resistance and fitness cost. *Pest Manag. Sci. 62*, 746–751.

Grosman, D.M., Upton, W.W., **2006**. Efficacy of systemic insecticides for protection of loblolly pine against southern pine engraver beetles (Coleoptera: Curculionidae: Scolytinae) and wood borers Coleoptera: Cerambycidae). *J. Econ. Entomol. 99*, 94–101.

Grosman, D.M., Upton, W.W., McCook, F.A., Billings, R.F., **2002**. Systemic insecticide injections for control of cone and seed insects in loblolly pine seed orchards – 2 year results. *South. J. Appl. For. 26*, 146–152.

He, L., Gao, X., Wang, J., Zhao, Z., Liu, N., **2009**. Genetic analysis of abamectin resistance in *Tetranychus urticae*. *Pest. Biochem. Physiol. 95*, 147–151.

Hu, J., Liang, P., Shi, X., Gao, X., **2008**. Effects of insecticides on the fluidity of mitochondrial membranes of the diamondback moth, *Plutella xylostella*, resistant and susceptible to avermectin. *J. Insect Sci. 8*, Article 3 p. 9. www.insectscience.org/8.03.

Lees, F., Baille, M., Gettinby, G., Revie, C.W., **2008**. The efficacy of emamectin benzoate against infestations of *Lepeophtheirus salmonis* on farmed Atlantic salmon (*Salmo salar*) in Scotland, 2002–2006. *PLoS ONE 3*(2), e1549. doi:10.1371/journal.pone.00001549.

McCavera, S., Walsh, T.K., Wolstenholme, A.J., **2007**. Nematode ligand-gated chloride channels: an appraisal of their involvement in macrocyclic lactone resistance and prospects for developing molecular markers. *Parasitology 134*, 1111–1121.

McCavera, S., Rogers, A.T., Yates, D.M., Woods, D.J., Wolstenholme, A.J., **2009**. An ivermectin-sensitive glutamate-gated chloride channel from the parasitic nematode *Haemonchus contortus*. *Mol. Parasitol. 75,* 1347–1355.

Prichard, R.K., Roulet, A., 2007. ABC transporters and β-tubulin in macrocyclic lactone resistance: prospects for marker development. *Parasitology 134,* 1123–1132.

Qian, L., Cao, G., Song, J., Yin, Q., Han, Z., 2008. Biochemical mechanisms conferring cross-resistance between tebufenozide and abamectin in *Plutella xylostella*. *Pest. Biochem. Physiol. 91,* 175–179.

Rossiter, L., Gunning, R., McKenzie, F., 2008. Silver Anniversary of resistance management in the Australian cotton industry. An overview and the current situation for *Helicoverpa armigera*. Australian Cotton Conference 2008 p. 8. http://australiancottonconference. com.au/index.php?url=/cottons/showpage/3/179/198.

Sato, M.E., daSilva, M.Z., Raga, A., de Souza Filho, M.F., 2005. Abamectin resistance in *Tetranychus urticae* Koch (Acari:Tetranychidae): selection, cross-resistance and stability of resistance. *Neotrop. Entomol. 34,* 991–998.

Stavrinides, M.C., Hadjistylli, M., 2009. Two-spotted spider mite in Cyprus: ineffective acaricides, causes and considerations. *J. Pest Sci. 82,* 123–128.

Wolstenholme, A.J., Rogers, A.T., 2005. Glutamate-gated chloride channels and the mode of action of the avermectin/milbemycin anthelmintics. *Parasitology 131,* S85–S95.

Zhao, J.-Z., Collins, H.L., Li, Y.-X., Mau, R.F.L., Thompson, G.D., Hertlein, M., Andaloro, J.T., Boykin, R., Shelton, A.M., 2006. Monitoring of diamondback moth (Lepidoptera: Plutellidae) resistance to spinosad, indoxacarb, and emamectin benzoate. *J. Econ. Entomol. 99,* 176–181.

4 Spider Toxins and their Potential for Insect Control

**F Maggio, B L Sollod, H W Tedford, and
G F King**, University of Connecticut Health
Center, Farmington, CT, USA

© 2005, Elsevier BV. All Rights Reserved.

4.1. Introduction

Insects are the most diverse and successful animals on the planet, with the number of extant species estimated to be ~5 million (Novotny *et al.*, 2002). However, the ability of insects to inhabit a wide variety of ecological niches has inevitably brought them into conflict with humans. Although only a small minority of insects are classified as pests, they nevertheless destroy 20–30% of the world's food supply (Oerke, 1994) and transmit a diverse array of human and animal pathogens.

In terms of agronomic importance, lepidopterans are the most pernicious insects, with ~40% of all chemical insecticides directed against heliothine species (Brooks and Hines, 1999). However, from a global human health perspective, mosquitoes are the most problematic arthropods, being responsible for the transmission of malaria, filariasis, and numerous arboviruses. Although malaria is unquestionably the most devastating insect-borne disease, causing an estimated 2 million deaths per year (Gubler, 1998), the increasing incidence of epidemics caused by mosquito-borne arboviruses (e.g., dengue, West Nile, Japanese encephalitis, yellow fever, and Rift Valley fever) is a mounting public health issue (Gubler, 2002). Other insects of significant public health importance include sandflies, tsetse flies, fleas, and triatomid bugs, which transmit the causative agents of leishmaniasis, trypanosomiasis (African sleeping sickness), plague, and Chagas disease, respectively (Gubler, 1998; Gratz, 1999).

4.2. Chemical Control of Insect Pests

Humans have been engaged in internecine chemical warfare with insects ever since Paris Green (copper acetoarsenite) was introduced as an insecticide by French grape growers in the 1870s (Winston, 1997; Dent, 2000). However, our chemical assault on insects dramatically intensified with the introduction of dichlorodiphenyltrichloroethane (DDT), the first synthetic organic pesticide, during World War II (Casida and Quistad, 1998). DDT was effective not only in the agricultural sector, but formed the basis of initially successful malaria eradication programs (Attaran *et al.*, 2000). The subsequent introduction of organophosphate and carbamate insecticides in the 1960s (Casida and Quistad, 1998) encouraged the prevailing view that chemical insecticides would prove to be a universal remedy for the control of insect pests.

This optimistic outlook was unwarranted for several reasons. First, chemical control subjects the insect population to Darwinian selection and the development of resistance is as inevitable as the evolution of bacterial resistance to overprescribed antibiotics. Moreover, the fact that the vast majority of insecticides act on one of just four nervous system targets – the voltage-gated sodium channel, the nicotinic acetylcholine receptor (nAChR), the γ-aminobutyric acid (GABA)-gated chloride channel, and acetylcholinesterase – promoted the development of cross-resistance to different families of insecticides (Brogdon and McAllister, 1998). As of 1992, more than 500 species of insects and mites, including 56 anopheline and 39 culicine mosquitoes, had developed resistance to one or more classes of insecticides (Feyereisen, 1995; Brogdon and McAllister, 1998). Second, most chemical insecticides are relatively broad-spectrum, and indiscriminate use can lead to catastrophic environmental, ecological, and human health-related problems. Because these chemicals kill nontarget arthropods, including natural enemies of the targeted pest, there are several instances where insecticide application has exacerbated rather than ameliorated the targeted insect-pest problem (Winston, 1997; Dent, 2000).

The publication in 1962 of Rachel Carson's landmark book *Silent Spring* (Carson, 1962) substantially raised public awareness of the potential adverse impacts of synthetic insecticides and ultimately led to a paradigm shift in the development and use of these chemicals. Deregistration of DDT and other organochlorines in the 1970s in the USA stimulated the agrochemical industry to develop a new generation of insecticides that have increased biological activity (leading to reduced application rates/amounts), enhanced specificity, reduced environmental persistence, and decreased mammalian toxicity (Dent, 2000). Nevertheless, several classes of widely used insecticides will soon be lost from the marketplace due to either resistance development or deregistration by regulatory authorities following retrospective review. Combined with more demanding registration requirements for new agrochemicals, this is likely to decrease the pool of effective chemical insecticides in the near future (Rose, 2001; Zaim and Guillet, 2002). Thus, there is a pressing need to isolate safe insecticidal compounds and to develop viable alternatives to chemical insecticides. In this chapter, the potential of natural insecticidal toxins derived from spider venoms to fuel developments in each of these areas will be reviewed.

4.3. Spiders and Their Venoms: A Brief Overview

Excluding insects, which are their primary prey, spiders are the most successful terrestrial invertebrates (Rash and Hodgson, 2002). There are about 38 000 described species (Platnick, 1997), with at least a similar number undescribed. The vast majority belong to the two major suborders Araneomorphae and Mygalomorphae. Araneomorphs (modern spiders) represent ~90% of the world's spiders. The primitive spiders, or mygalomorphs, use their silk in a rudimentary manner compared to modern web-weavers and they prey primarily on nonflying arthropods. This suborder includes most of the large spiders that have inspired popular myths and Hollywood movies, including the tarantulas (theraphosids) and trapdoors, which are both harmless to humans (Escoubas and Rash, 2004; Isbister and White, 2004), and the highly venomous Australian funnel-web spiders (Nicholson and Graudins, 2002; Isbister and White, 2004).

As for other venomous creatures, such as scorpions and cone snails, the venoms of most spiders are a heterogeneous mixture of inorganic salts, low molecular mass organic molecules (<1 kDa), small linear peptides (typically 2–4 kDa), disulfide-reticulated polypeptides (typically 3–8 kDa), and high molecular mass proteins including enzymes (>10 kDa) (Escoubas et al., 2000; Gomez et al., 2002; Rash and Hodgson, 2002; Kuhn-Nentwig et al., 2004; Tedford et al., 2004). These complex chemical cocktails have evolved for the primary purpose of killing or paralyzing arthropod prey, although some compounds may play an additional role in predigestion of the injected victim (Rash and Hodgson, 2002). In the following sections, a brief overview of the insecticidal peptide and polyamine components that have been isolated from spider venoms are provided.

4.4. Insecticidal Acylpolyamine Toxins from Spider Venom

Polyamine toxins are low molecular weight secondary metabolites found in the venom glands of spiders and wasps (Strømgaard et al., 2001). They appear to be a common component of both araneomorph (Aramaki et al., 1986; Grishin et al., 1989; Quistad et al., 1991) and mygalomorph (Fisher and Bohn, 1957; Gilbo and Coles, 1964; Skinner et al., 1990) venoms, where their primary function is to induce rapid paralysis of envenomated prey. Although first identified in spider venoms more than 40 years ago, they were not structurally and functionally characterized until the mid-1980s (Strømgaard et al., 2001; Rash and Hodgson, 2002; Mellor and Usherwood, 2004).

Acylpolyamine toxins typically contain an unbranched polyamine chain composed of two to four secondary amino groups separated from each other by two to eight methylene groups (Strømgaard *et al.*, 2001). One end of the chain usually terminates with a primary amino or guanidino group, while an aromatic moiety (such as a 2,4-dihydroxyphenylacetyl group or an indol-3-acetyl group) is typically attached to the other end via an amide bond (**Figure 1a**). Polyamine toxins are capable of low affinity, nonspecific interactions with a variety of proteins, and as such their pharmacology is complex and not amenable to description with a few generalizations. Although some polyamine toxins from wasp venom antagonize nAChRs, and argiotoxin-636 from the orb-weaving spider *Argiope lobata*

inhibits voltage-gated Na^+ and K^+ currents in vertebrate neurons, these toxins are thought to have evolved primarily as antagonists of arthropod ionotropic glutamate receptors (Strømgaard *et al.*, 2001; Mellor and Usherwood, 2004). While these toxins have proved to be valuable pharmacological tools, they have never been seriously exploited as insecticide leads, possibly due to undesirable characteristics such as poor phyletic specificity, undesirable pharmacokinetics, and difficulties with synthesis and purification (Strømgaard *et al.*, 2001). However, none of these characteristics absolutely rules out a possible success for insecticide discovery strategies that

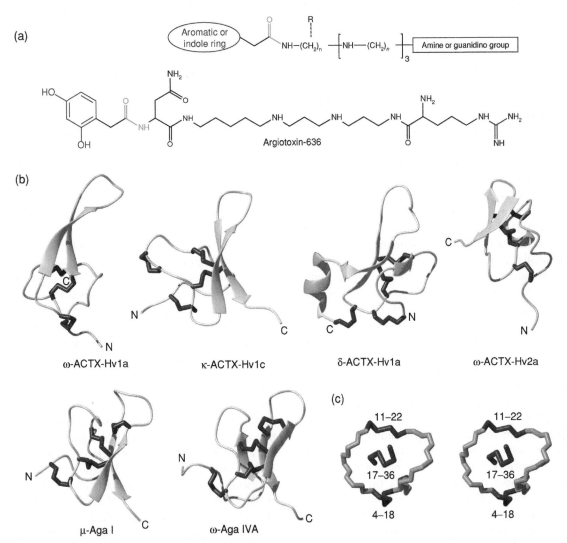

Figure 1 (a) The structure of argiotoxin-636 from the orb-weaving spider *Argiope lobata* below a generalized outline of the structure of spider venom acylpolyamines. (b) Schematic representations of the three-dimensional structure of the only six insecticidal spider toxins whose tertiary structure has been determined. The atracotoxins (ACTX) are from the Australian funnel-web spider *Hadronyche versuta* and the agatoxins (Aga) are from the unrelated American funnel-web spider *Agelenopsis aperta*. The β strands and α helices are shown in green and orange, respectively, and disulfide bridges are drawn as magenta tubes. The N- and C-termini of each toxin are labeled. Protein Data Bank accession codes are given in **Table 1**. (c) Stereo view of the cystine knot motif of ω-ACTX-Hv1a. The Cys17–Cys36 disulfide bond (magenta) is threaded through a closed loop formed by the Cys4–Cys18 and Cys11–Cys22 disulfide bridges (red) and the intervening sections of polypeptide backbone (blue).

exploit acylpolyamines either as rational design templates or as tracers in competitive ligand binding assays.

4.5. Insecticidal Peptide Toxins from Spider Venom

4.5.1. Spider Venom Polypeptides: An Underexploited Pharmacological Reservoir

Peptide toxins are expressed in the venom gland of spiders in a combinatorial fashion, and hence they tend to be the principal venom constituents (Escoubas et al., 2000; Escoubas and Rash, 2004; Kuhn-Nentwig et al., 2004; Tedford et al., 2004) (see Section 4.5.2). However, few spider venoms have been studied in sufficient detail to enable the extent and diversity of these peptidic components to be reliably estimated. A recent mass spectrometric analysis of 55 different tarantula venoms led to a conservative estimate of ~50 peptides per venom (Escoubas and Rash, 2004), and a similar level of peptide diversity has been reported in the venoms of the primitive hunting spider *Plectreurys tristis* (Quistad and Skinner, 1994), the Australian funnel-web spider *Hadronyche versuta* (Wang et al., 1999; Tedford et al., 2004), and the Central American wandering spider *Cupiennius salei* (Kuhn-Nentwig et al., 2004). If one assumes a total of ~80 000 spider species, then a conservative estimate of 20 pharmacologically distinct peptides per species leads to an estimated total of ~1.5 million spider venom polypeptides, which is a much larger pharmacological repertoire than the 50 000–100 000 peptides estimated to be present in the venoms of cone snails (Olivera and Cruz, 2001) and scorpions (Possani et al., 1999). It was recently estimated, based on available journal and patent literature, that <0.01% of the pharmacological diversity of spider venoms has been tapped thus far (Tedford et al., 2004).

Apart from studies of acylpolyamines, most early investigations of spider venoms focused on isolation of polypeptides with vertebrate neurotoxic activity. Some of these peptides, such as the ω-agatoxins (Olivera et al., 1994), have become invaluable tools in vertebrate pharmacology because of their selectivity for specific subtypes of vertebrate voltage-gated ion channels. Although genes encoding these peptides might be suitable for engineering recombinant insect viruses (see Section 4.6), their broad-spectrum toxicity makes them unsuitable as leads for chemical insecticides. However, the overwhelming majority of spiders are harmless to humans (Isbister and White, 2004), and therefore it is not surprising that appropriately designed venom screens have led to the discovery of insect-specific spider venom polypeptides (Skinner et al., 1992; Krapcho et al., 1995a, 1995b, 1997; Fletcher et al., 1997b; Johnson and Kral, 1997; Johnson et al., 1997, 1998; Atkinson et al., 1998; Balaji et al., 2000; Corzo et al., 2000; Wang et al., 2000, 2001; de Figueiredo et al., 2001; Lipkin et al., 2002; Zhang et al., 2003).

4.5.2. Expression and Evolution of Spider Venom Polypeptides: Rolling the Dice

Toxic polypeptides are found in the venoms of numerous animals, including scorpions, snakes, and cone snails, as well as the nematocysts of sea anemones. While these peptides are invariably expressed as longer precursors that are posttranslationally processed to yield the mature toxin, the nature of the precursor depends on the animal. In the aquatic cone snails and sea anemones, toxic polypeptides are initially translated as prepropeptides that are posttranslationally processed to yield the mature toxin (Olivera et al., 1999; Anderluh et al., 2000). In contrast, the precursors of most peptide toxins from snakes and scorpions lack a propeptide region (Goudet et al., 2002; Fujimi et al., 2003). Araneomorph and mygalomorph spiders generally conform to the prepropeptide paradigm (Wang et al., 2001; Gomez et al., 2002; Corzo et al., 2003). The precursor typically comprises an N-terminal signal sequence (16–25 residues) followed by a propeptide region of highly variable length (13–58 residues) which precedes a single downstream copy of the mature toxin sequence. For reasons that have yet to be determined, the propeptide region of most spider toxin precursors is rich in acidic residues (Santos et al., 1992; Diniz et al., 1993; Krapcho et al., 1995b; Wang et al., 2001; Liang, 2004; Ostrow et al., 2003).

Spider toxins are typically not produced as "one-offs." Rather, each toxin is generally expressed in the form of a small library of variants that can differ significantly or by as little as one amino acid in the mature toxin sequence (Skinner et al., 1989, 1992; Stapleton et al., 1990; Quistad and Skinner, 1994; Krapcho et al., 1995b; Sanguinetti et al., 1997; Johnson et al., 1998; Kalapothakis et al., 1998; Wang et al., 1999, 2000; Corzo et al., 2000, 2003; Lipkin et al., 2002; Kuhn-Nentwig et al., 2004; Liang, 2004). Recent comprehensive analyses of venom gland cDNA libraries from several species of Australian funnel-web spider indicate that a single species typically expresses three to ten variants of each peptide toxin (Sollod et al., 2003). Interspecies comparisons reveal that the mature toxin region has been poorly conserved during speciation whereas the signal sequences of the variants are nearly identical. This dichotomy between a highly conserved signal sequence and highly variable mature toxin sequence suggests that spiders, like cone snails (Duda and Palumbi, 1999; Olivera *et al.*, 1999; Conticello *et al.*, 2001), have diversified their toxin pool by developing a mechanism for accelerated evolution (hypermutation) of the mature toxin region relative to the signal sequence. Remarkably, the cysteine residues, which are critical for determining the tertiary structure of the toxins, are strictly conserved within the otherwise hypervariable mature toxin region.

Thus, spiders are master combinatorial chemists; for hundreds of millions of years they have been synthesizing combinatorial peptide libraries using various cystine frameworks as a structural scaffold. Hypermutation of the mature toxin sequence has facilitated the evolution and optimization of toxins with new functionalities, while maintenance in the venom of a minilibrary of toxin variants presumably enables the spider to target slightly altered versions of the same receptor in different insects, or even differentially edited/spliced versions of the same receptor within a single insect.

4.5.3. Three-Dimensional Structure of Spider Venom Polypeptides: A Tale of Ropes and Knots

While it is difficult to make structural generalizations about spider venom polypeptides based on the small number of toxins that have been studied, some trends are apparent. The peptides that have been isolated thus far can be grouped into three categories:

1. Linear amphipathic peptides, typically of mass 2–4 kDa, with no disulfide bonds.
2. Medium-sized neurotoxic peptides, typically of mass 3–6 kDa and reticulated with three to five disulfide bonds.
3. Larger polypeptides (6–9 kDa) with four or more disulfide bonds.

There are few characterizations of the highly reticulated peptides with mass >6 kDa, and no three-dimensional structures have been determined for any member of this group. An extensive mass spectrometric analysis of 55 tarantula venoms (see figure 2b in Escoubas and Rash, 2004), as well as less detailed mass spectrometric analyses of Australian funnel-web spiders (Wang, 2000; Wilson, 2001), suggests that the medium-sized neurotoxic peptides predominate in mygalomorph venoms. However, venom from the araneomorph *C. salei* appears to be markedly different: small amphipathic peptides appear to be the principal constituents of this venom, which also contains a higher proportion of high-mass peptides and less of the medium-sized

neurotoxic peptides than mygalomorph venoms (Kuhn-Nentwig *et al.*, 2004).

The small linear peptides expressed in spider venoms have antimicrobial activity and in this respect they are similar to the cecropin-like peptides that are found in the hemolymph of numerous arthropods, where they form part of the innate immune system that protects against invasion by pathogenic organisms (Boman, 2003). However, the peptides found in the venoms of spiders and other invertebrates, including bees, wasps, ants, and scorpions, differ from the cecropins in that their antimicrobial activity appears to be a consequence of relatively nonspecific cytolytic activity against prokaryotic and eukaryotic cells. While these peptides appear to form amphipathic α helices in membrane-mimetic solvents or when bound to biological membranes, they exist in flexible "rope-like" conformations in the absence of lipid bilayers (Corzo *et al.*, 2001, 2002; Kuhn-Nentwig *et al.*, 2004). Some of these peptides are neurotoxic to insects, but their primary function appears to be to synergize the neurotoxic activity of the larger disulfide-reticulated peptides by optimizing their access to the insect nervous system (Corzo *et al.*, 2002; Kuhn-Nentwig, 2003; Kuhn-Nentwig *et al.*, 2004). The broad-spectrum activity of the linear cytolytic peptides makes them unsuitable as insecticide leads, although it has been suggested that genes encoding these toxins might synergize with neurotoxin transgenes in enhancing the efficacy of insect-specific viruses (Corzo *et al.*, 2002; Kuhn-Nentwig, 2003).

The majority of insect-specific toxins isolated from spider venom are medium-sized neurotoxic polypeptides (typically 30–50 amino acid residues) with three to five disulfide bonds (summarized in **Table 1**). Most of these toxins act presynaptically at insect neuromuscular junctions, either by enhancing or inhibiting neurotransmitter release, although several act at sites within the insect central nervous system (Johnson *et al.*, 1998; Bloomquist, 2003; Tedford *et al.*, 2004). Three-dimensional structures have been determined for six of these toxins (**Figure 1b**), and they all contain a structural motif known as the inhibitor cystine knot (ICK) (Pallaghy *et al.*, 1994; Norton and Pallaghy, 1998; Craik *et al.*, 2001). The cystine knot comprises a ring formed by two disulfides and the intervening sections of polypeptide backbone, with a third disulfide piercing the ring to create a pseudo-knot (see **Figure 1c**). Except for the special case of cyclic ICK peptides, cystine knots are not true topological knots since, at least theoretically, they can be untied by a nonbond-breaking geometrical transformation (Craik *et al.*, 2001). Nevertheless, the cystine knot

Table 1 Overview of neurotoxic insecticidal peptides isolated from spider venom

Spider	Name of toxin[a]	Number of residues	Number of disulfide bonds	Molecular target[b]	Discovery reference	Insect selective?	Three-dimensional fold[c]	Structure reference	Insectophore mapped?
Agelenopsis aperta	μ-Aga I[d]	36	4	VGSC	1	Yes/No	1EIT	2	No
	ω-Aga IA	66 + 3[e]	5	VGCC	3	No	No		No
	ω-Aga IIA	?[f]	?[f]	VGCC	3	No	No		No
	ω-Aga IVA	48	4	VGCC	4	No	1OAW	5	No
Aptostichus schlingeri	Aptotoxin I	74	4	Unknown	6	Unknown	No		No
	Aptotoxin III	38	4	Unknown	6	Unknown	No		No
	Aptotoxin VII	33	3	Unknown	6	Unknown	No		No
Calisoga sp.	Peptide A	39	4	Unknown	7	Yes	No		No
Coremiocnemis validus	Covalitoxin-II	31	3	Unknown	8	Yes	No		No
Cupiennius salei	CSTX-1	74	4	VGCC	9	No	No		No
Diguetia canities	DTX 9.2	56	4	VGSC	10, 11	Yes	No		No
Hadronyche versuta	ω-ACTX-Hv1a	37	3	VGCC	12	Yes	1AXH	12	Yes[g]
	ω-ACTX-Hv2a	45	3	VGCC	13	Yes	1G9P	13	No
	κ-ACTX-Hv1c	37	4	K$^+$ channel	14, 15	Yes	1DL0	14	No
	δ-ACTX-Hv1a[i]	42	4	VGSC	16, 17	No	1VTX	18	Yes[h]
Macrotheles gigas	Magi-2	40	3	VGSC	19	Yes	No		No
	Magi-3	46	5	VGSC	19	Yes	No		No
	Magi-5	29	3	VGSC	19	No	No		No
	Magi-6	36	4	Unknown	19	No	No		No
Oxyopes kitabensis	Oxytoxin 1	69	5	VGSC	20	Yes	No		No
Phoneutria nigriventer	Tx4(6-1)	48	5	VGSC	21, 22	Yes	No		No
Plectreurys tristis	Plectoxin II	44[j]	5	VGCC	23	Unknown	No		No
	Plectoxin V	46	5	Unknown	24	Unknown	No		No
	Plectoxin IX	46	4	Unknown	24	Unknown	No		No
	Plectoxin X	49	5	Unknown	24	Unknown	No		No
Segestria florentina	SFI1[k]	46	4	Unknown	25	Unknown	No		No
Selenocosmia huwena[l]	Huwentoxin-V	35	3	Unknown	26	Yes	No		No
Tegenaria agrestis	TalTX-1	50	3	Unknown CNS target	27	Yes	No		No

[a]If a venom contains several variants of a toxin, only the protoypic family member is listed.

[b]Abbreviations used: VGSC, voltage-gated sodium channel; VGCC, voltage-gated calcium channel; CNS, central nervous system.

[c]The Protein Data Bank accession number is given for toxins whose three-dimensional structure has been determined.

[d]Agelenopsis aperta contains several variants of this toxin (μ-Aga II to VI) (Skinner et al., 1989). Similar toxins are found in the venom of the related American funnel-web spider Hololena curta (Stapleton et al., 1990) and Paracoeletes luctuosus (Corzo et al., 2000). Some of these toxins appear to be insect-selective, while others are neurotoxic to mice (Corzo et al., 2000).

[e]This toxin is a heterodimer in which 66- and 3-residue peptides are linked by an intermolecular disulfide bond (Santos *et al.*, 1992).

[f]This 11 kDa toxin has never been fully sequenced (Adams *et al.*, 1990).

[g]Tedford *et al.* (2001, 2004).

[h]Maggio and King (2002).

[i]A similar toxin (Magi 4) is found in the venom of *Macrotheles gigas* (Corzo *et al.*, 2003).

[j]The C-terminal Thr residue in PLTX-II is amidated and *O*-palmitoylated (Branton *et al.*, 1993).

[k]A toxin with similar N-terminal sequence was isolated from *Segestria* sp. many years earlier and shown to be insect-selective, but the complete amino acid sequence was not reported (Johnson and Kral, 1997).

[l]Reclassified as *Omithoctonus huwena* (Wang) in a recent taxonomic revision.

[1]Adams *et al.* (1989).

[2]Omecinsky *et al.* (1996).

[3]Adams *et al.* (1990).

[4]Mintz *et al.* (1992).

[5]Kim *et al.* (1995).

[6]Skinner *et al.* (1992).

[7]Johnson *et al.* (1997).

[8]Balaji *et al.* (2000).

[9]Kuhn-Nentwig *et al.* (2004).

[10]Krapcho *et al.* (1995b).

[11]Bloomquist *et al.* (1996).

[12]Fletcher *et al.* (1997b).

[13]Wang *et al.* (2001).

[14]Wang *et al.* (2000).

[15]Gunning, Maggio, King, and Nicholson (unpublished data).

[16]Brown *et al.* (1988).

[17]Nicholson *et al.* (1996).

[18]Fletcher *et al.* (1997a).

[19]Corzo *et al.* (2003).

[20]Corzo *et al.* (2002).

[21]Figueiredo *et al.* (1995).

[22]de Lima *et al.* (2002).

[23]Branton *et al.* (1993).

[24]Quistad and Skinner (1994).

[25]Lipkin *et al.* (2002).

[26]Zhang *et al.* (2003).

[27]Johnson *et al.* (1998).

provides peptidic neurotoxins with tremendous stability and resistance to proteases, thereby ensuring that the injected toxin is not degraded before reaching its nervous system target.

4.6. Toxin-Based Target Validation and Screening

Surprisingly, the molecular target of about 50% of the neurotoxins listed in **Table 1** has not been determined. However, it is striking that all spider venom-derived peptide neurotoxins with known molecular targets are directed against ion channels. Ion channels are well-validated insecticide targets (Bloomquist, 1996), with the vast majority of synthetic insecticides acting on voltage-gated sodium channels (the target of DDT, pyrethroids, oxadiazines, dihydropyrazoles, veratrum alkaloids, and N-alkylamides), nicotinic acetylcholine receptors (the target of neonicotinoids and spinosyns), or GABA/glutamate-gated chloride channels (the target of dieldrin, lindane, endosulfan, avermectins, and phenylpyrazoles). Of the major insecticide classes, only the organophosphate and carbamate insecticides (which target acetylcholinesterase) are not directed against this group of ligand- and voltage-gated ion channels.

Clearly, in order to minimize the potential for cross-resistance, it would be highly advantageous for new insecticides to act on targets outside of this established ion channel triumvirate. While the majority of well-characterized spider toxins appear to act on voltage-gated sodium channels (VGSCs), the overutilization of this target makes these toxins the least interesting in terms of developing high-throughput screens (HTSs) for small-molecule insecticide leads. It should be noted, however, that at least some of these toxins (e.g., δ-ACTX-Hv1a (Nicholson et al., 1996) and Magi 2, 3, and 5 (Corzo et al., 2003)) bind VGSCs at sites distinct from current-generation insecticides. Thus, the high penetrance in many insect populations of alleles conferring resistance to certain VGSC insecticides may not compromise newly developed chemical control agents that utilize different sodium channel binding sites (Bloomquist, 1996). In fact, insects with these resistance alleles might have increased susceptibility to such chemicals since insects with pyrethroid-resistance alleles are significantly more susceptible to both N-alkylamides and scorpion toxins that bind VGSCs at sites different from the pyrethroids (Zlotkin et al., 2000).

It is apparent from **Table 1** that many of the most potent spider venom toxins target neuronal voltage-gated calcium channels (VGCCs). This is not particularly surprising since insect VGCCs play a critical role in modulating cellular excitability and neurotransmitter release, and severe loss-of-function mutations in *Drosophila melanogaster* genes encoding the pore-forming α_1 subunits of both the L-type (*Dmca1D*) and P/Q/N-type (*Dmca1A*) VGCCs are embryonic lethal (Eberl et al., 1998; Smith et al., 1998; Kawasaki et al., 2002) (note that the designation of L- and P/Q/N-type channels is based solely on gene homologies with vertebrate VGCCs, not electrophysiological or pharmacological criteria). It is therefore remarkable that no commercial insecticides have ever been targeted against VGCCs. The lack of phyletic specificity of the ω-agatoxins possibly encouraged the misconception that insect VGCCs could not be selectively targeted. However, the highly insect-selective ω-atracotoxins (King et al., 2002; Tedford et al., 2004) demonstrate that such specificity is indeed possible. Hence, as suggested previously (Hall et al., 1994), insect VGCCs appear to be excellent HTS targets. The major impediment to implementing such HTSs will be development of heterologous expression systems for the relevant insect VGCCs. Thus far, there are no examples in the scientific or patent literature of heterologous expression of a functional arthropod VGCC.

An exciting recent development is the observation that the κ-atracotoxins (previously known as the Janus-faced atracotoxins) (Wang et al., 2000; King et al., 2002; Maggio and King, 2002) are selective for insect K$^+$ channels (Tedford et al., 2004). As far as we are aware, no synthetic chemical insecticides have ever been developed that target K$^+$ channels. Isolation of a lethal invertebrate-specific K$^+$ channel blocker was somewhat unexpected, since all loss-of-function K$^+$ channel alleles in *Drosophila* are viable (Wicher et al., 2001). Nevertheless, induction of a κ-ACTX-Hv1c transgene in *D. melanogaster* for just 10–15 min induces a lethal phenotype (Maggio, King, and Reenan, unpublished data), which strongly validates this ion channel as a viable insecticide target. In contrast with insect VGCCs, heterologous expression systems are available for numerous invertebrate K$^+$ channels (Wicher et al., 2001), and many HTS formats have been developed for K$^+$ channel targets (Kirsch et al., 2001; Ford et al., 2002).

An important consideration in implementing high-throughput insecticide screens with these ion channels is that one would like to mimic the phyletic specificity as well as the potency of the toxin used to validate the target. This contrasts with pharmaceutical development, where one can simply look for potent channel blockers (or activators) without regard for phyletic specificity. Ion channels are typically highly promiscuous with respect to available binding sites, as exemplified by VGSCs which display at least nine topologically distinct binding sites for toxins and insecticides (Zlotkin *et al.*, 2000). Thus, a screen designed to simply select for potent channel blockers has a high probability of yielding phyletically indiscriminate lead compounds that bind the channel at a location distinct from the binding site for the toxin used to validate the target. The probability of selecting lead compounds that are insect-selective is likely to be significantly increased if the primary screen is developed to identify molecules that compete with the toxin for its specific channel binding site; alternatively, a noncompetitive primary screen could be followed by a secondary screen that selects for such competitors.

4.7. Toxin-Based Bioinsecticides: A New Era of Biological Control

4.7.1. A Brief History of Conventional Biological Control Methods

Biological control methods were introduced well before the advent of synthetic insecticides, and in many instances were extremely successful. Indeed, the first use of biological control in North America was a spectacular success (Caltagirone and Doutt, 1989; Winston, 1997). In the 1880s, a massive outbreak of the accidentally introduced cottony-cushion scale (*Icerya purchasi*) threatened to decimate the infant Californian citrus industry. Over the winter of 1888–89, at a cost of only $1500, the US Department of Agriculture released the Australian ladybird beetle, *Rodolia cardinalis*, a voracious and specific predator of the cottony-cushion scale. Within a matter of months, the problematic scale was completely controlled within the introduced areas. Unfortunately, the introduction of broad-spectrum synthetic insecticides in the 1940s substantially diminished the success rate in subsequent predator/parasite release programs, largely because the introduced predators and parasites were just as susceptible to insecticides as the targeted pest insect. Ironically, cottony-cushion scale reappeared as a problem in California in the 1940s when citrus growers began experimenting with DDT and other insecticides that failed to discriminate between *Rodolia* and scale insects (Winston, 1997).

Another biological control method that has been used with variable success is the release of sterile insects. In this approach, male insects are sterilized, usually by irradiation, and released in large excess over the native male population. A textbook case was the total eradication of screwworm, a major pest of goats, from the Venezuelan island of Curaçao

in the 1950s (Winston, 1997). However, Curaçao is small and well separated from the mainland and neighbouring islands, and very few sterile insect release (SIR) programs have worked when applied to less ecologically isolated areas. In part, this is because the cost of traditional SIR programs are astronomical compared with predator release programs due to the massive rearing facilities that need to be built and maintained to enable sterile males to be produced and released in requisite excess over the native population (Winston, 1997).

Other biological control methods include the release of insect pathogens (especially viruses, which are discussed in more detail below) (see Section 4.7.2 and **Chapter 5**), insect pheromones, and entomopathogenic fungi (reviews: Hall and Menn, 1999; Copping and Menn, 2000).

4.7.2. Biological Control in the New Millennium

The genomics era has ushered in new methods of biological control. We now have the capacity to genetically manipulate insects and insect pathogens to have desired traits (Atkinson, 2002). The potential benefits for insect release programs have been reviewed recently (Handler, 2002; Benedict and Robinson, 2003). The ability to engineer male sterility by genetic methods should increase success in future SIR programs by substantially improving the fitness level of released males. Moreover, genetic sexing techniques (Robinson, 2002) should enable facile elimination of females from the release population, which will enhance SIR by increasing the number of matings between sterile males and wild females.

An even more exciting development is the possibility of releasing insects carrying either a conditional lethal trait or female-killing allele (Handler, 2002). Both of these techniques, under optimal conditions, should be far superior to traditional SIR (Schliekelman and Gould, 2000a, 2000b). A conditional lethal trait that has been tested in *Drosophila*, but which has the advantage of being applicable to virtually all insect species that can be genetically manipulated, is temperature-sensitive alleles of genes encoding either diphtheria or ricin toxins. The insects are engineered to express only the A subunit of the toxin; in the absence of the B subunit that normally allows transmembrane movement, toxicity is cell-specific, thus preventing damage to predators. In principle, genes encoding insect-specific spider toxins could be used in a similar manner. It has been shown that the ω- and κ-atracotoxins can be expressed under heat shock control in *Drosophila*; insects are phenotypically normal at 18 °C but a heat shock to 37 °C for just 10–15 min kills all flies (Tedford, Maggio, King, and Reenan, unpublished data). Since the toxins are not orally active, insects bearing these alleles would pose no threat to predatory organisms.

An alternative to SIR and parasite/predator release is incorporation of a transgene encoding a specific insecticidal component into the genome of plants or insect-specific pathogens. One recently introduced, and thus far highly successful, approach is the planting of transgenic crops that express insecticidal toxins, such as engineered corn and cotton crops that express δ-endotoxins from the soil bacterium *Bacillus thuringiensis* (Peferoen, 1997; Jenkins, 1999). In addition to minimizing effects on nontarget organisms, *Bt* plants have generated increases in crop yields and significantly reduced insecticide use on many farms (Huang *et al.*, 2002; Pray *et al.*, 2002; Toenniessen *et al.*, 2003). However, this technology has several potential drawbacks. First, constitutive expression of an insect toxin in transgenic plants is likely to expedite the development of resistance (Ferré and Van Rie, 2002; Morin *et al.*, 2003), thus limiting the long-term benefit of the toxin transgene and compromising the effectiveness of *Bt* spores used widely by organic farmers. Second, some consumers are uncomfortable with, or philosophically opposed to, the introduction of nonplant proteins into their food (for a discussion of the emotive "Frankenfood" pejorative, see Dale (1999a, 1999b)). An additional concern associated with this technology is the possibility of transgene introgression into wild relatives of the engineered plant (Arnaud *et al.*, 2003).

Bt endotoxins are uniquely suited to incorporation into transgenic plants as their activation is dependent upon the alkaline environment found in the gut of many insects, which ensures that these toxins are not active in humans and other vertebrates. In principle, the remarkable phyletic specificity of many spider-derived peptide neurotoxins makes them ideal for incorporation into transgenic plants. It was recently shown, for example, that rice transformed with a transgene encoding an insecticidal spider toxin was significantly less susceptible to infestation by the rice leaf-folder *Cnaphalocrocis medinalis* and striped stemborer *Chilo suppressalis* (Huang *et al.*, 2001). However, most spider-derived polypeptide neurotoxins are not active when fed to insects, although it is possible that oral activity could be engineered by cofeeding with an agent that increases insect gut permeability, such as cytolytic peptides or even *Bt* endotoxins.

An alternative bioinsecticide strategy that is more specific, and which obviates the problem of introducing a foreign protein into the food supply, is the release of insect-specific viruses. Although there are many different types of insect viruses, all studies of this technology to date have focused on arthropod-specific baculoviruses, partly because of their proven inability to infect vertebrates (Gröner, 1986; Black *et al.*, 1997; Kost and Condreay, 2002; Herniou *et al.*, 2003). For most members of the nucleopolyhedrovirus (NPV) subgroup, infectivity is restricted to a few closely related lepidopterans, although some viruses only replicate efficiently in a single species (Black *et al.*, 1997; Federici, 1999) (see **Chapter 5**). Thus, toxin exposure is limited to the targeted insect, while inculpable populations that are susceptible to chemical pesticides, including predators/parasites of the targeted insect, are unaffected.

Wild-type NPVs have become a key component of integrated pest management programs in several countries. They are used extensively in Brazil to protect soybean crops from the velvetbean caterpillar *Anticarsia gemmatalis* (Moscardi, 1999), and in China to safeguard cotton from the highly insecticide-resistant *Helicoverpa armigera* bollworm (Sun *et al.*, 2002). In most situations, however, wild-type NPVs are not competitive with chemical pesticides because they typically take 5–10 days to kill their hosts, during which time the insect continues to feed and cause crop damage (Bonning and Hammock, 1996; Black *et al.*, 1997). This shortcoming has been addressed by engineering recombinant baculoviruses that express heterologous insecticidal neurotoxins (Cory *et al.*, 1994; Treacy, 1999; Inceoglu *et al.*, 2001). Introduction of transgenes encoding peptidic neurotoxins from mites (Tomalski and Miller, 1991; Popham *et al.*, 1997), scorpions (McCutchen *et al.*, 1991; Stewart *et al.*, 1991; Gershburg *et al.*, 1998; Sun *et al.*, 2002), spiders (Prikhod'ko *et al.*, 1996, 1998; Hughes *et al.*, 1997), and sea anemones (Prikhod'ko *et al.*, 1996, 1998) have all been demonstrated to reduce the time interval between virus application and cessation of feeding or death (**Figure 2**). Importantly, introduction of a toxin transgene does not appear to alter the intrinsic infectivity of the virus or its natural host range (Black *et al.*, 1997). (For more in depth discussion of naturally occurring and genetically modified baculoviruses see **Chapter 5**.)

Few attempts have been made to enhance the insecticidal efficacy of baculoviruses by incorporation of spider toxin transgenes. This is rather surprising since, as outlined in **Figure 2**, the most dramatic improvement obtained thus far in the insecticidal activity of *Autographa californica* NPV (AcNPV) resulted from incorporation of a transgene encoding μ-agatoxin IV from the American funnel-web spider

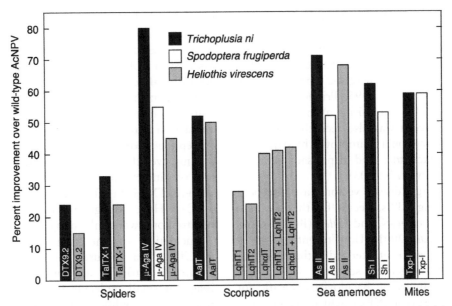

Figure 2 Histogram displaying the percent improvement (relative to wild-type virus) in the time required for 50% of a test population to be killed/paralyzed after feeding on *Autographa californica* nucleopolyhedrosis virus (AcNPV) engineered to contain transgenes encoding various arachnid or sea anemone toxins. It was difficult to make valid comparisons between studies because of differences in the manner of transgene insertion, choice of promoter driving the transgene, toxin signal sequence, and method of bioassay. Thus, this comparison is restricted to studies that used similar AcNPV constructs and which tested the virus on tobacco budworm (*Heliothis virescens*, light gray bars), fall armyworm (*Spodoptera frugiperda*, white bars), or cabbage looper (*Trichoplusia ni*, black bars). In studies where a number of different constructs of the same transgene were tested, the construct that gave the best improvement over wild-type virus is reported here. Toxins are grouped taxonomically and their names are indicated on the histogram. (Data are from Lu, A., Seshagiri, S., Miller, L.K., **1996**. Signal sequence and promoter effects on the efficacy of toxin-expressing baculoviruses as biopesticides. *Biol. Control 7*, 320–332; Hughes, P.R., Wood, H.A., Breen, J.P., Simpson, S.F., Duggan, A.J., *et al.*, **1997**. Enhanced bioactivity of recombinant baculoviruses expressing insect-specific spider toxins in lepidopteran crop pests. *J. Invertebr. Pathol. 69*, 112–118; Popham, H.J.R., Prikhod'ko, G.G., Felcetto, T.J., Ostlind, D.A., Warmke, J.W., *et al.*, **1998**. Effect of deltamethrin treatment on lepidopteran larvae infected with baculoviruses expressing insect-selective toxins μ-Aga-IV, As II, or Sh I. *Biol. Control 12*, 79–87; Prikhod'ko, G.G., Popham, H.J.R., Felcetto, T.J., Ostlind, D.A., Warren, V.A., *et al.*, **1998**. Effects of simultaneous expression of two sodium channel toxin genes on the properties of baculoviruses as biopesticides. *Biol. Control 12*, 66–78; and Regev, A., Rivkin, H., Inceoglu, B., Gershburg, E., Hammock, B.D., *et al.*, **2003**. Further enhancement of baculovirus insecticidal efficacy with scorpion toxins that interact cooperatively. *FEBS Lett. 537*, 106–110.)

Agelenopsis aperta. However, a complicating factor in comparing outcomes from different studies (apart from different bioassay strategies, as discussed in Treacy (1999)) is that the efficiency with which a particular toxin is expressed and folded in virus-infected insects is generally not determined. A highly potent insecticidal neurotoxin could be ineffective as a baculoviral transgene because the toxin fails to be properly processed and/or folded in virus-infected insects. This is a critical issue because all of the arachnid and sea anemone toxins used to enhance baculoviral efficacy contain three to seven disulfide bonds, and the number of possible disulfide isomers increases dramatically as the number of cysteine residues increases; for example, six cysteines can be paired in 15 different ways to form three disulfide bonds, whereas eight cysteines can be covalently linked in 105 different ways to form four disulfide bridges. Thus, if a toxin with four disulfide bonds folded in a completely stochastic manner, the yield of correctly

folded toxin would be <1%. In this respect, it is worth noting that heterologous expression systems have not been developed for any of the spider toxin genes previously used to generate recombinant baculoviruses, and hence it is unknown whether significant amounts of properly folded, bioactive spider toxin are produced in virus-infected insects.

The ω- and κ-atracotoxins appear to be excellent prospects for engineering viral bioinsecticides. They are the only insect-specific spider toxins for which heterologous expression systems have been developed (Tedford *et al.*, 2001; Maggio and King, 2002), they fold efficiently even in the absence of cellular folding enzymes (Wang *et al.*, 2000, 2001; Tedford *et al.*, 2001), and bioactive toxin is expressed when transgenes encoding these toxins are introduced into *D. melanogaster* (Tedford, Maggio, King, and Reenan, unpublished data). Our preliminary studies indicate that AcNPV expressing either ω-ACTX-Hv1a or κ-ACTX-Hv1c transgenes kills *Heliothis virescens* more quickly than wild-type virus; moreover, potency can be significantly improved by expressing both toxins simultaneously in the pest insect (Maggio, Mukherjee, and King, unpublished data). Coexpression of synergistic toxin combinations (Herrmann *et al.*, 1995; Prikhod'ko *et al.*, 1998) appears to be a promising avenue for future investigations since it was also recently noted that baculoviral efficacy could be improved by synergistic expression of excitatory and depressant scorpion neurotoxins (Regev *et al.*, 2003).

Viruses that target disease vectors have also been isolated, including NPVs specific for the mosquito genera *Aedes*, *Anopheles*, and *Culex* (Becnel *et al.*, 2001; Moser *et al.*, 2001). Until recently, none of these NPVs had been characterized at the molecular level. However, the genome sequence was recently determined for an exotic baculovirus specific for *Culex* mosquitoes (CuniNPV), which are key vectors of St. Louis encephalitis, Eastern equine encephalitis, and West Nile virus in the United States (Afonso *et al.*, 2001). This work opens up the possibility of incorporating spider toxin transgenes into CuniNPV and related mosquito viruses in order to improve their efficacy.

4.8. Spider Toxins as Templates for Mimetic Design

In general, polypeptides are not good drugs because of their poor pharmacokinetic properties (Latham, 1999). Linear peptides are typically not even good drug leads because their interconversion between a plethora of different solution conformations makes it difficult to elucidate meaningful structure–activity relationships (Kieber-Emmons *et al.*, 1997). In contrast, conformationally constrained peptides and peptidomimetic libraries have emerged as valuable tools in the drug discovery process. The cystine knot toxins found in spider venoms are more akin to these libraries with one exception: the spider toxins come preoptimized for the desired target. Cystine knot toxins essentially comprise a robust structural scaffold (the cystine framework) onto which four loops are tethered, each bounded by half-cystines (Tedford *et al.*, 2001). As outlined above (see Section 4.5.2), spiders have been performing combinatorial chemistry with this framework on an evolutionary timescale, randomly changing residues in the intercystine loops and selecting those peptides that are most effective in helping to kill or paralyze their insect prey.

Thus, in principle, the long evolutionary history of ICK spider toxins should ensure that they are better leads than those resulting from screens of peptide or peptidomimetic libraries. Why then have they not been utilized for rational insecticide design? The answer is obvious from a quick glance at **Table 1**. For rational design, one needs to know not just the three-dimensional structure of the toxin, but also the spatial arrangement of residues that are critical for interaction with the toxin target. Although dozens of peptide toxins have been isolated from spider venoms over the past 15 years, the target binding site, or "insectophore," has only been elucidated for ω-ACTX-Hv1a (Tedford *et al.*, 2001, 2004) and κ-ACTX-Hv1c (Maggio and King, 2002), and these insectophores were only mapped in the past 2 years.

Elucidation of the target binding sites for ω-ACTX-Hv1a (**Figure 3a**) and κ-ACTX-Hv1c (**Figure 3b**) provides an opportunity for rational design of small molecule mimetics based on the mapped insectophores, a process refered to as "insectophore cloning." Several recent examples have provided encouraging validation of this concept (review: Norton *et al.*, 2003). Particularly relevant are three recent studies that involved cloning the pharmacophores of peptide toxins that block vertebrate calcium or potassium channels (Menzler *et al.*, 2000; Baell *et al.*, 2001, 2002). In each case, an attempt was made to rationally design nonpeptidic analogs that topologically mimicked the functional moieties corresponding to three key amino acid hits from mutagenesis studies. In the most successful study, an attempt was made to clone the pharmacophore of the cone snail ICK peptide ω-conotoxin MVIIA, a specific blocker of vertebrate N-type VGCCs and a lead for the development of novel analgesics. An alkylphenyl backbone was decorated with chemical groups corresponding to the key functional residues Arg10, Leu11, and Tyr13 (**Figure 3c**). Even though other ω-conotoxin MVIIA residues are involved in the toxin–channel interaction, the rationally designed mimics nevertheless had IC_{50} values of ~3 μM against human N-type VGCCs (Menzler *et al.*, 2000).

Thus, it should be possible to clone topologically restricted spider toxin insectophores to produce nonpeptidic lead compounds with low micromolar potency against the desired insect targets. What remains to be determined is whether a rationally designed mimic can maintain the desired phyletic specificity of the parent toxin. An obvious caveat is that residues deemed not to be functionally important according to mutagenesis studies might nevertheless provide a steric impediment to binding to the vertebrate counterpart of the targeted insect channel. Failure to take this into account might lead to phyletically indiscriminate mimics.

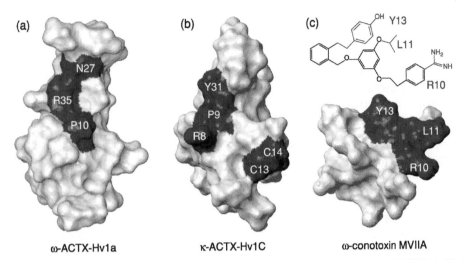

Figure 3 Molecular surface representation of (a) ω-ACTX-Hv1a (Protein Data Bank file 1AXH), (b) κ-ACTX-Hv1c (Protein Data Bank file 1DL0), and (c) ω-conotoxin MVIIA (Protein Data Bank file 1OMG) showing the channel binding surfaces (red) as mapped by mutagenesis studies (Nadasdi *et al.*, 1995; Tedford *et al.*, 2001, 2004; Maggio and King, 2002). Also shown in (c) is a rationally designed mimetic of the ω-conotoxin MVIIA pharmacophore, which blocks cloned human N-type VGCCs with IC$_{50}$ ∼ 3 μM (Menzler *et al.*, 2000).

4.9. Future Directions

Following a decade in which research on spider venoms was focused primarily on vertebrate-active compounds, there has been an acceleration of interest in recent years in the isolation and characterization of insect-specific toxins. So far, however, this immense reservoir of insecticidal compounds has been sparsely sampled, and it can be expected that many more insect-specific spider toxins will be isolated over the next few years. What remains to be seen is whether this basic research can be translated into viable insect control methods. In too few cases has there been a concerted effort to uncover the molecular target of these toxins, determine their three-dimensional structure, and elucidate detailed structure–function relationships. It is striking that the target binding site has been mapped in detail for only two spider toxins. Nevertheless, with appropriate focus on these aspects, it is likely that peptidic neurotoxins from spider venoms will facilitate discovery of new insecticide targets and provide molecular templates for rational design of novel chemical insecticides that act at these sites. Moreover, it should be possible to employ these spider toxins as bioinsecticides by incorporating genes encoding these toxins into insect-specific viruses, or by using such genes as conditional lethal alleles in transgenic insects.

Acknowledgments

We would like to acknowledge our many collaborators, especially Roger Drinkwater, Simon Gunning, Graham Nicholson, Robert Reenan, and David Wilson for allowing us to discuss unpublished collaborative results. Thanks to Susan Rowland and Scott Chouinard for critical appraisal of the manuscript, and to US National Science Foundation and the NIH National Institute of Allergy and Infectious Diseases for financial support.

References

Adams, M.E., Herold, E.E., Venema, V.J., **1989**. Two classes of channel-specific toxins from funnel web spider venom. *J. Comp. Physiol. A 164*, 333–342.

Adams, M.E., Bindokas, V.P., Hasegawa, L., Venema, V.J., **1990**. ω-Agatoxins: novel calcium channel antagonists of two subtypes from funnel web spider (*Agelenopsis aperta*) venom. *J. Biol. Chem. 265*, 861–867.

Afonso, C.L., Tulman, E.R., Lu, Z., Balinsky, C.A., Moser, B.A., *et al.*, **2001**. Genome sequence of a baculovirus pathogenic for *Culex nigripalpus*. *J. Virol. 75*, 11157–11165.

Anderluh, G., Podlesek, Z., Macek, P., **2000**. A common motif in proparts of cnidarian toxins and nematocyst collagens and its putative role. *Biochim. Biophys. Acta 1476*, 372–376.

Aramaki, Y., Yasuhara, T., Higashijima, T., Yoshioka, M., Miwa, A., *et al.*, **1986**. Chemical characterization of spider toxin, JSTX and NSTX. *Proc. Jpn. Acad. Ser. B 62*, 359–362.

Arnaud, J.F., Viard, F., Delescluse, M., Cuguen, J., **2003**. Evidence for gene flow via seed dispersal from crop to wild relatives in *Beta vulgaris* (Chenopodiaceae): consequences for the release of genetically modified crop species with weedy lineages. *Proc. Roy. Soc. B 270*, 1565–1571.

Atkinson, P.W., 2002. Genetic engineering in insects of agricultural importance. *Insect Biochem. Mol. Biol.* 32, 1237–1242.

Atkinson, R.K., Howden, M.E.H., Tyler, M.I., Vonarx, E.J., 1998. Insecticidal toxins derived from funnel web (*Atrax* or *Hadronyche*) spiders. US Patent No. 5 763 568.

Attaran, A., Roberts, D.R., Curtis, C.F., Kilama, W.L., 2000. Balancing risks on the backs of the poor. *Nature Med.* 6, 729–731.

Baell, J.B., Forsyth, S.A., Gable, R.W., Norton, R.S., Mulder, R.J., 2001. Design and Synthesis of Type-III Mimetics of ω-Conotoxin GVIA. *J. Comput. Aided Mol. Des.* 15, 1119–1136.

Baell, J.B., Harvey, A.J., Norton, R.S., 2002. Design and synthesis of Type-III mimetics of ShK toxin. *J. Comput. Aided Mol. Des.* 16, 245–262.

Balaji, R.A., Sasaki, T., Gopalakrishnakone, P., Sato, K., Kini, R.M., et al., 2000. Purification, structure determination and synthesis of covalitoxin-II, a short insect-specific neurotoxic peptide from the venom of the *Coremiocnemis validus* (Singapore Tarantula). *FEBS Lett.* 474, 208–212.

Becnel, J.J., White, S.E., Moser, B.A., Fukuda, T., Rotstsein, M.J., et al., 2001. Epizootiology and transmission of a newly discovered baculovirus from the mosquitoes *Culex nigripalpus* and *C. quinquefasciatus*. *J. Gen. Virol.* 82, 275–282.

Benedict, M.Q., Robinson, A.S., 2003. The first releases of transgenic mosquitoes: an argument for the sterile insect technique. *Trends Parasitol.* 19, 349–355.

Black, B.C., Brennan, L.A., Dierks, P.M., Gard, I.E., 1997. Commercialization of baculoviral insecticides. In: Miller, L.K. (Ed.), The Baculoviruses. Plenum, New York, pp. 341–387.

Bloomquist, J.R., 1996. Ion channels as targets for insecticides. *Annu. Rev. Entomol.* 41, 163–190.

Bloomquist, J.R., 2003. Mode of action of atracotoxin at central and peripheral synapses of insects. *Invertebr. Neurosci.* 5, 45–50.

Bloomquist, J.R., Kinne, L.P., Deutsch, V., Simpson, S.F., 1996. Mode of action of an insecticidal peptide toxin from the venom of a weaving spider (*Diguetia canities*). *Toxicon* 34, 1072–1075.

Boman, H.G., 2003. Antibacterial peptides: basic facts and emerging concepts. *J. Intern. Med.* 254, 197–215.

Bonning, B.C., Hammock, B.D., 1996. Development of recombinant baculoviruses for insect control. *Annu. Rev. Entomol.* 41, 191–210.

Branton, W.D., Rudnick, M.S., Zhou, Y., Eccleston, E.D., Fields, G.B., et al., 1993. Fatty acylated toxin structure. *Nature* 365, 496–497.

Brogdon, W.G., McAllister, J.C., 1998. Insecticide resistance and vector control. *Emerg. Infect. Dis.* 4, 605–613.

Brooks, E., Hines, E., 1999. Viral biopesticides for heliothine control: fact or fiction? *Today's Life Sci.* January/February, 38–44.

Brown, M.R., Sheumack, D.D., Tyler, M.I., Howden, M.E.H., 1988. Amino acid sequence of versutoxin,

a lethal neurotoxin from the venom of the funnel-web spider *Atrax versutus*. *Biochem. J.* 250, 401–405.

Caltagirone, L.E., Doutt, R.L., 1989. The history of the *Vedalia* beetle importation to California and its impact on the development of biological control. *Annu. Rev. Entomol.* 34, 1–6.

Carson, R., 1962. Silent Spring. Houghton Mifflin, Boston, MA.

Casida, J.E., Quistad, G.B., 1998. Golden age of insecticide research: past, present, or future? *Annu. Rev. Entomol.* 43, 1–16.

Conticello, S.G., Gilad, Y., Avidan, N., Ben-Asher, E., Levy, Z., et al., 2001. Mechanisms for evolving hypervariability: the case of conopeptides. *Mol. Biol. Evol.* 18, 120–131.

Copping, L.G., Menn, J.J., 2000. Biopesticides: a review of their action, applications and efficacy. *Pest Mgt. Sci.* 56, 651–676.

Cory, J.S., Hirst, M.L., Williams, T., Hails, R.S., Goulson, D., et al., 1994. Field trial of a genetically improved baculovirus insecticide. *Nature* 370, 138–140.

Corzo, G., Escoubas, P., Stankiewicz, M., Pelhate, M., Kristensen, C.P., et al., 2000. Isolation, synthesis and pharmacological characterization of δ-palutoxins IT, novel insecticidal toxins from the spider *Paracoelotes luctuosus* (Amaurobiidae). *Eur. J. Biochem.* 267, 5783–5795.

Corzo, G., Escoubas, P., Villegas, E., Barnham, K.J., He, W., et al., 2001. Characterization of unique amphipathic antimicrobial peptides from venom of the scorpion *Pandinus imperator*. *Biochem. J.* 359, 35–45.

Corzo, G., Gilles, N., Satake, H., Villegas, E., Dai, L., et al., 2003. Distinct primary structures of the major peptide toxins from the venom of the spider *Macrothele gigas* that bind to sites 3 and 4 in the sodium channel. *FEBS Lett.* 547, 43–50.

Corzo, G., Villegas, E., Gomez-Lagunas, F., Possani, L.D., Belokoneva, O.S., et al., 2002. Oxyopinins, large amphipathic peptides isolated from the venom of the wolf spider *Oxyopes kitabensis* with cytolytic properties and positive insecticidal cooperativity with spider neurotoxins. *J. Biol. Chem.* 277, 23627–23637.

Craik, D.J., Daly, N.L., Waine, C., 2001. The cystine knot motif in toxins and implications for drug design. *Toxicon* 39, 43–60.

Dale, P.J., 1999a. Public reactions and scientific responses to transgenic crops. *Curr. Opin. Biotechnol.* 10, 203–208.

Dale, P.J., 1999b. Public concerns over transgenic crops. *Genome Res.* 9, 1159–1162.

de Figueiredo, S.G., de Lima, M.E., Cordeiro, M.N., Diniz, C.R., Patten, D., et al., 2001. Purification and amino acid sequence of a highly insecticidal toxin from the venom of the brazilian spider *Phoneutria nigriventer* which inhibits NMDA-evoked currents in rat hippocampal neurones. *Toxicon* 39, 309–317.

de Lima, M.E., Stankiewicz, M., Hamon, A., de Figueiredo, S.G., Cordeiro, M.N., et al., 2002. The toxin Tx4(6-1) from the spider *Phoneutria nigriventer* slows down Na+

current inactivation in insect CNS via binding to receptor Site 3. *J. Insect Physiol. 48*, 53–61.

Dent, D., 2000. Insect Pest Management. CAB International, Wallingford.

Diniz, M.R., Paine, M.J., Diniz, C.R., Theakston, R.D., Crampton, J.M., 1993. Sequence of the cDNA coding for the lethal neurotoxin Tx1 from the Brazilian "armed" spider *Phoneutria nigriventer* predicts the synthesis and processing of a preprotoxin. *J. Biol. Chem. 268*, 15340–15342.

Duda, T.F., Jr., Palumbi, S.R., 1999. Developmental shifts and species selection in gastropods. *Proc. Natl Acad. Sci. USA 96*, 10272–10277.

Eberl, D.F., Ren, D., Feng, G., Lorenz, L.J., Vactor, D.V., et al., 1998. Genetic and developmental characterization of *Dmca1D*, a calcium channel α_1 subunit gene in *Drosophila melanogaster*. *Genetics 148*, 1159–1169.

Escoubas, P., Diochot, S., Corzo, G., 2000. Structure and pharmacology of spider venom neurotoxins. *Biochimie 82*, 893–907.

Escoubas, P., Rash, L., 2004. Tarantulas: eight-legged pharmacists and combinatorial chemists. *Toxicon 43*, 555–574.

Federici, B.A., 1999. Naturally occurring baculoviruses for insect pest control. In: Hall, F.R., Menn, J.J. (Eds.), Biopesticides: Use and Delivery. Humana Press, Totowa, NJ, pp. 301–320.

Ferré, J., Van Rie, J., 2002. Biochemistry and genetics of insect resistance to *Bacillus thuringiensis*. *Annu. Rev. Entomol. 47*, 501–533.

Feyereisen, R., 1995. Molecular biology of insecticide resistance. *Toxicol. Lett. 82–83*, 83–90.

Figueiredo, S.G., Garcia, M.E., Valentim, A.C., Cordeiro, M.N., Diniz, C.R., et al., 1995. Purification and amino acid sequence of the insecticidal neurotoxin Tx4(6-1) from the venom of the "armed" spider *Phoneutria nigriventer* (Keys). *Toxicon 33*, 83–93.

Fisher, F.G., Bohn, H., 1957. Die Giftsekrete der Volgenspinnen. *Liebigs Ann. Chem. 603*, 232–250.

Fletcher, J.I., Chapman, B.E., Mackay, J.P., Howden, M.E.H., King, G.F., 1997a. The structure of versutoxin (δ-atracotoxin-Hv1) provides insights into the binding of site 3 neurotoxins to the voltage-gated sodium channel. *Structure 5*, 1525–1535.

Fletcher, J.I., Smith, R., O'Donoghue, S.I., Nilges, M., Connor, M., et al., 1997b. The structure of a novel insecticidal neurotoxin, ω-atracotoxin-Hv1, from the venom of an Australian funnel web spider. *Nature Struct. Biol. 4*, 559–566.

Ford, J.W., Stevens, E.B., Treherne, J.M., Packer, J., Bushfield, M., 2002. Potassium channels: gene family, therapeutic relevance, high-throughput screening technologies and drug discovery. *Progr. Drug Res. 58*, 133–168.

Fujimi, T.J., Nakajyo, T., Nishimura, E., Ogura, E., Tsuchiya, T., et al., 2003. Molecular evolution and diversification of snake toxin genes, revealed by analysis of intron sequences. *Gene 313*, 111–118.

Gershburg, E., Stockholm, D., Froy, O., Rashi, S., Gurevitz, M., et al., 1998. Baculovirus-mediated expression of a scorpion depressant toxin improves the insecticidal efficacy achieved with excitatory toxins. *FEBS Lett. 422*, 132–136.

Gilbo, C.M., Coles, N.W., 1964. An investigation of certain components of the venom of the female Sydney funnel web spider *Atrax robustus* Cambr. *Austr. J. Biol. Sci. 17*, 758–763.

Gomez, M.V., Kalapothakis, E., Guatimosim, C., Prado, M.A.M., 2002. *Phoneutria nigriventer* venom: a cocktail of toxins that affect ion channels. *Cell. Mol. Neurobiol. 22*, 579–588.

Goudet, C., Chi, C.W., Tytgat, J., 2002. An overview of toxins and genes from the venom of the Asian scorpion *Buthus martensi* Karsch. *Toxicon 40*, 1239–1258.

Gratz, N.G., 1999. Emerging and resurging vector-borne diseases. *Annu. Rev. Entomol. 44*, 51–75.

Grishin, E.V., Volkova, T.M., Arseniev, A.S., 1989. Isolation and structure analysis of components from venom of the spider *Argiope lobata*. *Toxicon 27*, 541–549.

Gröner, A., 1986. Specificity and safety of baculoviruses. In: Granados, R.R., Federici, B.A. (Eds.), The Biology of Baculoviruses. CRC Press, Boca Raton, FL, pp. 177–202.

Gubler, D.J., 1998. Resurgent vector-borne diseases as a global health problem. *Emerg. Infect. Dis. 4*, 442–450.

Gubler, D.J., 2002. The global emergence/resurgence of arboviral diseases as public health problems. *Arch. Med. Res. 33*, 330–342.

Hall, F.R., Menn, J.J., 1999. Biopesticides: Use and Delivery. Humana Press, Totowa, NJ.

Hall, L.M., Ren, D., Feng, G., Eberl, D.F., Dubald, M., et al., 1994. The calcium channel as a new potential target for insecticides. In: Clark, J.M. (Ed.), Molecular Action of Insecticides on Ion Channels. American Chemical Society, Washington, DC, pp. 162–172.

Handler, A.M., 2002. Prospects for using genetic transformation for improved SIT and new biocontrol methods. *Genetica 116*, 137–149.

Herniou, E.A., Olszewski, J.A., Cory, J.S., O'Reilly, D.R., 2003. The genome sequence and evolution of baculoviruses. *Annu. Rev. Entomol. 48*, 211–234.

Herrmann, R., Moskowitz, H., Zlotkin, E., Hammock, B.D., 1995. Positive cooperativity among insecticidal scorpion neurotoxins. *Toxicon 33*, 1099–1102.

Huang, J., Rozelle, S., Pray, C., Wang, Q., 2002. Plant biotechnology in China. *Science 295*, 674–676.

Huang, J.Q., Wel, Z.M., An, H.L., Zhu, Y.X., 2001. *Agrobacterium tumefaciens*-mediated transformation of rice with the spider insecticidal gene conferring resistance to leaffolder and striped stem borer. *Cell Res. 11*, 149–155.

Hughes, P.R., Wood, H.A., Breen, J.P., Simpson, S.F., Duggan, A.J., et al., 1997. Enhanced bioactivity of recombinant baculoviruses expressing insect-specific spider toxins in lepidopteran crop pests. *J. Invertebr. Pathol. 69*, 112–118.

Inceoglu, A.B., Kamita, S.G., Hinton, A.C., Huang, Q., Severson, T.F., et al., 2001. Recombinant baculoviruses for insect control. Pest. Mgt. Sci. 57, 981–987.

Isbister, G.K., White, J., 2004. Clinical consequences of spider bite: recent advances in our understanding. Toxicon 43, 477–492.

Jenkins, J.N., 1999. Transgenic plants expressing toxins from Bacillus thuringiensis. In: Hall, F.R., Menn, F.R. (Eds.), Biopesticides: Use and Delivery. Humana Press, Totowa, NJ, pp. 211–232.

Johnson, J.H., Bloomquist, J.R., Krapcho, K.J., Kral, R.M., Trovalto, R., et al., 1998. Novel insecticidal peptides from Tegenaria agrestis spider venom may have a direct effect on the insect central nervous system. Arch. Insect Biochem. Physiol. 38, 19–31.

Johnson, J.H., Kral, R.M., Jr., 1997. Insecticidal peptides from Segestria sp. spider venom. US Patent No. 6 674 846.

Johnson, J.H., Kral, R.M., Jr., Krapcho, K. 1997. Insecticidal peptides from spider venom. US Patent No. 5 688 764.

Kalapothakis, E., Penaforte, C.L., Beirão, P.S.L., Romano-Silva, M.A., Cruz, J.S., et al., 1998. Cloning of cDNAs encoding neurotoxic peptides from the spider Phoneutria nigriventer. Toxicon 36, 1843–1850.

Kawasaki, F., Collins, S.C., Ordway, R.W., 2002. Synaptic calcium-channel function in Drosophila: analysis and transformation rescue of temperature-sensitive paralytic and lethal mutations of cacophony. J. Neurosci. 22, 5856–5864.

Kieber-Emmons, T., Murali, R., Greene, M.I., 1997. Therapeutic peptides and peptidomimetics. Curr. Opin. Biotechnol. 8, 435–441.

Kim, J.I., Konishi, S., Iwai, H., Kohno, T., Gouda, H., et al., 1995. Three-dimensional solution structure of the calcium channel antagonist ω-agatoxin IVA: consensus molecular folding of calcium channel blockers. J. Mol. Biol. 250, 659–671.

King, G.F., Tedford, H.W., Maggio, F., 2002. Structure and function of insecticidal neurotoxins from Australian funnel-web spiders. J. Toxicol. Toxin Rev. 21, 359–389.

Kirsch, D.R., Heinrich, J.N., Pausch, M.H., Silverman, S., Baumbach, W.M.L., et al., 2001. Molecular genetic screen design for agricultural and pharmaceutical product discovery. In: Seethala, R., Fernandes, P.B. (Eds.), Handbook of Drug Screening. Dekker, New York, pp. 153–188.

Kost, T.A., Condreay, J.P., 2002. Recombinant baculoviruses as mammalian cell gene-delivery vectors. Trends Biotechnol. 20, 173–180.

Krapcho, K.J., Jackson, J.R.H., Johnson, J.H., DelMar, E.G., Kral, R.M., Jr., 1995a. Insecticidally effective peptides. US Patent No. 5 441 934.

Krapcho, K.J., Jackson, J.R.H., VanWagenen, B.C., Kral, R.M., Jr., 1997. Insecticidally effective peptides. US Patent No. 5 658 781.

Krapcho, K.J., Kral, R.M., Jr., VanWagenen, B.C., Eppler, K.G., Morgan, T.K., 1995b. Characterization and

cloning of insecticidal peptides from the primitive weaving spider Diguetia canities. Insect Biochem. Mol. Biol. 25, 991–1000.

Kuhn-Nentwig, L., 2003. Antimicrobial and cytolytic peptides of venomous arthropods. Cell. Mol. Life Sci. 60, 2651–2668.

Kuhn-Nentwig, L., Schaller, J., Nentwig, W., 2004. Biochemistry, toxicology and ecology of the venom of the spider Cupiennius Salei (Ctenidae). Toxicon 43, 543–553.

Latham, P.W., 1999. Therapeutic peptides revisited. Nature Biotechnol. 17, 755–757.

Liang, S., 2004. An overview of peptide toxins from the venom of the Chinese bird spider Selenocosmia huwena Wang [Ornithoctonus huwena (Wang)]. Toxicon 43, 575–585.

Lipkin, A., Kozlov, S., Nosyreva, E., Blake, A., Windass, J.D., et al., 2002. Novel insecticidal toxins from the venom of the spider Segestria florentina. Toxicon 40, 125–130.

Lu, A., Seshagiri, S., Miller, L.K., 1996. Signal sequence and promoter effects on the efficacy of toxin-expressing baculoviruses as biopesticides. Biol. Control 7, 320–332.

Maggio, F., King, G.F., 2002. Scanning mutagenesis of a Janus-faced atracotoxin reveals a bipartite surface patch that is essential for neurotoxic function. J. Biol. Chem. 277, 22806–22813.

McCutchen, B.F., Choudary, P.V., Crenshaw, R., Maddox, D., Kamita, S.G., et al., 1991. Development of a recombinant baculovirus expressing an insect-selective neurotoxin: potential for pest control. Biotechnology 9, 848–852.

Mellor, I.R., Usherwood, P.N.R., 2004. Targeting ionotropic receptors with polyamine-containing toxins. Toxicon 43, 493–508.

Menzler, S., Bikker, J.A., Suman-Chauhan, N., Horwell, D.C., 2000. Design and biological evaluation of non-peptide analogues of ω-conotoxin MVIIA. Bioorg. Med. Chem. Lett. 10, 345–347.

Mintz, I.M., Venema, V.J., Swiderek, K.M., Lee, T.D., Bean, B.P., et al., 1992. P-type calcium channels blocked by the spider toxin ω-Aga-IVA. Nature 355, 827–829.

Morin, S., Biggs, R.W., Sisterson, M.S., Shriver, L., Ellers-Kirk, C., et al., 2003. Three cadherin alleles associated with resistance to Bacillus thuringiensis in pink bollworm. Proc. Natl Acad. Sci. USA 100, 5004–5009.

Moscardi, F., 1999. Assessment of the application of baculoviruses for control of Lepidoptera. Annu. Rev. Entomol. 44, 257–289.

Moser, B., Becnel, J., White, S., Afonso, C., Kutish, G., et al., 2001. Morphological and molecular evidence that Culex nigripalpus baculovirus is an unusual member of the family Baculoviridae. J. Gen. Virol. 82, 283–297.

Nadasdi, L., Yamashiro, D., Chung, D., Tarczy-Hornoch, K., Adriaenssens, P., et al., 1995. Structure–activity analysis of a Conus peptide blocker of N-type neuronal calcium channels. Biochemistry 34, 8076–8081.

Nicholson, G.M., Graudins, A., 2002. Spiders of medical importance in the Asia–Pacific: atracotoxin, latrotoxin, and related spider neurotoxins. *Clin. Exp. Pharmacol. Physiol. 29*, 785–794.

Nicholson, G.M., Little, M.J., Tyler, M., Narahashi, T., **1996**. Selective alteration of sodium channel gating by Australian funnel-web spider toxins. *Toxicon 34*, 1443–1453.

Norton, R.S., Baell, J.B., Angus, J.A., **2003**. Calcium channel blocking polypeptides: structure, function, and molecular mimicry. In: McDonough, S.I. (Ed.), Calcium Channel Pharmacology. Kluwer, Amsterdam, pp. 143–179.

Norton, R.S., Pallaghy, P.K., **1998**. The cystine knot structure of ion channel toxins and related polypeptides. *Toxicon 36*, 1573–1583.

Novotny, V., Basset, Y., Miller, S.E., Weiblen, G.D., Bremer, B., *et al.*, **2002**. Low host specificity of herbivorous insects in a tropical forest. *Nature 416*, 841–844.

Oerke, E.-C., **1994**. Estimated crop losses due to pathogens, animal pests and weeds. In: Oerke, E.-C., Dehne, H.-W., Schönbeck, F., Weber, A. (Eds.), Crop Production and Crop Protection: Estimated Losses in Major Food and Cash Crops. Elsevier, Amsterdam, pp. 72–78.

Olivera, B.M., Cruz, L.J., **2001**. Conotoxins, in retrospect. *Toxicon 39*, 7–14.

Olivera, B.M., Miljanich, G.P., Ramachandran, J., Adams, M.E., **1994**. Calcium channel diversity and neurotransmitter release: the ω-conotoxins and ω-agatoxins. *Annu. Rev. Biochem. 63*, 823–867.

Olivera, B.M., Walker, C., Cartier, G.E., Hooper, D., Santos, A.D., *et al.*, **1999**. Speciation of cone snails and interspecific hyperdivergence of their venom peptides: potential evolutionary significance of introns. *Ann. New York Acad. Sci. 870*, 223–237.

Omecinsky, D.O., Holub, K.E., Adams, M.E., Reily, M.D., **1996**. Three-dimensional structure analysis of μ-agatoxins: further evidence for common motifs among neurotoxins with diverse ion channel specificities. *Biochemistry 35*, 2836–2844.

Ostrow, K.L., Mammoser, A., Suchyna, T., Sachs, F., Oswald, R., *et al.*, **2003**. cDNA sequence and *in vitro* folding of GsMTx4, a specific peptide inhibitor of mechanosensitive channels. *Toxicon 42*, 263–274.

Pallaghy, P.K., Nielsen, K.J., Craik, D.J., Norton, R.S., **1994**. A common structural motif incorporating a cystine knot and a triple-stranded β-sheet in toxic and inhibitory polypeptides. *Protein Sci. 3*, 1833–1839.

Peferoen, M., **1997**. Progress and prospects for field use of *Bt* genes in crops. *Trends Biotechnol. 15*, 173–177.

Platnick, N.I., **1997**. Advances in Spider Taxonomy, 1992–1995: With Redescriptions 1940–1980. New York Entomological Society and The American Museum of Natural History, New York. (Updated online version available at http://research.amnh.org/entomology/spiders/catalog81-87/).

Popham, H.J.R., Li, Y., Miller, L.K., **1997**. Genetic improvement of a *Helicoverpa zea* nuclear polyhedrosis virus as a biopesticide. *Biol. Control 10*, 83–91.

Popham, H.J.R., Prikhod'ko, G.G., Felcetto, T.J., Ostlind, D.A., Warmke, J.W., *et al.*, **1998**. Effect of deltamethrin

treatment on lepidopteran larvae infected with baculoviruses expressing insect-selective toxins μ-Aga-IV, As II, or Sh I. *Biol. Control 12*, 79–87.

Possani, L.D., Becerril, B., Delepierre, M., Tytgat, J., **1999**. Scorpion toxins specific for Na$^+$ channels. *Eur. J. Biochem. 264*, 287–300.

Pray, C.E., Huang, J., Hu, R., Rozelle, S., **2002**. Five years of *Bt* cotton in china – the benefits continue. *Plant J. 31*, 423–430.

Prikhod'ko, G.G., Popham, H.J.R., Felcetto, T.J., Ostlind, D.A., Warren, V.A., *et al.*, **1998**. Effects of simultaneous expression of two sodium channel toxin genes on the properties of baculoviruses as biopesticides. *Biol. Control 12*, 66–78.

Prikhod'ko, G.G., Robson, M., Warmke, J.W., Cohen, C.J., Smith, M.M., *et al.*, **1996**. Properties of three baculovirus-expressing genes that encode insect-selective toxins: μ-Aga-IV, As II, and Sh I. *Biol. Control 7*, 236–244.

Quistad, G.B., Reuter, C.C., Skinner, W.S., Dennis, P.A., Suwanrumpha, S., *et al.*, **1991**. Paralytic and insecticidal toxins from the funnel web spider, *Hololena curta*. *Toxicon 29*, 329–336.

Quistad, G.B., Skinner, W.S., **1994**. Isolation and sequencing of insecticidal peptides from the primitive hunting spider, *Plectreurys tristis* (Simon). *J. Biol. Chem. 269*, 11098–11101.

Rash, L.D., Hodgson, W.C., **2002**. Pharmacology and biochemistry of spider venoms. *Toxicon 40*, 225–254.

Regev, A., Rivkin, H., Inceoglu, B., Gershburg, E., Hammock, B.D., *et al.*, **2003**. Further enhancement of baculovirus insecticidal efficacy with scorpion toxins that interact cooperatively. *FEBS Lett. 537*, 106–110.

Robinson, A.S., **2002**. Genetic sexing strains in medfly, *Ceratitis capitata*, sterile insect technique programmes. *Genetica 116*, 5–13.

Rose, R.I., **2001**. Pesticides and public health: integrated methods of mosquito management. *Emerg. Infect. Dis. 7*, 17–23.

Sanguinetti, M.C., Johnson, J.H., Hammerland, L.G., Kelbaugh, P.R., Volkmann, R.A., *et al.*, **1997**. Heteropodatoxins: peptides isolated from spider venom that block K$_v$4.2 potassium channels. *Mol. Pharmacol. 51*, 491–498.

Santos, A.D., Imperial, J.S., Chaudhary, T., Beavis, R.C., Chait, B.T., *et al.*, **1992**. Heterodimeric structure of the spider toxin ω-Agatoxin IA revealed by precursor analysis and mass spectrometry. *J. Biol. Chem. 267*, 20701–20705.

Schliekelman, P., Gould, F., **2000a**. Pest control by the release of insects carrying a female-killing allele on multiple loci. *J. Econ. Entomol. 93*, 1566–1579.

Schliekelman, P., Gould, F., **2000b**. Pest control by the introduction of a conditional lethal trait on multiple loci: potential, limitations, and optimal strategies. *J. Econ. Entomol. 93*, 1543–1565.

Skinner, W.S., Adams, M.E., Quistad, G.B., Kataoka, H., Cesarin, B.J., *et al.*, **1989**. Purification and characterization of two classes of neurotoxins from the funnel web spider, *Agelenopsis aperta*. *J. Biol. Chem.* 264, 2150–2155.

Skinner, W.S., Dennis, P.A., Lui, A., Carney, R.L., Quistad, G.B., 1990. Chemical characterization of acyl-polyamine toxins from venom of a trap-door spider and two tarantulas. *Toxicon* 28, 541–546.

Skinner, W.S., Dennis, P.A., Li, J.P., Quistad, G.B., 1992. Identification of insecticidal peptides from venom of the trap-door spider, *Aptostichus schlingeri* (Ctenizidae). *Toxicon* 30, 1043–1050.

Smith, L.A., Peixoto, A.A., Kramer, E.M., Villella, A., Hall, J.C., 1998. Courtship and visual defects of *cacophony* mutants reveal functional complexity of a calcium-channel α1 subunit in *Drosophila*. *Genetics* 149, 1407–1426.

Sollod, B.L., Wilson, D., Drinkwater, R., King, G.F., 2003. Atracotoxin expression and evolution. In: Proc. 14th World Congr. Int. Soc. Toxinol. Adelaide, Australia, p. 23.

Stapleton, A., Blankenship, D.T., Ackermann, B.L., Chen, T.M., Gorder, G.W., et al., 1990. Curtatoxins: neurotoxic insecticidal polypeptides isolated from the funnel-web spider *Hololena curta*. *J. Biol. Chem.* 265, 2054–2059.

Stewart, L.M.D., Hirst, M., Ferber, M.L., Merryweather, A.T., Cayley, P.A., et al., 1991. Construction of an improved baculovirus insecticide containing an insect-specific toxin gene. *Nature* 352, 85–88.

Strømgaard, K., Andersen, K., Krogsgaard-Larsen, P., Jaroszewski, J.W., 2001. Recent advances in the medicinal chemistry of polyamine toxins. *Mini Rev. Med. Chem.* 1, 317–338.

Sun, X., Chen, X., Zhang, Z., Wang, H., Bianchi, F.J., et al., 2002. Bollworm responses to release of genetically modified *Helicoverpa armigera* nucleopolyhedroviruses in cotton. *J. Invertebr. Pathol.* 81, 63–69.

Tedford, H.W., Fletcher, J.I., King, G.F., 2001. Functional significance of the β-hairpin in the insecticidal neurotoxin ω-atracotoxin-Hv1a. *J. Biol. Chem.* 276, 26568–26576.

Tedford, H.W., Sollod, B.L., Maggio, F., King, G.F., 2004. Australian funnel-web spiders: master insecticide chemists. *Toxicon* 43, 601–618.

Toenniessen, G.H., O'Toole, J.C., DeVries, J., 2003. Advances in plant biotechnology and its adoption in developing countries. *Curr. Opin. Plant Biol.* 6, 191–198.

Tomalski, M.D., Miller, L.K., 1991. Insect paralysis by baculovirus-mediated expression of a mite neurotoxin gene. *Nature* 352, 82–85.

Treacy, M.F., 1999. Recombinant baculoviruses. In: Hall, F.R., Menn, J.J. (Eds.), Biopesticides: Uses and Delivery. Humana Press, Totowa, NJ, pp. 321–340.

Wang, X.-H., 2000. Discovery and characterization of insecticidal neurotoxins from Australian funnel-web spider venom. Ph.D. thesis, Department of Biochemistry, University of Sydney.

Wang, X.-H., Connor, M., Smith, R., Maciejewski, M.W., Howden, M.E.H., et al., 2000. Discovery and characterization of a family of insecticidal neurotoxins with a rare vicinal disulfide bond. *Nature Struct. Biol.* 7, 505–513.

Wang, X.-H., Connor, M., Wilson, D., Wilson, H.I., Nicholson, G.M., et al., 2001. Discovery and structure of a potent and highly specific blocker of insect calcium channels. *J. Biol. Chem.* 276, 40806–40812.

Wang, X.-H., Smith, R., Fletcher, J.I., Wilson, H., Wood, C.J., et al., 1999. Structure–function studies of ω-atracotoxin, a potent antagonist of insect voltage-gated calcium channels. *Eur. J. Biochem.* 264, 488–494.

Wicher, D., Walther, C., Wicher, C., 2001. Non-synaptic ion channels in insects: basic properties of currents and their modulation in neurons and skeletal muscles. *Progr. Neurobiol.* 64, 431–525.

Wilson, D.T.R., 2001. The identification and characterization of Australian funnel-web spider venom. Ph.D. thesis, Institute for Molecular Biosciences, University of Queensland.

Winston, M.L., 1997. Nature Wars: People vs. Pests. Harvard University Press, Cambridge, MA.

Zaim, M., Guillet, P., 2002. Alternative insecticides: an urgent need. *Trends Parasitol.* 18, 161–163.

Zhang, P.F., Chen, P., Hu, W.J., Liang, S.P., 2003. Huwen-toxin-V, a novel insecticidal peptide toxin from the spider *Selenocosmia huwena*, and a natural mutant of the toxin indicates the key amino acid residues related to the biological activity. *Toxicon* 42, 15–20.

Zlotkin, E., Fishman, Y., Elazar, M., 2000. AaIT: from neurotoxin to insecticide. *Biochimie* 82, 869–881.

A4   Addendum: Spider Toxins and Their Potential for Insect Control

V Herzig and G F King, Institute for Molecular
Bioscience, The University of Queensland,
Queensland, Australia

© 2010, Elsevier BV. All Rights Reserved.

Evolution of insecticide resistance combined with deregistration of key insecticides due to their adverse environmental and health impacts (Tedford et al., 2004; Nicholson, 2007b) has created an unmet demand for a new generation of environmentally friendly insecticides. Spiders are efficient insect killers and they are proving to be a rich source of novel, insect-active compounds with promise as bioinsecticides.

ArachnoServer, the recently introduced spider toxin database (Wood et al., 2009), contains 162 toxins with proven insecticidal activity (see **Table A1**), which accounts for 31% of all 521 listed toxins. About two thirds of these toxins are peptides comprising only 30–50 amino acid residues (**Figure A1**). There are very few examples of large insecticidal proteins such as the latroinsectotoxins from black widow spiders (Rohou et al., 2007). The main targets of spider toxins appear to be voltage-gated sodium channels (reviewed in King, 2007) and voltage-gated calcium channels (reviewed in Nicholson, 2007a), which account for 18 and 21%, respectively, of the insecticidal toxins for which the molecular target has been determined.

Assuming that the percentage of currently listed insecticidal toxins is representative of all toxins that have not yet been assigned a phyletic activity, we conclude that ~70% of all spider toxins have insecticidal activity. This estimate makes sense in terms of spider ecology, with insects being their main prey. While the number of currently described insecticidal spider toxins is rather limited, the possibility of discovery of new ones is immense and still increasing. Despite the mega-diversity of spiders, with at least 41,000 extant species, toxins currently listed in ArachnoServer were isolated from venoms of only 69 species. It was recently suggested that the venoms of spiders might contain more than 16 million biologically active peptides (Escoubas et al., 2006) but even this might be an underestimation since recent work has demonstrated that some spider venoms contain more than 1000 different peptides (Escoubas et al., 2006). Moreover, due to refinement of both

genomic and proteomic techniques, the size of spiders that can be used for drug and insecticide discovery has decreased dramatically (Escoubas and King, 2009). Overall, there are likely to be millions of insecticidal spider toxins yet to be discovered, which makes spider venoms a virtually limitless resource for the discovery of novel bioinsecticides.

One of the main hurdles for harnessing the potential of venom-derived peptide toxins for the purpose of insect control is efficient delivery to target insects (Whetstone and Hammock, 2007). Peptides have generally been considered unlikely to be useful insecticides because of their presumed instability in gut and hemolymph, and their inability to traverse the insect gut epithelium. However, recent studies have seriously challenged this paradigm. First, ω-HXTX-Hv1a, a 37-residue peptide from the Australian funnel-web spider *Hadronyche versuta*, acts in the central nervous system of insects, indicating that it is capable of breaching the insect blood–brain barrier and must be stable for significant periods of time in insect hemolymph (Bloomquist, 2003). Second, there is only a marginal reduction in the toxicity of ω-HXTX-Hv1a when it is fed to the tick *Amblyomma americanum* compared to when it is injected (Mukherjee et al., 2006). Third, ω-HXTX-Hv1a is orally active against the cotton bollworm *Helicoverpa armigera* and the Egyptian cotton leafworm *Spodoptera littoralis* when expressed in transgenic tobacco (Khan et al., 2006). Despite these promising discoveries, it is unlikely that ω-HXTX-Hv1a is orally active against all pest species. In addition, these findings do not necessarily imply that all spider peptide toxins are orally active in insects. Hence, other strategies are required for delivering spider toxins to target insects.

One way of increasing oral activity of an insecticidal peptide toxin is by fusing it to a carrier protein that assists in toxin delivery across the midgut epithelium of a target insect. An example of such a carrier protein is a mannose-specific lectin from the snowdrop plant *Galanthus nivalis* (GNA). When fed to insects, GNA binds glycoproteins on

Table A1 Overview of insecticidal toxins isolated from spider venom[a]

Genus	Species	ID[b]	Toxin name[c]	Common synonyms	Number of isoforms	Number of residues	Number of disulfide bonds	Molecular target
Agelena	orientalis	287	U$_2$-AGTX-Ao1a	Agelenin	3	35	3	Unknown
Agelenopsis	aperta	375	μ-AGTX-Aa1a	μ-agatoxin I	6	36	4	VGSC[d]
		175	ω-AGTX-Aa1a	ω-agatoxin IA	2	66	4	VGCC[e]
		177	ω-AGTX-Aa2a	ω-agatoxin IIA	1	(28)*	?	VGCC
		178	ω-AGTX-Aa3a	ω-agatoxin IIIA	7	76	6	VGCC
		182	ω-AGTX-Aa4a	ω-agatoxin IVA	3	48	4	VGCC
Allagelena	opulenta	286	U$_2$-AGTX-Aop1a	Agelenin	1	35	3	Unknown
Aphonopelma	sp.	222	U$_1$-TRTX-Asp1a	ESTX (isoform 1)	2	39	3	Unknown
Apomastus	schlingeri	396	U$_1$-CUTX-As1a	Aptotoxin-1	4	74	4	Unknown
		403	U$_2$-CUTX-As1a	Aptotoxin-3	1	37	4	Unknown
		408	U$_3$-CUTX-As1a	Aptotoxin-7	1	32	3	Unknown
Araneus	ventricosus	114	U$_1$-AATX-Av1a	Toxin 1	1	80	3	Unknown
		113	U$_2$-AATX-Av1a	Toxin 2	1	64	3	Unknown
Atrax	robustus	305	δ-HXTX-Ar1a	δ-ACTX-Ar1a	1	42	4	VGSC
		20	ω-HXTX-Ar1a	ω-ACTX-Ar1a	7	37	3	VGCC
Brachypelma	smithi	753	U$_1$-TRTX-Bs1b	Brachypelma smithi toxin 5	1	39	3	Unknown
		648	U$_3$-TRTX-Bs1a	Bs1	1	41	3	Unknown
Calisoga	sp.	28	U$_1$-NETX-Csp1a	Peptide A	3	39	4	Unknown
Coremiocnemis	valida	422	U$_1$-TRTX-Cv1a	Covalitoxin-2	1	31	3	Unknown
Cupiennius	salei	289	M-CNTX-Cs1a	Cupiennin 1a	4	35	0	Cytolytic
		299	U$_1$-CNTX-Cs1a	CSTX-9	1	68	4	Unknown
		210	U$_1$-CNTX-Cs2a	CSTX-13	1	63	4	Unknown
		292	ω-CNTX-Cs1a	CSTX-1	1	74	4	VGCC
Diguetia	canities	352	μ-DGTX-Dg1a	DTX9.2	3	56	4	VGSC
Hadronyche	formidabilis	171	κ-HXTX-Hf1a	κ-ACTX-Hf1a, J-ACTX-Hf1a	1	37	4	K$_{Ca}$[f]
Hadronyche	versuta	307	δ-HXTX-Hv1a	δ-ACTX-Hv1a, Versutoxin	2	42	4	VGSC
		172	κ-HXTX-Hv1a	κ-ACTX-Hv1a, J-ACTX-Hv1a	5	36	4	K$_{Ca}$
		193	ω-HXTX-Hv1a	ω-ACTX-Hv1a	8	37	3	VGCC
		204	ω-HXTX-Hv2a	ω-ACTX-Hv2a	1	45	3	VGCC
Haplopelma	huwenum	328	U$_1$-TRTX-Hh1a	Huwentoxin-2 (form 1)	7	37	3	Unknown
		331	U$_2$-TRTX-Hh1a	Huwentoxin-3	1	33	3	Unknown
		333	ω-TRTX-Hh2a	Huwentoxin-5	1	35	3	VGCC
Hogna	carolinensis	65	M-LCTX-Hc1a	Lycotoxin-1	1	25	0	Cytolytic
Hololena	curta	294	μ-AGTX-Hc1a	Curtatoxin I	3	36	4	VGSC

Genus	Species	ID[b]	Recommended name[c]	Toxin name				Target
Lachesana	*tarabaevi*	615	M-ZDTX-Lt8a	Cyto-insectotoxin 1a	16	69	0	Cytolytic
Latrodectus	*tredecimguttatus*	62	α-LIT-Lt1a	α-Latroinsectotoxin	1	1170	5	Presynaptic[g]
		63	δ-LIT-Lt1a	δ-Latroinsectotoxin	1	991	2	Presynaptic
Loxosceles	*intermedia*	225	U$_1$-SCRTX-Li1a	LiTx1	3	66	5	Unknown
		273	U$_2$-SCRTX-Li1a	LiTx3	1	53	5	Unknown
Macrothele	*gigas*	378	U$_3$-HXTX-Mg1b	Magi-2	3	40	3	VGSC
		380	U$_6$-HXTX-Mg1a	Magi-3	1	46	5	VGSC
		386	U$_7$-HXTX-Mg1a	Magi-6	1	36	4	Unknown
		384	β-HXTX-Mg1a	Magi-5	1	29	3	VGSC
		382	δ-HXTX-Mg1a	Magi-4	2	43	4	VGSC
Macrothele	*raveni*	417	β-HXTX-Mr1a	Raventoxin III	1	29	3	VGSC
Oxyopes	*lineatus*	207	ω-OXTX-Ol1a	Oxytoxin-1	2	69	5	VGCC
Oxyopes	*takobius*	185	M-OXTX-Ot1a	Oxyopinin-1	1	48	0	Cytolytic
		186	M-OXTX-Ot2a	Oxyopinin-2a	4	37	0	Cytolytic
Phoneutria	*keyserlingi*	246	U$_4$-CNTX-Pk1a	PKTx28C4	1	38	4	Unknown
Phoneutria	*nigriventer*	234	U$_2$-CNTX-Pn1a	PNTx2-1	2	53	5	Unknown
		242	U$_4$-CNTX-Pn1a	PNTx2-5	2	49	5	Unknown
		247	U$_5$-CNTX-Pn1a	PNTx2-9	1	32	3	Unknown
		279	Γ-CNTX-Pn1a	Tx4(5–5), Pn4A	1	47	5	NMDA[h]
		280	δ-CNTX-Pn1a	PNTx4(6-1)	2	48	5	VGSC
		243	δ-CNTX-Pn2a	PNTx2-6	1	48	5	VGSC
Pireneitega	*luctuosa*	301	δ-AMATX-Pl1a	δ-palutoxin IT1	5	37	4	VGSC
Plectreurys	*tristis*	405	U$_1$-PLTX-Pt1a	Plectoxin-V/VI	6	46	5	Unknown
		411	U$_2$-PLTX-Pt1a	Plectoxin-IX	1	46	4	Unknown
		391	U$_3$-PLTX-Pt1a	Plectoxin-X	1	49	5	Unknown
		412	ω-PLTX-Pt1a	α-PLTX-II, Plectreurys toxin II	1	44	5	VGCC
Segestria	*florentina*	125	U$_1$-SGTX-Sf1a	Insectotoxin SIT	1	35	4	Unknown
		116	U$_2$-SGTX-Sf1a	Toxin SFI1, Toxin F5.6	9	46	4	Unknown
Tegenaria	*agrestis*	349	U$_1$-AGTX-Ta1a	TalTX-1	3	50	3	Unknown

[a] Data extracted from ArachnoServer (www.arachnoserver.org).
[b] ArachnoServer ID numbers.
[c] Recommended names based on recently devised rational nomenclature for spider-venom toxins (King et al., 2008).
[d] Voltage-gated sodium channel.
[e] Voltage-gated calcium channel.
[f] Calcium-activated potassium channel.
[g] Causes massive neurotransmitter release from insect presynaptic nerve terminals.
[h] NMDA (N-methyl-d-aspartate) ionotropic glutamate receptor.
* The asterisk indicates that only the N-terminal sequence is known for this toxin.

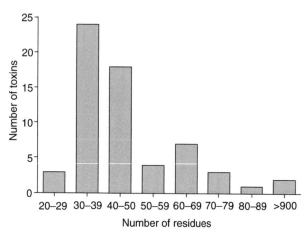

Figure A1 Size distribution of insecticidal toxins isolated from spider venoms. Data were compiled using ArachnoServer (www.arachnosever.org).

the lumenal surface of intestinal microvilli, leading to its uptake (via an unknown mechanism) and subsequent accumulation in Malpighian tubules and hemolymph (Fitches *et al.*, 2001). In a series of seminal papers, Gatehouse and Fitches demonstrated use of this property of GNA to illicitly transport peptides into the insect hemocoel (Fitches *et al.*, 2001). They showed that the oral activity of several arachnid toxins could be significantly enhanced by fusing them to GNA (Fitches *et al.*, 2004, 2010; Down *et al.*, 2006; Pham Trung *et al.*, 2006).

An alternative method of delivery is heterologous expression of toxins in plants, insect-specific viruses (as discussed in the original version of this chapter), or entomopathogenic fungi. As demonstrated by Khan *et al.* (2006), expression of ω-HXTX-Hv1a in tobacco provided complete protection against two lepidopteran pests. Similarly, expression in tobacco of the insecticidal peptide Magi-6 from the spider *Macrothele gigas* conferred resistance against *Spodoptera frugidera* (Hernandez-Campuzano *et al.*, 2009). Another genetic approach yet to be attempted with spider toxins is the introduction of a toxin transgene into entomopathogenic fungi or bacteria. However, expression of an insecticidal scorpion toxin in the entomopathogenic fungus *Metarhizium anisopliae* significantly increased its toxicity against tobacco hornworm (*Manduca sexta*) and *Aedes aegypti* mosquitoes (Wang and St Leger, 2007).

In conclusion, due to their pharmacological diversity, high potency, and both molecular and phyletic specificity, spider toxins provide extraordinary potential for the control of insect pests. A key step in their commercial deployment will be the development of efficient methods for delivering these peptide toxins to target insects.

Acknowledgments

We thank David Wood for extracting data from ArachnoServer and the Australian Research Council for financial support (Discovery Grant DP0878450).

References

Bloomquist, J.R., 2003. Mode of action of atracotoxin at central and peripheral synapses of insects. *Invertebr. Neurosci.* 5, 45–50.

Down, R.E., Fitches, E.C., Wiles, D.P., Corti, P., Bell, H.A., et al., 2006. Insecticidal spider venom toxin fused to snowdrop lectin is toxic to the peach-potato aphid, *Myzus persicae* (Hemiptera: Aphididae) and the rice brown planthopper, *Nilaparvata lugens* (Hemiptera: Delphacidae). *Pest Manag. Sci.* 62, 77–85.

Escoubas, P., King, G.F., 2009. Venomics as a drug discovery platform. *Expert Rev. Proteomics* 6, 221–224.

Escoubas, P., Sollod, B.L., King, G.F., 2006. Venom landscapes: mining the complexity of spider venoms via a combined cDNA and mass spectrometric approach. *Toxicon* 47, 650–663.

Fitches, E., Woodhouse, S.D., Edwards, J.P., Gatehouse, J.A., 2001. In vitro and in vivo binding of snowdrop (*Galanthus nivalis agglutinin*; GNA) and jackbean (*Canavalia ensiformis*; Con A) lectins within tomato moth (*Lacanobia oleracea*) larvae; mechanisms of insecticidal action. *J. Insect Physiol.* 47, 777–787.

Fitches, E., Edwards, M.G., Mee, C., Grishin, E., Gatehouse, A.M., et al., 2004. Fusion proteins containing insect-specific toxins as pest control agents: snowdrop lectin delivers fused insecticidal spider venom toxin to insect haemolymph following oral ingestion. *J. Insect Physiol.* 50, 61–71.

Fitches, E.C., Bell, H.A., Powell, M.E., Back, E., Sargiotti, C., et al., 2010. Insecticidal activity of scorpion toxin (ButaIT) and snowdrop lectin (GNA) containing fusion proteins towards pest species of different orders. *Pest Manag. Sci.* 66, 74–83.

Hernandez-Campuzano, B., Suarez, R., Lina, L., Hernandez, V., Villegas, E., et al., 2009. Expression of a spider venom peptide in transgenic tobacco confers insect resistance. *Toxicon* 53, 122–128.

Khan, S.A., Zafar, Y., Briddon, R.W., Malik, K.A., Mukhtar, Z., 2006. Spider venom toxin protects plants from insect attack. *Transgenic Res* 15, 349–357.

King, G.F., 2007. Modulation of insect Ca_V channels by peptidic spider toxins. *Toxicon* 49, 513–530.

King, G.F., Gentz, M.C., Escoubas, P., Nicholson, G.M., 2008. A rational nomenclature for naming peptide toxins from spiders and other venomous animals. *Toxicon* 52, 264–276.

Mukherjee, A.K., Sollod, B.L., Wikel, S.K., King, G.F., 2006. Orally active acaricidal peptide toxins from spider venom. *Toxicon* 47, 182–187.

Nicholson, G.M., 2007a. Insect-selective spider toxins targeting voltage-gated sodium channels. *Toxicon* 49, 490–512.

Nicholson, G.M., 2007b. Fighting the global pest problem: preface to the special Toxicon issue on insecticidal toxins and their potential for insect pest control. *Toxicon* 49, 413–422.

Pham Trung, N., Fitches, E., Gatehouse, J.A., 2006. A fusion protein containing a lepidopteran-specific toxin from the South Indian red scorpion (*Mesobuthus tamulus*) and snowdrop lectin shows oral toxicity to target insects. *BMC Biotechnol 6*, 18.

Rohou, A., Nield, J., Ushkaryov, Y.A., 2007. Insecticidal toxins from black widow spider venom. *Toxicon 49*, 531–549.

Tedford, H.W., Sollod, B.L., Maggio, F., King, G.F., 2004. Australian funnel-web spiders: master insecticide chemists. *Toxicon 43*, 601–618.

Wang, C., St Leger, R.J., 2007. A scorpion neurotoxin increases the potency of a fungal insecticide. *Nat. Biotechnol. 25*, 1455–1456.

Whetstone, P.A., Hammock, B.D., 2007. Delivery methods for peptide and protein toxins in insect control. *Toxicon 49*, 576–596.

Wood, D.L., Miljenovic, T., Cai, S., Raven, R.J., Kaas, Q., et al., 2009. ArachnoServer: a database of protein toxins from spiders. *BMC Genomics 10*, 375.

5 Baculoviruses: Biology, Biochemistry, and Molecular Biology

B C Bonning, Iowa State University, Ames, IA, USA

© 2005, Elsevier BV. All Rights Reserved.

5.1. Introduction

Initial interest in baculoviruses stemmed from their potential use in insect pest management (Moscardi, 1999). Modern baculovirology is driven by genetic enhancement of their insecticidal properties (van Beek and Hughes, 1998; Bonning *et al.*, 2002), their use for the study of fundamental biological processes (Clem, 2001; Manji and Friesen, 2001), for protein expression (Jarvis, 1997), and for gene therapy (Ghosh *et al.*, 2002; Kost and Condreay, 2002). Addressing their limitations for these purposes allows for increased understanding of baculovirus biology. As a result of the various applications of baculoviruses, they represent the best studied of the invertebrate viruses.

Autographa californica multiple nucleopolyhedrovirus (AcMNPV), the type species of the *Nucleopolyhedrovirus* genus (Blissard *et al.*, 2000), has been particularly extensively studied both in terms of molecular biology and host range. Hence much of this review will focus on AcMNPV with particular emphasis on recent advances in baculovirus research. Relatively little molecular, genetic, and biochemical research has been conducted on granuloviruses (GV), in part because of the difficulty of establishing cell culture systems for GV propagation (Winstanley and Crook, 1993). Certain aspects of baculovirus biology such as baculovirus ecology (Cory *et al.*, 1997), epizootiology (Fuxa and Tanada, 1987), baculoviruses of nonlepidopteran hosts (Couch, 1991; Federici, 1997), and tritrophic interactions (Duffey *et al.*, 1995) are described elsewhere.

5.1.1. Taxonomy

Baculoviruses belong to the family Baculoviridae, a large family of bacilliform, occluded viruses that infect arthropods, primarily the Lepidoptera (Adams and McClintock, 1991). Baculoviruses also have been isolated from members of the Diptera and Hymenoptera (Blissard *et al.*, 2000). Members of the two genera within the Baculoviridae, *Nucleopolyhedrovirus* (NPV) and *Granulovirus* (GV), are distinguished by occlusion body (OB) morphology and the cytopathology that they cause (Blissard *et al.*, 2000). NPVs produce large polyhedra (typically 1–5 μm in diameter) with many virions or occlusion-derived virus (ODV), while GV produce smaller occlusion bodies or granules (150 nm in diameter by 400–600 nm in length) with a single virion (**Figure 1**). NPVs are designated as SNPV or MNPV according to whether the ODV contains single (S) or multiple (M) nucleocapsids (**Figure 1**). There are many more described MNPVs than SNPVs (Volkman, 1997). Because all of the MNPV characterized to date infect species within the order Lepidoptera, while SNPV have been isolated from species within the Lepidoptera, Hymenoptera, and Diptera, the progenitor NPV is believed to be an SNPV (Rohrmann, 1996). Despite morphological similarity to baculoviruses, at least one of the shrimp baculoviruses, the white spot syndrome virus, is unrelated at the amino acid level (Yang *et al.*, 2001) and has been reclassified in the family Nimaviridae, genus *Whispovirus*.

The NPVs are divided into two groups on the basis of phylogenetic relatedness (Zanotto *et al.*, 1993; Herniou *et al.*, 2001). Group I NPVs have 17 genes that are absent from group II NPVs (Herniou *et al.*, 2001). The two groups have distinct budded

virus (BV) envelope fusion proteins, with GP64 in group I NPVs, and F (fusion) proteins in group II NPVs (Ijkel *et al.*, 1999; Pearson *et al.*, 2000). Homologs of GP64 are present in the thogotoviruses (Orthomyxoviridae) (Morse *et al.*, 1992), while proteins related to F proteins are present in the genus *Errantivirus* (Metaviridae) (Malik *et al.*, 2000; Rohrmann and Karplus, 2001; Pearson and Rohrmann, 2002).

GVs are classified according to tissue tropism: type 1 GVs such as *Trichoplusia ni* GV enter the host via the midgut epithelium but infect only the fat body. Because other important tissues are not affected, infection may be prolonged with lethargy only a day or two before death. For type 2 GV, tissue tropism is similar to that of NPVs affecting most tissues, while type 3 GVs are restricted to the midgut epithelium. *Harrisina brillians* GV is the only type 3 GV characterized to date and causes acute disease (Federici, 1997). The pathogenesis of NPV diseases has been studied in much greater detail compared to GV disease. A fundamental difference between the cytopathology of NPV and GV infections is that the nuclear membrane breaks down during replication of GV but not NPV, resulting in mixing of cytoplasm and nucleoplasm (Winstanley and Crook, 1993). GV-infected cells separate from each other and from the basement membrane, resulting in proliferation of regenerative cells adjacent to the basement membrane, which become infected as they differentiate.

5.1.2. Nomenclature

Baculoviruses are named according to viral genus based on morphological characteristics, and the Latin name of the host from which they were first isolated. This system can be misleading because many baculoviruses infect more than one host species. AcMNPV for example, although first isolated from *A. californica*, infects at least 40 other species within the Lepidoptera (Gröner, 1986). Complexities also arise when the same virus is isolated from different host species (e.g., *Rachiplusia ou* MNPV and *Anagrapha falcifera* MNPV) (Hostetter and Puttler, 1991; Harrison and Bonning, 1999), and when different viruses are isolated from the same host species (e.g., *Mamestra configurata* MNPV A and B) (Li *et al.*, 2002a, 2002b).

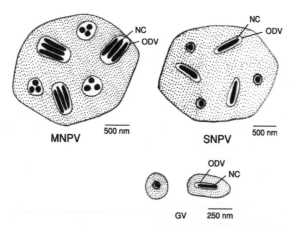

Figure 1 Schematic representation of occlusion bodies of multiple and single nucleopolyhedroviruses (MNPV and SNPV, respectively) and granuloviruses (GVs). The figure illustrates envelopment of multiple or single rod-shaped nucleocapsids (NC) within occlusion-derived virus (ODV) of the polyhedra of MNPV and SNPV, respectively. Each granule of the GV contains a virion with a single nucleocapsid. The relative sizes of the GV and NPV are indicated.

5.2. Structure

Baculoviruses exist as two phenotypes, BV and occlusion-derived virus (ODV), which have a common nucleocapsid structure and carry the same genetic information (Blissard, 1996; Rohrmann, 1992). ODV, contained within the occlusion bodies

(OB: polyhedra or granules), are responsible for transmission between host insects, while BV effect systemic transmission within the host, and virus propagation in cell culture.

5.2.1. Nucleocapsids

Nucleocapsids are tubular in shape with cap structures at each end (**Figure 2**). The genome is condensed to about 100-fold within the inner nucleoprotein core (Bud and Kelly, 1980). The supercoiled, circular genome of double-stranded DNA is complexed with a highly basic protein, P6.9 in NPV and VP12 in GV (Tweeten *et al.*, 1980; Wilson *et al.*, 1987). Binding of these arginine-rich molecules to the DNA produces a compact, insoluble DNA complex. The viral genomes appear to be prepackaged within the

virogenic stroma (an electron-dense structure produced within the nucleus at the onset of viral DNA synthesis), before incorporation into the capsid shells (Fraser, 1986) (**Figure 3**). The end structures of nucleocapsids consist of a flat disk at the basal end and nipple structures at the apical end (Federici, 1986). Nucleocapsids attach to membranes at the apical end of the capsid. VP39 is the major component of the nucleocapsid in BV and OB of AcMNPV (Pearson *et al.*, 1988; Thiem and Miller, 1989). P80 and P24 are also associated with the capsid and PP78/83 is associated with the basal end complexed with EC27 and C42 (Braunagel *et al.*, 2001) (**Figure 2**). VP1054 and VP91 are associated with both BV and OB (Olszewski and Miller, 1997a; Russell and Rohrmann, 1997).

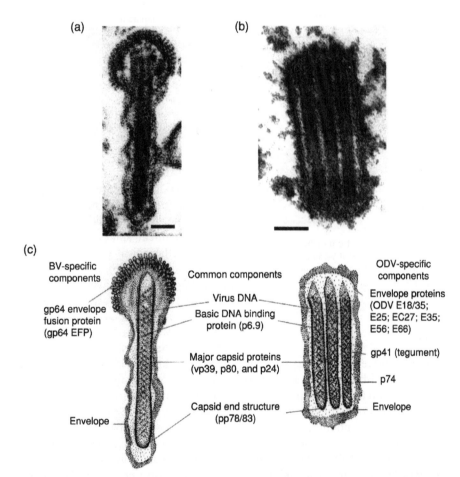

Figure 2 Structure of budded and occlusion-derived virus. (a) Transmission electron micrograph (TEM) of budded virus (BV) of *Lymantria dispar* MNPV showing glycoprotein spikes, or peplomers at the anterior end. Scale bar = 50 nm. (Reproduced with permission from Adams, J.R., Goodwin, R.H., Wilcox, T.A., **1977**. Electron microscopic investigations on invasion and replication of insect baculoviruses *in vivo* and *in vitro*. *Biol. Cellulaire 28*, 261–268.) (b) TEM of occlusion-derived virus (ODV) of *Agrotis ipsilon* MNPV (AgipMNPV) in a hemocyte of *Agrotis ipsilon*. Scale bar 100 nm. (c) Schematic representation of BV and ODV indicating virus-encoded structural components common to both or unique to each structure (after Funk *et al.*, 1997). References for structural proteins: P6.9 (Wilson *et al.*, 1987); VP39 (Thiem and Miller, 1989); P80 (Lu and Carstens, 1992); PP78/83 (Vialard and Richardson, 1993); ODV-E25 (Russell and Rohrmann, 1993); ODV-E66 (Hong *et al.*, 1994); ODV-E56 (Braunagel *et al.*, 1996a; Theilmann *et al.*, 1996); ODV-E18, -E35, and -E27 (Braunagel *et al.*, 1996b); GP41 (Whitford and Faulkner, 1992); p74 (Kuzio *et al.*, 1989); GP64 (group I NPVs: Whitford *et al.*, 1989) or F protein (group II NPVs and GVs: Pearson *et al.*, 2001).

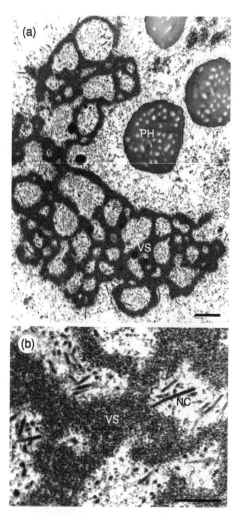

Figure 3 (a) Detail of the virogenic stroma (VS) and polyhedra (PH) within the nucleus of an AgipMNPV-infected hemocyte of *Agrotis ipsilon*. Scale bar = 1 μm. (b) Rod-shaped nucleocapsids (NC) produced within the virogenic stroma (VS). Scale bar = 500 nm.

5.2.2. Budded Virus

Nucleocapsids initially move out of the nucleus for exit at the plasma membrane, and appear to move in vesicles derived from the nuclear membrane in association with filaments of actin. The presence of multivesicular aggregates containing many nucleocapsids suggests that fusion of vesicles may occur (Williams and Faulkner, 1997). The nuclear membrane-derived vesicle is lost in transit and unenveloped nucleocapsids align at the apical end with the plasma membrane, which is modified by the viral-encoded glycoprotein, GP64 (**Figures 4** and **5**). The GP64 envelope fusion protein is the major virus-encoded protein associated with the BV envelope. GP64 is expressed during the early and late phases (Blissard and Rohrmann, 1989; Whitford *et al.*, 1989; Monsma and Blissard, 1995; Monsma *et al.*,

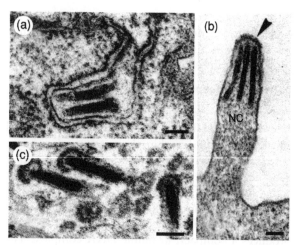

Figure 4 Production of budded virus. (a) Nucleocapsids become encased in a vesicle of nuclear membrane as they move out of the nucleus. Nucleocapsids are subsequently released into the cytoplasm as the vesicle breaks down. (*Autographa californica* MNPV in *Spodoptera exigua* fat body cell.) (b) Nucleocapsids move to and bud through regions of the plasma membrane that have been modified by virus-produced GP64 (arrow head). (*Helicoverpa zea* SNPV in *H. zea* cell line.) (c) Budded virus with distinctive peplomer structures at the anterior end. (*Trichoplusia ni* SNPV in *T. ni*.) Scale bars = 100 nm. (Reproduced with permission from Adams, J.R., Goodwin, R.H., Wilcox, T.A., **1977**. Electron microscopic investigations on invasion and replication of insect baculoviruses *in vivo* and *in vitro*. Biol. Cellulaire 28, 261–268.)

1996) and accumulates at the plasma membrane (Volkman, 1986). On budding through the plasma membrane, nucleocapsids acquire a loose envelope with prominent spikes or peplomers consisting of GP64 at the apical end (Volkman and Goldsmith, 1984; Volkman, 1986). Ubiquitin is localized to the inner surface of the envelope (Guarino *et al.*, 1995). GP64 is a membrane fusion protein that is activated at low pH and is involved in BV entry into the cell by endocytosis (Blissard and Wenz, 1992; Chernomordik *et al.*, 1995; Monsma and Blissard, 1995). The BV envelope fuses with host endosomal membranes thereby releasing nucleocapsids into the cytoplasm. GP64 is essential for cell-to-cell movement of virus (Monsma *et al.*, 1996). Baculovirus GP64 is related to the envelope glycoproteins of two orthomyxovirus-like arboviruses that are vectored by ticks. These proteins mediate membrane fusion and hemagglutination (Morse *et al.*, 1992). GP64 is absent from ODV (Blissard and Rohrmann, 1989; Jarvis and Garcia, 1994) and also from the two variants of the single NPV of the S phenotype for which the genome has been completely sequenced (Chen *et al.*, 2001, 2002).

GP64 is absent from group II NPVs, which have F proteins that serve a similar functional role. However, group I NPVs also encode F protein homologs that

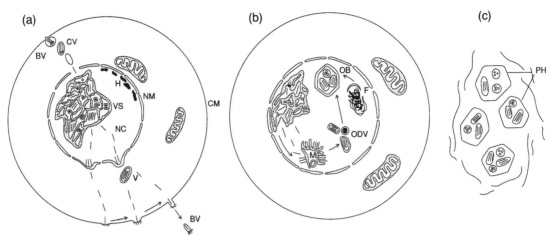

Figure 5 Production of budded virus and occlusion bodies during the late and very late phases of NPV infection, followed by cell lysis. (a) A budded virus (BV) binds to a putative receptor at the cell membrane (CM) and enters the cell by adsoprtive endocytosis within clathrin-coated vesicles (CV). The BV envelope fuses with host endosomal membranes thereby releasing nucleocapsids into the cytoplasm. Nucleocapsids uncoat at the nuclear membrane. Heterochromatin (H) becomes marginated and the virogenic stroma (VS) is produced at the onset of virus replication and the late phase of infection. Nucleocapsids (NC) produced within the virogenic stroma move out of the nucleus initially in vesicles (V) that appear to be derived form the nuclear membrane (NM). The vesicles dissociate and nucleocapsids bud through the GP64-modified regions of the plasma membrane to acquire a loose envelope with peplomers at the apical end. (b) A switch occurs at about 24 h after infection of the cell from BV to OB production, marking the onset of the very late phase of infection. Intranuclear vesicles dissociate to produce membranes (M) within the ring zone of the infected nucleus, which then envelope nucleocapsids to produce occlusion-derived virus (ODV). Fibrillar bodies (F) are produced that associate with polyhedral envelope precursor structures. ODV become embedded in the polyhedrin matrix, which is then surrounded by the polyhedral envelope. (c) Polyhedra (PH) are released on lysis of the infected cell.

are more divergent than those found in group II NPVs and GVs. The F protein homologs in group I NPVs appear to lack a fusion function (Pearson *et al.*, 2000; Lung *et al.*, 2003). There is overlap between the distribution of GP64 and the F protein homologs in the budded virus, but biochemical fractionation studies suggest that the F protein homologs are also associated with BV nucleocapsids. This localization suggests that the F protein homologs may play a role in the interaction between the budded virus envelope and the nucleocapsid potentially mediated by the cytoplasmic tail domain of the protein (Pearson *et al.*, 2001). However, the F protein homolog of AcMNPV is not essential for virus replication in cell culture or in insects. The AcMNPV F protein homolog has been shown to function as a viral pathogenicity factor that accelerates the mortality of infected hosts (Lung *et al.*, 2003).

The baculovirus F protein is related to the envelope protein of insect retroviruses in the genus *Errantivirus* (family Metaviridae). Evidence suggests that a *gypsy*-like noninfectious retrotransposon acquired a baculovirus F progenitor resulting in formation of a retrovirus (Malik *et al.*, 2000; Rohrmann and Karplus, 2001; Pearson and Rohrmann, 2002). A *Drosophila melanogaster* F homolog, which is not part of a retrovirus-like element, has also been identified, but it is unclear whether this is an insect

gene, or a remnant of an integrated retrotransposon (Malik *et al.*, 2000; Rohrmann and Karplus, 2001). Hence the evolutionary events leading to the presence of F protein homologs in insects, baculoviruses, and insect retroviruses remain unclear. It is possible that the insect F homolog may represent the progenitor of both the baculovirus F protein and the envelope proteins of insect retroviruses.

Both viral and host-derived ubiquitin with an unusual phospholipid anchor have been found on the inner surface of the BV envelope (Guarino, 1990; Guarino *et al.*, 1995). Free ubiquitin is also present between the nucleocapsid and envelope and is almost as abundant as GP64, accounting for 2% of the weight of the virion. Ubiquitin is one of the most highly conserved of known proteins, but diverged considerably since acquisition by the baculoviruses. Baculovirus ubiquitin is the most divergent ubiquitin known with only 76% identity to animal ubiquitin (Guarino, 1990). A primary function of ubiquitin is to signal degradation of proteins by the 26S proteosome by covalent attachment of ubiquitin molecules to the protein. However, the viral protein supports protein degradation at only 40% the rate of the eukaryotic ubiquitin in *in vitro* assays, and multiubiquitination of some proteins was inhibited (Haas *et al.*, 1996). This suggests that virus ubiquitin may prevent degradation of selected proteins during infection. Analysis of ubiquitin mutants of AcMNPV showed that ubiquitin may function in BV assembly although *ubi* is not an essential gene (Fraser *et al.*, 1995; Reilly and Guarino, 1996). The baculovirus proteins IAP2, IE2, and PE38, which have RING finger domains, may function as ubiquitin ligase (E3) enzymes during baculovirus infection (Imai *et al.*, 2003).

There are various other proteins that may be associated with BV including IE1 (Theilmann and Stewart, 1993) and actin (Lanier *et al.*, 1996). The viral cathepsin V-Cath is associated with both the nucleocapsids and the envelopes of BV, and degrades actin (Lanier *et al.*, 1996). This viral cathepsin L-like cysteine protease is required for liquefaction of the host insect during the final stages of virus infection. In the absence of this protease, tissues retain their integrity (Slack *et al.*, 1995).

5.2.3. Occlusion-Derived Virus

At about 24 h post infection (p.i.), a switch occurs from BV to OB production, although the switch is not absolute (Faulkner, 1981) (**Figure 5**). The FP25K protein is a potential regulator of the switch from BV to OB production during baculovirus infection (Jarvis *et al.*, 1992). During the very late phase, the nucleocapsids remain in the nucleus when sufficient FP25K protein is produced. The FP25K protein, which is expressed during the late phase, may be involved in the retention of the nucleocapsids within the nucleus (Harrison and Summers, 1995). FP25K is predominantly found in the cytoplasm of baculovirus-infected cells (Braunagel *et al.*, 1999). Disruption of FP25K

expression that occurs frequently on repeated passaging of virus in cell culture, leads to the few polyhedra (FP) phenotype with (1) enhanced BV production, (2) decreased polyhedra production, (3) increased synthesis of some BV structural proteins, (4) reduced levels of some ODV envelop proteins, (5) decreased polyhedrin protein synthesis, and (6) a block in the postmortem liquefaction of the host (Harrison and Summers, 1995; Harrison *et al.*, 1996; Katsuma *et al.*, 1999; Rosas-Acosta *et al.*, 2001).

During the very late phase of infection, trilamellate membrane profiles appear within the ring zone of the nucleus and envelope nucleocapsids to form ODV (Stoltz *et al.*, 1973; Fraser, 1986; Williams and Faulkner, 1997). These membranes may derive from invaginations of the inner nuclear membrane that produce microvesicles within the infected nucleus (Fraser, 1986; Hong *et al.*, 1994) (**Figures 5** and **6**). Regulation of the ODV-specific components in the viral envelope may be important for the switch from BV to OB production (Braunagel and Summers, 1994; Braunagel *et al.*, 1996b; Williams and Faulkner, 1997).

Nomenclature for some viral envelope proteins includes designation of protein source (BV or ODV), followed by location, E (envelope) or C (capsid), and the apparent molecular mass (Braunagel *et al.*, 1996a). ODV-E66 (Hong *et al.*, 1994, 1997) and ODV-E25 (Russell and Rohrmann, 1993) are both envelope proteins found in the virus-induced microvesicles in the cell nucleus. ODV-E56 is present in ODV envelopes, virus-induced intranuclear microvesicles, and the inner and outer nuclear membranes (Braunagel *et al.*,

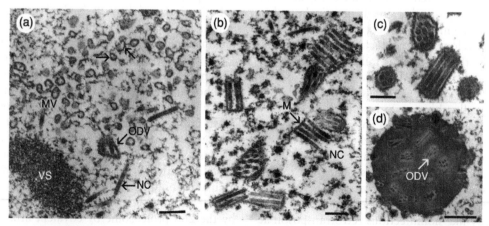

Figure 6   Production of occluded virus of AgipMNPV: intranuclear envelopment and occlusion of nucleocapsids. During the very late phase of infection, nucleocapsids (NC) are enveloped as they leave the virogenic stroma (VS). (a) Membrane profiles (arrows) appear to break down to produce microvesicles (MV), the occlusion-derived virus (ODV) envelope precursors in the ring zone of NPV-infected cells. (b) These microvesicles appear to produce trilamellar membrane profiles (M) distal to the nuclear membrane. Nucleocapsids (NC) align with these membranes, and (c) are enveloped to produce ODV. Transverse and longitudinal sections illustrate the bundling of multiple nucleocapsids within the ODV of MNPV. (d) The ODV are then occluded within the polyhedrin matrix. (a) and (d) fat body, (b) and (c) hemocyte of *Agrotis ipsilon*. Scale bar = 200 nm (a, b, and c), 500 nm (d).

1996a; Theilmann *et al.*, 1996). ODV-E35 shares the same N-terminus with ODV-E18 and may be translated from a transcript that includes both AcMNPV ORF143 (ODV-E18) and 144 (ODV-EC27) derived by ribosomal frameshifting at a step-loop structure (Braunagel *et al.*, 1996b). ODV-EC27 is present in ODV nucleocapsids and envelopes, and is also present in a modified form in BV (Braunagel *et al.*, 1996b). ODV-EC27 functions as a multifunctional viral cyclin with roles that may include arrest of the host cell cycle at G_2/M phase, or regulation of viral/cellular DNA replication (Belyavskyi *et al.*, 1998). P74 is associated with the ODV envelope and is essential for oral infectivity. It is hypothesized to function in attachment to or fusion with midgut epithelial cells (Kuzio *et al.*, 1989; Hill *et al.*, 1993; Faulkner *et al.*, 1997; Slack *et al.*, 2001). The *p74* gene lacks a conventional late promoter sequence, but is expressed at low levels during the late phase of infection.

The tegument protein GP41 is present between the envelope and the nucleocapsid of ODV (Whitford and Faulkner, 1992). GP41 is required for movement of the nucleocapsids out of the infected cell nucleus to produce BV (Olszewski and Miller, 1997b). Protein tyrosine phosphatase (PTP) has also been found associated with both BV and ODV (Li and Miller, 1995).

The lipid composition of the BV and ODV envelopes differs and ODV envelopes contain a greater density of protein (Braunagel and Summers, 1994). For AcMNPV BV derived from Sf9 cells, the major phospholipid constituent of the envelope is phosphatidylserine, comprising about 50% of the phospholipid content. For the ODV envelope, phosphatidylcholine and phosphatidylethanolamine are the predominant constituents. The BV envelope appears to be more fluid than that of ODV, which may be important for the different roles of these phenotypes in the baculovirus life cycle (Braunagel and Summers, 1994).

5.2.4. Occlusion Bodies

Large amounts of the occlusion matrix protein polyhedrin (or granulin) and a nonstructural protein P10 are produced during the very late phase of infection. These proteins accumulate in the cytoplasm (detectable at 10–12 and 14–16 h p.i.), and after about 20 h p.i. are also located in the nucleus. Polyhedrin becomes primarily intranuclear while P10 forms fibrillar bodies in both the cytoplasm and the nucleus (Vlak *et al.*, 1988; Williams *et al.*, 1989) (**Figure 7**). Large amounts of polyhedrin are required for production of OB. The ODV are embedded in the occlusion matrix of polyhedrin or granulin (Rohrmann,

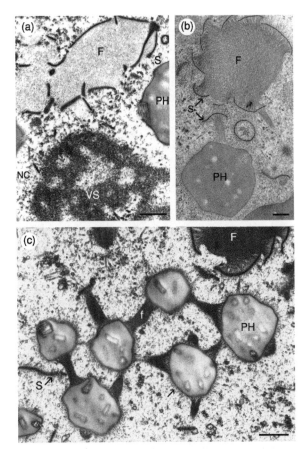

Figure 7   Maturation of the occlusion body. (a, b) Bilamellar fibrous sheets (S), which are calyx (polyhedral envelope) precursor structures, associate with the P10-rich fibrillar body (F). These sheets appear to form near to the nuclear membrane before incorporation into the fibrillar body. Nucleocapsids (NC), virogenic stroma (VS) and polyhedra (PH) are indicated. (c) The fibrillar bodies (F) and fibrillar structures (f) associated with the maturing polyhedra (PH) form a shell around the surface of the polyhedra prior to or during addition of the calyx (arrow). Addition of the calyx is the final step in OB maturation. (a), AgipMNPV, (b, c) *Rachiplusia ou* MNPV (RoMNPV). Scale bar = 500 nm.

1986, 1992; Funk *et al.*, 1997) (**Figure 6**). Polyhedrin is highly conserved (25–33 kDa) and related to the GV matrix protein granulin (with about 50% amino acid identity) that produces a paracrystalline lattice and accounts for 95% of the OB mass (Faulkner, 1981; Vlak and Rohrmann, 1985; Rohrmann, 1992). Point mutations in conserved domains of the polyhedrin gene have resulted in (1) a single cubic OB per cell with a paracrystalline matrix that occludes few virions (Brown *et al.*, 1980; Jarvis *et al.*, 1991), (2) small polyhedrin condensations that lack the lattice structure and do not occlude virions (Duncan *et al.*, 1983), and (3) large amorphous condensations of polyhedrin (Carstens *et al.*, 1992).

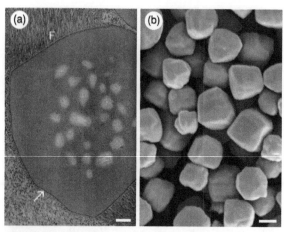

Figure 8 Occlusion bodies. (a) Mature polyhedron of RoMNPV showing polyhedral envelope or calyx (arrow) surrounding the polyhedron. The adjacent fibrillar body (F) is also indicated. Scale bar = 200 nm. (b) Scanning electron micrograph showing the polyhedral structure of occluded virus of RoMNPV derived from larvae of *Heliothis virescens*. Scale bar = 1 μm.

OB production is completed by addition of the polyhedron envelope or calyx (**Figure 8**), which consists primarily of carbohydrate and the phosphoprotein PP34, also known as the polyhedral envelope protein, PEP (Minion *et al.*, 1979; Gombart *et al.*, 1989). The precursor structures referred to as bilamellar fibrous sheets or electron-dense spacers form in the ring zone of the infected cell in association with P10 fibrillar bodies (Russell *et al.*, 1991) (**Figure 7**), although calyx formation is not dependent on the expression of P10 (van Lent *et al.*, 1990; Oers and Vlak, 1997). Thiol bonds link PP34 molecules to each other and to polyhedrin (Whitt and Manning, 1988). Mutants lacking the calyx show increased sensitivity to alkali and have pitted surfaces, suggesting possible loss of virions from the OB matrix (Zuidema *et al.*, 1989; Gross, 1994).

Occlusion bodies are resistant to desiccation and freezing (Jaques, 1985), and to putrefaction and various chemical treatments (Benz, 1986), but do not protect ODV well against the harmful effects of ultraviolet light (Ignoffo *et al.*, 1995, 1997). Baculoviruses lose most of their activity within 24 h as a result of damage to DNA when exposed to direct sunlight (Young and Yearian, 1974). On exposure to ultraviolet light, pyrimidine dimers are produced. Baculovirus expression of a pyrimidine dimer-specific glycosylase conferred some protection to BV but not to OB (Petrik *et al.*, 2003). It has been postulated that the baculovirus superoxide dismutase gene (*sod*) may function to protect OB from superoxide radicals generated on exposure to sunlight (Tomalski *et al.*, 1991; Ignoffo and Garcia, 1994).

5.3. Life Cycle

5.3.1. Infection

Baculoviruses infect the larval stages of insect hosts. Caterpillars typically ingest occlusion bodies (OB) (polyhedra or granules) on feeding, although NPVs can be transmitted via the egg. The polyhedrin or granulin matrix of the occlusion bodies dissolves rapidly in the alkaline conditions of the caterpillar midgut (pH 9.5–11.5), releasing the infectious ODV (**Figure 9**). Occlusion dissolution is also dependent on ionic strength (Kawanishi and Paschke, 1970) and is facilitated by midgut proteases (Granados and Williams, 1986). The ODV must then pass through the peritrophic membrane, an extracellular fibrous matrix of chitin, glycosaminoglycans, and protein that lines the midgut (Richards and Richards, 1977; Brandt *et al.*, 1978; Wang and Granados, 2001). GV and some NPV occlusions contain a viral enhancing factor, a metalloprotease called enhancin. Enhancin, which is released into the midgut on occlusion dissolution, facilitates penetration of the peritrophic membrane by degrading mucin (Derksen, 1988; Gallo *et al.*, 1991; Wang *et al.*, 1994; Gijzen *et al.*, 1995; Lepore *et al.*, 1996; Bischoff and Slavicek, 1997; Wang and Granados, 1997, 1998; Popham *et al.*, 2001). The chitin-binding agent calcofluor (M2R) inhibits formation of the peritrophic membrane (Wang and Granados, 2000) and increases the susceptibility of larvae to baculovirus infection. Calcofluor also the enhances the retention of infected midgut cells (Washburn *et al.*, 1998) and the relative contributions of these two effects to calcofluor-enhanced susceptibility of the host insect are unclear (Washburn *et al.*, 1998; Wang and Granados, 2001). It is also unclear how viruses that lack enhancin penetrate the peritrophic membrane. Possibilities include that virus infection occurs before the peritrophic membrane is fully formed, particularly in insects that produce the peritrophic membrane in response to feeding (Tellam, 1996), or through variation in peritrophic membrane structure in different species of Lepidoptera (Hopkins and Harper, 2001). Alternatively, there may be other virus proteins with peritrophic membrane-degrading activity.

Virions cross the peritrophic membrane and bind to the brush border microvilli of epithelial cells (**Figure 9**). ODV are highly infectious to midgut epithelial cells and less infectious to other tissues (Volkman and Summers, 1977). The primary target cells are the columnar cells of the midgut epithelium (Granados and Williams, 1986). ODV enter the cells by direct fusion of the viral and cellular membranes (Granados and Lawler, 1981), and the entry is likely

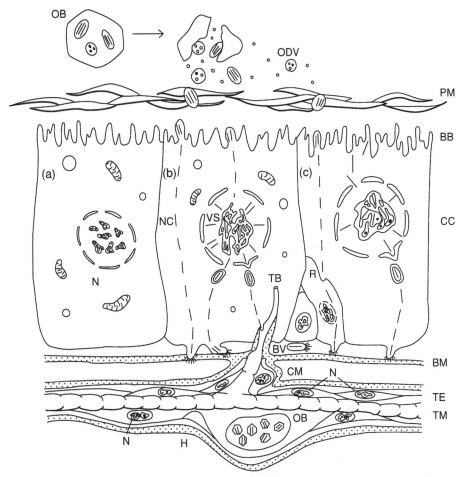

Figure 9 Infection of the midgut epithelium of the host insect by MNPV. Occlusion bodies (OB) dissolve in the midgut releasing ODV, which move through the peritrophic membrane (PM). (a) Uninfected columnar cell (CC) showing the normal distribution of heterochromatin in the nucleus (N); (b) columnar cell showing entry of virus at the brush boarder membrane (BB). Some nucleocapsids (NC) move to the nucleus, while other nucleocapsids undergo trancytosis. Infection of tracheoblasts (TB) closely associated with the midgut epithelium and subsequent infection of the tracheal epithelium (TE), as indicated by the presence of OB, is also shown. (c) Direct infection of regenerative cells (R), by transcytosis as for *Spodoptera exigua*, in addition to infection of columnar cell (CC). Budded virus acquires an envelope as it buds through the virus-modified plasma membrane. BV may enter the hemocoel (H) by budding through the basement membrane (BM) or via the tracheal epithelium (TE). Enlargement of the nucleus on infection of the cell is shown in (b) and (c). VS, virogenic stroma; CM, circular muscle; TM, tracheal matrix.

to involve a highly abundant or relatively non-specific receptor site (Horton and Burand, 1993). ODV entry is insensitive to inhibitors of adsorptive endocytosis.

Nucleocapsids move to the nucleus and uncoat, or may pass directly through the cell to infect other cells, or enter the hemocoel (Granados and Lawler, 1981; Keddie *et al.*, 1989) (**Figure 9**). The mechanism for movement of nucleocapsids within microvilli is unknown in the absence of microtubules (Volkman, 1997). BV bud from the basal side of the midgut epithelium and infectious virus can be detected in the hemolymph a few hours after infection (Granados and Lawler, 1981). When multiple nucleocapsids infect a single columnar epithelial

cell, some nucleocapsids derived from ODV can be converted directly into BV without replication in the nucleus, taking advantage of GP64 that is expressed as an early gene on entry of other nucleocapsids into the nucleus (Washburn *et al.*, 2003a). Infection with the M or S phenotype of NPV results in infection of primary tissues with multiple or single nucleocapsids, respectively. Infection of secondary tissues is more rapid for the M phenotype supporting the hypothesis that ODV nucleocapsids can be repackaged as budded virus, while systemic infections of the S phenotype require virus replication in the primary target cells (Washburn *et al.*, 2003a). Direct acquisition of the envelope protein GP64 by ODV-derived nucleocapsids of the M phenotype allows rapid movement of the virus through the midgut epithelial cells for further dissemination within the host insect (Washburn *et al.*, 2003b). In the case of *Spodoptera exigua*, ODV-derived nucleocapsids of AcMNPV infect both the columnar cells and the underlying regenerative cells (Flipsen *et al.*, 1995b) (**Figure 9**). Some lepidopteran larvae clear AcMNPV-infected cells from the midgut with almost full recovery of the tissue by 60 h p.i. (Keddie *et al.*, 1989; Washburn *et al.*, 2003b). The direct movement of virus through the midgut cells would provide an advantage to the virus in larvae that shed infected midgut epithelial cells and may have provided the driving force for evolution of the MNPV from the SNPV. The clearing of infected cells from the midgut may benefit MNPV by allowing the host insect to continue to feed (Volkman and Keddie, 1990).

Infection results in rapid formation of filamentous actin (F-actin) bundles, which may facilitate transport of nucleocapsids to the nucleus (Charlton and Volkman, 1993; Lanier and Volkman, 1998). Nucleocapids migrate to the nucleus and uncoat at the nuclear pore complex (GV) or just inside the nuclear membrane (Bilimoria, 1991). A capsid-associated protein kinase associated with NPV and GV virions may function in unpackaging of the viral genome by phosphorylation of the basic core protein P6.9 (Miller *et al.*, 1983; Wilson and Consigli, 1985; Funk and Consigli, 1993). Transcription of immediate early genes rapidly follows with virus-specific RNA detectable within 30 min of infection (Chisholm and Henner, 1988). At this stage, F-actin aggregates form at the plasma membrane mediated by an early viral gene product, actin rearrangement-inducing factor-1 (ARIF-1) which colocalizes with the F-actin aggregates (Dreschers *et al.*, 2001).

Numerous morphological changes occur in the cell and the nucleus during the course of infection.

Within the cytoplasm, there are alterations to the cytoskeleton and endomembrane systems, and large fibrillar structures develop. Within the first hour after infection, thick microfilament cables form from nucleocapsid-induced rearrangement of cellular actin (Charlton and Volkman, 1991). These cables, which are associated with nucleocapsids prior to viral gene expression, are only present at 1–4 h p.i. The capsid proteins P39 and P78/83 appear to interact with actin and induce actin polymerization, thereby enabling movement of nucleocapsids to and/or into the nucleus (Lanier and Volkman, 1998). Within the first 6–8 h p.i., the nucleus swells, the cell becomes rounded, and the distribution of heterochromatin and morphology of the nucleolus change. Rounding of the cell appears to be the result of virus-mediated rearrangement of the microtubules and is the first cytopathic effect seen by light microscopy following infection (Volkman and Zaal, 1990) (**Figure 10**). Heterochromatin becomes marginated along the inner nuclear membrane. The transition from early to late gene expression marks the onset of viral replication, and at this time the virogenic stroma, or viral replication center, is produced in the nucleus (Knudson and Harrap, 1976; Young *et al.*, 1993). The virogenic stroma is electron dense, contains intrastromal spaces, and has a convoluted periphery (**Figure 3**). Formation of the virogenic stroma is dependent on viral DNA synthesis and late gene expression (Carstens *et al.*, 1994). During the late phase, F-actin appears in the virogenic stroma and in the surrounding ring zone next to the inner nuclear membrane (Charlton and Volkman, 1991) (**Figure 11**), associated with P78/83 and VP39 (Charlton and Volkman, 1991; Lanier and Volkman, 1998). Nuclear F-actin is required for production of nucleocapsids (Ohkawa and Volkman, 1999). Six early viral genes, including *ie-1*, have been implicated

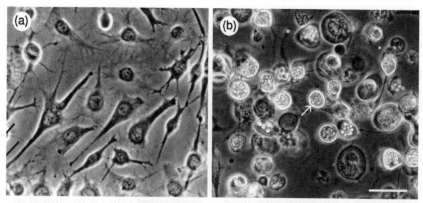

Figure 10   Cells of the *Manduca sexta* cell line GV1 illustrating (a) the normal fibroblast-like morphology with elongate extensions, and (b) rounding of cells on infection with AcMNPV. Refractile polyhedra are apparent within the distended nuclei by 3 days post infection (arrow). Rounding of the cells is the first response to baculovirus infection as seen by light microscopy. Photographs were taken under phase contrast. Scale bar = 50 μm.

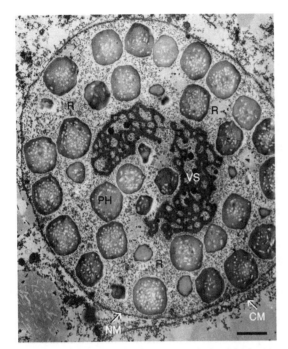

Figure 11 Hemocyte of *Agrotis ipsilon* infected with *Agrotis ipsilon* MNPV. The nucleus of the cell is shown with multiple polyhedra (PH) and the central, electron dense virogenic stroma (VS). The ring zone (R), nuclear membrane (NM), and cell membrane (CM) are indicated. Scale bar = 2 μm.

Table 1 Temporal classes of baculovirus gene expression; examples are provided to illustrate the progression of baculovirus gene expression from transregulatory genes, to those involved with DNA replication and production of budded and occluded virus in the late and very late stages of infection

Gene category	Early gene	Late genes	Very late genes
Transregulatory	ie0, ie1, ie2	ie1	
DNA replication	dnapol	lef3	
	p143		
Structural			
Nucleocapsid		p6.9	
		vp39	
		p80	
		p24	
		pp78/83	
		Tegument gp41	
BV	gp64	gp64	
		ubi	
ODV		odv-e66, odv-e25	
OB		pep	polh
Auxiliary	egt	sod	p10
	p35	ubq	
		cathepsin	
		ctl	
		pcna	
		ptp	
		protein kinase	

BV, budded virus; OB, occlusion bodies; ODV, occlusion-derived virus.

in recruitment of the monomeric G-actin to the nucleus (Ohkawa *et al.*, 2002). Progeny nucleocapsids appear within the virogenic stroma at about 10 h p.i. and synthesis of viral DNA continues for at least 12 h. As the virogenic stroma matures, the ring zone, a peristromal compartment of nucleoplasm appears (**Figure 11**). Intranuclear envelopment of virions and OB formation occurs in this compartment. Budded virus is produced during the late phase, and occluded virus is produced during the very late phase of virus infection (**Table 1**). The diameter of the nuclei of infected cells increases to five to ten times that of the nucleus of an uninfected cell (Federici, 1997). Baculovirus infection results in arrest of the cell cycle at the G_2/M phase, a phenomenon that may be mediated by viral proteins (Braunagel *et al.*, 1998).

Progeny nucleocapsids produced within the virogenic stroma initially bud through the plasma membrane modified by the virus-encoded protein GP64, to acquire a BV envelope (see Section 5.2.2). Later in infection nucleocapsids are enveloped and incorporated into the matrix of OB (see Section 5.2.3), while BV production is curtailed. Mature OB are released when the infected cell lyses. Large cytoplasmic vacuoles associate with both the nuclear and the outer plasma membranes and the nucleus disintegrates (van Oers *et al.*, 1993; Williams

and Faulkner, 1997). The P10 protein, cellular microfilaments and virus-produced cathepsin and chitinase are required for OB release from the host insect (Williams *et al.*, 1989; Hawtin *et al.*, 1997; Thomas *et al.*, 2000).

Production of occlusion bodies in the midgut epithelial cells does not occur for most lepidopteran insects. The high levels of actin associated with the microvilli of the columnar epithelial cells may have an inhibitory effect on the production of occluded virus (Volkman *et al.*, 1996). The infected cells are sloughed off when the larva molts and the midgut epithelium is regenerated (Flipsen *et al.*, 1993).

5.3.2. Dissemination within the Host

With the exception of lepidopteran NPVs and an NPV of the cranefly *Tipula paludosa*, all other NPVs infect only the midgut epithelium (Federici, 1993). In contrast, lepidopteran NPVs transiently infect the midgut and then spread to virtually all other tissues. With the exception of the GV of *H. brillians*, which is restricted to the midgut, all other GVs transiently infect the midgut before invading other tissues. For lepidopteran NPVs, systemic spread may occur through direct penetration of the basement membrane into the hemocoel (Granados and Lawler,

1981; Flipsen *et al.*, 1995b; Federici, 1997) and via the tracheal system (Engelhard *et al.*, 1994) although there is debate over whether one route predominates the other (Federici, 1997). Following infection of the midgut epithelial cells, viruses infect the tracheolar cells that service the midgut (Washburn *et al.*, 2001). The tracheoles often penetrate the basement membrane to supply tissues with oxygen, and thereby provide a means for the virus to bypass the basement membrane barrier. However, because tracheolar cells also are covered by basement membrane, it is not clear how BV would then enter the hemocoel without crossing a basement membrane. Given that the pores in the basement membrane are too small to allow physical passage of BV (Reddy and Locke, 1990), penetration of the basement membrane by BV may be an enzymatic process or may be possible when the basement membrane is thin, in early instars and surrounding tracheoles that are suspended in the hemolymph (Federici, 1997; Volkman, 1997). Because the midgut epithelial cells contribute to production of the overlying basement membrane, it is also possible that infection of these cells disrupts maintenance of the basement membrane (Knebel-Mörsdorf *et al.*, 1996), or that enlargement of infected cells alters basement membrane structure thereby allowing passage of budded virus. Following infection of *Heliothis virescens* and *Helicoverpa zea* with AcMNPV, infection of tracheal cells occurred 6–10 h prior to the appearance of BV in the hemolymph, suggesting that BV entering the hemocoel may be derived from the tracheal cells (Trudeau *et al.*, 2001).

BV enter cells by adsorptive endocytosis (Volkman and Goldsmith, 1985; Blissard and Wenz, 1992) and are transported in clathrin-coated vesicles to the endosome where they fuse with the endosome membrane thereby releasing nucleocapsids into the cytoplasm (Blissard, 1996). Fusion is mediated by GP64, the major envelope glycoprotein as the endosomal pH decreases (Monsma and Blissard, 1995).

The infection spreads within a tissue by cell-to-cell transmission. The tissue tropism varies from mono-organotrophic (e.g., *Neodiprion sertifer* NPV and *Culex nigripalpus* NPV that infect only the midgut epithelium) to polyorganotrophic with infection of most tissues (Granados and Williams, 1986; Moser *et al.*, 2001). Once the tracheal matrix and hemocytes become infected in susceptible polyorganotrophic hosts, infection spreads rapidly to the epithelial and fat body tissues (Keddie *et al.*, 1989; Engelhard *et al.*, 1994; Flipsen *et al.*, 1995b). Infection then proceeds to nerve, muscle, pericardial cells, and reproductive and glandular tissues. Some tissues such as the Malpighian tubules and salivary glands are not well infected with only immediate early genes

expressed suggesting that tissue tropism may be restricted by the availability of appropriate host factors to support virus replication (Knebel-Mörsdorf *et al.*, 1996).

The ability of the hemocytes of a host to support virus replication appears to be a key factor determining the susceptibility of some species to virus infection, through both virus amplification within the hemocoel, and the ability of the hemocytes to encapsulate sites of infection. In permissive hosts such as *H. virescens* and *T. ni*, the hemocytes play a primary role in amplification of virus in the hemocoel for initiation of new sites of infection. In some semipermissive hosts such as *H. zea*, although nucleocapsids of AcMNPV reach the nuclei of hemocytes, these cells do not support virus replication, and hence effectively remove virus from the hemocoel. The hemocytes of *H. zea* and *Manduca sexta* also appear to block virus dissemination by encapsulating infected tracheal elements possibly in response to ruptured basement membrane (Washburn *et al.*, 1996, 2000). Infection of hemocytes of the permissive host *H. virescens* compromises their ability to encapsulate (Trudeau *et al.*, 2001). There are contradictory reports on the role of hemocytes in virus dissemination within *Spodoptera frugiperda*, which is semipermissive to infection by AcMNPV. Clarke and Clem (2002) found lack of infection of hemocytes in this species, suggesting that the tracheal system may serve as the primary conduit for virus dissemination within this host. In contrast, Haas-Stapleton *et al.* (2003) found that hemocytes were infected by BV within the hemocoel. The opposing conclusions of these two studies may result from differential dispersal of virus from the sites of BV injection, between the first and second abdominal segment such that virus was injected to the rear of the abdomen (Clarke and Clem, 2002), versus through the planta of one of the prolegs (Haas-Stapleton *et al.*, 2003).

The response of the host insect to baculovirus infection varies dramatically according to the host–virus combination, which larval instar becomes infected, the virus dose acquired, and environmental conditions. As an example, about 4 days after infection of third instar *H. virescens* (a permissive host) with AcMNPV, host larvae become less responsive with reduced feeding activity (**Figure 12**). The cuticle becomes swollen and glossy in appearance as a result of hypertrophy of infected cells and tissues. In species that are lightly pigmented, the larvae take on a whitish coloration resulting from the polyhedra in the nuclei of fat body and epidermal cells. The baculovirus gene *ctl* encodes a small peptide with sequence similar to the conotoxins of predatory

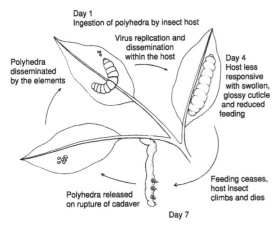

Day 1
Ingestion of polyhedra by insect host

Virus replication and
dissemination
within the host

Day 4
Host less
responsive
with swollen,
glossy cuticle
and reduced
feeding

Polyhedra
disseminated
by the elements

Feeding ceases,
host insect
climbs and dies

Polyhedra released
on rupture of cadaver

Day 7

Figure 12 Effects of baculovirus infection on a susceptible host insect. The timing of the various stages varies according to the virus dose, larval instar, and environmental conditions. The timing of events would be typical for AcMNPV infection of third instar larvae of *Heliothis virescens*.

marine snails and aptotoxins of spiders (Eldridge *et al.*, 1992; Skinner *et al.*, 1992) (see **Chapter 4**). Conotoxins are calcium channel antagonists, while the aptotoxins are insecticidal causing rapid paralysis and death (Skinner *et al.*, 1992). While the function of baculovirus CTL remains to be determined, it may function to paralyze the insect during late stages of infection thereby promoting viral use of diminishing host resources (Eldridge *et al.*, 1992).

5.3.3. Dissemination from the Host

By 6–7 days after infection, feeding ceases. OB are released on lysis of infected cells. As the cells lyse, millions of polyhedra accumulate and the basement membrane degrades. Prior to death, many species exhibit negative geotropism and climb to the top of the vegetation, thereby facilitating broad dispersal of progeny virus. The cuticle becomes fragile and the dead insect liquefies. The baculovirus chitinase (*chiA*) and cathepsin (*v-cath*) genes, which are present in members of both NPV and GV, act in concert to facilitate release of occlusion bodies from the insect cadaver by degrading the cuticle and internal tissues (Hawtin *et al.*, 1995, 1997; Slack *et al.*, 1995; Kang *et al.*, 1998).

The chitinase of AcMNPV has both exo- and endo-chitinase activity against a range of substrates and is active from pH 3 to 10 (Hawtin *et al.*, 1995; Thomas *et al.*, 2000). It accumulates within the endoplasmic reticulum as a result of a C-terminal ER retention motif (KDEL). The chitinase is released as infected cells lyse (Thomas *et al.*, 1998). Removal of the ER retention motif and consequent secretion of the enzyme results in more rapid host liquefaction (Saville

et al., 2002). V-Cath is a cysteine protease that accumulates in the cell as an inactive proenzyme, proV-Cath (Bromme and Okamoto, 1995; Hom *et al.*, 2002). The viral cathepsin, which has a pH optimum of 5.0–5.5 is also active at pH 7 (Ohkawa *et al.*, 1994). ProV-Cath activation occurs on death of the infected cell possibly as a result of lysosomal protease release, and appears to contribute to death of the cell (Hom *et al.*, 2002). ChiA is required for the processing of V-Cath (Hom and Volkman, 2000).

A single dead insect may release as many as 10^{10} polyhedra (Entwistle and Evans, 1985) with the fat body, epidermis, and tracheal matrix producing the largest quantities of polyhedra (Aizawa, 1963). The amount of virus produced depends in part on the speed of kill of the host insect. In theory, the longer the insect survives, the more time the virus has for replication and production of progeny. For example, infection of the semipermissive host *Mamestra brassicae* with AcMNPV resulted in a higher virus yield than infection of the permissive host *T. ni* that succumbed more rapidly to infection (Hernandez-Crespo *et al.*, 2001). However, the yield of polyhedra may vary with species according to the physiological bases for the semipermissive character.

Liquefaction of the host cadaver increases transmission of the virus to conspecifics and facilitates dissemination in the environment. Polyhedra can survive outside the host in soil, in crevices of plants, or other refugia for years (Jaques, 1975), and are ubiquitous. Hence, in contrast to other viruses for which death of the host represents an evolutionary dead end, death of the baculovirus-infected host represents a successful replication strategy.

Predatory arthropods, scavengers, and parasitoids feeding on, or parasitizing baculovirus-infected lepidopteran larvae may become contaminated with virus and play an important role in dissemination of baculoviruses under field conditions (Ruberson *et al.*, 1991; Fuxa *et al.*, 1993; Chittihunsa and Sikorowski, 1995; Sait *et al.*, 1996; Vasconcelos *et al.*, 1996). Grazing mammals and birds also contribute to baculovirus dissemination (Entwistle and Evans, 1985; Fuxa, 1991).

The most common route for baculovirus transmission between the infected and healthy individuals is via horizontal transmission when a larva ingests occlusion bodies from the environment. Vertical transmission (from parent to offspring) is common for nucleopolyhedroviruses (Smits and Vlak, 1988; Kukan, 1999; Fuxa *et al.*, 2002). Artificial selection resulted in an increased rate of vertical transmission of *S. frugiperda* NPV suggesting a genetic basis for this characteristic (Fuxa and

Richter, 1991). Vertical transmission of baculoviruses may be important for species such as *S. frugiperda* that disperse or migrate such that successive generations are located at different geographic locations.

5.4. Virus Replication

Regulation of more than 100 baculovirus genes is complex and requires coordinated expression of temporally regulated early, late, and very late genes (Rohel and Faulkner, 1984). Each stage of virus replication is dependent on expression of genes in the preceding stage. Baculovirus gene expression appears to be regulated at the level of transcription, and includes a unique switch from transcription of early genes by host RNA polymerase II, to late and very late genes by a virus-encoded RNA polymerase (Fuchs *et al.*, 1983; Guarino *et al.*, 1998a, 1998b). *Trans*-acting protein factors (IE0, IE1, IE2, and PE38) interact with *cis*-acting DNA elements (upstream activation regions or homologous regions) in the virus genome to initiate early gene expression (Friesen, 1997).

5.4.1. Early Gene Expression

Early genes (transcribed before initiation of virus DNA replication) are distributed throughout the genome and transcription often results in overlapping RNAs that differ at the 5' end. RNA transcripts are detectable within the first 15 min indicating that transcription likely begins on uncoating of the DNA in the nucleus. Early gene transcription peaks at 6–12 h p.i. in *S. frugiperda* cells. Downregulation of early gene transcription may result from expression of late genes or viral DNA replication. The expression of some early genes, such as *p35*, *gp64*, *39K*, and *ie1*, continues into the late phase as a result of the presence of both early and late promoter motifs. The early promoters, which are responsive to the host RNA polymerase II and host transcription factors, are readily distinguished from late promoters on the basis of nucleotide sequence. The early promoters consist of core transcription elements and *cis*-acting elements either in close proximity or within transcriptional enhancers. Most early promoters are TATA-containing (TATA+), initiator-containing (INR+), or composite (TATA+, INR+). The TATA element regulates the rate of transcription initiation and defines the start site for transcription. The INR motif (e.g., ATCA(G/T)T(C/T)) overlaps the RNA start site, contributes to basal promoter activity, and determines the start site in the absence of the TATA motif. The tetranucleotide CATG is the most conserved

of the INR-like motifs of baculovirus early genes and is also prevalent in the promoters of arthropod genes (Cherbas and Cherbas, 1993). Both INR and TATA bind specific factors that recruit the TFIID transcription complex at the promoter. The presence of both motifs may enhance recruitment of the required factors to ensure adequate expression of early genes, and use of universal host transcription factors may facilitate gene expression in a broad range of tissue types. An upstream activation region (UAR), often present for early baculovirus promoters, potentiates transcription from the early promoter. The activity of the UAR may vary in a host-specific manner, which is consistent with the hypothesis that host transcription factors act via the UAR to potentiate basal promoter activity. Transient expression assays suggest that other early baculovirus genes including *dnapol* that do not have the conventional INR or TATA promoter motifs may be more dependent on virus-encoded transactivators (Ohresser *et al.*, 1994). Downstream activating region (DAR) motifs contribute to basal promoter activity and are likely to stabilize protein interactions at the INR.

The homologous regions (*hrs*) are repetitive and highly conserved sequences that occur in multiple copies in the baculovirus genome. They are also unique to baculovirus genomes. *hrs* consist of repeated sequences encompassing direct repeats of imperfect palindromic sequences. These elements function as transcriptional enhancers of *cis*-linked early viral genes, and experimental data also indicate that they serve as origins of virus replication (Pearson *et al.*, 1992; Rodems and Friesen, 1993). *hrs* may also contribute to late gene transcription (Lu and Miller, 1995b). Enhancement of transcription is most likely through enhancement of transcription initiation. The *hrs* vary in size according to the number of palindromic repeats and in the number of copies in different baculovirus genomes. Although host factors can mediate *hr*-stimulated transcription from viral promoters in the absence of viral proteins, *hr*-mediated enhancement of transcription is dramatically increased by IE1, which binds as a dimer directly or indirectly to the *hr* (Friesen, 1997). *hrs* also function as origins of DNA replication in transient replication assays (Ahrens *et al.*, 1995b; Kool *et al.*, 1995).

On initial infection when DNA template concentrations are low, baculoviruses express transregulatory proteins that stimulate early gene expression. The immediate early transregulator IE1 plays a key role in regulation of gene expression. IE1 activates expression of early genes (*p35*, *gp64*, *39K*, *p143*, *dnapol*, *pe38*, *lef-1*, *lef-2*, *lef-3*, and *ie1*) and is required for DNA replication and transcription of late genes. IE1 dramatically enhances expression of genes that are *cis*-linked to *hr* enhancers. IE1 down-regulates transcription of *pe38* and *ie2*. The *ie1* promoter along with associated upstream sequences is active in uninfected cells, and can be used to direct foreign gene expression in transformed lepidopteran cells (Jarvis *et al.*, 1990). The steady state level of IE1 increases during infection although the function of this protein during virus replication remains to be determined.

IE0 is an immediate early transregulator identical to IE1 except for an additional 54 N-terminal residues. IE0 results from splicing of the upstream *ie0* exon and fusion to the *ie1* exon. RNA splicing is a rare event in baculovirus gene expression and probably only occurs during the early stages of infection (Kovacs *et al.*, 1991). IE0 transactivates the *hr*-linked promoters of *ie-1* and *39K*. IE2 stimulates transcription of several promoters, but to a lesser extent than IE1. IE2 and PE38 contain motifs typical of transcription regulators, including a leucine zipper and a RING finger. PE38 stimulates expression from the *p143* helicase promoter.

5.4.2. DNA Replication

For *S. frugiperda* cells infected by AcMNPV, viral DNA replication occurs from 6 to 18 h p.i., and then declines. Viral DNA synthesis is dependent on production of early viral proteins. *Cis*-acting elements that may be involved in initiation of viral DNA synthesis have been identified by analysis of defective virus particles generated on repeated passage of AcMNPV in cell culture. These defective viruses lack large regions of the AcMNPV genome with four regions that become amplified. Two of these regions contain *hrs*, and both the *hr* and non-*hr* regions can support transient replication of plasmids (Kool *et al.*, 1993b, 1994a). The efficiency of replication decreases with reduction of the number of palindromes present in the *hr* (Leisy *et al.*, 1995). The presence of multiple *hrs* dispersed throughout the genome may provide redundancy in the event that an *hr* is deleted. Loss of a single *hr* element from *Bombyx mori* NPV or AcMNPV has no impact on virus replication (Majima *et al.*, 1993; Rodems and Friesen, 1993). However, demonstration that *hrs* function *in vivo* as origins of replication is lacking. Non-*hr* origins of replication have been identified in AcMNPV and OpMNPV (Kool *et al.*, 1993a; Pearson *et al.*, 1993).

The isolation of conditional-lethal mutants that are defective in DNA replication has facilitated identification of virus genes that are involved in virus DNA replication (Lee and Miller, 1978; Brown

et al., 1979; Gordon, 1984; Lu and Carstens, 1991; Ribeiro *et al.*, 1994). Transient replication assays were used to identify regions of the baculovirus genome that are able to support replication of cotransfected origin-containing plasmid DNA (Kool *et al.*, 1994a, 1994b; Ahrens *et al.*, 1995a; Ahrens and Rohrmann, 1995a, 1995b; Lu and Miller, 1995b). Five genes were classified as essential (*p143*, *ie1*, *lef-1*, *lef-2*, and *lef-3*) and five as stimulatory (*dnapol*, *p35*, *ie-2*, *lef-7*, and *pe38*) for replication of AcMNPV DNA (Kool *et al.*, 1994b; Lu and Miller, 1995b). The *p143* gene encodes a helicase, which acts as a host range determinant. Incorporation of a fragment of *B. mori* NPV (BmNPV) helicase into AcMNPV (with few amino acids differing between the AcMNPV and BmNPV sequences) allowed AcMNPV to infect *B. mori* cells and larvae (Maeda *et al.*, 1993; Croizier *et al.*, 1994; Argaud *et al.*, 1998). *lef-1* is an early gene important for late and very late gene expression (Passarelli and Miller, 1993), and essential for DNA replication (Lu and Miller, 1995b). Conserved domains indicate that Lef-1 may be a baculovirus primase (Barrett *et al.*, 1996). Lef-2 is required for late gene expression and DNA replication. Lef-2 interacts with Lef-1 suggesting that these two proteins may function in replication as a heterodimer (Evans *et al.*, 1997). Lef-3 is a single-stranded DNA binding protein that is essential for DNA replication (Hang *et al.*, 1995). Although Lu and Miller (1995b) classified *dnapol* as stimulatory in transient assays, other reports have shown that *dnapol* is essential for virus replication (Pearson *et al.*, 1993; Kool *et al.*, 1994a, 1994b; Ahrens and Rohrmann, 1995a). The host DNA polymerase, in conjunction with virus genes essential for replication, may be able to replicate viral DNA at a suboptimal level. P35 is an apoptotic suppressor of *S. frugiperda* cells infected with AcMNPV (Clem *et al.*, 1991), and stimulates DNA replication by preventing apoptosis possibly induced by expression of late gene products or DNA replication (Clem and Miller, 1994). Under some circumstances IE1 has been reported to trigger apoptosis (Prikhod'ko and Miller, 1996). However, IE1 also induces expression of P35. The genes *ie-2* and *pe38* function as stimulators of DNA replication probably through activation of genes involved in replication. Both genes encode transactivators of early gene expression (Carson *et al.*, 1988; Lu and Carstens, 1993). PE38 activates *p143*, the baculovirus helicase homolog, while IE-2 stimulates production of PE38 and IE1. The effects of IE-2 are cell line specific (Lu and Miller, 1995a). Lef-7 is a putative single-stranded DNA binding protein and, along with Hcf-1, is a cell line-specific regulator of

DNA replication (Lu and Miller, 1995a, 1996). In addition to the genes identified as being required for replication by transient replication assays, Lin and Blissard (2002b) found that *lef-11* is necessary for viral DNA replication in the context of virus infection.

Because alteration of only one residue in AcMNPV helicase enables this virus to replicate in the otherwise nonpermissive BmN cell line, it is likely that a host protein interacts with the viral helicase, and that this interaction has virus and cell line specificity. Hence, host factors may be involved in replication of baculovirus DNA in addition to the viral genes described here. The host cell topoisomerase would also be required to relieve torsional stress as the duplex DNA is unwound by P143.

Given that infection-dependent replication of origin-containing plasmids results in production of concatemers of plasmid DNA, baculovirus DNA may be amplified by a rolling circle mode of replication. The detection of multiple unit-length genome fragments from replicating viral DNA also supports this hypothesis (Leisy and Rohrmann, 1993; Oppenheimer and Volkman, 1997). The mechanism for production of unit length genomes *in vivo* however remains to be established. Alternatively, replication may involve the theta mechanism initially, rolling circle synthesis, and recombination at the late stage, as for herpesviruses (Mikhailov, 2003).

5.4.3. Late and Very Late Gene Expression

Expression of late and very late genes begins following initiation of viral DNA replication, and is dependent on viral DNA replication. The late phase genes, which are transcribed at about 6–24 h p.i., include those encoding the structural proteins of the nucleocapsid, such as the major capsid protein VP39 and the basic core protein P6.9. The very late genes, transcribed at 18–72 h p.i., encode polyhedrin or granulin. The P10 protein, which affects disintegration of the nucleus, is expressed during the very late phase (Roelvink *et al.*, 1992). Production of the late and very late viral genes coincides with a dramatic decline in the host RNAs (6–18 h p.i.) (Ooi and Miller, 1988; Okano *et al.*, 2001) and host protein synthesis (Carstens *et al.*, 1979). The decline in host RNAs appears to result from one or more late viral genes and is concurrent with margination of host cell chromatin (Ooi and Miller, 1988). However, the amount of host TATA-binding protein (TBP), a transcription factor that localizes to sites of viral DNA replication, increases, possibly as a result of viral disruption of TBP turnover (Quadt *et al.*, 2002). Transcription of mitochondrial encoded genes is also unaffected by baculovirus infection (Okano *et al.*, 2001).

A unique four-subunit RNA polymerase is responsible for the switch from early to late and very late gene expression (Beniya *et al.*, 1996). All four subunits are encoded by the viral genome, specifically by *lef-4*, *lef-8*, *lef-9*, and *p47* (Guarino *et al.*, 1998a). Lef-4 is hypothesized to provide a capping function for RNA transcribed by the polymerase (Jin *et al.*, 1998). Two separate enzyme activities can be distinguished that differ in their ability to transcribe the very late genes (Xu *et al.*, 1995). This α-amanitin resistant polymerase requires the promoter element TAAG. Late promoters initiate from (A/G/T)TAAG, while the motif (T/A)ATAAGNA (T/A/C)T(T/A)T is conserved for promoters of *polyhedrin (polh)*, *granulin*, and *p10* genes (Rohrmann, 1986). The late and very late genes are distinguished by an additional promoter element, the burst element, in the untranslated leader of very late genes (Ooi *et al.*, 1989).

Nineteen AcMNPV genes, referred to as late expression factor (LEF) genes, are required for optimal expression of late and very late proteins in transient expression assays (Lu and Miller, 1995b; Todd *et al.*, 1995, 1996; Rapp *et al.*, 1998; Li *et al.*, 1999; Lin and Blissard, 2002a) (**Table 2**). A subset of these genes is required for the viral DNA replication (Kool *et al.*, 1994a; Lu and Miller, 1995b; Lin *et al.*, 2001; Guarino *et al.*, 2002b; Lin and Blissard, 2002b) (**Table 2**). Among the LEF genes are *ie-1* and *ie-2*, which transactivate early gene expression and may

Table 2 Late expression factors required for optimal expression of late and very late viral proteins in transient expression assays

Function	Late expression factors (LEFs)
Transregulator	IE1
	IE2
DNA replication	DNAPOL (polymerase)
	LEF-1 (primase)
	LEF-2
	LEF-3
	P143 (helicase)
	P35 (antiapoptosis)
	LEF-7
	PE38
	LEF-11
Late gene transcription	LEF-4 (subunit of RNA polymerase)
	LEF-8 (subunit of RNA polymerase)
	LEF-9 (subunit of RNA polymerase)
	P47 (subunit of RNA polymerase)
	LEF-5 (initiation factor)
	LEF-6
	LEF-10
	LEF-12
	VLF-1

5: Baculoviruses: Biology, Biochemistry, and Molecular Biology 141

transactivate other *lefs* including the replication-associated *lefs* such as *dnapol*, *lef-3*, and *p143*. Many of the LEF genes are expressed during the early phase of gene expression. LEFs include the four genes *lef-4*, *lef-8*, *lef-9*, and *p47*, which encode subunits of the virus-encoded RNA polymerase (Lu and Miller, 1997; Gross and Shuman, 1998; Guarino *et al.*, 1998a, 1998b). Lef-5 is an initiation factor and increases transcription from both late and very late promoters (Guarino *et al.*, 2002a). Very late expression factor-1 (Vlf-1) appears to regulate transcription of very late genes (McLachlin and Miller, 1994). Vlf-1 has sequence similarity to integrase/resolvases and DNA-binding activity without integrase activity (McLachlin and Miller, 1994). It interacts with the burst element of very late gene promoters, thereby directing the viral-specific RNA polymerase to these sites (Yang and Miller, 1999). No other genes that regulate very late gene expression have been identified (Todd *et al.*, 1996).

Expression of late and very late genes is regulated primarily at the level of transcription. Many late transcripts have mini-open reading frames, or are di- or polycistronic, which may provide mechanisms for translational regulation of downstream ORFs. There is also evidence of *cis*-acting transcriptional regulation of flanking genes, which may be common given that genes of different temporal classes are dispersed throughout the genome (Ooi and Miller, 1990; Gross and Rohrmann, 1993).

5.5. Effects on the Host

5.5.1. Virulence

Virulence is defined as the relative ability of a microorganism to overcome host defenses, or the degree of pathogenicity within a group or species (Poulin and Combes, 1999). Insect species are classified as permissive, semipermissive, or nonpermissive to virus infection, according to the degree of susceptibility (Bishop *et al.*, 1995). Enhanced virulence against

semipermissive species will result in host range expansion, and hence virulence and host range are intrinsically linked. The LD_{50} of a virus in a given host (the dose of virus required to kill 50% of the population) provides an indicator of the efficiency with which the baculovirus can lethally infect that host, and hence provides a measure of adaptation to the host. Various studies have been conducted to select for increased baculovirus virulence against a particular species under laboratory conditions (**Table 3**). The protocol typically involves repeated passaging of a baculovirus through the insect host. The mechanistic bases for the increased virulence in such cases remain to be determined however.

For fatal infection to occur, the baculovirus must overcome a number of barriers or defenses (that are distinct from mechanisms of resistance) (see Section 5.5.5) that may be presented by the host insect. The first potential barrier to baculovirus infection is the requirement for an alkaline environment in the midgut for dissolution of OB and release of ODV. The high alkalinity within the midgut of some Lepidoptera is hypothesized to protect larvae from toxic tannins in the diet (Berenbaum, 1980). Organisms without this characteristic would not be exposed to ODV, which are required for infection of the midgut epithelium. ODV must then negotiate the peritrophic membrane, a process that is poorly understood and likely varies with species according to temporal and structural properties of the peritrophic membrane (Tellam, 1996). Having negotiated the peritrophic membrane, various mechanisms may prevent virus replication within the cell, such as requirements for viral attachment, fusion and entry into the cell, apoptosis (Clem and Miller, 1993), shut down of protein synthesis (Du and Thiem, 1997), and incompatibility between the host cell and virus replication machinery (Croizier *et al.*, 1994). Most viruses would not replicate in nonpermissive organisms because of restrictions at the molecular level. Studies on infection of cell lines with

Table 3 Examples of selection of baculoviruses for increased virulence by repeated passaging in the insect host

Virus	Host insect	Number of passages	Increase in virulence (-fold)	Reference
HzSNPV	*Helicoverpa zea*	5	1.8	Shapiro and Ignoffo (1970)
OpMNPV	*Trichoplusia ni*	7	10	Martignoni and Iwai (1986)
CfMNPV	*T. ni*	1		Stairs *et al.* (1981)
	Galleria mellonella			
OpMNPV	*Estigmene acrea*	6		Shapiro *et al.* (1982)
CfMNPV		9	25–250	
LdMNPV		9		
AcNPV	*T. ni*		12–15	Tompkins *et al.* (1981, 1988)
	Spodoptera exigua			
AcMNPV	*Plutella xylostella*		15	Kolodny Hirsch and Beek (1997)

BV show that most semi- and nonpermissive insect cells support expression from baculovirus early gene promoters, but not from late or very late promoters (Morris and Miller, 1992, 1993; Liu and Carstens, 1993). Expression from late promoters in some cases suggests that the block likely occurs during or after viral DNA replication. Interestingly, baculoviruses are able to enter mammalian cells and mammalian promoters are used to drive gene expression for gene therapy purposes (Van Loo et al., 2001; Ghosh et al., 2002; Kost and Condreay, 2002; Huser and Hofmann, 2003). Baculoviruses do not replicate in mammalian cell lines.

Even if the baculovirus is able to replicate in the midgut cells, Lepidopteran larvae are able to slough midgut cells that are infected by ODV (Briese, 1986b; Engelhard and Volkman, 1995; Washburn et al., 1995, 1998). Infected midgut epithelial cells are also shed during molting. Evasion of the shedding of infected cells is hypothesized to have led to evolution of the M phenotype of NPV, and the ability of ODV-derived nucleocapsids to cross the midgut epithelial cells and rapidly produce budded virus (Granados and Lawler, 1981; Washburn et al., 1999, 2003a) (see Section 5.3.1 and **Figure 9**).

Baculoviruses encounter developmental resistance, with infection of later instars requiring a higher viral inoculum. Larvae also become increasingly resistant to fatal baculovirus infection as they age within each instar (Evans, 1986; Teakle et al., 1986). In addition, baculovirus infection of some semipermissive species is curtailed by the inability of hemocytes to support virus replication, and by melanization of infected tissues (Trudeau et al., 2001).

The arms race between the host defenses and evolution of baculoviruses to counter those defenses contributes to the virulence of a baculovirus against a specific host. Genes encoding enhancin for degradation of the peritrophic membrane, ecdysteroid UDP-glucosyl transferase for inactivation of ecdysteroids, and suppressors of apoptosis provide particularly nice examples of baculovirus weapons in this arms race. Understanding of the barriers to baculovirus infection will facilitate genetic manipulation to enhance the virulence of baculovirus insecticides.

5.5.2. Host Range

NPVs have been isolated from more than 500 insect species in several orders (Lepidoptera, Hymenoptera, Diptera, Coleoptera, Thysanura, Trichoptera). GVs have only been identified from lepidopteran hosts and have been isolated from more than 100 species. The NPVs tend to infect only members of the genus, or in some cases the family of the host insect species from which the virus was originally isolated. Few detailed studies of baculovirus host range have been conducted, and there is a wide variation in bioassay methodology, which makes comparison between studies difficult. One potential shortcoming of laboratory-based host range studies is that the heterogeneity observed in baculoviruses isolated from the field (Lee and Miller, 1978; Smith and Summers, 1979) may be more representative for studies of host range than laboratory clones, and indeed may be important for virus survival (Weitzmann et al., 1992).

Lymantriid NPVs such as those from *Lymantria dispar* and *Orgyia antiqua* tend to have a narrow host range (Barber et al., 1993). Some viruses of noctuid species infect few species, while other viruses isolated from noctuids, such as *A. californica*, *A. falcifera*, and *M. brassicae* (Payne, 1986; Doyle et al., 1990; Hostetter and Puttler, 1991), infect 30–40 species from multiple families. AcMNPV infects 43 species of Lepidoptera within 11 families (Payne, 1986). The GVs and hymenopteran NPVs appear to have narrower host ranges than the lepidopteran NPVs.

Current research on baculovirus host range is driven in part by the requirement for assessment of baculovirus host range for registration of baculovirus insecticides. This task has become more important with the registration of recombinant baculovirus insecticides for assessment of potential risks from recombinant baculoviruses to nontarget organisms. Understanding the genetic bases for host range may allow manipulation of baculovirus host range characteristics for pest management purposes.

5.5.2.1. Host range genes   There are eight baculovirus genes that are known to encode host range determinants (Miller and Lu, 1997) (**Table 4**), most of which have been identified by analysis of restriction of baculovirus replication to specific cell lines. Insertion or alteration of a single viral gene can influence host range. Alteration of a single residue in the *p143* gene, which encodes a DNA helicase homolog, was sufficient to enable AcMNPV to infect cell lines of *B. mori*, while alteration of two residues allowed AcMNPV to infect and kill larvae of *B. mori* (Croizier et al., 1994; Kamita and Maeda, 1997; Argaud et al., 1998).

Although TnGV and AcMNPV both replicate in *T. ni*, the DNA helicase of TnGV could not substitute for that of AcMNPV. This result suggests that specific protein–protein or protein–DNA interactions are necessary for a functional replication complex (Bideshi and Federici, 2000). Introduction of the host range factor 1 (*hrf-1*) gene of LdMNPV into AcMNPV enabled AcMNPV to infect *L. dispar* cell

Table 4   Nucleopolyhedrovirus genes implicated in host range or virulence

Gene	Origin of gene	Function	Manipulation	Effect on replication of AcMNPV in:				Reference
				Bombyx mori	Trichoplusia ni	Spodoptera spp.	Lymantria dispar	
p143	BmNPV	helicase	Hybrid gene	+ve				Croizier et al. (1994), Kamita and Maeda (1997), Argaud et al. (1998)
hrf-1	LdNPV	rescues tRNA	Insertion				+ve	Chen et al. (1998), Mazzacano et al. (1999)
hcf-1	AcMNPV	Unknown	Deletion		−ve			Lu and Miller (1996)
lef-7	AcMNPV	Unknown	Deletion			−ve		Chen and Thiem (1997)
p35	AcMNPV	Antiapoptosis	Deletion			−ve		Clem et al. (1991, 1994), Clem and Miller (1993), Chen and Thiem (1997), Zhang et al. (2002), Clarke and Clem (2003)
orf603	AcMNPV	Unknown	Deletion			−ve		Popham et al. (1998)
p94	AcMNPV	Unknown	Deletion					Clem et al. (1994)
ie0	AcMNPV	Transcriptional regulator	Deletion			+ve		Lu et al. (2003)

lines and larvae. The presence of *hrf*-1 in the AcMNPV genome resulted in a slight enhancement of AcMNPV virulence against *H. zea* (Chen *et al.*, 1998). Hrf-1 overcomes an apparent inhibition of tRNA synthesis in *L. dispar* cells infected with AcMNPV (Mazzacano *et al.*, 1999). Host cell-specific factor 1 (*hcf*-1) is required for replication of AcMNPV in *T. ni* cell lines but not for replication of AcMNPV in a *S. frugiperda* cell line (Lu and Miller, 1996). Hcf-1 was necessary for infection of *T. ni* larvae by injection, and disruption of *hcf*-1 resulted in a 20–30% increase in time taken by the virus to kill the host insect following oral infection. The effect of *hcf*-1 on virulence of AcMNPV in *T. ni* larvae would confer a significant advantage in the field. AcMNPV *lef*-7 (late expression factor 7) is required for efficient infection of a *S. frugiperda* and a *S. exigua* cell line, but not a *T. ni* cell line (Chen and Thiem, 1997). AcMNPV ORF 603 is important for infection of *S. frugiperda* larvae but not for infection of *T. ni* larvae (Popham *et al.*, 1998). Disruption of expression of IE0, and concomitant increased expression of the apoptosis suppressor P35, enables AcMNPV to replicate in *S. littoralis* cells that undergo apoptosis on infection with wild-type AcMNPV (Lu *et al.*, 2003). The baculovirus antiapoptotic gene *p35* allows AcMNPV to replicate in a *S. frugiperda* cell line and in larvae, but is not required for replication in *T. ni* cells and larvae (Clem *et al.*, 1991, 1994; Clem and Miller, 1993). The

presence in AcMNPV of at least one gene that specifically facilitates replication in *T. ni* larvae (*hcf*-1) and at least one gene that specifically facilitates replication in *S. frugiperda* larvae (*lef*-7) indicates that the ability to infect multiple species provides a selective advantage to the virus.

A DNA microarray of the AcMNPV genome has been used to identify six AcMNPV genes (*orf150*, *p10*, *pk2*, *lef-3*, *p35*, and *lef-6*) that are differentially regulated in cell lines derived from *S. frugiperda* (Sf9) and *T. ni* (High Five®) (Yamagishi *et al.*, 2003). Although the functional significance of these genes (with the exception of *p35*) in the two cell lines is unclear, these results are indicative of host cell-specific regulation of AcMNPV genes, suggesting differential viral gene expression for optimal replication in different cell types.

5.5.2.2. Antiapoptotic genes   P35 is a 35 kDa suppressor of apoptosis that inhibits CED-3/ICE-like proteases produced in response to virus infection and involved in apoptosis of infected insect cells (Clem, 1997). Viruses lacking a functional *p35* gene induce apoptosis of the infected cell, with characteristic blebbing of the plasma membrane, nuclear condensation, and intranucleosomal cleavage of cellular DNA. Baculovirus-induced apoptosis is correlated with reduced infectivity, replication, and dissemination within the host insect (Zhang *et al.*, 2002; Clarke and Clem, 2003). P35 is expressed during the early and late stages of infection to suppress apoptosis that occurs during the transition from the early to the late stage of infection. A reduction in OB production and lack of liquefaction of cadavers is observed in larvae of *T. ni* infected with *p35* mutants (Herschberger *et al.*, 1992; Clem *et al.*, 1994). P35 is the most broadly acting antiapoptotic protein known and is able to block apoptosis in a wide range of organisms. Hence it is of interest that P35 does not prevent apoptosis in cell lines from *S. littoralis* and *Choristoneura fumiferana* (Chejanovsky and Gershburg, 1995; Palli *et al.*, 1996).

A second family of baculovirus antiapoptotic genes, called *iap* (inhibitor of apoptosis) has been found in all baculoviruses except CuniNPV (Hughes, 2002). There are three *iap* genes in baculoviruses, but only EppoMNPV and OpMNPV have all three. AcMNPV has *iap1* and *iap2* in addition to *p35*, but the function of AcMNPV *iap* genes has not yet been demonstrated (Clem and Miller, 1994; Griffiths *et al.*, 1999). These Iap proteins may serve an auxiliary role in antiapoptotic activity (Viliplana and O'Reilly, 2003), act by blocking signal transduction pathways stimulated by various death stimuli (Clem, 1997), or may only function as inhibitors of apoptosis in certain hosts.

The protein P94 may induce apoptosis in tissues that are important for oral infection (Clem *et al.*, 1994). P35 is not required to inhibit apoptosis if P94 is not expressed (Kamita *et al.*, 1993; Clem, 1997). P94 is not required for infection of larvae of *S. frugiperda* or *T. ni*, but may be important for other species. If P94 induces apoptosis, there would be strong selection to acquire both of these genes. In line with this hypothesis, P35 in BmNPV is not required for infection of *B. mori* and most of the *p94* gene in this virus has been deleted (Kamita *et al.*, 1993; Clem, 1997).

5.5.3. Survival Time and Yield

The amount of time taken by the virus to kill the host varies widely, according to the host–virus combination, virus dose, larval instar, and environmental conditions, temperature in particular. In general, an increase in the dose of virus ingested results in a decrease in the survival time of the host insect (van Beek *et al.*, 1988). The survival time of 50% of the infected population (ST_{50}) and the LD_{50} both provide measures of pathogen virulence. The survival time of the host is also linked to the virus yield, with longer survival times allowing for additional rounds of virus replication, resources permitting. To promote increased survival time and virus yield, baculoviruses have acquired a gene that encodes ecdysteroid UDP-glucosyltransferase (*egt*).

Expression of the baculovirus *egt* gene has a complex effect on the host larva. By inactivating ecdysteroids, Egt allows the insect to live longer and grow larger thereby allowing increased production of virus. *egt* is expressed during the early phase of infection, and is secreted from infected cells rather than being retained in the endoplasmic reticulum as in the case of related mammalian proteins (O'Reilly, 1995). Egt catalyzes the conjugation of ecdysteroids with UDP-glucose or UDP-galactose and is specific for ecdysteroids (O'Reilly and Miller, 1989; O'Reilly *et al.*, 1992; Kelly *et al.*, 1995; Clark *et al.*, 1996). This sugar conjugation and inactivation of ecdysteroids suppresses host molting, prevents pupation, and prevents the conversion of permissive larvae into nonpermissive hosts. A critical level of Egt is required for the arrest of host development and hence the timing of arrest depends on the dose of virus received and also the timing of infection during the larval instar (O'Reilly, 1997). Suppression of molting causes infected larvae to continue to feed when they would otherwise cease feeding in preparation for the molt, and the yield of OB is increased (O'Reilly and Miller, 1991). Deletion of the *egt* gene from the AcMNPV genome results in a 20–30% reduction in the time taken by the virus to kill the host and an associated reduction in virus yield (O'Reilly and Miller, 1991). In *S. exigua*, Egt did not arrest development but Egt served to maintain the integrity of the Malpighian tubules of infected larvae (Flipsen *et al.*, 1995a). Hence, the role of Egt may not be dependent solely on inhibition of molting. Other evidence suggests that high titers of ecdysteroids are detrimental to virus replication and Egt may serve to mitigate these effects (O'Reilly *et al.*, 1995; O'Reilly, 1997).

5.5.4. Sublethal Effects and Latent Infections

The impact of baculoviruses on host populations extends beyond mortality of infected larvae, because of physiological effects that result from infection at sublethal doses of virus. Effects typical of sublethal infections include change in development time and sex ratio, and reduced fecundity and egg viability (Rothman and Myers, 1996a). The mechanisms of sublethal effects are poorly understood. Virus-induced alteration of hormonal and enzyme levels may affect host development or production of viable eggs or sperm (Subrahmanyam and Ramakrishnan, 1980; O'Reilly and Miller, 1989; Burand and Park, 1992). Latent viruses do not induce symptoms until activated by various stressors (Hughes *et al.*, 1993, 1997). Host insects that harbor latent viruses do not transmit virus horizontally, but can transmit virus vertically.

5.5.5. Resistance

Resistance is defined as physiological alteration in response to selection pressure from exposure to baculoviruses, resulting in decreased susceptibility of the host insect to the baculovirus. This is distinct from developmental resistance described above (see Section 5.5.1).

The development of resistance to baculoviruses is a key issue for use of baculoviruses in pest management but has received relatively little attention (Briese, 1986a, 1986b; Fuxa, 1993). Insects of the same species collected from different locations often vary in their susceptibility to a particular baculovirus, presumably because prior exposure in some localities has been selected for resistant populations (Fuxa, 1987). Resistance to and virulence of baculoviruses are likely to vary geographically according to frequency and levels of exposure. In Brazil, the resistance of the velvet bean caterpillar to *Anticarsia gemmatalis* MNPV increased according to the number of years of treatment at different sites (Abot *et al.*, 1995). Several studies have shown that it is possible to select for resistant hosts under selection pressure from different baculoviruses, although not for all virus–host combinations (Kaomini and Roush, 1988). The cabbage looper, *T. ni*, has been shown to evolve up to 22-fold resistance to the virus TnSNPV (Milks and Myers, 2000), which was not associated with a detectable fitness cost and was stable in the absence of virus exposure (Milks *et al.*, 2002). In contrast, resistance to other NPVs or a GV was associated with reduced pupal weight, developmental time, and/or egg production or hatch in *A. gemmatalis* (Fuxa and Richter, 1998), *S. frugiperda* (Fuxa *et al.*, 1988), and *Plodia interpunctella* (Boots and Begon, 1993). *Spodoptera frugiperda* acquired 4.5-fold resistance to SfNPV following selection for seven generations (Fuxa *et al.*, 1988); *P. interpunctella* acquired twofold resistance following exposure to *P. interpunctella* GV for 2 years (Boots and Begon, 1993); *A. gemmatalis* acquired fivefold resistance in eight generations in the laboratory, and >1000-fold resistance in 13 generations in the field in Brazil (Fuxa and Richter, 1998). As expected for resistance associated with fitness cost, resistance was rapidly lost on removal of virus selection from *S. frugiperda* and *A. gemmatalis* (Fuxa and Richter, 1989, 1998). There is also evidence that resistance to *S. frugiperda* NPV confers cross-resistance to other baculoviruses (Fuxa and Richter, 1990).

The mechanistic bases for resistance to baculoviruses have not been addressed. An additional complexity is that sublethal infection can cause the same effects as resistance such as altered development time and reduced fecundity. Assessment of field populations of the Western tent caterpillar *Malacosoma californicum pluviale*, for example, showed that effects were more consistent with sublethal infection than disease-induced resistance (Rothman and Myers, 1996b).

5.6. Baculovirus Genomics

5.6.1. General Properties of Baculovirus Genomes

At the time of writing, the genomes of 19 baculoviruses have been sequenced (**Table 5**). The sizes of the genomes range from approximately 80 to 180 kbp, due to the presence of unique genes, and gene duplications including the "baculovirus repeat ORFs" or *bro* genes (Gomi *et al.*, 1999; Hayakawa *et al.*, 1999; Kang *et al.*, 2003). The genomes vary as a result of deletions, gene insertions, inversions, and duplications. Transposable elements have contributed to some of these modifications (Friesen, 1993; Jehle, 1996; Jehle *et al.*, 1997). The organization of genes in NPV genomes is highly conserved, with rearrangements of genomic segments and presence or absence of auxiliary genes contributing to their diversity. The G+C content is highly variable, ranging from 28% to 59% of the genome.

In addition to the essential genes that are required for virus propagation (including genes required for virus gene expression, replication, virus assembly, and transmission), the large size of baculovirus genomes allows for inclusion of a second class of genes, the auxiliary genes (**Table 6**). These genes are not essential but confer some selective advantage to the virus, and may act at the cellular or organismal level (O'Reilly, 1997). These genes include *ubi*, *egt*, and *ctl*. Many baculovirus genes have yet to be characterized and the functions of many genes that have been studied remain unknown. Extensive analysis of 13 baculovirus genomes revealed a core of 30 common genes, including 10 of unknown function, and an additional 32 genes that are present in all of the lepidopteran baculoviruses (Herniou *et al.*, 2003). In addition, 27 genes are specific to GVs and 14 to lepidopteran NPVs suggesting a role for these genes in the differences in the biology of these two groups. There are 17 genes that are specific to group I NPVs, including *gp64*.

There are numerous examples of baculovirus genes with homology to genes from eukaryotic or prokaryotic organisms (Possee and Rohrmann, 1997) (**Table 6**). The high frequency of homologous recombination during viral replication, combined with movement of genetic material by transposable

Table 5   Insect baculovirus genomes sequenced

Baculovirus	Reference
Nucleopolyhedroviruses (NPV)	
Adoxophyes honmai NPV	Nakai, Goto, Kang, Shikata, and Kunimi (unpublished data)
Autographa californica MNPV[a]	Ayres et al. (1994)
Bombyx mori NPV[a]	Gomi et al. (1999)
Choristoneura fumiferana MNPV	de Jong, Dominy, Lauzon, Arif, Carstens, and Krell (unpublished data)
Culex nigripalpus NPV	Afonso et al. (2001)
Epiphyas postvittana MNPV	Hyink et al. (2002)
Helicoverpa armigera SNPV[b]	Chen et al. (2001)
Helicoverpa zea SNPV[b]	Chen et al. (2002)
Lymantria dispar MNPV	Kuzio et al. (1999)
Mamestra configurata NPV A	Li et al. (2002b)
Mamestra configurata NPV B	Li et al. (2002a)
Orygia pseudotsugata MNPV	Ahrens et al. (1997)
Rachiplusia ou MNPV[a] (*Anagrapha falcifera* MNPV)	Harrison and Bonning (2003)
Spodoptera litura MNPV	Pang et al. (2001)
Spodoptera exigua MNPV	Ijkel et al. (1999)
Granuloviruses (GV)	
Cydia pomonella GV	Luque et al. (2001)
Phthorimaea operculella GV	Croizier, Taha, Croizier, and Lopez Ferber (unpublished data)
Plutella xylostella GV	Hashimoto et al. (2000)
Xestia c-nigrum GV	Hayakawa et al. (1999)

[a,b]Viruses with the same superscript letter are considered to be variants of the same virus (Blissard et al., 2000).

elements (Miller and Miller, 1982; Fraser et al., 1983; Friesen, 1993), provide possible mechanisms for genetic exchange between disparate organisms. Some of the genes such as *egt*, the ecdysteroid UDP-glucosyl transferase gene, and *sod*, the superoxide dismutase gene homolog, are most likely of insect origin. Indeed, a genome-wide survey revealed six genes likely to have been horizontally transferred into baculoviruses, namely *DNA ligase*, *rr1*, *gta*, *iap*, *chi*, and *egt*, of which *egt* and *iap* are derived from insects (Hughes, 2002; Hughes and Friedman, 2003). The AcMNPV chitinase gene *chiA* has the highest similarity to bacterial chitinase (Hawtin et al., 1995). The higher G+C content of this gene is also more typical of the bacterial gene than other baculovirus genes. The F protein genes are also likely to have been acquired from insect hosts (Malik et al., 2000; Rohrmann and Karplus, 2001).

The sequencing of multiple baculovirus genomes has allowed inference of baculovirus evolution and genetic diversity (Herniou et al., 2003), horizontal gene transfer (Hughes and Friedman, 2003), identification of genes that have evolved in response to changing selection pressures (Harrison and Bonning, 2003, 2004), and identification of genes involved in virus–host insect interactions (Dall et al., 2001).

5.6.2. Baculovirus Evolution

Comprehensive phylogenetic analyses of 30 genes common to 13 baculovirus genomes highlighted the division of baculoviruses into four groups: GVs, group I NPVs, group II NPVs, and the dipteran NPV CuniNPV (Herniou et al., 2003). Phylogenetic analysis of *dnapol* from NPV and GV provides an example of this division (**Figure 13**). It is likely that the Baculoviridae will be reclassified to accommodate CuniNPV, which clearly does not belong with the lepidopteran NPVs.

Gene content mapping highlighted the fluid nature of baculovirus genomes including examples of repeated gene acquisitions and losses. Thirty-three genes were identified that have been acquired separately on two or even three occasions (Herniou et al., 2003; Hughes and Friedman, 2003). Baculoviruses have sufficient flexibility with incorporation of additional genetic material to allow "sampling" for beneficial genes from the environment (Herniou et al., 2001). Recombination between baculoviruses may occur when different baculoviruses infect the same cell. This is most likely to occur in the field between related viruses that share common hosts (Smith and Summers, 1980; Croizier and Ribeiro, 1992). Gene loss and acquisition may also result from the activity of transposable elements, several of which have been observed on culture of baculoviruses in cell lines. Sites favored by transposons include the *fp25K* gene resulting in several mutant viruses with the few polyhedra (FP) phenotype. *Trichoplusia ni* cell-derived transposable elements include IFP2 ("piggyBac") (Cary et al., 1989), TFP3 ("tagalong") (Wang et al., 1989), *hitchhiker* (Bauser et al., 1996), and TED (Miller and Miller, 1982). For Sf cells, the elements IFP1.6 and IFP2.2 have been characterized (Beames and Summers, 1988, 1990), and a Tc1-like transposon has been identified in the genome of CpGV (Jehle et al., 1995, 1997, 1998).

Shared protein phylogenetic profiles indicate the presence of viral proteins specific for infection of lymandtriid hosts. Only OpMNPV and LdMNPV that infect the lymantriids *L. dispar* and *Orygia pseudotsugata* encode *hrf1*, *ctl1*, and *ctl2*. Other baculoviruses encode one or other of the *ctl* genes,

Table 6 Baculovirus genes with homologs in other organisms

Gene	Homolog	Other locations
Virus propagation[a]		
iap	Inhibitor of apoptosis	Mammals, insects
fgf	Fibroblast growth factor	Mammals
gta	General transcription activator	Mammals
dnapol	DNA polymerase	Eukaryotes, prokaryotes, other virus families
vlf-1	Integrase/resolvase	Yeast, prokaryotes
pnk/pnl	Polynucleotide kinase and ligase	Phage, yeast, plants
alk-exo	Alkaline exonuclease	Eukaryotes, prokaryotes
gp64	Viral glycoprotein	Thogoto virus
gp37	Spindlin	Entomopoxviruses
dUTPase	dUTPase	Eukaryotes, prokaryotes, other virus families
rr 1, 2	Ribonucleotide reductase subunits	Eukaryotes, prokaryotes, other virus families
f	Envelope fusion protein	*Drosophila melanogaster*
Auxiliary genes[b]		
ptp	Protein tyrosine phosphatase	Eukaryotes, prokaryotes, poxviruses
ctl	Conotoxin	Cone snails, spiders
egt	Ecdysteroid UDP-glucosyl transferase	Eukaryotes, prokaryotes
pk1	Protein kinase	Eukaryotes, prokaryotes, other virus families
pk2	Protein kinase, eiF2-α-like	Eukaryotes
pcna	Proliferating cell nuclear antigen	Eukaryotes
sod	Superoxide dismutase	Eukaryotes, prokaryotes, poxviruses
ubi	Ubiquitin	Eukaryotes, prokaryotes
chiA	Chitinase	Eukaryotes, prokaryotes
cath	Cathepsin	Eukaryotes
p10	Filament-associated late protein	Entomopoxviruses

[a]Genes required for virus propagation including those encoding structural proteins, those required for virus gene expression, replication, production of progeny virus.
[b]Genes not required for virus replication but providing some selective advantage.
After Possee and Rohrmann (1997).

but not both, and *hrf1* is only found in OpMNPV and LdMNPV. The link between the presence of all three genes and infection of a particular lepidopteran family suggests that these proteins may function in specific interactions with this host family (Herniou *et al.*, 2003).

5.6.3. Identification of Genes Involved in Virus–Insect Interactions

Dall *et al.* compared the genes of representative viruses from three different families that include viruses that infect insects, namely the Poxviridae, Baculoviridae, and Iridoviridae (Dall *et al.*, 2001). Six groups of genes were identified that were present in all three families, present in viruses with insect hosts, and absent in members with vertebrate hosts. Four of the six groups of genes are found in baculoviruses, and include the fusolin/gp37 group, which encodes proteins that facilitate infection (Xu and Hukuhara, 1992; Mitsuhashi *et al.*, 1998; Hukuhara *et al.*, 1999). Analysis of the other three groups found in baculoviruses may provide leads for genetic improvement of baculovirus insecticides by modification of host range or virulence.

5.6.4. Identification of Genes that Have Undergone Adaptive Molecular Evolution

An alternative approach for identifying genes underlying host range or virulence is to compare the genomes of closely related baculoviruses that exhibit differential virulence to host species. The closely related viruses *Rachiplusia ou* MNPV (*A. falcifera* MNPV) (Harrison and Bonning, 1999) and AcMNPV have broad, overlapping host ranges but there are significant differences in the ability of these two viruses to infect at least six species. These species include the corn earworm *H. zea*, the European corn borer *Ostrinia nubilalis*, and the navel orangeworm *Amyelois transitella* (Lewis and Johnson, 1982; Hostetter and Puttler, 1991; Vail *et al.*, 1993; Cardenas *et al.*, 1997; Harrison and Bonning, 1999). On comparison of the genomes of these two viruses, it was found that the *hrs* of RoMNPV contain fewer palindromic repeats than in AcMNPV, homologs of five AcMNPV genes including *ctl* are missing from RoMNPV, and the composition of promoter motifs differed in 23 cases (Harrison and Bonning, 2003). On the basis that the primary selection pressure on a virus comes from its

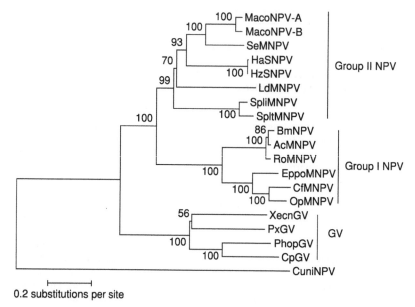

Figure 13 Phylogenetic analysis of predicted amino acid sequences of *dnapol* illustrating the separation of groups I and II NPVs from GVs and from the dipteran NPV derived from *Culex nigripalpus*. Phylogram produced by minimum evolution (ME) analysis of baculovirus DNA polymerase predicted amino acid sequences. Bootstrap values >50% (*n* = 1000 replicates) are shown at interior branches where they occur. The Group I and Group II NPV clades are indicated. AcMNPV, *Autographa californica* multiple nucleopolyhedrovirus (Ayres *et al.*, 1994); BmNPV, *Bombyx mori* nucleopolyhedovirus (Gomi *et al.*, 1999); CfMNPV, *Choristoneura fumiferana* multiple nucleopolyhedrovirus (Liu and Carstens, 1995); CpGV, *Cydia pomonella* granulovirus (Luque *et al.*, 2001); CuniNPV, *Culex nigripalpus* nucleopolyhedrovirus (Afonso *et al.*, 2001); EppoMNPV, *Epiphyas postvittana* multiple nucleopolyhedrovirus (Hyink *et al.*, 2002); HaSNPV, *Helicoverpa armigera* single nucleopolyhedrovirus (Chen *et al.*, 2001); HzSNPV, *Helicoverpa zea* single nucleopolyhedrovirus (Chen *et al.*, 2002); LdMNPV, *Lymantria dispar* multiple nucleopolyhedrovirus (Kuzio *et al.*, 1999); MacoNPV-A, *Mamestra configurata* nucleopolyhedrovirus strain 90/2 (Li *et al.*, 2002b); MacoNPV-B, *Mamestra configurata* nucleopolyhedrovirus strain 96B (Li *et al.*, 2002a); OpMNPV, *Orgyia pseudotsugata* multiple nucleopolyhedrovirus (Ahrens *et al.*, 1997); PxGV, *Plutella xylostella* granulovirus (Hashimoto *et al.*, 2000); PhopGV, *Phthorimaea operculella* granulovirus (accession number NC_004062); *Rachiplusia ou* multiple nucleopolyhedrovirus (Harrison and Bonning, 2003); SeMNPV, *Spodoptera exigua* multiple nucleopolyhedrovirus (Ijkel *et al.*, 1999); SpliMNPV, *Spodoptera littoralis* multiple nucleopolyhedrovirus (accession number AF215639); SpltMNPV, *Spodoptera litura* multiple nucleopolyhedrovirus (Pang *et al.*, 2001); XecnGV, *Xestia c-nigrum* granulovirus (Hayakawa *et al.*, 1999).

host, and that genes that have adaptively evolved may be involved in virulence, positive selection analysis was conducted on the genes of AcMNPV and RoMNPV (Yang *et al.*, 2000; Harrison and Bonning, 2003). Genes in which nonsynonymous (amino acid changing) substitutions are fixed at a higher rate than synonymous (silent) substitutions are identified as having undergone adaptive molecular evolution, or positive selection. This maximum likelihood analysis resulted in the identification of two genes, *arif-1* and *odv-e18*, predicted to have undergone positive selection. Odv-E18 is present in the envelope of occlusion-derived virus (Braunagel *et al.*, 1996b) and hence in an ideal location as a host range determinant for initial infection of midgut cells of the host insect. Arif-1 (actin-rearrangement-inducing factor) is involved with dissociation of the actin network of the host cell and the actin cables that form early during virus infection, and in production of actin aggregates at the plasma membrane (Dreschers *et al.*, 2001).

The strength of the positive selection analysis increases with increasing number of data sets. The positive selection analyses have been extended to include genes from multiple NPV genomes, and from this analysis nine genes predicted to have undergone positive selection have been identified (Harrison and Bonning, 2004). Alignments of 83 group I NPV genes were fitted to models of codon substitution that allow for varying selection intensity among codon sites. Genes predicted to have undergone positive selection include *ac38*, *lef-10*, *lef-12*, *odv-e18*, and *odv-e56*. NPV genes that have undergone positive selection may modulate the ability of different NPVs to replicate efficiently in cells (*lef-10*, *lef-12*) or establish primary infection of the midgut (*odv-e18*, *odv-e56*) of different host species.

5.7. Conclusion

Baculoviruses will continue to play an important role in the management of insect pests, as model systems for studies of fundamental biological processes, as tools for recombinant protein expression, and as candidate vectors for gene therapy. The potential of mosquito baculoviruses such as CuniNPV for management of mosquito-borne disease will also be of particular interest. Sequencing of additional baculovirus genomes and comparative genomics will provide further insights into baculovirus biology and evolution. Data acquired from the comparative genomics studies such as identification of baculovirus genes that have undergone adaptive molecular evolution (Harrison and Bonning, 2003) and those potentially involved in interaction with the insect host (Dall *et al.*, 2001) provide particularly useful leads for elucidation of how baculoviruses have evolved with their hosts. DNA microarrays of baculovirus genomes (Yamagishi *et al.*, 2003), expressed sequence tag (EST) libraries (Okano *et al.*, 2001), and RNA interference (Means *et al.*, 2003) will be valuable tools for future study of baculovirus gene expression and virus–host cell interactions. Increased understanding of baculovirus biology will facilitate further optimization of baculovirus-related technologies.

Acknowledgments

Thanks to Drs Robert Harrison, Loy Volkman and Gary Blissard for critical reading of this chapter. Funded in part by the Cooperative State Research, Education, and Extension Service, US Department of Agriculture, under Agreement Numbers 00-39210-9772 and 2003-35302-13558, as well as Hatch Act and State of Iowa funds.

References

Abot, A.R., Moscardi, F., Fuxa, J.R., Sosa-Gómez, D.R., Richter, A.R., **1995**. Susceptibility of populations of *Anticarsia gemmatalis* (Lepidoptera: Noctuidae) from Brazil and the United States to a nuclear polyhedrosis virus. *J. Entomol. Sci. 30*, 62–69.

Adams, J.R., McClintock, J.T., **1991**. Baculoviridae. Nuclear polyhedrosis viruses, part 1: Nuclear polyhedrosis viruses of insects. In: Adams, J.R., Bonami, J.R. (Eds.), Atlas of Invertebrate Viruses. CRC Press, Boca Raton, FL, pp. 87–204.

Adams, J.R., Goodwin, R.H., Wilcox, T.A., **1977**. Electron microscopic investigations on invasion and replication of insect baculoviruses *in vivo* and *in vitro*. *Biol. Cellulaire 28*, 261–268.

Afonso, C.L., Tulman, E.R., Lu, Z., Balinsky, C.A., Moser, B.A., *et al.*, **2001**. Genome sequence of a baculovirus pathogenic for *Culex nigripalpus*. *J. Virol. 75*, 11157–11165.

Ahrens, C.H., Carlson, C., Rohrmann, G.F., **1995a**. Identification, sequence and transcriptional analysis of *lef-3*, a gene essential for *Orgyia pseudotsugata* baculovirus DNA replication. *Virology 210*, 372–382.

Ahrens, C.H., Pearson, M.N., Rohrmann, G.F., **1995b**. Identification and characterization of a second putative origin of DNA replication in a baculovirus of *Orgyia pseudotsugata*. *Virology 207*, 572–576.

Ahrens, C.H., Rohrmann, G.F., **1995a**. Identification of essential *trans*-acting regions required for DNA replication of the *Orgyia pseudotsugata* multinucleocapsid nuclear polyhedrosis virus: *lef-1* is an essential replication gene. *Virology 207*, 417–428.

Ahrens, C.H., Rohrmann, G.F., **1995b**. Replication of *Orgyia pseudotsugata* baculovirus DNA: *ie-1* and *lef-2* are essential and *ie-2*, *p34* and *Op-iap* are stimulatory genes. *Virology 212*, 650–662.

Ahrens, C.H., Russell, R.L.Q., Funk, C.J., Evans, J.T., Harwood, S.H., *et al.*, **1997**. The sequence of the *Orgyia pseudotsugata* multinucleocapsid nuclear polyhedrosis virus genome. *Virology 229*, 381–399.

Aizawa, K., **1963**. The nature of infections caused by nuclear-polyhedrosis viruses. In: Steinhaus, E.A. (Ed.), Insect Pathology: An Advanced Treatise, vol. 1. Academic Press, New York, pp. 381–412.

Argaud, O., Croizier, L., López-Ferber, M., Croizier, G., **1998**. Two key mutations in the host-range specificity domain of the *p143* gene of *Autographa californica* nucleopolyhedrovirus are required to kill *Bombyx mori* larvae. *J. Gen. Virol. 79*, 931–935.

Ayres, M.D., Howard, S.C., Kuzio, J., Lopez-Ferber, M., Possee, R.D., **1994**. The complete DNA sequence of *Autographa californica* nuclear polyhedrosis virus. *Virology 202*, 586–605.

Barber, K.N., Kaupp, W.J., Holmes, S.B., **1993**. Specificity testing of the nuclear polyhedrosis virus of the gypsy moth, *Lymantria dispar* (L.) (Lepidoptera: Lymantriidae). *Can. Entomol. 125*, 1055–1066.

Barrett, J.W., Lauzon, H.A.M., Mercuri, P.S., Krell, P., Sohi, S.S., *et al.*, **1996**. The putative LEF-1 proteins from two distinct *Choristoneura fumiferana* multiple nucleopolyhedroviruses share domain homology to eukaryotic primases. *Virus Genes 13*, 229–237.

Bauser, C.A., Elick, T.A., Fraser, M.J., **1996**. Characterization of *hitchhiker*, a transposon insertion frequently associated with baculovirus FP mutants derived upon passage in the TN-368 cell line. *Virology 216*, 235–237.

Beames, B., Summers, M.D., **1988**. Comparisons of host cell DNA insertions and altered transcription at the site of insertions in few polyhedra baculovirus mutants. *Virology 162*, 206–220.

Beames, B., Summers, M.D., **1990**. Sequence comparison of cellular and viral copies of host cell DNA insertions found in *Autographa californica* nuclear polyhedrosis virus. *Virology 174*, 354–363.

Belyavskyi, M., Braunagel, S.C., Summers, M.D., **1998**. The structural protein ODV-EC27 of *Autographa californica* nucleopolyhedrovirus is a multifunctional viral cyclin. *Proc. Natl Acad. Sci. USA 95*, 11205–11210.

Beniya, H., Funk, C.J., Rohrmann, G.F., Weaver, R.F., **1996**. Purification of a virus-induced RNA polymerase from *Autographa californica* nuclear polyhedrosis virus-infected *Spodoptera frugiperda* cells that accurately initiates late and very late transcription *in vitro*. *Virology 216*, 12–19.

Benz, G.A., **1986**. Introduction: historical perspectives. In: Granados, R., Federici, B.A. (Eds.), The Biology of Baculoviruses. CRC Press, Boca Raton, FL, pp. 1–35.

Berenbaum, M., **1980**. Adaptive significance of midgut pH in larval Lepidoptera. *Am. Nat. 115*, 138–146.

Bideshi, D.K., Federici, B.A., **2000**. The *Trichoplusia ni* granulovirus helicase is unable to support replication of *Autographa californica* multicapsid nucleopolyhedrovirus in cells and larvae of *T ni*. *J. Gen. Virol. 81*, 1593–1599.

Bilimoria, S.L., **1991**. The biology of nuclear polyhedrosis viruses. In: Kurstak, E. (Ed.), Viruses of Invertebrates. Dekker, New York, pp. 1–72.

Bischoff, D.S., Slavicek, J.M., **1997**. Molecular analysis of an *enhancin* gene in *Lymantria dispar* nuclear polyhedrosis virus. *J. Virol. 71*, 8133–8140.

Bishop, D.H.L., Hirst, M.L., Possee, R.D., Cory, J.S., **1995**. Genetic engineering of microbes: virus insecticides – a case study. In: Darby, G.K., Hunter, P.A., Russell, A.D. (Eds.), 50 Years of Microbials. Cambridge University Press, Cambridge, pp. 249–277.

Blissard, G.W., **1996**. Baculovirus–insect cell interactions. *Cytotechnology 20*, 73–93.

Blissard, G.W., Black, B., Crook, N., Keddie, B.A., Possee, R., *et al.*, **2000**. Family Baculoviridae. In: Regenmortel, M.H.V.V., Fauquet, C.M., Bishop, D.H.L., Carstens, E.B., Estes, M.K., Lemon, S.M., Maniloff, J., Mayo, M.A., McGeoch, D.J., Pringle, C.R., Wickner, R.B. (Eds.), Virus Taxonomy, 7th Report of the International Committee on the Taxonomy of Viruses. Springer, New York, pp. 195–202.

Blissard, G.W., Rohrmann, G.F., **1989**. Location, sequence, transcriptional mapping, and temporal expression of gp64 envelope glycoprotein gene of the *Orgyia pseudotsugata* multicapsid nuclear polyhedrosis virus. *Virology 170*, 537–555.

Blissard, G.W., Wenz, J.R., **1992**. Baculovirus gp64 envelope glycoprotein is sufficient to mediate pH dependent membrane fusion. *J. Virol. 66*, 6829–6835.

Bonning, B.C., Boughton, A.J., Jin, H., Harrison, R.L., **2002**. Genetic enhancement of baculovirus insecticides. In: Upadhyay, R.K. (Ed.), Advances in Microbial Control of Insect Pests. Kluwer Academic, London, pp. 109–125.

Boots, M., Begon, M., **1993**. Trade-offs with resistance to a granulosis virus in the Indian meal moth, examined by a laboratory evolution experiment. *Func. Ecol. 7*, 528–534.

Brandt, C.R., Adang, M.J., Spence, K.D., **1978**. The peritrophic membrane: ultrastructural analysis and function as a mechanical barrier to microbial infection in *Orgyia pseudotsugata*. *J. Invertebr. Pathol. 32*, 12–24.

Braunagel, S.C., Burks, J.K., Rosas-Acosta, G., Harrison, R.L., Ma, H., *et al.*, **1999**. Mutations within the *Autographa californica* nucleopolyhedrovirus FP25K gene decrease the accumulation of ODV-E26 and alter its intranuclear transport. *J. Virol. 73*, 8559–8570.

Braunagel, S.C., Elton, D.M., Ma, H., Summers, M.D., **1996a**. Identification and analysis of an *Autographa californica* nuclear polyhedrosis virus structural protein of the occlusion-derived virus envelope: ODV-E56. *Virology 217*, 97–110.

Braunagel, S.C., Guidry, P.A., Rosas-Acosta, G., Engelking, L., Summers, M.D., **2001**. Identification of BV/ODV-C42, an *Autographa californica* nucleopolyhedrovirus orf101-encoded structural protein detected in infected-cell complexes with ODV-EC27 and p78/83. *J. Virol. 75*, 12331–12338.

Braunagel, S.C., He, H., Ramamurthy, P., Summers, M.D., **1996b**. Transcription, translation, and cellular localization of three *Autographa californica* nuclear polyhedrosis virus structural proteins: ODV-E18, ODV-E35, and ODV-EC27. *Virology 222*, 100–114.

Braunagel, S.C., Parr, R., Belyavskyi, M., Summers, M.D., **1998**. *Autographa californica* nucleopolyhedrovirus infection results in Sf9 cell cycle arrest at G$_2$/M phase. *Virology 244*, 195–211.

Braunagel, S.C., Summers, M.D., **1994**. *Autographa californica* nuclear polyhedrosis virus, PDV, and ECV viral envelopes and nucleocapsids: structural proteins, antigens, lipid and fatty acid profiles. *Virology 202*, 315–328.

Briese, D.T., **1986a**. Host resistance to microbial control agents. In: Franz, J.M. (Ed.), Biological Plant and Health Protection. Fischer Verlag, Stuttgart, pp. 233–256.

Briese, D.T., **1986b**. Insect resistance to baculoviruses. In: Federici, B.A., Granados, R. (Eds.), The Biology of Baculoviruses, vol. 2. CRC Press, Boca Raton, FL, pp. 237–263.

Bromme, D., Okamoto, K., **1995**. The baculovirus cysteine protease has a cathepsin B-like S2 subsite specificity. *Biol. Chem. Hoppe-Seyler 376*, 611–615.

Brown, M., Crawford, A.M., Faulkner, P.F., **1979**. Genetic analysis of a baculovirus *Autographa californica* nuclear polyhedrosis virus: isolation of temperature-sensitive mutants and assortment into complementation groups. *J. Virol. 31*, 190–198.

Brown, M., Faulkner, P., Cochran, M.A., Chung, K.L., **1980**. Characterization of two morphology mutants of *Autographa californica* nuclear polyhedrosis virus with large cuboidal inclusion bodies. *J. Gen. Virol. 50*, 309–316.

Bud, H.M., Kelly, D.C., **1980**. An electron microscope study of partially lysed baculovirus nucleocapsids: the intranucleocapsid packaging of viral DNA. *J. Ultrastruct. Res. 73*, 361–368.

Burand, J.P., Park, E.J., **1992**. Effect of a nuclear polyhedrosis virus infection on the development and pupation of gypsy moth larvae. *J. Invertebr. Pathol.* 60, 171–175.

Cardenas, F.A., Vail, P.V., Hoffmann, D.F., Tebbets, J.S., Schreiber, F.E., **1997**. Infectivity of celery looper (Lepidoptera: Noctuidae) multiple nucleocapsid polyhedrosis virus to navel orangeworm (Lepidoptera: Pyralidae). *Environ. Entomol.* 26, 131–134.

Carson, D.D., Guarino, L.A., Summers, M.D., **1988**. Functional mapping of an AcNPV immediate early gene which augments expression of the IE-1 *trans*-activated 39k gene. *Virology* 162, 444–451.

Carstens, E.B., Chan, H., Yu, H., Williams, G.V., Casselman, R., **1994**. Genetic analysis of temperature-sensitive mutations in baculovirus late expression factors. *Virology* 204, 323–337.

Carstens, E.B., Tjia, S.T., Doerfler, W., **1979**. Infection of *Spodoptera frugiperda* cells with *Autographa californica* nuclear polyhedrosis virus. I. Synthesis of intracellular proteins after virus infection. *Virology* 99, 386–398.

Carstens, E.B., Williams, G.V., Faulkner, P., Partington, S., **1992**. Analysis of polyhedra morphology mutants of *Autographa californica* nuclear polyhedrosis virus: molecular and ultrastructural features. *J. Gen. Virol.* 71, 3035–3040.

Cary, L.C., Goebel, M., Corsaro, B.G., Wang, H.G., Rosen, E., *et al.*, **1989**. Transposon mutagenesis of baculoviruses: analysis of *Trichoplusia ni* transposon IFP2 insertions within the FP-locus of nuclear polyhedrosis viurses. *Virology* 172, 156–169.

Charlton, C.A., Volkman, L.E., **1991**. Sequential rearrangement and nuclear polymerization of actin in baculovirus-infected *Spodoptera frugiperda* cells. *J. Virol.* 65, 1219–1227.

Charlton, C.A., Volkman, L.E., **1993**. Penetration of *Autographa californica* nuclear polyhedrosis virus nucleocapsids into IPBL Sf21 cells induces actin cable formation. *Virology* 197, 245–254.

Chejanovsky, N., Gershburg, E., **1995**. The wild-type *Autographa californica* nuclear polyhedrosis virus induces apoptosis of *Spodoptera littoralis* cells. *Virology* 209, 519–525.

Chen, C.-J., Quentin, M.E., Brennan, L.A., Kukel, C., Thiem, S.M., **1998**. *Lymantria dispar* nucleopolyhedrovirus *hrf-1* expands the larval host range of *Autographa californica* nucleopolyhedrovirus. *J. Virol.* 72, 2526–2531.

Chen, C.-J., Thiem, S.M., **1997**. Differential infectivity of two *Autographa californica* nucleopolyhedrovirus mutants on three permissive cell lines is the result of *lef-7* deletion. *Virology* 227, 88–95.

Chen, X., Ijkel, W.F.J., Tarchini, R., Sun, X., Sandbrink, H., *et al.*, **2001**. The sequence of the *Helicoverpa armigera* single nucleocapsid nucleopolyhedrovirus genome. *J. Gen. Virol.* 82, 241–257.

Chen, X., Zhang, W.-J., Wong, J., Chun, G., Lu, A., *et al.*, **2002**. Comparative analysis of the complete genome sequences of *Helicoverpa zea* and *Helicoverpa armigera* single-nucleocapsid nucleopolyhedroviruses. *J. Gen. Virol.* 83, 673–684.

Cherbas, L., Cherbas, P., **1993**. The arthropod initiator: the capsite consensus plays an important role in transcription. *Insect Biochem. Mol. Biol.* 23, 81–90.

Chernomordik, L., Leikina, E., Cho, M.-S., Zimmerberg, J., **1995**. Control of baculovirus gp64-induced syncitium formation by membrane lipid composition. *J. Virol.* 69, 3049–3058.

Chisholm, G.E., Henner, D.J., **1988**. Multiple early transcripts and splicing of the *Autographa californica* nuclear polyhedrosis virus IE-1 gene. *J. Virol.* 62, 3193–3200.

Chittihunsa, T., Sikorowski, P.P., **1995**. Effects of nonoccluded baculovirus (Baculoviridae) infection on *Microplitis croceipes* (Hymenoptera: Braconidae). *Biol. Control* 24, 1708–1712.

Clark, E.E., Tristem, M., Cory, J.S., O'Reilly, D.R., **1996**. Characterization of the ecdysteroid UDP-glucosyltransferase gene from *Mamestra brassicae* nucleopolyhedrosis virus. *J. Gen. Virol.* 77, 2865–2871.

Clarke, T.E., Clem, R.J., **2002**. Lack of involvement of haemocytes in the establishment and spread of infection in *Spodoptera frugiperda* larvae infected with the baculovirus *Autographa californica* M nucleopolyhedrovirus by intrahemocoelic injection. *J. Gen. Virol.* 83, 1565–1572.

Clarke, T.E., Clem, R.J., **2003**. *In vivo* induction of apoptosis correlating with reduced infectivity during baculovirus infection. *J. Virol.* 77, 2227–2232.

Clem, R.J., **1997**. Regulation of programmed cell death by baculoviruses. In: Miller, L.K. (Ed.), The Baculoviruses. Plenum, New York, pp. 237–266.

Clem, R.J., **2001**. Baculoviruses and apoptosis: the good, the bad, and the ugly. *Cell Death Differ.* 8, 137–143.

Clem, R.J., Fechheimer, M., Miller, L.K., **1991**. Prevention of apoptosis by a baculovirus gene during infection of insect cells. *Science* 254, 1388–1389.

Clem, R.J., Miller, L.K., **1993**. Apoptosis reduces both the *in vitro* replication and the *in vivo* infectivity of a baculovirus. *J. Virol.* 67, 3730–3738.

Clem, R.J., Miller, L.K., **1994**. Control of programmed cell death by the baculovirus genes *p35* and *iap*. *Mol. Cell. Biol.* 14, 5212–5222.

Clem, R.J., Robson, M., Miller, L.K., **1994**. Influence of infection route on the infectivity of baculovirus mutants lacking the apoptosis-inhibitory gene *p35* and the adjacent gene *p94*. *J. Virol.* 68, 6759–6762.

Cory, J.S., Hails, R.S., Sait, S.M., **1997**. Baculovirus ecology. In: Miller, L.K. (Ed.), The Baculoviruses. Plenum, New York, pp. 301–339.

Couch, J.A., **1991**. Nuclear polyhedrosis viruses of invertebrates other than insects. In: Adams, J.R., Bonami, J.R. (Eds.), Atlas of Invertebrate Viruses. CRC Press, Boca Raton, FL, pp. 205–226.

Croizier, G., Croizier, L., Argaud, O., Poudevigne, D., **1994**. Extension of *Autographa californica* nuclear polyhedrosis virus host range by interspecific replacement of a short DNA sequence in the p143 helicase gene. *Proc. Natl Acad. Sci. USA* 91, 48–52.

Croizier, G., Ribeiro, H.C.T., **1992**. Recombination as a possible major cause of genetic heterogeneity in *Anticarsia gemmatalis* nuclear polyhedrosis virus populations. *Virus Res. 26*, 183–196.

Dall, D., Luque, T., O'Reilly, D., **2001**. Insect–virus relationships: sifting by informatics. *BioEssays 23*, 184–193.

Derksen, A.C.G., **1988**. Alteration of a lepidopteran peritrophic membrane by baculoviruses and enhancement of viral infectivity. *Virology 167*, 242–250.

Doyle, C.J., Hirst, M.L., Cory, J.S., Entwistle, P.E., **1990**. Risk assessment studies: detailed host range testing of wild-type cabbage moth, *Mamestra brassicae* (Lepidoptera: Noctuidae), nuclear polyhedrosis virus. *Appl. Environ. Microbiol. 56*, 2704–2710.

Dreschers, D., Roncarati, R., Knebel-Morsdorf, D., **2001**. Actin rearrangement-inducing factor of baculoviruses is tyrosine phosphorylated and colocalizes with F-actin at the plasma membrane. *J. Virol. 75*, 3771–3778.

Du, X., Thiem, S.M., **1997**. Responses of insect cells to baculovirus infection: protein synthesis shutdown and apoptosis. *J. Virol. 71*, 7866–7872.

Duffey, S.S., Hoover, K., Bonning, B.C., Hammock, B.D., **1995**. The impact of host plant on the efficacy of baculoviruses. *Rev. Pestic Toxicol. 3*, 137–275.

Duncan, R., Chung, K.L., Faulkner, P., **1983**. Analysis of a mutant of *Autographa californica* nuclear polyhedrosis virus with a defect in the morphogenesis of the occlusion body macromolecular lattice. *J. Gen. Virol. 64*, 1531–1542.

Eldridge, R., Li, Y., Miller, L.K., **1992**. Characterization of a baculovirus gene encoding a small conotoxin-like polypeptide. *J. Virol. 66*, 6563–6571.

Engelhard, E.K., Kam-Morgan, L.N.W., Washburn, J.O., Volkman, L.E., **1994**. The insect tracheal system: a conduit for the systemic spread of *Autographa californica* M nuclear polyhedrosis virus. *Proc. Natl Acad. Sci. USA 91*, 3224–3227.

Engelhard, E.K., Volkman, L.E., **1995**. Developmental resistance in fourth instar *Trichoplusia ni* orally inoculated with *Autographa californica* M nuclear polyhedrosis virus. *Virology 209*, 384–389.

Entwistle, P.F., Evans, H.F., **1985**. Viral control. In: Gilbert, L.I., Kerkut, G.A. (Eds.), Comprehensive Insect Physiology, Biochemistry, and Pharmacology, vol. 12. Pergamon, Oxford, pp. 347–412.

Evans, H.F., **1986**. Ecology and epizootiology of baculoviruses. In: Granados, R.R., Federici, B.A. (Eds.), The Biology of Baculoviruses, vol. 2. CRC Press, Boca Raton, FL, pp. 89–132.

Evans, J.T., Leisy, D.J., Rohrmann, G.F., **1997**. Characterization of the interaction between the baculovirus replication factors, LEF-1 and LEF-2. *J. Virol. 71*, 3114–3119.

Faulkner, P., **1981**. Baculovirus. In: Davidson, E.W. (Ed.), Pathogenesis of Invertebrate Microbial Diseases. Allanheld, Osmun, Totowa, NJ, pp. 3–33.

Faulkner, P., Kuzio, J., Williams, G.V., Wilson, J.A., **1997**. Analysis of p74, a PDV envelope protein of *Autographa californica* nucleopolyhedrovirus required for occlusion body infectivity *in vivo*. *J. Gen. Virol. 78*, 3091–3100.

Federici, B.A., **1986**. Ultrastructure of baculoviruses. In: Granados, R.R., Federici, B.A. (Eds.), The Biology of Baculoviruses, vol. 1. CRC Press, Boca Raton, FL, pp. 61–88.

Federici, B.A., **1993**. Viral pathobiology in relation to insect control. In: Beckage, N.E., Thompson, S.N., Federici, B.A. (Eds.), Parasites and Pathogens of Insects, vol. 2. Academic Press, New York, pp. 81–101.

Federici, B.A., **1997**. Baculovirus pathogenesis. In: Miller, L.K. (Ed.), The Baculoviruses. Plenum, New York, pp. 33–60.

Flipsen, J.T.M., Lent, J.W.M.V., Goldbach, R.W., Vlak, J.M., **1993**. Expression of polyhedrin and p10 in the midgut of AcMNPV-infected *Spodoptera exigua* larvae: an immunoelectron microscopic investigation. *J. Invertebr. Pathol. 61*, 17–23.

Flipsen, J.T.M., Mans, R.M.W., Kleefsman, A.W., Knebel-Morsdorf, D., Vlak, J.M., **1995a**. Deletion of the baculovirus UDP-glucosyltransferase gene induces early degeneration of Malpighian tubules in infected insects. *J. Virol. 69*, 4529–4532.

Flipsen, J.T.M., Martens, J.W.M., Oers, M.M.V., Vlak, J.M., Lent, J.W.M.V., **1995b**. Passage of *Autographa californica* nuclear polyhedrosis virus through the midgut epithelium of *Spodoptera exigua* larvae. *Virology 208*, 328–335.

Fraser, M.J., **1986**. Ultrastructural observations of virion maturation in *Autographa californica* nuclear polyhedrosis virus infected *Spodoptera frugiperda* cell cultures. *J. Ultrastruct. Mol. Struct. Res. 95*, 189–195.

Fraser, M.J., Cary, L.C., Boonvisudhi, K., Wang, H.G.H., **1995**. Assay for movement of lepidopteran transposon IFP2 in insect cells using a baculovirus genome as a target DNA. *Virology 211*, 397–407.

Fraser, M.J., Smith, G.E., Summers, M.D., **1983**. Acquisition of host cell DNA sequences by baculoviruses: relationship between host DNA insertions and FP mutants of *Autographa californica* and *Galleria mellonella* nuclear polyhedrosis viruses. *J. Virol. 47*, 287–300.

Friesen, P.D., **1993**. Invertebrate transposable elements in the baculovirus genome: characterization and significance. In: Beckage, N.E., Thompson, S.N., Federici, B.A. (Eds.), Parasites and Pathogens of Insects: Parasites, vol. 2. Academic Press, London, pp. 147–178.

Friesen, P.D., **1997**. Regulation of baculovirus early gene expression. In: Miller, L.K. (Ed.), The Baculoviruses. Plenum, New York, pp. 141–170.

Fuchs, L.Y., Woods, M.S., Weaver, R.F., **1983**. Viral transcription during *Autographa californica* nuclear polyhedrosis virus infection: a novel RNA polymerase induced in infected *Spodoptera frugiperda* cells. *J. Virol. 43*, 641–646.

Funk, C.J., Braunagel, S.C., Rohrmann, G.F., **1997**. Baculovirus structure. In: Miller, L.K. (Ed.), The Baculoviruses. Plenum, New York, pp. 7–32.

Funk, C.J., Consigli, R.A., **1993**. Phosphate cycling on the basic protein of *Plodia interpunctella* granulosis virus. *Virology 193*, 396–402.

Fuxa, J.R., **1987**. *Spodoptera frugiperda* susceptibility to nuclear polyhedrosis virus isolates with reference to insect migration. *Environ. Entomol. 16*, 218–223.

Fuxa, J.R., **1991**. Release and transport of entomopathogenic microorganisms. In: Levin, M.A., Strauss, H.S. (Eds.), Risk Assessment in Genetic Engineering. McGraw Hill, New York, pp. 83–113.

Fuxa, J.R., **1993**. Insect resistance to viruses. In: Beckage, N.E., Thompson, S.N., Federici, B.A. (Eds.), Parasites and Pathogens of Insects, vol. 2. Academic Press, New York, pp. 197–209.

Fuxa, J.R., Mitchell, F.L., Richter, A.R., **1988**. Resistance of *Spodoptera frugiperda* (Lep.: Noctuidae) to a nuclear polyhedrosis virus in the field and laboratory. *Entomophaga 33*, 55–63.

Fuxa, J.R., Richter, A.R., **1989**. Reversion of resistance by *Spodoptera frugiperda* to nuclear polyhedrosis virus. *J. Invertebr. Pathol. 53*, 52–56.

Fuxa, J.R., Richter, A.R., **1990**. Response of nuclear polyhedrosis virus-resistant *Spodoptera frugiperda* larvae to other pathogens and to chemical insecticides. *J. Invertebr. Pathol. 55*, 272–277.

Fuxa, J.R., Richter, A.R., **1991**. Selection for an increased rate of vertical transmission of *Spodoptera frugiperda* (Lepidoptera: Noctuidae). *Environ. Entomol. 20*, 603–609.

Fuxa, J.R., Richter, A.R., **1998**. Repeated reversion of resistance to nucleopolyhedrovirus by *Anticarsia gemmatalis*. *J. Invertebr. Pathol. 71*, 159–164.

Fuxa, J.R., Richter, A.R., Ameen, A.O., Hammock, B.D., **2002**. Vertical transmission of TnSNPV, TnCPV, AcMNPV, and possibly recombinant NPV in *Trichoplusia ni*. *J. Invertebr. Pathol. 79*, 44–50.

Fuxa, J.R., Richter, A.R., Strother, M.S., **1993**. Detection of *Anticarsia gemmatalis* nuclear polyhedrosis virus in predatory arthropods and parasitoids after viral release in Louisiana soybean. *J. Entomol. Sci. 28*, 51–60.

Fuxa, J.R., Tanada, Y., **1987**. Epizootiology of Insect Diseases. Wiley, New York.

Gallo, L.G., Corsaro, B.G., Hughes, P.R., Granados, R.R., **1991**. *In vivo* enhancement of baculovirus infection by the viral enhancing factor of a granulosis virus of the cabbage looper, *Trichoplusia ni* (Lepidoptera: Noctuidae). *J. Invertebr. Pathol. 58*, 203–210.

Ghosh, S., Parvez, M.K., Banerjee, K., Sarin, S.K., Hasnain, S.E., **2002**. Baculovirus as mammalian cell expression vector for gene therapy: an emerging strategy. *Mol. Ther. 6*, 5–11.

Gijzen, M., Roelvink, P., Granados, R., **1995**. Characterization of viral enhancing activity from *Trichoplusia ni* granulosis virus. *J. Invertebr. Pathol. 65*, 289–294.

Gombart, A.F., Pearson, M.N., Rohrmann, G.F., Beaudreau, G.S., **1989**. A baculovirus PE-associated protein: genetic location, nucleotide sequence and immunocytochemical characterization. *Virology 169*, 182–193.

Gomi, S., Majima, K., Maeda, S., **1999**. Sequence analysis of the genome of *Bombyx mori* nucleopolyhedrovirus. *J. Gen. Virol. 80*, 1323–1337.

Gordon, J.D., **1984**. Phenotypic characterization and physical mapping of a temperature-sensitive mutant of *Autographa californica* nuclear polyhedrosis virus defective in DNA synthesis. *Virology 138*, 69–81.

Granados, R.R., Lawler, K.A., **1981**. *In vivo* pathway of *Autographa californica* baculovirus invasion and infection. *Virology 108*, 297–308.

Granados, R.R., Williams, K.A., **1986**. *In vivo* infection and replication of baculoviruses. In: Granados, R.R., Federici, B.A. (Eds.), The Biology of Baculoviruses, vol. 1. CRC Press, Boca Raton, FL, pp. 89–108.

Griffiths, C.M., Barnett, A.L., Ayres, M.D., Windass, J., King, L.A., et al., **1999**. *In vitro* host range of *Autographa californica* nucleopolyhedrovirus recombinants lacking functional *p35*, *iap1* or *iap2*. *J. Gen. Virol. 80*, 1055–1066.

Gröner, A., **1986**. Specificity and safety of baculoviruses. In: Granados, R.R., Federici, B.A. (Eds.), The Biology of Baculoviruses, vol. 1. CRC Press, Boca Raton, FL, pp. 177–202.

Gross, C.H., **1994**. The *Orgyia pseudotsugata* baculovirus p10 and polyhedrin envelope protein genes: analysis of their relative expression levels and role in polyhedron structure. *J. Gen. Virol. 75*, 1115–1123.

Gross, C.H., Rohrmann, G.F., **1993**. Analysis of the role of 5' promoter elements and 3' flanking sequences on the expression of a baculovirus polyhedron envelope protein gene. *Virology 92*, 273–281.

Gross, C.H., Shuman, S., **1998**. RNA 5'-triphosphatase, nucleoside triphosphatase, and guanylyltransferase activities of baculovirus LEF-4 protein. *J. Virol. 72*, 10020–10028.

Guarino, L.A., **1990**. Identification of a viral gene encoding a ubiquitin-like protein. *Proc. Natl Acad. Sci. USA 87*, 409–413.

Guarino, L.A., Dong, W., Jin, J., **2002a**. *In vitro* activity of the baculovirus late expression factor LEF-5. *J. Virol. 76*, 12663–12675.

Guarino, L.A., Jin, J., Dong, W., **1998a**. Guanylyltransferase activity of the LEF-4 subunit of baculovirus RNA polymerase. *J. Virol. 72*, 10003–10010.

Guarino, L.A., Mistretta, T., Dong, W., **2002b**. Baculovirus *lef-12* is not required for viral replication. *J. Virol. 76*, 12032–12043.

Guarino, L.A., Smith, G.E., Dong, W., **1995**. Ubiquitin is attached to membranes of baculovirus particles by a novel type of phopholipid anchor. *Cell 80*, 301–309.

Guarino, L.A., Xu, B., Jin, J., Dong, W., **1998b**. A virus-encoded RNA polymerase purified from baculovirus-infected cells. *J. Virol. 72*, 7985–7991.

Haas, A.L., Katzung, D.J., Reback, P.M., Guarino, L.A., **1996**. Functional characterization of the ubiquitin variant encoded by the baculovirus *Autographa californica*. *Biochemistry 35*, 5385–5394.

Haas-Stapleton, E., Washburn, J.O., Volkman, L.E., **2003**. Pathogenesis of *Autographa californica*

M nucleopolyhedrovirus in fifth instar *Spodoptera frugiperda*. *J. Gen. Virol. 84*, 2033–2040.

Hang, X., Dong, W., Guarino, L.A., **1995**. The *lef-3* gene of *Autographa californica* nuclear polyhedrosis virus encodes a single-stranded DNA-binding protein. *J. Virol. 69*, 3924–3928.

Harrison, R.L., Bonning, B.C., **1999**. The nucleopolyhedroviruses of *Rachiplusia ou* and *Anagrapha falcifera* are isolates of the same virus. *J. Gen. Virol. 80*, 2793–2798.

Harrison, R.L., Bonning, B.C., **2003**. Comparative analysis of the genomes of *Rachiplusia ou* and *Autographa californica* multiple nucleopolyhedrovirus. *J. Gen. Virol. 84*, 1827–1842.

Harrison, R.L., Bonning, B.C., **2004**. Application of maximum likelihood models to selection pressure analysis of group I nucleopolyhedrovirus genes. *J. Gen. Virol. 85*, 197–210.

Harrison, R.L., Jarvis, D.L., Summers, M.D., **1996**. The role of the AcMNPV *25K* gene "*FP25*," in baculovirus *polh* and *p10* expression. *Virology 226*, 34–46.

Harrison, R.L., Summers, M.D., **1995**. Mutations in the *Autographa californica* multinucleocapsid nuclear polyhedrosis virus 25 kDa protein gene result in reduced virion occlusion, altered intranuclear envelopment and enhanced virus production. *J. Gen. Virol. 76*, 1451–1459.

Hashimoto, Y., Hayakawa, T., Ueno, Y., Fujita, T., Sano, Y., *et al.*, **2000**. Sequence analysis of the *Plutella xylostella* granulovirus genome. *Virology 275*, 358–372.

Hawtin, R.E., Arnold, K., Ayres, M.D., Zanotto, P.M., Howard, S.C., *et al.*, **1995**. Identification and preliminary characterization of a chitinase gene in the *Autographa californica* nuclear polyhedrosis virus genome. *Virology 212*, 673–685.

Hawtin, R.E., Zarkowska, T., Arnold, K., Thomas, C.J., Gooday, G.W., *et al.*, **1997**. Liquefaction of *Autographa californica* nucleopolyhedrovirus-infected insects is dependent on the integrity of virus-encoded chitinase and cathepsin genes. *Virology 238*, 243–253.

Hayakawa, T., Ko, R., Okano, K., Seong, S., Goto, C., *et al.*, **1999**. Sequence analysis of the *Xestia c-nigrum* granulovirus genome. *Virology 262*, 277–297.

Hernandez-Crespo, P., Sait, S.M., Hails, R.S., Cory, J.S., **2001**. Behavior of recombinant baculovirus in lepidopteran hosts with different susceptibilities. *Appl. Environ. Microbiol. 67*, 1140–1146.

Herniou, E.A., Luque, T., Chen, X., Vlak, J.M., Winstanley, D., *et al.*, **2001**. Use of whole genome sequence data to infer baculovirus phylogeny. *J. Virol. 75*, 8117–8126.

Herniou, E.A., Olszewski, J.A., Cory, J.S., O'Reilly, D.R., **2003**. The genome sequence and evolution of baculoviruses. *Annu. Rev. Entomol. 48*, 211–234.

Herschberger, P.A., Dickson, J.A., Friesen, P.D., **1992**. Site-specific mutagenesis of the 35 kilodalton protein gene encoded by *Autographa californica* nuclear

polyhedrosis virus: cell line-specific effects on virus replication. *J. Virol. 66*, 5525–5533.

Hill, J.E., Kuzio, J., Wilson, J.A., Mackinnon, E.A., Faulkner, P., **1993**. Nucleotide sequence of the *p74* gene of a baculovirus pathogenic to the spruce budworm *Choristoneura fumiferana* multicapsid nuclear polyhedrosis virus. *Biochim. Biophys. Acta 1172*, 187–189.

Hom, L.G., Ohkawa, T., Trudeau, D., Volkman, L.E., **2002**. *Autographa californica* M nucleopolyhedrovirus ProV-CATH is activated during infected cell death. *Virology 296*, 212–218.

Hom, L.G., Volkman, L.E., **2000**. *Autographa californica* M nucleopolyhedrovirus chiA is required for processing of V-CATH. *Virology 277*, 178–183.

Hong, T., Braunagel, S.C., Summers, M.D., **1994**. Transcription, translation and cellular localization of PDV-E66: a structural protein of the PDV envelope of *Autographa californica* nuclear polyhedrosis virus. *Virology 204*, 210–222.

Hong, T., Summers, M.D., Braunagel, S.C., **1997**. N-terminal sequences from *Autographa californica* nuclear polyhedrosis virus envelope proteins ODV-E66 and ODV-E25 are sufficient to direct reporter proteins to the nuclear envelope, intranuclear microvesicles, and the envelope of occlusion derived virus. *Proc. Natl Acad. Sci. USA 94*, 4050–4055.

Hopkins, T.L., Harper, M.S., **2001**. Lepidopteran peritrophic membranes and effects of dietary wheat germ agglutinin on their formation and structure. *Arch. Insect Biochem. Physiol. 47*, 100–109.

Horton, H.M., Burand, J.P., **1993**. Saturable attachment sites for polyhedron-derived baculovirus on insect cells and evidence for entry via direct membrane fusion. *J. Virol. 67*, 1860–1868.

Hostetter, D.L., Puttler, B., **1991**. A new broad host spectrum nuclear polyhedrosis virus isolated from a celery looper, *Anagrapha falcifera* (Kirby) (Lepidoptera: Noctuidae). *Environ. Entomol. 20*, 1480–1488.

Hughes, A.L., **2002**. Evolution of inhibitors of apoptosis in baculoviruses and their insect hosts. *Infect. Genet. Evol. 2*, 3–10.

Hughes, A.L., Friedman, R., **2003**. Genome-wide survey for genes horizontally transferred from cellular organisms to baculoviruses. *Mol. Biol. Evol. 20*, 979–987.

Hughes, D.S., Possee, R.D., King, L.A., **1993**. Activation and detection of a latent baculovirus resembling *Mamestra brassicae* nuclear polyhedrosis virus in *M. brassicae* insects. *Virology 194*, 608–615.

Hughes, D.S., Possee, R.D., King, L.A., **1997**. Evidence for the presence of a low-level, persistent baculovirus infection of *Mamestra brassicae* insects. *J. Gen. Virol. 78*, 1801–1805.

Hukuhara, T., Hayakawa, T., Wijonarko, A., **1999**. Increased baculovirus susceptibility of armyworm larvae feeding on transgenic rice plants expressing an entomopoxvirus gene. *Nature Biotechnol. 17*, 1122–1124.

Huser, A., Hofmann, C., 2003. Baculovirus vectors: novel mammalian cell gene-delivery vehicles and their applications. *Am. J. Pharmacogenomics 3*, 53–63.

Hyink, O., Dellow, R.A., Olsen, M.J., Caradoc-Davies, K.M.B., Drake, K., *et al.*, 2002. Whole genome analysis of the *Epiphyas postvittana* nucleopolyhedrovirus. *J. Gen. Virol. 83*, 957–971.

Ignoffo, C.M., Garcia, A., 1994. Antioxidant and oxidant enzyme effects on the inactivation of inclusion bodies of the *Heliothis* baculovirus by simulated sunlight UV. *Environ. Entomol. 23*, 1025–1029.

Ignoffo, C.M., Garcia, C., Saathoff, S.G., 1997. Sunlight stability and rain-fastness of formulations of *Baculovirus heliothis*. *Environ. Entomol. 26*, 1470–1474.

Ignoffo, G.M., Garcia, C., Zuidema, D., Vlak, J.M., 1995. Relative *in vivo* activity and simulated sunlight-UV stability of inclusion bodies of a wild-type and an engineered polyhedra envelope-negative isolate of the nucleopolyhedrosis virus of *Autographa californica*. *J. Invertebr. Pathol. 66*, 212–213.

Ijkel, W.F.J., Strien, E.A. v., Heldens, J.G.M., Broer, R., Zuidema, D., *et al.*, 1999. Sequence and organization of the *Spodoptera exigua* multicapsid nucleopolyhedrovirus genome. *J. Gen. Virol. 80*, 3289–3304.

Imai, N., Matsuda, N., Tanaka, K., Nakano, A., Matsumoto, S., *et al.*, 2003. Ubiquitin ligase activities of *Bombyx mori* nucleopolyhedrovirus RING finger proteins. *J. Virol. 77*, 923–930.

Jaques, R.P., 1975. Persistence, accumulation and denaturation of nuclear polyhedrosis and granulosis viruses. In: Summers, M.D., Engler, R., Falcon, L.A., Vail, P.V. (Eds.), Baculoviruses for Insect Pest Control: Safety Considerations. American Society of Microbiology, Washington DC, pp. 55–67.

Jaques, R.P., 1985. Stability of insect viruses in the environment. In: Maramorosch, K., Sherman, K.E. (Eds.), Viral Insecticides for Biological Control. Academic Press, Orlando, FL, pp. 289–360.

Jarvis, D.L., 1997. Baculovirus expression vectors. In: Miller, L.K. (Ed.), The Baculoviruses. Plenum, New York, pp. 389–431.

Jarvis, D.L., Bohlmeyer, D.A., Garcia, A., 1991. Requirements for nuclear localization and supramolecular assembly of a baculovirus polyhedrin protein. *Virology 185*, 795–810.

Jarvis, D.L., Bohlmeyer, D.A., Garcia, J.A., 1992. Enhancement of polyhedrin nuclear localization during baculovirus infection. *J. Virol. 66*, 6903–6911.

Jarvis, D.L., Fleming, J.D.W., Kovacs, G.R., Summers, M.D., Guarino, L.A., 1990. Use of early baculovirus promoters for continuous expression and efficient processing of foreign gene products in stably transformed lepidopteran cells. *Bio/Technol. 8*, 950–955.

Jarvis, D.L., Garcia, A., 1994. Biosynthesis and processing of the *Autographa californica* nuclear polyhedrosis virus gp64 protein. *Virology 205*, 300–313.

Jehle, J.A., 1996. Transmission of insect transposons in baculovirus genomes: an unusual host–pathogen interaction. In: Tomiuk, J., Wohrmann, K., Sentker, A. (Eds.), Transgenic Organisms: Biological and Social Implications. Birlhauser Verlag, Basel, pp. 81–97.

Jehle, J.A., Fritsch, E.F., Nickel, A., Huber, J., Backhaus, H., 1995. TC14.7: a novel lepidopteran transposon found in *Cydia pomonella* granulosis virus. *Virology 207*, 369–379.

Jehle, J.A., Linden, I.F. van der, Backhaus, H., Vlak, J.M., 1997. Identification and sequence analysis of the integration site of transposon TCp3.2 in the genome of *Cydia pomonella* granulovirus. *Virus Res. 50*, 151–157.

Jehle, J.A., Nickel, A., Vlak, J.M., Backhaus, H., 1998. Horizontal escape of the novel Tc10-like lepidopteran transposon TCp3.2 into *Cydia pomonella* granulovirus. *J. Mol. Evol. 46*, 215–224.

Jin, J., Dong, W., Guarino, L.A., 1998. The LEF-4 subunit of baculovirus RNA polymerase has RNA 5'-triphosphatase and ATPase activities. *J. Virol. 72*, 10011–10019.

Kamita, S.G., Maeda, S., 1997. Sequencing of the putative DNA helicase-encoding gene of the *Bombyx mori* nuclear polyhedrosis virus and fine-mapping of a region involved in host range expansion. *Gene 190*, 173–179.

Kamita, S.G., Majima, K., Maeda, S., 1993. Identification and characterization of the *p35* gene of *Bombyx mori* nuclear polyhedrosis virus that prevents virus-induced apoptosis. *J. Virol. 67*, 455–463.

Kang, W., Tristem, M.N., Maeda, S., Crook, N.E., O'Reilly, D.R., 1998. Identification and characterization of the *Cydia pomonella* granulovirus cathepsin and chitinase genes. *J. Gen. Virol. 79*, 2283–2292.

Kang, W.K., Imai, N., Iwanaga, M., Matsumoto, S., Zemskov, E.A., 2003. Interaction of *Bombyx mori* nucleopolyhedrovirus BRO-A and host cell protein laminin. *Arch. Virol. 148*, 99–113.

Kaomini, M., Roush, R.T., 1988. Absence of response to selection for resistance to nucleopolyhedrosis virus in *Heliothis virescens* (F.) (Lepidoptera: Noctuidae). *J. Entomol. Sci. 23*, 379–382.

Katsuma, S., Noguchi, Y., Zhou, C.L.E., Kobayashi, M., Maeda, S., 1999. Characterization of the 25K FP gene of the baculovirus *Bombyx mori* nucleopolyhedrovirus: implications for post-mortem host degradation. *J. Gen. Virol. 80*, 783–791.

Kawanishi, C.Y., Paschke, J.D., 1970. The relationship of buffer pH and ionic strength on the yield of virions and nucleocapsids obtained by the dissolution of *Rachiplusia ou* nuclear polyhedra. *4th Int. Colloqu. Insect Pathol.* 127–146.

Keddie, B.A., Aponte, G.W., Volkman, L.E., 1989. The pathway of infection of *Autographa californica* nuclear polyhedrosis virus in an insect host. *Science 243*, 1728–1730.

Kelly, T.J., Park, E.J., Masler, C.A., Burand, J.P., 1995. Characterization of the glycosylated ecdysteroids in the hemolymph of baculovirus-infected gypsy moth larvae and cells in culture. *Eur. J. Entomol. 92*, 51–61.

Knebel-Mörsdorf, D., Flipsen, J.T.M., Roncarati, R., Jahnel, F., Kleefsman, A.W.F., *et al.*, 1996. Baculovirus infection of *Spodoptera exigua* larvae: *lacZ* expression driven by promoters of early genes *pe38* and *me53* in larval tissue. *J. Gen. Virol.* 77, 815–824.

Knudson, D.L., Harrap, K.A., **1976**. Replication of a nuclear polyhedrosis virus in a continuous cell culture of *Spodoptera frugiperda*: microscopy study of the sequence of events of the virus infection. *J. Virol.* 17, 254–268.

Kolodny Hirsch, D.M., van Beek, N.A.M., **1997**. Selection of a morphological variant of *Autographa californica* nuclear polyhedrosis virus with increased virulence following serial passage in *Plutella xylostella*. *J. Invertebr. Pathol.* 69, 205–211.

Kool, M., Ahrens, C., Goldbach, R.W., Rohrmann, G.F., Vlak, J.M., **1994a**. Identification of genes involved in DNA replication of the *Autographa californica* nuclear polyhedrosis virus. *Proc. Natl Acad. Sci. USA* 91, 11212–11216.

Kool, M., Ahrens, C.H., Vlak, J.M., Rohrmann, G.F., **1995**. Replication of baculovirus DNA. *J. Gen. Virol.* 76, 2103–2118.

Kool, M., Goldbach, R.W., Vlak, J.M., **1993a**. A putative non-hr origin of DNA replication of the *Hind*III-K fragment of *Autographa californica* nuclear polyhedrosis virus. *J. Gen. Virol.* 75, 3345–3352.

Kool, M., van den Berg, P.M., Tramper, J., Goldbach, R.W., Vlak, J.M., **1993b**. Location of two putative origins of DNA replication of *Autographa californica* nuclear polyhedrosis virus. *Virology* 192, 94–101.

Kool, M., Voeten, J.T.M., Vlak, J.M., **1994b**. Functional mapping of regions of the *Autographa californica* nuclear polyhedrosis viral genome required for DNA replication. *Virology* 198, 680–689.

Kost, T.A., Condreay, J.P., **2002**. Recombinant baculoviruses as mammalian cell gene-delivery vectors. *Trends Biotechnol.* 20, 173–180.

Kovacs, G.R., Choi, J., Guarino, L.A., Summers, M.D., **1991**. Identification of spliced baculovirus RNAs expressed late in infection. *Virology* 185, 633–643.

Kukan, B., **1999**. Vertical transmission of nucleopolyhedrovirus in insects. *J. Invertebr. Pathol.* 74, 103–111.

Kuzio, J., Jaques, R., Faulkner, P., **1989**. Identification of *p74*, a gene essential for virulence of baculovirus occlusion bodies. *Virology* 173, 759–763.

Kuzio, J., Pearson, M.N., Harwood, S.H., Funk, C.J., Evans, J.T., *et al.*, **1999**. Sequence and analysis of the genome of a baculovirus pathogenic for *Lymantria dispar*. *Virology* 253, 17–34.

Lanier, L.M., Slack, J.M., Volkman, L.E., **1996**. Actin binding and proteolysis by the baculovirus AcMNPV: the role of virion-associated V-CATH. *Virology* 216, 380–388.

Lanier, L.M., Volkman, L.E., **1998**. Actin binding and nucleation by *Autographa californica* M nucleopolyhedrovirus. *Virology* 243, 167–177.

Lee, H.H., Miller, L.K., **1978**. Isolation of genotypic variants of *Autographa californica* nuclear polyhedrosis virus. *J. Virol.* 27, 754–767.

Leisy, D.J., Rasmussen, C., Kim, H.-T., Rohrmann, G.F., **1995**. The *Autographa californica* nuclear polyhedrosis virus homologous region 1a: identical sequences are essential for DNA replication activity and transcriptional enhancer function. *Virology* 208, 742–752.

Leisy, D.J., Rohrmann, G.F., **1993**. Characterization of the replication of plasmids containing hr sequences in baculovirus-infected *Spodoptera frugiperda* cells. *Virology* 196, 722–730.

Lepore, L.S., Roelvink, P.R., Granados, R.R., **1996**. Enhancin, the granulosis virus protein that facilitates nucleopolyhedrovirus (NPV) infections, is a metalloprotease. *J. Invertebr. Pathol.* 68, 131–140.

Lewis, L.C., Johnson, T.B., **1982**. Efficacy of two nuclear polyhedrosis viruses against *Ostrinia nubilalis* [Lep.: Pyralidae] in the laboratory and field. *Entomophaga* 27, 33–38.

Li, L., Donly, C., Li, Q., Willis, L.G., Keddie, B.A., *et al.*, **2002a**. Identification and genomic analysis of a second species of nucleopolyhedrovirus isolated from *Mamestra configurata*. *Virology* 297, 226–244.

Li, L., Harwood, S.H., Rohrmann, G.F., **1999**. Identification of additional genes that influence baculovirus late gene expression. *Virology* 255, 9–19.

Li, Q., Donly, C., Li, L., Willis, L.G., Theilmann, D.A., *et al.*, **2002b**. Sequence and organization of the *Mamestra configurata* nucleopolyhedrovirus genome. *Virology* 294, 106–121.

Li, Y., Miller, L.K., **1995**. Properties of a baculovirus mutant defective in the protein phosphatase gene. *J. Virol.* 69, 4533–4537.

Lin, G., Blissard, G.W., **2002a**. Analysis of an *Autographa californica* multicapsid nucleopolyhedrovirus *lef-6*-null virus: LEF-6 is not essential for viral replication but appears to accelerate late gene transcription. *J. Virol.* 76, 5503–5514.

Lin, G., Blissard, G.W., **2002b**. Analysis of an *Autographa californica* nucleopolyhedrovirus *lef-11* knockout: LEF-11 is essential for viral DNA replication. *J. Virol.* 76, 2770–2779.

Lin, G., Slack, J.M., Blissard, G.W., **2001**. Expression and localization of LEF-11 in *Autographa californica* nucleopolyhedrovirus infected Sf9 cells. *J. Gen. Virol.* 82, 2289–2294.

Liu, J.J., Carstens, E.B., **1993**. Infection of *Spodoptera frugiperda* and *Choristoneura fumiferana* cell lines with the baculovirus *Choristoneura fumiferana* nuclear polyhedrosis virus. *Can. J. Microbiol.* 39, 932–940.

Liu, J.J., Carstens, E.B., **1995**. Identification, localization, transcription, and sequence analysis of the *Choristoneura fumiferana* nuclear polyhedrosis virus DNA polymerase gene. *Virology* 209, 538–549.

Lu, A., Carstens, E.B., **1991**. Nucleotide sequence of a gene essential for viral DNA replication in the baculovirus *Autographa californica* nuclear polyhedrosis virus. *Virology* 181, 336–347.

Lu, A., Carstens, E.B., **1992**. Nucleotide sequence and transcriptional analysis of the *p80* gene of *Autographa californica* nuclear polyhedrosis virus: a homologue of the *Orgyia pseudotsugata* nuclear polyhedrosis virus capsid-associated gene. *Virology 190*, 201–209.

Lu, A., Carstens, E.B., **1993**. Immediate early baculovirus genes transactivate the *p143* gene promoter of *Autographa californica* nuclear polyhedrosis virus. *Virology 195*, 710–718.

Lu, A., Miller, L.K., **1995a**. Differential requirements for baculovirus late expression factor genes in two cell lines. *J. Virol. 69*, 6265–6272.

Lu, A., Miller, L.K., **1995b**. The roles of eighteen baculovirus late expression factor genes in transcription and DNA replication. *J. Virol. 69*, 975–982.

Lu, A., Miller, L.K., **1996**. Species-specific effects of the *hcf-1* gene on baculovirus virulence. *J. Virol. 70*, 5123–5130.

Lu, A., Miller, L.K., **1997**. Regulation of baculovirus late and very late gene expression. In: Miller, L.K. (Ed.), The Baculoviruses. Plenum, New York, pp. 217–235.

Lu, L., Du, Q., Chejanovsky, N., **2003**. Reduced expression of the immediate-early protein IE0 enables efficient replication of *Autographa californica* multiple nucleopolyhedrovirus in poorly permissive *Spodoptera littoralis* cells. *J. Virol. 77*, 535–545.

Lung, O.Y., Cruz-Alvarez, M., Blissard, G.W., **2003**. Ac23, an envelope fusion protein homolog in the baculovirus *Autographa californica* multicapsid nucleopolyhedrovirus, is a viral pathogenicity factor. *J. Virol. 77*, 328–339.

Luque, T., Finch, R., Crook, N., O'Reilly, D.R., Winstanley, D., **2001**. The complete sequence of the *Cydia pomonella* granulovirus genome. *J. Gen. Virol. 82*, 2531–2547.

Maeda, S., Kamita, S.G., Kondo, A., **1993**. Host range expansion of *Autographa californica* nuclear polyhedrosis virus (NPV) following recombination of a 0.6-kilobase-pair DNA fragment originating from *Bombyx mori* NPV. *J. Virol. 67*, 6234–6238.

Majima, K., Kobara, R., Maeda, S., **1993**. Divergence and evolution of homologous regions of *Bombyx mori* nuclear polyhedrosis virus. *J. Virol. 67*, 7513–7521.

Malik, H.S., Henikoff, S., Eickbush, T.H., **2000**. Poised for contagion: evolutionary origins of the infectious abilities of invertebrate retroviruses. *Genome Res. 10*, 1307–1318.

Manji, G.A., Friesen, P.D., **2001**. Apoptosis in motion: an apical, P35-insensitive caspase mediates programmed cell death in insect cells. *J. Biol. Chem. 276*, 16704–16710.

Martignoni, M.E., Iwai, P.J., **1986**. Propagation of multicapsid nuclear polyhedrosis virus of *Orgyia pseudotsugata* in larvae of *Trichoplusia ni*. *J. Invertebr. Pathol. 47*, 32–41.

Mazzacano, C.A., Du, X., Thiem, S.M., **1999**. Global protein synthesis shutdown in *Autographa californica* nucleopolyhedrovirus-infected Ld652Y cells is rescued by tRNA from uninfected cells. *Virology 260*, 222–231.

McLachlin, J.R., Miller, L.K., **1994**. Identification and characterization of *vlf-1*, a baculovirus gene involved in very late gene expression. *J. Virol. 68*, 7746–7756.

Means, J.C., Muro, I., Clem, R.J., **2003**. Silencing of the baculovirus Op-*iap3* gene by RNA interference reveals that it is required for prevention of apoptosis during *Orgyia pseudotsugata* M nucleopolyhedrovirus infection of Ld652Y cells. *J. Virol. 77*, 4481–4488.

Mikhailov, V.S., **2003**. Replication of the baculovirus genome. *Mol. Biol. 37*, 250–259.

Milks, M.L., Myers, J.H., **2000**. The development of larval resistance to a nucleopolyhedrovirus in not accompanied by an increased virulence in the virus. *Evol. Ecol. 14*, 645–664.

Milks, M.L., Myers, J.H., Leptich, M.K., **2002**. Costs and stability of cabbage looper resistance to a nucleopolyhedrovirus. *Evol. Ecol. 16*, 369–385.

Miller, D.W., Miller, L.K., **1982**. A virus mutant with an insertion of a *copia*-like transposable element. *Nature 299*, 562–564.

Miller, L.K., Adang, M.J., Browne, D., **1983**. Protein kinase activity associated with the extracellular and occluded forms of the baculovirus *Autographa californica* NPV. *J. Virol. 46*, 275–278.

Miller, L.K., Lu, A., **1997**. The molecular basis of baculovirus host range. In: Miller, L.K. (Ed.), The Baculoviruses. Plenum, New York, pp. 217–235.

Minion, F.C., Coons, L.B., Broome, J.R., **1979**. Characterization of the PE of the nuclear polyhedrosis virus of *Heliothis virescens*. *J. Invertebr. Pathol. 34*, 303–307.

Mitsuhashi, W., Furuta, Y., Sato, M., **1998**. The spindles of an entomopoxvirus of Coleoptera (*Anomala cuprea*) strongly enhance the infectivity of a nucleopolyhedrovirus in Lepidoptera (*Bombyx mori*). *J. Invertebr. Pathol. 71*, 186–188.

Monsma, S.A., Blissard, G.W., **1995**. Identification of a membrane fusion domain and an oligomerization domain in the baculovirus GP64 envelope fusion protein. *J. Virol. 69*, 2583–2595.

Monsma, S.A., Oomens, A.G.P., Blissard, G.W., **1996**. The gp64 envelope fusion protein is an essential baculovirus protein required for cell-to-cell transmission of infection. *J. Virol. 70*, 4607–4616.

Morris, T.D., Miller, L.K., **1992**. Promoter influence on baculovirus-mediated gene expression in permissive and nonpermissive insect cell lines. *J. Virol. 66*, 7397–7405.

Morris, T.D., Miller, L.K., **1993**. Characterization of productive and non-productive AcMNPV infection in selected insect cell lines. *Virology 197*, 339–348.

Morse, M.A., Marriott, A.C., Nuttall, P.A., **1992**. The glycoprotein of Thogoto virus (a tick-borne orthomyxo-like virus) is related to the baculovirus glycoprotein gp64. *Virology 186*, 640–646.

Moscardi, F., **1999**. Assessment of the application of baculoviruses for control of Lepidoptera. *Annu. Rev. Entomol. 44*, 257–289.

Moser, B.A., Becnel, J.J., White, S.E., Afonso, C., Kutish, G., *et al.*, **2001**. Morphological and molecular evidence that *Culex nigripalpus* baculovirus is an unusual member of the family Baculoviridae. *J. Gen. Virol. 82*, 283–297.

Oers, M.M.V., Vlak, J.M., 1997. The baculovirus 10-kDa protein. *J. Invertebr. Pathol.* 70, 1–17.

Ohkawa, T., Majima, K., Maeda, S., 1994. A cysteine protease encoded by the baculovirus *Bombyx mori* nuclear polyhedrosis virus. *J. Virol.* 68, 6619–6625.

Ohkawa, T., Rowe, A.R., Volkman, L.E., 2002. Identification of six *Autographa californica* multicapsid nucleopolyhedrovirus early genes that mediate nuclear localization of G-actin. *J. Virol.* 76, 12281–12289.

Ohkawa, T., Volkman, L.E., 1999. Nuclear F-actin is required for AcMNPV nucleocapsid morphogenesis. *Virology* 264, 1–4.

Ohresser, M., Morin, N., Cerruti, M., Delsert, C., 1994. Temporal regulation of a complex and unconventional promoter by viral products. *J. Virol.* 68, 2589–2597.

Okano, K., Shimada, T., Mita, K., Maeda, S., 2001. Comparative expressed-sequence-tag analysis of differential gene expression profiles in BmNPV-infected BmN cells. *Virology* 282, 348–356.

Olszewski, J., Miller, L.K., 1997a. Identification and characterization of a baculovirus structural protein, VP1054, required for nucleocapsid formation. *J. Virol.* 71, 5040–5050.

Olszewski, J., Miller, L.K., 1997b. A role for baculovirus GP41 in budded virus production. *Virology* 233, 292–301.

Ooi, B.G., Miller, L.K., 1988. Regulation of host RNA levels during baculovirus infection. *Virology* 166, 515–523.

Ooi, B.G., Miller, L.K., 1990. Transcription of the baculovirus polyhedrin gene reduces the levels of an antisense transcript initiated downstream. *J. Virol.* 64, 3126–3129.

Ooi, B.G., Rankin, C., Miller, L.K., 1989. Downstream sequence augments transcription from the essential initiation site of a baculovirus polyhedrin gene. *J. Mol. Biol.* 210, 721–736.

Oppenheimer, D.I., Volkman, L.E., 1997. Evidence for rolling circle replication of *Autographa californica* M nucleopolyhedrovirus genomic DNA. *Arch. Virol.* 142, 2107–2113.

O'Reilly, D., 1995. Baculovirus-encoded ecdysteroid UDP-glucosyltransferases. *Insect Biochem. Mol. Biol.* 25, 541–550.

O'Reilly, D., 1997. Auxiliary genes. In: Miller, L.K. (Ed.), The Baculoviruses. Plenum, New York, pp. 267–300.

O'Reilly, D.R., Brown, M.R., Miller, L.K., 1992. Alteration of ecdysteroid metabolism due to baculovirus infection of the fall armyworm *Spodoptera frugiperda*: host ecdysteroids are conjugated with galactose. *Insect Biochem. Mol. Biol.* 22, 313–320.

O'Reilly, D.R., Kelly, T.J., Masler, E.P., Thyagaraja, B.S., Rohson, R.M., et al., 1995. Overexpression of *Bombyx mori* prothoracicotropic hormone using baculovirus vectors. *Insect Biochem. Mol. Biol.* 25, 475–485.

O'Reilly, D.R., Miller, L.K., 1989. A baculovirus blocks insect molting by producing ecdysteroid UDP-glucosyl transferase. *Science* 245, 1110–1112.

O'Reilly, D.R., Miller, L.K., 1991. Improvement of a baculovirus pesticide by deletion of the EGT gene. *Bio/Technol.* 9, 1086–1089.

Palli, S.R., Caputo, G.F., Sohi, S.S., Brownright, A.J., Ladd, T.R., et al., 1996. CfMNPV blocks AcMNPV-induced apoptosis in a continuous midgut cell line. *Virology* 222, 201–213.

Pang, Y., Yu, J., Wang, L., Hu, X., Bao, W., et al., 2001. Sequence analysis of the *Spodoptera litura* multicapsid nucleopolyhedrovirus genome. *Virology* 287, 391–404.

Passarelli, A.L., Miller, L.K., 1993. Identification and characterization of *lef-1*, a baculovirus gene involved in late and very late gene expression. *J. Virol.* 67, 3481–3488.

Payne, C.C., 1986. Insect pathogenic viruses as pest control agents. In: Franz, J.M. (Ed.), Biological Plant and Health Protection. Georg Fischer Verlag, Stuttgart, pp. 183–200.

Pearson, M.N., Bjornson, R.M., Ahrens, C.H., Rohrmann, G.F., 1993. Identification and characterization of a putative origin of DNA replication in the genome of a baculovirus pathogenic for *Orgyia pseudotsugata*. *Virology* 197, 715–725.

Pearson, M.N., Bjornson, R.M., Pearson, G., Rohrmann, G.F., 1992. The *Autographa californica* baculovirus genome: evidence for multiple replication origins. *Science* 257, 1382–1384.

Pearson, M.N., Groten, C., Rohrmann, G.F., 2000. Identification of the *Lymantria dispar* nucleopolyhedrovirus envelope fusion protein provides evidence for a phylogenetic division of the Baculoviridae. *J. Virol.* 74, 6126–6131.

Pearson, M.N., Rohrmann, G.F., 2002. Transfer, incorporation, and substitution of envelope fusion proteins among members of the Baculoviridae, Orthomyxoviridae, and Metaviridae (insect retrovirus) families. *J. Virol.* 76, 5301–5304.

Pearson, M.N., Russell, R., Rohrmann, G.F., 2001. Characterization of a baculovirus encoded protein that is associated with infected-cell membranes and budded virus. *Virology* 291, 22–31.

Pearson, M.N., Russell, R.L.Q., Rohrmann, G.F., Beaudreau, G.S., 1988. P39, a major baculovirus structural protein: immunocytochemical characterization and genetic location. *Virology* 167, 407–413.

Petrik, D.T., Iseli, A., Montelone, B.A., Van Etten, J.L., Clem, R.J., 2003. Improving baculovirus resistance to UV inactivation: increased virulence resulting from expression of a DNA repair enzyme. *J. Invertebr. Pathol.* 82, 50–56.

Popham, H.J.R., Bishop, D.H.L., Slavicek, J.M., 2001. Both *Lymantria dispar* nucleopolyhedrovirus enhancin genes contribute to viral potency. *J. Virol.* 75, 8639–8648.

Popham, H.J.R., Pellock, B.J., Robson, M., Dierks, P.M., Miller, L.K., 1998. Characterization of a variant of *Autographa californica* nuclear polyhedrosis virus with a nonfunctional ORF 603. *Biol. Control* 12, 223–230.

Possee, R.D., Rohrmann, G.F., 1997. Baculovirus genome organization and evolution. In: Miller, L.K. (Ed.), The Baculoviruses. Plenum, New York, pp. 109–140.

Poulin, R., Combes, C., **1999**. The concept of virulence: interpretations and implications. *Parasitol. Today 15,* 474–475.

Prikhod'ko, E.A., Miller, L.K., **1996**. Induction of apoptosis by baculovirus transactivator IE1. *J. Virol. 70,* 7116–7124.

Quadt, I., Mainz, D., Mans, R., Kremer, A., Knebel-Morsdorf, D., **2002**. Baculovirus infection raises the level of TATA-binding protein that colocalizes with viral replication sites. *J. Virol. 76,* 11123–11127.

Rapp, J.C., Wilson, J.A., Miller, L.K., **1998**. Nineteen baculovirus open reading frames, including *LEF-12,* support late gene expression. *J. Virol. 72,* 10197–10206.

Reddy, J.T., Locke, M., **1990**. The size limited penetration of gold particles through insect basal laminae. *J. Insect Physiol. 36,* 397–407.

Reilly, L.M., Guarino, L.A., **1996**. The viral ubiquitin gene of *Autographa californica* nuclear polyhedrosis virus is not essential for viral replication. *Virology 218,* 243–247.

Ribeiro, B.M., Hutchinson, K., Miller, L.K., **1994**. A mutant baculovirus with a temperature-sensitive IE-1 transregulatory protein. *J. Virol. 68,* 1075–1084.

Richards, A.G., Richards, P.A., **1977**. The peritrophic membrane of insects. *Annu. Rev. Entomol. 22,* 219–240.

Rodems, S.M., Friesen, P.D., **1993**. The hr5 transcriptional enhancer stimulates early expression from the *Autographa californica* nuclear polyhedrosis virus genome but is not required for virus replication. *J. Virol. 67,* 5776–5785.

Roelvink, P.W., Van Meer, M.M., De Kort, C.A.D., Possee, R.D., Hammock, B.D., *et al.,* **1992**. Dissimilar expression of *Autographa californica* multiple nucleocapsid nuclear polyhedrosis virus polyhedrin and p10 genes. *J. Gen. Virol. 73,* 1481–1489.

Rohel, D.Z., Faulkner, P., **1984**. Time course analysis and mapping of *Autographa californica* nuclear polyhedrosis virus transcripts. *J. Virol. 50,* 739–747.

Rohrmann, G.F., **1986**. Polyhedrin structure. *J. Gen. Virol. 67,* 1499–1513.

Rohrmann, G.F., **1992**. Baculovirus structural proteins. *J. Gen. Virol. 73,* 749–761.

Rohrmann, G.F., **1996**. Evolution of occluded baculoviruses. In: Granados, R.R., Federici, B.A. (Eds.), The Biology of Baculoviruses, vol. 1. CRC Press, Boca Raton, FL, pp. 203–215.

Rohrmann, G.F., Karplus, P.A., **2001**. Relatedness of baculovirus and gypsy retrotransposon envelope proteins. *BMC Evol. Biol. 1,* 1–9.

Rosas-Acosta, G., Braunagel, S.C., Summers, M.D., **2001**. Effects of deletion and overexpression of the *Autographa californica* nuclear polyhedrosis virus FP25K gene on synthesis of two occlusion-derived virus envelope proteins and their transport into virus-induced intranuclear membranes. *J. Virol. 75,* 10829–10842.

Rothman, L.D., Myers, J.H., **1996a**. Debilitating effects of viral diseases on host Lepidoptera. *J. Invertebr. Pathol. 67,* 1–10.

Rothman, L.D., Myers, J.H., **1996b**. Is fecundity correlated with resistance to viral disease in the western tent caterpillar? *Ecol. Entomol. 21,* 396–398.

Ruberson, J.R., Young, S.Y., Kring, T.J., **1991**. Suitability of prey infected by nuclear polyhedrosis virus for development, survival, and reproduction of the predator *Nabis roseipennis* (Heteroptera: Nabidae). *Environ. Entomol. 20,* 1475–1479.

Russell, R.L.Q., Pearson, M.N., Rohrmann, G.F., **1991**. Immunoelectron microscopic examination of *Orgyia pseudotsugata* multicapsid nuclear polyhedrosis virus-infected *Lymantria dispar* cells: time course and localization of major polyhedron-associated proteins. *J. Gen. Virol. 72,* 275–283.

Russell, R.L.Q., Rohrmann, G.F., **1993**. A 25 kilodalton protein is associated with the envelopes of occluded baculovirus virions. *Virology 195,* 532–540.

Russell, R.L.Q., Rohrmann, G.F., **1997**. Characterization of P91, a protein associated with virions of an *Orgyia pseudotsugata* baculovirus. *Virology 233,* 210–233.

Sait, S.M., Begon, M., Thompson, D.J., Harvey, J.A., **1996**. Parasitism of baculovirus-infected *Plodia interpunctella* by *Venturia canescens* and subsequent virus transmission. *Funct. Ecol. 10,* 586–591.

Saville, G.P., Thomas, C.J., Possee, R.D., King, L.A., **2002**. Partial redistribution of the *Autographa californica* nucleopolyhedrovirus chitinase in virus-infected cells accompanies mutation of the carboxy-terminal KDEL ER-retention motif. *J. Gen. Virol. 83,* 685–694.

Shapiro, M., Ignoffo, C.M., **1970**. Nucleopolyhedrosis of *Heliothis*: activity of isolates from *Heliothis zea. J. Invertebr. Pathol. 16,* 107–111.

Shapiro, M., Martignoni, M.E., Cunningham, J.C., Goodwin, R.H., **1982**. Potential use of the saltmarsh caterpillar as a production host for nucleoployhedrosis viruses. *J. Econ. Entomol. 75,* 69–71.

Skinner, W.S., Dennis, P.A., Li, J.P., Quistad, G.B., **1992**. Identification of insecticidal peptides from venom of the trap-door spider, *Aptostichus ichlingeri* (Ctenizidae). *Toxicon 30,* 1043–1050.

Slack, J.M., Dougherty, E.M., Lawrence, S.D., **2001**. A study of the *Autographa californica* multiple nucleopolyhedrovirus ODV envelope protein p74 using a GFP tag. *J. Gen. Virol. 82,* 2279–2287.

Slack, J.M., Kuzio, J., Faulkner, P., **1995**. Characterization of *v-cath*, a cathepsin L-like proteinase expressed by the baculovirus *Autographa californica* multiple nuclear polyhedrosis virus. *J. Gen. Virol. 76,* 1091–1098.

Smith, G.E., Summers, M.D., **1979**. Restriction maps of five *Autographa californica* MNPV variants, *Trichoplusia ni* MNPV, and *Galleria mellonella* MNPV DNAs with endonucleases *Sma*I, *Kpn*I, *Bam*HI, *Sac*I, *Xho*I, and *Eco*RI. *J. Virol. 30,* 828–838.

Smith, G.E., Summers, M.D., **1980**. Restriction map of *Rachiplusia ou* and *Rachiplusia ou–Autographa californica* baculovirus recombinants. *J. Gen. Virol. 33,* 311–319.

Smits, P.H., Vlak, J.M., **1988**. Biological activity of *Spodoptera exigua* nuclear polyhedrosis virus against *S. exigua* larvae. *J. Invertebr. Pathol. 51,* 107–114.

Stairs, G.R., Fraser, T., Fraser, M., 1981. Changes in growth and virulence of a nuclear polyhedrosis virus from *Choristoneura fumiferana* after passage in *Trichoplusia ni* and *Galleria mellonella*. *J. Invertebr. Pathol.* 38, 230–235.

Stoltz, D.B., Pavan, C., DaCunha, A., 1973. Nuclear polyhedrosis virus: a possible example of *de novo* intranuclear membrane morphogenesis. *J. Gen. Virol.* 19, 145–150.

Subrahmanyam, B., Ramakrishnan, N., 1980. The alteration of juvenile hormone titer in the haemolymph of *Spodoptera litura* (F.) due to baculovirus infection. *Experientia 36*, 471–472.

Teakle, R.E., Jensen, J.M., Giles, J.E., 1986. Age-related susceptibility of *Heliothis punctiger* to a commercial formulation of nuclear polyhedrosis disease. *J. Invertebr. Pathol.* 47, 82–92.

Tellam, R.L., 1996. The peritrophic membrane. In: Lehane, M.J., Billingsley, B.F. (Eds.), Biology of the Insect Midgut. Chapman and Hall, New York, pp. 86–114.

Theilmann, D.A., Chantler, J.K., Stewart, S., Flipsen, H.T.M., Vlak, J.M., et al., 1996. Characterization of a highly conserved baculovirus structural protein that is specific for occlusion-derived virions. *Virology 218*, 148–158.

Theilmann, D.A., Stewart, S., 1993. Analysis of the *Orgyia pseudotsugata* multicapsid nuclear polyhedrosis virus trans-activators IE-1 and IE-2 using monoclonal antibodies. *J. Gen. Virol.* 74, 1819–1826.

Thiem, S.M., Miller, L.K., 1989. Identification, sequence, and transcriptional mapping of the major capsid protein gene of the baculovirus *Autographa californica* nuclear polyhedrosis virus. *J. Virol.* 63, 2008–2018.

Thomas, C.J., Brown, H.L., Hawes, C.R., Lee, B.Y., Min, M.K., et al., 1998. Localization of a baculovirus-induced chitinase in the insect cell endoplasmic reticulum. *J. Virol.* 72, 10207–10212.

Thomas, C.J., Gooday, G.W., King, L.A., Possee, R.D., 2000. Mutagenesis of the active coding region of the *Autographa californica* nucleopolyhedrovirus *chiA* gene. *J. Gen. Virol.* 81, 1403–1411.

Todd, J.W., Passarelli, A.L., Lu, A., Miller, L.K., 1996. Factors regulating baculovirus late and very late gene expression in transient-expression assays. *J. Virol.* 70, 2307–2317.

Todd, J.W., Passarelli, A.L., Miller, L.K., 1995. Eighteen baculovirus genes, including *lef-11*, *p35*, 39K and *p47*, support late gene expression. *J. Virol.* 69, 968–974.

Tomalski, M.D., Eldridge, R., Miller, L.K., 1991. A baculovirus homolog of a Cu/Zn superoxide dismutase gene. *Virology 184*, 149–161.

Tompkins, G.J., Dougherty, E.M., Adams, J.R., Diggs, D., 1988. Changes in the virulence of nuclear polyhedrosis virus when propagated in altenative noctuid (Lepidoptera: Noctuidae) cell lines and hosts. *J. Econ. Entomol.* 81, 1027–1032.

Tompkins, G.J., Vaughn, J.L., Adams, J.R., Reichelderfer, C.F., 1981. Effects of propagating *Autographa californica* nuclear polyhedrosis virus and its *Trichoplusia*

ni variant in different hosts. *Environ. Entomol. 10*, 801–806.

Trudeau, D., Washburn, J.O., Volkman, L.E., 2001. Central role of hemocytes in *Autographa californica* M nucleopolyhedrovirus pathogenesis in *Heliothis virescens* and *Helicoverpa zea*. *J. Gen. Virol.* 75, 996–1003.

Tweeten, K.A., Bulla, L.A., Consigli, R.A., 1980. Characterization of an extremely basic protein derived from granosis virus nucleocapsids. *J. Virol.* 33, 866–876.

Vail, P.V., Hoffmann, D.F., Streett, D.A., Manning, J.S., Tebbets, J.S., 1993. Infectivity of a nuclear polyhedrosis virus isolated from *Anagrapha falcifera* (Lepidoptera: Noctuidae) against production and postharvest pests and homologous cell lines. *Environ. Entomol.* 22, 1140–1145.

van Beek, N.A.M., Hughes, P.R., 1998. The response time of insect larvae infected with recombinant baculoviruses. *J. Invertebr. Pathol.* 72, 338–347.

van Beek, N.A.M., Wood, H.A., Hughes, P.R., 1988. Quantitative aspects of nuclear polyhedrosis virus infections in lepidopterous larvae: the dose-survival time relationship. *J. Invertebr. Pathol.* 51, 58–63.

van Lent, J.W.M., Goenen, J.T.M., Klinge-Roode, E.C., Rohrmann, G.F., Zuidema, D., et al., 1990. Localization of the 34 kDa polyhedron envelope protein in *Spodoptera frugiperda* cells infected with *Autographa californica* nuclear polyhedrosis virus. *Arch. Virol. 111*, 103–114.

van Loo, N.-D., Fortunati, E., Ehlert, E., Rabelink, M., Grosveld, F., et al., 2001. Baculovirus infection of non-dividing mammalian cells: mechanisms of entry and nuclear transport of capsids. *J. Virol.* 75, 961–970.

van Oers, M.M., Flipsen, J.T.M., Reusken, C.B.E.M., Sliwinsky, E.L., Goldbach, R.W., et al., 1993. Functional domains of the p10 protein of *Autographa californica* nuclear polyhedrosis virus. *J. Gen. Virol.* 74, 563–574.

Vasconcelos, S.D., Williams, T., Hails, R.S., Cory, J.S., 1996. Prey selection and baculovirus dissemination by carabid predators of Lepidoptera. *Ecol. Entomol. 21*, 98–104.

Vialard, J.E., Richardson, C.D., 1993. The 1629-nucleotide open reading frame located downstream of the *Autographa californica* nuclear polyhedrosis virus polyhedrin gene encodes a nucleocapsid-associated phosphoprotein. *J. Virol.* 67, 5859–5866.

Viliplana, L., O'Reilly, D., 2003. Functional interaction between *Cydia pomonella* granulovirus IAP proteins. *Virus Res. 92*, 107–111.

Vlak, J.M., Klinkenberg, F.A., Zaal, K.J.M., Usmany, M., Klingeroode, E.C., et al., 1988. Functional studies on the *p10* gene of *Autographa californica* nuclear polyhedrosis virus using a recombinant expressing a p10-β-galactosidase fusion gene. *J. Gen. Virol.* 69, 765–776.

Vlak, J.M., Rohrmann, G.F., 1985. The nature of polyhedrin. In: Maramorosch, K., Sherman, K.E. (Eds.), Viral Insecticides for Biological Control. Academic Press, New York, pp. 489–542.

Volkman, L.E., **1986**. The 64K envelope protein of budded *Autographa californica* nuclear polyhedrosis virus. *Curr. Topics Microbiol. Immunol. 131*, 103–118.

Volkman, L.E., **1997**. Nucleopolyhedrovirus interactions with their insect hosts. *Adv. Virus Res. 48*, 313–348.

Volkman, L.E., Goldsmith, P.A., **1984**. Budded *Autographa californica* NPV 64K protein: further biochemical analysis and effects of postimmunoprecipitation sample preparation conditions. *Virology 139*, 295–302.

Volkman, L.E., Goldsmith, P.A., **1985**. Mechanism of neutralization of budded *Autographa californica* nuclear polyhedrosis virus by a monoclonal antibody: inhibition of entry by adsorptive endocytosis. *Virology 143*, 185–195.

Volkman, L.E., Keddie, B.A., **1990**. Nuclear polyhedrosis virus pathogenesis. *Semin. Virol. 1*, 249–256.

Volkman, L.E., Storm, K., Aivazachvili, V., Oppenheimer, D., **1996**. Overexpression of actin in AcMNPV-infected cells interferes with polyhedrin synthesis and polyhedra formation. *Virology 225*, 369–376.

Volkman, L.E., Summers, M.D., **1977**. *Autographa californica* nuclear polyhedrosis virus: comparative infectivity of the occluded, alkali-liberated, and nonoccluded forms. *J. Invertebr. Pathol. 30*, 102–103.

Volkman, L.E., Zaal, K.J.M., **1990**. *Autographa californica* M nuclear polyhedrosis virus: microtubules, p10, and replication. *Virology 175*, 292–302.

Wang, H.H., Fraser, M.J., Cary, L.C., **1989**. Transposon mutagenesis of baculoviruses: analysis of TFP3 lepidopteran transposon insertions at the FP locus of nuclear polyhedrosis viruses. *Gene 81*, 97–108.

Wang, P., Granados, R.R., **1997**. An intestinal mucin is the target substrate for a baculovirus enhancin. *Proc. Natl Acad. Sci. USA 94*, 6977–6982.

Wang, P., Granados, R.R., **1998**. Observations on the presence of the peritrophic membrane in larval *Trichoplusia ni* and its role in limiting baculovirus infection. *J. Invertebr. Pathol. 72*, 57–62.

Wang, P., Granados, R.R., **2000**. Calcofluor disrupts the midgut defense system of insects. *Insect Biochem. Mol. Biol. 30*, 135–143.

Wang, P., Granados, R.R., **2001**. Molecular structure of the peritrophic membrane (PM): identification of potential PM target sites for insect control. *Arch. Insect Biochem. Physiol. 47*, 110–118.

Wang, P., Hammer, D.A., Granados, R.R., **1994**. Interaction of *Trichoplusia ni* granulosis virus-encoded enhancin with the midgut epithelium and peritrophic membrane of four lepidopteran insects. *J. Gen. Virol. 75*, 1961–1967.

Washburn, J.O., Chan, E.Y., Volkman, L.E., Aumiller, J.J., Jarvis, D.L., **2003a**. Early synthesis of budded virus envelope fusion protein GP64 enhances *Autographa californica* multicapsid nucleopolyhedrovirus virulence in orally infected *Heliothis virescens*. *J. Virol. 77*, 280–290.

Washburn, J.O., Chen, E.Y., Volkman, L.E., Aumiller, J.J., Jarvis, D.L., **2001**. Comparative pathogenesis of *Helicoverpa zea* S nucleopolyhedrovirus in noctuid larvae. *J. Gen. Virol. 82*, 1777–1784.

Washburn, J.O., Haas-Stapleton, E.J., Tan, F.F., Beckage, N.E., Volkman, L.E., **2000**. Co-infection of *Manduca sexta* larvae with polydnavirus from *Cotesia congregata* increases susceptibility to fatal infection by *Autographa californica* M. *nucleopolyhedrovirus*. *J. Insect Physiol. 46*, 179–190.

Washburn, J.O., Kirkpatrick, B.A., Haas-Stapleton, E., Volkman, L.E., **1998**. Evidence that the stilbene-derived optical brightener M2R enhances *Autographa californica* M nucleopolyhedrovirus infection of *Trichoplusia ni* and *Heliothis virescens* by preventing sloughing of infected midgut epithelial cells. *Biol. Control 11*, 58–69.

Washburn, J.O., Kirkpatrick, B.A., Volkman, L.E., **1995**. Comparative pathogenesis of *Autographa californica* M nuclear polyhedrosis virus in larvae of *Trichoplusia ni* and *Heliothis virescens*. *Virology 209*, 561–568.

Washburn, J.O., Kirkpatrick, B.A., Volkman, L.E., **1996**. Insect protection against viruses. *Nature 383*, 767.

Washburn, J.O., Lyons, E.H., Haas-Stapleton, E.J., Volkman, L.E., **1999**. Multiple nucleocapsid packaging of *Autographa californica* nucleopolyhedrovirus accelerates the onset of systemic infection in *Trichoplusia ni*. *J. Virol. 73*, 411–416.

Washburn, J.O., Trudeau, D., Wong, J.F., Volkman, L.E., **2003b**. Early pathogenesis of *Autographa californica* multiple nucleopolyhedrovirus and *Helicoverpa zea* single nucleopolyhedrovirus in *Heliothis virescens*: a comparison of the 'M' and 'S' strategies for establishing fatal infection. *J. Gen. Virol. 84*, 343–351.

Weitzmann, M., Possee, R.D., King, L.A., **1992**. Characterization of two variants of *Panolis flammea* multiple nucleocapsid nuclear polyhedrosis virus. *J. Gen. Virol. 73*, 1881–1886.

Whitford, M., Faulkner, P., **1992**. Nucleotide sequence and transcriptional analysis of a gene encoding gp41, a structural glycoprotein of the baculovirus *Autographa californica* nuclear polyhedrosis virus. *J. Virol. 66*, 4763–4768.

Whitford, M., Stewart, S., Kuzio, J., Faulkner, P., **1989**. Identification and sequence analysis of a gene encoding gp67 an abundant glycoprotein of the baculovirus, *Autographa californica* nuclear polyhedrosis virus. *J. Virol. 63*, 1393–1399.

Whitt, M.A., Manning, J.S., **1988**. A phosphorylated 34-kDa protein and a subpopulation of polyhedrin are thiol linked to the carbohydrate layer surrounding a baculovirus occlusion body. *Virology 164*, 33–42.

Williams, G.V., Faulkner, P., **1997**. Cytological changes and viral morphogenesis during baculovirus infection. In: Miller, L.K. (Ed.), The Baculoviruses. Plenum, New York, pp. 61–107.

Williams, G.V., Rohel, D.Z., Kuzio, J., Faulkner, P., **1989**. A cytopathological investigation of *Autographa californica* nuclear polyhedrosis virus p10 gene function using insertion/deletion mutants. *J. Gen. Virol. 70*, 187–202.

Wilson, M.E., Consigli, R.A., **1985**. Functions of a protein kinase activity associated with purified capsids of the granulosis virus infecting *Plodia interpunctella*. *Virology 143*, 526–535.

Wilson, M.E., Mainprize, T.H., Friesen, P.D., Miller, L.K., **1987**. Location, transcription and sequence of a baculovirus gene encoding a small arginine-rich polypeptide. *J. Virol. 61*, 661–666.

Winstanley, D., Crook, N., **1993**. Replication of *Cydia pomonella* granulosis virus in cell cultures. *J. Gen. Virol. 74*, 1599–1609.

Xu, B., Yoo, S., Guarino, L.A., **1995**. Differential transcription of baculovirus late and very late promoters: fractionation of nuclear extracts by phosphocellulose chromatography. *J. Virol. 69*, 2912–2917.

Xu, J., Hukuhara, T., **1992**. Enhanced infection of a nuclear polyhedrosis virus in larvae of the armyworm, *Pseudoletia separata*, by a factor in the spheroids of an entomopoxvirus. *J. Invertebr. Pathol. 60*, 259–264.

Yamagishi, J., Osobe, R., Takebuchi, T., Bando, H., **2003**. DNA microarrays of baculovirus genomes: differential expression of viral genes in two susceptible insect cell lines. *Arch. Virol. 148*, 587–597.

Yang, F., He, J., Lin, X., Li, Q., Pan, D., *et al.*, **2001**. Complete genome sequence of the shrimp white spot bacilliform virus. *J. Virol. 75*, 11811–11820.

Yang, S., Miller, L.K., **1999**. Activation of very late promoters by interaction with very late factor 1. *J. Virol. 73*, 3404–3409.

Yang, Z., Nielsen, R., Goldman, N., Pedersen, A.-M.K., **2000**. Codon-substitution models for heterogeneous selection pressure at amino acid sites. *Genetics 155*, 431–449.

Young, J.H.C., MacKinnon, E.A., Faulkner, P., **1993**. The architecture of the virogenic stroma in isolated nuclei of *Spodoptera frugiperda* cells *in vitro* infected by *Autographa californica* nuclear polyhedrosis virus. *J. Struct. Biol. 110*, 141–153.

Young, S.Y., Yearian, W.C., **1974**. Persistence of *Heliothis* NPV on foliage of cotton, soybean, and tomato. *Environ. Entomol. 3*, 253–260.

Zanotto, P.M., Kessing, B.D., Maruniak, J.E., **1993**. Phylogenetic interrelationships among baculoviruses: evolutionary rates and host associations. *J. Invertebr. Pathol. 62*, 147–164.

Zhang, P., Yang, K., Xiaojiang, D., Su, D., **2002**. Infection of wild-type *Autographa californica* multicapsid nucleopolyhedrovirus induces *in vivo* apoptosis of *Spodoptera litura* larvae. *J. Gen. Virol. 83*, 3003–3011.

Zuidema, D., Klinge-Roode, E.C., van Lent, J.W.M., Vlak, J.M., **1989**. Construction and analysis of an *Autographa californica* nuclear polyhedrosis virus mutant lacking the polyhedral envelope. *Virology 173*, 98–108.

A5 Addendum: Baculoviruses: Biology, Biochemistry, and Molecular Biology

R L Harrison, Invasive Insect Biocontrol and
Behavior Laboratory, USDA Agricultural
Research Service, Plant Sciences Institute, MD, USA

© 2010, Elsevier BV. All Rights Reserved.

A5.1. Introduction

Since the publication of the *Comprehensive Molecular Insect Science* series, research on baculoviruses has continued at an unabated pace. The following addendum reviews developments in baculovirology since this chapter was written.

A5.1.1. Taxonomy

Based on genome sequences, phylogenetic analyses and biological/morphological traits of dipteran (mosquito) and hymenopteran (sawfly) baculoviruses (Afonso *et al.*, 2001; Duffy *et al.*, 2006; Lauzon *et al.*, 2004, 2006), family Baculoviridae has been reorganized into four genera: *Alphabaculovirus* (lepidopteran-specific nucleopolyhedroviruses/NPVs), *Betabaculovirus* (lepidopteran-specific granuloviruses/GVs), *Gammabaculovirus* (hymenopteran-specific NPVs) and *Deltabaculovirus* (dipteran-specific NPVs). A criterion for baculovirus species demarcation based on the genetic distances of partial *polh/gran*, *lef-8*, and *lef-9* sequences has been proposed (Jehle *et al.*, 2006) and used by researchers (for example, see Mukawa and Goto, 2008).

A5.2. Structure

Proteomics analyses have determined the total protein compositions of occlusion-derived virus (ODV) from three baculoviruses (Braunagel *et al.*, 2003; Deng *et al.*, 2007; Perera *et al.*, 2007). In some cases, these analyses detected viral transcription and replication proteins and also the presence of putative host proteins, though it is uncertain whether these proteins are genuine components of the virion or copurifying contaminants.

A5.2.1. Nucleocapsids

Additional nucleocapsid structural proteins have been identified, including proteins encoded by the genes *vlf-1* (Vanarsdall *et al.*, 2006), *ac141* (*exon0*) (Fang *et al.*, 2007) and *ac142* (McCarthy *et al.*, 2008) as well as *ac109*, the gene product of which occurs in both nucleocapsid and envelope protein fractions (Fang *et al.*, 2003, 2009).

A5.2.2. Budded Virus

Further studies of AcMNPV GP64 have defined domains and amino acids involved in receptor binding (Zhou and Blissard, 2008b) as well as targeting of GP64 to the virion and virion budding (Zhou and Blissard, 2008a). Features of the GP64 transmembrane domain important for GP64 translocation to/anchoring in the plasma membrane and for GP64's fusion activity have also been identified (Li and Blissard, 2009). GP64 associates with lipid rafts and colocalizes with filamentous actin in AcMNPV-infected Sf9 cells (Haines *et al.*, 2009).

While proteomics computational analysis and the crystal structure of the postfusion conformation of GP64 have revealed it to be a class III fusion protein, the F protein is a class I fusion protein (Garry and Garry, 2008; Kadlec *et al.*, 2008). Experiments with inactivated AcMNPV budded virus pseudotyped with either GP64 or F protein indicate that the two types of fusion protein bind to different receptors on tissue culture cells (Westenberg *et al.*, 2007). The cytoplasmic tail domain of F protein is required for infectivity, which is not the case for GP64 (Long *et al.*, 2006b).

A5.2.3. Occlusion-Derived Virus

A pathway for sorting ODV envelope proteins from the endoplasmic reticulum to the inner nuclear membrane has been elucidated and involves the viral proteins FP25K and BV/ODV-E26 and the cellular protein importin-α-16 (Braunagel *et al.*, 2004, 2009; Burks *et al.*, 2007; Saksena *et al.*, 2004, 2006).

ODV-E18, now acknowledged to be encoded by a core gene, is required for BV production (McCarthy and Theilmann, 2008). The *per os* infectivity factors PIF-2 (*ac22*) and PIF-3 (*ac115*), along with P74 and PIF-1 (*ac119*), are ODV envelope proteins (Song *et al.*, 2008). Both enhancins of *Lymantria dispar* MNPV are also located in the ODV envelope (Slavicek and Popham, 2005).

A5.2.4. Occlusion Bodies

Occlusion body proteins of *Culex nigripalpus* (mosquito) NPV and *Penaeus monodon* (shrimp) NPV are unrelated to, and significantly larger than, other baculovirus polyhedrins and granulins (Chaivisuthangkura *et al.*, 2008; Perera *et al.*, 2006). The function of P10 and its role in occlusion morphogenesis are still not fully understood, but immunofluorescence laser scanning confocal microscopy has revealed that it forms three distinct structures: (1) cytoplasmic filaments associated with and dependent upon host cell microtubules; (2) a cage of tubules surrounding the nucleus that is not associated with microtubules; and (3) intranuclear structures associated with occlusion bodies (Carpentier *et al.*, 2008; Patmanidi *et al.*, 2003).

A5.3. Life Cycle

A5.3.1. Infection

Of the four PIF (*per os* infectivity factor) proteins identified, P74, PIF1, and PIF2 directly mediate binding and infection of midgut cells (Haas-Stapleton *et al.*, 2004; Ohkawa *et al.*, 2005; Pijlman *et al.*, 2003). Trypsin cleavage of P74 is required for its role in *per os* infectivity (Slack *et al.*, 2008).

During baculovirus infection, the host cell nucleus partitions into two discrete compartments defined by the virogenic stroma and the peristromal region (where ODV envelopement occurs) (Kawasaki *et al.*, 2004). Marginalization of host heterochromatin during infection is driven by expansion of both the virogenic stroma and the peristromal region, and appears to require *hr* elements and expression of *ie1*, *lef3*, and *p143/dnahel* (Nagamine *et al.*, 2008).

Induction of actin polymerization by the PP78/83 capsid protein is mediated through activation of the host cell Arp2/3 complex, which in turn nucleates actin and assembles actin filaments within the nucleus of infected cells (Goley *et al.*, 2006; Wang *et al.*, 2007).

A5.3.2. Dissemination Within the Host

Treatment of AcMNPV-infected cells with an inhibitor of clathrin-mediated endocytosis has confirmed that secondary infection by BV occurs by entry through clathrin-coated pits (Long *et al.*, 2006a). Efficient dissemination within the host appears to require the *vfgf* gene product, which stimulates cell motility and influences both survival time and the proportion of infected hemocytes in larvae (Detvisitsakun *et al.*, 2005, 2006, 2007; Katsuma *et al.*, 2006, 2008). These results suggest that VFGF may promote systemic spread of baculovirus infection from the midgut epithelium by promoting branching and migration of tracheal cells or attracting uninfected hemocytes to infected tracheae.

A5.3.3. Dissemination from the Host

The *Bombyx mori* (Bm)NPV protein tyrosine phosphate (*ptp*) gene product induces light-activated enhanced locomotory activity in infected larvae, which likely increases dissemination and transmission of progeny virus (Kamita *et al.*, 2005). Research with BmNPV-infected silkworm moths has shown that dissemination of baculovirus infection by vertical transmission can occur by both transovarial and venereal routes, possibly involving infected spermatocytes and oocytes in adult moths that have survived baculovirus infection (Khurad *et al.*, 2004).

A5.4. Virus Replication

Experiments with inhibitors of signaling pathway kinases, phosphopeptide antibodies, and RNAi have established that signaling pathways involving extracellular signal-regulated kinase (ERK), c-Jun NH2-terminal kinase (JNK), and phosphatidylinositol 3-kinase/protein kinase B (PI3K-Akt) contribute to the optimization of progeny virus production during baculovirus infection (Katsuma *et al.*, 2007; Xiao *et al.*, 2009).

A5.4.1. Early Gene Expression

Deletion of the PE38 early gene transcription factor causes a delay in early gene expression, along with significant reductions in DNA replication, budded virus production, and oral infectivity toward larvae (Jiang *et al.*, 2006; Milks *et al.*, 2003). Either IE1 or IE0 is essential for infection, and both are required for optimal levels of BV and occlusion bodies (Stewart *et al.*, 2005).

No single *hr* of AcMNPV is essential for early gene expression or DNA replication (Carstens and Wu, 2007). Conserved bZIP family transcription factor binding sites are present in the interpalindromic regions of lepidopteran NPV *hr*s, and these sites promote reporter gene expression in uninfected insect cells (Landais *et al.*, 2006).

A5.4.2. DNA Replication

Research with a *dnapol* deletion mutant confirmed that baculovirus DNA polymerase is required for baculovirus DNA replication and that host DNA polymerases cannot substitute for the *dnapol* gene product (Vanarsdall *et al.*, 2005). Although AcMNPV alkaline nuclease deletion mutants replicate DNA at normal levels, the viral DNA produced consists of subgenomic fragments (Okano *et al.*, 2007). The deletion of the gene encoding DBP (*ac25*), which like LEF-3 has DNA unwinding and annealing activity (Mikhailov *et al.*, 2008), also results in accumulation of subgenomic viral DNA fragments (Vanarsdall *et al.*, 2007). These subgenomic fragments may be intermediates in a baculovirus genome assembly process involving recombination of subgenomic products of replication into intact genomes (Rohrmann, 2008). LEF-3 may play a role in such recombination events, as it promotes strand exchange between donor and acceptor DNA molecules (Mikhailov *et al.*, 2006).

A5.4.3. Late and Very Late Gene Expression

The *ie-0* gene product stimulates late gene expression along with the other 19 late expression factors (LEFs) (Huijskens *et al.*, 2004). The *Bombyx mori* NPV *bm34* gene, a homolog of AcMNPV *ac43*, is required for optimal *polh* and *p10* gene expression, possibly through a mechanism involving *vlf-1* and *fp25k* (Katsuma and Shimada, 2009).

The *pk-1*-encoded protein kinase is associated with a very late gene transcription complex, where it phosphorylates a 102 kDa protein that could be LEF-8 (Mishra *et al.*, 2008). It is still unclear how baculoviruses form 7-methylguanosine caps on their transcripts, as the RNA triphosphatase activities of neither AcMNPV LEF-4 nor PTP are required for formation of the baculovirus mRNA cap structures (Li and Guarino, 2008).

A5.5. Effects on the Host

A5.5.1. Virulence

Examination of ODV infections of the larval midgut of the fall armyworm (*Spodopera frugiperda*) suggests that susceptibility to primary infection is affected not only by the quantity of ODV bound to the midgut, but also by the quality of binding (i.e., whether the binding leads to productive ODV fusion, entry, and infection; Haas-Stapleton *et al.*, 2005). Furthermore, research with isolates of *Cydia pomonella* (Cp)GV that differ in their virulence and replication characteristics suggests that the virulence parameters of LD_{50}, median survival time (ST_{50}), and progeny virus production do not necessarily dictate the ability of different virus isolates to compete with each other in mixed infections of the sort that are likely to occur in the field (Arends *et al.*, 2005).

A5.5.2. Host Range

A5.5.2.1. Host Range Genes   In addition to other genes involved in determining host range, the 11K gene family members *ac145* and *ac150* have also been found to influence virulence in a species-specific fashion (Lapointe *et al.*, 2004; Zhang *et al.*, 2005). Host range factor-1 (*hrf-1*) is required for replication in *L. dispar* cells by three additional NPVs in addition to AcMNPV (Ishikawa *et al.*, 2004). Host cell factor-1 (*hcf-1*) localizes to foci within the nuclei of infected cells and contains a RING-finger domain that is required for its ability to promote AcMNPV replication in *Trichoplusia ni* cells (Wilson *et al.*, 2005).

A5.5.2.2. Antiapoptotic Genes

IE-1 is required for AcMNPV-triggered apoptosis and appears to counteract its own induction of apoptosis by upregulating expression of *p35* (Schultz *et al.*, 2009). NPV antiapoptotic genes cause the translational arrest arising from AcMNPV infection of *L. dispar* cells, but in a way that does not involve suppressing caspase activation (Thiem and Chejanovsky, 2004). Deletion of the AcMNPV *p35* gene diminishes the ability of AcMNPV to infect both *S. frugiperda* and *Anticarsia gemmatalis* larvae (Clarke and Clem, 2003; da Silveira *et al.*, 2005). The *p49* gene product, which is also a caspase inhibitor, was found to function as a homodimer that binds two caspase dimers each; altering the caspase recognition motif of P49 affected the specificity of caspases that it inhibits (Guy and Friesen, 2008).

A5.5.3. Survival Time and Yield

A number of genes have been identified whose deletion primarily affects survival time of infected larvae, including *vfgf* (Detvisitsakun *et al.*, 2007; Katsuma *et al.*, 2006), *ac30/bm21* (Huang *et al.*, 2008), and *ac68* (Li *et al.*, 2008).

A5.5.4. Sublethal Effects and Latent Infections

Asymptomatic, vertically transmitted persistent infections from which overt infections can be triggered, which had previously been reported only in laboratory cultures of virus hosts, have been discovered in several field populations of *Mamestra brassicae* by PCR analysis (Burden *et al.*, 2003).

A5.5.5. Resistance

Populations of codling moth (*C. pomonella*) with high levels of dominant, stable sex-linked resistance to CpGV appeared in Germany in 2003–2005, likely because of the overuse of a single strain of CpGV for control (Asser-Kaiser *et al.*, 2007). Elevated levels of larval plasma phenoloxidase, plasma selenium, midgut NADPH oxidoreductase, and a thicker peritrophic matrix are correlated with resistance to baculovirus infection in other lepidopteran species, suggesting several mechanisms for host resistance (Levy *et al.*, 2007; Selot *et al.*, 2007; Shelby and Popham, 2006, 2007).

Studies looking at host gene expression in baculovirus-infected cells and larvae have identified a heat shock cognate gene (*hsc70*) (Nobiron *et al.*, 2003), a paralytic peptide-binding protein (*PP-BP*) gene (Hu *et al.*, 2006), and a gene family termed *Response to Pathogen* (REPAT; Herrero *et al.*, 2007) that are upregulated in response to baculovirus infection. Recombinant AcMNPV expressing REPAT-1 exhibits reduced infectivity against *Spodoptera exigua* larvae (Herrero *et al.*, 2007). The *B. mori suppressor of profiling 2* gene is expressed at a higher level in a BmNPV-resistant silkworm moth strain than in more susceptible strains (Xu *et al.*, 2005).

A5.6. Baculovirus Genomics

A5.6.1. General Properties of Baculovirus Genomes

As of October 2009, 53 complete sequences of baculovirus genomes have been determined and deposited in GenBank. The characteristics of baculovirus genomes as inferred from 43 of these sequences were recently reviewed by van Oers and Vlak (2007).

A5.6.2. Baculovirus Evolution

Phylogenetic analysis of sequences from lepidopteran, hymenopteran, and dipteran baculoviruses suggests that baculoviruses have coevolved with their insect hosts, with either (a) baculoviruses originating with the first ancestral insects and coevolving with insects as they diverged into the current insect orders, or (b) baculoviruses originating with one already-established order of insects and cross-infecting other orders very early during their evolution (Herniou *et al.*, 2004). Analysis of baculovirus genomes suggests a role for *bro* genes and *hr*s in rearrangement events leading to gene loss and acquisition (de Jong *et al.*, 2005; Harrison and Popham, 2008). Instances of homologous recombination and allelic replacement have also been documented among isolates and variants of AcMNPV (Harrison, 2009; Harrison and Lynn, 2007; Jehle, 2004).

Acknowledgments

The author thanks Bryony Bonning for reviewing this addendum and apologizes to all the researchers whose work could not be mentioned because of space considerations.

References

Afonso, C.L., Tulman, E.R., Lu, Z., Balinsky, C.A., Moser, B.A., Becnel, J.J., Rock, D.L., Kutish, G.F., **2001**. Genome sequence of a baculovirus pathogenic for *Culex nigripalpus*. *J. Virol.* 75, 11157–11165.

Arends, H.M., Winstanley, D., Jehle, J.A., 2005. Virulence and competitiveness of *Cydia pomonella* granulovirus mutants: parameters that do not match. *J. Gen. Virol.* 86, 2731–2738.

Asser-Kaiser, S., Fritsch, E., Undorf-Spahn, K., Kienzle, J., Eberle, K.E., Gund, N.A., Reineke, A., Zebitz, C.P., Heckel, D.G., Huber, J., Jehle, J.A., 2007. Rapid emergence of baculovirus resistance in codling moth due to dominant, sex-linked inheritance. *Science 317*, 1916–1918.

Braunagel, S.C., Russell, W.K., Rosas-Acosta, G., Russell, D.H., Summers, M.D., 2003. Determination of the protein composition of the occlusion-derived virus of *Autographa californica* nucleopolyhedrovirus. *Proc. Natl. Acad. Sci. USA 100*, 9797–9802.

Braunagel, S.C., Williamson, S.T., Saksena, S., Zhong, Z., Russell, W.K., Russell, D.H., Summers, M.D., 2004. Trafficking of ODV-E66 is mediated via a sorting motif and other viral proteins: facilitated trafficking to the inner nuclear membrane. *Proc. Natl. Acad. Sci. USA 101*, 8372–8377.

Braunagel, S.C., Cox, V., Summers, M.D., 2009. Baculovirus data suggest a common but multifaceted pathway for sorting proteins to the inner nuclear membrane. *J. Virol. 83*, 1280–1288.

Burden, J.P., Nixon, C.P., Hodgkinson, A.E., Possee, R.D., Sait, S.M., King, L.A., Hails, R.S., 2003. Covert infections as a mechanism for long-term persistence of baculoviruses. *Ecol. Lett. 6*, 524–531.

Burks, J.K., Summers, M.D., Braunagel, S.C., 2007. BV/ODV-E26: a palmitoylated, multifunctional structural protein of *Autographa californica* nucleopolyhedrovirus. *Virology 361*, 194–203.

Carpentier, D.C., Griffiths, C.M., King, L.A., 2008. The baculovirus P10 protein of *Autographa californica* nucleopolyhedrovirus forms two distinct cytoskeletal-like structures and associates with polyhedral occlusion bodies during infection. *Virology 371*, 278–291.

Carstens, E.B., Wu, Y., 2007. No single homologous repeat region is essential for DNA replication of the baculovirus *Autographa californica* multiple nucleopolyhedrovirus. *J. Gen. Virol. 88*, 114–122.

Chaivisuthangkura, P., Tawilert, C., Tejangkura, T., Rukpratanporn, S., Longyant, S., Sithigorngul, W., Sithigorngul, P., 2008. Molecular isolation and characterization of a novel occlusion body protein gene from *Penaeus monodon* nucleopolyhedrovirus. *Virology 381*, 261–267.

Clarke, T.E., Clem, R.J., 2003. In vivo induction of apoptosis correlating with reduced infectivity during baculovirus infection. *J. Virol. 77*, 2227–2232.

da Silveira, E.B., Cordeiro, B.A., Ribeiro, B.M., Bao, S.N., 2005. *In vivo* apoptosis induction and reduction of infectivity by an *Autographa californica* multiple nucleopolyhedrovirus *p35*(-) recombinant in hemocytes from the velvet bean caterpillar *Anticarsia gemmatalis* (Hubner) (Lepidoptera: Noctuidae). *Res. Microbiol. 156*, 1014–1025.

de Jong, J.G., Lauzon, H.A., Dominy, C., Poloumienko, A., Carstens, E.B., Arif, B.M., Krell, P.J., 2005. Analysis of the *Choristoneura fumiferana* nucleopolyhedrovirus genome. *J. Gen. Virol. 86*, 929–943.

Deng, F., Wang, R., Fang, M., Jiang, Y., Xu, X., Wang, H., Chen, X., Arif, B.M., Guo, L., Hu, Z., 2007. Proteomics analysis of *Helicoverpa armig*era single nucleocapsid nucleopolyhedrovirus identified two new occlusion-derived virus-associated proteins, HA44 and HA100. *J. Virol. 81*, 9377–9385.

Detvisitsakun, C., Berretta, M.F., Lehiy, C., Passarelli, A.L., 2005. Stimulation of cell motility by a viral fibroblast growth factor homolog: proposal for a role in viral pathogenesis. *Virology 336*, 308–317.

Detvisitsakun, C., Hutfless, E.L., Berretta, M.F., Passarelli, A.L., 2006. Analysis of a baculovirus lacking a functional viral fibroblast growth factor homolog. *Virology 346*, 258–265.

Detvisitsakun, C., Cain, E.L., Passarelli, A.L., 2007. The *Autographa californica* M nucleopolyhedrovirus fibroblast growth factor accelerates host mortality. *Virology 365*, 70–78.

Duffy, S.P., Young, A.M., Morin, B., Lucarotti, C.J., Koop, B.F., Levin, D.B., 2006. Sequence analysis and organization of the *Neodiprion abietis* nucleopolyhedrovirus genome. *J. Virol. 80*, 6952–6963.

Fang, M., Wang, H., Yuan, L., Chen, X., Vlak, J.M., Hu, Z., 2003. Open reading frame 94 of *Helicoverpa armigera* single nucleocapsid nucleopolyhedrovirus encodes a novel conserved occlusion-derived virion protein, ODV-EC43. *J. Gen. Virol. 84*, 3021–3027.

Fang, M., Dai, X., Theilmann, D.A., 2007. *Autographa californica* multiple nucleopolyhedrovirus EXON0 (ORF141) is required for efficient egress of nucleocapsids from the nucleus. *J. Virol. 81*, 9859–9869.

Fang, M., Nie, Y., Theilmann, D.A., 2009. Deletion of the AcMNPV core gene ac109 results in budded virions that are non-infectious. *Virology 389*, 66–74.

Garry, C.E., Garry, R.F., 2008. Proteomics computational analyses suggest that baculovirus GP64 superfamily proteins are class III penetrenes. *Virol. J. 5*, 28.

Goley, E.D., Ohkawa, T., Mancuso, J., Woodruff, J.B., D'Alessio, J.A., Cande, W.Z., Volkman, L.E., Welch, M.D., 2006. Dynamic nuclear actin assembly by Arp2/3 complex and a baculovirus WASP-like protein. *Science 314*, 464–467.

Guy, M.P., Friesen, P.D., 2008. Reactive-site cleavage residues confer target specificity to baculovirus P49, a dimeric member of the P35 family of caspase inhibitors. *J. Virol. 82*, 7504–7514.

Haas-Stapleton, E.J., Washburn, J.O., Volkman, L.E., 2004. P74 mediates specific binding of *Autographa californica* M nucleopolyhedrovirus occlusion-derived virus to primary cellular targets in the midgut epithelia of *Heliothis virescens* larvae. *J. Virol. 78*, 6786–6791.

Haas-Stapleton, E.J., Washburn, J.O., Volkman, L.E., 2005. *Spodoptera frugiperda* resistance to oral infection by *Autographa californica* multiple nucleopolyhedrovirus linked to aberrant occlusion-derived virus binding in the midgut. *J. Gen. Virol. 86*, 1349–1355.

Haines, F.J., Griffiths, C.M., Possee, R.D., Hawes, C.R., King, L.A., 2009. Involvement of lipid rafts and cellular actin in AcMNPV GP64 distribution and virus budding. *Virol. Sin. 24*, 333–349.

Harrison, R.L., 2009. Structural divergence among genomes of closely related baculoviruses and its implications for baculovirus evolution. *J. Invertebr. Pathol. 101*, 181–186.

Harrison, R.L., Lynn, D.E., 2007. Genomic sequence analysis of a nucleopolyhedrovirus isolated from the diamondback moth, *Plutella xylostella. Virus Genes 35*, 857–873.

Harrison, R.L., Popham, H.J., 2008. Genomic sequence analysis of a granulovirus isolated from the Old World bollworm, *Helicoverpa armigera. Virus Genes 36*, 565–581.

Herniou, E.A., Olszewski, J.A., O'Reilly, D.R., Cory, J.S., 2004. Ancient coevolution of baculoviruses and their insect hosts. *J. Virol. 78*, 3244–3251.

Herrero, S., Ansems, M., Van Oers, M.M., Vlak, J.M., Bakker, P.L., de Maagd, R.A., 2007. REPAT, a new family of proteins induced by bacterial toxins and baculovirus infection in *Spodoptera exigua. Insect. Biochem. Mol. Biol. 37*, 1109–1118.

Hu, Z.G., Chen, K.P., Yao, Q., Gao, G.T., Xu, J.P., Chen, H.Q., 2006. Cloning and characterization of *Bombyx mori* PP-BP a gene induced by viral infection. *Yi Chuan Xue Bao 33*, 975–983.

Huang, J., Hao, B., Deng, F., Sun, X., Wang, H., Hu, Z., 2008. Open reading frame Bm21 of *Bombyx mori* nucleopolyhedrovirus is not essential for virus replication in vitro, but its deletion extends the median survival time of infected larvae. *J. Gen. Virol. 89*, 922–930.

Huijskens, I., Li, L., Willis, L.G., Theilmann, D.A., 2004. Role of AcMNPV IE0 in baculovirus very late gene activation. *Virology 323*, 120–130.

Ishikawa, H., Ikeda, M., Alves, C.A., Thiem, S.M., Kobayashi, M., 2004. Host range factor 1 from *Lymantria dispar* nucleopolyhedrovirus (NPV) is an essential viral factor required for productive infection of NPVs in IPLB-Ld652Y cells derived from L. dispar. *J. Virol. 78*, 12703–12708.

Jehle, J.A., 2004. The mosaic structure of the polyhedrin gene of the *Autographa californica* nucleopolyhedrovirus (AcMNPV). *Virus Genes 29*, 5–8.

Jehle, J.A., Lange, M., Wang, H., Hu, Z., Wang, Y., Hauschild, R., 2006. Molecular identification and phylogenetic analysis of baculoviruses from Lepidoptera. *Virology 346*, 180–193.

Jiang, S.S., Chang, I.S., Huang, L.W., Chen, P.C., Wen, C.C., Liu, S.C., Chien, L.C., Lin, C.Y., Hsiung, C.A., Juang, J.L., 2006. Temporal transcription program of recombinant *Autographa californica* multiple nucleopolyhedrosis virus. *J. Virol. 80*, 8989–8999.

Kadlec, J., Loureiro, S., Abrescia, N.G., Stuart, D.I., Jones, I.M., 2008. The postfusion structure of baculovirus gp64 supports a unified view of viral fusion machines. *Nat. Struct. Mol. Biol. 15*, 1024–1030.

Kamita, S.G., Nagasaka, K., Chua, J.W., Shimada, T., Mita, K., Kobayashi, M., Maeda, S., Hammock, B.D., 2005. A baculovirus-encoded protein tyrosine phosphatase gene induces enhanced locomotory activity in a lepidopteran host. *Proc. Natl. Acad. Sci. USA 102*, 2584–2589.

Katsuma, S., Shimada, T., 2009. *Bombyx mori* nucleopolyhedrovirus ORF34 is required for efficient transcription of late and very late genes. *Virology 392*, 230–237.

Katsuma, S., Horie, S., Daimon, T., Iwanaga, M., Shimada, T., 2006. *In vivo* and *in vitro* analyses of a *Bombyx mori* nucleopolyhedrovirus mutant lacking functional *vfgf. Virology 355*, 62–70.

Katsuma, S., Mita, K., Shimada, T., 2007. ERK- and JNK-dependent signaling pathways contribute to *Bombyx mori* nucleopolyhedrovirus infection. *J. Virol. 81*, 13700–13709.

Katsuma, S., Horie, S., Shimada, T., 2008. The fibroblast growth factor homolog of *Bombyx mori* nucleopolyhedrovirus enhances systemic virus propagation in B. mori larvae. *Virus. Res. 137*, 80–85.

Kawasaki, Y., Matsumoto, S., Nagamine, T., 2004. Analysis of baculovirus IE1 in living cells: dynamics and spatial relationships to viral structural proteins. *J. Gen. Virol. 85*, 3575–3583.

Khurad, A.M., Mahulikar, A., Rathod, M.K., Rai, M.M., Kanginakudru, S., Nagaraju, J., 2004. Vertical transmission of nucleopolyhedrovirus in the silkworm, *Bombyx mori* L. *J. Invertebr. Pathol. 87*, 8–15.

Landais, I., Vincent, R., Bouton, M., Devauchelle, G., Duonor-Cerutti, M., Ogliastro, M., 2006. Functional analysis of evolutionary conserved clustering of bZIP binding sites in the baculovirus homologous regions (*hr*s) suggests a cooperativity between host and viral transcription factors. *Virology 344*, 421–431.

Lapointe, R., Popham, H.J., Straschil, U., Goulding, D., O'Reilly, D.R., Olszewski, J.A., 2004. Characterization of two *Autographa californica* nucleopolyhedrovirus proteins, Ac145 and Ac150, which affect oral infectivity in a host-dependent manner. *J. Virol. 78*, 6439–6448.

Lauzon, H.A., Lucarotti, C.J., Krell, P.J., Feng, Q., Retnakaran, A., Arif, B.M., 2004. Sequence and organization of the *Neodiprion lecontei* nucleopolyhedrovirus genome. *J. Virol. 78*, 7023–7035.

Lauzon, H.A., Garcia-Maruniak, A., Zanotto, P.M., Clemente, J.C., Herniou, E.A., Lucarotti, C.J., Arif, B.M., Maruniak, J.E., 2006. Genomic comparison of *Neodiprion sertifer* and *Neodiprion lecontei* nucleopolyhedroviruses and identification of potential hymenopteran baculovirus-specific open reading frames. *J. Gen. Virol. 87*, 1477–1489.

Levy, S.M., Falleiros, A.M., Moscardi, F., Gregorio, E.A., 2007. Susceptibility/resistance of *Anticarsia gemmatalis* larvae to its nucleopolyhedrovirus (AgMNPV): Structural study of the peritrophic membrane. *J. Invertebr. Pathol. 96*, 183–186.

Li, Y., Guarino, L.A., 2008. Roles of LEF-4 and PTP/BVP RNA triphosphatases in processing of baculovirus late mRNAs. *J. Virol. 82*, 5573–5583.

Li, Z., Blissard, G.W., 2009. The *Autographa californica* multicapsid nucleopolyhedrovirus GP64 protein: analysis of transmembrane domain length and sequence requirements. *J. Virol. 83*, 4447–4461.

Li, G., Wang, J., Deng, R., Wang, X., 2008. Characterization of AcMNPV with a deletion of *ac68* gene. *Virus Genes 37*, 119–127.

Long, G., Pan, X., Kormelink, R., Vlak, J.M., 2006a. Functional entry of baculovirus into insect and mammalian cells is dependent on clathrin-mediated endocytosis. *J. Virol. 80*, 8830–8833.

Long, G., Pan, X., Westenberg, M., Vlak, J.M., 2006b. Functional role of the cytoplasmic tail domain of the major envelope fusion protein of group II baculoviruses. *J. Virol. 80*, 11226–11234.

McCarthy, C.B., Theilmann, D.A., 2008. AcMNPV ac143 (odv-e18) is essential for mediating budded virus production and is the 30th baculovirus core gene. *Virology 375*, 277–291.

McCarthy, C.B., Dai, X., Donly, C., Theilmann, D.A., 2008. *Autographa californica* multiple nucleopolyhedrovirus ac142, a core gene that is essential for BV production and ODV envelopment. *Virology 372*, 325–339.

Mikhailov, V.S., Okano, K., Rohrmann, G.F., 2006. Structural and functional analysis of the baculovirus single-stranded DNA-binding protein LEF-3. *Virology 346*, 469–478.

Mikhailov, V.S., Vanarsdall, A.L., Rohrmann, G.F., 2008. Isolation and characterization of the DNA-binding protein (DBP) of the *Autographa californica* multiple nucleopolyhedrovirus. *Virology 370*, 415–429.

Milks, M.L., Washburn, J.O., Willis, L.G., Volkman, L.E., Theilmann, D.A., 2003. Deletion of pe38 attenuates AcMNPV genome replication, budded virus production, and virulence in *Heliothis virescens*. *Virology 310*, 224–234.

Mishra, G., Chadha, P., Das, R.H., 2008. Serine/threonine kinase (pk-1) is a component of *Autographa californica* multiple nucleopolyhedrovirus (AcMNPV) very late gene transcription complex and it phosphorylates a 102 kDa polypeptide of the complex. *Virus Res. 137*, 147–149.

Mukawa, S., Goto, C., 2008. *In vivo* characterization of two granuloviruses in larvae of *Mythimna separata* (Lepidoptera: Noctuidae). *J. Gen. Virol. 89*, 915–921.

Nagamine, T., Kawasaki, Y., Abe, A., Matsumoto, S., 2008. Nuclear marginalization of host cell chromatin associated with expansion of two discrete virus-induced subnuclear compartments during baculovirus infection. *J. Virol. 82*, 6409–6418.

Nobiron, I., O'Reilly, D.R., Olszewski, J.A., 2003. *Autographa californica* nucleopolyhedrovirus infection of *Spodoptera frugiperda* cells: a global analysis of host gene regulation during infection, using a differential display approach. *J. Gen. Virol. 84*, 3029–3039.

Ohkawa, T., Washburn, J.O., Sitapara, R., Sid, E., Volkman, L.E., 2005. Specific binding of *Autographa californica* M nucleopolyhedrovirus occlusion-derived virus to midgut cells of *Heliothis virescens* larvae is mediated by products of *pif* genes Ac119 and Ac022 but not by Ac115. *J. Virol. 79*, 15258–15264.

Okano, K., Vanarsdall, A.L., Rohrmann, G.F., 2007. A baculovirus alkaline nuclease knockout construct produces fragmented DNA and aberrant capsids. *Virology 359*, 46–54.

Patmanidi, A.L., Possee, R.D., King, L.A., 2003. Formation of P10 tubular structures during AcMNPV infection depends on the integrity of host-cell microtubules. *Virology 317*, 308–320.

Perera, O.P., Valles, S.M., Green, T.B., White, S., Strong, C.A., Becnel, J.J., 2006. Molecular analysis of an occlusion body protein from *Culex nigripalpus* nucleopolyhedrovirus (CuniNPV). *J. Invertebr. Pathol. 91*, 35–42.

Perera, O., Green, T.B., Stevens, S.M.Jr., White, S., Becnel, J.J., 2007. Proteins associated with *Culex nigripalpus* nucleopolyhedrovirus occluded virions. *J. Virol. 81*, 4585–4590.

Pijlman, G.P., Pruijssers, A.J., Vlak, J.M., 2003. Identification of *pif-2*, a third conserved baculovirus gene required for *per os* infection of insects. *J. Gen. Virol. 84*, 2041–2049.

Rohrmann, G.F., 2008. DNA replication and genome processing. In: Baculovirus Molecular Biology. National Library of Medicine (US), National Center for Biotechnology Information, Bethesda (MD), pp. 55–70.

Saksena, S., Shao, Y., Braunagel, S.C., Summers, M.D., Johnson, A.E., 2004. Cotranslational integration and initial sorting at the endoplasmic reticulum translocon of proteins destined for the inner nuclear membrane. *Proc. Natl. Acad. Sci. USA 101*, 12537–12542.

Saksena, S., Summers, M.D., Burks, J.K., Johnson, A.E., Braunagel, S.C., 2006. Importin-alpha-16 is a translocon-associated protein involved in sorting membrane proteins to the nuclear envelope. *Nat. Struct. Mol. Biol. 13*, 500–508.

Schultz, K.L., Wetter, J.A., Fiore, D.C., Friesen, P.D., 2009. Transactivator IE1 is required for baculovirus early replication events that trigger apoptosis in permissive and nonpermissive cells. *J. Virol. 83*, 262–272.

Selot, R., Kumar, V., Shukla, S., Chandrakuntal, K., Brahmaraju, M., Dandin, S.B., Laloraya, M., Kumar, P.G., 2007. Identification of a soluble NADPH oxidoreductase (BmNOX) with antiviral activities in the gut juice of *Bombyx mori*. *Biosci. Biotechnol. Biochem. 71*, 200–205.

Shelby, K.S., Popham, H.J., 2006. Plasma phenoloxidase of the larval tobacco budworm, *Heliothis virescens*, is virucidal. *J. Insect Sci. 6*, 1–12.

Shelby, K.S., Popham, H.J., 2007. Increased plasma selenium levels correlate with elevated resistance of *Heliothis virescens* larvae against baculovirus infection. *J. Invertebr. Pathol. 95*, 77–83.

Slack, J.M., Lawrence, S.D., Krell, P.J., Arif, B.M., 2008. Trypsin cleavage of the baculovirus occlusion-derived virus attachment protein P74 is prerequisite in per os infection. *J. Gen. Virol. 89*, 2388–2397.

Slavicek, J.M., Popham, H.J., 2005. The *Lymantria dispar* nucleopolyhedrovirus enhancins are components of occlusion-derived virus. *J. Virol. 79*, 10578–10588.

Song, J., Wang, R., Deng, F., Wang, H., Hu, Z., 2008. Functional studies of per os infectivity factors of *Helicoverpa armigera* single nucleocapsid nucleopolyhedrovirus. *J. Gen. Virol. 89*, 2331–2338.

Stewart, T.M., Huijskens, I., Willis, L.G., Theilmann, D.A., 2005. The *Autographa californica* multiple nucleopolyhedrovirus ie0-ie1 gene complex is essential for wild-type virus replication, but either IE0 or IE1 can support virus growth. *J. Virol. 79*, 4619–4629.

Thiem, S.M., Chejanovsky, N., 2004. The role of baculovirus apoptotic suppressors in AcMNPV-mediated translation arrest in Ld652Y cells. *Virology 319*, 292–305.

van Oers, M.M., Vlak, J.M., 2007. Baculovirus genomics. *Curr. Drug Targets 8*, 1051–1068.

Vanarsdall, A.L., Okano, K., Rohrmann, G.F., 2005. Characterization of the replication of a baculovirus mutant lacking the DNA polymerase gene. *Virology 331*, 175–180.

Vanarsdall, A.L., Okano, K., Rohrmann, G.F., 2006. Characterization of the role of very late expression factor 1 in baculovirus capsid structure and DNA processing. *J. Virol. 80*, 1724–1733.

Vanarsdall, A.L., Mikhailov, V.S., Rohrmann, G.F., 2007. Characterization of a baculovirus lacking the DBP (DNA-binding protein) gene. *Virology 364*, 475–485.

Wang, Q., Liang, C., Song, J., Chen, X., 2007. HA2 from the *Helicoverpa armigera* nucleopolyhedrovirus: a WASP-related protein that activates Arp2/3-induced actin filament formation. *Virus Res. 127*, 81–87.

Westenberg, M., Uijtdewilligen, P., Vlak, J.M., 2007. Baculovirus envelope fusion proteins F and GP64 exploit distinct receptors to gain entry into cultured insect cells. *J. Gen. Virol. 88*, 3302–3306.

Wilson, J.A., Forney, S.D., Ricci, A.M., Allen, E.G., Hefferon, K.L., Miller, L.K., 2005. Expression and mutational analysis of *Autographa californica* nucleopolyhedrovirus HCF-1: functional requirements for cysteine residues. *J. Virol. 79*, 13900–13914.

Xiao, W., Yang, Y., Weng, Q., Lin, T., Yuan, M., Yang, K., Pang, Y., 2009. The role of the PI3K-Akt signal transduction pathway in *Autographa californica* multiple nucleopolyhedrovirus infection of *Spodoptera frugiperda* cells. *Virology 391*, 83–89.

Xu, J.P., Chen, K.P., Yao, Q., Lin, M.H., Gao, G.T., Zhao, Y., 2005. Identification and characterization of an NPV infection-related gene *Bmsop2* in *Bombyx mori* L. *J. Appl. Entomol. 129*, 425–431.

Zhang, J.H., Ohkawa, T., Washburn, J.O., Volkman, L.E., 2005. Effects of Ac150 on virulence and pathogenesis of *Autographa californica* multiple nucleopolyhedrovirus in noctuid hosts. *J. Gen. Virol. 86*, 1619–1627.

Zhou, J., Blissard, G.W., 2008a. Display of heterologous proteins on gp64null baculovirus virions and enhanced budding mediated by a vesicular stomatitis virus G-stem construct. *J. Virol. 82*, 1368–1377.

Zhou, J., Blissard, G.W., 2008b. Identification of a GP64 subdomain involved in receptor binding by budded virions of the baculovirus *Autographica californica* multicapsid nucleopolyhedrovirus. *J. Virol. 82*, 4449–4460.

6   Amino Acid and Neurotransmitter Transporters

D Y Boudko, University of Florida,
St. Augustine, FL, USA
B C Donly, Agriculture and Agri-Food Canada,
London, ON, Canada
B R Stevens, University of Florida, Gainesville,
FL, USA
W R Harvey, University of Florida,
St. Augustine, FL, USA

© 2005, Elsevier BV. All Rights Reserved.

6.1. Introduction

6.1.1. Role of Amino Acid Transporters in Insects

Insects use amino acids as substrates for protein synthesis, as essential cofactors for intermediary metabolism, as intracellular messengers between distinct metabolic pathways, and as intercellular messengers – neurotransmitters. Amino acids are precursors in a variety of catabolic and anabolic pathways in which amino acid transporters serve as rate-limiting components (Saier, 2000a); e.g., glutamate transporters regulate the synthesis of GABA (Mathews and Diamond, 2003). In addition some insects use amino acids rather than glucose as an energy source (Balboni, 1978; Gade and Auerswald, 2002) and as osmolytes (Patrick and Bradley, 2000).

6.1.2. Uniporters, Symporters, and Antiporters

The movement of amino acids across membranes is catalyzed by transporters, which are polypeptides arranged as monomers or multimers. The collective interaction of these transporters with specific substrates and electrochemical driving forces generated by primary ATP-driven pumps and secondary mineral ion transporters is termed a "transport system" (Christensen, 1964; Gerencser and Stevens, 1994). Amino acid transporters fall into three groups on the basis of their electrochemical profiles: uniporters, symporters, and antiporters. Uniporters facilitate diffusion of amino acids down their own electrochemical gradients. Thermodynamically, uniporters are equivalent to channels but differ kinetically in possessing specific amino acid binding sites and lower conductances than inorganic ion channels. In contrast, amino acid:ion symporters (cotransporters) use electrochemical motive forces of inorganic ions for the translocation of amino acids, stoichiometrically coupled to the ions, usually inorganic cations or anions. A cotransporter can move an amino acid up its electrochemical gradient while the cotransported ion moves down its electrochemical gradient (the amino acid undergoes secondary active transport). Under specific conditions the coupling stoichiometry between amino acid and ion may change, resulting in a so-called "molecular slip" (Nelson et al., 2002). In contrast to symporters, antiporters move thermodynamically coupled solutes in opposite directions and utilize different electrochemical mechanisms.

6.1.3. Energization and Electrochemistry of Amino Acid Uptake

The energization and electrochemical mechanisms of amino acid transporters vary greatly, depending upon the available electrochemical conditions (summarized in **Figure 1**). For example, classically characterized reuptake of neurotransmitters by high-affinity neurotransmitter transporters (NTTs) utilizes a steady-state Na^+ electrochemical gradient across neuronal membranes. The Na^+ gradient is thought to be generated by a Na^+/K^+ ATPase and the amino acid uptake is inhibited by agents, such as ouabain or vanadate, that inhibit Na^+/K^+ ATPase activity (Clark and Amara, 1993). More recently, it has become apparent that a proton-translocating, vacuolar-type ATPase (H^+ V-ATPase) has a dominant role in energization of membranes interfacing low sodium ion domains (Wieczorek et al., 1999a). For example, it energizes processes in intracellular membranes (endomembranes), e.g., vesicular and lysosomal amino acid carriers. It serves as an essential energization component in synaptic vesicles, where it contributes to the hyperpolarization of the vesicular membrane and therefore facilitates Na^+-coupled reuptake of neurotransmitters (Nelson and Lill, 1994). Amino acid uptake mediated by transporters is also energized by V-ATPases in plasma membranes of cells exposed to media with low Na^+ concentrations, as in the midgut lumen of caterpillars and mosquito larvae (Harvey et al., 1998). Nanomolar concentrations of bafilomycin or concanamycin inhibit the V-ATPase selectively (Drose et al., 2001), providing a valuable pharmacological tool for the identification of V-ATPase-dependent carrier mechanisms.

Whether an amino acid is anionic $(-)$, zwitterionic $(+/-)$, or cationic $(+)$ depends on the local pH and on amino acid dissociation constants. Other conditions being constant, the lower the pH the more protonated is the amino acid and the more positive is its charge. Conversely, the higher the pH the less protonated and the more negative is the amino acid. This dependence of charge transition upon pH value is especially important in mosquito larvae and caterpillars where amino acid uptake occurs through membranes with large electrostatic and pH gradients. Although the electrochemical interaction between H^+ V-ATPases and amino acid transporters in insect midgut is poorly explored, a tentative experimental model may be suggested by analogy with inorganic ion transporters, which mediate lumen alkalinization of caterpillar and mosquito larval midgut. One such component recruited in alkalinization is H^+ V-ATPase-energized Cl^-/HCO_3^- exchange, which was found in larval midgut of *Aedes aegypti* (Boudko et al., 2001a). Alkalinization of anterior midgut and related pH gradients along the midgut lumen appear to be chiefly, if not exclusively, generated by transepithelial exchange of

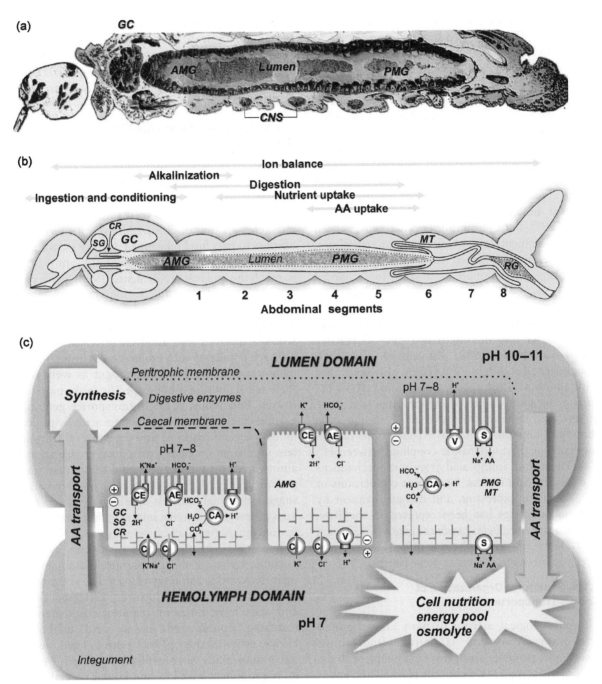

Figure 1 Energization and electrochemical mechanisms of amino acid uptake in insects (example from midgut of mosquito larvae). (a) Low magnification micrograph of longitudinal section showing general cytology of the larval midgut of 4th instar *Aedes aegypti*; (b) diagram illustrating distribution of specific physiological functions in alimentary canal of mosquito larvae; (c) simplified diagram showing putative electrochemical mechanisms that are involved in amino acid uptake in mosquito larvae. Abbreviations: AA, amino acids; AMG, anterior region of midgut; CR, cardia; CNS, central nervous system (ventral nerve cord ganglia); GC, gastric caeca; MT, Malpighian tubules; RG, rectal gland; PMG, posterior midgut; SG, salivary glands; S, symporter, V, H$^+$V-ATPase; C, channel; CE, cation exchanger; AE, anion exchanger; CA, carbonic anhydrase.

strong Cl$^-$ and weak HCO$_3^-$ anions that is energized by H$^+$ V-ATPases. So far no evidence for a significant role of Na$^+$/K$^+$ ATPases in energization of mosquito larval midgut cells has been obtained, but by analogy with known anionic pathways, a spatial exchange of strong inorganic cations and weak organic cations may mimic the role of Na$^+$/K$^+$ ATPases in generating the transmembrane Na$^+$ concentration gradients that are necessary for amino acid uptake. Alternatively, different motive forces, e.g., for H^+, K^+, Cl^- as well as one for substrate itself can be used by nutrient amino acid transporters.

Since H^+ V-ATPase constantly moves protons from the cytosol to the exterior it renders the inside of its resident membrane negative to the outside, i.e., it strongly hyperpolarizes the resident membrane (Harvey, 1992). Due to electrical integrity of the insect midgut epithelia, both apical and basal membranes are hyperpolarized (Beyenbach, 2001). For example, even though the V-ATPase is resident on the apical plasma membranes of specific cells in posterior midgut of *Ae. aegypti* larvae (Zhuang *et al.*, 1999), both membranes in such cells, as well as entire segments of posterior midgut epithelium, would be hyperpolarized relative to the lumen, implicating electrophoretic forces for charged amino acids and inorganic ions. Primary absorption of nutrient amino acids in posterior midgut of mosquito larvae (Clements, 1992) colocalizes with the residence and high activity of H^+ V-ATPases (Filippova *et al.*, 1998; Zhuang *et al.*, 1999; Boudko *et al.*, 2001b). V-ATPases also colocalize with amino acid transporters in certain cells of the gastric caeca and cardia, where they appear to drive amino acid uptake from hemocoel to the cells (Boudko, unpublished data). These observations suggest that thermodynamic coupling between H^+ V-ATPases and amino acid transport mechanisms is widespread in insects. Significant correlations of H^+ V-ATPase expression and midgut invasion by malaria ookinetes has been reported in the Ross cells of adult *Anopheles gambiae* midgut (Cociancich *et al.*, 1999). However, the role of V-ATPases in the infection mechanism remains to be clarified.

6.1.4. Structural Organization of Known Transporters

The physiological activity of most known amino acid transport systems is derived from transmembrane proteins that are monomers or form multimeric complexes within the plasma- and endomembranes. These proteins typically integrate 8–14 alpha-helical transmembrane spanning domains (TMDs) and a certain number of submembrane loops, which bear sites for interaction with the cytoskeleton and with other membrane and cytosolic proteins. Molecular identities of transport proteins have been explored mostly in mammals and, to some extent, in model insects (**Table 1**). In addition to monomeric transporters, there are several groups of unique transporters that act as obligatory heterodimers. For example, HAT heterodimers consist of a light chain, a catalytic subunit with 12–14 putative TMDs (SLC7), which is associated with a

glycosylated globular heavy chain subunit that has only a single membrane spanning domain (SLC3) (see Section 6.7.2). Monomeric and heterodimeric transport systems appear in both nonpolarized cells and epithelial cells; however, they play different and most often complementary roles.

6.1.5. How Many Amino Acid Transporters Are Required?

Analysis of recently available genomes suggests that complex organisms require many transporters to regulate local amino acid concentrations in different tissues and cells. Due to their large size and ionic character, all amino acids require specific molecular carriers to pass lipid bilayers of cellular membranes. Mosquito larvae require 10 essential L-amino acids to reach the 2nd instar (Clements, 1992): three basic ones (arginine, lysine, and histidine) and seven neutral ones (valine, leucine, isoleucine, phenylalanine, tryptophan, threonine, and methionine). Thus, one would expect to find a subset of transporters for primary uptake of such amino acids through apical membranes in the posterior midgut as well as for their redistribution among specific body parts and tissues. Potentially, a single transporter gene may encode carriers for several amino acids with common structural features; however, several transporters would be necessary to reduce competition of amino acids for binding sites and to increase transport efficiency. A growing number of observations suggest that high-affinity uptake is required for a high degree of selectivity, implying that a large number of transporter phenotypes may exist. Nonessential amino acids, i.e., glycine, alanine, tyrosine, cysteine, serine, proline, aspartate, glutamate, and their derivatives, can be synthesized *de novo* within the insect organism. However, different tissues and cells have different requirements, often not matching nonessential amino acid anabolism. Hence, essential roles of transporters for nonessential amino acids have evolved and have become more prominent with increased tissue specialization and histological complexity of modern organisms. In addition to the population of plasma membrane amino acid transporters, a large number of specific transporters are necessary for translocation of amino acids between different intracellular domains. Moreover, insects undergo radical physiological and ecological transitions between larval and adult lifestyles. These transitions imply dramatic differences in amino acid requirements, which, together with certain requirements for genomic and phenotypic plasticity of amino acid transport mechanisms, may result in an increase of amino acid transporters being encoded. They may remain

Table 1 Classification of insect nervous system amino acid and neurotransmitter transporters

Transporter family	TC number (Saier, 2000b)	Solute carrier orthology	Common names	Cloned insect genes[a]	Accession number	Substrates
Vesicular monoamine transporter (VMOAT)	2.A.1.22	SLC18	H+ dependent vesicular monoamine transporter/vesicular acetylcholine transporter family	drmVAChT	AAB86609	Monoamines, acetylcholine
Solute: sodium symporter (SSS)	2.A.21	SLC5	Na+ dependent glucose transporter family	trnCHT	AY629593	Choline
Neurotransmitter:sodium symporter (NSS)	2.A.22	SLC6	Na+/Cl− dependent neurotransmitter transporter family	masGAT	S65673	GABA, monoamines, proline
				trnGAT	AAF70819	
				drmSERT	AAD10615	
				masSERT	AAN59781	
				drmDAT	AAF76882	
				trnDAT	AAN52844	
				trnOAT	AAL09578	
				masPROT	AF454914	
				drmINE	AAC47292	
				masINE	AAF21642	
				drmBLOT	CAB53640	
Dicarboxylate/amino acid:cation (Na+ or H+)	2.A.23	SLC1	Na+/K+ dependent excitatory amino acid transporter family	trnEAAT1	AAB84380	Glutamate, aspartate
				dipEAAT1	AAF71701	
				drmEAAT1	AAD09142	
				drmEAAT2	AAD47830	
				apmEAAT1	AAD34586	
				aeaEAAT	AAP76304	
Anion:cation symporter (ACS; VGLUT) family	2.A.1.14	SLC17	H+ dependent vesicular glutamate transporter			Glutamate
Amino acid/auxin permease (AAAP) family	2.A.18	SLC32	H+ dependent vesicular inhibitory amino acid transporters			GABA, glycine

[a]Additionally, some putative transporters identified by phylogenetic proximity with characterized neurotransmitter transporter are discussed in Section 6.5.

inactive in insect genomes until activated, usually by a hormone, at a particular developmental stage.

6.1.6. Neurotransmitter Transporters and Nutrient Amino Acid Transporters

Amino acid transporters can be divided into two major categories with respect to their physiological roles: neurotransmitter transporters (NTTs) and nutrient amino acid transporters (NATs). Many phylogenetically distinct groups of amino acid transporters include both categories.

NTT denotes a carrier of neurotransmitters, which may be amino acids or their derivatives. Classical examples are plasma membrane insect NTTs for the high-affinity reuptake of glutamate, aspartate, glycine, and γ-aminobutyric acid (GABA), as well as the monoamines, dopamine, octopamine, 5-hydroxytryptamine (5-HT or serotonin), and histamine. Plasma membrane NTTs transport neurotransmitters from synaptic clefts into glial cells and presynaptic neurons. Intracellular NTTs serve for accumulation of neurotransmitters into synaptic vesicles, implicating them as one of the most critical components of presynaptic maturation for chemical signaling.

NAT denotes a carrier of one or more of the 20 natural α-amino acids with no apparent involvement in signaling processes, although there is weak background for such discrimination. Classical examples of insect NATs are those that absorb amino acids from the midgut or Malpighian tubule lumen through the apical membrane of specialized epithelial cells. NATs in the basal membrane export amino acids from epithelial cells into the hemolymph, comprising electrochemically different transport systems. Different NAT types are expressed in the basal cell membranes of the mosquito larval cardia, gastric caeca, and salivary glands, where they take up specific amino acids from the hemolymph and supply substrates for the synthesis of peritrophic and other secretory peptides. NATs in membranes of internal organelles participate in intermediary metabolism by supplying amino acids for metabolic pathways. Although glutamate, aspartate, and glycine transporters can be classified as NTTs or NATs, depending upon their role in specific cell types, in this review they will be discussed along with GABA transporters in Section 6.3.

A great deal is known about NATs in prokaryotes, yeast, and humans (Palacin et al., 1998; Wipf et al., 2002), but what little is known about them in insects is restricted mainly to caterpillars, fruit flies, and mosquitoes (discussed in Section 6.8.5.2). Much more is known about neurotransmitter amino acid transporters of insects (discussed immediately below and in Section 6.3).

6.2. Physiology of Insect NTTs

6.2.1. Synaptic Neurotransmission

The nervous system is responsible for the acquisition, integration, and storage of information. Neurons are specialized cells that enable electrochemical communication with each other, sensory cells, and target cells. This intercellular communication occurs predominantly at synapses, through the exocytotic release of chemical transmitters from the presynaptic neuron (Pennetta et al., 1999; Caveney and Donly, 2002; Watson and Schurmann, 2002). Released transmitters diffuse across the synaptic space and bind to receptors on the postsynaptic membrane, resulting in either depolarization or other effects on the target cell (**Figure 2**). Transmitters are stored in the presynaptic terminal in vesicles, which are stocked with these chemicals by the action of transporters located in the vesicular membranes. The necessary stores of transmitter are acquired both through the biosynthesis of new molecules and the reuptake of released ones. Recovery of released transmitter is accomplished by transporters located in the plasma membranes of the various cells impinging on the synaptic space, both neuronal and glial cells (**Figure 2**). Thus, transport of neurotransmitters is important for the functioning of the synapse at two levels: for the packaging of transmitters in presynaptic vesicles prior to their release by exocytosis and for their reuptake from the extracellular space into the cytoplasm (Krantz et al., 1999).

6.2.2. Role of Plasma Membrane NTTs in Synaptic Transmission

The action of released neurotransmitters is terminated by three major mechanisms: diffusion away from the synapse, buffering by the receptor and transporter ligand binding sites, and removal via uptake by high-affinity transporters present on neighboring membranes of presynaptic, postsynaptic, and glial cells (**Figure 2**). Much of the progress in understanding transmitter clearance has been carried out on various glutamatergic synapses in the mammalian CNS. Initially, diffusion was believed to play the predominant role in determining the concentration profile of glutamate in the synaptic cleft. In this view, plasma membrane transporters only act externally to the cleft – to lower extracellular glutamate levels, to contain the transmitter to the

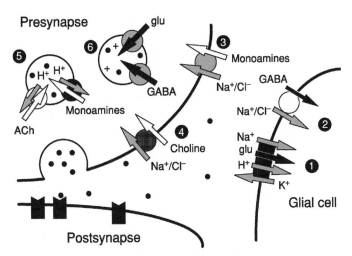

Figure 2 Neurotransmitters involved in insect synaptic signaling. Vesicles in the presynaptic neuron fuse with the presynaptic membrane and release neurotransmitters (●) that diffuse across the synaptic cleft and bind to receptors on the postsynaptic membrane. Unbound transmitters are taken up by various transport proteins in the plasma membranes of the presynaptic axon and adjacent glial cells: excitatory amino acid transporters (EAATs) that mediate Na^+ and K^+ dependent uptake of glutamate and aspartate into glutamatergic neurons or glial cells (1); and GABA and various monoamine transporters (SNFs) that mediate Na^+ and Cl^- dependent uptake of these ligands into glial cells (2) or back into neurons (3). A breakdown product of acetylcholine, choline, is transported into neurons by a different type of Na^+ and Cl^- dependent transporter that is related to glucose transporters (4). Neurotransmitters in neurons and glial cells are packaged into vesicles by transporters in the vesicular membrane, which are energized by proton/voltage gradients generated by the vacuolar H^+-ATPase. Transport of monoamines and acetylcholine into the vesicles is driven by a pH gradient (5), whereas transport of glutamate is driven by the electrical gradient (6). Vesicular GABA uptake is dependent on both gradients. Packaged neurotransmitters in neurons derived from this reuptake and from *de novo* synthesis are ready to be released when activated by an incoming voltage signal.

synapse, and to maintain a steep concentration gradient that promotes rapid diffusion (Krantz *et al.*, 1999). Transporters were thought to play a limited role in the clearance of transmitter from the synapse. Moreover, because they cycle at very low speeds relative to receptor inactivation times, transporters were thought to affect only the slow components of glutamate removal (Otis *et al.*, 1996). However, new data suggest that transporters play a dynamic role in the regulation of neurotransmission. For example, glutamate transporters may contribute to neurotransmitter inactivation on a rapid timescale. When present in high density, transporters may serve as binding sites without actually clearing the bound substrates (Seal and Amara, 1999). Also, Auger and Attwell (2000) found that glutamate removal by postsynaptic transporters at the climbing fiber synapse of rat Purkinje cells is very rapid, implying that translocation of glutamate may occur very early in the transport cycle. These observations suggest an active role for transporters in synaptic homeostasis, in which most of the presynaptically released transmitter is taken up rapidly and directly from the synapse. Other considerations affecting the relative roles of diffusion and uptake in regulating glutamate removal include the anatomy of individual synapses, with synapse diameter and

the presence or absence of transporter-rich glia affecting the termination of postsynaptic excitation (Danbolt, 2001). Thus, the effects of diffusion and uptake on neurotransmitter concentration in the synaptic cleft depend upon the geometry of the cleft and adjacent spaces.

Analysis of amine transporter knockouts in mice also reveals a direct influence of transporters on the volume transmission of monoamines. Elimination of the relevant transporters revealed that by limiting neurotransmitter removal to diffusion alone, the extracellular transmitter concentration increased fivefold for dopamine, sixfold for serotonin, and twofold for norepinephrine (Gainetdinov and Caron, 2002; Torres *et al.*, 2003). The persistence of dopamine in the extracellular space also increased 300-fold, relative to that in wild-type mice. Furthermore, intracellular transmitter concentrations were severely reduced, indicating reuptake is also crucial for the maintenance of the stores of these compounds. Therefore, the action of plasma membrane transporters is necessary for presynaptic homeostasis both for maintaining clearance kinetics and for providing adequate substrate levels.

Rapid removal of neurotransmitter from the synaptic cleft raises the prospect that transporter activity may influence the characteristics of postsynaptic potentials, with broad implications for behavior and learning. However, this prospect has scarcely been explored.

6.2.3. Study of Insect NTTs

Much current knowledge on transporter structure and function has been derived from studies of vertebrate (particularly mammalian) examples. In general, transporters from insects are less well characterized than their vertebrate counterparts. However, several groups and even particular species of insect transporters are clearly orthologous to vertebrate transporters. Much of what is known of many of the specific categories of NTTs (from both a structural and functional perspective) have been inferred from discoveries in the mammalian CNS (reviews: Deken et al., 2002; Kanai et al., 2002; Torres et al., 2003). Commensurate with overall genome size, there appears to be less specialization in NTT complements in insects than in their mammalian equivalents. This relative simplicity may render NTTs promising targets for pest control as the insect nervous system possesses relatively less molecular redundancy to be overcome. The simplicity should also prove to make them tractable model systems for genetic and molecular dissection. Indeed, characterizing membrane transporters in the model organism Drosophila melanogaster has progressed rapidly and there has been significant progress in analyzing NTT transporters in Lepidoptera, as described below.

Insects may nevertheless provide a rich resource of NTT diversity, based on their extensive physiological variation and adaptation to diverse environments and lifestyles. In addition to the examples already identified by functional studies, the completion of the fruit fly and mosquito genomes has now allowed the entire complements of transporters from these two species to be categorized, as discussed below (Sections 6.5–6.9). Initial annotation of the Drosophila genome according to the gene ontology (GO) system resulted in the assignment of 665 transcripts within the "function" category as transporters (Adams et al., 2000). Comparison with the recently completed mosquito genome (Holt et al., 2002) has yielded further novel insights into insect transporters as well as a wealth of inferential insights based on comparison with the growing catalog of genomic databases (see Section 6.5).

6.2.4. Potential for Use of NTTs as Insecticide Targets

The importance of neurotransmitter reuptake at the synapse suggests that insect NTTs may serve as targets for insecticidal action. Many potent drugs that affect behavior by acting as "selective reuptake blockers" of plasma membrane NTTs have been identified in the mammalian brain (Torres et al., 2003). Gene knockouts (and antisense knockdowns) of various transporters in rodents have also demonstrated significant physiological consequences resulting from disruption of transporter function (Rothstein et al., 1996; Torres et al., 2003). In insects, the natural plant compound, cocaine, may exert behavioral and insecticidal effects through antagonism of the reuptake of amines in the nervous system, in particular that of octopamine (Nathanson et al., 1993). The role of GABAergic uptake activity in modulating behavior in D. melanogaster has been examined using a pharmacological approach (Leal and Neckameyer, 2002). Blocking GABA transporter function with reuptake inhibitors revealed specific effects on behavior including reduction of locomotor activity, geotactic abilities, and loss of the righting reflex. The specificity of the pharmacological effect to GABAergic activity was shown by the restoration of normal locomotor activity when inhibitors were applied in the presence of a GABA receptor antagonist. Thus, there is good reason to think that agents that block transporter activity may prove effective in perturbing pest behavior (by antifeedant or other insecticidal activities).

6.3. Molecular Biology of Insect NTTs

6.3.1. Classification of NTTs

The increasing availability of genomic information now allows a holistic approach to the analysis of the structure and function of specific transporters. By comparing complete genome complements of protein components in diverse organisms, it is possible to develop hierarchies based on structural conservation reflecting biological themes that have evolved in response to functional requirements. For membrane proteins that transport amino acids, a comparison of three sequenced eukaryotic genomes (Saccharomyces cerevisiae, Arabidopsis thaliana, and Homo sapiens) distinguished five different transporter superfamilies based on phylogeny, substrate spectrum, transport mechanism, and cell specificity (Wipf et al., 2002). An even more comprehensive scheme has been devised by Saier (2000b) to classify all permeases based on both phylogeny and function. The resulting transporter classification (TC) system uses five criteria to group individual transporters within each family. Transporters in the nervous system belong to six separate families in the TC system (**Table 1**). The primary distinction demarcating the various nervous system transporters is the site at which the transport system is active. Plasma membrane transporters transfer released neurotransmitter from the synaptic space back into the cytoplasm of neurons and glial cells using the inwardly directed Na^+ gradient across the plasma membrane (Masson et al., 1999). Vesicular transporters located in neuronal synaptic vesicles further concentrate transmitters from the cytoplasm inside the vesicles using proton electrochemical gradients generated by the vacuolar type H^+-ATPase (Yelin and Schuldiner, 2002).

6.3.2. Plasma Membrane Neurotransmitter Transporters

It has long been recognized that nervous tissues are capable of taking up a variety of endogenous substrates by way of specific, high-affinity, Na^+-dependent plasma membrane transport. The energy for this uptake derives from coupling the inward transport of neurotransmitter to the flow of sodium ions, using the inwardly directed electrochemical Na^+ gradient generated by the Na^+/K^+-ATPase (see Section 6.1.3). The molecular identities of the membrane proteins responsible have become known over the last 10–15 years, revealing structural similarities to at least three families of transporters (**Table 1**; Krantz et al., 1999). Members of the excitatory amino acid transporter (EAAT) family (for glutamate and aspartate) are distinguished by the counter-transport of potassium ions, and this group includes other members that transport neutral amino acids as substrate. Proteins in the sodium neurotransmitter symporter family (SNF) are characterized by 12 transmembrane domains and depend upon chloride for uptake activity. This family is made up of four subfamilies, encompassing a variety of substrates ranging from various amino acids to amines. Finally, the most recently characterized family is the glucose transporter family, which includes carriers for choline. All three of these families also include many other more distantly related prokaryotic transporters. A new classification scheme of insect SNF transporters is proposed in Section 6.8.5.

6.3.3. Excitatory Amino Acid Transporters (EAAT; SLC1)

6.3.3.1. **Roles of glutamate and EAATs in insects** L-Glutamate has been implicated in many important physiological processes in mammals including neuronal development, plasticity, and long-term potentiation, as well as in pathological conditions including epilepsy, cerebral ischemia, amyotrophic lateral sclerosis, Alzheimer's disease, Parkinson's disease, and schizophrenia (Danbolt, 2001). L-Glutamate also is the most abundant amino acid in the insect CNS, acting as both an excitatory and an inhibitory neurotransmitter (Bicker et al., 1988; Usherwood, 1994) via high-affinity binding/activation of postsynaptic receptors, which are classified as metabotropic and ionotropic subtypes (Monaghan et al., 1989; Gasic and Hollmann, 1992). It also acts as the principal excitatory transmitter at neuromuscular junctions (Jan and Jan, 1976a, 1976b) binding to specific receptors on somatic muscles (Schuster et al., 1991; Petersen et al., 1997). Whereas, essential roles of glutamate in the insect CNS have been deduced from extensive immunohistochemical labeling of the ligand, its receptors, and transporters, its physiological roles, as well as their central mechanisms, remain poorly understood. Apparently, brain-specific glutamate in insects mimics processes known in the mammalian CNS, including afferent signaling and memory; e.g., it appears to mediate central mechanisms that are associated with long-term olfactory memory in the honeybee (Maleszka et al., 2000), and to mediate N-methyl-D-aspartate (NMDA) type glutamate receptor-coupled regulatory pathways in the biosynthesis of juvenile hormone (Chiang et al., 2002). Moreover, little is known about the physiology of glutamatergic transmission and glutamate/aspartate excitotoxicity in insects.

6.3.3.2. **Insect EAATs** Transporters of glutamate in the nervous system have recently been cloned and characterized from several insect species. The isolation of these genes was based on sequence similarities with vertebrate EAATs, using PCR with degenerate primers: sequence comparisons showed them to be members of this distinct family of transporters (dicarboxylate/amino acid:cation (Na^+ or H^+) symporter family by Saier, 2000b (TC 2.A.23); **Table 1**). This group corresponds to the mammalian solute ligand carrier group 1 (SLC1). Cloned insect EAATs comprise two from fruit flies (Seal et al., 1998; Besson et al., 1999), one from caterpillars (Donly et al., 1997), one from cockroaches (Donly et al., 2000), one from yellow fever mosquitoes (Umesh et al., 2003), and one from honeybees (Kucharski et al., 2000). They share up to 45% protein sequence identity and common secondary structures, which are characterized by six putative alpha-helical transmembrane domains and 2–4 beta-pleated domains resembling the beta-barrel structures of pore-forming ion channels (Seal et al., 1998). Demonstrated substrates for these insect EAATs include glutamate and also aspartate, which may sometimes act as a cotransmitter with glutamate. Biophysical studies with vertebrate transporters have shown that the transport of these substrates is electrogenic, involving the uptake of one amino acid$^-$ anion coupled to the cotransport of three Na$^+$ cations, one H$^+$ cation, and the countertransport of one K$^+$ cation (net inward flux of two positive charges per cycle). A novel property of vertebrate EAATs is the possession of a substrate-gated anion channel activity, as suggested by the presence of channel-like structures in the C-terminal portion of the protein. *Drosophila* EAAT1 protein expressed in frog oocytes induces a chloride flux that is consistent with this postulated channel-like activity (Seal *et al.*, 1998).

In situ hybridization detected the expression of both dmEAAT1 and dmEAAT2 in neuronal cells, however with distinct patterns for each transcript (Besson *et al.*, 1999). Similar extensive expression of EAAT has been detected in adult brain of the honeybee *Apis mellifera*, where it was associated with the mushroom bodies in which glial cells are rare (Kucharski *et al.*, 2000). In contrast, *in situ* hybridization, using *Trichoplusia ni* EAAT cDNA probes on whole mount caterpillar ganglia, identified a population of interfacial glial cells (Caveney and Donly, 2002). An immunohistochemical analysis on EAAT localization using antipeptide antibodies raised against the *T. ni* EAAT amino acid sequence produced selective staining in the ganglionic neuropil, as well as extensions of glial sheath cells wrapped around nerve processes that innervate skeletal muscles (Gardiner *et al.*, 2002). In contrast, an antipeptide antibody raised against the *A. aegypti* EAAT stained only the thoracic ganglia and not the muscles (Umesh *et al.*, 2003).

Comparisons between insect and vertebrate EAATs, based on structure, pharmacology, or expression, have failed to indicate clearly to which classes of vertebrate EAATs the insect transporters belong (Donly *et al.*, 1997; Seal *et al.*, 1998; Kucharski *et al.*, 2000). Most unexpected is the strong preference of fruit fly EAAT2 for aspartate as a substrate (Besson *et al.*, 2000). Sequence comparison among insect EAATs (**Figure 3**), reveals a cluster of putative aspartate transporters with similarity to the fruit fly EAAT2. Among the remaining glutamate transporters in the tree, there is no pattern relating to localization of expression, as there are examples of expression in both glial cells and neurons (drmEAAT1 is neuronal whereas trnEAAT is glial). Several putative EAATs (also compared in **Figure 3**) are derived from the recently completed mosquito (*A. gambiae*) genome and an early

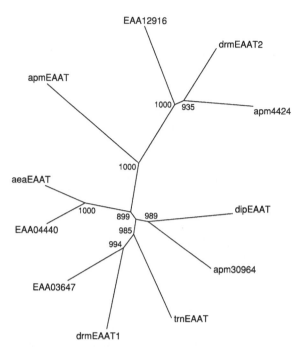

Figure 3 Dendrogram/bootstrap analysis of insect excitatory amino acid transporters. Sequence similarity was assessed using amino acid alignments in Clustal X 1.81 (Thompson *et al.*, 1997) and an unrooted tree was calculated by the "neighbor joining" method. Confidence values were derived by "bootstrapping" the dataset, using 1000 replicates and a generator seed value of 333 (Clustal X 1.81). The alignment was visualized using Treeview 1.6.6 (Page, 1996). The variable N- and C-terminal regions of the amino acid sequences were trimmed so that the region used for comparison extended from the first to the last predicted transmembrane domains. The accession numbers for the characterized transporters are found in **Table 1**. The three putative *Anopheles gambiae* EAAT sequences have accession numbers EAA12916, EAA04440, and EAA03647 as marked. The two putative honeybee EAAT sequences (in addition to the known apmEAAT) were deduced from contigs 30964 and 4424 from the initial assembly of the *A. mellifera* genome (Baylor Human Genome Sequencing Center).

assembly of the honeybee (*Ap. mellifera*) genome. Both genomes contain evidence for at least three potential EAATs. In contrast, the fruit fly genome contains only two EAATs. Based on this difference in EAAT complement between two related dipteran species, it seems possible that significant variability in EAAT complements may exist across the breadth of the Insecta.

6.3.4. Na$^+$/Cl$^-$-Driven Neurotransmitter Symporters (SNF; SLC6)

The essential role of neurotransmitter uptake in synaptic transmission and integrative nerve function was established long before the cloning of the first transporter (Levi and Raiteri, 1974). The existence of a family of plasma membrane transporters that is responsible for Na$^+$ and Cl$^-$-dependent transport in the nervous system first became apparent when similarity was observed between two mammalian transporter cDNAs that were isolated based on protein purification and sequencing (Guastella et al., 1990; Liu et al., 1992a) and on expression cloning (Pacholczyk et al., 1991). The similarity between these two transporters, one for GABA and the other for norepinephrine, was subsequently used to isolate a variety of other transporters in many organisms (including insects), which transport diverse substrates in the nervous system and other tissues. These transporters make up the Sodium Neurotransmitter Symporter family (SNF), which corresponds to the mammalian solute carrier group, SLC6 (TC 2.A.22; **Table 1**). All such transporters possess common features of structure and function, but nevertheless exhibit remarkable differences in their interaction with transmembrane electrochemical gradients for Na^+, K^+, and Cl^- as well as substrate selectivity, affinity, and carrier capacity. Protein sequence hydrophobicity and conformation profiles predict quite similar 12-TMD structures for these transporters, with cytosolic N- and C-terminals and an extended extracellular loop between TMDs 3 and 4 having a variable number of N-glycosylation sites (see Section 6.8.5).

Sequence comparison among vertebrate neurotransmitter transporters reveals five orthologous clusters among family members that can be defined according to substrate specificity as: (1) inhibitory neurotransmitter transporters for GABA and glycine; (2) aminergic neurotransmitter transporters for norepinephrine, serotonin, and dopamine; (3) the osmolyte transporters for betaine and taurine; (4) metabolic intermediate creatine transporters; and (5) amino acid transporters for proline and glycine (B^{0+} system). In addition, mammalian members of this family include several groups of "orphan" transporters for which no substrates and functions are yet known (Chen et al., 2003; Lill and Nelson, 1998). Comparison of the amino acid sequences of insect examples also reveals orthologous clustering, however differences from the vertebrate results are evident (**Figures 4, 6,** and **10**). Phylogenetic analyses show very clear differences between two large clusters of insect SNF members (**Figures 6** and **10**), which can be defined as Neurotransmitter Transporters (NTTs) and nutrient amino acid transporters (NATs) based also on their functional characteristics, which have been determined for many cloned insect transporters (see discussion below and Section 6.8.5.2, respectively). Due to their clear functional and phylogenetic specificity, these clusters represent a first and most significant divergence within the insect SNF; these clusters can reasonably be defined as NTT and NAT subfamilies. In addition, most insect NTTs form orthologous pairings with mammalian complements mentioned above, as well as sharing substrate and physiological roles in neuronal signaling, each of which has been confirmed for at least one cloned and characterized insect transporter. Despite their phylogenetic proximity to B^{0+} transporters, insect NATs encompass no substantial correlations in ligand specificity to such mammalian transporters. In contrast, they represent a large and lineage-specific (paralogous) group of NATs, which apparently are absent in the mammals and some other lineages (see discussion in Section 6.8.5.2).

6.3.4.1. GABA transporters   As in the vertebrate CNS, GABA serves as the primary inhibitory neurotransmitter in the insect nervous system (Callec, 1985). Immunocytochemical studies of GABAergic transmission sites using antibodies against GABA, glutamic acid decarboxylase (the essential enzyme for GABA synthesis), and various GABA receptor subunits have revealed an extensive pattern of GABA activity in the nervous systems of a variety of insect species. This pattern implies roles for GABA in diverse processes including visual and olfactory processing, as well as learning and memory (reviews: in Caveney and Donly, 2002; Umesh and Gill, 2002). The main mechanism by which GABA activity is terminated on the postsynaptic membrane is through reuptake into neurons and glial cells by high-affinity transporters (Callec, 1985). High-affinity insect GABA transporters (GATs) have been cloned and characterized from the hawkmoth, *Manduca sexta* (Mbungu et al., 1995) and the cabbage looper, *T. ni* (Gao et al., 1999). In each species, Northern analysis was used to localize expression of a single observed mRNA transcript to the nervous system. The hawkmoth masGAT protein has also been further localized by immunocytochemistry (Umesh and Gill, 2002). The results indicated that transporter immunoreactivity coincides closely with previously determined patterns of GABAergic activity, including staining of brain areas involved in processing visual and olfactory information.

The two characterized lepidopteran GATs show ionic stoichiometry and kinetics similar to those of the vertebrate GAT1 subtype (Mbungu et al., 1995; Gao et al., 1999). However, the pharmacological profiles of the lepidopteran transporters are distinct from any known vertebrate subtype, particularly in their insensitivity to several nipecotic acid derivatives, which are potent inhibitors of GAT1. Only one GAT has been found in each moth species; in contrast, vertebrates exhibit diversity at two levels. First, the vertebrate GATs, which in the rat are enumerated as GATs 1–3, display unique cellular expression patterns and affinities for GABA. Second, various related transporters have been identified in vertebrates, showing affinities for other substrates including taurine, creatine, and betaine. The presence of multiple GATs has been reported in *Drosophila* (Neckameyer, 1998), which is in contrast to the apparently unified phenotype studied in two lepidopteran models (Mbungu *et al.*, 1995; Gao *et al.*, 1999). Analysis of the fruit fly and mosquito genomes shows that each contains only a single GAT locus (at least of the known Na$^+$- and Cl$^-$-dependent type). Consequently, any diversity in GAT transcripts or proteins must result from alternative splicing of a single gene. In addition, no genes or transcripts that appear to encode taurine or creatine transporters have been identified in insects to date.

6.3.4.2. Monoamine transporters

Monoamines that serve as neurotransmitters in insects include serotonin (5-HT), dopamine, octopamine, and histamine (Monastirioti, 1999). The neuroactive amines in insects differ from those of vertebrates in that octopamine appears to substitute functionally for norepinephrine in insects. Tyramine, the immediate precursor of octopamine, may also have a role in neurotransmission but its status has remained controversial. Recently, tyramine was shown to be the effective ligand of the fruit fly amine receptor (TyrR) in modifying larval neuromuscular junction potentials, producing new evidence for a role for tyramine as a neuromodulator (Nagaya *et al.*, 2002). Monoamines play diverse roles in insect behavior and physiology, acting in the nervous system as well as in many peripheral tissues. In the nervous system, serotonin, dopamine, and octopamine are slow-acting neurotransmitters that bind to G protein-coupled receptors and modulate various second messengers, such as cyclic nucleotides and Ca^{2+}. After being released, they are taken up by high-affinity transporters for reuse or inactivation by N-acetylation. Transporters for all three of these monoamines are present in insects, clustering according to their transport substrate in sequence comparisons of insect Na$^+$/ Cl$^-$-dependent transporters (**Figures 4, 6,** and **10**).

6.3.4.2.1. Serotonin transporter (SERT; 5HTT)

Indolamine uptake activity in the insect nervous system was first demonstrated in nerve fibers of the locust brain (Klemm and Schneider, 1975), followed by the finding of high-affinity serotonin

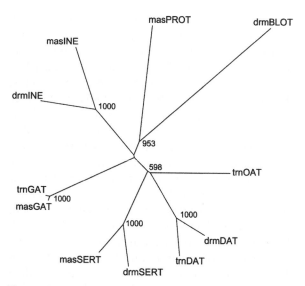

Figure 4 Dendrogram/bootstrap analysis of insect nervous system Na$^+$/Cl$^-$ dependent neurotransmitter transporters. The sequences used were limited to those for which there is a characterized insect example published (accession numbers are listed in **Table 1**). The comparison was performed as in **Figure 3**, using amino acid sequences trimmed to the first and last predicted transmembrane domains.

transport activity in many other insect species. In insects serotonin is also involved in neuromuscular transmission (McDonald, 1975) and epithelial transport processes, such as fluid balance (Maddrell, 1962; O'Donnell and Maddrell, 1984; Brown and Nestler, 1985), as well as in modulation of acid/base balance and midgut alkalinization (D. Boudko, unpublished data). It also serves in the integration of complex behavior patterns in insects, including aggression and defense (Baier *et al.*, 2002; Kostowski *et al.*, 1975), feeding (Novak and Rowley, 1994) and mating arousal (Yellman *et al.*, 1997), circadian rhythms (Muszynska-Pytel and Cymboroski, 1978a; Muzynska-Pytel and Cymborowski, 1978b), and memory (Carew, 1996). The first insect serotonin transporter cDNA (dSERT) to be cloned and characterized was that from *D. melanogaster* (Corey *et al.*, 1994; Demchyshyn *et al.*, 1994). The fruit fly transporter has pharmacological properties that are distinct from those of mammalian SERTs, such as a much lower sensitivity to antidepressant antagonists. A SERT that was recently cloned and characterized from the hawkmoth, *M. sexta*, was less sensitive to cocaine binding than other SERTs (Sandhu *et al.*, 2002b). Structural modification of the cocaine-sensitive human SERT, by using chimeras that mimic fragments of the insect molecule, revealed important elements involved in the interaction with cocaine (Sandhu *et al.*, 2002b). This finding is relevant for human health and important for understanding serotonin transporter biology; it also illustrates the value of arthropod models in uncovering structure–function relationships of complex transport proteins through comparative studies. Serotonin transporters are represented by a single gene in the *D. melanogaster* genome and apparently in the *An. gambiae* genome as well. Two conceptual translation entries seen in NCBI (EAA05837 and EAA03384) are identical by nucleotide sequence and have yet to be mapped, which together suggest that they may be fragments of a single gene copy (**Figure 6**).

6.3.4.2.2. Dopamine transporter (DAT)

Dopamine is the most abundant catecholamine in the insect nervous system, with dopamine-containing neurons being widely distributed throughout a variety of species (Osborne, 1996). In contrast, only trace quantities of the vertebrate catecholamine transmitter, norepinephrine, can be detected in the insect nervous system. Dopaminergic neurons modulate both visceral and skeletal muscle activity, with dopamine affecting flight motor patterns in *M. sexta* and the escape circuit in cockroaches (Osborne, 1996). Dopamine release from peripheral nerve endings also stimulates salivary gland secretions in many species. The first insect dopamine transporter (DAT) was cloned and characterized from the fruit fly (Porzgen *et al.*, 2001). This transporter exhibited substrate selectivity similar to vertebrate DATs (dopamine and tyramine are preferred substrates), but showed a distinct pharmacological profile that is reminiscent of mammalian norepinephrine transporters (with high affinities for tricyclic antidepressants). A second DAT, isolated from the cabbage looper moth, has similar properties (Gallant *et al.*, 2003). These transporters cluster closely, along with another putative DAT sequence that is present in the mosquito genome (**Figures 4, 6, and 10**). It has been proposed that these invertebrate DATs represent conserved copies of a primordial catecholamine transporter that existed before the divergence of mammalian catecholamine carriers into subtypes responsible for dopamine and norepinephrine uptake (Porzgen *et al.*, 2001).

6.3.4.2.3. Octopamine transporter (OAT)

Octopamine is the monohydroxylic analog of norepinephrine in insects. It is present in high concentrations in a range of insect tissues including the nervous system, where it may act as a neurotransmitter, neurohormone, and neuromodulator (Roeder, 1999). It modulates afferent and efferent signaling in locusts (Homberg, 2002; Leitch *et al.*, 2003). Na^+-dependent (and independent) octopamine

uptake has been reported in various cockroach tissues (Sinakevitch *et al.*, 1994) and firefly lantern (Robertson and Carlson, 1976; Copeland and Robertson, 1982). In the honeybee, it mediates consolidation of olfactory memory (Farooqui *et al.*, 2003) and hygienic behavior (Spivak *et al.*, 2003) as well as complex social behavior, including labor establishment (Schulz *et al.*, 2002). Despite abundant neurotransmitter roles for octopamine, examination of the completed dipteran genomes reveals no alternative genes that may encode a Na^+/Cl^--dependent transporter for octopamine uptake. In these insects, it has been proposed that octopamine may be dispersed by an alternate mechanism, or perhaps taken up by a member of an unrelated transporter family (Porzgen *et al.*, 2001). Nevertheless, a novel transporter cDNA that was isolated and cloned from the lepidopteran *T. ni*, using RT-PCR with degenerate primers, showed remarkable sequence similarity to DAT transporters (Malutan *et al.*, 2002). However, this transporter is distinct from known DATs in that it has a high affinity for octopamine along with tyramine and dopamine; in addition it is expressed in a different subset of neuronal cells in the caterpillar nervous system than DAT. This octopamine transporter (OAT) showed relaxed ionic dependency relative to other insect monoamine transporters and was generally resistant to pharmacological inhibitors of monoamine transporters. Comparison of the OAT sequence with other Na^+/Cl^--dependent transporters shows that it clusters with mammalian norepinephrine transporters but not other insect transporters, including those for monoamines (**Figures 6 and 10**). OAT emerges almost equidistant from the serotonin and dopamine transporter branches in a high-resolution dendrogram (**Figure 4**). The extent to which this novel type of transporter is distributed among insect orders remains to be determined.

Finally, most of the histamine in the insect nervous system is in the retina and optic lamina, where it functions as an inhibitory neurotransmitter (Stuart, 1999). However, searches for a histamine transporter based on expected homology to other monoamine transporters have detected no putative candidates. This failure may indicate that the histamine transporter is a member of a novel or unrelated transporter family, as in the case of the delayed discovery of transporters for choline in the nervous system (see Section 6.3.5). Alternatively, functional screening may yet identify a histamine transporter.

6.3.4.3. Glycine and proline transporters

In vertebrates, a distinct subfamily of Na^+/Cl^--dependent NTTs transports the amino acids glycine (GLYT) and proline (PROT) (Deken *et al.*, 2002). Multiple subtypes of transporters for the inhibitory amino acid glycine are expressed both in the CNS and in peripheral tissues. Although arthropods apparently lack receptors for glycine (Witte *et al.*, 2002), sequence comparison with vertebrate GLYTs reveals potential Gly type transporters in both the *D. melanogaster* and *An. gambiae* genomes (CG5549 or NP_611836 and EAA07088, respectively). Vertebrates have only a single high-affinity proline transporter type (it is brain specific), which is thought to be associated with modulating activity at glutamatergic synapses (Deken *et al.*, 2002). In insects, proline is an alternative energy source and major osmolyte, but its role in the nervous system remains unproven. Nevertheless, a Na^+/Cl^--dependent proline transporter was recently cloned from *M. Sexta*, where it is predominantly expressed in the nervous system (Sandhu *et al.*, 2002a). This protein transported proline exclusively when heterologously expressed in *Xenopus* oocytes. Sequence comparison shows that the hawkmoth PROT is related to known vertebrate glycine and proline transporters (Sandhu *et al.*, 2002a). The *D. melanogaster* and *An. gambiae* genomes also each contain a related sequence (CG7075 or NP_723303 and EAA12299, respectively) representing putative orthologs. Moreover, two gene products that apparently represent alternative splicoforms of a single gene in the 2L chromosome are present in *D. melanogaster* expressed sequence tags (ESTs for NP_723303 and NP_609135, **Figure 6**).

6.3.4.4. Orphan inebriated (ine) and bloated tubules (blot) transporters

The identification of Na^+/Cl^--dependent transporter family members based on structural similarity has resulted in the discovery of many putative transporters for which the substrates are not known. Many of these so-called "orphan" transporters are larger than other Na^+/Cl^--dependent transporters, with extended fourth and sixth extracellular loops and a larger C-terminus. The expression patterns of these orphan transporters are often diverse, with extensive nervous system and peripheral distributions. Insect examples include the *Drosophila* blot (bloated tubules) and *ine* (inebriated) proteins, and the *M. sexta ine* protein (ms*ine*). The "inebriated" locus, relating to a characteristic "inebriated" behavior phenotype (Stern and Ganetzky, 1992) was initially cloned from *D. melanogaster* and its expression was mapped by *in situ* hybridization in the posterior hindgut, Malpighian tubules, anal plate, garland cells, and a subset of cells in the CNS of developing embryos (Soehnge *et al.*, 1996). It has been proposed by

analogy with other NTTs that the neuronal hyperexcitability characteristic of the *ine* phenotype is a result of failure in some neurotransmitter uptake due to *ine* mutations (Soehnge *et al.*, 1996); however, the identity of any such ligands remains elusive. Additional transgenic characterization led to the hypothesis that a neuronal dm*ine* isoform mediates downregulation of sodium channels via a neurotransmitter uptake (Huang and Stern, 2002) whereas an epithelial dm*ine* isoform compensates osmotic load in the hindgut and Malpighian tubules via an osmolyte carrier (Huang *et al.*, 2002). Substantial progress in characterization of *ine* has been made by the heterologous expression of an ine-orthologous transporter from *M. sexta* in *Xenopus* oocytes (Chiu *et al.*, 2000). That study revealed that *ine* responds to hyper-osmotic stimuli via the inositol triphosphate-coupled release of intracellular Ca^{2+}, which subsequently activates Ca^{2+}-dependent K^+ channels. The phylogenetic analysis indicates a single *ine* gene in *An. gambiae* and two splicoforms of a single *ine* gene in *D. melanogaster* (CG15444, located on the L2 chromosome). The insect *ine* group, with undefined ligand and transient functions, represents an evolutionary root of the NTTs and apparently provides primordial genetic sources in the differentiation of GLYT and GAT and monoamine neurotransmitter transporters. Therefore, the reasons why substrates for these putative transporters have remained elusive may not only be technical, but biological as well. The primary function of some of these proteins may be to regulate transmembrane ion fluxes in different tissues including the CNS.

Another orphan gene, named by similarity with a mammalian orphan group, has been extensively analyzed in *Drosophila* (Johnson *et al.*, 1999). This gene, named bloated tubules (*blot*), exhibits complex and dynamic expression patterns during embryogenesis and supports processes that are essential for cell division, differentiation, and epithelial morphogenesis. Insect *ine* and *blot* do not cluster in high-resolution phylogenetic analyses. Nevertheless, their broader phylogenetic proximity with other uncharacterized "orphan type" sequences, as well as the related nutrient-type transporters, msKAAT1 and ms CAATCH1, from *M. sexta* and aeAAT1 and agAAT8 from mosquito (**Figures 6 and 10**; Section 6.8.5.2) indicates that several discrete functional groups may eventually emerge from under the "orphan" umbrella. The analysis showed five putative orphan genes in *An. gambiae* and four genes encoding seven presently known isoforms in *D. melanogaster*, which are in phylogenetic proximity with *blot* and represent a transition between the SNF

(Na^+/Cl^--dependent transporter) and PAT (proton amino acid transporter) families (**Figure 6**).

6.3.5. Choline Transporters (CHTs)

Acetylcholine (ACh) is the principal excitatory neurotransmitter in the insect CNS, although it has not been shown to have any role in transmission at the insect neuromuscular junction (Callec, 1985). ACh released from neurons activates postsynaptic receptors of both the ionotropic and metabotropic type and is then eliminated exclusively by enzymatic and nonenzymatic degradation, rather than a reuptake mechanism, as with other transmitters. Extracellular cholinesterases, widely exploited as insecticide targets, catalyze hydrolysis of ACh to choline and acetate; then specific transport systems remove the choline from the synaptic cleft via neuronal high-affinity, Na^+-driven reuptake. The resulting neuronal choline uptake represents a rate-limiting initial step in the ACh synthesis pathway in insects (Breer and Knipper, 1990).

An insect choline transporter (CHT) was isolated first from locust synaptosomes, using monoclonal antibodies that specifically blocked choline uptake *in vivo* (Knipper *et al.*, 1989). The transporter was shown to be sensitive to the choline uptake inhibitor hemicholinium-3 and was localized by immunohistochemistry in ganglionic neuropils in the locust CNS. Initial attempts to identify CHT cDNAs in mammals using homology-based PCR were unsuccessful until 2000 when a first draft of the *C. elegans* genomic project became available. Using this information Okuda and colleagues cloned a nematode CHT (cho-1) and almost immediately thereafter the high-affinity hemicholinium-3 sensitive rat CHT1 (Okuda and Haga, 2000). All currently known CHTs are characterized by a 13-transmembrane domain configuration and are members of the sodium ion-dependent glucose transporter family, which corresponds to mammalian SLC5, with a SLC5A7 membership tag assigned to CHTs. This group includes transporters for diverse substrates, including glucose and galactose (urea and water), I^- (ClO_4^-, SCN^-, NO_3^-, Br^-), biotin, lipoate and pantothenate, as well as choline. In addition, SLC5 members may act as glucose *myo*-inositol (glucose) Na^+ (H^+) channels. This family of transporters is present in species from cyanobacteria to humans. They are placed in the Solute:Sodium Symporter (SSS) family (Saier, 2000b; TC 2.A.21), which comprises one of the largest groups of solute carriers in mammals (SLC5; **Table 2**). A putative insect CHT recently cloned from the cabbage looper (NCBI; trnCHT: AY629593; see **Table 1**) shows

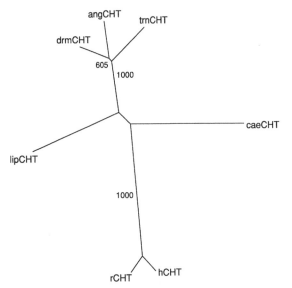

Figure 5 Dendrogram/bootstrap analysis of the amino acid sequences of insect choline transporters. Putative insect CHT sequences are included in the analysis, resulting in a discrete clade separate from the other invertebrate CHTs. The comparison was performed as in **Figure 3**, using amino acid sequences trimmed to the first and last predicted transmembrane domains. Accession numbers for the sequences are: human (hCHT; NP_068587), rat (rCHT; NP_445973), horseshoe crab (lipCHT; AAG41055), nematode (caeCHT; NP_502539), fruit fly (drmCHT; NP_650743), mosquito (angCHT; EAA07459), and cabbage looper (trnCHT; AY629593).

strong similarity to the known CHTs, and forms an orthologous cluster with putative CHT sequences from the fruit fly and mosquito genomes (**Figure 5**). The pharmacological effects of blocking choline uptake by this CHT may be different from blocking other neurotransmitter transporters since ACh synthesis should be suppressed without transmitter concentration building up in the synaptic space. It remains to be seen whether CHT may serve as a useful target for insecticide activity.

6.3.6. Vesicular Membrane Transporters

Neurotransmitters are taken up from the synaptic cleft by plasma membrane transporters and are sequestered in synaptic vesicles, thereby protecting the cell from potential toxicity and protecting the transmitters from intraneuronal metabolism (Masson *et al.*, 1999). Low-affinity transporters, which are located in the vesicular membrane, transfer the transmitters into the vesicle lumen. Accumulation of transmitter in vesicles also facilitates uptake of transmitter into the cell by plasma membrane transporters, since it lowers the concentration gradient across the neuronal membrane. Vesicular transport can employ H^+ and ligand-specific electrochemical motive forces that are induced by the activity of H^+ V-ATPases located on the vesicular membrane. Typically, such transporters exchange one or more protons inside the vesicle with a neurotransmitter molecule in the cytoplasm. There is less molecular information on these transporters than on their plasma membrane counterparts; however, it has become clear from sequence comparisons that the vesicular transporters characterized to date fall into three separate families; these are discussed immediately below.

6.3.6.1. Vesicular inhibitory amino acid transporters (vIAATs or vGATs)

A vesicular GABA transporter was initially identified in *C. elegans* as a product of the *unc-47* gene that mimics a GABA-defective behavior phenotype (McIntire *et al.*, 1993). This gene encodes a 10-TMD protein that colocalizes with synaptic vesicles of GABAergic neuronal terminals, where it mediates low-affinity vesicular accumulation of inhibitory neurotransmitters via an H^+/GABA(Gly) antiporter mechanism ($K_m^{GABA(Gly)} = 5(25)$ mM; McIntire *et al.*, 1997). Subsequently, when mammalian orthologs were cloned it was recognized that glycine may also be a physiological substrate, leading to the term vesicular inhibitory amino acid transporter (vIAAT, more commonly vGAT). These transporters are distantly related to other transporters that resemble SLC36 and SLC38 in mammals and bear some resemblance to a group of permeases that place them in the Amino Acid/Auxin Permease (AAAP) family of transporters (TC 2.A.18) in the TC classification scheme (Saier, 2000b; see also phylogenetic analysis in Section 6.8.1). Functional characterization of rodent orthologous transporters revealed that they contribute to GABA- and glycine-ergic signaling and glycine-dependent vesicular accumulation of GABA (Sagne *et al.*, 1997). However, no insect vIAAT has been characterized as yet.

6.3.6.2. Vesicular monoamine and acetylcholine transporters (vMATs and vAChTs)

The first recognized and most thoroughly studied vesicular transporters make up the Vesicular Neurotransmitter Transporter (VNT) family (TC 2.A.1.22), a group classified within the Major Facilitator superfamily (Saier, 2000b). They include the vesicular transporters for monoamines (vMATs) and for acetylcholine (vAChT). These proteins have 12 predicted TMDs, which differ from the vIAATs (above) whose sequences predict only 10 (Masson *et al.*, 1999). Many vMATs and vAChT related transporters have been cloned from a variety of organisms, but the only insect vesicular transporter that has been

studied to date is the *D. melanogaster* vAChT (Kitamoto *et al.*, 1998; Kitamoto *et al.*, 2000a; Yasuyama *et al.*, 2002). Additional data and analysis of phylogenetic proximity of dipteran vMATs to *H. sapiens* vMATs are included in Sections 6.5 and 6.9.1. Like the gene for nematode and vertebrate transporters, the gene for the fruit fly vAChT nests completely within the first intron of the choline acetyl transferase gene, encoding the enzyme that synthesizes ACh (Kitamoto *et al.*, 1998). Mutation analysis of the fly vAChT gene demonstrated that the function of the transporter is essential for viability, because mutants are recessive lethal (Kitamoto *et al.*, 2000b). VAChT is involved in cholinergic transmission since decreased activity in heterozygous mutants resulted in defects in the giant fiber pathway, a neural circuit known to contain cholinergic synapses. Functional studies of the fruit fly vAChT in an isolated cell system should yield information on its kinetics.

6.3.6.3. Vesicular glutamate transporters (vGLUTs)

Vesicular transporters for glutamate were only recently recognized when a human brain-specific inorganic phosphate transporter (BNPI) was found to also transport glutamate (Bellocchio *et al.*, 2000). Three subtypes of mammalian vesicular glutamate transporter are now known (SLC17; see Section 6.9.4), but no insect examples have yet been studied. Based upon their structure, these transporters also belong to the Major Facilitator superfamily (see above), but within a different family, the Anion:Cation Symporter (ACS) family (TC 2.A1.14; **Table 1**).

6.4. Nutrient Amino Acid Transport in Insects

6.4.1. Early Studies of Insect Amino Acid Uptake

Clements (1992) reviewed the work on amino acid uptake in Malpighian tubules of living insects by Wigglesworth, Ramsay, Phillips, and others. Giordana *et al.* (1989), Castagna *et al.* (1997), and Wolfersberger (2000) reviewed the physiology and biochemistry of amino acid uptake in isolated midguts. In the early 1970s, Signe Nedergaard (1972) reported the transport of alpha-aminoisobutyric acid from lumen to blood-side of the isolated caterpillar midgut by a voltage-dependent process that requires oxygen and K^+ but is independent of active K^+ transport. Studying brush border membrane vesicles from caterpillar midgut, Hanozet, Giordana, and Sacchi (Hanozet *et al.*, 1980) showed that phenylalanine is cotransported with K^+ by a voltage-dependent process. Since the K^+ pump is electrogenic (Harvey et al., 1968) and oxygen dependent (Harvey et al., 1967), it appeared that it uses energy from ATP hydrolysis to charge the apical plasma membrane (lumen positive to cell) and that the voltage drives a positively charged K^+ amino acid complex across the membrane, carrying both solutes into the cell. The oxygen dependence arises because the midgut has little anaerobic reserve, ATP levels dropping as oxygen pressure drops (Mandel et al., 1980). After the K^+ pump-containing, goblet cell apical membrane was isolated (Cioffi and Wolfersberger, 1983; Harvey et al., 1983), the K^+ pump was shown to consist of an H^+ translocating ATPase (Wieczorek et al., 1989) that charges the membrane and is thought to drive secondary $K^+/2H^+$ exchange (Wieczorek et al., 1991). The membrane charging by the V-ATPase has been confirmed many times (Wieczorek et al., 1999b) but the nature of the secondary K^+/H^+ exchange remains unresolved. The working hypothesis for amino acid uptake is that the H^+ V-ATPase charges the apical membrane (lumen positive to cell) and that the voltage drives amino acid:K^+ cotransport from lumen to cell.

6.4.2. Amino Acid Uptake by Brush Border Membrane Vesicles

Several major amino acid transport systems have been characterized in brush border membrane vesicles (BBMV) from lepidopteran midgut. Uptake has been studied in *Philosamia cynthia* (Giordana and Parenti, 1994), *M. sexta* (Hennigan et al., 1993a, 1993b; Reuveni and Dunn, 1994; Bader et al., 1995; Liu and Harvey, 1996a, 1996b), *Bombyx mori* (Parenti et al., 2000; Leonardi et al., 2001a, 2001b) and *Pieris brassicae* (Reuveni and Dunn, 1994; Giordana et al., 1989; Wolfersberger, 2000). Facilitative transport systems are widespread (Wolfersberger, 2000) along with separate K^+- or Na^+-dependent neutral and cationic amino acid transporters, which are driven mainly by the membrane potential (Giordana and Parenti, 1994; Castagna et al., 1997). Some evidence suggests that leucine uptake is partially modulated by a hormonal and nutrient-related mechanism (Leonardi et al., 2001b). Nutrient glutamate transport appears to be anomalous (Xie et al., 1994). In summary, four principal transport systems were characterized: (1) a broad spectrum neutral amino acid:K^+ cotransport system; (2) an arginine K^+ cotransport system; (3) an acidic amino acid system with anomalous properties; and (4) a glycine or proline transporter. Extensive kinetics studies on BBMV showed that the transport systems had low affinity for amino acids (K_m around 1 mM) and K^+

(K_m around 30 mM). Presteady state kinetic studies on BBMV suggested a symmetric on/off mechanism (Parenti et al., 1992). The kinetic studies also suggested that there might be more than one transporter for each amino acid and that several amino acids use the same transporter. Hence, the kinetic parameters from BBMV studies represent average values rather than values from single transporters.

Surprisingly, there are no available references on amino acid uptake by isolated brush border membrane vesicles from larval midgut of mosquitoes. There is a single report of uptake of radiolabeled amino acids in intact midgut of adult *Anopheles stephensi* (Schneider et al., 1986). There is a published method for vesicle preparation from midgut preparations, but it was used only for *Bacillus sphaericus* binding studies (Nielsen-LeRoux and Charles, 1992). Therefore, most existing insect vesicle transport data are from caterpillars, taking advantage of their large size in preparation of large tissue samples. Nevertheless, dipteran larvae of *An. gambiae* and *D. melanogaster* serve in the postgenomic era as more convenient and comprehensive models for molecular, cellular, and integrative studies of transport biology, including explorations of amino acid transport metabolons. BBMV studies on caterpillars confirmed voltage-driven, amino acid transport.

6.5. Postgenomic Analysis of Insect AATs

In the postgenomic era, there is a rapid advance toward the identification of virtually all transport proteins in a growing number of genomic models. This advance facilitates the accumulation and systematization of molecular and physiological data, which makes it possible to deduce the function of particular transporters by their phylogenetic proximity, which in turn enables one to identify new species and groups of transporters with previously unknown functions. This review takes advantage of this new opportunity. A matrix of dipteran transporters, together with reference sequences of characterized amino acid transporters, was used to establish phylogenetic reciprocity within comprehensive genomic populations of insect amino acid transporters. The resulting phylogram (**Figure 6**) incorporates the best reciprocal matches from predicted gene products of *D. melanogaster* (Adams et al., 2000) and *An. gambiae* (Holt et al., 2002), along with characterized metazoan amino acid transporters, which are deposited in the National Center for Bioinformatics (NCBI) databases. The properties of insect amino acid transporters will be discussed in the framework of the classical divisions

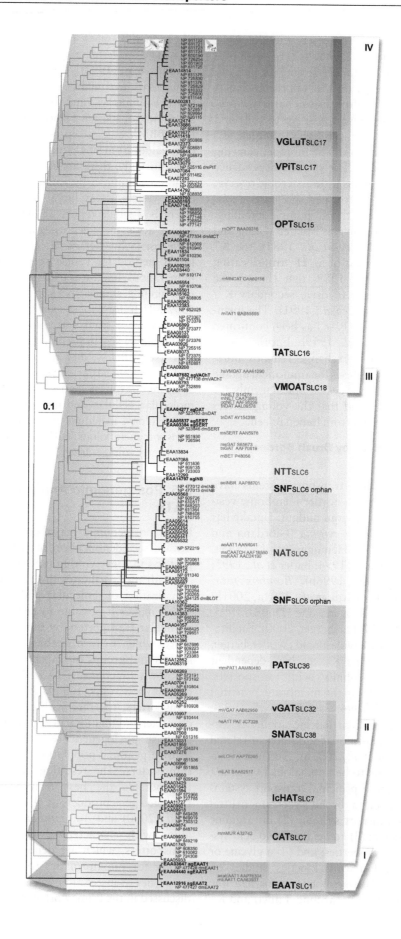

of neurotransmitter phenotypes and a novel division of nutrient amino acid phenotypes. Then this description is integrated into a new scheme that discloses phylogenetic relationships of such transporters in the order in which they are self-clustered in the low-resolution, unrooted phylogram (**Figure 6**), which was constructed by using ClustalX software for progressive multiple sequence alignment and comparison (Thompson *et al.*, 1997). Since there are an indefinite number of phenotypic transitions and putative substrates within several phylogenetic groups, an entire group of amino acid transporters was included if at least one of its members, or its reciprocal sequence homolog, serves as a carrier of amino acids or related substrates. The final phylogenetic tree (**Figure 6**) is a product of multiple iterations of comprehensive subsets of sequences. It includes 85 gene products of *An. gambiae* (blue font color, proximal column) and 108 gene products of *D. melanogaster*, which are encoded by 88 genes, but some are represented by several established splicoforms (red font color, middle column). Included also are 25 transporters, which were cloned and characterized from other insect species (green font color, distal column). Several *An. gambiae* gene products, which are definitely orthologs of cloned and characterized transporters, are indicated in bold. Six families that differ substantially in their electrochemical transport mechanisms are represented by different background colors. In addition to the integrated low-resolution tree in **Figure 6**, analyses of phylogenetic proximity of mammalian and dipteran transporters for each particular family are included, which are shown here as ClustalX dendrograms, phylograms, and/or cladograms, depending upon the view that is best suited for representing peculiarities within a particular group. Due to the lack of clear consensus in transport protein classification, we will follow the most natural solute carrier nomenclature (SLC), which rests on a comprehensive subset of characterized *H. sapiens* and other mammalian transporters. Four large clusters that exhibit phylogenetic proximity between different families of insect amino acid transporters are introduced in **Figure 6** as panels with a Roman integer.

Special emphasis is placed on the insect Nutrient Amino Acid Transporter (NAT) subfamily which is introduced here for the first time. It resides within the SNF, and apparently represents a group of lineage-specific, essential nutrient amino acid transporters, which appear to be good targets for insect population control (see Section 6.4).

Figure 6 Comprehensive low-resolution phylogram (blue tree) and cladogram (black tree) of dipteran amino acid transporters. The unrooted phylogenetic trees were constructed using Clustal X software (Thompson *et al.*, 1997), which utilizes a neighbor-joining algorithm for tree reconstruction (Saitou and Nei, 1987), converted to Windows metafile by TreeView(Win32) (Page, 1996) and decorated with CorelDraw software. This phylogenetic analysis includes 193 dipteran sequences (Smith-Waterman all-against-all e-value cutoff $<10^{-5}$), which represent putative orthologs of 10 solute carrier families (SoLute Carrier numbers; SLC with Arabic integers) and 15 identified[1] phenotypes of amino acid transporters (large font on the transparent ribbon). Sequences are represented by nonredundant subsets of gene product accession numbers, which can be used to retrieve these sequences from NCBI protein databases (National Center for Biotechnology Information (**NCBI**)[2]). First row from the left (blue font) are 85 putative gene products identified in *An. gambiae*, each of which is represented by a single splicoform; second row (red font) 108 gene products found in *D. melanogaster*, which are encoded by 88 genes, but some are represented by several established splicoforms; third row (green font) represent 25 characterized transporters from various species that are used to reference tree clusters[3]. Common transporter phenotype abbreviations are used, e.g.: EAAT, excitatory amino acid transporter; CAT, cationic amino acid transporter; lcHAT, light chain subunit of heterodimeric amino acid transporter; SNAT, sodium neutral amino acid transporter; vGAT, vesicular GABA transporter; PAT, proton amino acid transporter; SNF, sodium neurotransmitter transporter family; NAT, nutrient amino acid transporter; NTT, neurotransmitter transporter; vMAT, vesicular monoamine transporter; TAT, T-system amino acid transporter; OPT, oligopeptide transporter; P$_i$T, inorganic phospate (P$_i$) transporter; vGluT, vesicular glutamate transporter. Sequences from different lineages are shown in different font color, e.g: *Ae. aegypti*, green; *An. gambiae*, blue; *D. melanogaster*, red; *H. sapiens*, black. Accession numbers from first draft of *An. gambiae* genome annotation (EAA numbers) are duplicated by accession numbers for recently reviewed sequences (XP_#). Tree reconstruction was done from the unaltered alignment of sequences as they were deposited in the NCBI database, so be aware that some *An. gambiae* sequences may have incorrect composition of open reading frame as well as prediction of 3′ and 5′ ends, which could produce tree distortion. Scale bar are evolutionary distances in mutations per site. [All families in **Figure 6** are shown in amplified form in appropriate sections of this chapter.]

[1]Not all transport phenotypes are shown in this phylogram; see other figures for details.
[2]Low dash, which links NP and numerical integers, are missing due to file conversion and must be inserted to avoid accession errors, e.g. NP 477427 represents NP_477427. Accession numbers starting with EAA integer are from first draft of *An. gambiae* genome annotation; since a few of them will disappear during future revisions of genomic databases they are duplicated by new accession numbers started with XP_ in the high resolution phylogram (**Figures 8–10** and **13–17**).
[3]Each referencing sequence is represented by a transporter phenotype name following by an accession number, e.g., msCAATCH AAF18560 is *M. sexta* cation amino acid transporter channel; other species abbreviations here and on other figures are: ae, *Ae. aegypti*; dm, *D. melanogaster*; gg, *Gallus gallus*; hs, *Homo sapiens*; rn, *Rattus norvegicus*; mm, Mus musculus; tn, *Trichoplusia ni*.

6.6. AAT Cluster I (SLC1)

This type of transporter, abundant in metazoan lineages and present in bacteria and Archaea, is absent in plants and fungi (**Table 2**). Mammalian SLC1 transporters are extensively characterized, comprising two ASCT (alanine, serine, and cysteine transporters) and five EAAT genes in *Homo sapiens*. Phylogenetically, EAAT is a progeny of the ASCT group which together comprise only one SLC1 (solute carrier type 1) family in mammals. It is devoted to small nutrient and excitatory (glutamate and aspartate) amino acid transporters. Mammalian SLC1 genes encode two classically characterized carrier systems (**Table 2**). System ASC (ASCT1 and 2) serves as a facultative sodium ion driven substrate provider of glutamine, alanine, serine, cysteine, and asparagine in tissues with a low capacity or absence of anabolic pathways for the production of such amino acids (Arriza *et al.*, 1993; Utsunomiya-Tate *et al.*, 1996; Furuya and Watanabe, 2003; Wolfgang *et al.*, 2003) and in proliferating tissues, including tumors (Kekuda *et al.*, 1996; Li *et al.*, 2003). Intriguingly, some transporters of this group have been identified as retroviral receptors (Rasko *et al.*, 1999); for example, SLC5A1 (ASCT2) mammalian transporters are utilized as a surface binding receptor and an intracellular gateway by a diverse group of retroviruses (Marin *et al.*, 2003).

System $X^{A,G}$ (EAAT) maintains local glutamate/aspartate concentrations, including those in central excitatory insect synapses (Donly *et al.*, 2000), glial tissue (Kanai and Hediger, 1992; Kanai *et al.*, 1993; Kirschner *et al.*, 1994; Suchak *et al.*, 2003), visual and olfactory afferents (Arriza *et al.*, 1997; Besson *et al.*, 1999; Kucharski *et al.*, 2000), macrophages (Rimaniol *et al.*, 2000), and osteocytes (Huggett *et al.*, 2002). Mammalian EAAT genes appear to have diverged through adaptation for particular physiological conditions (Arriza *et al.*, 1994; Amara and Fontana, 2002) and specifically may expand their functions beyond the tuning of excitatory amino acids uptake at membranes that surround a synaptic cleft. For example, EAATs may serve as facultative providers of cysteine for glutathione synthesis (Chen and Swanson, 2003) or act as leak channels and/or ligand-gated chloride channels (Fairman *et al.*, 1995; Arriza *et al.*, 1997).

Apparently, the ASCT group is absent from dipteran genomes, which also appear to have few EAAT genes. EAAT comprises two established genes in *D. melanogaster* and three genes in *An. gambiae*, which split into two clusters in the dipterans (**Figure 6**), but have no obvious orthologs among the five mammalian transporters. A brief description of cloned EAATs from several insects is given above in Section 6.3.3.

6.7. AAT Cluster II (SLC7)

Like Cluster I, Cluster II corresponds to one solute carrier family in mammals (SLC7) that integrates two large functionally and structurally different groups of transporters, the cationic amino acid transporter (CAT) subfamily and the heterodimeric amino acid transporter (HAT) subfamily, based on phylogenetic, structural, and physiological properties (Verrey *et al.*, 2003). SLC7-related transporter heterogeneity has been partially explored in insects and one representative HAT-homologous sequence (AeaLAT) has recently been cloned from *Ae. aegypti* (Jin *et al.*, 2003). Phylogenetic analysis of insect transporters revealed three specific clusters. Two large and distal clusters apparently comprise insect versions of CAT and HAT subfamilies, respectively. The third small cluster is parental to the CAT/HAT cluster (**Figures 6** and **7**), which by its phylogenetic loci represents transporters similar to mammalian orphan SLC7A4, with an as yet indefinite function (Wolf *et al.*, 2002). In mammals SLC7 genes encode a broad variety of transport systems including sodium independent b^+, y^+, y^+L (for cationic AA) as well as $b^{0,+}$ and a sodium-dependent system $B^{0,+}$; y^+L (for neutral AAs) (Closs, 2002; Verrey *et al.*, 2003). Since very few insect transporters in this phylogenetic group are characterized, a brief description of characterized mammalian transporters is included below, along with a description of the phylogenetic pattern of related dipteran amino acid transporters (**Figure 7**).

6.7.1. Cationic Amino Acid Transporters (CATs)

The cationic amino acid transporter (CAT, SLC7A1–4) subfamily comprises 14 TMD transporters that are typically located on the plasma membrane (Wolf *et al.*, 2002). Most of these transporters regulate local concentrations of cationic amino acids through (1) facilitated diffusion and/or (2) differential amino acid exchange between relatively saturated physiological domains in organisms, e.g., growing and/or actively metabolizing cells, as well as nutrient and secretory epithelia (Verrey *et al.*, 2003). The first CAT was identified as the receptor for murine ecotropic leukaemia viruses in the mouse, MuLVR or mmCAT1, which mimics a y^+-transport system (sodium ion-independent L-arginine, L-lysine, and L-ornithine uptake) phenotype in *Xenopus* oocytes (Albritton *et al.*, 1989;

Table 2 Classification properties and taxonomic proximity of insect amino acid transporters (last five columns indicate taxonomic proximity by Blast LINK, NCBI; value is protein number, cutoff max 200; database 13.10.2003)

Type	Mammalian SLC members (protein name, respectively)	Insect protein	Transport systems (some aliases)	Substrate selectivity (weak substrate)	TMDs (other prediction)	Carrier type	Apparent affinity	Stoichiometry ↓absorption ↑emission (uncoupled)	Representative H. sapiens	Orthology (proximity) An. gambiae	Archaea	Bacteria	Metazoa	Fungi	Plant
Cluster I															
ASCT	SLC1A4,5 (ASCT1,2)	Absent	B0,ATB0,SA AT	A,S,C,(T,Q)	8(6,10)	S	L,M	↓Na^+Sb(Cl^-)	NP_003029	**Figure 6**	7	61	132		6
EAAT	SLC1A1,2 (EAAT3,2)	EAAT2 (dEAAT2)	X_{AG}^-	E,D(K_m^O<<K_m^E in dEAAT2)	8(6,10)	E	L,M	↓Na^+Sb↑Sb	NP_004161	[EAA12916]	5	62	132		17
	SLC1A3,6,7(EAAT1,4,5)	Absent	X_{AG}^-	E,D	8(6,10)	S	H	↓$3Na^+$Sb$^-H^+$↑K^+	NP_004163		7	58	135		21
		EAAT1(dEAAT1)	X_{AG}^-	E,D	8(6,10)	S	H	↓$3Na^+$Sb$^-H^+$↑K^+	**Figure 6**	EAA03647	7	64	129		19
		EAAT3(AeaEAAT)	X_{AG}^-	E,D	8(6,10)	S	H	↓2-$3Na^+$Sb$^-H^+$↑K^+		EAA04440	7	63	130		20
Cluster II															
CAT	Insect specific				14					EAA05933		13	39		25
	Insect specific				14					EAA01745		56	125		20
	SLC7A1(CAT1)		y^+	R,K,H	14	U,E	L,M	↓Sb; ↓Sb↑Sb^+	NP_003036	**Figure 7**		112	59		19
	SLC7A2(CAT2)		y^+	R,K,H	14	U	L	↓Sb	NP_003037			110	59		20
	SLC7A3(CAT3)		y^+	R,K	14	U	L	↓Sb	NP_116192			112	62		6
	SLC7A4(CAT4)		Orphan		14				NP_004164	EAA09935		113	52		7
					14					EAA09874		97	63		3
					14					EAA09921		106	65		6
					14					EAA09918		107	64		5
HAT	SLC7A6		y^+L, y^+LAT2	K,R,Q,H,M,L,A,C	12+AP[1]	E	M,H	↓Sb^+↑Sb; ↓Na^+Sb↑Sb^{+0}	NP_003974		5	59	123	7	6
	SLC7A7		y^+L, y^+LAT1	K,R,Q,H,M,L	12+AP[1]	E	M,H	↓Na^+Sb↑Sb^+	NP_003973	EAA07276	7	62	117	7	7
					12+AP[1]					EAA13031	7	70	114	6	3
		AeaHAT			12+AP[1]				**Figure 7**	EAA00996	11	64	109	7	6
	SLC7A5		L,LAT1	H,M,L,I,V,F,Y,W,(Q)	12+AP[1]	E	↓H↑L	↓Sb↑Sb	NP_003477	[EAA01959]	10	64	113	6	5
	SLC7A8		L, LAT2	A,S,C,T,N,Q,R,H, M,L,I,V,F,Y,W	12+AP[1]	E	M	↓Sb↑Sb	NP_036376		5	77	106	7	5
	SLC7A10		Asc1	G,A,S,C,T (D & L isoforms)	12+AP[1]	E		↓Sb↑Sb	NP_062823		10	63	119	6	1
					12+AP[1]						7	69	118	5	
					12+AP[1]					EAA03429	7	69	109	6	
										EAA10600	6	73	103	10	
	SLC7A11		x_c^-(xCT)	E, C-(substrate for GSH)	12+AP[1]	E	H	↓C↑E	NP_055146		7	67	119	7	
	SLC7A9		rBAT, b^{0+}	K,R,A,S,C,T,N, Q,H,M,I,L,V,F,Y,W	12+AP[1]	E	M	↓Sb^{+0}↑Sb	NP_055085		6	60	124	7	3

Continued

Table 2 Continued

Type	Mammalian SLC members (protein name, respectively)	Insect protein	Transport systems (some aliases)	Substrate selectivity (weak substrate)	TMDs (other prediction)	Carrier type	Apparent affinity	Stoichiometry ↓absorption ↑emission (uncoupled)	Representative H. sapiens	Orthology (proximity) An. gambiae	Archaea	Bacteria	Metazoa	Fungi	Plant
Cluster III															
SNAT	SCL38A3(SNAT3)		N1, SN1	Q,N,H	11	SE	L	↓Na$^+$↑Sb↑H$^+$	NP_006832				104	18	51
	SCL38A5(SNAT5)		N2, SN2	Q,N,H,S,G	11	SE	L	↓Na$^+$↑Sb↑H$^+$	NP_277053	Figure 8			109	18	47
	SCL38A1(SNAT1)		A1, ATA1	Q,N,H,S,G,A,C,M	11	S	L	↓Na$^+$↑Sb	NP_109599				108	17	46
	SCL38A2(SNAT2)		A2, ATA2	Q,N,H,S,G,A,C,M,P	11	S	L	↓Na$^+$↑Sb	NP_061849				105	17	49
	SCL38A4(SNAT4)		A3, ATA3	N,S,G,A,C,M,P	11	S	L	↓Na$^+$↑Sb	NP_060488				100	17	50
	SCL38A6		Orphan		11				NP_722518	[EAA10907]			112	17	42
	SCL38A7		Orphan		11				NP_612637	EAA00995		1	111	12	42
	SCL38A8		Orphan		11				NP_775785	[EAA07508]		1	86	4	30
vGAT	SLC32		VGABAT(VIAAT)	GAVA, G	10	E	L	↓Sb↑H$^+$(ves)	NP_542119	EAA05252			89	9	87
PAT	SLC36A3,4		Orphan		11(9)	S			NP_861439			2	118	15	53
	SLC36A1		IMINO(LYAA T1)	P,G,A,GABA	11(9)	S	L	↑SbH$^+$(lys); ↓SbH$^+$(ep)	NP_510968	Figure 8			110	15	68
	SLC36 A2		Tramdorin 1	P,G,A	11(9)	S	M	↑SbH$^+$(lys); ↓SbH$^+$(ep)	NP_861441	EAA10362	1	2	134	12	34
SNF$^+$	Earlier insect specific	NAT, dBLOT	orphan (blot)		12								190		
NAT	Earlier insect specific		orphan		12					EAA00597			185		
NAT	Insect specific	aeAAT1,KAAT1, CAATCH1	NAT1(BO$^+$)	F,C,H,A,S,M,I,Y,T,G, N,P,L-DOPA	12	S	M	↓2Na$^+$↑Sb[Cl−]; ↓2K$^+$↑Sb[Cl−]		[EAA05435]			200		
NAT	Insect specific	agAAT6	NAT2(BO$^+$)	W,Y,5-HT,F,A	12	S	M	↓2Na$^+$↑Sb[Cl−]		EAA05614			200		
NAT	Insect specific	agAAT8	NAT3(BO$^+$)	Y,F,L-DOPA,W,5-HT	12	S	M	↓2Na$^+$↑Sb[Cl−]	Figure 10	EAA05568			200		
NAT	Insect specific		NAT4(BO$^+$)		12	S				EAA05532			200		
NAT	Insect specific		NAT5(BO$^+$)		12	S				EAA05604			200		
NAT	Insect specific		NAT6(BO$^+$)		12	S				EAA05441			200		
NAT	Insect specific		NAT7(BO$^+$)		12	S				EAA05529			200		
Cluster III															
NNT	Earlier insect specific	NNT, dINE, msINE	Orphan(ine)		12			K$^+$(Ca$^+$) channel		EAA14797			200		
NNT	SLC6A5,7,9	GlyT, ProT,	G,P	12	S	H	↓2-3Na$^+$↑Sb[Cl−]	See Figure 10	[EAA12299]			200			
NNT	SLC6A14	BO$^+$(ATBO$^+$)	K,R,A,S,C,T,N,O,H,M, I,L,V,F,Y,W	12	S	M	↓2Na$^+$↑Sb[Cl−]	NP_009162	[EAA07088]			200			
NNT	SLS6A8,6,11,12,13	GABA/Bet, CT	GABA, betaine, creatine	12	S	H	↓2Na$^+$↑Sb[Cl−]	Figure 10				200			
NNT	SLC6A1	GABAT	GABA, neuronal high affinity	12	S	H	2-3Na$^+$↑GABA[Cl]	NP_003033	EAA13834			200			
NNT	SLC6A3,2	tnOAT, tnDAT, dDAT	DAT, NET	DA, OA, NE	12	S	H	↓2Na$^+$↑Sb↑K$^+$ [Cl−]	Figure 10	EAA04277			200			
NNT	SLC6A4	msSERT, dSERT	SERT	5HT	12	S	H	↓2Na$^+$↑5HT↑K$^+$ [Cl−]	NP_001036	EAA05837			200			

Family	Name	Transporter	Ligands	Predicted TM	S/E/U	H/M/L	Coupling	RefSeq	Accession				
Cluster VI													
VMOAT	SLC18A1	MAT, MOAT	DA, 5HT, [HA, EP, NE, OA]	12(10)	E		↓Sb+↑H+	NP_003044	[EAA09208]	7	110	61	1
	SLC18A2	MAT, MOAT	DA, 5HT, HA, EP, NE	12(10)	E		↓Sb+↑H+	NP_003045	Figure 14	7	95	61	3
	SLC18A3	AChT	ACh [Oa]	12(10)	E		↓Sb+↑H+	NP_003046	EAA07682	3	122	46	1
	Insect specific			12(10)					EAA08793	6	127	61	1
	Insect specific			12(10)					EAA01169	5	105	65	2
MCT	SLC16A1,7,8,3(MCT1,2,3,4)	MCT	Lactate, pyruvate, ketone bodies	12+AP2	SE	LM	↓H+Sb; ↓Sb↑Sb	NP_003042	Figure 15	2	27	158	10
	SLC16A4,5,6,11,12,13	Orphan		12				NP_004686		1	26	126	5
	SLC9,14	Orphan		12				XP_166145	[EAA11834]	3	26	131	2
	Insect specific	Putative dmMCT		12				Figure 15	EAA06367			194	1
	SLC16A2	T3,T4	d, l aromatic amino acids	12	U	ML	↓Sb	NP_006508	EAA08073	1	15	139	5
	SLC16A10	TAT	d, l aromatic amino acids	12	U	ML	↓Sb	NP_061063	[EAA12383]	2	23	161	11
OPT	Insect specific			12	S	L	↓2H+Sb0; ↓1-2H+Sb+/-	Figure 15	EAA05501	3	70	122	2
	SLC15A1 (PEPT1)		Di-, tri-peptides	12	S	L	↓2H+Sb0; ↓1-2H+Sb+/-	NP_005064	Figure 16		11	67	6
	SLC15A2 (PEPT2)		di-, tri-peptides	12	S	H	↓2H+Sb0; ↓1-2H+Sb+/-	NP_066568			14	68	5
	SLC15A1,2 (PHT1,2)		histidine, di-, tri-peptides	12	S		↓2H+Sb0; c1-2H+Sb+/-	NP_057666			5	34	5
	Insect species specific			12					EAA00193		17	68	2
	Insect species specific			12					EAA07145		19	63	4
PIT(ANT)	SLC17A1 (NPT1)	P1T	P1T organic ions	12	S	M	↓Na+Pi, ↓Sb (?)	NP_005065	Figure 17	1	14	170	12
	SLC17A2,3,4 (NPT3,4) [Absent]	Orphan		9	S	M	↓H+Sb-	NP_005826			35	152	13
	SLC17A5 (AST) [Absent]	Sialin	Sialic acid	10(12,6)	S	ML	↓E↑H+(Cl-) ves	NP_036566			32	152	12
VGluT	SLC176,7,8(VGluT2,1,3)	DNPi;BNPi	E	9^SUSOI	E		↓E↑H+(Cl-) ves	NP_064705	EAA12373		42	145	12
VGluT	Insect specific		E	12^SUSOI	E		↓E↑H+(Cl-) ves		EAA11677		42	146	12
VGluT	Insect specific		E	12^SUSOI	E		↓E↑H+(Cl-) ves		EAA11418		42	146	12
	Insect specific			11^SUSOI					EAA14814		40	147	12
	Insect specific								EAA00281		38	149	12
	Insect specific								EAA09135				12
	Insect specific							Figure 17	EAA13079				12
	Insect specific								EAA07243				
	Insect specific								EAA07084				
	Insect specific								EAA13886				
	Insect specific								EAA12474				
	Insect specific								EAA14792	7^SUSOI	1	68	117	12

Notes, abbreviations, and special characters: + and −, positive and negative charges; AP1, ancillary protein (4F2hc/CD98 or rBAT); AP2, ancillary protein (CD147); U, uniporters (facilitated diffusion), E, exchanger (antiporters); S, symporters (cotransporter); C, channels (high throughput tunneling of inorganic ions); Sb, substrate; H, M, L, apparent substrate affinity high $<10^{-5}$, medium 10^{-3}–10^{-4}, and low $>10^{-3}$, respectively; lys, lysosomal membrane; ep, epithelial membrane; ves, vesicular membrane; na, not available. Ligands: specific amino acid are shown according to a single capital letter nomenclature; HA, histamine; DA, dopamine; 5HT, serotonin; NE, norepinephrine (noradrenalin); EP, epinephrine (adrenaline); ACh, acetylcholine; Ch, choline; GABA, gamma amino butyric acid; SUSOI, prediction from available open reading frame sequences using SUSOI algorithm and remote server (http://sosui.proteome.Bio.tuat.ac.jp/sosui_subunit.html) Mitaku Group, Department of Biotechnology, Tokyo University of Agriculture and Technology.

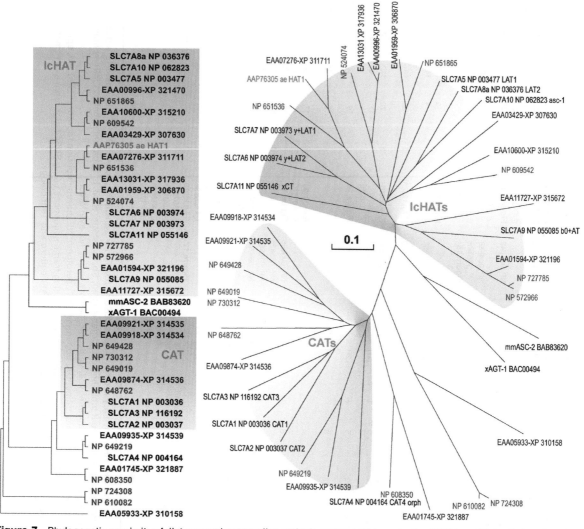

Figure 7 Phylogenetic proximity of dipteran and mammalian cationic amino acid transporter (CATs) and light chain subunits of heterodimeric amino acid transporter lcHATs (SLC7). Cladogram rooted with an EAA05933-XP_310158 *An. gambiae* sequence (left) and unrooted dendrogram (right) of putative dipteran CATs and lcHATs along with identified and partly characterized orthologs from *H. sapiens* genome.

Kim *et al.*, 1991). Now mammalian CAT subtypes 1–3 (SLC7A1–3, respectively) represent a large group of comparatively well-characterized Na^+-independent transporters. CATs in nonepithelial tissues predominantly function as y^+-system nutrient cationic amino acid providers. In epithelial tissues, CATs mediate basolateral transport which apparently is an essential component of transepithelial transport of amino acids. Thermodynamically, CAT types 2 and 3 mediate downhill diffusion (uniport) of selected cationic amino acids with extremely weak coupling or no coupling to other substrates. Since its substrates carry charge at physiological pH, CAT transport can be electrophoretic as well. In contrast, CAT type 1 is strongly trans-stimulated by intracellular substrates and may act as a thermodynamically

coupled exchanger of specific intracellular and extracellular amino acids. Neutral amino acids can be recognized by most CATs with rather low apparent affinity (Closs, 2002). In mammals, CAT subtypes exhibit abundant but subtype-specific spatial expression and immunolabeling patterns and appear to be genetic factors of several metabolic and immune diseases, including cystinuria and lysinuric protein intolerance (Simell, 2001). In addition to their relatively universal function in absorption and emission of cationic amino acids including arginine, CATs serve as rate-limiting substrate providers for enzymatic production of nitric oxide (NO) in endothelial cells and macrophages. CATs undergo complex regulation at transcriptional (Fernandez *et al.*, 2002) and posttranscriptional (Graf *et al.*,

2001) levels, which include modulation by substrates and nontransported ligands, as well as by cytoskeleton proteins (Zharikov et al., 2001).

6.7.2. Heterodimeric Amino Acid Transporters (HATs)

The heterodimeric amino acid transporter (HAT, mammalian SLC7A5–11) subfamily comprises 12-TMD transport proteins (also referred to as light chain catalytic subunit, ~40 kDa) whose functional expression requires association with its ancillary counterpart peptide, a heavy chain (~80 kDa) subunit (Verrey et al., 2003). The heavy chain subunits (mammalian SLC3) are currently represented by two proteins with apparent α-glucosidase homology: 4F2hc, the heavy chain of the lymphocyte activation antigen 4F2 (Hemler and Strominger, 1982) and rBAT, a type II membrane glycoprotein related to the $b^{0,+}$ amino acid transport phenotype (Bertran et al., 1992; Tate et al., 1992; Wells and Hediger, 1992). Heavy chain proteins have a single membrane spanning domain and an extracellular globular C-terminal domain in their secondary structure. Their interaction with light chain subunits is stabilized via 4F2hc-specific disulfide covalent binding (Chillaron et al., 2001) or rBAT-specific noncovalent interaction (Pfeiffer et al., 1998). The suggested role of the heavy chain SLC3 component is in coordinated delivery of the heterodimeric complex to the plasma membrane whereas the light chain SLC7 of HATs serves as a carrier mechanism (Palacin and Kanai, 2003; Verrey et al., 2003).

In contrast to facilitators and nonobligatory exchangers like CATs, the obligatory exchangers, HATs, are associated with a great variety of transport systems that are diverse in terms of distribution and substrate selectivity, including L system for large neutral amino acids; asc system for small neutral amino acids, e.g., Ala, Ser, Cys; x_c^- system for anionic amino acids; and $y^+L/b^{0,+}$ system that combines carriers for cationic and neutral amino acids (Verrey et al., 2003). Virtually all cloned HATs are obligatory exchangers of amino acids, except for the sodium ion driven y^+L transport of neutral amino acids by SLC7A6 or SLC7A7 and SLC3A3 HATs (Pfeiffer et al., 1999; Broer et al., 2000). The varied physiological roles of HATs have been extrapolated from their activity in heterologous expression systems, distribution, and earlier brush border vesicle analyses. For example, in nutritive epithelial tissues they serve as basolateral complements of apical B^0 and $b^{0,+}$ transporters, which are both essential for transepithelial absorption of amino acids (Fernandez et al., 2003). They also may serve as tertiary active transport mechanisms for epithelial

reabsorption of cysteine and dibasic amino acids (Chillaron et al., 1996). In nonpolarized cells, heterodimeric transporters serve in several unique mechanisms in which nutrient neutral amino acid uptake is thermodynamically coupled to emission of metabolized or redundant intracellular neutral amino acids (Meier et al., 2002). Notable examples of HAT function are neuronal and glial activity of a SLC7A11/xCT transporter, which is heterologously associated with a 4F2hc subunit (Sato et al., 1999) and mediates 1:1 exchange of extracellular cysteine for intracellular glutamate. This transporter, therefore, represents a rate-limiting component in GSH (glutathione) synthesis pathways. GSH is a tripeptide of glycine, glutamate, and cysteine that is crucial for cells and tissues that tolerate endogenous free-radical stress, e.g., activated macrophages, hypothalamic neurons, a variety of glial cells and pancreas, as well as several cell culture lines (Bassi et al., 2001; Bridges et al., 2001; Sato et al., 2002; Tomi et al., 2003; Wang et al., 2003).

6.7.2.1. SLC7 related genes in dipteran models
Transporters structurally similar to SLC7 have been reported throughout all life kingdoms; however, these have quite specific patterns of taxonomic proximity for particular species (**Table 2**). The insect CAT cluster, defined by phylogenetic reciprocity to mammalian CATs, includes six genes in both dipteran models (**Figures 4** and **7**). However, insect CAT genes do not form exact orthological matches with mammalian CATs, except for the SLC7A4 (NP_004164) gene, which clusters with insect NP_649219 and EAA09935 sequences (**Figure 7**). In contrast, insect CATs exhibit better interspecific homology by forming four orthologous clusters when they are analyzed within frames of dipteran lineages (**Figure 6**). It is most likely that proximal insect CATs are functional equivalents of mammalian CATs 1–3, which evolved via lineage-specific gene duplications; however, exact functions remain elusive, pending cloning and functional characterization of such gene transcripts. In addition to the four CATs characterized in mammalian models, both *Anopheles* and *Drosophila* genomes include two more, apparently parental, genes that are at a lower phylogenetic level relative to CATs and HATs.

Insect HATs comprise nine putative genes in *An. gambiae* but only five genes in *D. melanogaster*, which together occupy five apparently orthologous clusters (**Figures 6** and **7**). Only one such group forms a cluster with mammalian SLC7A9 (NP_055085) transporters (**Figure 7**). Several *An. gambiae* genes are not mapped, so the question about the exact number of HAT genes in this model remains to be clarified. One HAT gene cloned and characterized from *Ae. aegypti* (Jin *et al.*, 2003), AAP76305, appears to be orthologous to EAA07276 and NP_651536 genes in *An. gambiae* and *D. melanogaster*, respectively (see Section 6.7.2.3).

6.7.2.2. SLC3-related genes in dipteran models

SLC3 proteins (heavy chain subunit of HAT; **Figure 8**) serve as obligatory cofactors of SLC7 transport proteins, but do not appear to be phylogenetically related to some membrane solute carrier family. In contrast, this subunit is homologous to the α-amylase or glycosyl hydrolase Family 13, all of which bear a unique catalytic $(\beta/\alpha)_8$-barrel domain with the active site being at the C-terminal end of the barrel beta-strands. These proteins share a functional signature in that they catalyze a variety of glycoside bond cleavage/formation reactions (MacGregor *et al.*, 2001). The extracellular domain of SLC3 members has up to 30% to 40% peptide identity with insect maltases and bacterial α-glucosidases (Palacin and Kanai, 2003), which in

turn indicates a possible route of evolution common for α-amylase family proteins (Janecek, 1994). Nevertheless, rBAT (see below) lacks amylase-specific activity when heterologously expressed in *Xenopus* oocytes (Wells and Hediger, 1992); it appears to lack amylase-specific catalytic residues (Chillaron *et al.*, 2001) when compared to *Bacillus cereus* oligo 1,6-glucosidase (O1,6G), which is available in the Protein Data Bank (PDB) (Watanabe *et al.*, 1997).

At present, two members of mammalian SLC3 that form functional HATs are characterized (Palacin and Kanai, 2003): rBAT (SLC3A1) was initially identified by expression cloning in *Xenopus* oocytes (Bertran *et al.*, 1992; Tate *et al.*, 1992; Wells and Hediger, 1992); and 4F2hc (SLC3A2), was initially identified as a lymphocyte activation antigen (Hemler and Strominger, 1982). Both of these proteins appear to be widespread throughout the Metazoa. They also share remarkable homology with a number of putative genes from bacteria and fungi. However, the affiliation of these bacterial and fungal proteins with HAT forming proteins requires a detailed comparison of sequences and secondary

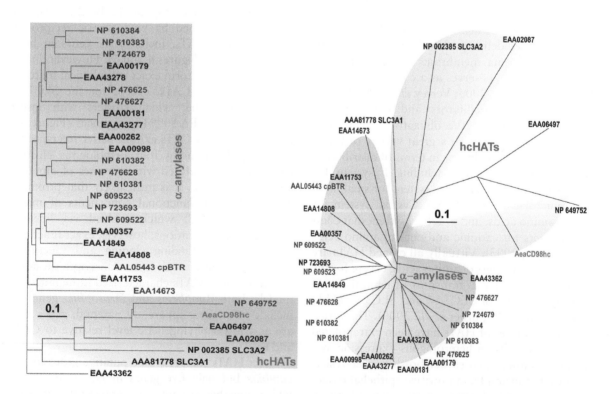

Figure 8 Phylogenetic proximity of mammalian HATs (hcHATs, SLC3) with dipteran orthologs of heavy chain subunits and with amylases. Unrooted phylogram (left) and dendrogram (right) of putative heavy chain subunit of HATs and amylases from dipteran genomes, along with known orthologs from *H. sapiens* genome (black font). Other font colors: green, heavy chain subunit of heterodimeric amino acid transporter from *Ae. aegypti*; blue, *An. gambiae* sequences; red, *D. melanogaster* sequences; pink, a binary toxin-binding alpha-glucosidase from *Culex pipiens*. Scale bar is evolutionary distances in mutations per site.

structures as well as functional assay, because of their apparent orthology with amylases. Intriguingly, SLC7 homologous proteins are present in plants and Archaea that lack SLC3 counterparts. This fact makes dialectic the obligatory requirement of SLC3 and CLC7 association at least in some lineages.

Two putative genes resembling the mammalian SLC3A2 heavy chain subunit are currently found in the *An. gambiae* genome (EAA06497, located on the X chromosome; and EAA02087, not yet mapped) whereas only one gene is present in *D. melanogaster* (NP_649753, located on the 3R chromosome). All other dipteran sequences with apparent SLC3 homology reside in an α-amylase specific gene cluster (**Figure 8**). Intriguingly, some products of these genes are recognized as toxin receptors, e.g., AAL05443, which encodes the receptor of the *B. sphaericus* binary toxin (BT) in *Culex pipiens* midgut (Darboux *et al.*, 2001). Another important finding from analysis of dipteran sequences is that heavy chain proteins apparently evolved not from α-amylases but from some common primordial ancestor (**Figure 6**).

6.7.2.3. Cloned insect HATs from *Ae. aegypti*; (AeaLAT + AeaCD98hc)

Two complementary subunits (AeaLAT and AeaCD98hc, light chain and heavy chain subunits, respectively) of an insect HAT have been cloned recently from the mosquito *Ae. aegypti*. Like mammalian HATs, insect transporters require expression of complementary subunits to produce a functional L-amino acid transporter phenotype in *Xenopus* oocytes. Heterologously expressed mosquito AeaLAT+AeaCD98hc subunits mediate uptake of large neutral and basic amino acids, partly resembling y⁺L system transporters. Small acidic and neutral amino acids appear to be poor substrates for this transporter (Jin *et al.*, 2003). Like mammalian HAT, these insect transporters can utilize a Na⁺ gradient (upregulated ~44%) for the uptake of neutral amino acids (L-leucine; $K_m^{leu} = 67\,\mu M$), but not basic amino acids (L-lysine), which is virtually Na⁺-independent at pH of 7.4 (Jin *et al.*, 2003). It has been suggested that the electrochemical gradient for amino acids at high extracellular concentrations appears to be the major factor that enables amino acid transport by AeaLAT+AeaCD98hc. AeaLAT showed high-level expression in the gastric caeca, Malpighian tubules, and hindgut of larvae. In caeca and hindgut, expression was in the apical cell membrane. However, in Malpighian tubules and midgut (which showed low-level expression), the transporter was detected in the basolateral membrane. This expression profile supports the conclusion that this AeaLAT + AeaCD98hc pair mediates a nutrient amino acid emission from posterior midgut epithelial cells to hemolymph (Jin *et al.*, 2003). A combination of similar basal y⁺L HATs with apical B⁰,⁺ transporters, such as msKAAT1, msCAATCH1, and (recently cloned from mosquito larvae midguts) aeAAT1 and agAAT8 may comprise a complete transepithelial transport metabolon for nutrient amino acid uptake. It has been stated that the apical location of AeaLAT in the gastric caeca implies its involvement in the absorption of amino acids from the lumen (Jin *et al.*, 2003). Similarly, its basal location in the midgut suggests that it may participate in the emission of amino acids or their derivatives to the hemolymph for circulation to the gastric caeca and cardia lumen where they are required for assembly of the peritrophic matrix.

6.8. AAT Cluster III (SLC6, SLC32, SLC36, SLC38)

6.8.1. SNATs and PATs (SLC38, SLC32, SLC36)

Sodium neutral amino acid transporters (SNATs), vesicular GABA Transporters (vGATs), and proton amino acid transporters (PATs), SLC38, SLC32, and SLC36, respectively, are grouped together because they represent phylogenetically related groups of transport proteins with a consecutive phylogenetic hierarchy as well as common motifs in their secondary structure, such as the presence of 10–12 TMDs. This clustering of insect transporters also corresponds to a previously identified, eukaryotic-specific, amino acid/auxin permease superfamily (Young *et al.*, 1999). Although the homology of these superfamilies of metazoan transporters is certain, to judge the exact orthology of particular insect transporters with known mammalian SNATs and PATs is problematic, because of extensive lineage-specific gene duplication/loss and lack of characterized insect transporters related to this superfamily.

6.8.2. Sodium Neutral Amino Acid Transporters, SNATs (SLC38)

This cluster represents the parental group in the SNAT + PAT superfamily, whose members fall into several consecutive phylogenetic levels in dipteran models (**Figures 6** and **9**). In mammals, characterized SNATs comprise System A and N transporters, each of which mediates a low-affinity transport $(0.3 < K_m^{substrate} < 2\,mM)$ of varied subsets of small,

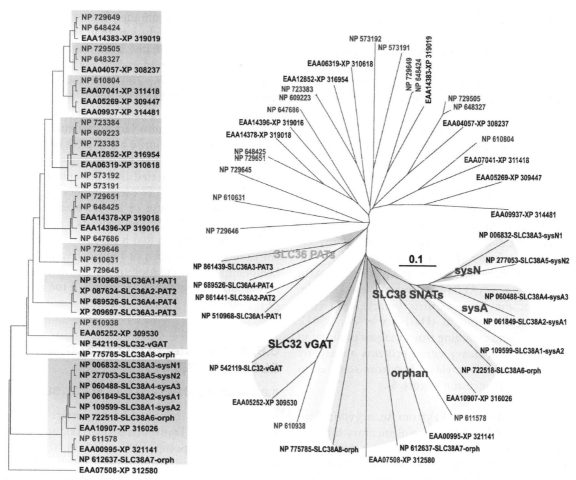

Figure 9 Phylogenetic transitions between dipteran and mammalian sodium neutral amino acid transporters (SNATs, SLC38), proton amino acid transporters (PATs, SLC36) and vesicular GABA transporter (vGAT, SLC32). Unrooted phylogram (left) and dendrogram (right). Font colors: blue, *An. gambiae* sequences; red, *D. melanogaster* sequences; black, *H. sapiens* sequences. Other abbreviations: sysA and sysN, systems A and system N transporters, respectively. Scale bar is evolutionary distances in mutations per site.

aliphatic amino acids (Mackenzie and Erickson, 2003). System A transporters appear to be broadly expressed in all mammalian tissues and to mediate Na$^+$-driven, electrophoretic, and pH sensitive neutral amino acid uptake through plasma membranes. In contrast, system N transporters have narrower substrate profiles and utilize proton motive forces, which may facilitate a reverse amino acid transport and consequently an active emission of catabolic amino acids (Chaudhry *et al.*, 2001; Broer *et al.*, 2002). Both systems are known to be tightly regulated via hormonal and intracellular signaling pathways and pH dependent allosteric modifications. SNATs are implicated in many physiological processes associated with reciprocal transcellular amino acid exchange, cellular nutrition, and metabolite secretion/detoxification pathways (Mackenzie and Erickson, 2003).

6.8.3. Vesicular GABA Transporters, vGATs (SLC32)

These transporters mediate vesicular accumulation of inhibitory neurotransmitters via an H$^+$/GABA(Gly) exchange (see Section 6.3.6.1). They presently comprise a single-member "family" in mammals designated SLC32 (Gasnier, 2003), which in insects represents a single level transition between A and N system transporters within the SLC38 family and proton amino acid transporters of the SLC36 family (**Figures 6** and **9**). A single vGAT gene encodes at least one splicoform in each dipteran model. The physiological role of vGATs is in vesicular accumulation of intracellular GABA, which is essential for inhibitory chemical signaling. Essential and universal throughout the Metazoa, inhibitory neurotransmission implicates a stabilizing selection that would conserve vGAT sequences and would make it reasonable to identify insect vGAT genes by phylogenetic relevance with mammalian genes (**Figures 6** and **9**). Hence, the function of *An. gambiae* EAAT05252 and *D. melanogaster* NP_610938 in vesicular GABA uptake can be considered.

6.8.4. Proton Amino Acid Transporters, PAT (SLC36)

Proton amino acid transporters (PATs) comprise a group of electrophoretic H$^+$/amino acid (1 : 1) symporters that mediate cytosolic acquisition of small neutral amino acids from intracellular (lysosomal) and extracellular proteolysis. In mammals, the PAT group is represented by four genes that encode two well-characterized transporters, PAT1/LYAAT-1 and PAT2, and two orphan transporters (Boll *et al.*, 2003a). The first orphan has been characterized as a rat brain-specific lysosomal amino acid transporter (rLYAAT-1) (Sagne *et al.*, 2001) and also as a mouse intestinal epithelium-specific plasma membrane proton/amino acid transporter (mPAT1), which induces H$^+$ driven, low affinity ($K_m^{G,A,P} = 7$, 7.5, 2.8 mM) transport of Gly, Ala, Pro, and several structurally related derivatives in *Xenopus* oocytes (Boll *et al.*, 2002). In contrast, PAT2 has a narrower substrate selectivity and higher affinity ($K_m^{G,A,P} = 0.59, 0.26, 0.12$ mM) (Boll *et al.*, 2002). PATs are differentially and broadly expressed in mammalian tissues except for the orphan PAT3, which so far has been found only in adult mice testis (Boll *et al.*, 2003b). Immunocytochemical studies show that rLYAAT-1 is associated with other cellular membranes, e.g., Golgi apparatus, ER, and occasionally with the plasma membrane (Agulhon *et al.*, 2003).

Dipteran genomes comprise nine putative genes in *An. gambiae* and 11 genes, encoding 15 identified transcripts, in *D. melanogaster*. These transporters split into five orthologous groups within the dipteran framework; one additional paralogous group is present in *D. melanogaster* (**Figure 9**). However, the entire population of these dipteran transporters is paralogous to mammalian PATs.

The phylogeny of SNATs and PATs apparently reflects transitions of electrochemical properties observed among characterized transporters in these families. Such transitions were suggested earlier as an evolutionary trend from an H$^+$ driven ancestral carrier, e.g., in PATs and vGAT, toward aquisition of a Na$^+$ driven mechanism, e.g., in SNATs (Boll *et al.*, 2003a; Chen *et al.*, 2003). An inverse evolutionary trend, from a Na$^+$ dependent ancestral mechanism toward a H$^+$ driven mechanism, is apparent from the phylogenetic analysis. Although H$^+$ driven mechanisms are indeed older; the SNATs and PATs population is self-rooted and apparently shares a common ancestor with the Na$^+$ driven SNF (SLC6) transporters (**Figure 6**). This fact means that Na$^+$ driven mechanisms were already established before the SNAT and PAT families diverged. In addition, a Na$^+$ driven mechanism is encoded in the older A system of SLC38 transporters that reside on the plasma membrane (**Figures 6** and **9**). The ability to modify properties under low intracellular Na$^+$ demand (N system SLC38) and directly utilize proton motive forces (SLC36) presumably was acquired later during the establishment of phagocytosis and lysosomal digestion in earlier eukaryotes. SLC36 members residing in various endomembranes correspond with lysosomal traffic for which the apical intestinal membrane is just a single possible position. Paralogous clustering in insect and mammalian SLC36 is not surprising, since this group is under the pressure of diverse nutrient requirements in different tissues and cell types. In contrast, vGAT (SLC 32) serving in inhibitory neurotransmission evolved under a stabilizing pressure, which is reflected by its trans-specific orthology. Despite structural similarity, SLC32 and SLC36 exhibit striking differences in their kinetic mechanisms: SLC32 vGAT is an exchanger (antiporter) utilizing H$^+$ motive forces from synergetic activity modulated by Cl$^-$ channels and V-ATPase for ligand accumulation (Gasnier, 2003), whereas SLC36 is a symporter utilizing both H$^+$ and ligand electrochemical motive forces for ligand secretion through lysosomal membranes and/or ligand uptake through plasma membranes (Boll *et al.*, 2003a).

6.8.5. Sodium Neurotransmitter Symporter Family (SNF; SLC6)

The title of this amino acid transporter family, which is present in many protein databases, reflects a historical, rather than a biological, perspective. It is misleading in four ways. First, the root of this family is located among primitive primordial carriers that have broad substrate selectivity and apparently share a common ancestor with archaeal and bacterial transporters (**Table 1**). Second, members of at least five other, phylogenetically remote, families participate directly in neurotransmitter transport, namely, EAATs (SLC1), PATs (SLC36), vGAT (SLC32), vGluTs (SLC17), and vMAT (SLC16). Third, the SNF family includes a large number of nutrient amino acid transporters (NATs) and undefined transporter functions (blot, ine) along with the neurotransmitter transporters (NTTs) (See **Figures 6** and **10; Table 1**). Fourth, electrochemical mechanisms encoded in these transporters can utilize transmembrane gradients for ions other than just Na$^+$ and/or Cl$^-$, e.g., K$^+$. Nevertheless, all SNF members share significant similarities in their sequences and secondary structures, which include 12 relatively conserved TMDs and a large extracellular loop between transmembrane domains 3 and 4, although prominent differences are present in this loop.

6.8.5.1. Neurotransmitter transporters, NTTs

The following subsets of insect Na$^+$/Cl$^-$ neurotransmitter transporters (SNF) are described above: GABA transporters (GAT; Section 6.3.4.1); dopamine transporters (DAT), octopamine transporters (OAT), and serotonin transporters (SERT or 5HTT; Section 6.3.4.2); glycine-proline transporters (Section 6.3.4.3); inebriated transporters (Ine; Section 6.3.4.4) (See also Sections 6.2 and 6.3).

6.8.5.2. Nutrient amino acid transporters (NATs)

NATs represent a large and phylogenetically separate cluster of carrier proteins with specific physiological roles in insects and apparently in many other organisms. Here a new scheme of SNF classification that includes two subfamilies, NTTs and NATs (**Figure 10**) is introduced. Phylogenetic analysis revealed six NAT genes in the *D. melanogaster* genome and seven genes in the *An. gambiae* genome (**Figures 6** and **10**). The properties of four currently cloned and characterized insect NATs, *M. sexta* KAAT1 and CAATCH1,

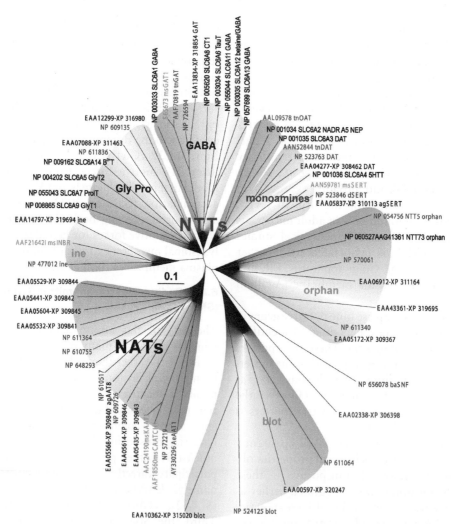

Figure 10 Unrooted neighbor-joining dendrogram of insect and mammalian neurotransmitter transporters (NTTs) and nutrient amino acid transporter (NATs) from the sodium neurotransmitter family (SNF, SLC6). Font colors: blue, *An. gambiae* sequences; red, *D. melanogaster* sequences; black, *H. sapiens* sequences; dark green, *T. ni*; light green, *M. sexta*; brown, bacterial SNF related transporter from *Bacillus anthracis* (NP_656078) and *An. gambiae* sequence, which apparently is symbiotic contamination (EAA02338). Scale bars are evolutionary distances in mutations per site.

Ae. aegypti AeAAT1 (Boudko *et al.*, in press), and *An. gambiae* agAAT8, are reviewed here. Cloned insect NATs share common 12-TMD's secondary structure and high homology in protein sequences (**Figure 11**).

msKAAT1, a potassium and sodium ion coupled amino acid transporter, was the first NAT in the SNF family to be cloned. Efforts to clone it based on phylogenic proximity to the NTT cluster were unsuccessful and it was cloned by extensive heterologous expression-screening of RNA fractions from posterior midgut of *M. sexta* (Castagna *et al.*,

1998). Because the msKAAT1 sequence is significantly similar to that of other SNF members, it is clear that the SNF category includes not only transporters for neurotransmitter uptake but also other transporters that mediate epithelial functions. *Xenopus* oocytes expressing KAAT1 induced uptake of neutral amino acids but not charged ones or GABA, resembling in this profile mammalian system B carriers (Castagna *et al.*, 1998) as well as the earlier reported amino acid uptake in brush border vesicles from midgut epithelium of *M. sexta* (Hennigan *et al.*, 1993a, 1993b). As expected from

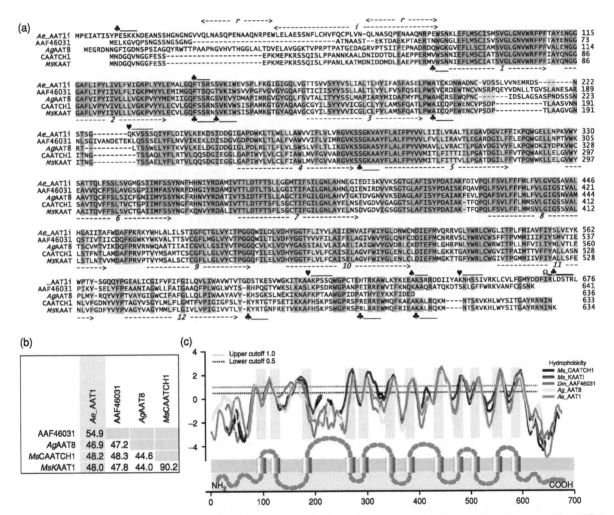

Figure 11 Molecular structure of insect NATs. (a) An open reading frame from the conceptual translation of *Ae. aegypti* Ae_AAT1i clone # 23 cDNA (AY330296) is aligned with the sequences of *M. sexta* KAAT1 (AAC241900) and CAATCH1 (AAF18560); *An. gambiae* AgAAT8 that was cloned in our laboratory (AAN40409) and a putatively homologous protein present in the *Drosophila* genome (AAF46031). (Sequence similarities are shown by the intensities of the green background.) Ae_AAT1 specific and universal homologous protein features are shown in upper and lower lines, respectively, as follows: ← # → similar transmembrane domains and their numbers; ← r →, internal sequence repeats; ← i →, insert that is absent in alternative Ae_AAT1 (AAN40410); ♣, protein kinase; C, phosphorylation sites ($n=10$); ♠, cGMP dependent protein kinase phosphorylation sites ($n=2$); ♥, myosin 1 heavy chain kinase phosphorylation sites ($n=3$); Ω, protein kinase C phosphorylation sites of all types ($n=1$). (b) Reciprocal amino acid sequence identity matrix. (c) Hydropathy plot and predicted transmembrane topology of all five transporters (TopPed II parameters were: GES-hydrophobicity scales; 1.0 upper and 0.5 lower cutoffs; 10 and 5, respectively; core and wage window sizes, 60; critical loop length, 2 for critical transmembrane spacer. Hydropathy values (vertical scale) and distribution of transmembrane spanning domain (insert at bottom) are aligned relative to amino acid position (horizontal scale).

the vesicle studies, msKAAT1 utilizes not only Na^+ but K^+ gradient as well. $K_m^{K^+}$ (1 mM Leu) $= 32 \pm 2.8$ mM, which was five times higher than $K_m^{Na^+}$. The high expression of msKAAT1 in absorptive columnar cells of the posterior midgut, which was confirmed by Northern blots and *in situ* hybridization, as well as the unusual mechanism that is adapted for K^+ rather than Na^+ motive forces for amino acid translocation, fits with the physiology of nutrition in caterpillar larvae, which have a low Na^+ and high K^+ plant diet. In addition, studies of pH dependency showed that msKAAT1 transport efficiency is maximal in the alkaline media that is physiologically normal for caterpillar midgut (Peres and Bossi, 2000).

Reciprocal ligand inhibition experiments show that msKAAT1 substrates require the presence of both the carboxylic group and the alpha-amino group, as well as the uncharged side chain (Vincenti *et al.*, 2000). The unique characteristics of msKAAT1 suggested that its inhibitors would represent a new type of lineage selective, and thus environmentally safe, insecticides and encouraged further analysis of its electrochemical, pharmacological, and structural properties. These studies led to several important findings, e.g., the ability of heterologous msKAAT1 to induce carrier coupled and uncoupled ion fluxes with specific ion selectivity profiles (Bossi *et al.*, 1999a, 1999b), the key role of the Y147 site in substrate selectivity and affinity (Liu *et al.*, 2003), the critical role of the E93 site in the functional folding of msKAAT1 (Sacchi *et al.*, 2003), and the irreversible pharmacological inhibition of msKAAT1 by modification of the R76 residue with an arginine-modifying reagent, phenylglyoxal (Castagna *et al.*, 2002).

msCAATCH1, a nutrient amino acid transporter that is also an ion channel, was cloned by a PCR based strategy using degenerate primers to conserved peptide motifs in the msSNF but avoiding msKAAT1-specific motives (Feldman *et al.*, 2000). In contrast to msKAAT1 and NTTs, the *Xenopus* oocyte expressed msCAATCH1 is Cl^- independent but combines a carrier associated current with a carrier independent leakage current, which can be blocked by methionine, alanine, leucine, histidine, glycine, and threonine. In addition, msCAATCH1 exhibits different substrate selectivity depending upon symported cations, e.g., preferring threonine in the presence of K^+ but preferring proline in the presence of Na^+ with apparent $K_m^{[Na+]Pro} = 330$ versus $K_m^{[Na+]Tre} = 35$ and $K_m^{[K+]Pro} = 1900$ versus $K_m^{[K+]Tre} = 235$ in μM, respectively. Simultaneous assay of isotope labeled amino acid uptake and voltage clamped currents revealed that transport and ligand gated channel mechanisms are thermodynamically uncoupled (Quick and Stevens, 2001). msCAATCH1 and msKAAT1 share 90.6% peptide sequence identity and could be alternative splicoforms of a single gene. In addition, they share the essential role of Y147 in the ligand affinity (Stevens *et al.*, 2002) along with the analogous role of Y140 in GAT1 (Bismuth *et al.*, 1997). Despite this striking structural similarity, the two msNATs encode transporters that are dramatically different in physiological and electrochemical properties. Hopefully, future studies of the relative expression and membrane location will clarify the physiological roles of these two transporters in the caterpillar midgut.

aeAAT1, the first NAT to be cloned from a mosquito larva, was cloned from a larval posterior midgut cDNA collection of the Yellow fever mosquito, *Ae. aegypti* (Boudko *et al.*, in press) using a PCR based similarity cloning approach (Feldman *et al.*, 2000) followed by rapid amplification of cDNA ends (RACE). Complete analysis revealed that aeAAT1 serves as a primary phenylalanine provider in mosquito larvae (Harvey *et al.*, 2003). Despite remarkable sequence similarity with msKAAT1 and msCAATCH1, the substrate and electrochemical profiles of aeAAT1 differ from the lepidopteran transporters. For example, the Cl^- dependency and ability to utilize both K^+ and Na^+ electrochemical gradients that are characteristic of msKAAT1 are absent in aeAAT1 (**Figure 12**). Additional screening of an entire-larval cDNA collection indicated the presence of two splicoforms, short (aeATT1) and long (aeAAT1i). Heterologous expression/characterization in *Xenopus* oocytes confirmed that long splicoform aeAAT1i is a low-affinity and high-throughput transporter with a specific preference for phenylalanine (substrate affinity profile $F \gg C > H > A \approx S > M > I > Y > T \approx G > N \approx P > $ L-DOPA $>$ GABA $>$ L $\gg$ DA $\approx$ Octopamine $\approx$ 5HT; $K_m^{Phe} = 0.45$ mM; $K_m^{Na+} = 38$ mM; $\eta Na^+/\eta Phe \approx 2/1$; **Figure 12**). Whole mount *in situ* hybridization confirmed a specific spatial transcription of aeAAT1 in the posterior midgut, salivary glands, cardia, and specific basal cells of gastric caeca (**Figure 13**).

Although aeAAT1 shares some properties with msNATs and the well-known mammalian B^{0+} system, for which the molecular identity so far consists of but one cloned transporter (Sloan and Mager, 1999), differences in substrate affinity and electrochemical profile, which apparently are a result of intensive adaptive divergence, make uncertain the possibility of defining a clear orthology between NAT species. Two other insect transporters that mediate aromatic amino acid uptake are members

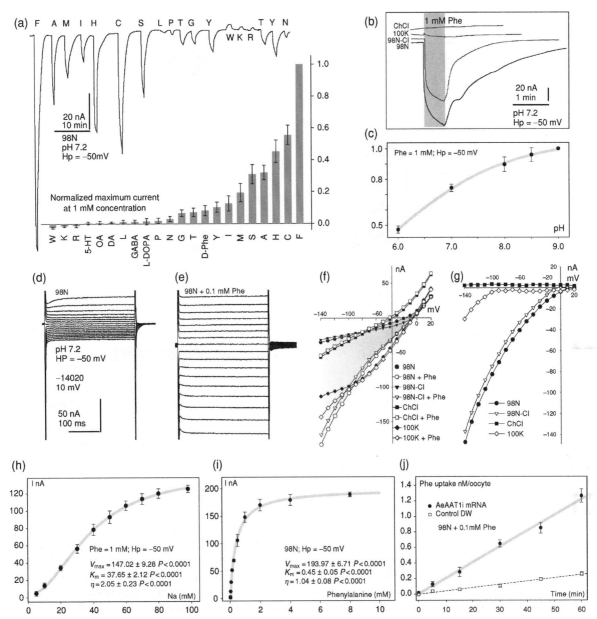

Figure 12 Electrochemical properties of Ae_AAT1i expressed in *Xenopus* oocytes. (a) Top: a trace representing amino acid induced currents in an oocyte injected with Ae_AAT1i mRNA (all amino acids were presented at 1 mM final concentration in 98 mM Na$^+$-containing saline at pH 7.2 for 30 s and washed until complete recovery of the background current). Bottom: amino acid and selected neurotransmitter sensitivity profiles in heterologously expressed Ae_ATT1i (bars are means of normalized responses ± SE; $n \geq 3$ for different oocytes). (b and c) Ion and pH dependency of phenylalanine induced currents. (d and e) Representative voltage-step induced currents in saline alone and after application of 0.1 mM phenylalanine. (f) I/V plots at different ionic conditions outside oocytes. Dots with similar shapes represent pairs of a control (no ligand) vs. 0.1 mM phenylalanine I/V plots. Gradual shading indicates a difference between I/V plots in control 98 saline and 98 saline with 0.1 mM phenylalanine. (g) Average I/V plot for the data from f. (h and i) Phenylalanine and sodium ion concentration dependency (means of current ± SE; $n = 3$ for each point); K_m, Michaelis-Menten constant and η, Hill coefficient values derived from these data, shown in graph; (j) Linear regression lines from phenylalanine uptake by RNA vs. water injected oocytes (concentration means ± SE, $n = 3$ or 2).

of the Na$^+$ independent TAT (Kim *et al.*, 2001) transport systems and heterodimeric LAT$^+$4F2hc group (Mastroberardino *et al.*, 1998; Rossier *et al.*, 1999), which are well characterized in mammals (SLC16 and SLC7$^+$SLC3 and SLC16 phylogenetic families, respectively). An increasing number of studies indicate that TAT and heterodimeric transport systems participate in the basolateral epithelial uptake/exchange of various organic solutes in growing cells (Verrey *et al.*, 2003). By contrast, aeAAT1

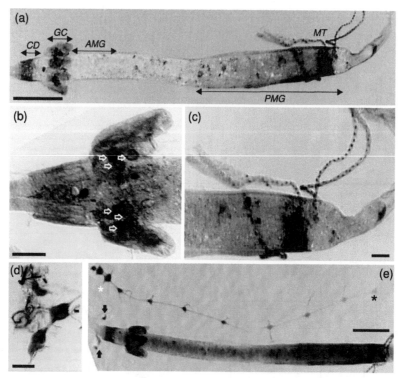

Figure 13   *In situ* hybridization of whole-mount early 4th instar *Ae. aegypti* larvae with Ae_ATT1i DIG mRNA probes. (a) Distribution of Ae_AAT1i (blue labeling). CD, cardia (esophageal invagination); GC, gastric caeca; AMG, anterior midgut; PMG, posterior midgut; MT, Malpighian tubules. (b and c) High magnification images of gastric caeca and posterior midgut, respectively. Large epithelial cells in the basal posterior part of GC are indicated by white outline arrows. (d) Image of cephalic ganglia isolated from the central nervous system (CNS). (e) Image from middle 4th instar larval preparation showing intensive and uniform Ae_AAT1 expression patterns in CD, GC, and PMG. Isolated ventral nerve cord showing different intensities of hybridization in anterior abdominal ganglia vs. posterior abdominal ganglia of the CNS (anterior and posterior ends of the cord, white and black asterisks, respectively). Black arrows point to salivary glands. Scale bar: (a, e) 1 mm; (b, c, d) 200 μm.

and other NATs appear to play a prominent role in the primary uptake of essential nutrient amino acids from the lumen to the epithelial cells through the apical brush border membranes in posterior midgut. aeAAT1 is also transcribed in the cardia and gastric caeca, where amino acid absorption is likely, because amino acids in these tissues are extensively depleted by the polymerization of the peritrophic matrix in the lumen of the cardia. These data support the hypothesis that aeAAT1 is a key transporter in phenylalanine-consuming pathways, such as the dopaminergic neurotransmission and ectoderm tanning pathways that are ubiquitous in the animal kingdom, as well as in insect-specific external cuticle sclerotization and polymerization of the peritrophic matrix.

agAAT8 (resembling predicted gene product EAAT05568) was cloned from larvae of *An. gambiae*, which serve as a genomic and tropical disease vector model. A novel, high throughput strategy was used for cloning and characterization of this transport protein (Boudko *et al.*, unpublished data). This strategy combines cyberanalysis to verify the correct 5′ and 3′ open reading frame (ORF) ends with PCR cloning from a high performance, tissue specific-cDNA collection (Matz, 2002) using long, ligation-ready, ORF-flanking primers and the efficient expression vector, pXOOM, which is designed for heterologous expression in *Xenopus* oocytes or mammalian cell lines (Jespersen *et al.*, 2002). In contrast to aeAAT1, agAAT8 has approximately equal apparent affinity for products of phenylalanine catabolism, including tyrosine and L-DOPA ($K_m^Y > 0.3$ mM). Although *in situ* hybridization revealed similar expression patterns for both mosquito transporters, some differences are prominent, e.g., the relative labeling intensity of aeAAT1 is significantly higher in the posterior midgut than in the gastric caeca, whereas labeling of agATT8 is approximately equal in the two regions. The transcript density of both transporters also varied depending upon developmental stage.

In summary, from the information currently available on insect NATs we conclude that this group comprises transporters that mediate the initial absorption of essential nutrient amino acids from the midgut lumen into the epithelial cells of the mosquito. Nutrient amino acid transporters do not possess the trans-specific orthology exhibited by neurotransmitter transporters (**Figures 6** and **10**), since they diverge through extensive gene duplication and losses during adaptation to particular environmental and nutrient niches. Most likely, similar paralogous groups of nutrient transporters will be defined in other insects and will be split into several, probably three, subclusters, each of which is adapted to transport an isomorphic subset of amino acids. In addition to encoding principal apical transporters, NATs may encode basolateral transporters through an alternative splicing of an anchoring motif. Owing to their crucial role in insect nutrition and phylogenetic specificity, NATs are considered to be excellent targets for the development of lineage-specific and environmentally safe mosquitocides.

6.8.5.3. Orphan transporters: transitions between SNF clusters

These transporters are essential for physiology and development but their substrates and transport mechanisms are currently undefined. Reasons for failure to characterize orphan transporters could vary, e.g., inappropriate heterologous system, incorrect protein folding, or lack of function enabling interactions. Phylogenetic analysis indicates that orphan species represent functional and presumably evolutionary transitions between groups with defined functions (**Figure 6**). For example, orphan *ine* transporters represent a transition between neurotransmitter transporters and nutrient amino acid transporter subfamilies. SNF orphans cannot be defined as a subfamily since they do not form a true separate cluster but instead represent several consecutive levels in the phylogenetic tree. One orphan gene, named by similarity with a mammalian orphan group, has been extensively analyzed in *Drosophila* (Johnson *et al.*, 1999). This gene, named bloated tubules (*blot*), exhibits a complex and dynamic expression pattern during embryogenesis of supporting processes that are essential for cell division, differentiation, and epithelial morphogenesis. The analysis showed five putative orphan genes in *An. gambiae* and four genes encoding seven presently known isoforms in *D. melanogaster*, which represent transitions between SNF and PAT (proton amino acid transporters) superfamilies (**Figure 3**).

6.9. AAT Cluster IV (SLC15, SLC16, SLC17, SLC18)

Cluster IV represents one of the most diverse groups of transporters, many of which do not transport amino acids but are phylogenetically close to those which do so (**Figure 6**). It corresponds to distantly related families of mammalian transporters, including: vesicular acetylcholine and monoamine transporters (vAChT and vMATs, SLC18, see also Section 6.3.3.2). It is the best characterized group of vesicular transporters. It includes monocarboxylate cotransporters (MCTs, SLC16), which comprise one characterized T-type amino acid transporter (TATs, SLC16A10), H$^+$/oligopeptide transporters (OPTs, PEPTs, SLC15), and vesicular glutamate transporters (vGluTs). The latter represent a family that is united with transporters of type I phosphate and other organic and inorganic anions (SLC17). Since these transporters have not been characterized in insect lineages we include the most comprehensive subset of predicted genes from dipteran genomes in the phylogenetic analysis.

6.9.1. Vesicular Monoamine and Acetylcholine Transporters, vMATs and vAChT (SLC18)

The SLC18 family comprises specific transporters mediating high-affinity vesicular accumulation of biogenic amines that are essential for cholinergic and aminergic transmission in CNS and neuromuscular synapses throughout the Metazoa (Eiden *et al.*, 2003). These transporters apparently are not related to amino acid carriers but are included here because they are functionally similar to vesicular amino acid carriers from other phylogenetic groups (vGAT, SLC32 and vGluT, SLC17) and show distant relationships to monocarboxylate transporters (MCT, SLC16), with at least one member serving as an amino acid carrier. More direct phylogenetic links with SLC18 have been defined with toxin-extruding antiporters (TEXANs) (Yelin and Schuldiner, 1995), which are abundant in bacteria, fungi, and Archaea (**Table 2**) (Paulsen and Skurray, 1993). In bacteria these transporters appear to be essential components of interspecific antibiotic tolerance and drug resistance (Schuldiner *et al.*, 1994; Schuldiner *et al.*, 1995). All metazoan transporters that are involved in neuronal signaling are highly conserved structurally, and SLC18 members are no exception. They share remarkable similarities, in sequences and electrochemical signatures, with different lineages of the Metazoa (Rand *et al.*, 2000). A different, though highly conserved, SLC18 member's electrochemical

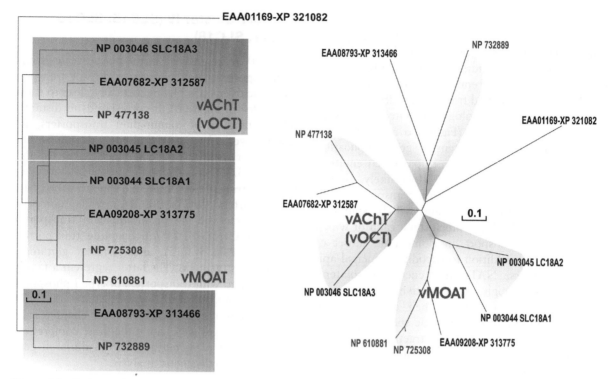

Figure 14 Phylogenetic proximity of dipteran and mammalian vesicular monoamine transporters (vMOATs; SLC18). Unrooted phylogram (left) and dendrogram (right). Font colors: blue, *An. gambiae* sequences; red, *D. melanogaster* sequences; black, *H. sapiens* sequences. Scale bars are evolutionary distances in mutations per site.

mechanism enables a high-affinity uptake of biogenic amines, which is thermodynamically coupled to H^+ V-ATPase generated proton motive forces. It exhibits a rigid stoichiometry of 1 ligand per $2H^+$ and is capable of vesicular accumulations to as high as $0.5\,M$ concentrations of particular ligands, exceeding up to 10^3-fold and 10^5-fold transmembrane gradients for ACh and monoamines, respectively (Parsons, 2000).

Although many mammalian SLC18 members have been analyzed since the first rodent vMAT1 and vMAT2 (SLC18A1 and SLC18A2, respectively) was cloned (Erickson *et al.*, 1992; Liu *et al.*, 1992b), only one SLC18 related insect transporter has been cloned from *D. melanogaster* (NP_477131 or AAB86609; Kitamoto *et al.*, 2000a). A 12-TMD structure for this transporter has been predicted. However, since experimental proof is lacking, an alternative 10-TMD model, arising from a different prediction algorithm, can be considered as well (Rost, 1996; Eiden *et al.*, 2003). Based on EM and immunolabeling, dmvAChT colocalized with acetylcholine transferase in the brain of the adult fly where it appears to mediate specific components of synaptic wiring in glomerula (Yasuyama *et al.*, 2002). vMATs and vAChTs are encoded by four putative genes in the *An. gambiae* and three genes

in the *D. melanogaster* genomes (**Figures 6** and **14**). vMATs and vAChTs form orthological clusters with mammalian SLC18 members. In contrast to the *H. sapiens* genome, only one gene encodes vMAT in dipteran genomes; however, two apparent splicoforms are encoded by dmvMAT (**Figure 14**). An additional dipteran specific group is absent in mammalian SLC18 (**Figure 14**).

6.9.2. Monocarboxylate Transporters and T System Amino Acid Transporters, MCTs and TATs (SLC16)

MCTs and TATs (SLC16) comprise a large family of H^+ coupled transporters and apparent uniporters (Bonen, 2000) that currently includes 14 members (Halestrap and Meredith, 2003) in *H. sapiens*, which, within the E value interval selected for these sequences, could be associated with 18 predicted gene products in NCBI and EMBL databases (**Figure 15**). Only five mammalian SLC16 members have been characterized (Halestrap and Meredith, 2003):

1. The first member was defined *in vivo* (Deuticke, 1982), subsequently cloned (Kim *et al.*, 1992; Garcia *et al.*, 1994), and designated MCT1 (SLC16A1); when heterologously expressed, it

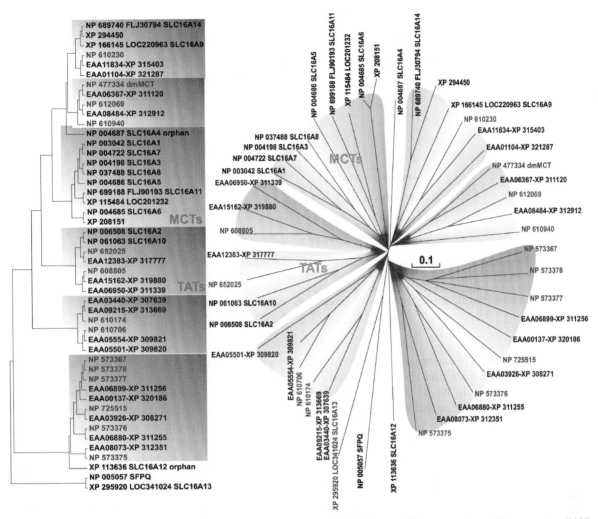

Figure 15 Phylogenetic distribution of dipteran and mammalian monocarboxilate and T-system amino acid transporters (MCTs and TATs; SLC16). Rooted cladogram (left) and unrooted dendrogram (right). Human orphan member of SLC16 (SLC16A13) and PTB associated splicing factor (NP_005057) are selected as outgroup. Font colors: blue, *An. gambiae* sequences; red, *D. melanogaster* sequences; black, *H. sapiens* sequences. Scale bars are evolutionary distances in mutations per site.

mediates a moderate affinity uptake of two to five carbon chain aliphatic monocarboxylates, e.g., $K_m^{pyruvate} = 0.7\,mM$, $K_m^{propionate} = 1.5\,mM$, and $K_m^{L\text{-}lactate} < 3\,mM$ (Broer *et al.*, 1998a; Manning Fox *et al.*, 2000).

2. Unlike MCT1, MCT2 (SLC16A2) expression intensity varies greatly in different mammalian models, e.g., its transcripts are weakly expressed in human (Price *et al.*, 1998) but are abundant in rodent (Garcia *et al.*, 1995; Jackson *et al.*, 1997) tissues where it induces a relatively high-affinity carrier; $K_m^{L\text{-}lactate} \approx 0.1\,mM$ and $K_m^{pyruvate} \approx 0.7\,mM$ (Lin *et al.*, 1998; Broer *et al.*, 1999).

3. MCT3 (SLC16A8), initially cloned from chick retinal pigment epithelium (Yoon *et al.*, 1997), mediates a low-affinity ($K_m^{L\text{-}lactate} > 6\,mM$) basal transport (Yoon *et al.*, 1997; Philp *et al.*, 2001), which together with an apical MCT1 (Bergersen

et al., 1999) comprises a transepithelial pathway in the retinal pigment epithelium and choroid plexus.

4. MCT4 (SLC16A3) is abundant in various tissues (Price *et al.*, 1998; Wilson *et al.*, 1998; Pilegaard *et al.*, 1999; Dimmer *et al.*, 2000; Bergersen *et al.*, 2002; Meredith *et al.*, 2002) and has the weakest substrate affinity among MCTs, e.g., measured $K_m^{lactate}$ and $K_m^{pyruvate}$ values with different assay techniques are in the tenths millimolar range (Dimmer *et al.*, 2000; Manning Fox *et al.*, 2000).

5. T system transporter TAT1 (SLC16A10) mediates H^+ and Na^+ uncoupled, submillimolar affinity, active transport or facilitated diffusion of aromatic amino acids and some related metabolites (L-phenylalanine, L-tryptophan, L-tyrosine, and L-DOPA) across basolateral membranes of intestinal and renal epithelial cells as well as

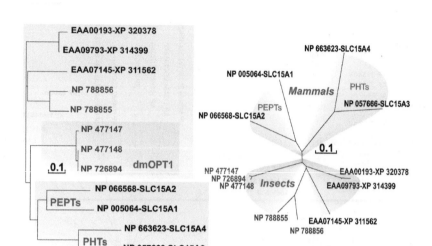

Figure 16 Paralogous distribution of dipteran and mammalian oligopeptide transporters (OPTs, SLC15). Unrooted phylogram (left) and dendrogram (right). Font colors correspond: blue, *An. gambiae* sequences; red, *D. melanogaster* sequences; black, *H. sapiens* sequences. Other abbreviations: dmOPTs, *D. melanogaster* oligopeptide transporters; PEPTs, *H. sapiens* peptide transporters; PHTs, *H. sapiens* peptide/histidine transporters. Scale bar is evolutionary distances in mutations per site.

plasma membranes of various nonpolarized cells, e.g., in placenta, liver, and muscle (Kim *et al.*, 2001, 2002; Kim do *et al.*, 2002). (SLC16A3 TATs should not be confused with sulfate transporter Tat1 (SLC23A11) (Toure *et al.*, 2001; Vincourt *et al.*, 2003) or bacteria/yeast specific TAT1 permeases (Bajmoczi *et al.*, 1998)).

The tissue distributions of a few other members of SLC16 have been analyzed (Price *et al.*, 1998) but no electrochemical and ligand data are yet available. A recent comprehensive review aids our SLC16 description by mentioning that heterologous expression of a TAT1 related SLC16A2 transporter in *Xenopus* oocytes revealed a thyroid hormone (T4 and T3) transporter phenotype, which has micromolar apparent substrate affinity and appears to be independent of sodium and proton motive forces (Halestrap and Meredith, 2003).

All characterized SLC16 members have a 12-TMD structure with a characteristic cytosolic loop between TMDs 6 and 7. At least some members require an auxiliary protein for trafficking and functional coordination in plasma membranes, e.g., MCT2 interacts with an unknown protein factor (Kirk *et al.*, 2000) and MCT1 and MCT4 interact with CD147 (also known as OX-47, extracellular matrix metalloproteinase inducer (EMMPRIN), HT7, or basigin) (Kirk *et al.*, 2000; Zhao *et al.*, 2001; Halestrap and Meredith, 2003), which was confirmed using a fluorescence resonance energy transfer (FRET) technique (Wilson *et al.*, 2002) and by reduction of transport by antisense silencing

of a CD147 ortholog in *Xenopus* oocytes (Meredith and Halestrap, 2000).

Phylogenetic analysis discloses 15 putative genes in *An. gambiae* and 14 genes, corresponding with the number of identified transcripts, in *D. melanogaster* (**Figures 6** and **15**). One member of *D. melanogaster* MCTs was identified before annotation of the genome (CAB42050); however, it does not cluster with mammalian MCT1 (SLC16A1) but instead fits into an apparently orthologous phylogenetic cluster with four other putative members of dipteran MCTs (**Figure 15**). In fact, most MCTs form mammalian or dipteran lineage specific, paralogous clusters, except for orphans SLC16A9 and SLC16A14 and the characterized aromatic amino acid transporter TAT (SLC16A10) and putative thyroid hormone transporter (SLC16A2), which show proximity with dipteran transporters and would reasonably be expected to have similar physiological roles (**Figure 16**).

6.9.3. Oligopeptide Transporters, OPTs (SLC15)

Oligopeptide transporters (OPTs), also known as peptide transporters (PTRs) (Steiner *et al.*, 1995) evolved as an efficient, energy-saving uptake mechanism complementing or replacing amino acid carriers in special cases, e.g., (1) they enable high throughput intracellular loading of nutrient amino acids in peptide form and (2) they remove specific and nonspecific oligopeptides, which may interfere with cellular signaling and cellular interactions from body fluids and subcellular media. In mammals,

OPTs mediate H$^+$ coupled uptake of small peptides or similar substrates into intestinal and renal epithelial cells, bile duct epithelium, glial and epithelia cells of the choroid plexus, lung, and mammary gland (Daniel and Kottra, 2003). OPTs mainly utilize transmembrane proton motive forces and appear to be reversible when heterologously expressed in HeLa cells or *Xenopus* oocytes (Fei *et al.*, 1994; Boll *et al.*, 1996) and *in vivo* (Tomita *et al.*, 1995).

The first proton-coupled peptide transporter PEPT1 (SLC15A1) was identified by expression-cloning from rabbit small intestine cDNA in the Hediger laboratory (Fei *et al.*, 1994). It mediates the uptake of small peptides, being independent of extracellular Na$^+$, K$^+$, and Cl$^-$ gradients and, surprisingly, independent of the membrane potential as well. PEPT1 is transcribed in intestine, kidney, and liver with small amounts in brain. Simultaneously, PEPT1 was cloned by Murer and colleagues in attempts to define the molecular basis for proton-dependent peptide transport in BBMV from rabbit small intestinal epithelia (Boll *et al.*, 1994). This group also cloned an orthologous transporter PEPT2 (SLC15A2) from kidney cortex epithelia which, in contrast to low-affinity high-throughput PepT1, mediates high-affinity low-throughput transport of a virtually identical subset of substrates (Boll *et al.*, 1996). A large number of PepT1 and PEPT2 related OPTs have been identified by homology cloning in subsequent years (Liang *et al.*, 1995; Saito *et al.*, 1996; Rubio-Aliaga *et al.*, 2000; Chen *et al.*, 2002; Verri *et al.*, 2003), including one from insects, i.e., *D. melanogaster* OPT-1 (Roman *et al.*, 1998). More recently, two other OPTs were identified in mammals: a brain-specific peptide histidine transporter 1 (rat PHT1 a.k.a. human PTR4, SLC15A4; Yamashita *et al.*, 1997) and peptide histidine transporter 2 (rPHT2 a.k.a. hPTR3, SLC15A3) isolated from the rat lymphatic tissue (Sakata *et al.*, 2001). In contrast to the broad substrate affinity of PEPT1 and 2, PHT1 and 2 have more specific sets of substrates including histidine and some model di- and tripeptides. These transporters apparently play different, more specific physiological roles, since they are strongly expressed in specialized tissues such as brain for PTH1 (Yamashita *et al.*, 1997) and lymphatic system, lung, spleen, and thymus for PHT 2 (Sakata *et al.*, 2001). OPTs range from 450 to 700 amino acid residues, most of the eukaryotic members having ~650 amino acid residues and 12 predicted TMDs with cytoplasmic N- and C-termini (Daniel and Kottra, 2003).

6.9.3.1. *Drosophila* oligopeptide transporters, OPTs

The only insect OPT characterized to date is from *D. melanogaster* (Roman *et al.*, 1998). The *opt* gene encodes a protein that mediates proton-dependent di- and tripeptide transport in intestinal and Malpighian tubular (renal) epithelia. Like mammalian OPTs, fruit fly OPT1 has 12 TMDs and a long (200 amino acid residue) fifth external loop. It is most similar to human renal PEPT2 and intestinal PEPT1 proteins, containing the highly conserved histidine residue, thought to be involved in peptide binding and transport, at position 88.

Up to 30% of total amino acids in the hemolymph of adult fruit flies are present in the form of di- or tripeptides, which serve in osmoregulation (Chen, 1966; Collett, 1976). Protein digestion in *Drosophila* occurs mainly in the midgut where resulting peptides and amino acids are rapidly taken up by the epithelium (Law *et al.*, 1977) and metabolized or transported to the hemolymph. Hemolymph amino acids are secreted into the Malpighian tubular lumen (equivalent to vertebrate glomerular filtration) but are reabsorbed in the hindgut. Compared to mammalian peptide transporters discussed above, little was known about the absorption of peptides in insects prior to the cloning of dmOPT1. The *opt1* gene encodes a high-affinity di- and tripeptide transporter that is proton dependent. It is transcribed in germinal and somatic tissues of both males and females where it is most highly expressed in nurse cells of the ovary. It is also expressed in epithelial cells of the midgut and rectum and is mildly expressed in neurons. Uptake was assayed in HeLa cells transfected with pCMVOPT1 or pCMV2 using lipofectamine. Tritiated L-alanylalanine uptake was measured in buffer with high NaCl and low KCl at pH 6.0. Evidence that the peptide transport is proton-coupled was a ten-fold decrease in uptake when extracellular pH was increased from 6 to 7 and a 20-fold decrease when it was increased from 6 to 8. Uptake was virtually abolished by FCCP (25 μM), which collapses the proton gradient. The K_m for OPT1-dependent alanylalanine transport at external pH = 6 was 48.8 μM (Roman *et al.*, 1998).

6.9.3.2. Phylogenetic characteristics of dipteran OPTs

OPT (SLC15) related transporters are present in most life kingdoms, except for the Archaea (**Table 2**), being especially abundant in plants where they are in phylogenetic proximity to nitrate transporters. OPTs comprise three putative genes in the *An. gambiae* genome and three genes in the *D. melanogaster* genome with three different splicoforms known for the single characterized example of a dmOPT1 gene (**Figure 16**). Within the framework of mammalian and dipteran lineages, OPTs form separate paralogous branches, indicating relatively recent gene duplication. Probably for the same reason, the proximity of different dipteran OPTs is difficult to deduce solely from the phylogenetic analysis.

6.9.4. Vesicular Glutamate and Phosphate Transporters, vGLUTs and P$_i$Ts (SLC17)

In mammals, the SLC17 family comprises two physiologically distinct groups of characterized transporters: type I inorganic phosphate transporters (P$_i$Ts) and vesicular glutamate transporters (vGLUTs). The P$_i$T family, which was initially restricted to transporters of inorganic phosphates across plasma membranes, has now been expanded to include transporters of various organic and inorganic anions across both plasma membranes and endomembranes. The vGluT family comprises extensively analyzed vesicular glutamate transporters, which are specialized to serve excitatory neurotransmission that enables endomembrane traffic and vesicular accumulation of excitatory amino acids. Since no SLC17-related insect transporter has yet been characterized, only a brief description of known SLC17 transporters was included from a recent comprehensive review (Reimer and Edwards, 2003).

6.9.4.1. Inorganic phosphate transporter, type I P$_i$Ts (SLC17A1)
The first member of mammalian SLC17A1 (NaP$_i$-1) was identified by expression cloning from a rabbit kidney cortex cDNA library, based on its ability to induce Na$^+$-dependent phosphate transport in *Xenopus* oocytes, with 750-fold higher apparent uptake than controls injected with total poly(A) RNA (Werner *et al.*, 1991). However, submillimolar substrate affinities of rabbit NaP$_i$-1 and the next-cloned ortholog from *H. sapiens* (Miyamoto *et al.*, 1995), as well as specific electrochemical properties of their heterologous expression in *Xenopus* oocytes, e.g., conductance for mineral and organic anions (Busch *et al.*, 1996a, 1996b; Broer *et al.*, 1998b) and low pH sensitivity (Busch *et al.*, 1996b), do not match characteristics of Na$^+$-dependent P$_i$ uptake in brush border membrane vesicles (Amstutz *et al.*, 1985) and do not support a primary role for these transporters in high-affinity phosphate transport (Murer *et al.*, 2000; Reimer and Edwards, 2003). Indeed, two different P$_i$T types have been identified (Magagnin *et al.*, 1993; Kavanaugh *et al.*, 1994) that are unrelated to SLC17 Na$^+$/P$_i^-$ cotransporters. One new type is a SLC34 family, type II P$_i$T, whose members mediate apical P$_i$ uptake in various polarized epithelia (Murer *et al.*, 2003). Transporters of the other new type were originally described as retroviral receptors (Olah *et al.*, 1994) but are now recognized as type III P$_i$Ts (SLC20), which appear to be basolateral components of P$_i$ homeostasis (Collins *et al.*, 2003). A recent revision of P$_i$Ts tends to support the earlier proposal that some members of the SLC17 family are facultative P$_i$Ts, with major roles in the rapid acquisition of electrochemical and pH balance via anion secretion (Amstutz *et al.*, 1985). In fact, they may utilize a variety of electrochemical mechanisms ranging from facilitated diffusion to electrophoretic Na$^+$ coupled symport (Reimer and Edwards, 2003).

Intriguingly, despite a crucial role of P$_i$ homeostasis in all organisms, and the presence of homologous proteins in other metazoa, as well as in bacteria, fungi, and plant lineages, there is no trace of SLC20 and SLC34 families in dipteran genomes and they appear to be absent in arthropods in general. Conversely, mammalian type I P$_i$Ts do not include any dipteran sequences (**Figure 17**), which leaves uncertain the presence of SLC17 related P$_i$Ts in dipteran genomes.

6.9.4.2. Sialin (SLC17A5)
Extensive genetic analysis and BBMV studies revealed a physiological role for another SLC17 member, sialin (SLC17A5), which apparently mediates lysosomal H$^+$ coupled (1 : 1) emission of amino carbohydrates (sialic acid) from lysosomal catabolism of glycosylated membrane proteins, as well as serving endomembrane traffic of some monocarboxylic acids (Reimer and Edwards, 2003).

6.9.4.3. vGlutT (SLC17A6, 7, 8)
Three members of mammalian SLC17 (SLC17A6, A7, and A8, a.k.a. vGluT1, 2 and 3) are recruited in excitatory neurotransmission, enabling vesicular storage of L-glutamate. Initially defined as plasma membrane, Na$^+$ coupled P$_i$Ts, vGluT1 (Ni *et al.*, 1994), and vGluT2 (Aihara *et al.*, 2000) transport phenotypes have since been shown to be H$^+$ coupled carriers that mediate accumulation of L-glutamate into synaptic vesicles of neurons (Bellocchio *et al.*, 2000) and glutamatergic endocrine cells (Hayashi *et al.*, 2001; Takamori *et al.*, 2001). All characterized vGluTs have been morphologically associated with glutamatergic neurons of the postnatal brain and especially abundant at membranes of synaptic vesicles, although spatial differences in subtype distribution in the brain are prominent (Bellocchio *et al.*, 1998; Fremeau *et al.*, 2002; Varoqui *et al.*, 2002b). The neuronal specificity of vGluTs makes them

useful markers of excitatory neuronal components, e.g., recent demonstration of the existence of peripheral glutamatergic components in the gastro-enteropancreatic system and testis (Hayashi *et al.*, 2003). The expression of vGluTs occurs very early in embryogenesis (Gleason *et al.*, 2003). High resolution immunoelectron microscopy also demonstrates a specific subcellular distribution of vGluT types, e.g., vGluT1 and vGluT2 are mainly associated with nerve terminals; in contrast, vGluT3 is present in vesicular structures of astrocytes and neuronal dendrites (Fremeau *et al.*, 2002).

Apparently, all vGluTs share highly conserved, pmf driven, electrochemical mechanisms with nearly millimolar apparent K_m^{Glu}. Like P_iTs, vGluTs enable Cl$^-$ and some other anionic conductances

in BBMV and heterologous models (Bellocchio *et al.*, 2000; Fremeau *et al.*, 2001; Varoqui *et al.*, 2002a). Vesicular glutamate accumulation has been shown to be inhibited by the broad specificity blocker of chloride conductance DIDS(4,4'-diisothiocya-natodihydrostilbene-2,2'disulfonic acid). Like all other endomembrane transporters, electrophoretic vGluTs are energized by electrogenic V-ATPase.

6.9.4.4. Phylogenetic characteristics of SLC17 transporters Before their molecular identities were established, P_iTs and vGluTs have been identified by comparative studies of vesicular glutamate uptake in BBMVs from neuronal tissues of several vertebrates and invertebrates (goldfish, frogs, turtles, pigeons, rats, fruit fly, and crayfish). Their

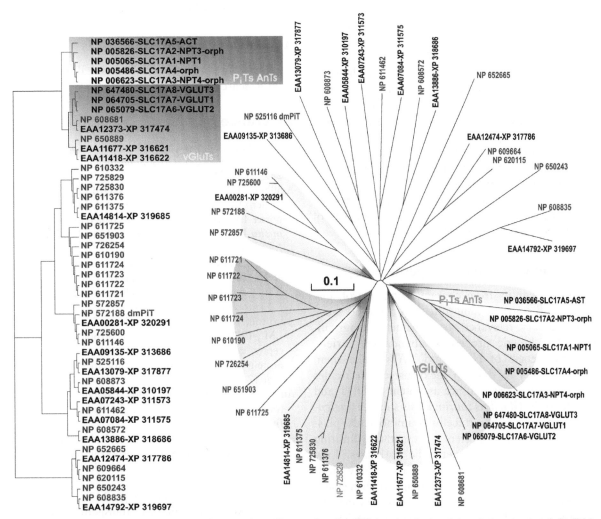

Figure 17 Phylogenetic proximity of putative organic and inorganic anion (P_iTs) and vesicular glutamate transporters (vGluTs) in dipteran and mammalian models (SLC17). Unrooted cladogram (left) and dendrogram (right). Other abbreviations: AST, anion/sugar transporter; NPT, sodium inorganic phosphate transporter[4]. Scale bars are evolutionary distances in mutations per site.

[4]NTP title reflects a historical drawback, since NTPs were initially cloned and characterized as inorganic phosphate transporter Type I. Later, the major role of this transporter phenotype in the transport of organic and inorganic anions was established.

properties, including millimolar Glu affinity, as well as V-ATPase energization and chloride stimulation, appear to be extremely conserved over 350–400 million years (Tabb and Ueda, 1991). Phylogenetic analysis indicates a single gene copy in *D. melanogaster* and *An. gambiae* genomes with close proximity to three, apparently paralogous, vGluT gene clusters in mammals (**Figure 17**). Two additional *An. gambiae* genes and one *D. melanogaster* gene are more distantly related to the vGluTs. No dipteran genes with proximity to P_iT transporters have been found to date. The earlier defined dmP_iT transporter gene (NP_572188) now appears to be phylogenetically separated from P_iT gene clusters; its function cannot be deduced from the phylogenetic pattern alone. Intriguingly, dipteran genomes include several new SLC17 gene clusters that are absent in mammals and have no matches with characterized genes in other lineages. Two of these clusters indicate remarkable gene duplications in *D. melanogaster* (**Figure 17**). In summary, relatively conserved VGluTs seem to have diverged through lineage-specific gene duplications that reflect increases in functional and morphological complexity of the CNS. Hence, a role for the genes that encode transporters related to the SLC17 family can be expected in insect vesicular glutamate homeostasis. In contrast, P_i absorption and homeostasis, which are essential for all organisms, appears to occur in insects through phylogenetically unrelated roots of transporters that are distinct from P_iTs of SLC20, SLC34, and even SLC17.

6.10. Summary and Perspectives

Practical benefits from characterizing neurotransmitter and nutrient amino acid transporters have extraordinary relevance to medicine and agriculture. The kinetic properties, abundance, variety, and uniqueness of NTTs and NATs make them prime targets to develop inexpensive insecticides and mosquitocides that are as potent as DDT and as safe as Bti (*Bacillus thuringiensis* subsp. *israelensis*). As providers of intracellular amino acids, plasma membrane transporters represent rate limiting components of crucial metabolic pathways. Low concentrations of high affinity or irreversibly binding inhibitors can block transport and abolish essential pathways, so such transport inhibitors may be potent insecticides. The abundance and variety of transport systems in insects provides a plethora of targets for potential amino acid transport inhibitors. This diversity may help to avoid the pitfalls of resistance that have doomed many previous insecticides such as DDT – before pests become resistant to one

transport inhibitor another one can be developed. In particular, the unique nature of mosquito NATs provides the potential for agents to be developed that will inhibit them without affecting vertebrate or other insect NATs and NTTs. The most likely class of potential mosquitocides would be amino acid derivatives and peptides. An almost infinite variety of such compounds can be synthesized and screened making use of new tools from recent advances in insect transport physiology and molecular biology. Vincent Wigglesworth invented Insect Physiology to help control tropical diseases. Today, integrative physiology and functional genomics of *An. gambiae* and other disease vectors can build on this basic science to provide rational strategies for impacting insect borne diseases through vector modification to reduce disease transmissibility and insecticide application to reduce vector population size.

Acknowledgments

Supported in part by The Whitney Laboratory and NIH Research grant R01 AI-30464.

References

Adams, M.D., Celniker, S.E., Holt, R.A., *et al.*, 2000. The genome sequence of *Drosophila melanogaster*. *Science* 287, 2185–2195.

Agulhon, C., Rostaing, P., Ravassard, P., Sagne, C., Triller, A., *et al.*, 2003. Lysosomal amino acid transporter LYAAT-1 in the rat central nervous system: an *in situ* hybridization and immunohistochemical study. *J. Comp. Neurol.* 462, 71–89.

Aihara, Y., Mashima, H., Onda, H., Hisano, S., Kasuya, H., *et al.*, 2000. Molecular cloning of a novel brain-type Na(+)-dependent inorganic phosphate cotransporter. *J. Neurochem.* 74, 2622–2625.

Albritton, L.M., Tseng, L., Scadden, D., Cunningham, J.M., 1989. A putative murine ecotropic retrovirus receptor gene encodes a multiple membrane-spanning protein and confers susceptibility to virus-infection. *Cell* 57, 659–666.

Amara, S.G., Fontana, A.C., 2002. Excitatory amino acid transporters: keeping up with glutamate. *Neurochem. Int.* 41, 313–318.

Amstutz, M., Mohrmann, M., Gmaj, P., Murer, H., 1985. Effect of pH on phosphate transport in rat renal brush border membrane vesicles. *Am. J. Physiol.* 248, F705–F710.

Arriza, J.L., Eliasof, S., Kavanaugh, M.P., Amara, S.G., 1997. Excitatory amino acid transporter 5, a retinal glutamate transporter coupled to a chloride conductance. *Proc. Natl Acad. Sci. USA* 94, 4155–4160.

Arriza, J.L., Fairman, W.A., Wadiche, J.I., Murdoch, G.H., Kavanaugh, M.P., *et al.*, 1994. Functional comparisons of three glutamate transporter subtypes cloned from human motor cortex. *J. Neurosci.* 14, 5559–5569.

Arriza, J.L., Kavanaugh, M.P., Fairman, W.A., Wu, Y.N., Murdoch, G.H., et al., 1993. Cloning and expression of a human neutral amino acid transporter with structural similarity to the glutamate transporter gene family. *J. Biol. Chem. 268*, 15329–15332.

Auger, C., Attwell, D., 2000. Fast removal of synaptic glutamate by postsynaptic transporters. *Neuron 28*, 547–558.

Bader, A.L., Parthasarathy, R., Harvey, W.R., 1995. A novel proline, glycine: K$^+$ symporter in midgut brush-border membrane vesicles from larval *Manduca sexta*. *J. Exp. Biol. 198*, 2599–2607.

Baier, A., Wittek, B., Brembs, B., 2002. *Drosophila* as a new model organism for the neurobiology of aggression? *J. Exp. Biol. 205*, 1233–1240.

Bajmoczi, M., Sneve, M., Eide, D.J., Drewes, L.R., 1998. TATI encodes a low-affinity histidine transporter in *Saccharomyces cerevisiae*. *Biochem. Biophys. Res. Commun. 243*, 205–209.

Balboni, E., 1978. A proline shuttle in insect flight muscle. *Biochem. Biophys. Res. Commun. 85*, 1090–1096.

Bassi, M.T., Gasol, E., Manzoni, M., Pineda, M., Riboni, M., et al., 2001. Identification and characterisation of human xCT that co-expresses, with 4F2 heavy chain, the amino acid transport activity system x(c) (-). *Pflugers Archiv. 442*, 286–296.

Bellocchio, E.E., Hu, H., Pohorille, A., Chan, J., Pickel, V.M., et al., 1998. The localization of the brain-specific inorganic phosphate transporter suggests a specific presynaptic role in glutamatergic transmission. *J. Neurosci. 18*, 8648–8659.

Bellocchio, E.E., Reimer, R.J., Fremeau, R.T., Jr., Edwards, R.H., 2000. Uptake of glutamate into synaptic vesicles by an inorganic phosphate transporter. *Science 289*, 957–960.

Bergersen, L., Johannsson, E., Veruki, M.L., Nagelhus, E.A., Halestrap, A., et al., 1999. Cellular and subcellular expression of monocarboxylate transporters in the pigment epithelium and retina of the rat. *Neuroscience 90*, 319–331.

Bergersen, L., Rafiki, A., Ottersen, O.P., 2002. Immunogold cytochemistry identifies specialized membrane domains for monocarboxylate transport in the central nervous system. *Neurochem. Res. 27*, 89–96.

Bertran, J., Werner, A., Moore, M.L., Stange, G., Markovich, D., et al., 1992. Expression cloning of a cDNA from rabbit kidney cortex that induces a single transport system for cystine and dibasic and neutral amino acids. *Proc. Natl Acad. Sci. USA 89*, 5601–5605.

Besson, M.T., Soustelle, L., Birman, S., 1999. Identification and structural characterization of two genes encoding glutamate transporter homologues differently expressed in the nervous system of *Drosophila melanogaster*. *FEBS Lett. 443*, 97–104.

Besson, M.T., Soustelle, L., Birman, S., 2000. Selective high-affinity transport of aspartate by a *Drosophila* homologue of the excitatory amino-acid transporters. *Curr. Biol. 10*, 207–210.

Beyenbach, K.W., 2001. Energizing epithelial transport with the vacuolar H(+)-ATPase. *News Physiol. Sci. 16*, 145–151.

Bicker, G., Schafer, S., Ottersen, O.P., Storm-Mathisen, J., 1988. Glutamate-like immunoreactivity in identified neuronal populations of insect nervous systems. *J. Neurosci. 8*, 2108–2122.

Bismuth, Y., Kavanaugh, M.P., Kanner, B.I., 1997. Tyrosine 140 of the gamma-aminobutyric acid transporter GAT-1 plays a critical role in neurotransmitter recognition. *J. Biol. Chem. 272*, 16096–16102.

Boll, M., Daniel, H., Gasnier, B., 2003a. The SLC36 family: proton-coupled transporters for the absorption of selected amino acids from extracellular and intracellular proteolysis. *Pflugers Arch.* 2003 May 14 (Epub ahead of print).

Boll, M., Foltz, M., Rubio-Aliaga, I., Daniel, H., 2003b. A cluster of proton/amino acid transporter genes in the human and mouse genomes. *Genomics 82*, 47–56.

Boll, M., Foltz, M., Rubio-Aliaga, I., Kottra, G., Daniel, H., 2002. Functional characterization of two novel mammalian electrogenic proton-dependent amino acid cotransporters. *J. Biol. Chem. 277*, 22966–22973.

Boll, M., Herget, M., Wagener, M., Weber, W.M., Markovich, D., et al., 1996. Expression cloning and functional characterization of the kidney cortex high-affinity proton-coupled peptide transporter. *Proc. Natl Acad. Sci. USA 93*, 284–289.

Boll, M., Markovich, D., Weber, W.M., Korte, H., Daniel, H., et al., 1994. Expression cloning of a cDNA from rabbit small-intestine related to proton-coupled transport of peptides, beta-lactam antibiotics and Ace-inhibitors. *Pflugers Arch. 429*, 146–149.

Bonen, A., 2000. Lactate transporters (MCT proteins) in heart and skeletal muscles. *Med. Sci. Sports Exercise 32*, 778–789.

Bossi, E., Centinaio, E., Castagna, M., Giovannardi, S., Vincenti, S., et al., 1999a. Ion binding and permeation through the lepidopteran amino acid transporter KAAT1 expressed in *Xenopus* oocytes. *J. Physiol. – Lond. 515*, 729–742.

Bossi, E., Sacchi, V.F., Peres, A., 1999b. Ionic selectivity of the coupled and uncoupled currents carried by the amino acid transporter KAAT1. *Pflugers Archiv. 438*, 788–796.

Boudko, D.Y., Kohn, A.B., Meleshkevitch, E.A., Dasher, M.K., Seron, T.J., et al., Ancestry and progency of nutrient amino acid transporters. *Proc. Natl Acad. Sci. USA* (in press).

Boudko, D.Y., Moroz, L.L., Harvey, W.R., Linser, P.J., 2001a. Alkalinization by chloride/bicarbonate pathway in larval mosquito midgut. *Proc. Natl Acad. Sci. USA 98*, 15354–15359.

Boudko, D.Y., Moroz, L.L., Linser, P.J., Trimarchi, J.R., Smith, P.J., et al., 2001b. *In situ* analysis of pH gradients in mosquito larvae using non-invasive, self-referencing, pH-sensitive microelectrodes. *J. Exp. Biol. 204*, 691–699.

Breer, H., Knipper, M., 1990. Regulation of high affinity choline uptake. *J. Neurobiol.* 21, 269–275.

Bridges, C.C., Kekuda, R., Wang, H.P., Prasad, P.D., Mehta, P., *et al.*, 2001. Structure, function, and regulation of human cystine/glutamate transporter in retinal pigment epithelial cells. *Invest. Ophthalmol. Vis. Sci.* 42, 47–54.

Broer, A., Albers, A., Setiawan, I., Edwards, R.H., Chaudhry, F.A., *et al.*, 2002. Regulation of the glutamine transporter SN1 by extracellular pH and intracellular sodium ions. *J. Physiol. –Lond.* 539, 3–14.

Broer, A., Wagner, C.A., Lang, F., Broer, S., 2000. The heterodimeric amino acid transporter 4F2hc/y+LAT2 mediates arginine efflux in exchange with glutamine. *Biochem. J.* 349, 787–795.

Broer, S., Broer, A., Schneider, H.P., Stegen, C., Halestrap, A.P., *et al.*, 1999. Characterization of the high-affinity monocarboxylate transporter MCT2 in *Xenopus laevis* oocytes. *Biochem. J.* 341, 529–535.

Broer, S., Schneider, H.P., Broer, A., Rahman, B., Hamprecht, B., *et al.*, 1998a. Characterization of the monocarboxylate transporter 1 expressed in *Xenopus laevis* oocytes by changes in cytosolic pH. *Biochem. J.* 333, 167–174.

Broer, S., Schuster, A., Wagner, C.A., Broer, A., Forster, I., *et al.*, 1998b. Chloride conductance and Pi transport are separate functions induced by the expression of NaPi-1 in *Xenopus* oocytes. *J. Membr. Biol.* 164, 71–77.

Brown, C.S., Nestler, C., 1985. Catecholamines and indolalkylamines. In: Kerkut, G.A., Gilbert, L.I. (Eds.), Comprehensive Insect Physiology, Biochemistry and Pharmacology, vol. 11. Pergamon Press, Oxford, pp. 436–497.

Busch, A.E., Biber, J., Murer, H., Lang, F., 1996a. Electrophysiological insights of type I and II Na/Pi transporters. *Kidney Int.* 49, 986–987.

Busch, A.E., Schuster, A., Waldegger, S., Wagner, C.A., Zempel, G., *et al.*, 1996b. Expression of a renal type I sodium/phosphate transporter (NaPi-1) induces a conductance in *Xenopus* oocytes permeable for organic and inorganic anions. *Proc. Natl Acad. Sci. USA* 93, 5347–5351.

Callec, J.J., 1985. Synaptic transmission in the central nervous system. In: Kerkut, G.A., Gilbert, L.I. (Eds.), Comprehensive Insect Physiology, Biochemistry and Pharmacology, vol. 5. Pergamon Press, Oxford, pp. 139–179.

Carew, T.J., 1996. Molecular enhancement of memory formation. *Neuron* 16, 5–8.

Castagna, M., Shayakul, C., Trotti, D., Sacchi, V.F., Harvey, W.R., *et al.*, 1997. Molecular characteristics of mammalian and insect amino acid transporters: implications for amino acid homeostasis. *J. Exp. Biol.* 200, 269–286.

Castagna, M., Shayakul, C., Trotti, D., Sacchi, V.F., Harvey, W.R., *et al.*, 1998. Cloning and characterization of a potassium-coupled amino acid transporter. *Proc. Natl Acad. Sci. USA* 95, 5395–5400.

Castagna, M., Vincenti, S., Marciani, P., Sacchi, V.F., 2002. Inhibition of the lepidopteran amino acid co-transporter KAAT1 by phenylglyoxal: role of arginine 76. *Insect Mol. Biol.* 11, 283–289.

Caveney, S., Donly, B.C., 2002. Neurotransmitter transporters in the insect nervous system. *Adv. Insect Physiol.* 29, 55–149.

Chaudhry, F.A., Krizaj, D., Larsson, P., Reimer, R.J., Wreden, C., *et al.*, 2001. Coupled and uncoupled proton movement by amino acid transport system N. *EMBO J.* 20, 7041–7051.

Chen, H., Pan, Y.X., Wong, E.A., Bloomquist, J.R., Webb, K.E., 2002. Molecular cloning and functional expression of a chicken intestinal peptide transporter (cPepT1) in *Xenopus* oocytes and Chinese hamster ovary cells. *J. Nutr.* 132, 387–393.

Chen, N.H., Reith, M.E., Quick, M.W., 2003. Synaptic uptake and beyond: the sodium- and chloride-dependent neurotransmitter transporter family SLC6. *Pflugers Arch.* 2003 Apr 29 (Epub ahead of print).

Chen, P.S., 1966. Amino acid and protein metabolism in insect development. *Adv. Insect Physiol.* 3, 53–132.

Chen, Y., Swanson, R.A., 2003. The glutamate transporters EAAT2 and EAAT3 mediate cysteine uptake in cortical neuron cultures. *J. Neurochem.* 84, 1332–1339.

Chen, Z., Fei, Y.J., Anderson, C.M., Wake, K.A., Miyauchi, S., *et al.*, 2003. Structure, function and immunolocalization of a proton-coupled amino acid transporter (hPAT1) in the human intestinal cell line Caco-2. *J. Physiol.* 546, 349–361.

Chiang, A.S., Lin, W.Y., Liu, H.P., Pszczolkowski, M.A., Fu, T.F., *et al.*, 2002. Insect NMDA receptors mediate juvenile hormone biosynthesis. *Proc. Natl Acad. Sci. USA* 99, 37–42.

Chillaron, J., Estevez, R., Mora, C., Wagner, C.A., Suessbrich, H., *et al.*, 1996. Obligatory amino acid exchange via systems bo,+-like and y+L-like. A tertiary active transport mechanism for renal reabsorption of cystine and dibasic amino acids. *J. Biol. Chem.* 271, 17761–17770.

Chillaron, J., Roca, R., Valencia, A., Zorzano, A., Palacin, M., 2001. Heteromeric amino acid transporters: biochemistry, genetics, and physiology. *Am. J. Physiol. Renal Physiol.* 281, F995–F1018.

Chiu, C.-S., Ross, L.S., Cohen, B.N., Lester, H.A., Gill, S.S., 2000. The transporter-like protein inebriated mediates hyperosmotic stimuli through intracellular signalling. *J. Exp. Biol.* 203, 3531–3546.

Christensen, H.N., 1964. Transport system serving for mono-+ diamino acids. *Proc. Natl Acad. Sci. USA* 51, 337–344.

Cioffi, M., Wolfersberger, M.G., 1983. Isolation of separate apical, lateral and basal plasma membrane from cells of an insect epithelium. A procedure based on tissue organization and ultrastructure. *Tissue Cell* 15, 781–803.

Clark, J.A., Amara, S.G., **1993**. Amino acid neurotransmitter transporters: structure, function, and molecular diversity. *Bioessays 15*, 323–332.

Clements, A.N., **1992**. The Biology of Mosquitoes. Chapman and Hall, London.

Closs, E.I., **2002**. Expression, regulation and function of carrier proteins for cationic amino acids. *Curr. Opin. Nephrol. Hypertens. 11*, 99–107.

Cociancich, S.O., Park, S.S., Fidock, D.A., Shahabuddin, M., **1999**. Vesicular ATPase-overexpressing cells determine the distribution of malaria parasite oocysts on the midguts of mosquitoes. *J. Biol. Chem. 274*, 12650–12655.

Collett, J.I., **1976**. Small peptides, a life-long store of amino acid in adult *Drosophila* and Calliphora. *J. Insect Physiol. 22*, 433–440.

Collins, J.F., Bai, L., Ghishan, F.K., **2003**. The SLC20 family of proteins: dual functions as sodium-phosphate cotransporters and viral receptors. *Pflugers Arch.* May 21 (Epub ahead of print).

Copeland, J., Robertson, H.A., **1982**. Octopamine as the transmitter at the firefly lantern – presence of an octopamine-sensitive and a dopamine-sensitive adenylate-cyclase. *Comp. Biochem. Physiol. C–Pharmacol. Toxicol. Endocrinol. 72*, 125–127.

Corey, J.L., Quick, M.W., Davidson, N., Lester, H.A., Guastella, J., **1994**. A cocaine-sensitive *Drosophila* serotonin transporter: cloning, expression, and electrophysiological characterization. *Proc. Natl Acad. Sci. USA 91*, 1188–1192.

Danbolt, N.C., **2001**. Glutamate uptake. *Prog. Neurobiol. 65*, 1–105.

Daniel, H., Kottra, G., **2003**. The proton oligopeptide cotransporter family SLC15 in physiology and pharmacology. *Pflugers Arch.* 2003 Aug 7 (Epub ahead of print).

Darboux, I., Nielsen-LeRoux, C., Charles, J.F., Pauron, D., **2001**. The receptor of *Bacillus sphaericus* binary toxin in *Culex pipiens* (Diptera: Culicidae) midgut: molecular cloning and expression. *Insect Biochem. Mol. Biol. 31*, 981–990.

Deken, S.L., Fremeau, R.T.J., Quick, M.W., **2002**. Family of sodium-coupled transporters for GABA, glycine, proline, betaine, taurine, and creatine. In: Reith, M.E.A. (Ed.), Neurotransmitter Transporters: Structure, Function and Regulation. Humana Press Inc., Totowa, New Jersey, pp. 193–234.

Demchyshyn, L.L., Pristupa, Z.B., Sugamori, K.S., Barker, E.L., Blakely, R.D., *et al.*, **1994**. Cloning, expression, and localization of a chloride-facilitated, cocaine-sensitive serotonin transporter from *Drosophila melanogaster. Proc. Natl Acad. Sci. USA 91*, 5158–5162.

Deuticke, B., **1982**. Monocarboxylate transport in erythrocytes. *J. Membr. Biol. 70*, 89–103.

Dimmer, K.S., Friedrich, B., Lang, F., Deitmer, J.W., Broer, S., **2000**. The low-affinity monocarboxylate transporter MCT4 is adapted to the export of lactate in highly glycolytic cells. *Biochem. J. 350*, 219–227.

Donly, B.C., Richman, A., Hawkins, E., McLean, H., Caveney, S., **1997**. Molecular cloning and functional expression of an insect high-affinity Na$^+$-dependent glutamate transporter. *Eur. J. Biochem. 248*, 535–542.

Donly, C., Jevnikar, J., McLean, H., Caveney, S., **2000**. Substrate-stereoselectivity of a high-affinity glutamate transporter cloned from the CNS of the cockroach *Diploptera punctata. Insect Biochem. Mol. Biol. 30*, 369–376.

Drose, S., Boddien, C., Gassel, M., Ingenhorst, G., Zeeck, A., *et al.*, **2001**. Semisynthetic derivatives of concanamycin A and C, as inhibitors of V- and P-type ATPases: structure-activity investigations and developments of photoaffinity probes. *Biochemistry 40*, 2816–2825.

Eiden, L.E., Schafer, M.K., Weihe, E., Schutz, B., **2003**. The vesicular amine transporter family (SLC18): amine/proton antiporters required for vesicular accumulation and regulated exocytotic secretion of monoamines and acetylcholine. *Pflugers Arch.* 2003 May 6 (Epub ahead of print).

Erickson, J.D., Eiden, L.E., Hoffman, B.J., **1992**. Expression cloning of a reserpine-sensitive vesicular monoamine transporter. *Proc. Natl Acad. Sci. USA 89*, 10993–10997.

Fairman, W.A., Vandenberg, R.J., Arriza, J.L., Kavanaugh, M.P., Amara, S.G., **1995**. An excitatory amino-acid transporter with properties of a ligand-gated chloride channel. *Nature 375*, 599–603.

Farooqui, T., Robinson, K., Vaessin, H., Smith, B.H., **2003**. Modulation of early olfactory processing by an octopaminergic reinforcement pathway in the honeybee. *J. Neurosci. 23*, 5370–5380.

Fei, Y.J., Kanai, Y., Nussberger, S., Ganapathy, V., Leibach, F.H., *et al.*, **1994**. Expression cloning of a mammalian proton-coupled oligopeptide transporter. *Nature 368*, 563–566.

Feldman, D.H., Harvey, W.R., Stevens, B.R., **2000**. A novel electrogenic amino acid transporter is activated by K$^+$ or Na$^+$, is alkaline pH-dependent, and is Cl$^-$-independent. *J. Biol. Chem. 275*, 24518–24526.

Fernandez, E., Torrents, D., Chillaron, J., Martin Del Rio, R., Zorzano, A., *et al.*, **2003**. Basolateral LAT-2 has a major role in the transepithelial flux of L-cystine in the renal proximal tubule cell line OK. *J. Am. Soc. Nephrol. 14*, 837–847.

Fernandez, J., Bode, B., Koromilas, A., Diehl, J.A., Krukovets, I., *et al.*, **2002**. Translation mediated by the internal ribosome entry Site of the cat-1 mRNA is regulated by glucose availability in a PERK kinase-dependent manner. *J. Biol. Chem. 277*, 11780–11787.

Filippova, M., Ross, L.S., Gill, S.S., **1998**. Cloning of the v-ATPase B subunit from *Culex Quinquefasciatus* and expression of the B and C subunits in mosquitoes. *Insect Mol. Biol. 7*, 223–232.

Fremeau, R.T., Jr., Burman, J., Qureshi, T., Tran, C.H., Proctor, J., *et al.*, **2002**. The identification of vesicular glutamate transporter 3 suggests novel modes of signaling by glutamate. *Proc. Natl Acad. Sci. USA 99*, 14488–14493.

Fremeau, R.T., Troyer, M.D., Pahner, I., Nygaard, G.O., Tran, C.H., *et al.*, **2001**. The expression of vesicular glutamate transporters defines two classes of excitatory synapse. *Neuron 31*, 247–260.

Furuya, S., Watanabe, M., 2003. Novel neuroglial and glioglial relationships mediated by L-serine metabolism. *Arch. Histol. Cytol. 66*, 109–121.

Gade, G., Auerswald, L., 2002. Beetles' choice – proline for energy output: control by AKHs. *Comp. Biochem. Physiol. B Biochem. Mol. Biol. 132*, 117–129.

Gainetdinov, R.R., Caron, M.G., 2002. Monoamine transporters. In: Reith, M.E.A. (Ed.), Neurotransmitter Transporters: Structure, Function and Regulation. Humana Press Inc., Totowa, New Jersey, pp. 171–192.

Gallant, P., Malutan, T., McLean, H., Verellen, L., Caveney, S., *et al.*, 2003. Functionally distinct dopamine and octopamine transporters in the CNS of the cabbage looper moth. *Eur. J. Biochem. 270*, 664–674.

Gao, X.J., McLean, H., Caveney, S., Donly, C., 1999. Molecular cloning and functional characterization of a GABA transporter from the CNS of the cabbage looper, *Trichoplusia ni. Insect Biochem. Mol. Biol. 29*, 609–623.

Garcia, C.K., Brown, M.S., Pathak, R.K., Goldstein, J.L., 1995. cDNA cloning of MCT2, a second monocarboxylate transporter expressed in different cells than MCT1. *J. Biol. Chem. 270*, 1843–1849.

Garcia, C.K., Li, X., Luna, J., Francke, U., 1994. cDNA cloning of the human monocarboxylate transporter-1 and chromosomal localization of the Slc16a1 locus to 1p13.2-P12. *Genomics 23*, 500–503.

Gardiner, R.B., Ullensvang, K., Danbolt, N.C., Caveney, S., Donly, B.C., 2002. Cellular distribution of a high-affinity glutamate transporter in the nervous system of the cabbage looper *Trichoplusia ni. J. Exp. Biol. 205*, 2605–2614.

Gasic, G.P., Hollmann, M., 1992. Molecular neurobiology of glutamate receptors. *Annu. Rev. Physiol. 54*, 507–536.

Gasnier, B., 2003. The SLC32 transporter, a key protein for the synaptic release of inhibitory amino acids. *Pflugers Arch.* 2003 May 16 (Epub ahead of print).

Gerencser, G.A., Stevens, B.R., 1994. Thermodynamics of symport and antiport catalyzed by cloned or native transporters. *J. Exp. Biol. 196*, 59–75.

Giordana, B., Parenti, P., 1994. Determinants for the activity of the neutral amino acid/K$^+$ symport in lepidopteran larval midgut. *J. Exp. Biol. 196*, 145–155.

Giordana, B., Sacchi, V.F., Parenti, P., Hanozet, G.M., 1989. Amino acid transport systems in intestinal brush-border membranes from lepidopteran larvae. *Am. J. Physiol. 257*, R494–R500.

Gleason, K.K., Dondeti, V.R., Hsia, H.L., Cochran, E.R., Gumulak-Smith, J., *et al.*, 2003. The vesicular glutamate transporter 1 (xVGlut1) is expressed in discrete regions of the developing *Xenopus laevis* nervous system. *Gene Expr. Patterns 3*, 503–507.

Graf, P., Forstermann, U., Closs, E. I., 2001. The transport activity of the human cationic amino acid transporter hCAT-1 is downregulated by activation of protein kinase C. *Br. J. Pharmacol. 132*, 1193–1200.

Guastella, J., Nelson, N., Nelson, H., Czyzyk, L., Keynan, S., *et al.*, 1990. Cloning and expression of a rat brain GABA transporter. *Science 249*, 1303–1306.

Halestrap, A.P., Meredith, D., 2003. The SLC16 gene family – from monocarboxylate transporters (MCTs) to aromatic amino acid transporters and beyond. *Pflugers Arch.* Aug 28 (Epub ahead of print).

Hanozet, G.M., Giordana, B., Sacchi, V.F., 1980. K$^+$-dependent phenylalanine uptake in membrane vesicles isolated from the midgut of *Philosamia cynthia* larvae. *Biochim. Biophys. Acta. 596*, 481–486.

Harvey, W.R., 1992. Physiology of V-ATPases. *J. Exp. Biol. 172*, 1–17.

Harvey, W.R., Cioffi, M., Dow, J.A., Wolfersberger, M.G., 1983. Potassium ion transport ATPase in insect epithelia. *J. Exp. Biol. 106*, 91–117.

Harvey, W.R., Haskell, J.A., Nedergaard, S., 1968. Active transport by the Cecropia midgut. III. Midgut potential generated directly by active K-transport. *J. Exp Biol. 48*, 1–12.

Harvey, W.R., Haskell, J.A., Zerahn, K., 1967. Active transport of potassium and oxygen consumption in the isolated midgut of *Hyalophora cecropia. J. Exp. Biol. 46*, 235–248.

Harvey, W.R., Maddrell, S.H.P., Telfer, W.H., Wieczorek, H., 1998. H$^+$ V-ATPases energize animal plasma membranes for secretion and absorption of ions and fluids. *Am. Zool. 38*, 426–441.

Harvey, W.R., Pung, L., Meleshkevitch, E.A., Kohn, A., Boudko, D.Y., 2003. Molecular and electrochemical integration of nutrient amino acid uptake in mosquito larvae. *Proc. 6th Int. Congr. Comp. Physiol. Biochem.* Mt. Buller, Australia.

Hayashi, M., Morimoto, R., Yamamoto, A., Moriyama, Y., 2003. Expression and localization of vesicular glutamate transporters in pancreatic islets, upper gastrointestinal tract, and testis. *J. Histochem. Cytochem. 51*, 1375–1390.

Hayashi, M., Otsuka, M., Morimoto, R., Hirota, S., Yatsushiro, S., *et al.*, 2001. Differentiation-associated Na$^+$-dependent inorganic phosphate cotransporter (DNPI) is a vesicular glutamate transporter in endocrine glutamatergic systems. *J. Biol. Chem. 276*, 43400–43406.

Hemler, M.E., Strominger, J.L., 1982. Characterization of antigen recognized by the monoclonal antibody (4F2): different molecular forms on human T and B lymphoblastoid cell lines. *J. Immunol. 129*, 623–628.

Hennigan, B.B., Wolfersberger, M.G., Harvey, W.R., 1993a. Neutral amino acid symport in larval *Manduca sexta* midgut brush-border membrane vesicles deduced from cation-dependent uptake of leucine, alanine, and phenylalanine. *Biochim. Biophys. Acta 1148*, 216–222.

Hennigan, B.B., Wolfersberger, M.G., Parthasarathy, R., Harvey, W.R., 1993b. Cation-dependent leucine, alanine, and phenylalanine uptake at pH 10 in brush-border membrane vesicles from larval *Manduca sexta* midgut. *Biochim. Biophys. Acta 1148*, 209–215.

Holt, R.A., Subramanian, G.M., Halpern, A., *et al.*, 2002. The genome sequence of the malaria mosquito *Anopheles gambiae*. *Science* 298, 129–149.

Homberg, U., 2002. Neurotransmitters and neuropeptides in the brain of the locust. *Microsc. Res. Tech. 56*, 189–209.

Huang, X., Huang, Y., Chinnappan, R., Bocchini, C., Gustin, M.C., *et al.*, 2002. The *Drosophila* inebriated-encoded neurotransmitter/osmolyte transporter: dual roles in the control of neuronal excitability and the osmotic stress response. *Genetics 160*, 561–569.

Huang, Y., Stern, M., 2002. *In vivo* properties of the *Drosophila* inebriated-encoded neurotransmitter transporter. *J. Neurosci. 22*, 1698–1708.

Huggett, J.F., Mustafa, A., O'Neal, L., Mason, D.J., 2002. The glutamate transporter GLAST-1 (EAAT-1) is expressed in the plasma membrane of osteocytes and is responsive to extracellular glutamate concentration. *Biochem. Soc. Trans. 30*, 890–893.

Jackson, V.N., Price, N.T., Carpenter, L., Halestrap, A.P., 1997. Cloning of the monocarboxylate transporter isoform MCT2 from rat testis provides evidence that expression in tissues is species-specific and may involve post-transcriptional regulation. *Biochem. J. 324*, 447–453.

Jan, L.Y., Jan, Y.N., 1976a. L-glutamate as an excitatory transmitter at the *Drosophila* larval neuromuscular junction. *J. Physiol. 262*, 215–236.

Jan, L.Y., Jan, Y.N., 1976b. Properties of the larval neuromuscular junction in *Drosophila melanogaster*. *J. Physiol. 262*, 189–214.

Janecek, S., 1994. Sequence similarities and evolutionary relationships of microbial, plant and animal alpha-amylases. *Eur. J. Biochem. 224*, 519–524.

Jespersen, T., Grunnet, M., Angelo, K., Klaerke, D.A., Olesen, S.P., 2002. Dual-function vector for protein expression in both mammalian cells and *Xenopus laevis* oocytes. *Biotechniques 32*, 536–538, 540.

Jin, X., Aimanova, K., Ross, L.S., Gill, S.S., 2003. Identification, functional characterization and expression of a LAT type amino acid transporter from the mosquito *Aedes aegypti*. *Insect Biochem. Mol. Biol. 33*, 815–827.

Johnson, K., Knust, E., Skaer, H., 1999. bloated tubules (blot) encodes a *Drosophila* member of the neurotransmitter transporter family required for organisation of the apical cytocortex. *Devel. Biol. 212*, 440–454.

Kanai, Y., Hediger, M.A., 1992. Primary structure and functional characterization of a high-affinity glutamate transporter. *Nature 360*, 467–471.

Kanai, Y., Smith, C.P., Hediger, M.A., 1993. The elusive transporters with a high affinity for glutamate. *Trends Neurosci. 16*, 365–370.

Kanai, Y., Trotti, D., Berger, U.V., 2002. The high-affinity glutamate and neutral amino-acid transporter family. In: Reith, M.E.A. (Ed.), Neurotransmitter Transporters: Structure, Function and Regulation. Humana Press Inc., Totowa, New Jersey, pp. 255–312.

Kavanaugh, M.P., Miller, D.G., Zhang, W., Law, W., Kozak, S.L., *et al.*, 1994. Cell-surface receptors for gibbon ape leukemia virus and amphotropic murine retrovirus are inducible sodium-dependent phosphate symporters. *Proc. Natl Acad. Sci. USA 91*, 7071–7075.

Kekuda, R., Prasad, P.D., Fei, Y.J., Torres-Zamorano, V., Sinha, S., *et al.*, 1996. Cloning of the sodium-dependent, broad-scope, neutral amino acid transporter Bo from a human placental choriocarcinoma cell line. *J. Biol. Chem. 271*, 18657–18661.

Kim do, K., Kanai, Y., Matsuo, H., Kim, J.Y., Chairoungdua, A., *et al.*, 2002. The human T-type amino acid transporter-1: characterization, gene organization, and chromosomal location. *Genomics 79*, 95–103.

Kim, C.M., Goldstein, J.L., Brown, M.S., 1992. Cdna Cloning of Mev, a mutant protein that facilitates cellular uptake of mevalonate, and identification of the point mutation responsible for its gain of function. *J. Biol. Chem. 267*, 23113–23121.

Kim, D.K., Kanai, Y., Chairoungdua, A., Enomoto, A., Inatomi, J., *et al.*, 2002. Identification and characterization of a novel epithelial aromatic amino acid transporter TAT1. *Jap. J. Pharmacol. 88*, 223P.

Kim, D.K., Kanai, Y., Chairoungdua, A., Matsuo, H., Cha, S.H., *et al.*, 2001. Expression cloning of a Na$^+$-independent aromatic amino acid transporter with structural similarity to H+/monocarboxylate transporters. *J. Biol. Chem. 276*, 17221–17228.

Kim, J.W., Closs, E.I., Albritton, L.M., Cunningham, J.M., 1991. Transport of cationic amino acids by the mouse ecotropic retrovirus receptor. *Nature 352*, 725–728.

Kirk, P., Wilson, M.C., Heddle, C., Brown, M.H., Barclay, A.N., *et al.*, 2000. CD147 is tightly associated with lactate transporters MCT1 and MCT4 and facilitates their cell surface expression. *EMBO J. 19*, 3896–3904.

Kirschner, M.A., Copeland, N.G., Gilbert, D.J., Jenkins, N.A., Amara, S.G., 1994. Mouse excitatory amino acid transporter EAAT2: isolation, characterization, and proximity to neuroexcitability loci on mouse chromosome 2. *Genomics 24*, 218–224.

Kitamoto, T., Wang, W., Salvaterra, P.M., 1998. Structure and organization of the *Drosophila* cholinergic locus. *J. Biol. Chem. 273*, 2706–2713.

Kitamoto, T., Xie, X., Wu, C. F., Salvaterra, P. M., 2000a. Isolation and characterization of mutants for the vesicular acetylcholine transporter gene in *Drosophila melanogaster*. *J. Neurobiol. 42*, 161–171.

Kitamoto, T., Xie, X., Wu, C.-F., Salvaterra, P.M., 2000b. Isolation and characterization of mutants for the vesicular acetylcholine transporter gene in *Drosophila melanogaster*. *J. Neurobiol. 42*, 161–171.

Klemm, N., Schneider, L., 1975. Selective uptake of indolamine into nervous fibers in brain of desert locust, Schistocerca-Gregaria Forskal (Insecta) – fluorescence and electron-microscopic investigation. *Comp. Biochem. Physiol. CPharmacol. Toxicol. Endocrinol. 50*, 177–182.

Knipper, M., Strotmann, J., Mädler, U., Kahle, C., Breer, H., 1989. Monoclonal antibodies against the high affinity choline transport system. *Neurochem. Int. 14*, 217–222.

Kostowski, W., Tarchalska-Krynska, B., Markowska, L., 1975. Aggressive behavior and brain serotonin and catecholamines in ants (*Formica rufa*). *Pharmacol. Biochem. Behav. 3*, 717–719.

Krantz, D.E., Chaudhry, F.A., Edwards, R.H., 1999. Neurotransmitter transporters. In: Bellen, H. (Ed.), Neurotransmitter Release. Oxford University Press, Oxford, pp. 145–207.

Kucharski, R., Ball, E.E., Hayward, D.C., Maleszka, R., 2000. Molecular cloning and expression analysis of a cDNA encoding a glutamate transporter in the honeybee brain. *Gene 242*, 399–405.

Law, J.H., Dunn, P.E., Kramer, K.J., 1977. Insect proteases and peptidases. *Adv. Enzymol. Relat. Areas Mol. Biol. 45*, 389–425.

Leal, S.M., Neckameyer, W.S., 2002. Pharmacological evidence for GABAergic regulation of specific behaviors in *Drosophila melanogaster*. *J. Neurobiol. 50*, 245–261.

Leitch, B., Judge, S., Pitman, R.M., 2003. Octopaminergic modulation of synaptic transmission between an identified sensory afferent and flight motorneuron in the locust. *J. Comp. Neurol. 462*, 55–70.

Leonardi, M.G., Casartelli, M., Fiandra, L., Parenti, P., Giordana, B., 2001a. Role of specific activators of intestinal amino acid transport in *Bombyx mori* larval growth and nutrition. *Arch. Insect Biochem. Physiol. 48*, 190–198.

Leonardi, M.G., Fiandra, L., Casartelli, M., Cappellozza, S., Giordana, B., 2001b. Modulation of leucine absorption in the larval midgut of *Bombyx mori* (Lepidoptera, Bombycidae). *Comp. Biochem. Physiol. A Mol. Integr. Physiol. 129*, 665–672.

Levi, G., Raiteri, M., 1974. Exchange of neurotransmitter amino-acid at nerve-endings can simulate high affinity uptake. *Nature 250*, 735–737.

Li, R., Younes, M., Frolov, A., Wheeler, T.M., Scardino, P., et al., 2003. Expression of neutral amino acid transporter ASCT2 in human prostate. *Anticancer Res. 23*, 3413–3418.

Liang, R., Fei, Y.J., Prasad, P.D., Ramamoorthy, S., Han, H., et al., 1995. Human intestinal H^+/peptide cotransporter. Cloning, functional expression, and chromosomal localization. *J. Biol. Chem. 270*, 6456–6463.

Lill, H., Nelson, N., 1998. Homologies and family relationships among Na^+/Cl^- neurotransmitter transporters. *Meth. Enzymol. 296*, 425–436.

Lin, R.Y., Vera, J.C., Chaganti, R.S., Golde, D.W., 1998. Human monocarboxylate transporter 2 (MCT2) is a high affinity pyruvate transporter. *J. Biol. Chem. 273*, 28959–28965.

Liu, Q.R., Mandiyan, S., Nelson, H., Nelson, N., 1992a. A family of genes encoding neurotransmitter transporters. *Proc. Natl Acad. Sci. USA 89*, 6639–6643.

Liu, Y., Peter, D., Roghani, A., Schuldiner, S., Prive, G.G., et al., 1992b. A cDNA that suppresses MPP^+ toxicity encodes a vesicular amine transporter. *Cell 70*, 539–551.

Liu, Z., Harvey, W.R., 1996a. Cationic lysine uptake by System R^+ and zwitterionic lysine uptake by System B in brush border membrane vesicles from larval *Manduca sexta* midgut. *Biochim. Biophys. Acta 1282*, 32–38.

Liu, Z., Harvey, W.R., 1996b. Arginine uptake through a novel cationic amino acid:K^+ symporter, System R^+, in brush border membrane vesicles from larval *Manduca sexta* midgut. *Biochim. Biophys. Acta 1282*, 25–31.

Liu, Z., Stevens, B.R., Feldman, D.H., Hediger, M.A., Harvey, W.R., 2003. K^+ amino acid transporter KAAT1 mutant Y147F has increased transport activity and altered substrate selectivity. *J. Exp. Biol. 206*, 245–254.

MacGregor, E.A., Janecek, S., Svensson, B., 2001. Relationship of sequence and structure to specificity in the alpha-amylase family of enzymes. *Biochim. Biophys. Acta 1546*, 1–20.

Mackenzie, B., Erickson, J.D., 2003. Sodium-coupled neutral amino acid (System N/A) transporters of the SLC38 gene family. *Pflugers Arch.* 2003 Jul 4 (Epub ahead of print).

Maddrell, S.H., 1962. A diuretic hormone in *Rhodnius Prolixus Stal. Nature 194*, 605–606.

Magagnin, S., Werner, A., Markovich, D., Sorribas, V., Stange, G., et al., 1993. Expression cloning of human and rat renal cortex Na/Pi cotransport. *Proc. Natl Acad. Sci. USA 90*, 5979–5983.

Maleszka, R., Helliwell, P., Kucharski, R., 2000. Pharmacological interference with glutamate re-uptake impairs long-term memory in the honeybee, *Apis mellifera*. *Behav. Brain Res. 115*, 49–53.

Malutan, T., McLean, H., Caveney, S., Donly, C., 2002. A high-affinity octopamine transporter cloned from the central nervous system of cabbage looper *Trichoplusia ni. Insect Biochem. Mol. Biol. 32*, 343–357.

Mandel, L.J., Moffett, D.F., Riddle, T.G., Grafton, M.M., 1980. Coupling between oxidative metabolism and active transport in the midgut of tobacco hornworm. *Am. J. Physiol. 238*, C1–C9.

Manning Fox, J.E., Meredith, D., Halestrap, A.P., 2000. Characterisation of human monocarboxylate transporter 4 substantiates its role in lactic acid efflux from skeletal muscle. *J. Physiol. 529*, 285–293.

Marin, M., Lavillette, D., Kelly, S.M., Kabat, D., 2003. N-linked glycosylation and sequence changes in a critical negative control region of the ASCT1 and ASCT2 neutral amino acid transporters determine their retroviral receptor functions. *J. Virol. 77*, 2936–2945.

Masson, J., Sagne, C., Hamon, M., El Mestikawy, S., 1999. Neurotransmitter transporters in the central nervous system. *Pharmacol. Rev. 51*, 439–464.

Mastroberardino, L., Spindler, B., Pfeiffer, R., Skelly, P.J., Loffing, J., et al., 1998. Amino-acid transport by heterodimers of 4F2hc/CD98 and members of a permease family. *Nature 395*, 288–291.

Mathews, G.C., Diamond, J.S., 2003. Neuronal glutamate uptake contributes to GABA synthesis and inhibitory synaptic strength. *J. Neurosci. 23*, 2040–2048.

Matz, M.V., 2002. Amplification of representative cDNA samples from microscopic amounts of invertebrate tissue to search for new genes. *Methods Mol. Biol. 183,* 3–18.

Mbungu, D., Ross, L.S., Gill, S.S., 1995. Cloning, functional expression, and pharmacology of a GABA transporter from *Manduca sexta. Arch. Biochem. Biophys. 318,* 489–497.

McDonald, T.J., 1975. Neuromuscular pharmacology of insects. *Ann. Rev. Entomol. 20,* 151–166.

McIntire, S.L., Jorgensen, E., Horvitz, H.R., 1993. Genes required for GABA function in *Caenorhabditis elegans. Nature 364,* 334–337.

McIntire, S.L., Reimer, R.J., Schuske, K., Edwards, R.H., Jorgensen, E.M., 1997. Identification and characterization of the vesicular GABA transporter. *Nature 389,* 870–876.

Meier, C., Ristic, Z., Klauser, S., Verrey, F., 2002. Activation of system L heterodimeric amino acid exchangers by intracellular substrates. *EMBO J. 21,* 580–589.

Meredith, D., Bell, P., McClure, B., Wilkins, R., 2002. Functional and molecular characterisation of lactic acid transport in bovine articular chondrocytes. *Cell. Physiol. Biochem. 12,* 227–234.

Meredith, D., Halestrap, A.P., 2000. OX47 (basigin) may act as a chaperone for expression of the monocarboxylate transporter MCT1 at the plasma membrane of *Xenopus laevis* oocytes. *J. Physiol. Lond. 526,* 23P–24P.

Miyamoto, K., Tatsumi, S., Sonoda, T., Yamamoto, H., Minami, H., *et al.,* 1995. Cloning and functional expression of a Na(+)-dependent phosphate co-transporter from human kidney: cDNA cloning and functional expression. *Biochem. J. 305,* 81–85.

Monaghan, D.T., Bridges, R.J., Cotman, C.W., 1989. The excitatory amino acid receptors: their classes, pharmacology, and distinct properties in the function of the central nervous system. *Annu. Rev. Pharmacol. Toxicol. 29,* 365–402.

Monastirioti, M., 1999. Biogenic amine systems in the fruit fly *Drosophila melanogaster. Microsc. Res. Tech. 45,* 106–121.

Murer, H., Forster, I., Biber, J., 2003. The sodium phosphate cotransporter family SLC34. *Pflugers Arch. 16,* 16.

Murer, H., Hernando, N., Forster, I., Biber, J., 2000. Proximal tubular phosphate reabsorption: molecular mechanisms. *Physiol. Rev. 80,* 1373–1409.

Muszynska-Pytel, M., Cymboroski, B., 1978a. The role of serotonin in regulation of the circadian rhythm of locomotor activity in the cricket (*Acheta domesticus* L.) II. Distribution of serotonin and variations in different brain structure. *Comp. Biochem. Physiol. C 59,* 17–20.

Muzynska-Pytel, M., Cymborowski, B., 1978b. The role of serotonin in regulation of the circadian rhythms of locomotor activity in the cricket (*Acheta domesticus* L.) I. Circadian variations in serotonin concentration in the brain and hemolymph. *Comp. Biochem. Physiol. C 59,* 13–15.

Nagaya, Y., Kutsukake, M., Chigusa, S.I., Komatsu, A., 2002. A trace amine, tyramine, functions as a neuromodulator in *Drosophila melanogaster. Neurosci. Lett. 329,* 324–328.

Nathanson, J.A., Hunnicutt, E.J., Kantham, L., Scavone, C., 1993. Cocaine as a naturally occurring insecticide. *Proc. Natl Acad. Sci. USA 90,* 9645–9648.

Neckameyer, W.S., 1998. Dopamine and mushroom bodies in *Drosophila*: experience-dependent and-independent aspects of sexual behavior. *Learn. Mem. 5,* 157–165.

Nedergaard, S., 1972. Active transport of alpha-aminoisobutyric acid by the isolated midgut of *Hyalophora cecropia. J. Exp. Biol. 56,* 167–172.

Nelson, N., Lill, H., 1994. Porters and neurotransmitter transporters. *J. Exp. Biol. 196,* 213–228.

Nelson, N., Sacher, A., Nelson, H., 2002. The significance of molecular slips in transport systems. *Nat. Rev. Mol. Cell. Biol. 3,* 876–881.

Ni, B., Rosteck, P.R., Jr., Nadi, N.S., Paul, S.M., 1994. Cloning and expression of a cDNA encoding a brain-specific Na(+)-dependent inorganic phosphate cotransporter. *Proc. Natl Acad. Sci. USA 91,* 5607–5611.

Nielsen-LeRoux, C., Charles, J.-F., 1992. Binding of *Bacillus sphaericus* binary toxin to a specific receptor on midgut brush-border membranes from mosquito larvae. *Eur. J. Biochem. 210,* 585–590.

Novak, M.G., Rowley, W.A., 1994. Serotonin depletion affects blood-feeding but not host-seeking ability in *Aedes triseriatus* (Diptera: Culicidae). *J. Med. Entomol. 31,* 600–606.

O'Donnell, M.J., Maddrell, S.H., 1984. Secretion by the Malpighian tubules of *Rhodnius prolixus stal*: electrical events. *J. Exp. Biol. 110,* 275–290.

Okuda, T., Haga, T., 2000. Functional characterization of the human high-affinity choline transporter. *FEBS Lett. 484,* 92–97.

Olah, Z., Lehel, C., Anderson, W.B., Eiden, M.V., Wilson, C.A., 1994. The cellular receptor for gibbon ape leukemia virus is a novel high affinity sodium-dependent phosphate transporter. *J. Biol. Chem. 269,* 25426–25431.

Osborne, R.H., 1996. Insect neurotransmission: neurotransmitters and their receptors. *Pharmacol. Ther. 69,* 117–142.

Otis, T.S., Wu, Y.C., Trussell, L.O., 1996. Delayed clearance of transmitter and the role of glutamate transporters at synapses with multiple release sites. *J. Neurosci. 16,* 1634–1644.

Pacholczyk, T., Blakely, R.D., Amara, S.D., 1991. Expression cloning of a cocaine- and antidepressant-sensitive human noradrenaline transporter. *Nature 350,* 350–354.

Page, R.D., 1996. TreeView: an application to display phylogenetic trees on personal computers. *Comput. Appl. Biosci. 12,* 357–358.

Palacin, M., Estevez, R., Bertran, J., Zorzano, A., 1998. Molecular biology of mammalian plasma membrane amino acid transporters. *Physiol. Rev. 78,* 969–1054.

Palacin, M., Kanai, Y., 2003. The ancillary proteins of HATs: SLC3 family of amino acid transporters. *Pflugers Arch.* 2003 Jun 11 (Epub ahead of print).

Parenti, P., Forcella, M., Pugliese, A., Casartelli, M., Giordana, B., et al., 2000. Substrate specificity of the brush border K⁺-leucine symport of *Bombyx mori* larval midgut. *Insect Biochem. Mol. Biol. 30*, 243–252.

Parenti, P., Villa, M., Hanozet, G.M., 1992. Kinetics of leucine transport in brush border membrane vesicles from lepidopteran larvae midgut. *J. Biol. Chem. 267*, 15391–15397.

Parsons, S.M., 2000. Transport mechanisms in acetylcholine and monoamine storage. *FASEB J. 14*, 2423–2434.

Patrick, M.L., Bradley, T.J., 2000. The physiology of salinity tolerance in larvae of two species of *Culex* mosquitoes: the role of compatible solutes. *J. Exp. Biol. 203*, 821–830.

Paulsen, I.T., Skurray, R.A., 1993. Topology, structure and evolution of 2 families of proteins involved in antibiotic and antiseptic resistance in Eukaryotes and Prokaryotes – an analysis. *Gene 124*, 1–11.

Pennetta, G., Wu, M., Bellen, H.J., 1999. Dissecting the molecular mechanisms of neurotransmitter release in Drosophila. In: Bellen, H. (Ed.), Neurotransmitter Release. Oxford University Press, Oxford, pp. 304–351.

Peres, A., Bossi, E., 2000. Effects of pH on the uncoupled, coupled and pre-steady-state currents at the amino acid transporter KAAT1 expressed in *Xenopus* oocytes. *J. Physiol. Lond. 525*, 83–89.

Petersen, S.A., Fetter, R.D., Noordermeer, J.N., Goodman, C.S., DiAntonio, A., 1997. Genetic analysis of glutamate receptors in *Drosophila* reveals a retrograde signal regulating presynaptic transmitter release. *Neuron 19*, 1237–1248.

Pfeiffer, R., Rossier, G., Spindler, B., Meier, C., Kuhn, L., et al., 1999. Amino acid transport of y+L-type by heterodimers of 4F2hc/CD98 and members of the glycoprotein-associated amino acid transporter family. *EMBO J. 18*, 49–57.

Pfeiffer, R., Spindler, B., Loffing, J., Skelly, P.J., Shoemaker, C.B., et al., 1998. Functional heterodimeric amino acid transporters lacking cysteine residues involved in disulfide bond. *FEBS Lett. 439*, 157–162.

Philp, N.J., Yoon, H.Y., Lombardi, L., 2001. Mouse MCT3 gene is expressed preferentially in retinal pigment and choroid plexus epithelia. *Am. J. Physiol. –Cell Physiol. 280*, C1319–C1326.

Pilegaard, H., Terzis, G., Halestrap, A., Juel, C., 1999. Distribution of the lactate/H⁺ transporter isoforms MCT1 and MCT4 in human skeletal muscle. *Am. J. Physiol. 276*, E843–E848.

Porzgen, P., Park, S.K., Hirsh, J., Sonders, M.S., Amara, S.G., 2001. The antidepressant-sensitive dopamine transporter in *Drosophila melanogaster*: a primordial carrier for catecholamines. *Mol. Pharmacol. 59*, 83–95.

Price, N.T., Jackson, V.N., Halestrap, A.P., 1998. Cloning and sequencing of four new mammalian monocarboxylate transporter (MCT) homologues confirms the existence of a transporter family with an ancient past. *Biochem. J. 329*, 321–328.

Quick, M., Stevens, B.R., 2001. Amino acid transporter CAATCH1 is also an amino acid-gated cation channel. *J. Biol. Chem. 276*, 33413–33418.

Rand, J.B., Duerr, J.S., Frisby, D.L., 2000. Neurogenetics of vesicular transporters in *C. elegans*. *FASEB J. 14*, 2414–2422.

Rasko, J.E.J., Battini, J.-L., Gottschalk, R.J., Mazo, I., Miller, A.D., 1999. The RD114/simian type D retrovirus receptor is a neutral amino acid transporter. *Proc. Natl Acad. Sci. USA 96*, 2129–2134.

Reimer, R.J., Edwards, R.H., 2003. Organic anion transport is the primary function of the SLC17/type I phosphate transporter family. *Pflugers Arch.* 2003 Jun 17 (Epub ahead of print).

Reuveni, M., Dunn, P.E., 1994. Proline transport into brush border membrane vesicles from the midgut of *Manduca sexta* larvae. *Comp. Biochem. Physiol. Comp. Physiol. 107*, 685–691.

Rimaniol, A.C., Haik, S., Martin, M., Le Grand, R., Boussin, F.D., et al., 2000. Na⁺-dependent high-affinity glutamate transport in macrophages. *J. Immunol. 164*, 5430–5438.

Robertson, H.A., Carlson, A.D., 1976. Octopamine – presence in firefly lantern suggests a transmitter role. *J. Exp. Zool. 195*, 159–164.

Roeder, T., 1999. Octopamine in invertebrates. *Progr. Neurobiol. 59*, 533–561.

Roman, G., Meller, V., Wu, K.H., Davis, R. L., 1998. The *opt1* gene of *Drosophila melanogaster* encodes a proton-dependent dipeptide transporter. *Am. J. Physiol. Cell Physiol. 44*, C857–C869.

Rossier, G., Meier, C., Bauch, C., Summa, V., Sordat, B., et al., 1999. LAT2, a new basolateral 4F2hc/CD98-associated amino acid transporter of kidney and intestine. *J. Biol. Chem. 274*, 34948–34954.

Rost, B., 1996. PHD: predicting one-dimensional protein structure by profile-based neural networks. *Methods Enzymol. 266*, 525–539.

Rothstein, J.D., Dykes-Hoberg, M., Pardo, C.A., Bristol, L.A., Jin, L., et al., 1996. Knockout of glutamate transporters reveals a major role for astroglial transport in excitotoxicity and clearance of glutamate. *Neuron 16*, 675–686.

Rubio-Aliaga, I., Boll, M., Daniel, H., 2000. Cloning and characterization of the gene encoding the mouse peptide transporter PEPT2. *Biochem. Biophys. Res. Commun. 276*, 734–741.

Sacchi, V.F., Castagna, M., Mari, S.A., Perego, C., Bossi, E., et al., 2003. Glutamate 59 is critical for transport function of the amino acid cotransporter KAAT1. *Am. J. Physiol. Cell Physiol. 285*, C623–C632.

Sagne, C., Agulhon, C., Ravassard, P., Darmon, M., Hamon, M., et al., 2001. Identification and characterization of a lysosomal transporter for small neutral amino acids. *Proc. Natl Acad. Sci. USA 98*, 7206–7211.

Sagne, C., El Mestikawy, S., Isambert, M.F., Hamon, M., Henry, J.P., et al., 1997. Cloning of a functional vesicular GABA and glycine transporter by screening of genome databases. *FEBS Lett. 417*, 177–183.

Saier, M.H., 2000a. Vectorial metabolism and the evolution of transport systems. *J. Bacteriol. 182*, 5029–5035.

Saier, M.H., Jr., 2000b. A functional-phylogenetic classification system for transmembrane solute transporters. *Microbiol. Mol. Biol. Rev. 64*, 354–411.

Saito, H., Terada, T., Okuda, M., Sasaki, S., Inui, K., 1996. Molecular cloning and tissue distribution of rat peptide transporter PEPT2. *Biochim. Biophys. Acta-Biomembr. 1280*, 173–177.

Saitou, N., Nei, M., 1987. The neighbor-joining method: a new method for reconstructing phylogenetic trees. *Mol. Biol. Evol. 4*, 406–425.

Sakata, K., Yamashita, T., Maeda, M., Moriyama, Y., Shimada, S., et al., 2001. Cloning of a lymphatic peptide/histidine transporter. *Biochem. J. 356*, 53–60.

Sandhu, S.K., Ross, L.S., Gill, S.S., 2002a. Molecular cloning and functional expression of a proline transporter from *Manduca sexta. Insect Biochem. Mol. Biol. 32*, 1391–1400.

Sandhu, S.K., Ross, L.S., Gill, S.S., 2002b. A cocaine insensitive chimeric insect serotonin transporter reveals domains critical for cocaine interaction. *Eur. J. Biochem. 269*, 3934–3944.

Sato, H., Tamba, M., Ishii, T., Bannai, S., 1999. Cloning and expression of a plasma membrane cystine/glutamate exchange transporter composed of two distinct proteins. *J. Biol. Chem. 274*, 11455–11458.

Sato, H., Tamba, M., Okuno, S., Sato, K., Keino-Masu, K., et al., 2002. Distribution of cystine/glutamate exchange transporter, system x(c) (-), in the mouse brain. *J. Neurosci. 22*, 8028–8033.

Schneider, M., Rudin, W., Hecker, H., 1986. Absorption and transport of radioactive-tracers in the midgut of the malaria mosquito, *Anopheles-Stephensi. J. Ultrastruct. Mol. Struct. Res. 97*, 50–63.

Schuldiner, S., Shirvan, A., Linial, M., 1995. Vesicular neurotransmitter transporters: from bacteria to humans. *Physiol. Rev. 75*, 369–392.

Schuldiner, S., Shirvan, A., Stern-Bach, Y., Steiner-Mordoch, S., Yelin, R., et al., 1994. From bacterial antibiotic resistance to neurotransmitter uptake. A common theme of cell survival. *Ann. N. Y. Acad. Sci. 733*, 174–184.

Schulz, D.J., Barron, A.B., Robinson, G.E., 2002. A role for octopamine in honey bee division of labor. *Brain Behav. Evol. 60*, 350–359.

Schuster, C.M., Ultsch, A., Schloss, P., Cox, J.A., Schmitt, B., et al., 1991. Molecular cloning of an invertebrate glutamate receptor subunit expression in *Drosophila* muscle. *Science 254*, 112–114.

Seal, R.P., Amara, S., 1999. Excitatory amino acid transporters: a family in flux. *Annu. Rev. Pharmacol. Toxicol. 39*, 431–456.

Seal, R.P., Daniels, G.M., Wolfgang, W.J., Forte, M., Amara, S., 1998. Identification and characterization of a cDNA encoding a neuronal glutamate transporter from *Drosophila melanogaster. Recept. Channels 6*, 51–64.

Simell, O., 2001. Lysinuric protein intolerance and other cationic amino acidurias. In: Scriver, C.R., Beaudet, S.W. (Eds.), Metabolic and Molecular Bases of Inherited Diseases. McGraw-Hill, New York, pp. 4933–4956.

Sinakevitch, I.G., Geffard, M., Pelhate, M., Lapied, B., 1994. Octopamine-like immunoreactivity in the dorsal unpaired median (Dum) neurons innervating the accessory-gland of the male cockroach *Periplaneta-Americana. Cell Tiss. Res. 276*, 15–21.

Sloan, J.L., Mager, S., 1999. Cloning and functional expression of a human Na^+ and Cl^--dependent neutral and cationic amino acid transporter B0+. *J. Biol. Chem. 274*, 23740–23745.

Soehnge, H., Huang, X., Becker, M., Whitley, P., Conover, D., et al., 1996. A neurotransmitter transporter encoded by the *Drosophila* inebriated gene. *Proc. Natl Acad. Sci. USA 93*, 13262–13267.

Spivak, M., Masterman, R., Ross, R., Mesce, K.A., 2003. Hygienic behavior in the honey bee (*Apis mellifera* L.) and the modulatory role of octopamine. *J. Neurobiol. 55*, 341–354.

Steiner, H.Y., Naider, F., Becker, J.M., 1995. The PTR family: a new group of peptide transporters. *Mol. Microbiol. 16*, 825–834.

Stern, M., Ganetzky, B., 1992. Identification and characterization of inebriated, a gene affecting neuronal excitability in *Drosophila. J. Neurogenet. 8*, 157–172.

Stevens, B.R., Feldman, D.H., Liu, Z., Harvey, W.R., 2002. Conserved tyrosine-147 plays a critical role in the ligand-gated current of the epithelial cation/amino acid transporter/channel CAATCH1. *J. Exp. Biol. 205*, 2545–2553.

Stuart, A.E., 1999. From fruit flies to barnacles, histamine is the neurotransmitter of arthropod photoreceptors. *Neuron 22*, 431–433.

Suchak, S.K., Baloyianni, N.V., Perkinton, M.S., Williams, R.J., Meldrum, B.S., et al., 2003. The 'glial' glutamate transporter, EAAT2 (Glt-1) accounts for high affinity glutamate uptake into adult rodent nerve endings. *J. Neurochem. 84*, 522–532.

Tabb, J.S., Ueda, T., 1991. Phylogenetic studies on the synaptic vesicle glutamate transport system. *J. Neurosci. 11*, 1822–1828.

Takamori, S., Rhee, J.S., Rosenmund, C., Jahn, R., 2001. Identification of differentiation-associated brain-specific phosphate transporter as a second vesicular glutamate transporter (VGLUT2). *J. Neurosci. 21*, RC182.

Tate, S.S., Yan, N., Udenfriend, S., 1992. Expression cloning of a Na(+)-independent neutral amino acid transporter from rat kidney. *Proc. Natl Acad. Sci. USA 89*, 1–5.

Thompson, J.D., Gibson, T.J., Plewniak, F., Jeanmougin, F., Higgins, D.G., 1997. The CLUSTAL_X windows interface: flexible strategies for multiple sequence alignment aided by quality analysis tools. *Nucleic Acids Res. 25*, 4876–4882.

Tomi, M., Funaki, T., Abukawa, H., Katayama, K., Kondo, T., et al., 2003. Expression and regulation of

L-cystine transporter, system x(c) (-), in the newly developed rat retinal Muller cell line (TR-MUL). *Glia 43*, 208–217.

Tomita, Y., Takano, M., Yasuhara, M., Hori, R., Inui, K.I., **1995**. Transport of oral cephalosporins by the H^+/dipeptide cotransporter and distribution of the transport activity in isolated rabbit intestinal epithelial cells. *J. Pharmacol. Exp. Ther. 272*, 63–69.

Torres, G.E., Gainetdinov, R.R., Caron, M.G., **2003**. Plasma membrane monoamine transporters: structure, regulation and function. *Nat. Rev. Neurosci. 4*, 13–25.

Toure, A., Morin, L., Pineau, C., Becq, F., Dorseuil, O., et al., **2001**. Tat1, a novel sulfate transporter specifically expressed in human male germ cells and potentially linked to RhoGTPase signaling. *J. Biol. Chem. 276*, 20309–20315.

Umesh, A., Cohen, B.N., Ross, L.S., Gill, S.S., **2003**. Functional characterization of a glutamate/aspartate transporter from the mosquito *Aedes aegypti*. *J. Exp. Biol. 206*, 2241–2255.

Umesh, A., Gill, S.S., **2002**. Immunocytochemical localization of a *Manduca sexta* gamma-aminobutyric acid transporter. *J. Comp. Neurol. 448*, 388–398.

Usherwood, P.N.R., **1994**. Insect glutamate receptors. *Adv. Insect. Physiol. 24*, 309–341.

Utsunomiya-Tate, N., Endou, H., Kanai, Y., **1996**. Cloning and functional characterization of a system ASC-like Na^+-dependent neutral amino acid transporter. *J. Biol. Chem. 271*, 14883–14890.

Varoqui, H., Schafer, M.K.H., Zhu, H.M., Weihe, E., Erickson, J.D., **2002a**. Identification of the differentiation-associated Na^+/P-I transporter as a novel vesicular glutamate transporter expressed in a distinct set of glutamatergic synapses. *J. Neurosci. 22*, 142–155.

Varoqui, H., Schafer, M.K., Zhu, H., Weihe, E., Erickson, J.D., **2002b**. Identification of the differentiation-associated Na^+/PI transporter as a novel vesicular glutamate transporter expressed in a distinct set of glutamatergic synapses. *J. Neurosci. 22*, 142–155.

Verrey, F., Closs, E.I., Wagner, C.A., Palacin, M., Endou, H., Kanai, Y., **2003**. CATs and HATs: the SLC7 family of amino acid transporters. *Pflugers Arch.* 2003. Jun 11 (Epub ahead of print).

Verri, T., Kottra, G., Romano, A., Tiso, N., Peric, M., et al., **2003**. Molecular and functional characterisation of the zebrafish (*Danio rerio*) PEPT1-type peptide transporter. *FEBS Lett. 549*, 115–122.

Vincenti, S., Castagna, M., Peres, A., Sacchi, V.F., **2000**. Substrate selectivity and pH dependence of KAAT1 expressed in *Xenopus laevis* oocytes. *J. Membr. Biol. 174*, 213–224.

Vincourt, J.B., Jullien, D., Amalric, F., Girard, J.P., **2003**. Molecular and functional characterization of SLC26A11, a sodium-independent sulfate transporter from high endothelial venules. *FASEB J. 17*, 890–892.

Wang, H.Y., Tamba, M., Kimata, M., Sakamoto, K., Bannai, S., et al., **2003**. Expression of the activity of cystine/glutamate exchange transporter, system x(c) (-),

by xCT and rBAT. *Biochem. Biophys. Res. Commun. 305*, 611–618.

Watanabe, K., Hata, Y., Kizaki, H., Katsube, Y., Suzuki, Y., **1997**. The refined crystal structure of *Bacillus cereus* oligo-1,6-glucosidase at 2.0 angstrom resolution: structural characterization of proline-substitution sites for protein thermostabilization. *J. Mol. Biol. 269*, 142–153.

Watson, A.H.D., Schurmann, F.-W., **2002**. Synaptic structure, distribution, and circuitry in the central nervous system of the locust and related insects. *Microsc. Res. Tech. 56*, 210–226.

Wells, R.G., Hediger, M.A., **1992**. Cloning of a rat kidney cDNA that stimulates dibasic and neutral amino acid transport and has sequence similarity to glucosidases. *Proc. Natl Acad. Sci. USA 89*, 5596–5600.

Werner, A., Moore, M.L., Mantei, N., Biber, J., Semenza, G., et al., **1991**. Cloning and expression of cDNA for a Na/Pi cotransport system of kidney cortex. *Proc. Natl Acad. Sci. USA 88*, 9608–9612.

Wieczorek, H., Brown, D., Grinstein, S., Ehrenfeld, J., Harvey, W.R., **1999a**. Animal plasma membrane energization by proton-motive V-ATPases. *Bioessays 21*, 637–648.

Wieczorek, H., Gruber, G., Harvey, W.R., Huss, M., Merzendorfer, H., **1999b**. The plasma membrane H^+-V-ATPase from tobacco hornworm midgut. *J. Bioenerg. Biomembr. 31*, 67–74.

Wieczorek, H., Putzenlechner, M., Zeiske, W., Klein, U., **1991**. A vacuolar-type proton pump energizes K^+/H^+ antiport in an animal plasma membrane. *J. Biol. Chem. 266*, 15340–15347.

Wieczorek, H., Weerth, S., Schindlbeck, M., Klein, U., **1989**. A vacuolar-type proton pump in a vesicle fraction enriched with potassium transporting plasma membranes from tobacco hornworm midgut. *J. Biol. Chem. 264*, 11143–11148.

Wilson, M.C., Jackson, V.N., Heddle, C., Price, N.T., Pilegaard, H., et al., **1998**. Lactic acid efflux from white skeletal muscle is catalyzed by the monocarboxylate transporter isoform MCT3. *J. Biol. Chem. 273*, 15920–15926.

Wilson, M.C., Meredith, D., Halestrap, A.P., **2002**. Fluorescence resonance energy transfer studies on the interaction between the lactate transporter MCT1 and CD147 provide information on the topology and stoichiometry of the complex in situ. *J. Biol. Chem. 277*, 3666–3672.

Wipf, D., Ludewig, U., Tegeder, M., Rentsch, D., Koch, W., et al., **2002**. Conservation of amino acid transporters in fungi, plants and animals. *Trends Biochem. Sci. 27*, 139–147.

Witte, I., Kreienkamp, H.-J., Gewecke, M., Roeder, T., **2002**. Putative histamine-gated chloride channel subunits of the insect visual system and thoracic ganglion. *J. Neurochem. 83*, 504–514.

Wolf, S., Janzen, A., Vekony, N., Martine, U., Strand, D., et al., **2002**. Expression of solute carrier 7A4 (SLC7A4) in the plasma membrane is not sufficient to mediate amino acid transport activity. *Biochem. J. 364,* 767–775.

Wolfersberger, M.G., 2000. Amino acid transport in insects. *Annu. Rev. Entomol. 45,* 111–120.

Wolfgang, C.L., Lin, C., Meng, Q., Karinch, A.M., Vary, T.C., et al., 2003. Epidermal growth factor activation of intestinal glutamine transport is mediated by mitogen-activated protein kinases. *J. Gastrointest. Surg. 7,* 149–156.

Xie, T., Parthasarathy, R., Wolfersberger, M.G., Harvey, W.R., 1994. Anomalous glutamate/alkali cation symport in larval *Manduca sexta* midgut. *J. Exp. Biol. 194,* 181–194.

Yamashita, T., Shimada, S., Guo, W., Sato, K., Kohmura, E., et al., 1997. Cloning and functional expression of a brain peptide/histidine transporter. *J. Biol. Chem. 272,* 10205–10211.

Yasuyama, K., Meinertzhagen, I.A., Schurmann, F.W., 2002. Synaptic organization of the mushroom body calyx in *Drosophila melanogaster. J. Comp. Neurol. 445,* 211–226.

Yelin, R., Schuldiner, S., 1995. The pharmacological profile of the vesicular monoamine transporter resembles that of multidrug transporters. *FEBS Lett. 377,* 201–207.

Yelin, R., Schuldiner, S., 2002. Vesicular neurotransmitter transporters. In: Reith, M.E.A. (Ed.), Neurotransmitter Transporters: Structure, Function and Regulation. Humana Press Inc., Totowa, New Jersey, pp. 313–355.

Yellman, C., Tao, H., He, B., Hirsh, J., 1997. Conserved and sexually dimorphic behavioral responses to biogenic amines in decapitated *Drosophila. Proc. Natl Acad. Sci. USA 94,* 4131–4136.

Yoon, H.Y., Fanelli, A., Grollman, E.F., Philp, N.J., 1997. Identification of a unique monocarboxylate transporter (MCT3) in retinal pigment epithelium. *Biochem. Biophys. Res. Commun. 234,* 90–94.

Young, G.B., Jack, D.L., Smith, D.W., Saier, M.H., Jr., 1999. The amino acid/auxin:proton symport permease family. *Biochim. Biophys. Acta 1415,* 306–322.

Zhao, C., Wilson, M.C., Schuit, F., Halestrap, A.P., Rutter, G.A., 2001. Expression and distribution of lactate/monocarboxylate transporter isoforms in pancreatic islets and the exocrine pancreas. *Diabetes 50,* 361–366.

Zharikov, S.I., Sigova, A.A., Chen, S., Bubb, M.R., Block, E.R., 2001. Cytoskeletal regulation of the L-arginine/NO pathway in pulmonary artery endothelial cells. *Am. J. Physiol. Lung Cell. Mol. Physiol. 280,* L465–L473.

Zhuang, Z., Linser, P.J., Harvey, W.R., 1999. Antibody to H(+) V-ATPase subunit E colocalizes with portasomes in alkaline larval midgut of a freshwater mosquito (*Aedes aegypti*). *J. Exp. Biol. 202,* 2449–2460.

Relevant Websites

http://sosui.proteome.bio.tuat.ac.jp – Proteome Bio Tuat.
http://www.ncbi.nlm.nih.gov – NCBI.

A6 Addendum: Amino Acid and Neurotransmitter Transporters

D Boudko, Rosalind Franklin University of Medicine and Science, North Chicago, IL, USA

© 2010, Elsevier BV. All Rights Reserved.

Insects represent the dominant class of terrestrial Metazoa by diversity of species and habitats (Engel and Grimaldi, 2004) as well as by their impact on human society, health, and economy (Berenbaum, 1995; Marquardt and Kondratieff, 2005; Scudder, 2009). A compilation of genome sequences from several insect species facilitates the study of practical and fundamental aspects of insect biology, promising major breakthroughs in the management of vector and pest arthropods, as well as in our understanding of the fundamental principles of genetics and evolution of animal functions. Amino acid and neurotransmitter transport comprise a complex of biological phenomena. In recent years, our systematic understanding of these phenomena has been improved by the acquisition and analysis of model insect genomes, leading to novel insect management possibilities via discovery of insect-specific transport mechanisms. The chapter "Amino Acid and Neurotransmitter Transporters" (Boudko et al., 2005c) originally published in Comprehensive Molecular Insect Science (Gilbert et al., 2005) has summarized the physiological and molecular study of insect transporters up to 2005. This addendum updates the chapter by briefly summarizing new key findings and ideas and providing new references.

The original review was focused on secondary transporters that perform transport of nutrient and signaling amino acids, including small derivatives of the amino acid metabolism monoamines and GABA neurotransmitters (Boudko et al., 2005c). With the emergence of a phylogenomic comparative analysis of transport proteins, several phylogenetic families of these transporters have been clearly identified in the growing diversity of metazoan genomes (Boudko, 2009a). Hence, it is clear now that the metazoan amino acid transport network for the alimentary absorption and systemic redistribution of canonical amino acids has remained conserved from the last universal metazoan ancestor (Boudko, 2009a). It recruits members from an identical phylogenetic subset of secondary transporters of ten SLC families in insects and 10 + 1 SLC families in mammals, resembling the SoLute Carrier family systematics. The SLC nomenclature proposed by the Human Genome Organization (HUGO) (Hediger et al., 2004) appears to be well suited to accommodate broader phylogenomic systematization of metazoan transporters and in particular, for classification of insect transporters (Boudko, 2009a). The evolutionary trend and phylogenetic specificity of metazoan amino acid transport is reflected in the more broad Transporter Classification Data Base (TCDB) (Busch and Saier, 2003), which also identifies 11 families of metazoan amino acid transporters that belong to six superfamilies. The molecular diversity of the mammalian amino acid transport network was summarized in a set of short reviews devoted to individual SLC families: 1 (Kanai and Hediger, 2004), 3 (Palacin and Kanai, 2003), 6 (Chen et al., 2003), 7 (Verrey et al., 2003), 15 [oligopeptide transporters family] (Daniel and Kottra, 2003), 16 (Halestrap and Meredith, 2003), 17 (Reimer and Edwards, 2003), 18 (Eiden et al., 2003), 32 (Gasnier, 2003), 36 (Boll et al., 2004), 38 (Mackenzie and Erickson, 2003), 43 (Bodoy et al., 2005).

Each of these families specializes in substrate selectivity and electrochemical coupling of transport mechanism, as well as expression site and time of actions. These specializations allow them to play specific, often unique roles in the amino acid transport network of organisms. Nearly every amino acid can be locally transported via more than one transporter, providing backup capacity for absorption in the case of mutational inactivation of a transport system (Broer, 2008). The mutual support between amino acid transporters usually occurs within a close phylogenetically related population of transporters, but rarely between phylogenetically distant convergent phenotypes. Phylogenomic analysis suggests that evolution of structural complicity correlates with metabolic specialization of cells and expansion of membrane transport mechanisms, rather than with the invention of new metabolic cascades (Boudko, 2009a). Evidently, amino acid transporters represent a very dynamic entity, with a remarkable degree of physiological and evolutionary plasticity (Boudko et al., 2005a, 2005b; Okech et al., 2008). Comparison of amino acid transporter families in the completed mammalian and insect genomes suggests that despite common phylogenetic routes, interspecific populations within the amino acid transport network may diverge dramatically, forming paralogous, lineage-specific phylogenetic clusters. The apparent protein sequence divergences usually coincide with notable differences in substrate profiles, relative selectivities, and affinities between such populations, which reflect adaptations of the transport mechanism to specific operational environments. One such adaptation is the early radiation of neurotransmitter transporters especially pronounced in the SLC6 family transporters (Caveney et al., 2006), of which several new representatives have been recently cloned from pest insects (Donly and Caveney, 2005; Donly et al., 2007). Another striking example of adaptive divergence is the identification of remarkable differences between mammalian and insect SLC6 transporters that mediate absorption and distribution of essential aromatic and aliphatic amino acids. A representative set of these transporters has been cloned from different mammalian models, revealing a molecular entity of B^0 system (Broad spectra ^{0}neutral substrates), which encompass a phylogenetically separate cluster of SLC6 transporters (A15 (Takanaga et al., 2005a; Broer et al., 2006), A16, A17 (Zaia and Reimer, 2009), A18 (Singer et al., 2009), A19 (Bohmer et al., 2005); +A20 (Kowalczuk et al., 2005; Takanaga et al., 2005b; Ristic et al., 2006) designated in IMINO, a proline preferred system). In contrast, only two insect transporters of the SLC6 family overlap with the mammalian B^0/IMINO cluster, while the majority represents paralogous expansion with 2–9 transporters in various insects (Boudko, 2009b). In addition to broad substrate spectra transporters (Boudko et al., 2005a, 2005b; Miller et al., 2008), the insect-specific cluster includes transporters with unusually narrow specificity to the most underrepresented essential amino acids: tryptophan (Meleshkevitch et al., 2009a), phenylalanine (Meleshkevitch et al., 2006), and methionine (Meleshkevitch et al., 2009b), and apparent roles in the amplification of absorption of the corresponding substrates (Boudko, 2009a, 2009b). Functional expression of the B^0 system undergoes auxiliary regulation by angiotensin-converting enzyme family members (Danilczyk et al., 2006; Broer, 2009). The auxiliary subunit MetS is also essential for the expression of a bacterial SLC6 member MetP in Corynebacterium glutamicum

(Trotschel et al., 2008). However, similar regulatory factors beyond vertebrate and prokaryote organisms remain undefined.

Astonishing progress has been made in unraveling structures of major amino acid transporter families. The crystal structures of prokaryotic representatives of SLC1 (Yernool et al., 2004; Boudker et al., 2007), SLC6 (Yamashita et al., 2005; Singh et al., 2007), and SLC7 (Shaffer et al., 2009) families have been solved. These families of transporters share significant protein sequence conservation between prokaryotic and eukaryotic members (pairwise identity 20–30%), allowing for rapid progress in 3D model-driven exploration of molecular architecture and mechanics of action in the corresponding eukaryotic transporters.

SLC7 family members enable the bulk of basolateral emission of nutrient amino acids in the metazoan alimentary canal. They act as high-throughput symporters and amino acid exchangers (Palacin et al., 2005; Closs et al., 2006). Furthermore, dipteran SLC7 members have been implicated in nutrient amino acid sensing cascades, as transmembrane transmitters and/or amplifiers of sub/intracellular amino acid concentration reference signals and as transceptors (Hundal and Taylor, 2009; Taylor, 2009). For example, minidisc, which represents a principal light subunit of the heterodimeric [SLC7 + SLC3 subfamilies] transporter (Martin et al., 2000), and slimfast (slif; CG1128) (Colombani et al., 2003) which encodes a single-subunit Cationic Amino acid Transporter [CAT subfamily of SLC7], are expressed in the fat body and modulate nutrition and growth of the fruit fly. Two CAT–SLC7s transporters, slimfast and iCAT2, are essential to the onset of blood meal-mediated TOR signaling in the females of yellow fever mosquitoes, Aedes aegypti, which subsequently triggers vitellogenic gene expression and oogenesis (Attardo et al., 2006). It has been shown that paired expression of heavy-chain subunit CG2791 is necessary to induce L-system amino acid transport by the Drosophila SLC7 transporter JhI-21 in Xenopus oocytes (Reynolds et al., 2009). Other nutrient amino acid transceptors, CG3424 and CG1139, were identified among the proton amino acid transporters of the Drosophila SLC36 family. These transporters mediate essential growth signaling via TOR- and other InR- coupled pathways without an apparent requirement in capacity transport of the signaling amino acids (Goberdhan et al., 2005). A new SLC36 member, AaePAT1 (AAEL007191), has been cloned and characterized from A. aegypti (Evans et al., 2009). It is expressed in the epithelial cell membranes of larval caeca and upregulated in the adult female midgut upon a blood meal, potentially playing a role as a transceptor. This implies a new role of APC-transporter superfamily members, including SLC7 and SLC36 families, in signaling the availability of amino acids to the target of the rapamycin (TOR) pathway, possibly through PI 3-kinase-dependent mechanisms (Hundal and Taylor, 2009).

References

Attardo, G.M., Hansen, I.A., Shiao, S.H., Raikhel, A.S., 2006. Identification of two cationic amino acid transporters required for nutritional signaling during mosquito reproduction. *J. Exp. Biol.* 209(Pt 16), 3071–3078.

Berenbaum, M., 1995. Bugs in the System: Insects and Their Impact on Human Affairs. Addison-Wesley, Reading, MA.

Bodoy, S., Martin, L., Zorzano, A., Palacin, M., Estevez, R., Bertran, J., 2005. Identification of LAT4, a novel amino acid transporter with system L activity. *J. Biol. Chem.* 280(12), 12002–12011.

Bohmer, C., Broer, A., Munzinger, M., Kowalczuk, S., Rasko, J.E.J., Lang, F., Broer, S., 2005. Characterization of mouse amino acid transporter B(0)AT1 (slc6a19). *Biochem. J.* 389, 745–751.

Boll, M., Daniel, H., Gasnier, B., 2004. The SLC36 family: proton-coupled transporters for the absorption of selected amino acids from extracellular and intracellular proteolysis. *Pflugers Arch.* 447(5), 776–779.

Boudker, O., Ryan, R.M., Yernool, D., Shimamoto, K., Gouaux, E., 2007. Coupling substrate and ion binding to extracellular gate of a sodium-dependent aspartate transporter. *Nature* 445(7126), 387–393.

Boudko, D.Y., 2009a. Molecular ontology of amino acid transport. In: Gerencser, G. (Ed.), Epithelial Transport Physiology. Humana-Springer Verlag, Springer, NY, pp. 379–472.

Boudko, D.Y., 2009b. Molecular ontology of essential amino acid transport. In: Lubec, G. (Ed.), 11th International Congress on Amino Acids, Peptides and Proteins, Vol. 37. Springer, Vienna, Austria, p. S92.

Boudko, D.Y., Kohn, A.B., Meleshkevitch, E.A., Dasher, M.K., Seron, T.J., Stevens, B.R., Harvey, W.R., 2005a. Ancestry and progeny of nutrient amino acid transporters. *Proc. Natl Acad. Sci. USA* 102(5), 1360–1365.

Boudko, D.Y., Meleshkevitch, E.A., Harvey, W.R., 2005b. Novel transport phenotypes in the sodium neurotransmitter symporter family. *FASEB J.* 19(4), A748.

Boudko, D.Y., Stevens, B.R., Donly, B.C., Harvey, W.R., 2005c. Nutrient amino acid and neurotransmitter transporters. In: Gilbert, L.I., Iatrou, K., Gill, S.S. (Eds.), Comprehensive Molecular Insect Science, Vol. 4. Elsevier, Amsterdam, pp. 255–309.

Broer, S., 2008. Amino acid transport across mammalian intestinal and renal epithelia. *Physiol. Rev.* 88(1), 249–286.

Broer, S., 2009. The role of the neutral amino acid transporter B0AT1 (SLC6A19) in Hartnup disorder and protein nutrition. *IUBMB Life* 61(6), 591–599.

Broer, A., Tietze, N., Kowalczuk, S., Chubb, S., Munzinger, M., Bak, L.K., Broer, S., 2006. The orphan transporter v7-3 (slc6a15) is a Na + -dependent neutral amino acid transporter (B0AT2). *Biochem. J.* 393(Pt 1), 421–430.

Busch, W., Saier, M.H.Jr., 2003. The IUBMB-endorsed transporter classification system. In: Yan, Q. (Ed.), Membrane Transporters: Methods and Protocols Methods in Molecular Biology. Humana Press, Totowa, NJ, pp. xii, 369.

Caveney, S., Cladman, W., Verellen, L., Donly, C., 2006. Ancestry of neuronal monoamine transporters in the metazoa. *J. Exp. Biol.* 209(Pt 24), 4858–4868.

Chen, N.H., Reith, M.E., Quick, M.W., 2003. Synaptic uptake and beyond: the sodium- and chloride-dependent neurotransmitter transporter family SLC6. *Pflugers Arch.* 29, 29.

Closs, E.I., Boissel, J.P., Habermeier, A., Rotmann, A., 2006. Structure and function of cationic amino acid transporters (CATs). *J. Membr. Biol.* 213(2), 67–77.

Colombani, J., Raisin, S., Pantalacci, S., Radimerski, T., Montagne, J., Leopold, P., 2003. A nutrient sensor mechanism controls Drosophila growth. *Cell* 114(6), 739–749.

Daniel, H., Kottra, G., 2003. The proton oligopeptide cotransporter family SLC15 in physiology and pharmacology. *Pflugers Arch.* 7, 7.

Danilczyk, U., Sarao, R., Remy, C., Benabbas, C., Stange, G., Richter, A., Arya, S., Pospisilik, J.A., Singer, D., Camargo, S.M., et al., 2006. Essential role for collectrin in renal amino acid transport. *Nature* 444(7122), 1088–1091.

Donly, B.C., Caveney, S., 2005. A transporter for phenolamine uptake in the arthropod CNS. *Arch. Insect Biochem. Physiol.* 59(3), 172–183.

Donly, C., Verellen, L., Cladman, W., Caveney, S., 2007. Functional comparison of full-length and N-terminal-truncated octopamine transporters from Lepidoptera. *Insect Biochem. Mol. Biol.* 37(9), 933–940.

Eiden, L.E., Schafer, M.K., Weihe, E., Schutz, B., 2003. The vesicular amine transporter family (SLC18): amine/proton antiporters required for vesicular accumulation and regulated exocytotic secretion of monoamines and acetylcholine. *Pflugers Arch.* 24, 24.

Engel, M.S., Grimaldi, D.A., 2004. New light shed on the oldest insect. *Nature* 427(6975), 627–630.

Evans, A.M., Aimanova, K.G., Gill, S.S., 2009. Characterization of a blood-meal-responsive proton-dependent amino acid transporter in the disease vector, *Aedes aegypti. J. Exp. Biol.* 212(Pt 20), 3263–3271.

Gasnier, B., 2003. The SLC32 transporter, a key protein for the synaptic release of inhibitory amino acids. *Pflugers Arch.* 16, 16.

Gilbert, L.I., Iatrou, K., Gill, S.S., 2005. Comprehensive Molecular Insect Science. Elsevier, Amsterdam, Boston.

Goberdhan, D.C., Meredith, D., Boyd, C.A., Wilson, C., 2005. PAT-related amino acid transporters regulate growth via a novel mechanism that does not require bulk transport of amino acids. *Development 132*(10), 2365–2375.

Halestrap, A.P., Meredith, D., 2003. The SLC16 gene family-from monocarboxylate transporters (MCTs) to aromatic amino acid transporters and beyond. *Pflugers Arch. 9, 9.*

Hediger, M.A., Romero, M.F., Peng, J.B., Rolfs, A., Takanaga, H., Bruford, E.A., 2004. The ABCs of solute carriers: physiological, pathological and therapeutic implications of human membrane transport proteins. Introduction. *Pflugers Arch. 447*(5), 465–468.

Hundal, H.S., Taylor, P.M., 2009. Amino acid transceptors: gate keepers of nutrient exchange and regulators of nutrient signaling. *Am. J. Physiol. Endocrinol. Metab. 296*(4), E603–E613.

Kanai, Y., Hediger, M.A., 2004. The glutamate/neutral amino acid transporter family SLC1: molecular, physiological and pharmacological aspects. *Pflugers Arch. 447*(5), 469–479.

Kowalczuk, S., Broer, A., Munzinger, M., Tietze, N., Klingel, K., Broer, S., 2005. Molecular cloning of the mouse IMINO system: an Na + -and Cl-dependent proline transporter. *Biochem. J. 386*, 417–422.

Mackenzie, B., Erickson, J.D., 2003. Sodium-coupled neutral amino acid (System N/A) transporters of the SLC38 gene family. *Pflugers Arch. 4, 4.*

Marquardt, W.C., Kondratieff, B.C., 2005. Biology of Disease Vectors. Elsevier Academic Press, Burlington, MA.

Martin, J.F., Hersperger, E., Simcox, A., Shearn, A., 2000. Minidiscs encodes a putative amino acid transporter subunit required non-autonomously for imaginal cell proliferation. *Mech. Dev. 92*(2), 155–167.

Meleshkevitch, E.A., Assis-Nascimento, P., Popova, L.B., Miller, M.M., Kohn, A.B., Phung, E.N., Mandal, A., Harvey, W.R., Boudko, D.Y., 2006. Molecular characterization of the first aromatic nutrient transporter from the sodium neurotransmitter symporter family. *J. Exp. Biol. 209*(Pt 16), 3183–3198.

Meleshkevitch, E.A., Robinson, M., Popova, L.B., Miller, M.M., Harvey, W.R., Boudko, D.Y., 2009a. Cloning and functional expression of the first eukaryotic Na + -tryptophan symporter, AgNAT6. *J. Exp. Biol. 212*(Pt 10), 1559–1567.

Meleshkevitch, E.A., Voronov, D., Miller, M.M., Fox, J., Popova, L.B., Boudko, D.Y., 2009b. Novel methionine-selective transporters from the neurotransmitter sodium symporter family. In: Lubec, G. (Ed.), 11th International Congress on Amino Acids, Peptides and Proteins, Vol. 37. Springer, Vienna, Austria, p. S84.

Miller, M.M., Popova, L.B., Meleshkevitch, E.A., Tran, P.V., Boudko, D.Y., 2008. The invertebrate B(0) system transporter, *D. melanogaster* NAT1, has unique d-amino acid affinity and mediates gut and brain functions. *Insect Biochem. Mol. Biol. 38*(10), 923–931.

Okech, B.A., Meleshkevitch, E.A., Miller, M.M., Popova, L.B., Harvey, W.R., Boudko, D.Y., 2008. Synergy and specificity of two Na + -aromatic amino acid symporters in the model alimentary canal of mosquito larvae. *J. Exp. Biol. 211*(Pt 10), 1594–1602.

Palacin, M., Kanai, Y., 2003. The ancillary proteins of HATs: SLC3 family of amino acid transporters. *Pflugers Arch. 6, 6.*

Palacin, M., Nunes, V., Font-Llitjos, M., Jimenez-Vidal, M., Fort, J., Gasol, E., Pineda, M., Feliubadalo, L., Chillaron, J., Zorzano, A., 2005. The genetics of heteromeric amino acid transporters. *Physiology (Bethesda) 20*, 112–124.

Reimer, R.J., Edwards, R.H., 2003. Organic anion transport is the primary function of the SLC17/type I phosphate transporter family. *Pflugers Arch. 17, 17.*

Reynolds, B., Roversi, P., Laynes, R., Kazi, S., Boyd, C.A., Goberdhan, D.C., 2009. *Drosophila* expresses a CD98 transporter with an evolutionarily conserved structure and amino acid-transport properties. *Biochem J. 420*(3), 363–372.

Ristic, Z., Camargo, S.M., Romeo, E., Bodoy, S., Bertran, J., Palacin, M., Makrides, V., Furrer, E.M., Verrey, F., 2006. Neutral amino acid transport mediated by ortholog of imino acid transporter SIT1/SLC6A20 in opossum kidney cells. *Am. J. Physiol. Renal Physiol. 290*(4), F880–F887.

Scudder, G.H., 2009. The importance of insects. In: Foottit, R., Adler, P.H. (Eds.), Insect Biodiversity: Science and Society. Wiley-Blackwell, Chichester, UK; Hoboken, NJ, pp. 7–32.

Shaffer, P.L., Goehring, A., Shankaranarayanan, A., Gouaux, E., 2009. Structure and mechanism of a Na + -independent amino acid transporter. *Science 325*(5943), 1010–1014.

Singer, D., Camargo, S.M., Huggel, K., Romeo, E., Danilczyk, U., Kuba, K., Chesnov, S., Caron, M.G., Penninger, J.M., Verrey, F., 2009. Orphan transporter SLC6A18 is renal neutral amino acid transporter B0AT3. *J. Biol. Chem. 284*(30), 19953–19960.

Singh, S.K., Yamashita, A., Gouaux, E., 2007. Antidepressant binding site in a bacterial homologue of neurotransmitter transporters. *Nature 448*(7156), 952–956.

Takanaga, H., Mackenzie, B., Peng, J.B., Hediger, M.A., 2005a. Characterization of a branched-chain amino-acid transporter SBAT1 (SLC6A15) that is expressed in human brain. *Biochem. Biophys. Res. Commun. 337*(3), 892–900.

Takanaga, H., Mackenzie, B., Suzuki, Y., Hediger, M.A., 2005b. Identification of mammalian proline transporter SIT1 (SLC6A20) with characteristics of classical system imino. *J. Biol. Chem. 280*(10), 8974–8984.

Taylor, P.M., 2009. Amino acid transporters: eminences grises of nutrient signalling mechanisms? *Biochem. Soc. Trans. 37*(Pt 1), 237–241.

Trotschel, C., Follmann, M., Nettekoven, J.A., Mohrbach, T., Forrest, L.R., Burkovski, A., Marin, K., Kramer, R., 2008. Methionine uptake in *Corynebacterium glutamicum* by MetQNI and by MetPS, a novel methionine and alanine importer of the NSS neuro-transmitter transporter family. *Biochemistry* 47, 12698–12709.

Verrey, F., Closs, E.I., Wagner, C.A., Palacin, M., Endou, H., Kanai, Y., 2003. CATs and HATs: the SLC7 family of amino acid transporters. *Pflugers Arch.* 11, 11.

Yamashita, A., Singh, S.K., Kawate, T., Jin, Y., Gouaux, E., 2005. Crystal structure of a bacterial homologue of Na +/Cl-dependent neurotransmitter transporters. *Nature* 437(7056), 215–223.

Yernool, D., Boudker, O., Jin, Y., Gouaux, E., 2004. Structure of a glutamate transporter homologue from *Pyrococcus horikoshii. Nature* 431(7010), 811–818.

Zaia, K.A., Reimer, R.J., 2009. Synaptic vesicle protein NTT4/XT1 (SLC6A17) catalyzes Na + -coupled neutral amino acid transport. *J. Biol. Chem.* 284(13), 8439–8448.

7 Biochemical Genetics and Genomics of Insect Esterases

J G Oakeshott, CSIRO Entomology, Canberra, Australia
C Claudianos, Australian National University, Canberra, Australia
P M Campbell, CSIRO Entomology, Canberra, Australia
R D Newcomb, HortResearch, Auckland, New Zealand
R J Russell, CSIRO Entomology, Canberra, Australia

© 2005, Elsevier BV. All Rights Reserved.

7.1. Introduction

7.1.1. Historical Perspective

Esterases have been a major part of insect biochemical research over the last 50 years. As an important component of insects' xenobiotic defence system, they have long been a focus for studies of chemical insecticide metabolism and resistance. Additionally, acetylcholinesterase (AChE) was a target for the early organophosphate (OP) and carbamate compounds. Other esterases have been implicated in the metabolism of specific hormones and pheromones and, consequently, in various aspects of insect development and behavior. Many esterases are easily scored by isozyme electrophoresis, with multiple isozymes and their high levels of genetic variation making them tractable, informative tools for a variety of biochemical and ecological genetic studies. For all these reasons esterase biochemistry featured prominently in the first edition of *Comprehensive Insect Biochemistry, Physiology and Pharmacology* in 1985.

At that time not a single insect esterase gene had been cloned but in the decade that followed several were isolated, particularly from *Drosophila*. The insect esterase genes cloned in this period all proved to derive from the one gene family, the carboxyl/cholinesterases (or just carboxylesterases in the Pfam nomenclature – see Pfam ID 00135). Members of this family were also cloned from various prokaryotes and vertebrates, albeit some of the noninsect esterase genes cloned in the period clearly belonged to different gene families. Alternatively, genes for several noncatalytic proteins involved in various signaling processes in insects and other eukaryotes were found to be members of the carboxyl/cholinesterase gene family.

Critically too, the tertiary structures of AChE and some sequence-related microbial lipases were resolved in this period and shown to lie within the one structurally defined superfamily of proteins. This defining structure is called the α/β hydrolase fold. Catalytically active members of this superfamily all use the same reaction mechanism, involving a catalytic triad of residues headed by a serine or sometimes cysteine residue as a nucleophil. While hydrolases (including some esterases) in some other superfamilies have the same reaction mechanism, they achieve this with otherwise unrelated structures, apparently as an outcome of convergent evolution.

The last decade has seen a few cases of esterase-based insecticide resistance resolved at a molecular level. All these cases still involve the carboxyl/cholinesterase gene family. All involve OP resistance. They fall into essentially three groups, first involving mutant AChE molecules that are less sensitive to inhibition by the insecticide, second involving many-fold amplifications of genes encoding carboxylesterases with limited ability to hydrolyze the insecticide but substantial scope to sequester it, and the third involving structural mutations in certain carboxylesterases that improve their kinetics for insecticide hydrolysis. Remarkably, some of the same mutations are found in widely different species, which might suggest very limited options for evolving resistance. Given the empirical data on the structure of AChE, and modeled structures of the carboxylesterases based on AChE and the lipases, it is becoming possible to understand the effects of the resistance mutations in precise structural terms.

Of course the other qualitative advance during the last decade has been the application of modern genomic and proteomic technologies to the study of insect esterases. This has greatly facilitated the isolation of esterase genes and improved our understanding of their functions and sequence relationships. It appears that insect species may each have about 30–60 different members of the carboxyl/cholinesterase gene family, about three-quarters of them catalytically active, and that these comprise a majority but not all of the esterase isozymes that can be visualized after native polyacrylamide gel electrophoresis (PAGE). The insect members of the family

fall into a small number of clades many of whose origins approach or, in a few cases, exceed the age of the class Insecta. Within some of these clades, however, there has been remarkably rapid evolution such that only a few orthologous genes can be identified except among quite closely related species.

This chapter reviews the biochemistry and genetics of insect esterases, focusing mainly on developments since the first edition of this series. It is written from a genomics perspective and the bulk of it is organized around the major clades of insect esterases as revealed by the comparative genomic analysis. First, however, some of the historical functional definitions and classifications applied to insect esterases are briefly revised.

7.1.2. Functional Definitions and Classifications

The esterase reaction, defined as the hydrolysis of an ester to its component alcohol and acid, encompasses hydrolysis of a diverse range of carboxylic, thio-, phospho-, and other ester substrates. It sits within a broader set of hydrolase reactions that also include glycosylases, proteases, amidases, and many others (Webb, 1992). Even more generally it can be considered a particular case, involving water, of the acylation of a nucleophil. Use of other nucleophils in otherwise similar reactions generates, for example, various dehalogenase activities (Ollis et al., 1992).

Most of the discussion in this review concerns enzymes hydrolyzing esters of carboxylic acids (E.C. 3.1.1). This is a valid and useful functional grouping insomuch as reaction energetics are significantly different for the hydrolysis of esters like thio- and phospho-esters, where there is a strong negative charge in the immediate environment of the ester bond (Aldridge and Reiner, 1972). Nevertheless, there is still great diversity even among the carboxylic acid ester hydrolases.

One major distinction separates the lipases from others in the group (Phythian, 1998; Arpigny and Jaeger, 1999; Fojan et al., 2000). Lipases have maximal activity for esters like triglycerides with simple alcohol groups and complex acid moieties, which are essentially insoluble in water. Most of the others have substrates with simple acid moieties but potentially complex alcohol moieties, which are at least partly soluble in water. It is this latter, nonlipase, group, which is the particular focus of this review.

Various classifications based on inhibitor sensitivities have been applied to subdivide the nonlipase group of carboxylic ester hydrolases. One early classification in use even before isozyme studies recognized A-, B-, and C-type esterases on the basis of their interaction with OPs. While A-type esterases could hydrolyze these molecules, the B-type enzymes were essentially irreversibly inhibited by them, and the C-type enzymes neither hydrolyzed nor were inhibited by them (Aldridge and Reiner, 1972). Most, but not all, insect esterases studied to date would be classified as B-type esterases on this scheme.

A subsequent elaboration of the Aldridge and Reiner (1972) scheme has been used to classify eukaryotic esterase isozymes detectable after native PAGE and staining with various artificial ester substrates. It relies on three classes of inhibitors, the OPs, sulfhydral reagents, and the carbamate eserine sulfate, from which can be discerned four classes of enzymes (Holmes and Masters, 1967; Coates et al., 1975; Healy et al., 1991). These are:

- acetylesterases, which are unaffected by any of these inhibitors and generally prefer substrates with acetyl acid groups and aromatic alcohol groups;
- arylesterases, which are only inhibited by the sulfhydral reagents and generally prefer substrates with aromatic alcohol groups;
- carboxylesterases, which are only inhibited by the OPs and prefer aliphatic esters, generally of longer acids than acetate; and
- cholinesterases, which are inhibited by OPs and eserine sulfate, and prefer substrates with charged alcohol moieties like choline esters over other aromatic or aliphatic esters.

A further classification that had been applied extensively to esterase isozymes, especially in *Drosophila* and mosquitoes, groups the enzymes according to their preferential hydrolysis of the isomeric artificial substrates, α- and β-naphthyl acetate (see Sections 7.4.3 and 7.5.1 below). This distinction is made because hydrolysis of the two isomers leads to differently colored bands when the reaction is coupled with several of the common diazo staining dyes. The classification has little value as a predictor of enzyme function and indeed allelic differences between the two preferences have been described for some isozymes. The α- and β-based nomenclatures are retained herein for some widely studied enzymes and enzyme clusters, noting, however, that in themselves they represent no broader biological distinctions.

A survey of 22 soluble *Drosophila melanogaster* esterase isozymes identified 10 carboxylesterases, six cholinesterases, and three acetylesterases, with no aryl esterases, and three isozymes not clearly classifiable on this scheme (Healy et al., 1991). Surveys of the major esterase isozymes of various mosquitoes and hemipterans yield broadly similar

results, albeit consistently recovering minority numbers of aryl as well as acetylesterases (VedBrat and Whitt, 1975; Casabé and Zerba, 1981; Narang and Seawright, 1982).

As noted earlier, all invertebrate carboxyl and cholinesterases that have been sequenced to date lie in the same gene family, which is therefore called the carboxyl/cholinesterase gene family (Oakeshott et al., 1993, 1995, 1999; Claudianos et al., 2001). No sequence information has yet been reported for invertebrate acetyl- or arylesterases, although data for a few from prokaryotes and vertebrates put them in unrelated families (Campbell et al., 2003). There are also a small number of functionally defined carboxylesterases from vertebrates that do not sit in the carboxyl/cholinesterase family (Zschunke et al., 1991).

Several different classes of mechanism have been described for hydrolase reactions, ranging across general acid/base, electrostatic, and covalent catalyses (Voet and Voet, 1990). They may be one or two step reactions. Some use metal ions in various capacities. However, the mechanism used by the carboxyl/cholinesterases entails a two-step reaction which is independent of metal ions and has an acyl-enzyme intermediate (**Figure 1**). The serine of the active site is made more nucleophilic than usual by interaction with a histidine residue, which itself is oriented correctly by an interaction with an acidic residue. These three residues constitute the so-called catalytic triad. Their functions are analogous to those of the well-known catalytic triad of serine proteases where the same mechanism occurs. Since the structures of the two types of protein are otherwise unrelated, their shared catalytic mechanism is assumed to reflect convergent evolution.

In the first step of this reaction mechanism the oxygen of the serine's side chain makes a nucleophilic attack on the carbonyl carbon atom of the substrate, displacing the alcohol product and forming a covalent linkage with the acyl moiety of the substrate. In the second step, the oxygen of a water molecule makes a similar attack on the carbonyl carbon, hydrolyzing the covalent acyl-enzyme link to yield free enzyme and the acid product of the reaction. It has generally been assumed that the water molecule is activated by the histidine and acid residues of the catalytic triad as described for the first, acylation step (**Figure 1**), but Thomas et al. (1999) and Behra et al. (2002) have proposed that a conserved "additional serine" might sometimes provide this function instead. Both steps require a tetrahedral transition state in which charge is moved to the carbonyl oxygen. As with serine proteases, the active site of carboxylesterases has certain

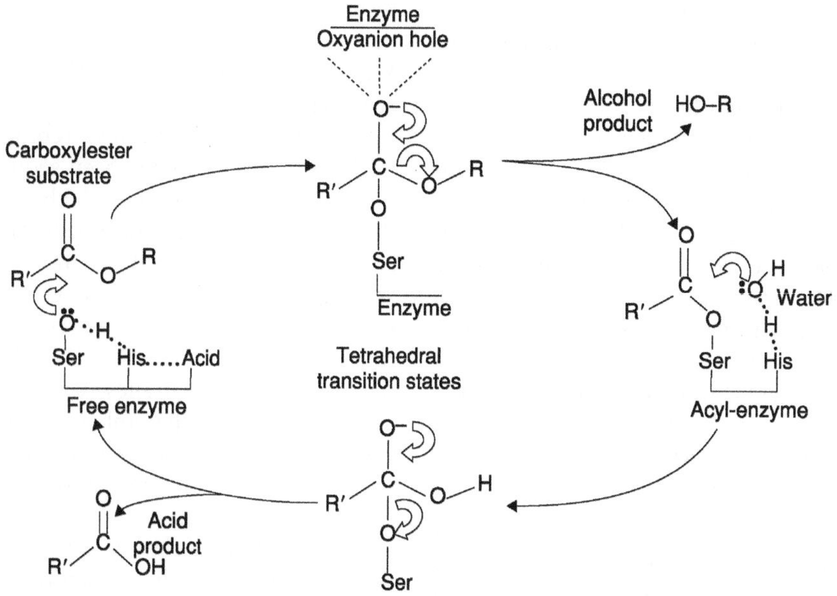

Figure 1 The carboxylesterase reaction mechanism. Carboxylesterases hydrolyze their substrates in two steps. First, the oxygen of a serine residue in the active site makes a nucleophilic attack on the carbonyl carbon of the substrate, displacing the alcohol product and forming a relatively stable acyl-enzyme. In the second step water makes a similar nucleophilic attack, this time displacing the serine residue to release the acid product of the reaction and regenerate the free enzyme. The serine of the active site is made more nucleophilic than usual by interaction with a histidine residue which is oriented correctly by an interaction with an acidic residue. During both nucleophilic substitution reactions there is a tetrahedral transition state the energy of which is reduced by an "oxyanion hole" in which the charge displaced onto the oxygen is accommodated by hydrogen bonding to the main-chain nitrogens of three amino acid residues around the active site. This mechanism is analogous to the mechanism of the serine proteases that also perform hydrolysis with a Ser–His–acid triad and main-chain nitrogens forming an oxyanion hole. The mechanism of serine proteases is frequently presented in detail in biochemistry textbooks with the similarity of carboxylesterases and serine proteases cited as an example of convergent evolution (e.g., Voet and Voet (1990), pp. 373–382).

conserved small residues, the main chain nitrogens of which are proposed to form an "oxyanion hole" to stabilize the transition states in both steps of the reaction by hydrogen bonding.

Hydrolysis of esters of carboxylic acids can be relatively undemanding energetically and examples of this reaction, including from the carboxyl/cholinesterases, have been described which approach diffusion limited kinetics (Taylor and Radic, 1994; Zera *et al.*, 2002; Devonshire *et al.*, 2003; Kamita *et al.*, 2003). Also, however, reactions of this type can often be relatively easily reversible, particularly in nonaqueous environments. The functional significance of the resultant esterification reactions has been well established for several bacterial "esterases," but the reverse reactions of eukaryotic esterases have received relatively little attention (Phythian, 1998).

7.2. Comparative Genomics of Insect Esterases

7.2.1. Overview

Consistent with their functional diversity, the sequences of the carboxyl/cholinesterases are also widely divergent. Even within the insect members there can be as little as about 20% amino acid identity between distant members of the family (Oakeshott *et al.*, 1999). In this and subsequent sections, it is shown how some of the sequence differences contribute to differences in the biochemistry of the proteins and their physiological functions.

Figure 2 presents a phylogeny of insect carboxyl/cholinesterase sequences. It includes all those identified in the fully sequenced *D. melanogaster* and *Anopheles gambiae* (African malaria mosquito) genomes (35 and 51, respectively), as well as 35 others for which there is both substantial sequence and some functional information from other insects. It therefore does not include the many apparently full length but functionally anonymous expressed sequence tag (EST) carboxyl/cholinesterase sequences in various insect EST databases. It also excludes several orthologs of *D. melanogaster* and *A. gambiae* sequences in other *Drosophila* and *Anopheles* species since these would simply appear in terminal bifurcations of the tree and provide little further information in terms of functional evolution. However, to give a broader context, all the lineages of *Caenorhabditis elegans* (nematode) carboxyl/ cholinesterases as identified by Oakeshott *et al.* (1999) that sit within the phylogeny (albeit for ease of tree building some nematode-specific radiations are simply represented by single sequences) are also included. Bacterial and vertebrate carboxyl/ cholinesterases are not discussed, although some of their relationships to the major insect clades are known (Oakeshott *et al.*, 1999). No lipase sequences have yet been identified within the carboxyl/cholinesterase family in insects; some may sit as yet anonymously within the sequences included. However, the limited information available for insects, and the precedents in bacteria and vertebrates, suggest that most insect lipases will sit outside the clades shown in **Figure 2**, albeit many are in other families in the α/β hydrolase fold superfamily (Pistillo *et al.*, 1998; Arpigny and Jaeger, 1999; Nardini and Dijkstra, 1999; Adams *et al.*, 2000; Arrese *et al.*, 2001).

One obvious feature of the phylogeny is the high degree of separation of the insect clades from the *C. elegans* genes. There are only three clear overlaps, in the AChE, gliotactin, and neuroligin radiations. By cross-referencing the trees constructed by Oakeshott *et al.* (1999), it is also told that only these clades would likely be shared with known vertebrate carboxyl/cholinesterases. All the other clades in the phylogeny are thus insect-specific. While further sequencing may reveal representatives from other eukaryotes in some of these clades, it nevertheless seems likely that most of the insect carboxyl/cholinesterase lineages have evolved since the insects diverged from both the nematodes and the vertebrates.

The 35 non-*D. melanogaster*, non-*A. gambiae*, insect proteins in the phylogeny include 11 from other dipterans, eight from various lepidopterans, and two each from the Hymenoptera and Coleoptera, all in the Holometabola, and, for the Hemimetabola, 12 from various hemipterans. Although there are no nondipteran proteins in a few of the major clades at this point it is expected that most of these will also prove to have members from several orders, and indeed from both the Holo- and Hemimetabola, as more sequences become available. Notable here are the branch orders among the major clades; even some of the most derived clades contain members from different orders. Thus most of the major clades are older than the split between the Holo- and Hemimetabola.

For the purposes of this review the radiations in the phylogeny containing insect sequences have been partitioned into 14 major clades, denoted A to N in **Figure 2**. In the rest of this section some major

sequence and structural distinctions among these clades are outlined and below (see Sections 7.3, 7.4, and 7.5) are reviewed what is known of the functions of proteins within each of them.

One of the major discriminants among the 14 clades is catalytic capability, as inferred from the presence of a consensus catalytic triad (see also **Figure 4** below). Significantly, there are several noncatalytic clades, including the neuroligin, gliotactin, and neurotactin lineages (clades L, M, and N), as well as two clades of uncharacterized proteins (I and K), on this criterion. As discussed later, all functional evidence from characterized members of these clades confirms their catalytic incompetence. These clades were generated at relatively early branch points in the phylogeny and, in the case of the neuroligins and gliotactins, are also known to have vertebrate orthologs (see Section 7.3.2). Whilst the evidence from prokaryote members of the family suggests that catalytic competence is the ancestral state (Oakeshott *et al.*, 1999), it is nevertheless clear that a few noncatalytic functions developed early in the evolution of higher eukaryotes and have been retained in diverse lineages since. Loss of catalytic activity in these clades is also associated with the acquisition of additional domains as either N- or C-terminal extensions (depending on the clade) to the single carboxyl/cholinesterase domain (**Figure 3**).

The only major clade which contains significant proportions of both presumptively catalytic and noncatalytic members is the glutactin lineage (clade H) (see Section 7.4.1). Functional data confirm a noncatalytic role for glutactin and, like most of the other noncatalytic proteins above, it also has extra C-terminal transmembrane and cytoplasmic domains (**Figure 3**). However, it has apparently arisen relatively recently in a lineage which is predominantly composed of catalytically active proteins and, consistent with this, it lacks vertebrate orthologs.

Interestingly, while catalytic competence is generally a good taxonomic character for the insect carboxyl/cholinesterases, the identity of the catalytic triad is not. All those with an identifiable triad use serine as a nucleophil and histidine as the base. However, the acid residue may be aspartate or glutamate, and there are several lineages within the tree where both can be found.

Cellular/subcellular localization is another taxonomically informative character, at least for much of the phylogeny. The presence of N-terminal signal peptides and/or, in many cases, direct functional information indicate that all the major noncatalytic clades contain secreted proteins, albeit those characterized are known to be tethered to the cell membrane by motifs in their N- or C-terminal

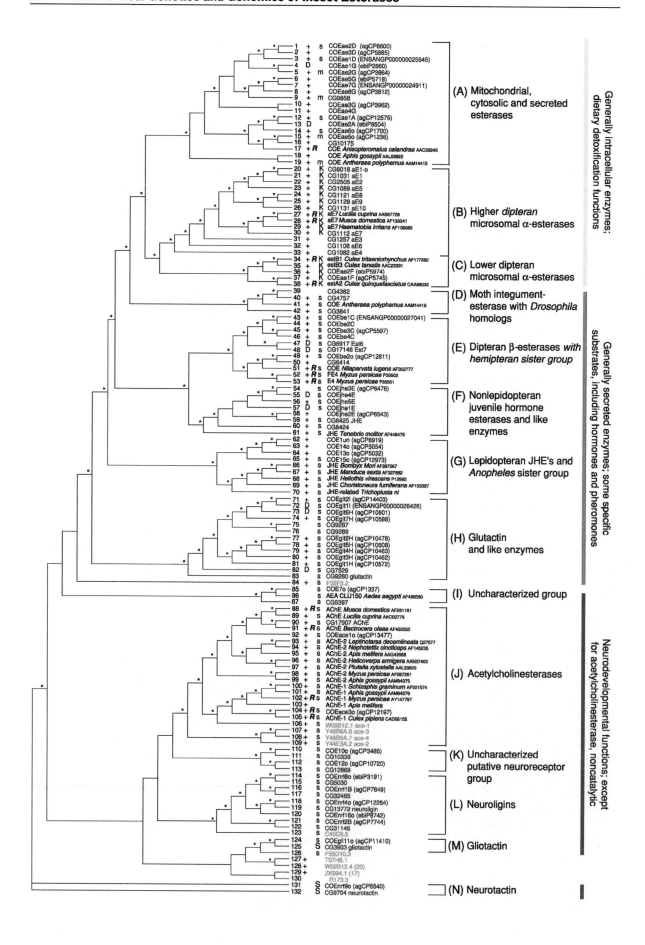

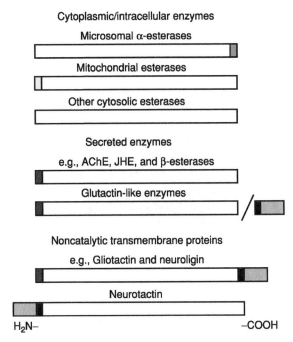

Figure 3 Schematic model (not to scale) of the consensus domain structure of insect carboxyl/cholinesterases in the various major clades indicated in **Figure 2**, showing N-terminal secretion (red) and mitochondrial targeting (yellow) and C-terminal endoplasmic retrieval (KDEL, green) signals for transmembrane proteins, the transmembrane (blue) and cytoplasmic (gray) domains. Slash symbol signifies that not all members of the glutactin clade have transmembrane domains (see Section 7.4.1). See also **Figure 2** regarding the secretion and endoplasmic reticulum retrieval signals.

extensions (**Figure 3**) (see Sections 7.3.2 and 7.4.1). There are also some radiations of mainly catalytically competent enzymes that are secreted (D–H, J), but with the exception of glutactin and the AChE clade (J) these are not known to be membrane-associated. Members of the other major clades (A–C) are essentially all catalytically active enzymes and mostly intracellular, albeit they fall into two subgroups on the basis of subcellular localization. Members of one subgroup (B, C) can be classified as at least largely microsomal on the basis of functional data and/or the presence of C-terminal KDEL targeting motifs (Pelham, 1996; Yamamoto *et al.*, 2003). Members of the other subgroup (A) lack the latter motifs but cover a range of other localizations. A few contain conventional amino terminal secretion signals and some contain mitochondrial targeting signals (Bannai *et al.*, 2002). The rest are presumptively cytosolic, although some use of as yet unidentified nonconventional targeting signals, or extracellular translocation following membrane insertion (Emanuelsson *et al.*, 2000; Chou, 2001; Kriek *et al.*, 2003; Lai, 2003), cannot be discounted.

At the highest level, the phylogeny of insect carboxyl/cholinesterases can thus be partitioned quite cleanly into three groups on the basis of catalytic competence and cellular/subcellular localization. The oldest group (I–N) contains clades of secreted proteins that are generally membrane associated and, except for AChE, catalytically incompetent.

Figure 2 Unrooted distance neighbor-joining tree showing a phylogeny of invertebrate carboxyl/cholinesterases. Sequences from the *Drosophila melanogaster* and *Anopheles gambiae* genomes (red and blue, respectively) are as annotated in Ranson *et al.* (2002). Sequences for various carboxyl/cholinesterases from other insects (black) were generally taken from the National Center for Biotechnology Information (NCBI) database; see text for inclusion criteria. The two exceptions were sequences 70 and 103, from Jones *et al.* (1994) and http://www.hgsc.bcm.tmc.edu. Sequences from the *Caenorhabditis elegans* genome (green) are as in Oakeshott *et al.* (1999) except that single representative sequences only were used for two nematode-specific lineages (see text and below). Sequences were aligned using CLUSTAL W (Thompson *et al.*, 1994) with the BLOSUM 45 scoring matrix, with gap opening and gap extension penalties of 10.0 and 0.1, respectively, followed by some minor manual corrections to conform to known structural features of carboxyl/cholinesterases. The tree was constructed with PAUP* (Swofford, 2000) using standard distances and mean character differences. Nodes with at least 50% resampling frequency in 1000 bootstrap replications are indicated with an asterisk. To the right of the tree, in order, are: sequence numbers (1–132) as used elsewhere in this paper; catalytic competence as assessed by presence of a consensus catalytic Ser–His–Glu (+) or Ser–His–Asp (D) triad; evidence of involvement in chemical insecticide resistance (R, see text), and presence of a secretion signal (s) as assessed by the iPSORT algorithm of Bannai *et al.* (2002) and, if not secreted, of a mitochondrial (m) or KDEL endoplasmic reticulum targeting signal (K) as defined by Pelham (1996) and Yamamoto *et al.* (2003). (The upper-case S for sequences 125, 131, and 132 signifies that they lack a conventional N-terminal secretion signal according to iPSORT but empirical data establish that they are presented on the extracellular side of cell membranes (see Section 7.3.2).) Then follow sequence names and NCBI, Celera Protein and ENSEMBL database accession numbers and, if not *D. melanogaster*, *A. gambiae*, or *C. elegans*, species names. Note that *C. elegans* sequences 128 and 129 are single representative sequences for radiations of 20 and 17 proteins, respectively. Finally, major clades of sequences recognized in the text are named (and for cross-referencing to **Figure 6** below, alphabetically identified) and three putative functional "super-clades" are also indicated. Note that some amino terminal sequence is missing for 18 of the proteins (4, 6, 7,13, 50, 58, 62–64, 78, 80, 103, 112, 114, 116, 120, 121, 126) and some C-terminal data are missing for 17 (6, 13, 54, 58, 60, 62–65, 77–80, 95, 103, 112, 118). Note also that the recently revised nomenclature of Weill *et al.* (2000, 2002, 2003) and Russell *et al.* (2004) for the AChE sequences have been followed (see also Section 7.3.1).

Members of the next group (D–H) are also almost all secreted but, except for glutactin, not known to be membrane-tethered and they are almost all catalytically competent. The third and most derived group (A–C) contains catalytically competent enzymes covering a wide range of cellular/subcellular localizations, one subgroup generally being microsomal and the remainder a mix of presumptively secreted, mitochondrial, and cytosolic enzymes.

As discussed below (see Sections 7.3, 7.4, and 7.5), the available functional data suggest that these three groups are also qualitatively distinct in terms of their physiological roles. Characterized members of the first group, containing the secreted, membrane-associated, noncatalytic proteins and the AChE clade, are almost all involved in neurodevelopmental processes. At least four clades in the second group, mainly composed of secreted catalytic proteins, include various hormone and pheromone esterases. Many members of the third group, comprising mainly intracellular catalytically active proteins, appear to have broad substrate specificities and more general dietary and/or detoxification functions. The mutant enzymes conferring insecticide resistance occur in all three groups: AChE enzymes resistant to OP and carbamate inhibition in the first group and pesticide detoxifying enzymes in the third and, at least for some sap-sucking insects, also in the second. First, however, the phylogeny is examined in more detail, with particular regard to some major features of tertiary structure.

7.2.2. The Structural Context

Figure 4 diagrams the tertiary structure of the carboxyl/cholinesterase domain of the mature AChE monomer from *D. melanogaster* (DmAChE) (Harel *et al.*, 2000) (see Section 7.3.1). This structure is used as a model in the following sections because it is the only insect member of the carboxyl/cholinesterase protein family whose structure has been reported to this point. There are published structures for AChEs from a few different vertebrates, for the closely related vertebrate butyrylcholinesterase (BuChE) and another vertebrate carboxylesterase, and from a few sequence-related lipases from lower eukaryotes (Schrag and Cygler, 1993; Bourne and Shindyalov, 1998; Nicolet *et al.*, 2003). The structure of the carboxyl/cholinesterase domain of DmAChE is very similar to these others in its overall organization, notwithstanding over 60% amino acid sequence divergence from them. The core of the structure is the α/β hydrolase fold, comprising a major β-sheet of 11 strands arranged in a characteristic order, plus 14 α-helices in loops situated

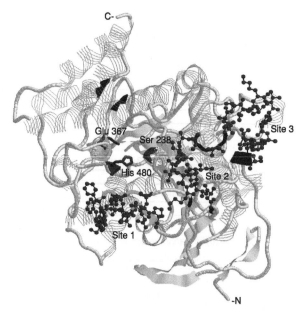

Figure 4   Three-dimensional cartoon structure of the carboxyl/cholinesterase domain of DmAChE (Harel *et al.*, 2000; PDB 1QO9; http://www.rcsb.org/pdb) generated by the RasMol program (http://www.openrasmol.org) looking down the catalytic gorge into the active site. Specifically indicated are residues contributing to the β-sheet (yellow; however, many are obscured behind the gorge and active site), several α-helices in the loops between β-strands (brown), six cysteines forming three disulfide bridges (black), the catalytic triad (Ser238, Glu367, His480, red) plus the candidate catalytic tetrad residue of Thomas *et al.* (1999) and Behra *et al.* (2002) (Ser264, green), 12 aromatic guidance residues lining the active site gorge (blue) and three peripheral substrate binding sites (purple). See also **Figure 6** for the relationship of these features to the primary sequence of the protein, and **Figure 5** for a schematic of the interior of the active site that cannot readily be shown in this figure.

between the strands. Also in the loops are the components of the active site and catalytic gorge and much of the external decoration for the structure. The structure is stabilized by three disulfide and four salt bridges.

There is quite detailed understanding of structure–function relationships in the regions of the AChE protein directly involved in catalysis. Many residues in these regions have been mutated *in vivo* in one or more of several cloned *ace* genes and the catalytic consequences for the protein determined (Doctor *et al.*, 1998; Ordentlich *et al.*, 1998; Villatte *et al.*, 2000a, 2000b; Boublik *et al.*, 2002; Pezzementi *et al.*, 2003 and references therein). **Figure 5** shows how the active site is built around the catalytic triad, with the nucleophilic serine protruding into the site on a conserved GXSXG pentapeptide termed the nucleophilic elbow. The rest of the site is largely composed of three subsites,

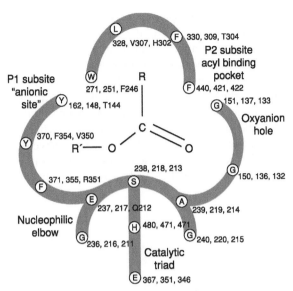

Figure 5 Schematic representation of the carboxyl/cholinesterase active site showing conserved amino acid residues. Each residue is shown with its single-letter code followed by its position in the sequences of DmAChE, LcαE7 (E3), and DmJHE, respectively. A letter is included where the residue in E3 or JHE differs from AChE. The residues of the catalytic triad, oxyanion hole, and nucleophilic elbow show strong conservation across active carboxyl/cholinesterases, while those constituting the subsites for the leaving and acyl groups of the substrate are more variable. These two subsites are described in greatest detail for AChE where they have often been called the P1 (or anionic) and P2 subsites, respectively. There is strong conservation of these sites across vertebrate and invertebrate AChEs (Harel *et al.*, 2000; Pezzementi *et al.*, 2003), but two of the indicated residues differ from AChE in E3 and seven of the eight differ in JHE.

which have variously been termed the leaving group pocket or P1 subsite (or in the particular case of AChE, anionic site), the acyl pocket or P2 subsite, and oxyanion hole. These accommodate the alcohol, acid, and oxyanion moieties of the substrate, respectively. Entry to the active site is via the catalytic gorge, which is over 20 Å long but only a few Å wide and is lined with several aromatic residues. Around the lip of the gorge are three "peripheral binding sites" which, when bound by substrate, trigger allosteric changes in the catalytic gorge that facilitate movement of substrate down the gorge. The end result is a particularly efficient enzyme with kinetics approaching diffusion limits (Taylor and Radic, 1994; Estrada-Mondaca and Fournier, 1998; Massoulie *et al.*, 1999).

Only two of the sequences in our phylogeny lack the additional serine (Ser264) which Thomas *et al.* (1999) and Behra *et al.* (2002) considered might, in combination with the established catalytic triad residues, form a "catalytic tetrad." One of these (33) is catalytically competent and the other (113) is not.

Given its retention in so many catalytically inactive proteins, and its apparent absence in at least one catalytically competent enzyme, it seems likely to us that Ser264 has some highly conserved structural function in the active site, but not one directly involved in catalysis.

Figure 6 shows a qualitative alignment of the sequences in the phylogeny annotated with some of the major structural and/or functional motifs outlined above (a full alignment is available from the authors on request). In general terms, the alignment is consistent with the conservation of the overall arrangement of β-sheets and α-helices specifying the α/β hydrolase fold structure. Nevertheless there are also numerous clade-specific insertions/deletions, particularly in regions corresponding to loops towards the C-terminus of AChE. Many of these loops provide the external decoration to the core α/β hydrolase fold structure. Perhaps unsurprisingly some of the most distinctive arrangements occur in the noncatalytic clades. Importantly, however, the other catalytic clades also differ from AChE in aspects including two of the disulfide bridges, two of the peripheral substrate binding sites, and many of the aromatic gorge residues.

It is difficult to assess the significance of these differences in the absence of empirical structural information for any of the catalytic clades except AChE. However, they would seem to imply some major differences in the molecular mechanisms underpinning interactions with substrates. Modeling of a lepidopteran juvenile hormone esterase (JHE) indeed suggests at least two novel structural features of its active site that are instrumental to its particular catalytic capabilities (Thomas *et al.*, 1999) (see Section 7.4.2). There are also intriguing subclade-specific differences, for example, in the aromatic gorge residues of the hemipteran AChE-1 subclade, which may well have implications for catalysis, and the interactions with OP and carbamate insecticides (see Section 7.3.1). And, as discussed, the biochemistries of some of the insecticide resistance mutations imply major, allele-specific differences in aspects of structure affecting catalytic processes in certain catalytic carboxylesterases (see Sections 7.5.2, 7.6.1–7.6.3). It may be that a major reason for the proliferation of carboxyl/cholinesterases in higher eukaryotes is the functional diversity that can be achieved by relatively minor sequence changes in their catalytic machinery. Empirical structural data for some of the non-AChE catalytic clades would clearly now add greatly to our understanding of the functional diversity that has evolved among insect carboxyl/cholinesterases.

7.3. Acetylcholinesterase and Noncatalytic Neurodevelopmental Clades

The six major clades (I–N) containing insect sequences diverging earliest in the phylogeny (**Figure 2**) involve AChE and five clades of noncatalytic secreted proteins. The AChE is discussed first (see Section 7.3.1) because of the wealth of structure–function data available for it, followed by five noncatalytic clades (see Section 7.3.2).

Before doing so, however, it is worth noting some interesting parallels with the convergently evolved serine protease family of proteins in respect of the evolution of noncatalytic clades. Although they belong to a different structurally defined superfamily of proteins, the serine proteases generally use the same catalytic mechanism, built around the same catalytic triad and oxyanion hole, as the carboxyl/cholinesterases. (Indeed a few vertebrate esterases have been found in the serine protease family; see for example, Zschunke *et al.* (1991).) And, intriguingly, there are

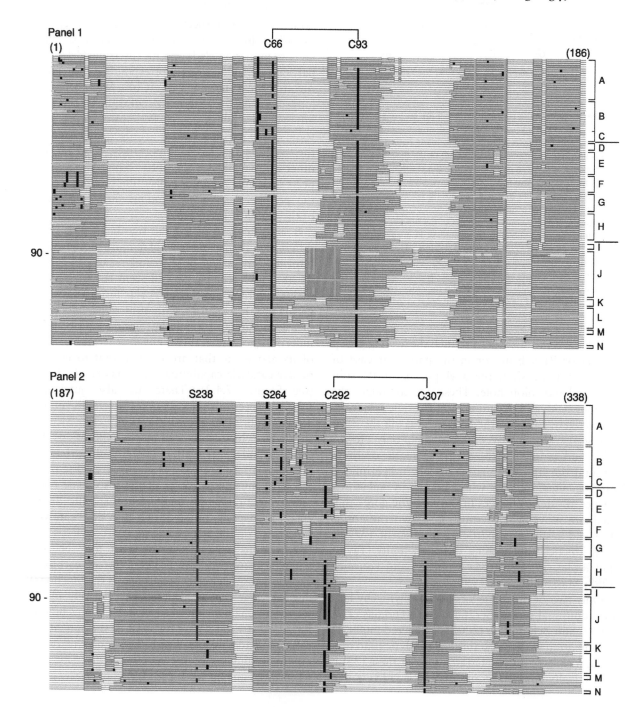

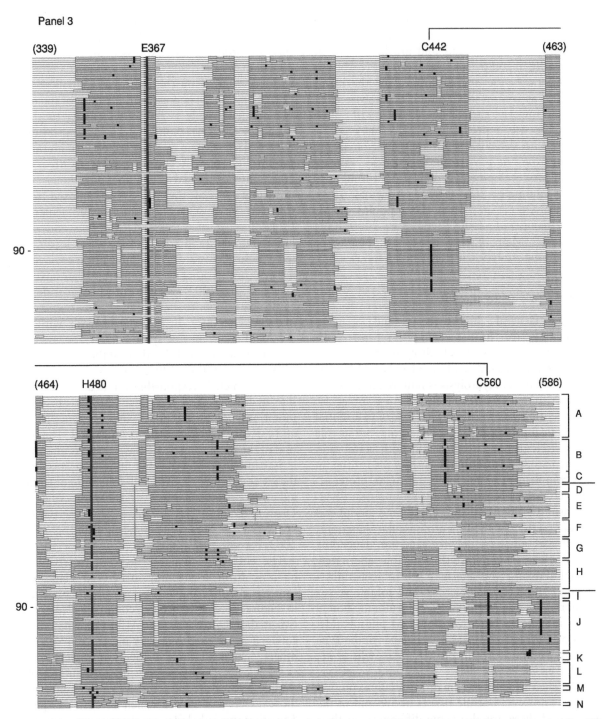

Figure 6 Block diagram generated by the GenDoc program (http://www.psc.edu) showing a qualitative alignment of the 132 carboxyl/cholinesterase amino acid sequences used to generate the phylogeny in **Figure 2**. Each sequence is simply represented as a series of blocks or lines to show the presence or absence of alignable residues. Sequences are ordered as in **Figure 2**, and their classification into the major clades recognized in **Figure 2** is shown on the right-hand side of each panel. The four panels cover sequences alignable to residues 1–186, 187–338, 339–463, and 464–585, respectively, in DmAChE (sequence 90 in the figure), which is used as a reference sequence because its tertiary structure is known (**Figure 4**). The β-strands and α-helices in DmAChE are shown in yellow and brown, respectively. Cysteines in all sequences are shown in black; those at residues 66 and 93, 292 and 307, and 442 and 560 in DmAChE are known to form three disulfide bonds. Aromatic residues corresponding to the aromatic gorge residues of DmAChE are shown in blue, conserved catalytic triad residues (Ser238, His480, and Glu367 in DmAChE) are shown in red, and the putative catalytic tetrad residue (Ser264 in DmAChE) of Thomas *et al.* (1999) and Behra *et al.* (2002) (which is present in all sequences covering the region except 33 and 113) in green. The known or presumptive peripheral substrate binding sites of all the AChE sequences are shown in pink. See **Figure 2** for the list of proteins lacking some amino or carboxy terminal sequence.

several lineages in the serine protease family, including in both vertebrates and invertebrates, that lack a consensus catalytic triad and are known to be catalytically incompetent. As with the carboxyl/cholinesterases, the ancestral state for the serine proteases, as seen in prokaryotes, is catalytic competence. However, early in the evolution of eukaryotes there was a proliferation of serine protease lineages with defective triads that took on ligand receptor roles in physiological processes ranging from neurodevelopmental regulation through to cellular immunity (Gabay and Almeida, 1993; Murugasu-Oei *et al.*, 1995; Dimopoulos *et al.*, 1997; Huang *et al.*, 2000; Kwon *et al.*, 2000; Kang *et al.*, 2002). Many of these themes recur among the clades of noncatalytic carboxyl/cholinesterases.

7.3.1. Acetylcholinesterase

7.3.1.1. General biochemistry and genetics Although it also has other important roles, the major function of AChE is hydrolysis of the neurotransmitter acetylcholine after it has bound at cholinergic synapses. Hydrolysis is required to reset the synapses so that they can receive the next molecules of the neurotransmitter. Acetylcholine is an essential neurotransmitter in all higher eukaryotes but its role in insects is distinct from that in both vertebrates and nematodes. Accordingly insects also differ in their complement of *ace* genes and AChE proteins.

In vertebrates acetylcholine is an important neurotransmitter in both the central and peripheral nervous system and at neuromuscular junctions. Vertebrates only have one *ace* gene but, depending on the species, it can have three or four alternately spliced 3′ exons. This alternate splicing generates a range of variously oligomerized and soluble versus membrane-associated forms. The membrane-bound forms may be anchored with glycosyl phosphatidyl inositol (GPI) or complexed with collagen or the Type 1 transmembrane protein PRiMA (Massoulie *et al.*, 1999; Perrier *et al.*, 2002; Scholl and Scheiffele, 2003). The various forms are differentially expressed in the various cholinergic tissues above and in the blood and there are also some differences in their ontogenetic specificities. Vertebrates also have the closely related BuChE gene/enzyme system but this system is not directly involved in acetylcholine hydrolysis and it has no ortholog in nematodes or insects (Nicolet *et al.*, 2003 and references therein).

Nematodes use acetylcholine as a neurotransmitter in both the central nervous system (CNS) and muscle tissue. *Caenorhabditis elegans* has four *ace* genes, but the products of two of these contribute the great majority of the organism's AChE activity

(Massoulie *et al.*, 1999, Combes *et al.*, 2000, 2003). These two genes each encode a single transcript and class of product, one expressed as GPI-anchored dimers in neural tissue, and the other as a complex with collagen in muscle tissue and a small proportion of neurons. The form and function of the other two *ace* gene products are not well understood.

Insects use acetylcholine as a neurotransmitter in their CNS but their muscles largely use glutamate for this function. Accordingly their AChE is predominantly expressed as GPI-anchored dimers, with no equivalent of the collagen-complexed forms found in the muscles of vertebrates and nematodes (Massoulie *et al.*, 1999). *Drosophila* and other higher Diptera (and possibly also a few lower Diptera) (Bourguet *et al.*, 1998) only appear to have one *ace* gene but all other insects studied to date have two (Weill *et al.*, 2002). Significantly, an acarine, the cattle tick *Boophilus microplus*, also has two (Baxter and Barker, 2002). In those arthropods with two *ace*/AChE gene/enzyme systems there appears to be just one system, *ace-1*/AChE-1 on the nomenclature of Weill *et al.* (2000, 2002, 2003) and Russell *et al.* (2004), which is responsible for neural acetylcholine hydrolysis, and it is also this one in which mutations conferring target site resistance to OP and carbamate insecticides may be found. The function of the other, *ace-2*/AChE-2, system is unknown; it is catalytically active but it is also speculated that it could have various cell–cell communication/adhesive properties, such as those described for *Drosophila* and vertebrate *ace*/AChE (Grisaru *et al.*, 1999; Behra *et al.*, 2002; Scholl and Scheiffele, 2003; Russell *et al.*, 2004).

The currently available invertebrate AChE sequences form a strongly structured clade with four clearly discernable subclades in the phylogeny in **Figure 2**. One subclade, comprising solely the two inabundant *C. elegans* AChEs and its muscle/collagen-complexed AChE, is an outgroup to all the others. The next most distinct subclade includes the *C. elegans* AChE responsible for neural acetylcholine hydrolysis, plus the AChE-1 sequences of those insects and the tick which have two *ace* genes. The other two subclades are sister groups, one comprising the AChE-2s of the species with two *ace* genes, and the other comprising the AChEs of the higher Diptera. It follows that the single-*ace* genotype found in the higher Diptera has evolved by loss of the *ace-1* gene in an ancestor with a twin-*ace* genotype. And the intriguing corollary is that the originally noncholinergic AChE-2 in this ancestor has then reverted to a cholinergic role to cover for the loss of the AChE-1 (Weill *et al.*, 2002; Russell *et al.*, 2004).

The mature GPI-linked dimers that are the predominant form of AChE in *D. melanogaster* are produced through a complex set of posttranslational modifications of a primary 70 kDa translation product (Mutero and Fournier, 1992; Estrada-Mondaca and Fournier, 1998). This primary product comprises a 15 kDa N-terminal domain (in addition to a secretion signal), a ~55 kDa central carboxyl/cholinesterase domain, and a short C-terminal hydrophobic peptide. Dimerization occurs through an intermolecular disulfide linkage between homologous cysteine residues towards the C-terminal of the carboxyl/cholinesterase domain (Cys577) (**Figure 6**). The 15 and 55 kDa domains are proteolytically cleaved, albeit they remain noncovalently associated. Then the C-terminal hydrophobic peptide is replaced with the GPI anchor to the cell membrane. Several glycans are also added to the 15 and 55 kDa domains during processing.

At least four catalytically efficient AChE isoforms are clearly resolvable on native PAGE of *D. melanogaster* extracts (Fournier *et al.*, 1988, 1992; Zador, 1989; Healy *et al.*, 1991). The predominant one is the mature GPI-linked AChE dimer above (often also called the amphiphilic dimer). The others include amphiphilic monomers and hydrophilic dimers and monomers. The amphiphilic forms are largely localized to the CNS but the soluble, hydrophilic forms are not. It is unclear to what extent the alternate forms result from incomplete posttranslational processing as above, or are the result of the programmed deconstruction of the amphiphilic dimers. However, at least some contribution from both processes has been suggested. The amphiphilic dimers clearly have the major role in the cleavage of acetylcholine at synapses but it is also suggested that the other forms have other, as yet only poorly understood functions (Grisaru *et al.*, 1999). Notwithstanding their extra *ace* gene, insects in other orders still produce an essentially similar array of AChE isoforms (Toutant, 1989; Bourguet *et al.*, 1998).

Data for several species suggest that insect AChEs are generally, like their vertebrate counterparts, highly efficient kinetically (Gnagey *et al.*, 1987; Devonshire *et al.*, 1998; Chen *et al.*, 2001 and references therein). Specificity constants, k_{cat}/K_m, are of the order of $10^8\,M^{-1}\,s^{-1}$ for vertebrate AChEs with acetylcholine or (as more commonly measured) acetylthiocholine (Gibney *et al.*, 1990; Taylor and Radic, 1994). These values are the outcome of particularly high turnover rates (k_{cat} values approaching $10^4\,s^{-1}$) but only modest substrate affinities (K_m values approaching $50\,\mu M$). The latter are rationalized in terms of the likely high local concentrations

of acetylcholine in the microenvironment of the synapse. Values for AChEs from the higher dipterans *D. melanogaster* and the Australian sheep blowfly *Lucilia cuprina* are three- to sixfold lower, due to slightly lower turnover values. Values for different isoforms of DmAChE are very similar to one another (Mutero and Fournier, 1992; Estrada-Mondaca and Fournier, 1998). There is one surprisingly low estimate of k_{cat} of around $2 \times 10^4\,min^{-1}$ for the Colorado potato beetle *Leptinotarsa decemlineata* (Zhu and Clark, 1995a) but the significance of this is unknown.

Two features of AChE biology in vertebrates that are generally associated with noncholinergic functions are also seen in insects, although evidence for the functional effects is still sparse for the insects (Darboux *et al.*, 1996; Scholl and Scheiffele, 2003). These features are the existence of nonneural soluble isoforms and the cell adhesive property. There are numerous studies showing effects of vertebrate AChE on the development of neuromuscular junctions and neurite outgrowth and many of these appear to depend on cell adhesive rather than catalytic properties of the enzyme (Sternfeld *et al.*, 1998; Grisaru *et al.*, 1999; Sharma *et al.*, 2001; Behra *et al.*, 2002, Scholl and Scheiffele, 2003). Soluble AChE is also overproduced in response to various stressors (Grisaru *et al.*, 1999; Lev-Lehman *et al.*, 2000). It is not known what parallels exist in insects but given the proliferation of closely related noncatalytic clades with clear functional effects they would seem well worth further investigation.

7.3.1.2. Target site insecticide resistance The activated or oxon forms of OP insecticides (see below) and carbamate insecticides act as essentially irreversible inhibitors, or suicide substrates, for AChE. This has proven true for all the well-characterized AChEs, although the insect AChEs seem particularly susceptible (Villatte *et al.*, 1998; Chen *et al.*, 2001; Boublik *et al.*, 2002). The inhibition occurs because the insecticides can serve as good substrates for the first step but poor substrates for the second step in the two-step reaction mechanism (**Figure 1**) (see Section 7.1.2). In other words they readily form a phosphorylated or carbamylated enzyme intermediate instead of the acyl-enzyme intermediate formed with carboxyl/choline ester substrates, but the phosphoryl- and carbamyl-enzyme intermediates are far more stable than the acyl-enzyme. The carbonyl carbon in the acyl enzyme intermediate is susceptible to nucleophilic attack by the oxygen of a water molecule, this oxygen being activated and oriented by residues in the active site. The slow to negligible hydrolysis of the phosphoryl- and carbamyl-enzymes is presumably due to the inappropriate geometry of the phosphate group or an interfering effect of the nitrogen atom adjacent to the carbonyl carbon, respectively, that prevents hydrolysis by the enzyme's usual mechanism.

Inhibition as above for AChE probably occurs to varying degrees for all catalytically competent carboxyl/cholinesterases with OPs. To a lesser degree it may also occur for some of these other enzymes with some of the carbamates. Whatever their effects on the other carboxyl/cholinesterases, however, it is the consequences of the pesticides' inhibition of AChE on the functioning of cholinergic nerves that are assumed to underlie much of their insecticidal effect. Insects or acarines exposed to the insecticides can survive until a surprisingly high proportion of their AChE molecules are inhibited (Smissaert et al., 1975; Hoffmann et al., 1992; Devonshire et al., 2003), but above this threshold signal transduction down cholinergic nerves fails and paralysis occurs.

Well over 20 cases of target site resistance to OPs and/or carbamates have now been described in various insects and acarines (reviews: Fournier and Mutero, 1994; Bourguet et al., 1998; Devonshire et al., 1998; Russell et al., 2004). The cases cover most of the major orders of insect pests although hemipteran and dipteran cases predominate and there are relatively few lepidopteran examples (Table 1). There is a little evidence that upregulation of the amount of AChE can augment tolerance to the insecticides (Fournier et al., 1992; Charpentier et al., 1998) but high-level resistance seems only to be achieved by amino acid substitutions affecting the catalytic machinery of the enzyme that make it less sensitive to inhibition by the insecticides.

Russell et al. (2004) have discerned two major patterns of target site resistance to OPs and/or carbamates among many of the cases that have been most thoroughly characterized so far (Table 1). Each pattern is reported from several orders, and Pattern Two also from acarines, and there are examples of both from the genus Culex. Pattern One resistance is characterized at a bioassay level by much greater resistance to carbamates than to OPs and, at a biochemical level, by a much greater reduction in the sensitivity of AChE to the carbamates than to the OPs. The target site resistance reported in the mosquito Culex pipiens, the cotton aphid Aphis gossypii, and the peach potato aphid Myzus persicae are well-characterized examples of Pattern One resistance. Pattern Two resistance is characterized by more equivalent levels of resistance and/or reductions in AChE sensitivity to the two classes of insecticide. Some well-characterized examples of Pattern Two resistance involve the mosquito Culex

tritaeniorhynchus, D. melanogaster, the housefly Musca domestica, the olive fruit fly Bactrocera oleae, and the silverleaf whitefly Bemisia tabaci. Importantly, the two types of pattern as defined by the bioassay and biochemical criteria also correspond to different sets of mutation in the catalytic machinery of AChE.

Molecular data on the resistance mutations are now available for 10 species in **Table 1**. A total of 18 mutations have been confidently associated with resistance in eight of the species, with four polymorphisms tentatively associated with resistance in the beetle L. decemlineata and the two-spotted spider mite Tetranychus urticae. Except for the L. decemlineata data, all cases involve the cholinergic form of AChE. As many as five mutations have been reported in each of D. melanogaster and M. domestica. However, the 22 differences involve just 11 DmAChE-equivalent sites. Mutations at five of the sites occur more than once in the data, sometimes involving the same substitutions in different species. Ten of the 11 sites lie in or near the active site/gorge.

At least on the current data, the options for Pattern One resistance mutations seem more constrained than those for Pattern Two resistance. A G151S mutation (on DmAChE numbering, which is used throughout this section) in the oxyanion hole has been associated with one clear case of Pattern One resistance, in C. pipiens, as well as two other anopheline cases almost certainly also of this pattern, in Anopheles gambiae and A. albimanus (Weill et al., 2002, 2003, 2004). It is unclear as yet how this relatively conservative mutation in the three lower dipterans reduces the sensitivity of the active site to inhibition by carbamates in particular. However, it is intriguing that a different mutation, to an aspartate, at the equivalent oxyanion hole residue of certain L. cuprina and M. domestica carboxylesterases actually confers metabolic resistance to OPs by allowing those enzymes to complete the second step of the reaction with OP substrates, albeit at the expense of their carboxylesterase activity (see Section 7.5.2).

However, Pattern One resistance in the hemipteran M. persicae is associated not with G151S but with a S371F change (Nabeshima et al., 2003). S371 is also close to the catalytic residues at the base of the gorge and has been implicated in substrate guidance and binding. While this is considered to be primarily with the alcohol portion of a carboxyl ester, its location (not obvious from the two-dimensional cartoon in **Figure 5**) is on the boundary between the anionic and acyl binding sites where it can also affect binding of the alkoxy group of OPs (Ordentlich et al., 1993, 1996; Millard et al.,

Table 1 Summary of cases of resistance to OPs and/or carbamates resulting from an insensitive AChE in various insect species; cases are included on the basis of bioassay and/or biochemical data implicating an insensitive AChE and, where possible, classified according to the two patterns of AChE insensitivity identified by Russell et al. (2004)[a]

Species	Resistance profile	Resistance mutations[b]	Reference
Pattern One: Carbamate resistance > OP resistance			
Thysanoptera			
Scirtothrips citri (citrus thrip)	AChE insensitivity is greater for propoxur than paraoxon in a resistant strain		Ferrari et al. (1993)
Hemiptera			
Aphis gossypii (cotton aphid)	Carbamate resistance (pirimicarb) is much greater than OP resistance (demeton-*S*-methyl, methamidiphos, monocrotophos, omethoate, pirimiphos-methyl)		Moores et al. (1996), Menozzi, cited in Weill et al. (2002), Li and Han (2002)
Myzus persicae (peach potato aphid)	Marked AChE insensitivity to pirimicarb (>100×) but no insensitivity to a range of other carbamates (e.g., methomyl, carbofuran, aldicarb, ethiofencarb) or OPs (demeton-*S*-methyl, dichlorvos, paraoxon)	S371F (gorge guidance residue near triad His/acyl pocket)	Moores et al. (1994), Williamson et al., cited in Villatte and Bachmann (2002); Nabeshima et al. (2003)
Nephotettix cincticeps of Iwata and Hama (green rice leafhopper)	43-fold resistance to propoxur, 43-fold to carbaryl, 6-fold to malathion, 18-fold to malaoxon		Iwata and Hama (1972)
Coleoptera			
Leptinotarsa decemlineata of Wierenga and Hollingworth (Colorado potato beetle)	>550-fold resistance to carbofuran and 9-fold resistance to azinphosmethyl		Wierenga and Hollingworth (1993)
Lepidoptera			
Helicoverpa armigera (cotton bollworm)	35-fold resistance to thiodicarb, 75-fold to methomyl and no cross-resistance to OPs		Gunning et al. (1996a)
Heliothis virescens (tobacco budworm)	k_i values for AChE inhibition are 10- to 100-fold less for propoxur than for several OPs		Brown and Bryson (1992)
Lower Diptera			
Anopheles albimanus of various authors	k_i values are 100-fold less for propoxur than paraoxon		Ayad and Georghiou (1975)
	k_i values are ~10-fold less for propoxur than for malaoxon		Hemingway et al. (1985)
	Resistance levels are >1000-fold for propoxur, 73-fold for carbaryl and 1.2–38-fold for OPs		Hemingway and Georghiou (1983)
Culex pipiens complex (house mosquito)	k_i values for three strains are 1000-fold less for propoxur than malaoxon, paraoxon, and aldicarb; resistance factors for one strain are 1600-fold for propoxur, 18-fold for chlorpyriphos, 140-fold for malaoxon and 5-fold for temephos	G151S (oxyanion hole) in all strains	Weill et al. (2002, 2003, 2004)
Pattern Two: Carbamate resistance ~ OP resistance			
Hemiptera			
Bemisia tabaci (silverleaf whitefly)	Similar k_i values for OPs and carbamates		Byrne and Devonshire (1993, 1997)
Coleoptera			
Diabrotica virgifera (western corn rootworm)	Slight increase in AChE insensitivity to paraoxon in a strain resistant to a range of OPs and carbaryl		Miota et al. (1998)
Lepidoptera			
Spodoptera frugiperda (fall armyworm)	Similar k_i values for OPs and carbamates		Yu (1991)

Continued

Table 1   Continued

Species	Resistance profile	Resistance mutations[b]	Reference
Lower Diptera			
Culex tritaeniorhynchus	AChE from a resistant strain has 870-fold decrease in OP sensitivity, and to a lesser extent to carbamates	F371W (gorge guidance residue near triad His/acyl pocket)	Nabeshima *et al.* (2004)
Higher Diptera			
Bactrocera oleae (olive fruit fly)	k_i data indicate insensitivity to omethoate and paraoxon, but not to propoxur	G436S (near triad Glu) I161V (gorge guidance residue near anionic site)	Vontas *et al.* (2001, 2002)
Drosophila melanogaster (vinegar fly)	Similar levels of resistance to OPs (paraoxon, malaoxon) and carbamates (carbaryl, propoxur)	F77S (anionic site) I161V/T (gorge guidance residue near anionic site) G265A (near triad Ser) F330Y (acyl pocket)	Mutero *et al.* (1994)
Musca domestica of Walsh *et al.* (housefly)	Similar k_i values for OPs and carbamates	V182L (little effect) G265A/V (near triad Ser) F330Y (acyl pocket) G368A (near triad Glu)	Walsh *et al.* (2001)
Acarina			
Amblyseius potentillae (predatory mite)	Similar k_i values for paraoxon and propoxur		Anber and Overmeer (1988)
Tetranychus kanzawai (Kanzawa spider mite)	OP resistant strains with low cross-resistance to carbamates		Kuwahara (1982)
Unknown			
Thysanoptera			
Frankliniella occidentalis (western flower thrip)	Ten-fold decrease in AChE sensitivity to diazoxon inhibition in a resistant strain		Zhao *et al.* (1994)
Homoptera			
Cacopsylla pyri (European pear sucker)	Resistance to monocrotophos and 70-fold decrease in k_i for monocrotophos		Berrada *et al.* (1994)
Hemiptera			
Lygus hesperus (plant bug)	34-fold decrease in AChE sensitivity to paraoxon inhibition in a resistant strain		Zhu and Brindley (1990a)
Nephotettix cincticeps of Tomita *et al.*	AChE inhibition by propoxur		Tomita *et al.* (2000)
Coleoptera			
Leptinotarsa decemlineata	Resistance to azinphosmethyl	S276G (peripheral binding site/active site)[c]	Zhu and Clark (1995a, 1995b), Zhu *et al.* (1996)
Lower Diptera			
Anopheles albimanus of Weill *et al.*	Resistance to carbamates and OPs based on AChE insensitivity but details not reported	G151S (oxyanion hole)	Weill *et al.* (2004)
Anopheles gambiae (African malaria mosquito)	Resistance to carbosulfan	G151S (oxyanion hole)	N'guessan *et al.* (2003); Weill *et al.* (2003, 2004)
Higher Diptera			
Musca domestica of Kozaki *et al.*	>40-fold decrease in AChE sensitivity to fenitroxon in resistant individuals	G265A (near triad Ser) F330Y (acyl pocket)	Kozaki *et al.* (2001)
Acarina			
Boophilus microplus (southern cattle tick)	Resistance to OPs		Baxter and Barker (2002)
Tetranychus urticae (two-spotted spider mite)	Multi-resistant strain has AChE that is more sensitive to OPs and carbamates (not less)	S/G151S (oxyanion hole) F371F/C (gorge guidance residue near triad His/acyl pocket) D/E160E (vicinity of gorge)	Anazawa *et al.* (2003)

1999). The clear inference from the *M. persicae* data is that it can also have profound effects on the binding of certain carbamates, at least in this hemipteran version of AChE-1.

In fact it is suspected that this *M. persicae* case may represent a subtype of Pattern One resistance that is specific for dimethyl carbamates like pirimicarb and the dimethyl carbamoyltriazole triazamate, and notably not including resistance to monomethyl carbamates such as propoxur that have been used to diagnose many of the mosquito and some of the other cases of Pattern One resistance above. Interestingly, the S371F change brings the *M. persicae* sequence into line with the status of this residue as a strongly conserved phenylalanine in most carboxyl/cholinesterases. Whatever benefit the serine might otherwise confer on susceptible aphids, it seems that it allows an accommodation of the slightly bulkier dimethyl carbamates at the base of the gorge that is hindered by the larger phenylalanine more usually found at this position in carboxyl/cholinesterases.

Another hemipteran, *A. gossypii*, also shows high resistance to pirimicarb, without cross-resistance to monomethyl carbamates, so it will be interesting to see if the same mutation occurs in its AChE-1. If so, it might suggest that the potency of the dimethyl carbamates as aphicides could at least in part be due to the unusual status of AChE-1 residue 371 in (susceptible) aphids. Interestingly, two other mutations to different residues which give very different resistance profiles have also been reported at residue 371 outside the Hemiptera and are discussed below.

A total of 12 mutations have been reported in three higher dipteran cases of known or presumptive Pattern Two resistance, involving *D. melanogaster*, *M. domestica*, and *B. oleae* (Mutero *et al.*, 1994; Kozaki *et al.*, 2001; Vontas *et al.*, 2001, 2002; Walsh *et al.*, 2001). These occur in just seven DmAChE-equivalent sites, with three of these sites (161, 265, 330) each used in two or three of the mutations. One of the 12 mutations, V182L, is located at some distance from the catalytic residues and apparently makes little contribution to resistance. However, the other 11 mutations all involve six sites at the base of the gorge, and most are substitutions to larger residues, which modeling suggests could act by constricting the space at the base of the gorge in such a way as to reduce the insecticides' access to the catalytic residues. Why they should effect Pattern Two as opposed to Pattern One resistance is as yet unclear.

Elegant kinetic analyses of AChE variants with different combinations of most of the 11 mutations indicate that most of them individually contribute towards the insensitivity phenotype. This is significant because various alleles in natural populations only contain a subset of the substitutions. Introduction of some of the same changes into the AChEs of *Aedes aegypti* and *L. cuprina* has a similar effect on the inhibition kinetics of baculovirus-expressed enzyme (Vaughan *et al.*, 1997b; ffrench-Constant *et al.*, 1998; Chen *et al.*, 2001); this supports the view that these or very similar mutations could underlie other cases of Pattern Two resistance.

Pattern Two resistance in *C. tritaeniorhynchus* involves a single change, F371W, not seen elsewhere as yet in this pattern (Nabeshima *et al.*, 2004). However, it does involve the same site at which other mutations have been associated with Pattern One resistance in *M. persicae* (above) and another very different resistance in the mite *T. urticae* (below). It fits the trend of changes to bulkier side chains evident in many of the mutations in both patterns. The fact that quite different insensitivity and resistance profiles can be obtained by the various substitutions at this site seen in different species may be an indication that the precise location of this residue around the boundary of the acyl and anionic sites differs among the AChEs of these species.

Russell *et al.* (2004) could not assign the case of target site resistance to azinphosmethyl in the beetle *L. decemlineata* to a pattern in the absence of toxicological or biochemical data on carbamate

---

[a]Cases were assigned to Pattern One or Two where bioassay and/or biochemical data were available for both OPs and carbamates. Pattern One resistance involves greater (target site) resistance factors and/or greater reductions in AChE sensitivity to carbamates than OPs. Pattern Two resistance involves more similar, or greater (target site) resistance or reductions in AChE sensitivity for the OPs compared to the carbamates. Cases for which an insensitive AChE has been reported, but for which the resistance profiles for both OPs and carbamates have not been reported, are classified as "unknown."

[b]All identified mutations except the S276G reported in *L. decemlineata* by Zhu *et al.* (1996) are in the cholinergic AChE, i.e., AChE in the higher Diptera and AChE-1 elsewhere (on the nomenclature of Weill *et al.* (2000), 2002, 2003 and Russell *et al.* (2004)). Residues are numbered according to the nomenclature for DmAChE (Harel *et al.*, 2000) with the residues in susceptible strains given before the numbers and those in resistant strains after them. A dash indicates that the mutation has yet to be identified. Bracketed comments on locations are taken from the authors or extrapolated from Harel *et al.* (2000).

[c]Weill *et al.* (2002) suggest that the association with resistance is not significant.

sensitivity. More importantly, Weill *et al.* (2003) have also expressed doubts about the significance of the resistance phenotype. This is important because molecular data (Zhu *et al.*, 1996) implicate an S276G change in the presumptively noncholinergic AChE-2 (Weill *et al.* (2003) and Russell *et al.* (2004) nomenclature) in this resistance. Neither this residue, nor this form of AChE have previously been implicated in resistance but modeling also puts the change in the vicinity of the active site, albeit not as close to the catalytic residues as most of those above. Zhu *et al.* (1996) suggest that the substitution could affect interactions with the pesticide either at the peripheral binding site or within the gorge. Clearly, further work is required to elucidate the biochemistry and molecular genetics of AChE-based resistance in this species.

The case of *T. urticae* is remarkable in that it apparently involves change(s) to AChE-1 that make it more, not less sensitive to inhibition by OPs and carbamates. No fixed differences have yet been reported in AChE-1 between susceptible and resistant strains. However, Anazawa *et al.* (2003) found that residue 151 in the oxyanion hole was polymorphic for Ser and Gly in susceptible strains but only Ser in resistant strains, residue 371 at the base of the gorge was polymorphic for Phe and Cys in resistant strains but only Phe in susceptibles, and residue 160, also near the active site, was polymorphic for Asp and Glu in susceptibles and only Glu in resistance. Residue 160 has not been implicated in resistance previously but both the other two have been. In both cases the precedent data (G151S, S371F, and F371W) associate changes to bulkier residues at these sites with reduced AChE sensitivity and target site resistance. In *T. urticae*, the F371F/C polymorphism associates a bulkier residue with lower AChE sensitivity but target site susceptibility, while the reverse is true for S/G151S. Whilst clearly difficult to interpret in detail, at face value the *T. urticae* results would seem to imply a very different AChE (in)sensitivity mechanism which nevertheless involves changes to some of the same residues implicated in the other AChE-based target site resistances so far resolved at a molecular level.

Some important insights into AChE-based target site resistance have also been gained from detailed analyses of the inhibition kinetics of various synthetic mutants of DmAChE. A series of studies (Villatte *et al.*, 1998, 2000a, 2000b; Boublik *et al.*, 2002) have examined a large number of substitutions at some of the sites recurrently implicated in natural resistance, as well as selected changes at over 20 other sites in the active site/gorge. DmAChE is already more sensitive than most characterized

noninsect AChEs to inhibition by OPs and carbamates (as indeed are other well-characterized insect AChEs) (Chen *et al.*, 2001) but a small number of mutations were found to make it even more sensitive. Most changes, however, decreased sensitivity, and most also reduced acetylcholine hydrolytic activity, although many still allowed levels of acetylcholine hydrolysis predicted to be viable *in vivo*. Some changes had generally greater effect on the interactions with OPs than with carbamates, others had more impact on carbamate interactions, and some had large effects on both. There was also significant, mutation-specific variation both among OPs and among carbamates in the size of the effects. Some mutations masked the effects of others. Several mutations at sites like I161 and F371 where natural resistance has been reported were found to have large effects, but several sites closer to the lip of the gorge and the peripheral binding sites were also found to cause significant sensitivity changes. Overall the results suggest that there may be several more options for Patterns One or Two resistance, or indeed other classes of target site resistance mutations, than have been documented thus far, with the "choice" among options depending both on the particular pesticides involved and on genetic constraints like pre-existing codon usage and variability at the key sites.

It will also not be surprising if there are phylogenetic differences in the mutations underlying the various insensitivity phenotypes. There are already indications of this in the data to date. For example, the three lower dipteran cases of known or presumptive Pattern One resistance all involve a G151S change, whereas, in *M. persicae* the causative change appears to be S371F. Likewise, none of the several changes seen in the three higher dipteran cases of Pattern Two is the same as the one found underlying this pattern in *C. tritaeniorhynchus*. It is suspected that such differences reflect phylogenetic differences in the details of the geometry and composition of the gorge; significant structural differences are indeed suggested by the alignment in **Figure 6**, the hemipteran AChE-1 subclade being but one example. The differences between the higher Diptera and other lineages reflecting their use of nonorthologous AChEs for cholinergic functions are also expected.

Both Pattern One and Pattern Two resistances seem to compromise the kinetics of acetylcholine hydrolysis to some degree (Fournier and Mutero, 1994; Devonshire *et al.*, 1998), which can lead to a fitness cost for resistant genotypes in the absence of the insecticide (Roush and McKenzie, 1987). The costs have been most thoroughly explored in two cases of Pattern One resistance.

The insensitive AChE in *M. persicae* affects fitness in two ways (Foster *et al.*, 2003). Aphids with the modified AChE reproduce more slowly and are more responsive to alarm pheromone. These effects have only recently been discerned as they are usually superimposed on fitness costs due to other resistance factors in clonal lines of this species (see Section 7.4.3.2). How they are mediated in terms of the physiology of AChE is as yet unclear.

A severe fitness cost has been described for the Pattern One resistance in *C. pipiens*; estimates by various means have put the fitness cost as high as about 60% (Chevillon *et al.*, 1997; Lenormand *et al.*, 1999). As a consequence there has been clinal variation in resistant allele frequencies away from geographic areas with heavy use of the insecticides towards those with much lighter use. There has also been regular seasonal variation in the frequency of the resistant allele in localities only seasonally exposed to the insecticide. Most recently a duplication of the *ace-1* locus has arisen and swept to high frequency in *C. pipiens* that combines both the susceptible wild-type and the resistant allele (Lenormand *et al.*, 1998a). This combination apparently provides both sufficient resistant AChE-1 protein to ensure survival in the presence of the insecticide, and sufficient wild-type AChE-1 protein to restore near-wild-type neurotransmission in the absence of the insecticide.

7.3.2. Noncatalytic Clades

The five clades of noncatalytic proteins considered here comprise the neurotactins (clade N in **Figure 2**), which constitute an outgroup to all the others, and then the gliotactins (M), putative neuroligins (L), and two functionally anonymous radiations (I, K) which are all more closely related to the AChEs (J). expectThe greater phylogenetic separation of the neurotactins is also borne out by other aspects of their biology. For example, the neurotactins have an N-terminal extension which specifies transmembrane and intracellular domains, whereas all the others appear to specify these domains in C-terminal extensions (Grisaru *et al.*, 1999). Nevertheless, as shown, there are some striking biochemical similarities between the neurotactins, gliotactins and AChE, and these are specified by their shared carboxyl/cholinesterase domains.

Drosophila melanogaster and *A. gambiae* each have a single neurotactin. In *D. melanogaster* it is only reported as a transmembrane glycoprotein in neuronal and epithelial tissue of embryos and larvae (Fremion *et al.*, 2000). It does not have a consensus N-terminal secretion signal, presumably relying on a nonconventional signal or another mechanism, such as translocation following membrane insertion (see Section 7.2.2). Its ligand is a secreted protein, Amalgam, which is a member of the immunoglobulin superfamily (Liebl *et al.*, 2003). However, there is as yet no evidence for a role in immunity in *Drosophila*. Instead, the Amalgam–neurotactin complex, aggregated with other as yet only partly ascertained proteins, acts as a heterophilic cell adhesion system in axon guidance processes during development (Hortsch *et al.*, 1990; Speicher *et al.*, 1998). Notably, the adhesive properties of neurotactin in *in vitro* systems can be mirrored by chimeric neurotactin proteins in which its carboxyl/cholinesterase domain is replaced with homologous sequences from the AChEs of *D. melanogaster* or the eel *Torpedo marmorata*, or *D. melanogaster* glutactin (Darboux *et al.*, 1996) (see Section 7.4.1). *Inter alia*, this bears out the evidence for the cell adhesive capabilities of the AChEs (see Section 7.3.1.2).

Significantly, both the Amalgam-binding and cell adhesive functions of *D. melanogaster* neurotactin have been mapped to a segment of about 140 residues corresponding to the N-terminal region of the carboxyl/cholinesterase domain of AChE (in fact most of panel 1 in **Figure 6**) (Darboux *et al.*, 1996; Fremion *et al.*, 2000). Botti *et al.* (1998) have also shown strong conservation of surface charge around the lip of the gorge among AChE, neurotactin, and some of the other noncatalytics. All this suggests that N-terminal regions in the noncatalytics corresponding to some of the peripheral substrate binding machinery and parts of the gorge in AChE are responsible for their ligand binding functions.

There is no known vertebrate ortholog of insect neurotactin (although unfortunately there is an unrelated vertebrate peptide which is also called neurotactin) (Jung *et al.*, 2000). There is another inactive carboxyl/cholinesterase in vertebrates, thyroglobulin, which is quite similar to insect neurotactin in its carboxyl/cholinesterase domain and which is also like insect neurotactin, and unlike the other noncatalytics has a membrane-spanning domain as an N-terminal extension (Grisaru *et al.*, 1999; Scholl and Scheifele, 2003). However, thyroglobulin apparently functions in thyroid hormone delivery which clearly has no close parallel with insect neurotactin.

Although there are no functional data on the putative insect neuroligins, as a group they show clear sequence orthology with the well-studied vertebrate neuroligins and they have therefore been tentatively identified as such (Ranson *et al.*, 2002). There are four neuroligin genes in the vertebrates studied to date, and these all express an array of alternately spliced transcripts, generating further diversity at the protein level (Bolliger *et al.*, 2001). Even greater diversity is seen in their ligands, the neurexins, which exist in hundreds of forms (Tabuchi and Sudhof, 2002). Neurexin diversity is again generated from about three genes, with alternate promoters and downstream splice variants generating the diversity. The well-studied neuroligin-1 group are postsynaptic membrane proteins that trigger synapse formation around axon contacts (Song *et al.*, 1999; Clarris *et al.*, 2002). Their extracellular carboxyl/cholinesterase domains attract and aggregate a subset of neurexins and certain other proteins, and their cytoplasmic domains then respond by assembling part of the exocytotic apparatus for the synapse (Dean *et al.*, 2003). It is assumed that the diversity of neuroligins and neurexins gives specificity in the formation and organization of the synapses (Ichtchenko *et al.*, 1996). There is some evidence for direct functional interactions between certain neuroligin-1 proteins and AChE in promoting neurite development (Grifman *et al.*, 1998; Grisaru *et al.*, 1999). At least in part this involves a level of functional overlap but how this is mediated in terms of ligands and specific sequence/structural motifs is unknown.

The insect neuroligin clade currently contains four pairs of *D. melanogaster* and *A. gambiae* orthologs, plus one additional *A. gambiae* sequence. EST and bioinformatic analyses also suggest at least some use of alternate splicing to generate additional diversity of gene products (Claudianos, unpublished data). Interestingly *D. melanogaster* only has one neurexin gene although, again, it apparently encodes at least a small number of different neurexin protein isoforms (Tabuchi and Sudhof, 2002; Schulte *et al.*, 2003). Nevertheless it seems unlikely that the insect neuroligin/neurexin system involves the same scale of protein diversity as seen in their vertebrate orthologs, so there has probably been significant divergence in their function(s).

There are just a single *D. melanogaster* sequence and its *A. gambiae* ortholog in the insect gliotactin clade at this point. The *D. melanogaster* gliotactin is a transiently expressed transmembrane protein in the peripheral nervous system of the developing embryo, where it is an essential component of a paracellular structure called the septate junction (Genova and Fehon, 2003). The *A. gambiae* sequence has a conventional N-terminal secretion signal but the *D. melanogaster* sequence does not; like neurotactin above, it may use a nonconsensus secretion signal or a membrane insertion–translocation mechanism for extracellular presentation. Gliotactin acts in part as a transmembrane

anchor for the septate junction structure, which also contains a small number of other proteins, notably including neurexin. There is no direct evidence that the neurexin acts as a ligand for the gliotactin although it might not be surprising, given the vertebrate data above indicating that neurexins are ligands for the related neuroligin receptors. There is evidence that the insect gliotactin and the septate junction function in part in axon guidance during development (Schulte *et al.*, 2003). However, the major role of the complex appears to be maintenance of the transepithelial nerve–hemolymph permeability barrier. In mutant embryos with defective gliotactin the integrity of this barrier is broken down, the peripheral motor axons are exposed to high concentrations of potassium ions from the hemolymph, action potentials therefore do not propagate, and substantive paralysis results (Auld *et al.*, 1995).

Consistent with the colocalization with neurexin in the septate junction, there are suggestions in the literature that *D. melanogaster* gliotactin is equivalent to the neuroligin-3 subset of vertebrate neuroligins, which are also found in ensheathing glia (Gilbert *et al.*, 2001). However, this correspondence does not reflect genuine orthology at a sequence level (Claudianos, unpublished data) suggesting either a functional shift or functional overlap between the two receptor radiations.

One of the two functionally anonymous noncatalytic clades (K in **Figure 2**), comprising just two pairs of *D. melanogaster* and *A. gambiae* orthologs, is phylogenetically closely related to the neuroligins and conceivably could represent additional members of this neurexin binding functional group. Like the neuroligins, their sequences show C-terminal extensions beyond their carboxyl/cholinesterase domains with clear consensus transmembrane domains and downstream, presumptively cytoplasmic, domains. The other functionally anonymous noncatalytic clade (I in **Figure 2**) includes a single orthologous sequence from each of *D. melanogaster*, *A. gambiae*, and *A. aegypti*. It is less closely related to the neuroligins, or AChEs. Although it does have a C-terminal extension it does not contain any good match with consensus membrane-spanning motifs. Functional analysis of both these clades using all the tools of modern *Drosophila* genetics would seem to be highly worthwhile from here.

7.4. Secreted Catalytic Clades

There are five major clades of secreted, generally catalytically active carboxyl/cholinesterases (**Figure 2**). One contains the CNS adhesive protein glutactin and its relatives, most apparently catalytically active (clade H) (see Section 7.4.1). Two others contain JHEs and similar enzymes albeit, interestingly, these two clades are not simply sister groups (clades F, G) (see Section 7.4.2). The fourth contains the well-studied dipteran β-esterases and their hemipteran relatives implicated in insecticide resistance (clade E) (see Section 7.4.3). Little is known about the fifth clade but it does include one enzyme putatively identified as a pheromone esterase (clade D) (see Section 7.4.4).

7.4.1. Glutactin and Related Proteins

This clade currently only contains *D. melanogaster* and *A. gambiae* sequences, four for the former and nine for the latter, which are arranged in two chromosomal clusters in each species. The only member of the clade for which there is currently functional information is *D. melanogaster* glutactin, which is an outgroup to all the others. This protein is catalytically inactive, as are two of the other *D. melanogaster* sequences. Conversely, the duplicate of *D. melanogaster* glutactin (82 in **Figure 2**) and all the *A. gambiae* sequences have consensus catalytic triads, and so are potentially catalytically competent. Glutactin is also one of only two proteins in this clade known to have additional carboxy terminal sequence specifying transmembrane and cytoplasmic domains, the other being another of the catalytically incompetent *D. melanogaster* proteins (76).

Little is yet known about the function of *D. melanogaster* glutactin (Olson *et al.*, 1990; Parker *et al.*, 1995; Darboux *et al.*, 1996; Grisaru *et al.*, 1999) and, as noted above, there are no known vertebrate orthologs to offer clues as to its function. The *D. melanogaster* protein is strongly anionic and heavily glycosylated. It is found in many basement membrane sites and is therefore inferred to be a component of the extracellular matrix. It is particularly abundant in the envelope of the CNS, so it fits the recurring theme of a nervous system cell surface adhesive protein seen above for AChE and the other noncatalytic members of the family. Nothing is known about possible ligands.

7.4.2. Juvenile Hormone Esterases

Juvenile hormone esterase (JHE) is one of the few insect esterases other than AChE to have a clearly defined physiological substrate. However, known insect JHEs fall into two distinct clades in the phylogeny in **Figure 2**. Known nonlepidopteran JHEs are found in a sister clade to the β-esterases (see Section 7.4.3), while known lepidopteran JHEs sit in a more distantly related radiation.

Juvenile hormones (JHs) are a family of sesquiterpenoid esters of methanol found throughout the Insecta (**Figure 7**). The ester moiety is stabilized by conjugation with a 2,3 double bond and it may be this feature of the substrate that requires a specific enzyme (Hammock, 1985). Hydrolysis of the conjugated methyl ester by JHE is traditionally regarded as inactivating the hormone, although mounting evidence suggests that the resultant JH acid also has a hormonal role in some circumstances, with or without re-esterification by a methyltransferase (Sparagana *et al.*, 1985; Ismail *et al.*, 1998, 2000). The most common form of JH is JHIII, which is found in all insect orders that have been investigated (Gilbert *et al.*, 2000). Other forms of JH have the same length of acid chain and generally also the 10(R),11 epoxide group. They differ principally in having various combinations of ethyl rather than methyl substituents on the acid chain (JH0, JHI, JHII), or by an additional 6,7 epoxide group (JHIII bisepoxide, JHB3). Some insects, notably the Lepidoptera and higher Diptera, may use mixtures of JHs, and the JH function may extend to other closely related compounds such as methyl farnesoate (JHIII without the epoxide) (Gilbert *et al.*, 2000).

JH has key roles in the regulation of insect development, metamorphosis, and reproduction (reviews: Gupta, 1990; Riddiford, 1996; Gilbert *et al.*, 2000; Truman and Riddiford, 2002). Certain developmental events require low JH titers and these often coincide with sharp peaks of JHE activity, suggesting that the role of JHE is to substantively clear the insect of JH (de Kort and Granger, 1981, 1996; Hammock, 1985; Roe and Venkatesh, 1990; Gilbert *et al.*, 2000). Some precisely timed experiments have shown peaks of JHE slightly after the JH titer had dropped, suggesting that the role of JHE could be to scavenge for traces of JH remaining after it has mostly been removed by other means,

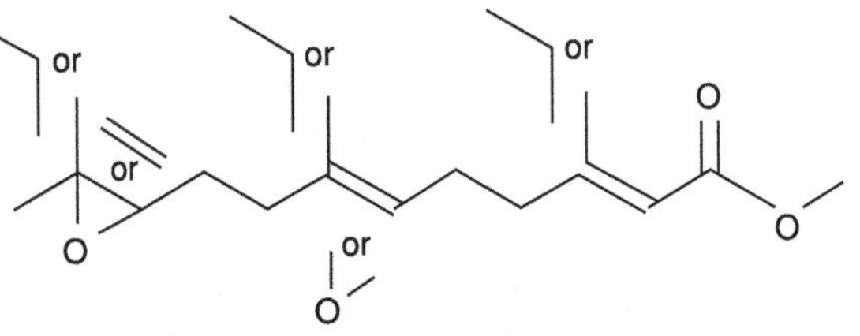

Figure 7 Juvenile hormone (JH) structures. The most common juvenile hormone (JHIII) is shown centrally. Other forms of JH have very similar structures. For example, JHIII bisepoxide has a second epoxide ring replacing the 6,7 double bond of JHIII while methylfarnesoate differs from JHIII by having a double bond in place of the 10,11 epoxide. These three forms of JH may act as a hormone blend in the higher Diptera (Yin *et al.*, 1995). Further forms of JH found particularly in Lepidoptera differ from JHIII by having ethyl side chains replacing one or more of the methyl side chains of JHIII (on C11 for JHII, on C11 and C7 for JHI, and on C11, C7, and C4 for JH0).

for example, by JH epoxide hydrolases (Baker *et al.*, 1987).

Genetic variation in JHE activity can lead to differences in JH titer, which in turn can have profound effects on development. For example, such variation acts as a genetic switch between short-winged, sedentary and long-winged, migratory forms of some crickets (Zera and Huang, 1999; Zera and Harshman, 2001). Further evidence for a morphogenic role of JHE comes from the disruption of morphogenesis by application of more-or-less specific inhibitors of JHE or interfering RNAs that downregulate JHE titers (Hajos *et al.*, 1999). The use of the interfering RNAs in final instars of the tobacco budworm *Heliothis virescens* led to aberrant morphogenesis, with the body organization remaining larvalike while the cuticle became pupalike.

The defining characteristics of a JH-specific esterase are a very high affinity for the hormone (typically a 20–200 nM K_m) combined with moderate turnover (k_{cat} of 0.6–$4.3\,s^{-1}$), yielding specificity constants (6–$50 \times 10^6\,M^{-1}\,s^{-1}$) approaching the diffusion limits (Zera *et al.*, 2002; Kamita *et al.*, 2003). This should be sufficient to provide efficient hydrolysis of realistic *in vivo* concentrations of the hormone (Hammock, 1985). At the same time it makes an interesting contrast with AChE (see Section 7.3.1), which has an even higher specificity constant for its substrate but achieves this through a very high turnover and only modest K_m.

Most *Jhe* genes have been isolated by the classical reverse-genetics route in which protein purification and amino acid sequencing have then allowed the design of suitable polymerase chain reaction (PCR) primers or probes. Of particular importance has been a trifluoromethylketone affinity technique used for the purification of various leptidopteran JHEs and a coleopteran JHE (Hinton and Hammock, 2003). However, this method has generally proven difficult when applied to nonlepidopteran JHEs, due to either poor binding to the ligand or poor stability during extensive dialysis steps (Campbell *et al.*, 1998b; Zera *et al.*, 2002), although Campbell *et al.* (2001) achieved good recovery of *D. melanogaster* JHE activity when a carrier protein was added. JHE from the cricket *Gryllus assimilis* is the only example of purification from a hemimetabolous insect and its N-terminal sequence has been used to isolate part of the cognate gene (Zera *et al.*, 2002).

The identification of the *Jhe* gene of *D. melanogaster* became possible when the genome of that species was completely sequenced. This was done by matching a peptide mass fingerprint from the purified protein with a predicted gene sequence in the genome sequence (Campbell *et al.*, 2001). It was then possible to identify an orthologous gene from *A. gambiae* once the genome of that species became available (**Figure 2**) (Ranson *et al.*, 2002), although it remains to be demonstrated directly that the latter encodes a functional JHE.

The availability of complete sequences of insect genomes allows us to address an outstanding question from the classical studies on JHE. Do insects have more than one *Jhe*/JHE gene/enzyme system? Multiple JHE activities with different isoelectric points or other physical differences are sometimes found in one species (Khlebodarova *et al.*, 1996; Zera *et al.*, 2002 and references therein). It is generally not clear whether the different JHEs are the products of different genes, differently processed products of the same gene, or allelic variants. Also, it is possible for an esterase to show some JHE activity *in vitro* without evidence for an *in vivo* role in JH metabolism (Campbell *et al.*, 1998b). No study has clearly shown more than one *Jhe* gene in any particular species, although Hinton and Hammock (2003) had evidence for two products from one *Jhe* gene. Nevertheless it is plausible that an insect might use different *Jhe*/JHE systems in different locations (say in tissues and in circulation) or with specificity for different forms of JH. The latter is particularly plausible in the Diptera and Lepidoptera where endogenous forms of JH might act as mixtures of components, which differ in potency in assays, binding affinity to carriers, or relative abundance at different developmental stages (Yin *et al.*, 1995; Gilbert *et al.*, 2000).

The complete genome of *A. gambiae* shows a clade of five esterase sequences that are monophyletic with a pair of esterase sequences in *D. melanogaster*, one of which is the identified JHE (**Figure 2**). The closest relative of these seven dipteran sequences is the identified JHE of a coleopteran, the yellow mealworm *Tenebrio molitor* (Hinton and Hammock, 2003). Thus it seems probable that these eight sequences descend from an ancestral JHE. However, it is also quite possible that some of them no longer function in JH metabolism. Only the *A. gambiae* sequence most closely related to *D. melanogaster* JHE shows the GQSAG version of the GXSXG nucleophilic elbow motif associated with verified JHEs (see below). The second *D. melanogaster* sequence is clearly a duplication of the identified JHE and is adjacent on the chromosome but it also lacks the GQSAG motif. Perhaps one gene in each dipteran has retained the ancestral JHE function, while one duplicated gene in *D. melanogaster* and four in *A. gambiae* have evolved to other functions.

The lepidopteran JHEs form a distinct clade from the dipteran and coleopteran JHEs yet they share the

GQSAG motif. Falling between these two clades are the β-esterases (see Section 7.4.3). However, the β-esterase sequences lack the GQSAG motif, there is no physiological evidence suggesting that they might function as JHEs, and one, EST6, has been purified and shown to lack JHE activity *in vitro* (Myers *et al.*, 1993). This suggests that perhaps a single ancestral JHE whose origin predates the separation of several insect orders has subsequently given rise to a non-JHE radiation. Alternatively, the JHE function could have evolved twice in distinct esterase radiations subsequent to the separation of these orders. Interestingly, the lepidopteran JHE clade also includes four *A. gambiae* genes but no *D. melanogaster* genes. This again raises the possibility of multiple JHEs, although none of the four *A. gambiae* genes contains the GQSAG motif that is present in all the lepidopteran JHEs. It also suggests a loss of these genes in the higher Diptera.

Further evidence suggesting multiple JHEs or JHEs diverging to other roles is an unusual protein decribed as "JHE-related" from the cabbage looper *Trichoplusia ni* (Kadono-Okuda *et al.*, 2000). Although this sequence (70 in **Figure 2**) falls within the lepidopteran JHE clade, *T. ni* has another JHE that more closely resembles other lepidopteran JHEs. The JHE-related enzyme hydrolyzes JHIII but its affinity is fourfold less than the more typical JHE from *T. ni* hemolymph. The enzyme shows much lower sensitivity to certain inhibitors and a different expression pattern from classical lepidopteran JHEs. Kadono-Okuda *et al.* (2000) suggest that its function might not be to hydrolyze JH but instead to break down some other, as yet unknown, JH-like compound.

Even more atypical are two putative JHE sequences isolated from *L. decemlineata* (Vermunt *et al.*, 1998). These show only about 17% identity with the JHEs described above and lack the GXSXG nucleophilic elbow motif around the catalytic serine that is characteristic of all known carboxylesterases. Alternatively, these sequences show 29% identity with a bacterial protein that shows oral toxicity against several mosquito species and one lepidopteran (Duchaud *et al.*, 2003). These results could imply an entirely unrelated group of JHE enzymes. Importantly, however, there is as yet no direct demonstration that these genes express JHE activity.

While presence of the above-mentioned GQSAG motif is strongly correlated with JHE sequences, it is not diagnostic for JHEs. The consensus nucleophilic elbow sequence around the catalytic serine residue of all carboxyl/cholinesterases is GXSXG, the first X being most often Glu and the second most often Ala. Gln in the first X position is rare overall but present in all the confidently identified JHEs (Campbell *et al.*, 2001; Hinton and Hammock, 2003). Thomas *et al.* (1999) modeled the structure of a lepidopteran JHE on the known tertiary structures of an AChE and a lipase, and proposed that Gln rather than Glu at this position indirectly shortens the hydrogen bond between the His and Glu residues of the catalytic triad. This might partially account for the very low K_m values of JHEs for their substrates. Aside from the known JHEs and the probable *A. gambiae* JHE, a Gln at this position in an active carboxyl/cholinesterase only occurs twice in the *A. gambiae* genome (14 and 15 in **Figure 2**) and once in the *D. melanogaster* genome (82). The above-mentioned *T. ni* JHE-related enzyme has GQSCG, with the Gln in common with the other JHEs but the Cys residue otherwise unique in the phylogeny.

A second structural feature associated with JHEs is a predicted amphipathic helix with basic residues along one face, first identified in the lepidopteran JHEs (Thomas *et al.*, 1999) and also predicted for *D. melanogaster* JHE (Lys184, Arg191, Arg195) (Campbell *et al.*, 2001). The basic residues form a positively charged patch on the opposite surface of the protein from the active site gorge. This patch might function in the recognition of JHE by specific cells that take up and then degrade the enzyme, thereby contributing to the regulation of its titer in hemolymph (Thomas *et al.*, 1999). This feature is strongly associated with both the lepidopteran and nonlepidopteran JHE clades but seldom seen in otherwise apparently catalytically competent enzymes elsewhere in the phylogeny. Outside the JHE clades there is a trend for substitutions (numbered as per DmJHE) of the second (Arg191) and third (Arg195) basic residues in the face of the amphipathic helix with Gln or the acidic Glu or Asp residues, while the first (Lys184) is generally conserved among clades A to D but not elsewhere.

Beyond the two sequence motifs mentioned above, JHEs must have specific residues that interact with JH molecules to confer their very low K_m values and high specificities. An homology-based model of the tertiary structure of the lepidopteran *H. virescens* JHE could only dock JH in the active site with its orientation reversed relative to acetylcholine in AChE, such that the small alcohol moiety of JH was accommodated by the region homologous to the smaller "acyl binding pocket" while the longer acyl group was accommodated by the larger "anionic site" (Thomas *et al.*, 1999). Activity series with artificial substrates and inhibitors are generally consistent with this reversed-orientation hypothesis (Kamita *et al.*, 2003). Studies with JH analogs show that almost any departure from the JH structure reduces binding (Campbell *et al.*, 1998b). Consistent with that observation, the *H. virescens* JHE model has JH making contacts with eight further residues (given below in HvJHE numbering) in addition to the residues of the catalytic triad (Thomas *et al.*, 1999). Two (Gly123, Gly124) are the generally conserved oxyanion hole residues. A further two (Pro295 and Phe127) occur often in the alignment. One (Ile449) is unusual but consistent with the generally small residues found adjacent to the His448 of the catalytic triad. The remaining three putative contact residues (Val75, Tyr77, and Phe338) show good conservation among the lepidopteran JHEs but no further and in particular they are not conserved among the nonlepidopteran JHEs. Instead they fall in regions of difficult alignment and indels in the overall esterase alignment (**Figure 6**), as one might expect if these are the regions that confer specificity for diverse substrates.

Much of the above speculation on JHE structure/function may soon be resolved. Kamita *et al.* (2003) announced the crystallization of JHE from the tobacco hornworm *Manduca sexta* with a JH analog, 3-octylthio-1,1,1-trifluoro-2-propanone (OTFP), and expect a high resolution structure to follow soon.

Jhe has attracted significant interest over several years as a potential transgene for insertion into recombinant insect baculoviruses (Hammock *et al.*, 1990). The objective has been to make the viruses commercially viable prospects as biological insecticides. A range of possible anti-JH effects might follow from an excess of virus-expressed JHE, which could be more effective than the JH-like effects of the chemical "juvenoid" insecticides. In larvae JH is a feeding stimulant so it was hoped that excess JHE would cause larvae to cease feeding while leaving the animals alive to allow further replication of the virus. Alternatively, an anti-JH effect at the time of a larval moult might have caused a precocious and fatal pupal molt.

In the first experiments with baculoviruses expressing wild-type HvJHE, first instar lepidopteran larvae were killed but older larvae were found to rapidly degrade the excess JHE (Hammock *et al.*, 1993; Bonning and Hammock, 1996). It appeared that the levels of circulating JHE were tightly regulated by specific uptake and degradation in pericardial cells. More recent work has therefore attempted to increase the efficacy of this JHE with mutations intended to interfere with its targeting for degradation, thereby allowing a greater increase of JHE activity in the insect (van Meer *et al.*, 2000; Shanmugavelu *et al.*, 2000 and references therein). When Lys29 and Lys524 in this JHE (not those in the amphipathic helix above) were replaced with arginine residues there was no reduction in the clearance of JHE from the hemolymph. However, less JHE was found in the lysosomes of pericardial cells, indicating that intracellular processing of the JHE taken up by the pericardial cells had been effected as intended. There was also less binding to a putative JHE binding protein and this correlated with improved insecticidal efficiency and reduced feeding damage and 17% of larvae showed contractile paralysis.

Curiously, more success was obtained with another mutation, which was actually expected to serve as a negative control. This mutation inactivated the enzyme by substituting a glycine for the catalytic serine residue. Viruses expressing this inactive enzyme killed the larvae more rapidly than did viruses expressing wild-type JHE and they also reduced larval feeding significantly. Some larvae (27%) died at the larval molt with extreme blackening of the cuticle. The remainder showed similar symptoms to larvae infected with the wild-type virus except that some also showed contractile paralysis. These pathologies differ from those seen with the first set of mutants, suggesting different mechanisms of action, but the mutations were not synergistic when combined in the same molecule (van Meer *et al.*, 2000). Enzyme-inactivating mutations of the other two residues of the catalytic triad resulted in viruses that were no more effective than the wild-type. There remains no satisfactory explanation for these results; at face value they could imply a noncatalytic function for the JHE used in these experiments. At the least they serve as dramatic reminders that the functions of JHEs and JHE-like enzymes are still very incompletely understood.

7.4.3. β-Esterases

This intensively studied clade (E) is the sister group to the nonlepidopteran JHEs. It comprises two subclades, each with *D. melanogaster* and *A. gambiae* representatives. One contains the so-called β-esterase cluster of *Drosophila*, and the other contains the hemipteran esterases implicated in metabolic resistance to OPs and some other insecticides.

7.4.3.1. The *Drosophila* β-esterase cluster   The β-esterase cluster was so named by Korochkin *et al.* (1987) because it includes the major esterase isozymes apparent in PAGE analysis of whole organism homogenates of most *Drosophila* species using β-naphthyl acetate as a substrate. As outlined below (see Section 7.5.2), those authors also named another, α-esterase cluster, on the basis that it included the major esterase isozymes evident after staining PAGE gels using the isomeric α-naphthyl acetate as substrate. In fact there are esterases that are polymorphic for α- and β-naphthyl acetate preferring forms (Zouros and van Delden, 1982), so the substrate preference does not necessarily indicate substantial sequence divergence. Nevertheless the names for the two clusters are now widely used and, with the caveat that there is imperfect correspondence with the substrate preference, the usage is retained here.

The β-esterase cluster has been characterized at a molecular and/or biochemical genetic level in many *Drosophila* species, mainly located in four species groups, covering both the subgenera *Sophophora* and *Drosophila* (reviews: Oakeshott *et al.*, 1990, 1993, 1995, 1999). In most species studied it contains two to four members arranged in a tightly linked cluster, with one member being an abundant hemolymph enzyme, another located in late larval integument and, in some cases, one being heavily expressed in male reproductive tissue (**Figure 8**). However, the roles of the different β-esterases in carrying out these functions vary greatly both among and within the species groups.

In the *melanogaster* species group in the subgenus *Sophophora*, the cluster comprises just two very tightly linked, tandemly arranged genes (Oakeshott *et al.*, 1995). In the ancestral state for this species group, as seen in *D. yakuba* and *D. erecta*, the upstream gene, *esterase6* (*Est6*), encodes a homodimer expressed mainly in the hemolymph of most life stages, while the product of the downstream gene, *esterase7* (*Est7*), also a homodimer, is essentially only seen in late larval integument (Dumancic *et al.*, 1997; Oakeshott *et al.*, 2001) (although EST analyses also reveal a brief pulse of expression in early embryos) (http://flygenome.yale.edu/Lifecycle/; see also Arbeitman *et al.*, 2002). However in the derived state, as seen in *D. melanogaster*, *D. simulans*, and *D. mauritiana*, all in the *melanogaster* subgroup of species, EST6 has become a monomer and acquired a major pulse of expression in the anterior sperm ejaculatory duct of the adult male (albeit still retaining strong hemolymph activity) (Oakeshott *et al.*, 2001). Furthermore, *D. melanogaster* at least shows a high frequency of null alleles of the *Est7* gene and it is postulated that it is becoming a functionless pseudogene in this species (Balakirev *et al.*, 2003).

Although little is known about the functional role of the ancestral hemolymph activity of EST6, there is abundant evidence that the acquired ejaculatory duct expression of EST6 in the melanogaster subgroup is important in reproductive biology (Meikle *et al.*, 1990; Richmond *et al.*, 1990). In these species the enzyme is transferred from the male to the female in the semen during mating. Most of it is

melanogaster subgroup species in the *melanogaster* group

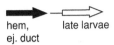
hem,
ej. duct
late larvae

D. yakuba and *D. erecta* in the *melanogaster* group

hem   late larvae

D. pseudoobscura in the *obscura* group

?   hem, eye   ?

D. virilis in the *virilis* group

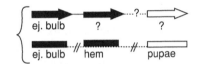
ej. bulb   ?   ?
ej. bulb   hem   pupae

D. buzzatii in the *repleta* group

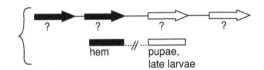
?   ?   ?   ?
hem   pupae,
late larvae

Figure 8 Organization of the β-esterase clusters in various *Drosophila* species as inferred from molecular and classical genetics (arrows and blocks respectively). *Est6* and its putative orthologs are shown in black, *Est7* and its putative orthologs are unshaded. For the molecular data, the arrows indicate directions of transcription. Major expression phenotypes are indicated below the arrows. Question marks indicate unknown linkage and orientation (between the arrows) or unknown expression (below the arrows). ····//··· indicates close linkage but precise relationship unknown. Hem, hemolymph; ej. duct, ejaculatory duct; ej. bulb, ejaculatory bulb. (Adapted from Oakeshott, J.G., Boyce, T.M., Russell, R.J., Healy, M.J., **1995**. Molecular insights into the evolution of an enzyme: esterase 6 in *Drosophila*. *Trends Ecol. Evol. 10*, 103–110; and Oakeshott, J.G., Claudianos, C., Russell, R.J., Robin, G.C., **1999**. Carboxyl/cholinesterases: a case study of the evolution of a successful multigene family. *BioEssays 21*, 1031–1042.)

not then retained in her reproductive tissue but rapidly translocated to her hemolymph, where traces of it survive for several days. It has also been suggested that the evolutionary shift of the enzyme to monomeric structure in the *melanogaster* subgroup has been to facilitate transport of this enzyme across the vaginal wall (Oakeshott *et al.*, 1995). How the male-donated enzyme in the female hemolyph impacts on the role of the female-produced EST6 in her hemolymph is a mystery. However, elegant experiments reviewed in Richmond *et al.* (1990) comparing females mated to males bearing wild-type versus laboratory null alleles of *Est6* have shown clear consequences of the transfer of active enzyme for the female's subsequent reproductive behavior. In general terms there is a stimulation of egg laying behavior and an inhibition of receptivity to remating. There are separable short (<24 h) and long (5–7 days) term effects on receptivity. Interestingly some similar effects have also been described for several small peptides found in the semen but there is no evidence as to whether or how the EST6 and small peptide effects are related (Chapman *et al.*, 2003; Liu and Kubli, 2003). EST6 shows activity against a broad range of substrates *in vitro*, plus some limited peptidase/amidase activity, but its *in vivo* substrate(s) is (are) unknown (Richmond *et al.*, 1990).

The localization of EST7 activity to the integument of late larvae/prepupae suggests a fairly specific function during this major metamorphic transition. Alternatively, the high frequency of null alleles in *D. melanogaster* suggests that the function is dispensable, without significant effects on fitness in this species at least.

Detailed comparative genomic analyses have given us some strong clues as to the genetic basis of the acquired reproductive tract expression of EST6 in the *melanogaster* subgroup of species. It appears that the first step was the insertion of a complex macrotransposon into the distal portion of the ancestral *Est6* promoter (Oakeshott *et al.*, 2001; C.W. Coppin, B.C. Morrish, and J.G. Oakeshott, unpublished data). This insertion created a rudimentary promoter element directing ejaculatory duct expression. There was then a burst of nucleotide change in and around this element to hone its function, plus another pulse of change in the *Est6* coding region to adapt the protein to its new function. About 10% of the residues in the protein changed in a period of only 2–6 million years, and many of the changes were clustered in regions that would correspond to the peripheral substrate binding sites of AChE and an inferred ancestral dimerization site.

There is also evidence for functional effects of polymorphic variation in the *Est6* promoter and coding region within *D. melanogaster*. Most populations of the species segregate for two haplotypic forms of the promoter, the two forms differing at 13 sites spread over a ~400 bp region encompassing the introduced ejaculatory duct motifs (Odgers *et al.*, 1995). One of these two forms directs two- to three-fold more expression of the gene in the ejaculatory duct than the other (Odgers *et al.*,

2002). In the laboratory at least, such differences in the males are associated with quantitative differences in the egg laying profiles and remating behaviors of their mates (Saad *et al.*, 1994). There is no direct evidence as to whether these effects carry over to field populations but there is strong evidence from statistical tests of frequency distributions that these haplotype differences are not selectively equivalent in the field (Balakirev *et al.*, 2002; Odgers *et al.*, 2002). Most field populations of *D. melanogaster* are also polymorphic for two amino acid differences in the region of the EST6 protein corresponding to the peripheral substrate binding sites of AChE (Cooke and Oakeshott, 1989). There are kinetic differences between the corresponding allozymes for artificial substrates (White *et al.*, 1988), although the biological significance of these is unclear, either in the laboratory or in the field. Again, however, there is clear evidence that the allozymes are not selectively equivalent in the field. In this case there are large-scale latitudinal clines in their respective frequencies that recur across continents and take complementary forms in the northern and southern hemispheres (Anderson and Oakeshott, 1984).

The other lineage of the subgenus *Sophophora* in which the β-esterase cluster has been well characterized is the *obscura* species group, specifically *D. pseudoobscura* (Brady *et al.*, 1990; Brady and Richmond, 1990, 1992). In this species the cluster comprises three genes (**Figure 8**). The 3′ gene, termed *Est5C*, is orthologous to the 3′ gene (*Est7*) in the *melanogaster* cluster, while the two upstream genes, *Est5A* and *Est5B* (the latter sometimes just termed *Est5*), are duplicates of their common ancestor with *Est6* in the *melanogaster* cluster. The EST5 protein shows the high hemolymph activity seen for EST6 in the *melanogaster* group, albeit EST5 also shows high activity of unknown physiological significance in the eye. Nothing is known of the physiological roles of EST5A and EST5C, although their sequences show all the motifs expected for catalytic activity. Interestingly, some allozymes of EST5 differ in monomer/dimer status, suggesting that the ancestral dimeric status can be abolished by one or a very few amino acid differences (Arnason and Chambers, 1987).

Albeit less thoroughly characterized, it is clear that the β-esterase cluster in the subgenus *Drosophila* has diverged substantially from those outlined above for the two sophophoran species groups (**Figure 8**). Classical genetic analyses show a cluster of three genes encoding catalytically active β-esterases in *D. virilis*, in the *virilis* species group (Korochkin *et al.*, 1987, 1990). One of these, termed EST2, is abundant in the hemolymph while another, ESTS, is abundant in the male reproductive tract. Unlike EST6, ESTS is produced in the ejaculatory bulb, not the duct, and, although transferred to the female, it is not transferred to hemolymph, but retained in the vagina. There are in fact reports of seminal fluid esterases that are retained in the vagina in various insects (Mikhailov and Torrado, 2000). Some of these have also been implicated in egg laying and remating behavior, with the mechanism suggested to involve the digestion of waxy matrices, which otherwise retain yet-to-be-used sperm in their sperm storage organs (Korochkin et al., 1976). This mechanism has been proposed to apply to D. virilis, in which mated females have a waxy plug at their entrance of their sperm storage organ whereas, notably, mated D. melanogaster females do not (Korochkin et al., 1976; Oakeshott et al., 1990; Healy et al., 1991).

Molecular work has also revealed a cluster of three β-esterase genes in D. virilis (Enikolopov et al., 1989; Korochkin et al., 1990, 1995; Sergeev et al., 1993, 1995) which map cytologically to the region where the classical genetics locates the three β-esterase isozyme loci. One of the cloned genes is heavily expressed in the ejaculatory bulb and was, therefore, originally nominated as the EstS gene. However, the sequence of the encoded protein shows a nonconservative substitution of the histidine in the catalytic triad which would almost certainly inactivate it. While it is tempting to put this down to a sequencing error, in fact (below) a parallel situation in the β-esterase cluster of D. buzzatii, in the repleta species group is seen. The alternative is to suggest that ESTS and perhaps also the other two β-esterase isozymes are encoded by as yet uncloned gene(s) at nearby chromosomal location(s). The corollary of this is that both an active and a catalytically inactive β-esterase are highly expressed in the ejaculatory bulb. Its abundant expression might suggest a function for the inactive protein, and as seen in the earlier sections there are many catalytically inactive carboxyl/cholinesterases, which have important ligand-binding functions. The D. virilis β-esterases clearly warrant further genetic and biochemical analyses.

Classical genetic analysis of D. mojavensis in the repleta species group shows a cluster of two β-esterase isozymes, one expressed in hemolymph and the other in late larvae (Zouros et al., 1982; Zouros and van Delden, 1982) and, although not mapped genetically, there is also evidence from several repleta species for a third β-esterase, which is highly expressed in male reproductive tissue (Kambysellis et al., 1968; Oakeshott et al., 1990). Molecular

analysis of D. buzzatii in this species group has identified a cluster of three carboxyl/cholinesterase genes, which sequence similarities suggest should be its β-esterase cluster (East et al., 1990; Gomez and Hasson, 2003). However, the sequences of two of these show a glycine in place of the catalytic serine, which would clearly render them catalytically inactive. Given that D. buzzatii does express active hemolymph and late larval β-esterases (East et al., 1990), it appears that there may be two clusters of β-esterase or like genes in this species. And as with D. virilis, one of them appears to encode noncatalytic products.

The A. gambiae genome has a four-gene cluster which is clearly the sister group for the Drosophila β-esterase cluster (Figure 2). All four genes appear to encode catalytically active enzymes but there appears to be no information on their expression profiles. Several esterase isozymes are expressed in hemolymph and male reproductive tract in this species and other anophelines, but their relationship to the sequences in question is uncertain (Narang and Seawright, 1982; Vernick et al., 1988).

7.4.3.2. Amplified hemipteran β-esterases
Also in part of the β-esterase clade (E) are a group of secreted esterases (which would actually be classified as α-esterases on the basis of their α/β-naphthyl acetate preferences) whose overexpression provides a major mechanism of resistance to OPs and carbamates in several hemipterans.

In M. persicae metabolic resistance to these insecticides is associated with a marked increase in total carboxylesterase activity (Needham and Sawicki, 1971). Four phenotypic classes have been identified; susceptible, moderately resistant (R1), highly resistant (R2), and extremely resistant (R3). Variously elevated levels of esterase activity are found in all three resistant classes, due to overexpression of either of two esterase isozymes, E4 and FE4 (Devonshire and Sawicki, 1979; Field et al., 1988, 1993). In some resistant lines the levels of the esterase protein reach 10 pM enzyme per aphid, or approximately 1% of the total soluble protein (Devonshire, 1989). The two enzymes differ slightly in molecular weight (E4 65 kDa, FE4 66 kDa) and they have slightly different catalytic activities. The genes encoding these enzymes are closely related (98% amino acid identity), genetically linked (19 kb apart), and likely to be recently derived by gene duplication (Field and Devonshire, 1998). Flanking sequences also differ between the two genes. The FE4 gene has a 1.7 kb deletion of promoter region compared with E4, resulting in a shorter 5' untranslated region (94 bp). It is also likely that there are two further esterases in the region, creating a cluster of at least four ester-ase genes even in susceptible *M. persicae* (Field and Devonshire, 1998).

Multiple mechanisms control the overexpression of the E4 and FE4 genes in resistant lines but gene amplification is the major one (Field *et al.*, 1999) (**Table 2**). Either gene can be amplified up to 80 times within a highly resistant R3 strain. Differences in resistance and esterase activity among the R1, R2, and R3 phenotypes largely correlate with the num-ber of copies of the esterase genes. Normally only one of E4 or FE4 is amplified within a single line. The unit of amplification is large, approximately 24 kb for E4 and 20 kb for FE4, with each amplicon containing just one copy of the respective esterase gene and no other open reading frames. Amplicons are generally arranged in a head-to-tail configura-tion. In many resistant lines the E4 amplifications are also associated with a particular chromosomal translocation, with the amplicons generally all located at a single site on one of the translocation chromosomes. In one resistant clone, however, sites of E4 amplification have been found on three sepa-rate chromosomes. The FE4 gene is not associated with any translocations but the amplicons are dis-tributed across the genome at from three to eight sites (Blackman *et al.*, 1995, 1999).

Restriction fragment length polymorphism (RFLP) patterns and 5′ sequence for E4 and FE4 genes from several resistant clones suggest that there was a single original amplification event for each of E4 and FE4 (Field *et al.*, 1994). Several subsequent mutations then generated the three major classes of resistance phenotype, the variation in the chromosomal sites for some of the E4 amplicons, and also some minor variation in gene copy number within the major phenotypic classes (Blackman *et al.*, 1995).

The levels of expression of E4 are also con-trolled by DNA methylation, with decreased methylation in some lines carrying amplifications resulting in reduced gene expression and reversion to a more susceptible phenotype (ffrench-Constant

Table 2 Esterase amplicons associated with insecticide resistance in *Myzus persicae* and *Culex pipiens*

Amplicon	Genetics and biochemistry	References
Myzus persicae		
E4	Secreted esterase with high affinity for OPs and carbamates, very slow turnover for OPs, and effectively none for carbamates. Up to 80-fold amplifications associated with a translocation (can be several sites) confers up to 500-fold resistance to OPs and 15-fold to carbamates. Quantitative variation in resistance correlates with amplification level. Reversion to susceptibility due to decreased flanking sequence methylation	Sawicki *et al.* (1978), Devonshire and Sawicki (1979), Devonshire and Moores (1982), Devonshire *et al.* (1986), ffrench-Constant and Devonshire (1988), Field *et al.* (1988, 1989, 1997), Blackman *et al.* (1995), Field and Devonshire (1998), Field (2000), Field and Blackman (2003)
FE4	As for E4 except for: slightly high turnover of OPs, amplification not associated with translocation, and no reversion due to altered methylation	As for E4
Culex pipiens complex		
Estα1	Low frequency in France, ≤70-fold amplification	Mouchès *et al.* (1987), Raymond *et al.* (1989), Chevillon *et al.* (1997), Lenormand *et al.* (1998b)
Estβ1[2]	Widespread in the Americas and China, ≤250-fold amplification, at least three sequence variants	Georghiou *et al.* (1980), Beyssat-Arnaouty *et al.* (1989), Mouchès *et al.* (1990), Cuany *et al.* (1993), Qiao and Raymond (1995), Vaughan *et al.* (1995)
Estβ6	Rare variant	Xu *et al.* (1994)
Estβ7	Rare variant	Xu *et al.* (1994)
Estβ8	Rare variant	Vaughan *et al.* (1995)
Estα2[1]Estβ2[1]	Common worldwide, partially replaced Estβ1 in North America, but now declining in frequency in some European populations, ≤60-fold amplification	Raymond *et al.* (1987, 1991), Wirth *et al.* (1990), Rivet *et al.* (1993), Karunaratne *et al.* (1995), Qiao and Raymond, (1995), Guillemaud *et al.* (1996, 1997) Chevillon *et al.* (1997), Lenormand *et al.* (1998b), Hemingway *et al.* (2000), Weill *et al.* (2000), Pasteur *et al.* (2001)
Estα3Estβ1	South/Central American amplicon, two Estβ1 sequence variants	De Silva *et al.* (1997)
Estα4Estβ4	Becoming more common in France, ≤50-fold amplification	Poirié *et al.* (1992), Chevillon *et al.* (1997), Lenormand *et al.* (1998b)
Estα5Estβ5	Becoming more common in Cyprus, ≤250-fold amplification	Poirié *et al.* (1992), Wirth and Georghiou (1996)

and Devonshire, 1988; Field *et al.*, 1989; Hick *et al.*, 1996; Field, 2000). No examples of such reversion have been described for the FE4 amplification suggesting that the phenomenon may be locus-specific. Methylation of the E4 gene in revertants is confined to CpG doublets within the coding region and is absent from CpG-rich regions in the 5′ and 3′ flanking DNA (Field, 2000). DNA methylation within a highly expressed gene is itself unusual in eukaryotes, methylation typically being associated with pretranscriptional gene silencing.

While the E4 and FE4 esterases are highly sensitive to inhibition by the insecticides they show poor hydrolytic activity against them. Thus the bimolecular rate constant (k_i) for the phosphorylation of E4 by paraoxon is about $133\,\mu M^{-1}\,min^{-1}$ (Devonshire, 1977). However, its maximum rate of hydrolysis (k_3) for dimethyl OPs is about $3\,h^{-1}$, for diethyl OPs it is less than $1\,h^{-1}$, and it is even slower for carbamates (Devonshire and Moores, 1982). FE4 hydrolyzes both dimethyl and diethyl OPs approximately 1.5 times faster than E4 (Devonshire, 1989). Because their kinetics are so poor, the enzymes are thought to protect the organism from OPs and carbamates by first sequestering the insecticide and then, at least for dimethyl OPs, very slowly degrading it (Devonshire and Moores, 1989).

Very similar elevated esterase phenotypes are also associated with OP resistance in populations of a tobacco-feeding form (Abdel-Aal *et al.*, 1990, 1992; Wolff *et al.*, 1994) originally classified as a distinct species, *Myzus nicotianae* (Blackman, 1987). It is now clear that the elevated E4/FE4 esterases and encoding amplicons in this form are identical to those of *M. persicae*, and that the amplicons were acquired by introgression through sexual hybridization rather than by independent evolution (Field *et al.*, 1994). Given this, and studies of RFLP and randomly amplified polymorphic DNA (RAPD) patterns, it is now considered that *M. persicae* and *M. nicotianae* are conspecific (Clements *et al.*, 2000).

A similar case of amplification-based resistance has also been described in the brown rice planthopper *Nilaparvata lugens*. In this species two esterase isozymes stain more intensely in OP and carbamate resistant strains (Chen and Sun, 1994; Karunaratne *et al.*, 1999; Small and Hemingway, 2000a). The two isozymes are differently glycosylated forms of the same protein (Small and Hemingway, 2000a). The encoding gene is amplified three to seven times in strains resistant to OPs and carbamates, and there is an eight- to tenfold increase in measured esterase activity (Small and Hemingway, 2000b; Vontas *et al.*, 2000). Detailed kinetic analyses have not been reported for this system, but it is assumed that sequestration is a major part of the resistance mechanism.

The situation is more complicated in the greenbug *Schizaphis graminum*, where there are two types of esterase-based metabolic resistance to OPs, each associated with elevated staining intensities of esterase isozymes (Siegfried and Ono, 1993a, 1993b; Shufran *et al.*, 1996; Siegfried *et al.*, 1997; Ono *et al.*, 1999). One well-characterized type involves overexpression of a single esterase, whose gene shares intron positions and shows strong sequence similarity to E4/FE4 from *M. persicae* (about 73% identity at the amino acid level across the partial sequence available). The encoded enzyme cross-reacts strongly with antibody raised against *M. persicae* E4. Four- to eightfold amplification of this esterase gene confers broad spectrum OP resistance, while reversion to susceptibility occurs despite retention of the amplified genes because of changes in methylation. As in *M. persicae*, the amplified genes in resistant strains are methylated at CCGG sites, but revertants do not show methylation in upstream sequences. And also as with *M. persicae*, the mechanism of resistance seems primarily to involve sequestration rather than hydrolysis of the pesticide.

The second resistance mechanism in *S. graminum* is associated with elevated staining intensities of several esterase isozymes, although one, not the same as that in the first type above, clearly predominates. Patterns of antibody cross-reactivity suggest that none of the more intensely staining esterases in this so-called type 2 resistance are as closely related to *M. persicae* E4 as the overexpressed esterase in Type 1 resistance. The major type 2 esterase from *S. graminum* has been partially purified and shows slightly different electrophoretic mobility in type 2 strains, while Western blots using a type 2 esterase antibody suggest similar levels of expression across susceptible type 1 and type 2 strains (Siegfried and Zera, 1994; Siegfried *et al.*, 1997). These data suggest that Type 2 resistance in *S. graminum* is, unusually among the Hemiptera, not simply a result of overexpression and gene amplification. The extent to which it involves sequestration versus hydrolysis is therefore also an open question.

One feature of the hemipteran esterases like E4/FE4 and their relatives in clade E which may predispose them to a role in insecticide resistance is that they all appear to be secreted enzymes. It may be particularly useful for a sucking insect to secrete the enzyme into the gut, or perhaps preferentially into the salivary secretions, which are pumped into the plant as they feed. Salivary secretions are characteristically replete with digestive hydrolases, and this may indeed be the ancestral function of the esterases recruited to a role in insecticide resistance.

Measurable fitness costs have been associated with the amplified esterases in at least some of the cases above (Dawson *et al.*, 1983; Foster *et al.*, 2003). These costs have been confounded in the case of *M. persicae* by the simultaneous presence in many clones of other resistance mechanisms with differing impacts on fitness (see Section 7.3.1.2). However the contributions from the various resistances are now being dissected, and the presence of the amplified esterase genes has been shown to slow the intrinsic rate of population increase (Foster *et al.*, 2003). With the possible exception of type 2 resistance in *S. graminum*, there are no known structural differences between the amplified esterases and their susceptible ancestors, so they would be expected still to perform the ancestral function. Nor does the metabolic cost associated with the production of such large amounts of an enzyme fully account for the fitness costs because, in *M. persicae* at least, revertants still appear to be less fit than susceptible aphids (Foster *et al.*, 1996).

7.4.4. Semiochemical Esterases

Clade D currently comprises one moth esterase and three *D. melanogaster* homologs. All appear catalytically active and three have good consensus secretion signals. While nothing is known of the functions of the *D. melanogaster* enzymes the moth esterase has been implicated in pheromone signaling. A clade A enzyme from the same species has also been tentatively implicated in this role but this section is used to review the topic of esterase involvement in pheromone and other semiochemically triggered signaling processes.

Sensitive and specific sensing of volatile chemical cues is crucial to the ability of insects to interpret their environment and signal to one another. Efficient degradation of the cues is an integral part of the signal reception and tranduction process and several specific odor degrading enzymes (ODEs) have now been identified in insects. Many of these ODEs are esterases because many of the semiochemical signals are esters. So far 231 esters, many of them C10–C16 acetates, have been identified as components of moth sex pheromone blends (http://www-pherolist.slu.se), while some of the 169 esters that have been identified in floral scents may be involved in the attraction of pollinators (Knudsen *et al.*, 1993). Little is known about esterases for other semiochemicals but there is quite detailed knowledge of some esterases involved in the reception of sex pheromones.

Sex pheromones are commonly found and mostly studied in Lepidoptera but they probably occur quite widely across insects. They are produced mainly by females to attract and court conspecific males. At least in the lepidopteran model systems studied, the males receive the signal in specialized sensillar hairs on their antennae. Inside these hollow sensillar hairs is an aqueous lymph where the external blend of volatiles is thought to be recreated for a subset of the molecules that can be bound by locally abundant pheromone binding proteins. The latter act to solubilize and transport the chemical signal to the dendrites of specific neurons that extend into the lymph. ODEs in the lymph compete to remove the signal, thereby maintaining the system's ability to respond to new stimuli. Receptors on the dendritic membrane, probably members of the G protein coupled receptor superfamily, specifically recognize the signal and trigger a signaling cascade that results in a neuronal impulse being sent to the antennal lobe and then the brain.

As well as the sensillar lymph, ODEs are also required at at least one and probably two other points in the process. One which is relatively well characterized involves cleaning pheromone off the external surface of the insect; most pheromone molecules are strongly hydrophobic and therefore prone to adhere to waxy surfaces. The males need to remove this material from their integument so that they do not become their own source of signal, while the females may need to remove it so that they can better control their signal (Vogt and Riddiford, 1981). The other point in the process where an esterase may be required is in the removal of bound pheromone off the receptor inside the olfactory neuron cells, if the signal remains bound to the receptor as it gets reset through vesicle cycling inside the cell.

The saturnid moth *Antheraea polyphemus* is one of the preferred models for studying the biochemistry of sex pheromone signal transduction, in large part because its large branched antennae and long sex pheromone-sensitive sensilla (sensillar tricoidea) are relatively amenable to study. Its sex pheromone is a 9 : 1 blend of the ester (E,Z)-6,11-hexadecadienyl acetate and the corresponding aldehyde (E,Z)-6,11-hexadecadienal (Kochansky *et al.*, 1975). Four esterase isozymes extracted from *A. polyphemus* may have functions in the processing of the ester component of its pheromone blend: one from its scales, two from its antennal integument, and one from the lymph inside its sensillar tricoidea (Vogt and Riddiford, 1981, 1986; Klein, 1987).

An isozyme able to degrade radiolabeled (E,Z)-6, 11-hexadecadienyl acetate has been found in the body scales of both sexes (Vogt and Riddiford, 1986). The enzyme appears to be a carboxylesterase, as is it inhibited by typical OP inhibitors (Vogt and Riddiford, 1986). It requires solubilization by detergents suggesting some membrane association. It migrates very slowly on native PAGE and has a presumptively dimeric molecular mass of 100 kDa (Klein, 1987). Closely related species using different pheromones were found to have scale esterases that did not have activity against (E,Z)-6,11-hexadecadienyl acetate (Vogt and Riddiford, 1986).

Two soluble, OP-inhibitable esterase isozymes have also been associated with the antennal integument of both male and female A. polyphemus (Klein, 1987). One, of molecular mass 65 kDa, resolves as two forms under IEF (pI 5.85 and 6.0). The other, of 90 kDa, resolves as a single band under isoelectric focussing (IEF) (pI 5.0). The latter may be a homodimer as the molecular weight was estimated under native conditions. Neither of these is found in sensillar lymph but one, probably the 65 kDa enzyme, is also found in various other cuticular tissues, including wing, head and abdominal cuticle, and trachea, albeit not the brain, ventral nerve cord, fat body, or hemolymph (Vogt and Riddiford, 1981); hence the name integument esterase. As with the membrane-associated scale esterase above, this soluble integument esterase can hydrolyze the pheromone ester to the corresponding alcohol (Vogt and Riddiford, 1981).

An esterase isozyme has been identified in the sensillar lymph of male but not female A. polyphemus. This sensillar enzyme has a molecular mass typical of the subunit sizes of other carboxyl/cholinesterases (55 kDa) but an unusually low pI (3.0), and is not inhibited by several of the usual OP and other carboxyl/cholinesterase inhibitors (Vogt et al., 1985; Klein, 1987). The pheromone analog 1,1,1-trifluoro-2-tetradecanone is a potent inhibitor (IC_{50} 5 nM) and kinetic analysis shows good affinity for the pheromone ($K_m = \sim 2 \times 10^{-6}$ M). Inhibition studies of the enzyme's ability to hydrolyze artificial substrates like naphthyl acetate with a range of pheromone-like compounds suggest that it has a high level of specificity for the native pheromone, including for the sites of unsaturation in its acyl chain backbone (Vogt et al., 1985). Although it is perhaps four orders of magnitude less abundant than the pheromone binding protein, Vogt et al. (1985) have estimated that its V_{max} should be sufficient to clear half of the pheromone inside the sensillar lymph in 15 ms.

No esterases or other ODEs have yet been reported from inside pheromone-receptive neuronal cells but they may well occur there. It is known that many neurotransmitter-receiving G protein coupled receptors are internalized into the cell once activated by ligand binding (McClintock and Sammeta, 2003). The sensitivity of the neuron to further stimulation is then controlled by the balance between recycling to the cell surface and degradation of the receptor within the endosomes. If pheromone receptors are endocytosed in this fashion, then the ligand may well also become internalized and require hydrolysis within the neuron.

Ishida and Leal (2002) have used a degerate PCR approach to clone an esterase gene from A. polyphemus cDNA that they suggest encodes the 65 kDa integument esterase (clade D, sequence 41 in **Figure 2**). This assignment is based on the similar predicted size (61.7 kDa) and broadly similar pI (7.49) of protein 41 to the 65 kDa integument isozyme, the presence of a predicted secretion signal in protein 41, and the expression of the 41 gene in cuticular tissues, including male legs, as well as in male and female antennae. Esterase genes similar to 41 are also present in other moth antennal EST collections. One from female antennal cDNA of M. sexta is 70% identical to 41 over 120 amino acids (GenBank Accession no. BE015493), while another from male antennae of the light brown apple moth Epiphyas postvittana is 56% identical over 234 amino acids (R.D. Newcomb, unpublished data). Although M. sexta uses a sex pheromone composed of aldehydes, mainly (E,E,Z)-10,12,14-hexadecatrienal (Tumlinson et al., 1989), E. postvittana uses a blend of two acetate esters (Bellas et al., 1983).

Ishida and Leal (2002) also cloned another esterase gene that is expressed in the antennae of A. polyphemus using their degenerate PCR approach. This gene (19 in **Figure 2**) sits not in clade D as above, but in clade A. The authors suggest that it may encode the sensillar esterase, since it is expressed in male but not female antennae. It is also predicted to have a secretion signal (which would be necessary for it to be found in the sensillar lymph) under the software used by Ishida and Leal (2002), albeit not by the packages that are used (**Figure 2**). Other possible anomalies in respect of this assignment are the slightly larger size of the predicted protein (60 versus 55 kDa) and its substantially higher predicted pI (6.6 versus 3.0). The link between 19 and the sensillar esterase therefore remains to be confirmed.

It is nevertheless noteworthy that 19, like 41, shows significant similarity to sequences in three antennal EST collections. In our phylogenetic analysis, 19 is most similar to an esterase from the cotton aphid *Aphis gossypii* (18) of no known function. However, esterase ESTs similar to 19 occur in male antennal cDNA libraries of *M. sexta* (GenBank Accession no. BF047018, 44% identical over 221 amino acids; GenBank Accession no. BF046933, 44% identical over 199 amino acids) and *E. postvittana* (42% identical over 321 amino acids) (R.D. Newcomb, unpublished data). Another esterase EST (GenBank Accession no. BP183087) very similar to 19 (70% identical over 148 amino acids) is also expressed in the pheromone gland of female silkmoth *Bombyx mori*. This moth uses an alcohol pheromone, bombycol ((*E,Z*)-10,12-hexadecadien-1-ol) (Butenandt *et al.*, 1959). Perhaps an ester is made in the gland but is hydrolyzed to the alcohol for use as the sex pheromone.

The three *D. melanogaster* esterases (39, 40, and 42) in clade D have yet to be characterized functionally and no mutants are available. However, sites of expression can at least be identified from EST data. No ESTs have been detected for 39 at the time of writing but ESTs corresponding to 42 have been recovered in adult testis, and to 40 in adult head and salivary gland cDNA libraries. Microarray life cycle data are also available for 40 (http://flygenome. yale.edu/Lifecycle/) showing a peak of expression in prepupae and adults. It remains to be determined whether these esterase genes play any role in hormone hydrolysis in *Drosophila*.

We are unaware of any reports of esterases for plant semiochemicals among insects as yet. However, the large number of volatile plant esters which elicit specific responses in insects suggest that they probably exist. Three examples serve to illustrate the diversity of plant ester cues and insect responses to them. First, the noctuid moth *Autographa gamma* is attracted to a number of flowering plants that have in common the floral scent compound benzyl benzoate. Wind tunnel and electrophysiological experiments indicate that the response is specific to this ester (Plepys *et al.*, 2002a, 2002b). Second, the volatile ester methyl salicylate is produced by many plants when they are attacked by herbivores (Shulaev *et al.*, 1997) and females of many parasitoids use this compound as a cue to locate a host (Poecke *et al.*, 2001). Finally, the ester ethyl (2*E*,4*Z*)-2,4-decadienoate produced by the Bartlett pear *Pyrus communis* is a powerful attractant for neonate larvae of the codling moth *Cydia pomonella* and is used by them to locate the fruit (Knight and Light, 2001). All these plant esters detected by insects will require rapid degradation by the insect to allow new incoming signal to be detected.

7.5. Intracellular Catalytic Clades

There are three major clades of mainly intracellular carboxyl/cholinesterases, all members so far identified apparently being catalytically active (clades A–C in **Figure 2**). Currently all but two of the sequences in these clades come from the Diptera. One clade is exclusively populated at the moment by an extensive radiation within the higher Diptera. Conversely, its sister clade, whilst not so extensively radiated, currently contains only lower dipteran sequences. These two clades, the lower and higher dipteran microsomal α-esterases (see Sections 7.5.1 and 7.5.2), are relatively intensively studied because of the involvement of certain members in metabolic resistance to various chemical insecticides. Less can be said as yet about the third major clade in this group, which appears to contain several cytosolic and a few secreted and mitochondrial enzymes. These enzymes are mainly from the lower Diptera, plus two from *D. melanogaster* and one each from a wasp, a moth, and an aphid. Interestingly the wasp esterase is also involved in insecticide resistance (see Section 7.5.3).

7.5.1. Lower Dipteran Microsomal α-Esterases

Both the anopheline and culicine mosquitoes studied to date contain a two-member cluster of genes encoding soluble/microsomal esterases. This two-member cluster in the mosquitoes constitutes clade C (**Figure 2**). In the culicines, where they are best studied, the two esterase genes are commonly designated *estα* and *estβ* because the corresponding enzymes, respectively, prefer α- and β-naphthyl acetate as substrates (Hemingway *et al.*, 2000). However, they also appear as *est3* and *est2*, respectively, in some literature (see, e.g., Pasteur and Georghiou, 1981; Pasteur *et al.*, 1981a). The *estα* and *estβ* genes share several intron positions and the amino acid sequences of the encoded Estα and Estβ proteins are about 50% identical (Vaughan *et al.*, 1995, 1997a; Ranson *et al.*, 2002). This compares with the 98% identity between the products of the duplicated aphid β-esterases implicated in OP and carbamate resistance (see Section 7.4.3.2), making duplication of the *estα* and *estβ* genes a much older event. Consistent with this, there is little immunological cross-reactivity between the two enzymes in *Culex pipiens* (Mouchès *et al.*, 1987; Karunaratne *et al.*, 1994).

The two genes are arranged head to head, with only about 600 bp between their coding regions in *A. gambiae* and about 1.7 kb between them in the wild-type (OP susceptible) *C. pipiens* (Vaughan *et al.*, 1997a; Ranson *et al.*, 2002). Although divergently transcribed, in *C. pipiens* at least it appears that the two genes may share some promoter elements (Hawkes and Hemingway, 2002). Expression levels are not particularly high in susceptible *C. pipiens* and the two enzymes can be difficult to detect in some electrophoretic assays, despite their relatively high activities for the naphthyl ester substrates (Karunaratne *et al.*, 1995). The *C. pipiens* enzymes separate mainly in the soluble fraction during purification but the presence of C-terminal KDEL motifs suggests at least a weak microsomal association (Fournier *et al.*, 1987; Merryweather *et al.*, 1990) (see Section 7.2.1). Estα and Estβ are highly polymorphic for isozyme variants in *C. pipiens*, with as much as 5% amino acid sequence differences reported among Estβ variants (Raymond *et al.*, 1991; Hemingway and Karunaratne, 1998). Estα behaves as a homodimer while Estβ behaves as a monomer (Fournier *et al.*, 1987). Monomer molecular weights of around 60 and 67 kDa for Estα and Estβ, respectively, are typical for carboxyl/cholinesterases (Fournier *et al.*, 1987).

Mutations of the *estα* and *estβ* genes have been associated with broad spectrum OP resistance in *C. pipiens* populations from diverse localities around the world. These populations cover both *C. pipiens pipiens* and *C. pipiens quinquefasciatus*, and indeed many of the same alleles are found in both subspecies (e.g., Beyssat-Arnaouty *et al.*, 1989; Hemingway and Karunaratne, 1998) (**Table 2**). In all cases characterized, resistance has been associated with overproduction of either or both of the Estα and Estβ proteins, and in almost all cases this in turn has been associated with amplification of either or both of the *estα* and *estβ* genes. One case of resistance apparently not due to amplification involves overproduction of the Estα1 variant by some other, as yet uncharacterized, mechanism (Rooker *et al.*, 1996).

At least nine different *estα* and/or *estβ* amplicons have now been described in *C. pipiens*, at least one involving just *estα*, four just *estβ*, and four involving both (Pasteur *et al.*, 1981b; Guillemaud *et al.*, 1998; Hemingway and Karunaratne, 1998; Hemingway *et al.*, 2000) (**Table 2**). These numbers are likely to be underestimates because several of the studies have only used isozyme mobilities to discriminate phenotypes and it is clear that these methods will miss many variants (Poirié *et al.*, 1992; Vaughan *et al.*, 1995; Karunaratne *et al.*, 1995). While many populations are polymorphic for several amplicons, two are widespread, *estβ1²* and *estα2¹/estβ2¹*, with the latter particularly common and apparently replacing several amplicons, including the former, in many populations around the world over time

(Raymond *et al.*, 1987; Qiao and Raymond, 1995). Intriguingly now in France and Cyprus the *estα4/estβ4* and *estα5/estβ5* amplicons, respectively, are in turn superceding the *estα2¹/estβ2¹* amplicon (Poirié *et al.*, 1992; Lenormand *et al.*, 1998b). The amplifications characterized to date map to broadly the same region of the chromosome, but they are adjacent insertions rather than allelic substitutions of the wild-type genes (Nance *et al.*, 1990; Wirth *et al.*, 1990; Karunaratne *et al.*, 1995; Hemingway *et al.*, 2000).

Quantitative PCR and dot blot methods have also revealed substantial quantitative variation in amplicon copy number among strains with the same type of amplicon (Mouchès *et al.*, 1985; Pasteur and Georghiou, 1989; Callaghan *et al.*, 1998; Weill *et al.*, 2000). This variation in turn underlies variation in levels of esterase activity and OP resistance (Villani *et al.*, 1983; El-Khatib and Georghiou, 1985). Copy numbers as high as 250 have been reported, associated with activity increases of 500-fold and resistance factors of 800-fold (Mouchès *et al.*, 1990; Poirié *et al.*, 1992). Copy number is stable over many generations in homozygous laboratory lines in the absence of selection (Raymond *et al.*, 1993). However, copy number has been readily increased in the laboratory by further OP selection (Wirth *et al.*, 1990). It is suggested that unequal crossing-over generates at least some of the variation in copy number (Weill *et al.*, 2000; Berticat *et al.*, 2001). Differences in DNA methylation as seen in the amplified aphid esterases (see Section 7.4.3.2) have not been reported for *C. pipiens estα* or *estβ*.

The resistance mechanism mediated by the *M. persicae* E4/FE4 enzymes involves predominantly sequestration but with some role for very slow hydrolysis for dimethyl OPs (Devonshire, 1977; Devonshire and Moores, 1982) (see Section 7.4.3.2). However, the mechanism mediated by the amplified *C. pipiens* esterases appears to depend even more strongly on sequestration. Thus the overexpressed Estα2 and Estβ2 isozymes characterized to date are all highly reactive with OPs, but have negligible turnover rates for them (Ketterman *et al.*, 1992, 1993; Karunaratne *et al.*, 1993, 1995). For example, estimates of bimolecular rate constants for the phosphorylation of Estα2¹ or Estβ2¹ with various OPs range from 0.02 to 155 $\mu M^{-1} min^{-1}$ (18 and 17, respectively, for paraoxon, cf. 133 $\mu M^{-1} min^{-1}$ for *M. persicae* E4) (Devonshire, 1977) (see Section 7.4.3.2). The corresponding turnover rates (k_3) for paraoxon are only about 0.06 h^{-1} compared to 0.6 h^{-1} for *M. persicae* E4 (Devonshire, 1977). Coupled with their abundance – as much as 0.4% of the total soluble protein of the organism (Karunaratne *et al.*, 1993) (cf. up to 1% for the *M. persicae* enzymes) (Devonshire, 1989) (see Section 7.4.3.2) – the overproduced enzymes can clearly provide the organism with an effective sink for insecticide.

There are several reports of significant differences in OP kinetics among the minority of the amplified Estα and Estβ isozymes that have been investigated in detail (Karunaratne *et al.*, 1993, 1995; Ketterman *et al.*, 1993; but see also Hemingway *et al.*, 2000). These include suggestions that the Estβ isozymes involved in amplification and resistance may have greater reactivity with OPs than Estβ isozymes that are not implicated in amplification and resistance. Further, thoroughgoing comparisons of the OP kinetics across amplified and nonamplified Estα and Estβ isozymes utilizing compounds used in the field might not only explain why particular isozymes are involved in resistance, but also elucidate the allele succession phenomena that are occurring among amplicons in the field.

Unlike the amplified aphid esterases, the amplified *C. pipiens* enzymes are only slightly reactive against carbamates biochemically (Cuany *et al.*, 1993; Karunaratne *et al.*, 1993) and do not confer carbamate resistance (Georghiou *et al.*, 1980; Priester *et al.*, 1981).

Two of the *C. pipiens* amplicons have been characterized intensively at a molecular level, one just involving the *estβ1²* allele and the other involving both *estα2¹* and *estβ2¹*. In the first case, typified by the TEMR strain, the amplicon is some 30 kb in length, containing a highly conserved internal 25 kb core region (Mouchès *et al.*, 1990; Vaughan *et al.*, 1995; Rooker *et al.*, 1996). The amplicons are arranged in a tandem array adjacent to the wild-type locus on chromosome 2; they create a noticeably longer chromosome cytologically (Nance *et al.*, 1990; Wirth *et al.*, 1990). This *estβ1²* amplicon was one of the first detected after the introduction of OPs, and has since declined in frequency in many localities where two-gene amplicons like *estα2¹/estβ2¹*, *estα4/estβ4*, and *estα5/estβ5* are proliferating (Raymond *et al.*, 1991; Poirié *et al.*, 1992; Rivet *et al.*, 1993; Lenormand *et al.*, 1998b).

The 25 kb *estα2¹/estβ2¹* amplicon as typified by the PelRR strain contains both esterase genes in a head-to-head configuration as per the wild-type unamplified genes (Villani *et al.*, 1983; Raymond *et al.*, 1987; Hemingway *et al.*, 1990). However, in PelRR the two genes are separated by 2.7 kb, whereas in susceptible strains like PelSS they are only separated by 1.7 kb, due to two insertions in the intergenic region in PelRR. There are also various single nucleotide substitutional differences between the intergenic regions of the two strains (Vaughan *et al.*, 1997a).

Although amplification of the two esterase genes in the PelRR strain results in a 1:1 ratio in gene number there is about threefold more Estβ2¹ protein in PelRR flies than Estα2¹ (Karunaratne, 1994, cited in Hawkes and Hemingway, 2002). Moreover, the two proteins differ in their tissue specificity (Pasteur *et al.*, 2001). Both are highly expressed in the alimentary canal, Malpighian tubules, and the oenocytes in the body cavity of larvae and adults. These are all tissues where a variety of detoxification functions are known to be carried out. However, Estβ2¹ is also expressed in the CNS, where its proximity to the target site, AChE, may confer an additional level of protection. Presumably these differences can be accounted for by differences in promoter function between *estα2¹* and *estβ1¹*. They may be encoded by the intergenic insertions noted above. Gel shift assays have indeed indicated altered binding of at least one transcription factor in Pel RR (Hawkes and Hemingway, 2002).

Another explanation suggested for the success of the *estα2¹/estβ2¹* amplicon involves the presence of an additional gene on the amplicon. A functional aldehyde oxidase gene (*ao1*) is also amplified in this strain (Hemingway *et al.*, 2000), resulting in mosquitoes which also have elevated aldehyde oxidase activity. It is suggested that this might be advantageous for the detoxification of pesticides or other xenobiotics containing aromatic groups. The *estβ1²* and *estα3/estβ1* amplicons were found not to contain intact functional copies of *ao1*.

Perhaps unsurprisingly, given the production of such large amounts of esterase protein, several of the amplification alleles are clearly associated with some fitness costs in the absence of insecticide. It appears that some resistant strains are not able to overwinter as well as susceptible strains (Gazave *et al.*, 2001), and that resistant males are not as successful in competing for mates as susceptible males (Berticat *et al.*, 2002). Amplicons with such fitness costs include *estα2¹/estβ1¹* but significantly, at least in terms of overwintering ability in French populations, apparently not *estα4/estβ4* (Chevillon *et al.*, 1997; Lenormand *et al.*, 1998b).

Intriguingly it seems that, along with insecticide resistance, the overexpressed *C. pipiens* esterases may also be associated with reduced levels of filarial parasites (McCarroll *et al.*, 2000). It will be of interest to see if any causal basis can be established for this correlation in the future. There were also earlier reports associating a particular esterase isozyme in *A. gambiae* with reduced transmission of the malarial parasite plasmodium (Collins *et al.*, 1986;

Vernick *et al.*, 1988; Vernick and Collins, 1989) but more detailed investigations subsequently argued against any causal connection with the esterase (Crews-Oyen *et al.*, 1993).

Although not yet fully characterized, amplifications of both *estα* and *estβ* genes have also been associated with broad spectrum OP resistance in *C. tritaeniorhynchus* and *C. tarsalis* (**Table 3**, below).

There are clearly fundamental similarities between the OP resistance mechanisms based on these amplified culicine α-esterases and the amplified hemipteran β-esterases reviewed above (see Section 7.4.3.2). In both cases high levels of resistance against a broad spectrum of OPs are principally achieved by a pesticide sequestration mechanism that uses massive gene amplification to dramatically overexpress a carboxylesterase isozyme with very high affinity for OPs. The versatility of this basic mechanism is borne out by some of the genetic and biochemical differences between the systems, for example, the relatively low levels of sequence similarity and the secreted versus intracellular locations of the enzymes. However in the following section we see how in the higher Diptera α-esterases with high affinity for OPs have also been recruited to a fundamentally different OP resistance mechanism. In this case a structural change in the enzyme converts it to an OP hydrolase, whilst qualitatively reducing its carboxylesterase activity in conventionally stained isozyme gels.

7.5.2. Higher Dipteran Microsomal α-Esterases

The *D. melanogaster* α-esterase radiation comprises 10 physically clustered genes plus one other located at a remote chromosomal location (Robin *et al.*, 1996, 2000a, 2000b; Russell *et al.*, 1996) (**Figure 9**). The 10 clustered α-esterase genes lie in a 60 kb segment that also contains four unrelated open reading frames and a closely related α-esterase pseudogene (αE4a, lying within an intron of αE6). All but one of the esterases in the cluster are oriented in the same direction. This α-esterase cluster is one of the largest clusters of sequence-related enzyme coding genes in the *D. melanogaster* genome (Adams *et al.*, 2000).

At the nucleotide level, all members of the higher dipteran α-esterase radiation show a distinctive pattern of introns (Oakeshott *et al.*, 1999). At the protein level, they show 37–66% amino acid identity, plus a distinctive pattern of small indels, a characteristic motif in a region corresponding to a peripheral binding site in AChE (plus some characteristic residues corresponding to the lining of the gorge in AChE) and the lack of an identifiable signal peptide, but almost all possess a carboxy terminal endoplasmic reticulum retention signal (Oakeshott *et al.*, 1999) (**Figure 6**); hence their assignment as microsomal α-esterases.

All the esterase genes in the α-cluster are expressed to some degree. Diverse and noncomplementary roles are suggested by considerable differences between the genes in developmental patterns, overall expression levels, and tissue and sex specificities (Campbell *et al.*, 2003). For example, three show varying degrees of male specificity, with one predominantly expressed in heads (*DmαE8*), another abundant in testes (*DmαE3*), and the third restricted to the germ line (*DmαE6*). Others are expressed broadly across feeding stages and this would be consistent with a role for at least some in the digestion of dietary or xenobiotic esters. A role in detoxification of xenobiotics is also implied by the role of several orthologs of one (αE7) in metabolic resistance to OPs (see below).

Figure 10 shows how some of the major events generating the *D. melanogaster* α-esterase radiation can be discerned by combining the evidence on the branch orders for their divergences in **Figure 2** with the information on their physical locations in **Figure 9**. Briefly, five major phases can be discerned. The first gave rise to four of the central genes in the extant cluster by a series of single gene duplications from the progenitor gene. The original progenitor then duplicated to generate one progenitor for the remainder of the genes on the left side of the cluster, and another progenitor for the remainder of those on its right side. Then successive single gene duplications of these latter two progenitors, respectively, generated the rest of the cluster, with one late-stage event also duplicating the left-terminal gene to generate the one remotely located member of the radiation. In general terms the radiation was thus generated from the inside outwards by a sequence of single gene duplications, mostly to adjacent locations and mostly in the same orientation.

This scenario can then be embellished by reference to available data on the composition of the α-esterase cluster in some other species, including fairly complete data for *D. pseudoobscura* and *D. buzzatii* and partial data for the *M. domestica* (**Figure 10**) (Oakeshott *et al.*, 1999; Robin *et al.*, 2000a; http://www.hgsc.bcm.tmc.edu). It is seen that most members of the cluster as now seen in *D. melanogaster* had arisen before the split of the drosophilids from the Muscidae (about 80 million years ago). However, it is also seen that the left end in particular remains a focus for ongoing change within the cluster, including both recurrent gene duplication and occasional gene loss. *Drosophila pseudoobscura*

Table 3 Summary of cases of esterases implicated in metabolic resistance to OPs (excluding malathion-specific resistance; see **Table 4**) and carbamates

Species	Biochemistry	Reference
Dictyoptera		
Blattella germanica (German cockroach)	Greater intensities of ≥3 OP- and carbamate-inhibitable esterase isozymes in OP and carbamate resistant strains	Prabhakaran and Kamble (1995), Scharf *et al.* (1997)
Thysanoptera		
Frankliniella occidentalis	Esterase activity in an OP resistant strain is lower and less sensitive to OP inhibition	Zhao *et al.* (1994)
Scirtothrips citri	Greater esterase activity in OP resistant strains	Ferrari *et al.* (1993)
Hemiptera		
Acyrthosiphon pisum (pea aphid)	Greater esterase activity in OP selected strains	Marullo *et al.* (1988)
Aphis gossypii	Greater esterase activity and ≥3 different or more intense esterase isozymes in OP and carbamate resistant strains	Furk *et al.* (1980), O'Brien *et al.* (1992)
Bemisia tabaci	Greater esterase activity in an OP-selected strain. However, the B-biotype $E_{0.14}$ esterase isozyme involved in SP resistance is not involved (see **Table 5**)	Bloch and Wool (1994), Byrne *et al.* (2000)
Brevicoryne brassicae (cabbage aphid)	Greater esterase activity in carbamate selected strains	Marullo *et al.* (1988)
Lygus hesperus	Greater OP-inhibitable esterase activity in strains whose OP resistance is also suppressed by carboxylesterase inhibitors	Zhu and Brindley (1990a, 1990b, 1992)
Myzus persicae	See **Table 2** and Section 7.4.3.2	
Myzus nicotianae[a] (tobacco aphid)	Greater esterase activity and more intense esterase isozymes in OP and carbamate resistant strains due to introgression of E4 and FE4 amplicons from *M. persicae* (see Section 7.4.3.2)	Abdel-Aal *et al.* (1990, 1992), Field *et al.* (1994), Wolff *et al.* (1994)
Nasonovia ribisnigri (currant-lettuce aphid)	Intense esterase isozyme in some carbamate resistant strains	Barber *et al.* (1999)
Nephotettix cincticeps	Greater esterase activity and ~4 more intense esterase isozymes in OP resistant strains	Hasui and Ozaki (1984), Motoyama *et al.* (1984), Chiang and Sun (1996)
Nilaparvata lugens (brown rice planthopper)	Greater esterase activity and two more intense isozymes (generated by different glycosylation) due to a three- to sevenfold amplification of an esterase gene in OP and carbamate resistant strains (see Section 7.4.3.2)	Karunaratne *et al.* (1999), Small and Hemingway (2000a, 2000b), Vontas *et al.* (2000)
Phorodon humuli (damson-hop aphid)	Greater esterase activity and greater intensity of ≥4 esterase isozymes in OP and carbamate resistant strains	Lewis and Madge (1984), Wachendorff and Zoebelein (1988)
Schizaphis graminum (greenbug)	Increased esterase activity and (mainly) two more intense isozyme zones in strains whose OP resistance is partially suppressed by carboxylesterase inhibitors. Elevation in one zone (type I) is due to four- to eightfold amplification of an esterase gene with strong similarity to *M. persicae* E4/FE4. Reversion to susceptibility due to reduced methylation as with E4. Elevation in the other zone (type 2) apparently due to structural differences in an isozyme produced in similar amount in susceptible and resistant strains (see Section 7.4.3.2)	Siegfried and Ono (1993a, 1993b), Siegfried and Zera (1994), Shufran *et al.* (1996), Siegfried *et al.* (1997), Ono *et al.* (1999)
Coleoptera		
Diabrotica virgifera	Greater intensity of at least one major esterase isozyme zone and greater carbamate and OP hydrolytic activity *in vivo* in strains whose OP and carbamate resistance is suppressed by carboxylesterase inhibitors	Miota *et al.* (1998), Scharf *et al.* (1999a, 1999b), Zhou *et al.* (2002)
Leptinotarsa decemlineata	Novel OP-inhibitable esterase isozyme in an OP resistant strain	Anspaugh *et al.* (1995)
Oryzaephilus surinamensis (saw-toothed grain beetle)	Greater esterase activity in a chlorpyrifos-methyl (CPM) and fenitrothion (FT) resistant strain is due to greater amounts of an esterase isozyme zone with greater sensitivity to OP inhibition (but no OP hydrolytic activity). A strain with high CPM but low FT resistance produces an alternate isozyme zone still highly sensitive to CPM inhibition but less sensitive to FT inhibition	Collins *et al.* (1992), Conyers *et al.* (1998), Rossiter *et al.* (2001)

Continued

Table 3 Continued

Species	Biochemistry	Reference
Lepidoptera		
Cydia pomonella (codling moth)	Reduced esterase activities in some OP resistant strains	Bush *et al.* (1993)
Epiphyas postvittana (light brown apple moth)	Greater esterase activity due to greater intensities of many esterase isozymes in OP resistant strains	Armstrong and Suckling (1988)
Heliothis virescens	Novel metal-activated esterase isozyme insensitive to OP inhibition but with OP hydrolytic activity isolated from an OP (methyl paraoxon) resistant strain	Konno *et al.* (1990), Kasai *et al.* (1992)
	Greater esterase activity due to three more intense esterase isozymes in a carbamate selected strain whose resistance is partially suppressed by carboxylesterase inhibitors and which also shows elevated cross-resistance to OPs	Goh *et al.* (1995), Zhao *et al.* (1996)
	Greater esterase activity due to greater intensities of several esterase isozymes, one in place of an alternate esterase isozyme, in OP (profenofos) resistant strains	Harold and Ottea (1997, 2000)
Platynota idaeusalis (tufted apple bud moth)	Greater esterase activity due to greater intensities of one major and several minor esterase isozymes and greater OP hydrolytic activity *in vivo* in strains whose OP resistance is suppressed by carboxylesterase inhibitors	Karoly *et al.* (1996)
Diptera		
Aedes aegypti (yellow fever mosquito)	Greater OP-inhibitable esterase activity in strains whose OP resistance is suppressed by carboxylesterase inhibitors	Mazzarri and Georghiou (1995)
Anopheles albimanus	Greater esterase activity and a more intense zone of esterase isozymes in a strain whose OP resistance is suppressed by carboxylesterase inhibitors	Brogdon and Barber (1990)
Culex pipiens complex	See **Table 2** and Section 7.5.1	
Culex tarsalis	Greater esterase activity and more intense isozyme zone due to ~30-fold amplification of orthologs of *Estα*/*Estβ* in *C. pipiens* in an OP resistant strain (see Section 7.5.1)	Mouchès *et al.* (1987), Beyssat-Arnaouty *et al.* (1989), Raymond *et al.* (1989)
Culex tritaenorhynchus	Greater esterase activity and more intense isozyme due to amplification of an orthologs of *C. pipiens Estα*/*Estβ* in an OP resistant strain (see Section 7.5.1)	Karunaratne *et al.* (1998), Karunaratne and Hemingway (2000, 2001)
Haematobia irritans (buffalo fly)	Greater esterase activity and greater intensity of an esterase isozyme is associated with increased expression of the LcαE7 ortholog in some OP resistant strains (see Section 7.5.2)	Guerrero *et al.* (2000)
Lucilia cuprina (Australian sheep blowfly)	Reduced general esterase activity and E3 isozyme staining intensity and significant OP hydrolase activity associated with Asp137 (and Ser/Leu251) variants of LcαE7 in strains with OP resistance partially suppressible by carboxylesterase inhibitors (see Section 7.5.2)	Hughes and Devonshire (1982), Hughes and Raftos (1985), Russell *et al.* (1990), McKenzie *et al.* (1992), Newcomb *et al.* (1997, 2004), Campbell *et al.* (1998a, 1998c), Devonshire *et al.* (2003)
Lucilia sericata (green bottle fly)	Asp137 (and Ser/Leu251) variants of the ortholog of LcαE7 in populations known to segregate for general OP resistance (see Section 7.5.2)	R.D. Newcomb, C.G. Young, and J.R. Stevens, unpublished data
Musca domestica	Reduced general esterase activity and Ali isozyme staining intensity associated with Asp137 (and Ser/Leu251) variants of MdαE7 in OP resistant strains (see Section 7.5.2)	Oppenoorth and van Asperen (1960), Campbell *et al.* (1997), Claudianos *et al.* (1999, 2001), Scott and Zhang (2003)
Simulium damnosum and *Simulium sanctipauli* (black flies)	Greater esterase activity and greater *in vivo* OP hydrolytic activity in strains whose OP resistance is suppressed by carboxylesterase inhibitors	Hemingway *et al.* (1989)
Acarina		
Amblyseius pontentillae	Less intense esterase in an OP resistant strain; the isozyme is highly reactive with OPs but has insufficient OP hydrolytic activity to be a major contributor to resistance	Anber and Oppenoorth (1989), Oppenoorth (1990)
Boophilus microplus	Greater intensities of several esterase isozymes in OP resistant strains	Rosario-Cruz *et al.* (1997)
Tetranychus kanzawai	Greater esterase activity in strains whose OP resistance is suppressed by carboxylesterase inhibitors	Kuwahara *et al.* (1982)

[a]Note that *M. nicotianae* is now recognized as conspecific with *M. persicae* (see Section 7.4.3.2).

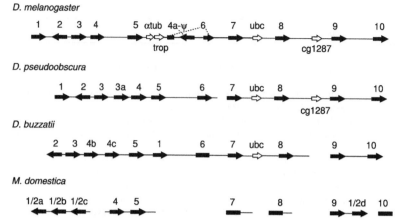

Figure 9 Organization of the α-esterase clusters as characterized to date from *D. melanogaster*, *D. pseudoobscura*, *D. buzzatii*, and *M. domestica* (Oakeshott *et al.*, 1999; Adams *et al.*, 2000; Robin *et al.*, 2000a; http://www.hgsc.bcm.tmc.edu). The whole of the cluster has been recovered from *D. melanogaster* and most of it at least from *D. pseudoobscura* and *D. buzzatii*, but it is only partly characterized in *M. domestica*. Nonesterase genes are unshaded (ubc, ubiquitin conjugating enzyme; trop, putative tropomyosin; αtub, α-tubulin; CG1287, functionally anonymous open reading frame). The contig in *D. melanogaster* is about 60 kb. Line breaks indicate uncharacterized gaps in *D. pseudoobscura*, *D. buzzatii*, and *M. domestica*. The orientations of some genes are unknown and accordingly shown as blocks rather than arrows. There is ambiguity as to the orthologs of the *Drosophila* αE1 and αE2 genes in *M. domestica* so the four candidates are shown as αE1/2a–d. Note that orthologs for six of the genes (αE1/2, αE5, αE7, αE8, αE9, αE10) have also been recovered from *Lucilia cuprina* but there is no information on their contig arrangement (Newcomb *et al.*, 1996).

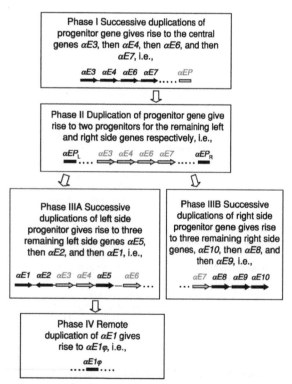

Figure 10 Schematic diagram showing a parsimonious reconstruction of the major phases in the evolution of the extant α-esterase cluster in *D. melanogaster*. Genes added in each phase are shown in bold, with pre-existing genes in gray. Genes of unknown orientation are shown as blocks rather than arrows. Distances are not to scale. αEP = progenitor α-esterase gene; αEP$_L$ and αEP$_R$ = progenitors for genes on the left and right side of the cluster, respectively.

lacks the αE4a pseudogene duplication and two of the nonesterase genes in the cluster, but has a tandem duplication of αE3. *Drosophila buzzatii* shows evidence of a different duplication of αE4 and has its αE1 in a different, central location within the cluster. Alternatively, *M. domestica* has two additional copies of αE1/αE2, including one at the other end of the cluster. It is not clear why the left end genes in particular should be a focus for change, but it may be relevant that they show relatively similar and broad expression profiles and lower levels of amino acid sequence conservation than the other genes.

Homologs of several of the α-esterases have also been identified in another higher dipteran, the calliphorid *L. cuprina* (**Figure 9**) (Oakeshott *et al.*, 1999) and, as with *M. domestica*, presumptive orthologs have been identified for several genes at the right end of the drosophilid cluster. The conservation of these orthologs across the Drosophilidae, Muscidae, and Calliphoridae clearly suggests the retention of important physiological functions, although as noted above there is as yet only indirect evidence as to the identities of these functions. The most highly conserved of the presumptive orthologs is αE7 (about 64% amino acid identities across the two lineages). This is also the one that has been implicated in metabolic resistance to OPs in a few muscid or calliphorid species involving what is often termed the mutant ali-esterase mechanism (Oppenoorth and van Asperen, 1960).

Organophosphates

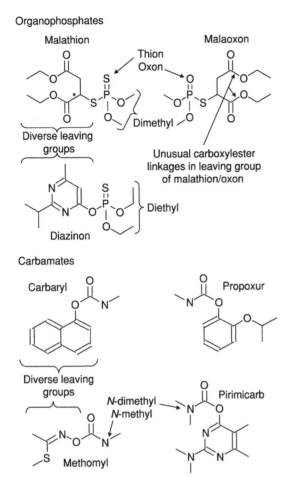

Figure 11 Structures of organophosphate (OP) and carbamate insecticides. OP insecticides all contain a central phosphorus atom. Most are produced in the thion form in which the phosphorus is linked by a double bond to a sulfur atom (phosphorothionates, exemplified here by diazinon, or phosphorothiolothionates when the leaving group is linked via another sulfur, exemplified here by malathion). Thion OPs are not the active forms of the insecticides but are converted *in vivo* by p450 oxidases to active oxon forms (phosphates, or phosphorothiolates when the leaving group is linked via a sulfur). Some OP insecticides (e.g., chlorfenvinphos) are produced in the oxon form and, therefore, do not require *in vivo* activation. The leaving groups are diverse electron-withdrawing groups that are displaced when oxon OPs react to phosphorylate the catalytic serine residue of a carboxyl/cholinesterase (**Figure 1**). They may be aromatic or aliphatic, may contain heteroatoms, and may be linked to the central phosphorus through an oxygen (most common), a sulfur or other atoms. The remaining two moieties attached to the central phosphorus are typically two methoxy groups (dimethyl) or two ethoxy groups (diethyl) although some other variations exist among commercial OPs. Malathion/oxon is unusual in that its leaving group contains two carboxylester linkages (and hydrolysis of either inactivates the OP). The optical center of malathion mentioned in the text is indicated with a asterisk. Carbamate insecticides resemble ordinary carboxylesters except that a nitrogen atom is adjacent to the carbonyl carbon of the carboxylester linkage. As with the OPs, the leaving groups are diverse, with varied sizes, aromatic or aliphatic groups, heteroatoms, etc. Where the leaving group is linked through a nitrogen atom rather than the more common carbon the carbamate is an oxime exemplified here by methomyl. The acyl group of most carbamate insecticides is *N*-methyl although there are variations, such as the *N*-dimethyl pirimicarb.

OP insecticides all contain a central phosphotriester moiety, which is crucial to their mode of action (**Figure 11**). It enables them to inhibit carboxyl/cholinesterases by acting as suicide substrates (see Section 7.3.1.2). Thus they rapidly phosphorylate the catalytic serine residue in an analogous manner to the first, acylation step in carboxylester hydrolysis. However, the second step (dephosphorylation for OPs rather than deacylation) proceeds at very slow to negligible rates, due to stereochemical differences between OP and carboxylester substrates. Carbamates (**Figure 11**) also inhibit by a similar mechanism. Commercial OP insecticides are mostly produced in an inactive "thion" form and must be activated to "oxons" by P450 enzymes in the insect before they inhibit as described above. Two of the three ester linkages around the central phosphorus in most commercial OPs involve methyl (dimethyl OPs) or ethyl groups (diethyl OPs), while the third is an electron-withdrawing "leaving group" (**Figure 11**). Hydrolysis of this third linkage detoxifies the insecticide. A few OP insecticides, most notably malathion, also contain carboxylester moieties within the leaving group, the hydrolysis of which also inactivates the insecticide. This leads to a special case of OP resistance, often termed malathion-specific resistance, which is discussed in detail below (see Section 7.6.2).

Three shared features of LcαE7 and MdαE7 have been suggested as predisposing them for a role in metabolic resistance to OPs (Campbell *et al.*, 1997). First, the wild-type enzymes are highly susceptible to inhibition by the activated, oxon forms of many OPs; indeed both interact more rapidly with many OPs than does the presumed target, AChE (Newcomb *et al.*, 1997; Chen *et al.*, 2001). For example, the k_i of LcαE7 with paraoxon is 6.3×10^7 $M^{-1} min^{-1}$ compared with $5.1 \times 10^6 M^{-1} min^{-1}$ for *L. cuprina* AChE (Newcomb *et al.*, 1997; Chen *et al.*, 2001). Second, the enzymes are expressed quite abundantly across the feeding life stages, larvae and adults, when the animals are likely to come in contact with insecticides (Parker *et al.*, 1991, 1996). Finally, they also show good kinetics (specificity constants $\sim 10^6 M^{-1} s^{-1}$) towards the carboxylester groups of malathion (malathion carboxylesterase activity, MCE), leading to the involvement of certain alleles in malathion-specific resistance.

Resistance to a broad spectrum of OPs in both *M. domestica* and *L. cuprina* has been conferred by amino acid substitutions at either of two positions in the active site of αE7. In both species a Gly137Asp substitution confers one class of broad-spectrum resistance while Trp251Ser/Leu substitutions confer a different class of broad-spectrum resistance, coupled with malathion-specific resistance (Campbell *et al.*, 1998a, 1998c). None of the substitutions confers cross-resistance to carbamates. In both species it appears that each resistance substitution initially occurred at least twice in different allelic backgrounds although only one allele encoding each type of resistance is now relatively common in each species (Claudianos *et al.*, 1999, 2001; Smyth *et al.*, 2000; Scott and Zhang, 2003; Newcomb *et al.*, 2004).

Alleles showing the Gly137Asp substitution now dominate contemporary populations of both *L. cuprina* and *M. domestica* (Newcomb *et al.*, 1997, 2004; Claudianos *et al.*, 1999; Smyth *et al.*, 2000; Scott and Zhang, 2003). The resulting mutant enzyme has improved ability to hydrolyze OPs, with a preference for diethyl (55-fold increase in hydrolytic ability) over dimethyl OPs (33-fold increase) (Devonshire *et al.*, 2003). While these increases represent very low turnover numbers (about three turnovers per hour per molecule of enzyme), they are sufficient to account for modest levels of resistance to dimethyl OPs (4–12-fold) and about twofold greater resistance to diethyl OPs (7–20-fold resistance in adult *L. cuprina*) (Campbell *et al.*, 1998a, 1998c; Devonshire *et al.*, 2003). Interestingly, the turnover numbers of Asp137-αE7 with OPs are very comparable to those that the *M. persicae* E4/FE4 enzymes achieve for the particular case of malathion, and larger than those for the *M. persicae* enzymes with dimethyl OPs. In both cases Devonshire and Moores (1989) have argued that hydrolysis plays a significant supplementary role to sequestration in the resistance mechanism (see Section 7.4.3.2).

Asp137 is a nonconservative substitution of the Gly137 whose main chain nitrogen atom contributes to the oxyanion hole within the active site (**Figure 5**). The increased ability of the Asp137 enzyme to hydrolyze OPs may be due to the ability of this residue to activate a water molecule in a suitable orientation for nucleophilic attack and dephosphorylation of the phosphoryl-enzyme intermediate (Newcomb *et al.*, 1997), in an analogous manner to the second, deacylation step of carboxylester hydrolysis (**Figure 1**) (see Section 7.1.2). Thus the OP is turned over as a normal substrate rather than blocking the active site as a suicide substrate. A similar mechanism is proposed to explain the increased OP hydrolytic activity caused by a substitution to a

histidine at a similar position in the oxyanion hole of human BuChE (Lockridge *et al.*, 1997). Forms of αE7 with Asp137 show drastically reduced carboxylesterase activities, possibly due to compromised functioning of the oxyanion hole for carboxylester hydrolysis (Newcomb *et al.*, 1997; Claudianos *et al.*, 1999). This then accounts for the association of OP resistance with apparent null phenotypes for the corresponding isozymes (E3 and Ali for *L. cuprina* and *M. domestica*, respectively) (Hughes and Raftos, 1985; Campbell *et al.*, 1997; Claudianos *et al.*, 1999) and reduced whole body homogenate activities for artificial substrates like naphthyl acetate or methyl butyrate (termed ali-esterase activity by Oppenoorth and van Asperen (1960)). It is this coincident gain of OP hydrolytic activity and loss of "ali-esterase" activity that defines the mutant ali-esterase mechanism.

Less abundant in *L. cuprina* and *M. domestica* populations are OP resistance alleles of αE7 in which Trp251 is substituted by Leu or Ser (Claudianos *et al.*, 1999; Smyth *et al.*, 2000). Flies with these active site substitutions also show reduced ali-esterase activities as this mutation interferes with some, although not all of the susceptible enzyme's carboxylesterase activities (Campbell *et al.*, 1998a). Like the Asp137 mutants, this enzyme has improved ability to hydrolyze OPs, except that in this case dimethyl OPs are preferred over diethyl OPs (34- and tenfold increases, respectively, over wild-type in OP turnover for the Leu251 enzyme from *L. cuprina*) (Devonshire *et al.*, 2003). This preference is reflected in two- to fivefold greater resistance to dimethyl OPs (5–27-fold resistance in adult *L. cuprina*) than their diethyl analogs (two to sixfold) (Campbell *et al.*, 1998c). Exceptional resistance levels are seen, however, for malathion (>130-fold) (see below). The effect of the Leu/Ser251 mutations is proposed to depend on replacement of a bulky aromatic group in the active site by smaller residues, which might provide less steric hindrance to the inversion about the tetrahedral phosphorus that must occur during hydrolysis of the serine–phosphorus bond (Campbell *et al.*, 1998a).

The exceptional resistance to malathion shown by *M. domestica* and *L. cuprina* with Leu or Ser251 variants of αE7 is due to the combination of MCE and OP hydrolase activities in these particular forms of the enzyme. Hydrolysis of the carboxylester groups of malathion accounts for the main products of detoxification of malathion in resistant flies, yet *in vitro* MCE activity seems poorly correlated with resistance (Smyth *et al.*, 2000). Flies with the Asp137 variant of αE7 show very low *in vitro* MCE activity (Asp137-αE7 loses nearly all carboxylesterase activity) but do not differ in their susceptibility to malathion from flies showing general OP susceptibility and far more *in vitro* MCE activity (having Gly137/Trp251-αE7). Alternatively, flies with Leu/Ser251-αE7 show modest increases of *in vitro* MCE activity over OP susceptible flies yet show very high level resistance to malathion. Here one needs to remember that malathion is converted *in vivo* to malaoxon, a potent dimethyl OP esterase inhibitor. Campbell *et al.* (1998c) showed that the OP susceptible (Gly137/Trp251) form of αE7 rapidly loses its MCE activity in the presence of malaoxon, modeling the rapid loss of detoxification capacity that would occur in an OP susceptible fly exposed to malathion. However, the Leu251 form of the enzyme was able to sustain its MCE activity in the presence of malaoxon. Devonshire *et al.* (2003) and Heidari *et al.* (2004a) demonstrated that the Leu251 and Ser251 enzymes' MCE kinetics are only slightly improved over those of the OP susceptible form but their abilities to hydrolyze dimethyl OPs are greatly enhanced as described above. Thus the Leu251 and Ser251 enzymes are able to reactivate due to their OP hydrolase activity if their catalytic serines become phosphorylated by malaoxon and, therefore, they can continue to contribute their MCE activity to the task of detoxification.

Consistent with the specificity of the OP hydrolase activity, malathion resistant *M. domestica* and *L. cuprina* show little resistance to higher (nondimethyl OP) analogs of malathion, even though these molecules are identical around the carboxylester groups that are hydrolyzed by MCE (Campbell *et al.*, 1997, 1998c and references therein). Detoxification of malathion analogs by carboxylesterase activity only occurs for those analogs where the enzyme's minor OP hydrolase activity allows reactivation of the oxon OP-inhibited enzyme.

Although confirmatory bioassay data are not yet available for all the sequenced isolates, it appears that parallel mutations at the 137 and 251 sites of the orthologous αE7 enzyme may also confer the two types of resistance on another blowfly, *Lucilia sericata*. Thus this species is known to show the two types of resistance. Recent allelic sequencing work has revealed one (presumptive diethyl OP resistant) Asp137 haplotype from New Zealand and three unrelated (presumptive dimethyl OP/malathion-specific resistant) Ser251 or Leu251 haplotypes from Europe, North America, and Africa, in addition to seven unique (presumptive OP susceptible) Gly137/Trp251 *LsαE7* haplotypes (R.D. Newcomb, C.G. Young, and J.R. Stevens, unpublished data).

In light of the recurrence of αE7-based OP resistance, Heidari *et al.* (2004a) explored the catalytic consequences of various rational substitutions around the LcαE7 enzyme's active site. Substitutions at residue 137 generally reduced carboxylesterase activities, but none was as effective as the naturally occurring Asp for enhancing OP turnover. Substitutions of Trp251 with smaller residues generally produced enzymes with qualitatively similar properties to the naturally occurring Leu or Ser substitutions. This is consistent with the idea that the substitutions remove a steric constraint on OP hydrolysis. Notably, two of the synthetic mutants, bearing W251T and W251A, had quantitatively better dimethyl OP hydrolytic activity than the naturally occurring mutants. The fact that these mutants have not been found in the field may be because they would both require a double mutation in the wild-type (susceptible) codon. The synthetic W251G mutation was interesting for a different reason. Compared with all the other substitutions at this site, Gly251 led to lesser improvements in OP turnover but 20-fold better MCE kinetics. It is noteworthy then that a glycine substitution at the homologous position occurs in an overexpressed esterase from a parasitoid wasp, *Anisopteromalus calandrae*, which shows high malathion-specific resistance (Zhu *et al.*, 1999a, 1999b) (see Section 7.5.3). Substitutions of various other residues around the active site were mostly detrimental to OP hydrolysis and MCE activities, while the natural mutations combined in one molecule (Asp137 + Leu251) caused no improvement over either mutation alone.

As noted earlier, the function of αE7 prior to selection with OPs is unknown. However, it is noteworthy that αE7-mediated resistance in *L. cuprina* was initially associated with a fitness cost and measurable fluctuating (bilateral) asymmetry in the absence of the insecticide (McKenzie *et al.*, 1982; Davies *et al.*, 1996). Resistance due to αE7 in fact did not rise to high frequency until the 1970s when a *modifier* gene arose at another locus which redressed the cost and asymmetry problems (Davies *et al.*, 1996). The cost in OP resistant flies was presumably due to compromised activity towards the enzyme's native substrate and, consistent with this, the Asp137 enzyme shows more comprehensive loss of carboxylesterase activities and was associated with more severe asymmetry than the Leu/Ser251 enzymes. The biochemical function of the *modifier* gene product is unknown, although it is now known not to be the ortholog of the developmental gene *Notch* in *D. melanogaster* (P.C. Batterham, personal communication). Clearly identification of the *modifier* gene function would yield important insights into the function of αE7 beyond detoxification of OPs.

Another set of mutations has apparently also occurred in *L. cuprina* since the spread of the original αE7 resistance alleles (Smyth *et al.*, 2000; Newcomb *et al.*, 2004). These are duplications of the chromosomal region containing αE7 which result in two copies of this gene and others from the α-esterase cluster being carried on one chromosome. So far three independent duplications have been observed at low frequencies. Importantly, these duplications result in dissimilar resistance alleles on the same chromosome, Asp137 alleles linked with Leu251 alleles. As a result strains homozygous for these duplications show both high levels of malathion resistance and slightly higher resistance levels to diazinon (a diethyl OP) than do nonduplicate Asp137 strains (Campbell *et al.*, 1998c).

An αE7 ortholog has also recently been implicated in OP resistance in the horn fly *Haemotobia irritans*, also in the Muscidae. In this case, however, resistance appears to have been achieved through overexpression of αE7 without amino acid changes (Guerrero, 2000). AChE is presumably protected because the OPs preferentially react with the very abundant HiαE7 in a fashion analogous to that of the overexpressed hemipteran esterases discussed earlier (see Section 7.4.3.2).

OP resistance has never been linked with αE7 in *D. melanogaster* and a possible reason is that the enzyme in this species tends to show lower rather than higher affinities for OPs when compared with its endogenous AChE (Parker *et al.*, 1991, 1996; Chen *et al.*, 2001). It is therefore suggested that little benefit would arise from mutations that might amplify the gene or enhance its turnover of OPs.

7.5.3. Nonmicrosomal Esterases

Although a relatively derived clade compared to others in **Figure 2**, this clade (A) contains sequences from both holo- and hemimetabolan orders, and it shows greater diversity in respect of cellular/subcellular localization than do any other clades. It contains a substantial proportion of the presumptively cytosolic enzymes in the phylogeny, plus the only presumptively mitochondrial enzymes and also three with good consensus secretion signals.

Little is yet known of the functions of clade A enzymes. We previously suggested one *D. melanogaster* member (9) may correspond to the developmental gene *cricklet* (Claudianos *et al.*, 2001), but it is now clear that it does not (http://flybase.bio.indiana.edu/). The only specific function now established for a clade A enzyme is the high level of malathion-specific resistance attributed to a mutant form of a cytosolic esterase (17) in the parasitoid *A. calandrae*

(Baker and Weaver, 1993; Baker *et al.*, 1998a, 1998b; Zhu *et al.*, 1999a, 1999b) (see Section 7.5.2).

Resistant strains of *A. calandrae* have a Trp to Gly substitution at a position equivalent to the Trp251Leu/Ser substitutions conferring malathion-specific resistance in *L. cuprina* and *M. domestica*. The effect is to improve the MCE kinetics of the enzyme. Heidari *et al.* (2004a) have since made a Trp251Gly change in the *L. cuprina* enzyme *in vitro* and expressed it in a baculovirus system, finding that it confers significantly greater benefits in respect of MCE activity than do the Trp251Leu/Ser mutations in this enzyme. The *A. calandrae* mutant thus provides strong support from a case of malathion-specific resistance in the field supporting the contention of Heidari *et al.* (2004a) that substitution of the Trp with any of several small residues might confer malathion-specific resistance.

Baker *et al.* (1998a) also found that the mutant *A. calandrae* enzyme is relatively unstable, at least *in vitro*. This is reminiscent of similar evidence for the Trp251Leu mutant in the blowfly enzyme (Smyth *et al.*, unpublished data). Interestingly, there is also 30-fold upregulation of the parasitoid enzyme in resistant strains (Zhu *et al.*, 1999b). This might compensate for any instability of the enzyme *in vivo*, or simply augment the resistance due to the structural mutation. There is at least some evidence for upregulation of the blowfly mutant as well (Whyard and Walker, 1994).

7.6. Other Esterases Implicated in Insecticide Resistance

As seen in the earlier sections that molecular/genomic approaches have yielded enormous insights into the mechanisms underlying certain cases of esterase-based insecticide resistance. However, these clearly cover just a small proportion of the ever-growing list of esterase-based resistance. Now attention can be turned to the great majority of these for which there is as yet no molecular information. Their biochemistry and (classical) genetics are summarized in an attempt to discern recurring similarities and differences compared to those now elucidated at a molecular level.

7.6.1. Resistance against Organophosphates and Carbamates

Table 3 lists over 20 cases involving insect species from six orders and three acarines in which changes in esterase activity have been associated with non-malathion specific OP resistance and/or carbamate resistance. About one-third of the cases involve hemipterans. Most involve OP resistance alone or OP plus carbamate resistance; only a few involve carbamate resistance alone. The most common finding is an association of resistance with increased esterase activity as measured spectrophotometrically or electrophoretically with artificial substrates like naphthyl acetate.

Only three cases of a decrease are found, all involving OP resistance, outside the higher dipteran cases (see Section 7.5.2). These involve the codling moth *Cydia pomonella*, the western flower thrip *Frankliniella occidentalis*, and the predacious mite *Amblyseius potentillae*. However, only the *A. potentillae* case has been characterized in any detail biochemically and in this case the esterase seems unlikely to be the major cause of resistance (Oppenoorth, 1990). This suggests that other instances of the mutant aliesterase mechanism may occur outside the higher Diptera, but are relatively uncommon.

The fact that the great majority of cases involve increases in esterase activity might suggest that sequestration as exemplified in the *Myzus* and *Culex* cases is the predominant mechanism for esterase-mediated metabolic resistance to OPs and carbamates. The data for several of the cases are indeed at least consistent with these paradigms but it is equally clear that others are not. Some that may involve sequestration are not readily interpreted in terms of the sort of gene amplification events seen in the *Myzus* and *Culex* cases. A few are most easily explained in terms of a hydrolytic mechanism, albeit not the mutant ali-esterase mechanism.

Clearly one key requirement of a sequestration mechanism is that the esterase involved has high binding affinity for the insecticide. This is demonstated directly in terms of the inhibition of the esterase activity by OPs *in vitro* in a few of the cases tabulated (the cockroach *Blattella germanica*, the beetles *Leptinotarsa decemlineata* and *Oryzaephilus surinamensis*, and the mosquito *A. aegypti*). There are also several others where it can reasonably be inferred because the resistance is suppressed *in vivo* in bioassays including OP reagents like *S,S,S*-tributylphosphorotrithioate (DEF). Most of these cases cover OP resistance but they do include a case of carbamate resistance, involving *H. virescens*. Such *in vitro* or *in vivo* inhibition is generally taken as evidence for involvement of a member of the carboxyl/cholinesterases, although it is worth noting that an esterase from the serine protease family would probably yield similar results.

Sequestration as seen in the earlier aphid and culicine examples relies principally on mutations enabling massive overexpression of the relevant enzymes. In a few of the cases in **Table 3**, most notably

the green rice leafhopper *Nephotettix cincticeps*, saw-toothed grain beetle *O. surinamensis*, and *B. microplus*, there is direct evidence for increased enzyme amounts. In a couple of others, it is indirectly inferred from the lack of difference in K_m or k_i estimates with various substrates or inhibitors between the relevant esterases from susceptible and resistant strains. In most cases, however, there is no evidence on this point and in one case, involving carbamate resistance in *H. virescens*, it actually appears not to be the case (see below).

In a significant proportion of the cases potentially involving sequestration which have been investigated electrophoretically, the greater level of spectrophotometric esterase activity in resistant lines appears to be largely due to the greater intensity of a single esterase isozyme. Cases like this include the currant lettuce aphid *Nasonovia ribisnigri*, the beetles *L. decemlineata* and *O. surinamensis*, and the tufted apple bud moth *Platynota idaeusalis*. However, there are in fact more cases where multiple isozymes show more intense staining in resistant strains. Some of the latter are probably just post-translationally modified variants of single gene products, while others involving just a small number of isozymes could reflect coamplification and overexpression of a small cluster of esterase-encoding genes. However, there are some cases, like *E. postvittana*, *H. virescens*, and *B. microplus*, where four to ten very different esterase isozymes are more intense in some resistant strains (Armstrong and Suckling, 1988; Rosario-Cruz *et al.*, 1997; Harold and Ottea, 2000). It is not obvious what the genetic basis for this resistance could be, although there have been suggestions that *trans*-acting regulatory "supergenes" (Sabourault *et al.*, 2001) could have such an effect.

For a few of the species in **Table 3** one or more of the elevated esterases associated with resistance has been purified and characterized to some degree biochemically. For the cases involving *B. germanica* (Prabhakaran and Kamble, 1995), *N. cincticeps* (Chiang and Sun, 1996), *L. decemlineata* (Anspaugh *et al.*, 1995), and *Anopheles albimanus* (Brogdon and Barber, 1990), soluble esterases with molecular masses (~50–60 kDa), pH optima, pI values, and inhibitor sensitivities typical of carboxylesterases have been described. Limited comparisons between the enzymes from susceptible and resistant strains have failed to find any differences in properties suggestive of catalytic or other structural differences, and tentative conclusions about sequestration mechanisms based on increased enzyme amounts have therefore been drawn. However, the following cases, involving *O. surinamensis*, the western corn rootworm *Diabrotica virgifera*, and *H. virescens*, all involve novel features suggestive of variously different mechanisms.

Two elevated esterase phenotypes are associated with OP resistance in *O. surinamensis*. One (HH) gives high level resistance to both chlorpyrifos-methyl and fenitrothion (both dimethyl thion OPs) (**Figure 11**), while the other (MH) gives high resistance to chlorpyrifos-methyl, but only moderate protection against fenitrothion. Isoelectric focusing resolves intense zones containing three and two isozymes, respectively, in HH and MH strains that are not apparent in susceptible individuals. None of these intense bands have the same mobility in HH and MH strains but all five have the same N-terminal sequence. It is therefore thought that the variation within each of the two phenotypes is posttranslational in origin, while the differences between phenotypes could be genetic, possibly due to allelic substitutions downstream of the sequenced N-terminal region.

The five isozymes all behave as conventional carboxylesterases: about 71 kDa in molecular mass, readily inhibitable by OPs, and without detectable OP hydrolase activity. Significantly, however, there are large quantitative differences in their sensitivities to OP inhibition. The isozymes in both HH and MH are more susceptible to chlorpyrifos-methyl and fenitrothion inhibition than extracts from susceptible strains. However, the HH isozymes are highly sensitive to inhibition by both OPs, whereas the MH isozymes are significantly less sensitive to fenitrothion. Given the consistency with the resistance data, the parsimonious interpretation is that there are structural differences between the HH and MH isozymes affecting the avidity of their binding to fenitrothion and hence their ability to sequester it. These structural differences may also be responsible for the greater naphthyl acetate staining intensities of the isozmyes compared to susceptible forms or, alternatively, the intensity differences could be due to a different mutation(s), involving gene amplification or upregulation of expression.

Parathion resistance in some populations of *D. virgifera* is associated with greatly increased intensities of (principally) two esterase isozymes (Scharf *et al.*, 1999a, 1999b; Zhou *et al.*, 2002). Little is known about the biochemistries of these isozymes, but the strong correlation in their intensities across individuals suggests that they are either posttranslational variants of a single gene product or that they are coamplified or coregulated. The fact that the resistance is partially suppressible by DEF suggests that they are carboxylesterases. The paradox is that there is also very good evidence that resistance is associated with *in vivo* hydrolysis of OPs (Miota *et al.*, 1998), whereas the mutant carboxylesterases conferring resistance through OP hydrolysis as per the mutant ali-esterase mechanism are associated with a loss, not a gain in carboxylesterase activity. Either the isozyme intensities and OP hydrolysis both reflect a single as yet undescribed mechanism, or they reflect two distinct mechanisms, one involving sequestration by the isozymes and the other involving hydrolysis by other, apparently electrophoretically cryptic enzymes. Significantly there are three other species where OP resistance is also both substantively suppressed by DEF and associated with OP hydrolytic activity *in vivo* (*P. idaeusalis* and two simulid species) (Hemingway *et al.*, 1989; Karoly *et al.*, 1996); all these cases might reflect a novel carboxylesterase mechanism involving enhanced hydrolysis for both OPs and the conventional staining substrates. Alternatively, there are at least two instances (*H. virescens* and the hemipteran *Triatoma infestans*; see below) where esteratic OP hydrolysis is clearly not related to carboxylesterases.

The biochemistries of two independent cases of OP resistance and one of carbamate resistance have been studied in *H. virescens*. In the case of the carbamate resistance three esterase isozymes resolvable by isoelectric focusing were found to be elevated in intensity in a thiodicarb-selected strain (Goh *et al.*, 1995; Zhao *et al.*, 1996). Two of these were unreactive against antibodies raised against either the amplified *M. persicae* esterase or the *M. domestica* αE7 esterase, but the third reacted to both these antibodies and amino terminal sequencing confirmed its similarity with these and other carboxylesterases. Similarity was greatest against the *Drosophila* and hemipteran β-esterases, which might suggest it belongs in clade E. Notably, immunoblotting did not show a clear difference in the amount of this isozyme between susceptible and resistant strains, suggesting that structural rather than expression differences could underlie its elevated esterase activity in the resistant strain. Unfortunately little information was obtained about the biochemistries of the other two isozymes showing elevated staining intensities in this strain. At this point neither sequestration nor hydrolysis can be discounted as the mechanism underlying this resistance.

Profenofos resistance in *H. virescens* has also been associated with increased staining intensities of multiple naphthyl acetate staining esterase isozymes (Harold and Ottea, 1997, 2000). Several different isozymes are involved and some are not always elevated in intensity. However, the intensity of one isozyme is always elevated in resistant individuals from both the field and laboratory populations analyzed, and the intensity of this band is negatively correlated with that of another, suggesting that the two may be allelic alternatives. This would argue for a structural difference between the two forms. It is not clear whether any of the isozymes implicated in OP resistance in this work correspond with those associated with carbamate resistance above; Harold and Ottea (2000) used native PAGE whereas the latter were resolved by isoelectric focusing. However it is notable that the carbamate selected line did also show substantial DEF-suppressible cross-resistance to OPs (Zhao *et al.*, 1996), which clearly supports the possibility of some overlap.

The other case of OP resistance in *H. virescens* does not involve carboxylesterases. It involves a parathion resistant line in which an arylesterase isozyme has been associated with OP hydrolysis both *in vitro* and *in vivo* (Konno *et al.*, 1990; Kasai *et al.*, 1992). The 120 kDa enzyme is concentrated in the soluble fraction, is activated by some divalent metal ions (Co^{2+} and Mn^{2+} but not Ca^{2+}) and has a pH optimum between 8 and 9. It has relatively low affinity for methyl paraoxon (K_m over $700\,\mu M$) but apparently good turnover rate ($V_{max} \sim 0.01\,\mu M\,min^{-1}\,mg^{-1}$ semipure protein), which contrasts directly with the mutant ali-esterases of *L. cuprina* and *M. domestica* (see Section 7.5.2). It is apparently inactive against parathion, but shows activity for paraoxon and a range of other oxon OPs. Some aspects of its biochemistry seem reminiscent of mammalian paraoxonases, but the latter are much smaller molecules (~ 45 kDa) that are preferentially activated by Ca^{2+} (Primo-Parmo *et al.*, 1996; La Du *et al.*, 1999; Claudianos *et al.*, 2001). Its activity was found to be many-fold higher in the resistant strain, compared to the susceptible control strain, but there was no evidence as to whether this reflected a difference in amount or activity of the molecule.

Further insights into OP hydrolytic activities not associated with carboxylesterases come from earlier work on OP hydrolysis by a strain of the kissing bug *Triatoma infestans* (Casabé and Zerba, 1981; De Malkenson *et al.*, 1984). The resistance status of this strain is not clear. However, analysis of the hydrolytic activity of native gel slices for various radiolabeled OPs found three distinct zones of activity from extracts of this strain. None corresponded with slices containing carboxylesterase isozymes although, as expected, some of these sequestered significant amounts of the OPs. One zone containing an aryl esterase isozyme could hydrolyze both oxon and thion OPs (**Figure 11**), another containing an acetylesterase isozyme could

hydrolyze a thion OP, and a third containing no detectable esterase isozyme could also hydrolyze the thion OP. The aryl esterase has some parallels with the *H. virescens* case elucidated by Konno *et al.* (1990) and Kasai *et al.* (1992), although the activity against the thion OP is a notable difference. More generally these data clearly indicate the potential for several esterase-based hydrolytic resistance mechanisms that do not involve carboxylesterases, or even detectable esterase isozymes.

7.6.2. The Special Case of Malathion Carboxylesterase

Malathion is unusual among OP insecticides in having two carboxyl ester linkages in addition to the phosphoester linkage common to all OPs (**Figure 11**). As well as the hydrolysis of the latter as with other OPs, malathion can also be detoxified by hydrolysis of either or both of its (α- and β-) carboxylester linkages (Matsumura and Hogendijk, 1964; Eto, 1974; Welling *et al.*, 1974; Raftos, 1986). We outlined above (see Sections 7.5.2 and 7.5.3) three specific cases, involving dipteran microsomal α-carboxylesterases and a cytosolic wasp carboxylesterase, where substitutions to smaller amino acids at a particular residue in their presumptive acyl pocket created an MCE capable of effective *in vivo* detoxification and high levels of malathion-specific resistance. In the case of the wasp, the amount of its mutant cytosolic MCE was also elevated, further boosting resistance. In this section the biochemistry of several other cases of resistance-conferring MCEs, which have not yet been elucidated at a molecular level is summarized.

However, it is first worth noting some particular features of the chemistry of the malathion molecule. One is that it has an asymmetric carbon in its diethyl succinate group (**Figure 11**) and commercial malathion insecticides are racemic mixes of the two stereo isomers generated (Talley *et al.*, 1998). Almost all the biochemistry of MCEs to date is likewise based on unresolved racemates. Also, malathion differs from most other commonly researched carboxylester substrates, like acetylcholine or naphthyl acetate, etc., which have simple acyl groups and more complex alcohol moieties. Malathion has simple alcohol moieties (both ethyl) and bulkier acyl groups (ethyl succinate linked to the organophosphorus moiety). In this it is reminiscent of juvenile hormone (**Figure 7**) (see Section 7.4.3). It also resembles the pyrethroids (see **Figure 12**, Section 7.6.3) in having the bulky acyl groups, albeit the latter have bulky alcohol moieties as well. This may explain occasional cases of esterase-mediated malathion and pyrethroid cross-resistance outlined below.

Reports of so-called malathion-specific resistance range across several insect orders, although about half involve the Diptera (**Table 4**). High levels of resistance, 50- to thousands-fold, are commonplace. With a few significant caveats, the resistance also shows a high level of specificity for malathion. One caveat is that in the few cases where it has been tested it also covers a small number of other OP insecticides like phenthoate, which also have carboxylester bonds in their acyl groups (Beeman and Schmidt, 1982; Scott and Georghiou, 1986; Raftos, 1986). Another is that while most relevant studies clearly show a lack of high levels of cross-resistance to noncarboxylester OPs, they could not preclude the possibility of low level resistance to such compounds, as seen occurs in the intensively characterized *L. cuprina* and *M. domestica* cases (see Section 7.5.2). Finally, there is some evidence for a degree of cross-resistance to pyrethroids, at least in the hemipterans *N. lugens* and *N. cincticeps* (Motoyama *et al.*, 1984; Chen and Sun, 1994; Karunaratne *et al.*, 1999).

Because the phenotype is generally so clear-cut, malathion-specific resistance has proven relatively amenable to classical genetic analysis. Essentially single-locus control has been reported for the higher dipterans *M. domestica*, *L. cuprina*, and *Chrysomya putoria* (Townsend and Busvine, 1969; Shono, 1983; Raftos and Hughes, 1986), the mosquitoes *A. stephensi*, *A. culicifacies*, *A. arabiensis*, *A. subpictus* (Hemingway, 1983, 1985; Malcolm and Boddington, 1989; Karunaratne and Hemingway, 2001) and *C. tarsalis* (Matsumura and Brown, 1963), the Indian meal moth *Plodia interpunctella* (Beeman and Schmidt, 1982), and the beetles *Tribolium cataneum* and *Cryptolestes ferrugineus* (White and Bell, 1988; Spencer *et al.*, 1998), amongst others. However, there can be two or more resistance alleles at the locus, including both multiple malathion-specific alleles (Picollo de Villar *et al.*, 1983), or also including one or more broad-spectrum OP resistance alleles (Campbell *et al.*, 1998c; Karunaratne *et al.*, 1999; Smyth *et al.*, 2000; Vontas *et al.*, 2000; and see below).

It has generally been found that malathion-specific resistance can be supressed by the use of the OP triphenyl phosphate (TPP) as a synergist (**Table 4**). This is often regarded as diagnostic of malathion-specific resistance and strongly implicates a carboxylesterase-based mechanism. Thin layer chromatography (TLC) indeed shows that malathion-specific resistance is generally associated with esteratic degradation at both the α- and β-carboxylester bonds of malathion (**Figure 11**), although the α-monoacid is generally the majority product and the diacid is generally less abundant than either monoacid (Matsumura and Voss, 1964; Kao *et al.*, 1985; Raftos, 1986; Malcolm and Boddington, 1989, and references therein). Quantitation of MCE activity as a V_{max} of whole body homogenates (as determined from an endpoint radiometric partition assay) also generally shows MCE activity to be elevated in malathion-specific resistant strains. Elevations of 50-fold or greater are commonly recorded, although notably they are often significantly less than the corresponding resistance factors (e.g., Beeman and Schmidt, 1982; Ziegler *et al.*, 1987; Smyth *et al.*, 1996; Spencer *et al.*, 1998). Characteristically the elevated MCE activities are potently inhibited by OPs, particularly TPP (e.g., Plapp and Eddy, 1961; Dyte and Rowlands, 1968; Townsend and Busvine, 1969; Hemingway, 1985; Spencer *et al.*, 1998).

However, there is a very variable relationship between elevated MCE activity and general carboxylesterase activity in malathion-specific resistance. As well as the well-characterized *L. cuprina* and *M. domestica* cases (see Section 7.5.2), those involving the blowfly *C. putoria*, the moth *P. interpunctella*, and the mite *Tetranychus urticae* are also associated with a reduced level of "general" esterase activity as assessed using methyl butyrate or various naphthyl esters as substrates. However, there are cases where such general esterase activities are essentially unchanged (e.g., in some of the anophelines) or substantially enhanced (e.g., *C. tarsalis* and some planthoppers and leafhoppers) in resistant strains. A similar range of results is found across the various isozyme studies. Elevated MCE activity is associated with reduced staining intensities of particular isozymes in *M. domestica* and *L. cuprina*, but with elevated staining intensities of particular isozymes in *N. cincticeps* and the brown planthopper *Laodelphax striatellus*. In *C. tarsalis* the majority of MCE is associated with a zymogram region lacking any detectable general esterase isozyme at all. The same is true of the bug *T. infestans*, although malathion resistance status was not reported in this case (Wood *et al.*, 1985).

Importantly it is clear that even susceptible strains often have significant levels of MCE activity. Notwithstanding large quantitative differences, at least some MCE is found to be widely distributed across tissues, life stages and subcellular fractions in both susceptible and resistant strains (Townsend and Busvine, 1969; Beeman and Schmidt, 1982; Motoyama *et al.*, 1984; Wood *et al.*, 1985) (**Table 4**). Studies of *N. lugens* and *N. cincticeps* found

Table 4   Summary of cases of esterases implicated in malathion-specific metabolic resistance

Species	Biochemistry	Reference
Hemiptera		
Laodelphax striatellus (brown planthopper)	Some immunologically related esterase isozymes with *in vitro* malathion carboxylesterase (MCE) activity and identical amino terminal sequences are elevated in amounts in malathion-resistant strains	Miyata and Saito (1976), Sakata and Miyata (1994)
Nephotettix cincticeps	Four isozymes with high esterase activity, three of which had malathion hydrolytic activity in three strains showing moderate levels of resistance to malathion albeit also several other OPs and some SPs	Miyata and Saito (1976), Motoyama *et al.* (1984), Chiang and Sun (1996)
Nilaparvata lugens	An isozyme with high MCE activity (and some SP hydrolytic activity) is more intense in strains with low level resistance to malathion (and some SPs) but not most OPs	Chang and Whalon (1987), Chen and Sun (1994)
Coleoptera		
Cryptolestes ferrugineus (rust red grain beetle)	Slightly lower general esterase activity but increased MCE activity *in vivo* and *in vitro* in a strain showing a high level of triphenyl phosphate (TPP)-suppressible malathion resistance	Spencer *et al.* (1998)
Tribolium castaneum (red flour beetle)	TPP-suppressible malathion-specific resistance	White and Bell (1988)
Lepidoptera		
Plodia interpunctella (Indian meal moth)	General esterase activity reduced but MCE activity elevated 33-fold in a strain showing high TPP-suppressible resistance to malathion and cross-resistance to phenthoate	Beeman and Schmidt (1982)
Hymenoptera		
Anisopteromalus calandrae	General esterase activity unchanged but (cytosolic) MCE activity 10–30-fold higher in a malathion resistant strain; W to G mutation at the same site as the W251L MCE mutation in *L. cuprina* and *M. domestica* contributes to malathion resistance, as does upregulation of expression	Baker *et al.* (1998a, 1998b), Zhu *et al.* (1999a, 1999b)
Diptera		
Anopheles arabiensis	General esterase activity unchanged and no elevated esterase isozymes but kinetically efficient MCE in strains with high levels of TPP-suppressible malathion specific resistance	Hemingway (1985)
Anopheles culicifacies	General esterase activity essentially unchanged and no elevated esterase isozyme but high levels of MCE activity and *in vivo* malathion metabolism in strains with high levels of TPP-suppressible malathion specific resistance	Herath *et al.* (1987), Malcolm and Boddington (1989), Karunaratne and Hemingway (2001)
Anopheles stephensi	General esterase activity essentially unchanged and no elevated esterase isozymes but three isozymes have MCE activity in strains resistant to malathion and phenthoate	Hemingway (1982, 1983), Scott and Georghiou (1986)
Anopheles subpictus	High levels of MCE activity and monoacid and diacid metabolites produced by mass homogenates of resistant field strain	Karunaratne and Hemingway (2001)
Chrysomya putoria (green blowfly)	General esterase activity is reduced in TPP-suppressible malathion-specific resistance; the alkyloxy methyl group of malathion is important for high level resistance	Townsend and Busvine (1969)
Culex tarsalis	General esterase activity enhanced in a strain with 150-fold resistance specifically to malathion that is synergised by TPP; MCE activity is associated with a zymogram region lacking esterase activity	Matsumura and Brown (1963), Ziegler *et al.* (1987), Whyard *et al.* (1994)
Lucilia cuprina	Reduced levels of general esterase activity, slightly higher levels of MCE activity and significant OP hydrolase activity associated with Ser/Leu251 variants of the LcαE7 (E3) isozyme in strains with high levels of TPP-suppressible resistance to malathion and phenthoate (see Section 7.5.2)	Hughes *et al.* (1984), Raftos (1986), Raftos and Hughes (1986), Smyth *et al.* (1994, 1996, 2000), Whyard and Walker (1994), Campbell *et al.* (1998a, 1998c), Newcomb *et al.* (2004)
Lucilia sericata	Ser/Leu251 variants found in LcαE7 ortholog in populations known to segregate for malathion specific resistance (see Section 7.5.2)	R.D. Newcomb, C.G. Young, and J.R. Stevens, unpublished data

Continued

Table 4   Continued

Species	Biochemistry	Reference
Musca domestica	Reduced levels of general esterase activity, slightly higher levels of MCE activity and Ser/Leu251 variants of the MdαE7 (Ali) isozyme in strains with high levels of TPP-suppressible resistance to malathion (see Section 7.5.2)	Plapp and Eddy (1961), Picollo de Villar *et al.* (1983), Shono (1983), Kao *et al.* (1984, 1985), Claudianos *et al.* (1999, 2001)
Acarina		
Tetranychus urticae	Reduced levels of general esterase activity, increased MCE activity and increased phosphatase activity in a strain with 60-fold malathion resistance and 10-fold parathion resistance	Matsumura and Voss (1964)

measurable levels of MCE activity associated with a majority of the esterase isozymes in both suscepti-ble and resistant strains (Chang and Whalon, 1987; Chen and Sun, 1994; Chiang and Sun, 1996). Together with the cases where MCE is associated with a nonstaining area of general esterase zymo-grams, this suggests that a range of MCEs is at least theoretically available from which mutants with improved activities might be selected to confer resistance.

As described above (see Section 7.5.2) one MCE mutation found in orthologous microsomal carbox-ylesterases from clade B (**Figure 2**) in both the higher dipterans *L. cuprina* and *M. domestica*. This active site mutation improves MCE kinetics very slightly (although with a specificity constant around $10^6 \, M^{-1} \, s^{-1}$ it is already high) but also confers sig-nificant OP hydrolase activity, so the MCE enzyme escapes from otherwise essentially irreversible inhi-bition if it binds at the phospho end of malaoxon (the activated form of malathion) rather than the carboxylester end of malaoxon or malathion. The mutant enzyme also has reduced activity for several of the conventional artificial substrates used to mea-sure general esterase activity. There is molecular evidence that a similar MCE mutation confers mal-athion specific resistance on the hymenopteran *A. calandrae*, albeit in this case it involves a soluble cytosolic esterase from clade A in **Figure 2**, and also appears to involve some upregulation in the amount of the esterase (see Section 7.5.3). Precise parallels in the biochemistry and toxicology of the phenotype have also led to suggestions that the malathion-specific resistance in another higher dipteran *C. putoria* and in the moth *P. interpunctella* would be due to the same mutation in a micro-somal clade B esterase (Campbell *et al.*, 1997, 1998c; Smyth *et al.*, 2000). Some of the better-characterized other cases of resistance to see if they also have biochemistries matching these cases are now considered.

Several cases of malathion-specific resistance in anopheline mosquitoes have been characterized to some degree biochemically, and none has proven associated with elevated staining of any esterase isozymes although all show the characteristic TPP-suppressible, MCE-based mechanism (Hemingway, 1985; Herath *et al.*, 1987; Malcolm and Boddington, 1989; Hemingway *et al.*, 1998; Karunaratne and Hemingway, 2001). Kinetic evidence in several cases indicates the presence of a highly efficient MCE enzyme in resistant strains that is apparently absent in susceptible strains (Hemingway, 1985; Hemingway *et al.*, 1998). *Anopheles stephensi* in Pakistan express at least three esterase isozymes with MCE activity but none of them is overex-pressed in resistant strains and indeed they are all inabundant in both susceptible and resistant strains (Jayawardena and Hemingway, cited in Hemingway *et al.*, 1998). All these data imply that a qualitative rather than quantitative change in an MCE confers malathion-specific resistance among these anophe-lines. The *A. gambiae* genome does not contain any sequences in clade B (although culicines do), but the biochemistry available for the anopheline MCEs does not yet suggest which clade they do belong to.

In the culicine mosquito *C. tarsalis*, malathion-specific resistant lines express in the mitochondria of their gut cells a kinetically efficient MCE, which is not detectable in susceptible strains (Ziegler *et al.*, 1987; Whyard *et al.*, 1994). There is also a cytosolic MCE in a broad range of tissues, but this is less efficient kinetically and equally expressed in both susceptible and resistant strains. It is thus suspected that the effect of the resistance mutation is to sub-stantially elevate the expression of a previously ina-bundant enzyme that is otherwise appropriate in its kinetics and expression profile for *in vivo* detoxifi-cation of malathion. The resistance-associated enzyme is not an ortholog of the clade B esterase amplified in broad-spectrum OP resistance in other culicines (Tittiger and Walker, 1997). The only clade in **Figure 2** containing sequences with mitochondrial targeting signals is clade A, although not all clade A enzymes have such signals. There are many precedents for mitochondrially located enzymes

performing detoxification functions in other organisms, including for pesticides like malathion (Bachurin *et al.*, 2003 and references therein).

An unusual variant of the secreted clade E esterase associated with broad-spectrum OP resistance in the planthopper *N. lugens* (see Section 7.4.3.2) has been described from Japan and the Philippines. This esterase provides at least a low level of resistance to malathion and certain synthetic pyrethroids but not to most OPs (Chang and Whalon, 1987; Chen and Sun, 1994) (see Section 7.6.3). While the variant associated with the broad-spectrum OP resistance has been shown to involve a three- to sevenfold amplification of the encoding gene, nothing has yet been reported of the molecular nature of the malathion/pyrethroid resistant variant (Karunaratne *et al.*, 1999; Small and Hemingway, 2000a, 2000b; Vontas *et al.*, 2000). While as many as eight different esterase isozymes have MCE activity in *N. lugens*, it is the E1 isozyme, which is more intense in the malathion/pyrethroid resistant strain, that appears to have the best MCE activity.

At least four esterase isozymes are strongly overexpressed in field strains of the leafhopper *N. cincticeps* showing varying levels of resistance to many OPs, including moderate resistance to malathion, and also some resistance to pyrethroids (Miyata and Saito, 1976; Motoyama *et al.*, 1984). The net effect is over 40-fold higher carboxylesterase activity in the resistant strains. At least three of the isozymes have some MCE and pyrethroid hydrolytic activities. It has been proposed these isozymes provide resistance to malathion and pyrethroids by hydrolytic detoxification, and to the other OPs by sequestration (Motoyama *et al.*, 1984) (see Section 7.6.1). Whilst not discounting this possibility, the latest data suggest that the contribution to malathion resistance due to the MCE activities of these enzymes may be relatively minor (Chiang and Sun, 1996). Even without this *N. cincticeps* case, it is apparent that the structural mutations described in *L. cuprina* and *M. domestica*, and inferred in *C. putoria* and *P. interpunctella*, are just one mechanism conferring MCE based malathion-specific resistance. Although not yet resolved at the molecular level there are clearly other mechanisms, involving different mutations, nonorthologous enzymes, and possibly different clades, which could confer MCE based resistance.

7.6.3. Resistance against Synthetic Pyrethroids

Most commercial synthetic pyrethroid (SP) insecticides are carboxylesters, with relatively large and highly hydrophobic acid and alcohol groups (Casida *et al.*, 1983) (**Figure 12**). The early, type 1, SPs are smaller than the more modern type 2 compounds, which differ in having an α-cyano substituent on their alcohol group and, in many cases, dibromo instead of dichloro substituents on their acyl groups. Most SPs also show high levels of racemic complexity, with as many as three chiral centers, one around the carbon to which the cyano moiety attaches in the alcohol group and two across the cyclopropane ring in the acyl group. The type 2 SPs generally have greater potency, in part because their greater bulk makes them less susceptible to hydrolytic detoxification by carboxylesterases (Elliott, 1977; Soderlund and Casida, 1977a, 1977b; Casida *et al.*, 1983). The various isomers also differ in potency against many insects, albeit not always in the same way across species. For example, it is generally but not always true that the *cis* isomers across the cyclopropane ring are more potent insecticides than their *trans* alternatives and, again, this is at least partly because they appear to be less susceptible to hydrolysis by carboxylesterases (see, e.g., Casida *et al.*, 1983; Ruigt, 1985; Chang and Whalon, 1986; Byrne *et al.*, 2000).

Table 5 summarizes about a dozen cases where elevated esterase levels have been associated with metabolic resistance to SPs. The cases tabulated range across several orders of insects and two acarines, although lepidopterans account for about half the cases. None of the cases tabulated has yet been fully elucidated at a molecular genetic or even biochemical level. The problems in studying SP resistance at a biochemical level stem from several refractory properties of the compounds themselves, including their isomeric complexity, difficulties in separating some isomers, and their very limited (low μM) aqueous solubilities. These difficulties have been compounded until recently by the unavailability of sensitive and convenient fluorometric or colorimetric assays, so kinetic analyses had to be carried out with expensive radiometric reagents in tedious partition assays. Fortunately Wheelock *et al.* (2003) have recently developed fluorometric analogs of some SPs, which should now greatly expedite progress in resolving the biochemistry of SP resistance.

At a phenotypic level, esterase based metabolic resistance to SPs is often reported in association with P450-based resistance. The contribution that the esterase mechanism makes to the overall metabolic resistance can be readily demonstrated experimentally by the use of respective inhibitors for the two enzyme systems, and the reliance of the P450 mechanism on NADP as a cofactor; however, it is difficult to quantify the relative contributions of the two mechanisms (Liu *et al.*, 1984; Ottea *et al.*,

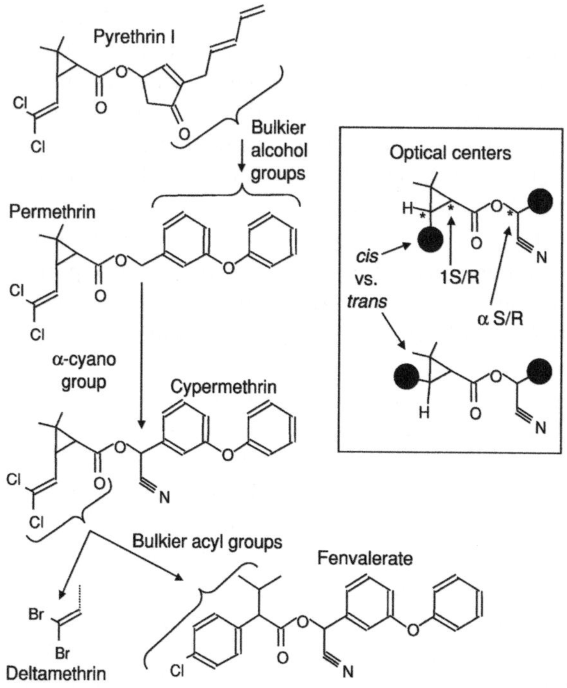

Figure 12 Pyrethroid structures. Pyrethrin I is shown first as an example of a natural pyrethroid. Synthetic pyrethroids (SPs) have shown increasing resistance to hydrolysis by esterases as the alcohol groups were made bulkier (e.g., permethrin), an α-cyano group was added (e.g., cypermethrin) and the acyl group was also made bulkier (e.g., fenvalerate), often by the replacement of the chlorine atoms with bromines (e.g., deltamethrin). Type 2 SPs contain the α-cyano moiety and often also the dibromo moiety. Pyrethroids generally have at least two and often three optical centers, one around the carbon to which the cyano moiety attaches and two across the cyclopropane ring, resulting in four or eight possible stereoisomers. Naming conventions describe stereo-isomers as α R or S with respect to the carbon to which the cyano group attaches and 1R or S for the optical center on the cylcopropane ring closer to the carboxylester linkage. However, the alternatives at the other optical center on the cyclopropane ring are not described directly. Instead the combined effect of the two optical centers either side of the cyclopropane ring is then described as *cis* or *trans* for whether the bulkier moieties either side of the cyclopropane ring are placed closer together or further from each other.

1995; Shan *et al.*, 1997; Shan and Ottea, 1998). Cross-resistance of the esterase component beyond SPs is unusual, although it has been reported for malathion in a small number of cases, e.g., *N. cincticeps* and *N. lugens* (**Table 5**), plus possibly also see *C. pipiens* (Bisset *et al.*, 1997). In these cases resistance may be mediated by a shared carboxylesterase-based mechanism. There is one report, involving *A. albimanus*, of cross-resistance to other OPs. The mechanism here is unknown but may involve amplification of an esterase that binds and sequesters both classes of compound (Brogdon and Barber, 1990). In a few cases multiple SPs have been tested; cross-resistance among SPs seems the rule, although (consistent with their heterogeneity in size and shape) resistance factors can vary substantially (Gunning *et al.*, 1999).

At a biochemical level, the usual finding is of elevated levels of esterase activity in *in vitro* assays of whole body extracts with compounds like naphthyl acetate as substrates. The increase at this level is generally not large, around two- to tenfold, although increases exceeding 50-fold have been reported, for example, in the cotton bollworm *Heliothis armigera*. However, the staining intensity of individual isozymes may be elevated over 100-fold in resistant strains, or in some cases they may be apparent in resistant strains but undetectable in susceptible strains. In some cases, like *Bemisia tabaci* and *B. microplus*, it essentially involves just a single band. More commonly it involves several bands; cases of this include *H. armigera*, the native budworm *Helicoverpa punctigera*, *H. virescens* and the tick *A. fallacis*. In *A. albimanus* the elevated intensity was resolved as a single band by native PAGE but separate bands by isoelectric focusing. There are no cases of SP resistance to our knowledge associated with reductions in staining intensity of esterase isozymes or the replacement of one isozyme band with another.

It is generally assumed that increased staining intensities of esterase isozymes associated with

Table 5 Summary of cases of esterases implicated in metabolic resistance to synthetic pyrethroids

Species	Biochemistry	Reference
Hemiptera		
Bemisia tabaci	Increased molar amount of the esterase isozyme $E_{0.14}$ is strongly associated with a marked increase in *in vivo* permethrin hydrolysis	Costa and Brown (1991), Byrne and Devonshire (1996), Byrne *et al.* (2000)
Nephotettix cincticeps	Increased amounts of four esterase isozymes with some SP (and malathion) hydrolytic activity tentatively associated with SP (and malathion) resistance; one isozyme with activity essentially only for fenvalerate. Activity greatest for *cis*-permethrin, then cypermethrin, then *trans*-permethrin (and then malathion)	Motoyama *et al.* (1984), Chiang and Sun (1996)
Nilaparvata lugens	Increased amount of the E1 isozyme in an SP and malathion resistant strain has more activity for *trans*-permethrin (and malathion) than *cis*-permethrin and cypermethrin. Several other isozymes also hydrolyse *cis*- and *trans*-permethrin (*trans* better than *cis*) and malathion	Chang and Whalon (1987), Chen and Sun (1994)
Coleoptera		
Leptinotarsa decemlineata	Increased *trans*-permethrin hydrolytic activity and general esterase activity in whole body homogenates of resistant individuals	Argentine *et al.* (1995)
Lepidoptera		
Helicoverpa armigera	Increased *in vivo* SP hydrolysis, general esterase activity and more intense staining of several esterase isozymes in strains whose resistance to multiple SPs is partially suppressible by nontoxic doses of OPs; increased staining intensity due to increased amount of enzyme	Gunning *et al.* (1996b, 1998, 1999), Campbell (2001)
Helicoverpa punctigera (native budworm)	Increased *in vivo* SP (fenvalerate) hydrolysis and more intense staining of several esterase isozymes in strains whose resistance is suppressible by DEF and propenofos	Gunning *et al.* (1997)
Helicoverpa zea (corn earworm)	Increased general esterase activity in resistant strains (not inducible by diet)	Muehleisen *et al.* (1989)
Heliothis virescens	Increased *in vivo* SP hydrolytic activity and general esterase activity and more intense staining of several esterase isozymes in resistant strains; increased staining intensity due to increased amount of enzyme; *trans*-permethrin hydrolyzed more rapidly than *cis*-permethrin	Dowd *et al.* (1987), Goh *et al.* (1995), Ottea *et al.* (1995), Shan and Ottea (1998)
Plutella xylostella (diamondback moth)	Esterase activity towards permethrin found in resistant strain	Liu *et al.* (1984)
Spodoptera exigua (beet armyworm)	Increased *in vivo* deltamethrin hydrolytic activity and general esterase activity in resistance strains	Delorme *et al.* (1988)
Spodoptera littoralis (Egyptian cotton leafworm)	Increased esterase activity associated with TPP-suppressible SP resistance	Riskallah (1983)
Diptera		
Anopheles albimanus	Increased intensity of an esterase isozyme zone in strains and individuals showing *S,S,S*-tributylphosphorotrithioate (DEF)-suppressible resistance to fenitrothion and deltamethrin	Brogdon and Barber (1990)
Acarina		
Amblyseius fallacis (predatory mite)	Many esterase isozymes in susceptible insects have some SP hydrolytic activity (preferentially *trans*- over *cis*-permethrin) but certain esterase isozymes from resistant strain have higher rates of SP hydrolysis	Chang and Whalon (1986)
Boophilus microplus	Increased *in vivo* SP hydrolytic activity and at least one unique esterase isozyme detected in strains whose SP resistance is at least partially suppressed by TPP	de Jersey *et al.* (1985), Jamroz *et al.* (2000)

resistance are due to increases in the amounts of the enzymes, and in some cases (*B. tabaci, N. cincticeps, N. lugens, H. armigera,* and *H. virescens*) immunochemical, tritium-labeling or other techniques have been used to validate this assumption. In *H. armigera* and the corn earworm *H. zea* it has been found that the elevated activity phenotype is stably expressed across a range of dietary conditions and is not the result of increased inducibility of the enzymes by the pyrethroids.

Although there may also be some protective effect due to the esterases binding and sequestering the pesticide (Goh *et al.*, 1995; Gunning *et al.*, 1999), there is good evidence from many studies indicating metabolism of SP by the esterases. In particular, increases in SP hydrolytic activity in whole body homogenates of resistant individuals have been recurrently reported. Examples of this include *B. tabaci*, *L. decemlineata*, *H. armigera*, *H. punctigera*, *H. virescens*, the beet armyworm *Spodoptera exigua*, *M. domestica*, and *B. microplus*. The generalities of greater hydrolytic activity for smaller compounds and *trans* isomers hold up in these extracts, but exceptions to both these generalities are also documented (Ishaaya and Casida, 1981) (**Table 5**).

As with malathion above, it appears that a substantial proportion of the esterase isozymes even in susceptible insects have some SP hydrolytic activity, at least *in vitro*. This has been mostly clearly demonstrated by Chang and Whalon (1986, 1987) who found that a substantial proportion of the esterase isozymes which they could resolve in *N. lugens* and *A. fallacis* had some SP hydrolytic activity; some of these were elevated in resistant strains but others were not. Evidence from other species is also consistent with these results (Abdel-Aal and Soderlund, 1980; Ishaaya and Casida, 1980; Yu, 1991) (**Table 5**).

Esterase isozymes with some SP hydrolytic activity have been purified and characterized to varying degrees from several species. Except for *B. tabaci*, evidence to implicate the particular enzymes studied in SP resistance is generally lacking (Byrne *et al.*, 2000). It is clear that isozymes with SP hydrolytic activity can be found from many tissues and developmental stages, and from both soluble and micrososmal fractions (Abdel-Aal and Soderlund, 1980; Ishaaya and Casida, 1980; Motoyama *et al.*, 1984). Evidence on inhibition by OPs, molecular mass, isoelectric points, pH optima, and immunological cross-reactivity are all consistent with the isozymes being in the carboxyl/cholinesterase family, without being sufficiently definitive to identify particular clades within the family (Abdel-Aal and Soderlund, 1980; Ishaaya and Casida, 1980; Motoyama *et al.*, 1984; de Jersey *et al.*, 1985; Chen and Sun, 1994; Goh *et al.*, 1995; Chiang and Sun, 1996; Byrne *et al.*, 2000). K_m estimates with SPs are generally in the low–mid μM range, with k_{cat} data generally lacking (Jao and Casida, 1974; Abdel-Aal and Soderlund, 1980; Yu, 1991). Interestingly, Abdel-Aal and Soderlund (1980) found quantitative differences in tissue distributions and inhibitor sensitivities between *cis* and *trans* permethrin preferring activities and therefore suggested that esterase-mediated metabolic resistance to SPs

would accrue through the combined effect of several different esterases of differing substrate preferences.

The most detailed kinetic analyses of SP hydrolysis by insect esterases have recently been carried out by Heidari *et al.* (unpublished data) for *in vitro* expressed forms of the *L. cuprina* αE7/E3 enzyme and various natural and synthetic mutants of it. LcαE7 is not involved in SP resistance but the wild-type enzyme shows high SP hydrolytic activity, including preferences for type 1 compounds, dichloro substituents, and *cis/trans* isomers across the cycopropane ring, with relatively little stereospecificity with respect to the α-cyano group. However, a few variants of the enzyme with substitutions in its presumptive acyl pocket were variously found to have reversed the *cis/trans* and *R/S* stereospecificities, and at least quantitatively altered preferences for the other structural features. These results support the earlier studies above indicating that most esterase isozymes are individually quite specific for the variable structural features of SP insecticides, and they further show that good hydrolytic activities against a variety of SP structures can be achieved across different (even very closely related) esterases. They also caution against the assumption that the changes in esterase biochemistry conferring SP resistance will always affect isozyme amounts; clearly there is scope for point mutations affecting enzyme structures to improve the suite of SP hydrolytic activities of the organism.

Two predictions can be made about the genetics of esterase-based SP resistance from the biochemical findings to date: they might differ considerably between species and sometimes can be polygenic. Whilst accepting the possibility of some structural changes in relevant enzymes, most of the data to date suggest that the changes will generally affect isozyme amounts. Gene amplification has been suggested as one possible mechanism for this (Gunning *et al.*, 1999) although it would probably need to involve multiple amplifications, or amplifications of multiple (closely linked) genes, to account for some of the cases of resistance involving increased staining of multiple isozymes. Regulatory changes are an alternative explanation, and certainly there is abundant evidence for regulatory variation affecting the amounts of esterase isozymes (Game and Oakeshott, 1990; Holloway *et al.*, 1997; Odgers *et al.*, 2002). Notably the best evidence for the coincident upregulation of multiple esterases appears to involve inducible systems (Salama *et al.*, 1992; Callaghan *et al.*, 1998; Campbell, 2001), which does not appear to apply to the esterases involved in SP resistance. It is therefore suspected that polygenic inheritance will sometimes be the case.

7.7. Concluding Remarks

The carboxyl/cholinesterase gene family has radiated extensively in the higher eukaryotes. Small numbers are found in a minority of prokaryotes and only a few occur in lower eukaryote genomes, but 30 or more have been found in each of the higher eukaryotic genomes annotated. In insects, it appears that this expansion corresponds at least in part with their recruitment to functions in neurodevelopmental processes and the transduction of hormonal and other signals, albeit the precise physiological functions of most remain largely, if not completely, unknown. The radiation in physiological function is paralleled at a biochemical level. Where known, the kinetics of individual enzymes for their particular physiological substrates can be highly efficient. However, the substrates for different members of the family differ quite widely in structure and, remarkably, a significant proportion of the family has moved into noncatalytic functions.

While a significant proportion of the major catalytic and noncatalytic clades appear to have origins as old as the Insecta, it is also clear from cases like glutactin, the noncatalytic β-esterases, and the evolution of insecticide resistance, that new radiations moving the family into new functions are still arising. In this respect it will be interesting to see the constitution of the family in the silkmoth *Bombyx mori* and honeybee *Apis mellifera,* whose genome sequences should be released before this article appears in print. Both species are clearly much more distantly related than *D. melanogaster* and *A. gambiae,* and an enormous diversity of esterase isozymes have been reported from several Lepidoptera – as many as 60 isozymes compared to just 20–30 in *D. melanogaster* (Campbell, 2001; Campbell *et al.,* 2003).

Notwithstanding the expansion of the carboxyl/cholinesterase family in higher eukaryotes, it is also clear that other protein families contribute significantly to the complement of esteratic activities of higher organisms. Vertebrate and prokaryote precedents suggest that insect aryl and acetyl esterases will derive from other families (Oakeshott *et al.,* 1993). There are also other insect esterase activities like the ortholog of vertebrate neuropathy target esterase (Borrás *et al.,* 2003) that clearly have other, nonserine dependent reaction mechanisms. And there are several vertebrate precedents for other enzyme families with serine based reaction mechanisms like the serine proteases that have generated good esteratic activities (Zschunke *et al.,* 1991). The carboxyl/cholinesterases probably contribute the majority of insect carboxylic ester

hydrolase activities, but the contributions of other protein families are as yet very poorly understood.

We will not understand why the carboxyl/cholinesterase family has been such a versatile platform for the evolution of new functions until we know a great deal more about structure, function, and structure–function relationships within the family. Fortunately, use of the modern tools of functional genomics and metabolomics could greatly improve our knowledge of their physiological and biochemical functions in the near term. The structural information, however, will be more difficult to expand. AChE is the only insect member of the family whose structure is currently known, albeit with a structure for a lepidopteran JHE also imminent. Even the modeled structure of JHE has suggested some qualitative differences from AChE directly concerned with catalysis, like the reverse orientation of the acyl and leaving group pockets. There is also particular priority to solve structures in some of the clades with noncatalytic functions and those from which the various metabolic resistances arise. In both cases ligand-bound structures will be of particular interest.

Perhaps the most spectacular demonstration of the evolutionary adaptability of the carboxyl/cholinesterases is their likely involvement in well over 50 cases of resistance to chemical insecticides over the last 50 years. Many involve the older OP and carbamate compounds where mutations of major biochemical effect are required to achieve either target site or metabolic resistance. Several also now involve the SPs although none of these has as yet been resolved at a molecular level. Esterase based SP resistance is also a notable feat of molecular evolution given the bulk and chiral complexity of SP insecticides. Nor are the more recent insecticides immune from esterase based resistance mechanisms – there are already reports of esterase involvement in resistance to various insect growth regulators (Ishaaya and Degheele, 1988; El Saidy *et al.,* 1989; Ishaaya and Klein, 1990) where at least one can rationalize a hydrolytic basis for detoxification. Intriguingly there are also reports of possible esterase involvement in metabolic resistance to imidacloprid (Wang *et al.,* 2002) and even the proteinaceous crystal toxins of *Bacillus thuringiensis* (Dang and Gunning, 2002), where a mechanism is at this time much less readily rationalized biochemically.

This chapter has taken a genomics perspective to the cases of esterase mediated resistance that have been resolved at a molecular level and then analyzed the biochemistries of about 70 cases of such resistances that have not yet been resolved at a molecular level. These analyses generally support the contention that the options for evolving resistance are limited. However, the biochemistry clearly indicates the existence of several more that have yet to be resolved at a molecular level in any species. Involvement of multiple clades of carboxyl/cholinesterases has already been observed, and it is clear that a few cases involve other protein families.

It has recently been suggested that there may be two broad classes of AChE insensitivity mutations conferring target site resistance to OPs and carbamates, one skewed towards OP resistance and the other to carbamate resistance. It also appears that there may be at least two subclasses within the carbamate resistant class, with one specific for dimethyl carbamates and possibly restricted to hemipterans. The validity of these classifications will only be tested by molecular work on several more independent cases of target site resistance. It also appears from biochemical analyses of synthetic mutants that there could be many more substitutional options available for conferring resistance than have been reported from natural populations to date. Understanding which are real options, and how they relate to the different patterns of insensitivity phenotype, will also require AChE structural data for additional taxa. Also important to our understanding of the evolution of AChE and its role in resistance will be much greater knowledge at a biochemical and molecular level of both the noncatalytic functions of AChE and of the partitioning of the catalytic and noncatalytic functions among the two AChE proteins found in most insects.

It appears that the gene amplification/sequestration mechanism, and possibly also the mutant aliesterase mechanism, found in the cases of metabolic resistance to OPs resolved at a molecular level so far will indeed recur in some other cases. However, the biochemistry clearly points to some significant variations on these themes, as well as some completely different mechanisms. The involvement of the metal activated aryl esterases in general OP resistance in *H. virescens* obviously constitutes at least one completely different mechanism. Whilst it seems that insects do not have homologs of the vertebrate paraoxonase subfamily of aryl esterases (Sorenson *et al.*, 1995), they appear to have homologs of at least some prokaryote OP hydrolytic enzymes (Hou *et al.*, 1996; Claudianos *et al.*, 2001; Horne *et al.*, 2002a, 2002b). Insect homologs of one of these, the metal-activated OPH (Pfam 02126), have been identified (Hou *et al.*, 1996; Claudianos *et al.*, 2001). *Spodoptera frugiperda* larvae infected with recombinant baculoviruses expressing high levels of a bacterial OPH show nearly 300-fold greater resistance to paraoxon than uninfected controls

(Dumas *et al.*, 1990) while *D. melanogaster* genetically transformed with this gene show over 20-fold greater paraoxon resistance than untransformed controls (Phillips *et al.*, 1990; Benedict *et al.*, 1994).

There are also diverse esterase biochemistries underlying malathion-specifc resistance, with the molecular biology only as yet resolved for the Leu/Ser251 versions of the mutant αE7/aliesterase mechanism.

Regulatory changes are clearly suggested by the increased staining intensities and abundances of some isozymes seen for general OP, malathion-specific and pyrethroid resistance. Some of these could be relatively straightforward changes to the promoters of the relevant esterase genes, as seems to be the case for at least one *C. pipiens* amplicon. However, one paradox is the finding of greater intensities for multiple isozymes in several cases of resistance. Some of these appear beyond the scope of explanations based on posttranslational enzyme modification or adjacent gene coamplifications. Alternative explanations involving *trans*-acting regulatory factors remain to be investigated.

With the notable exception of the carbamate–OP cross-resistance due to E4/FE4 sequestration in *M. persicae*, none of the esterase based metabolic resistances to carbamates and SPs have as yet been elucidated at a molecular level. A sequestration mechanism similar to the E4/FE4 model may underlie some of the many other cases of OP–carbamate cross-resistance reported. Some variant of it might also account for the occasional case of OP–SP cross-resistance. However hydrolysis could well underlie more of the cases not involving cross-resistance to OPs, which are now numerous for the SPs at least. It is clear that many insect esterases have significant hydrolytic activity against the carboxylester bonds in SPs *in vitro*. Little is known about hydrolysis of carbamates in insects but bacterial enzymes are known which have good hydrolytic activities against either the amide or carboxylester bonds in these compounds (Pohlenz *et al.*, 1992). The nature of the chemistry involved in hydrolysis of the carbamates and SPs is so different, both from that for OPs and from each other, that we would not expect hydrolytic mechanisms to confer cross-resistance among the three classes of compound.

One final perspective on resistance is that what is known at the moment is a snapshot of the "first generation" of insect responses to the insecticides. There is little insight as to what preceded resistance or what will follow the first generation response. What were the predisposing features of the relevant esterase systems? Did the mutations pre-date the use of the pesticides or has the breakdown of control waited on the occurrence of new mutations? Can insects develop "second generation" responses that substantively enhance their resistance or cross-resistance or avoid the costs of resistance? Has the repeated use of ester pesticides so depleted some insects' store of genetic variability in detoxifying enzymes that they will struggle to respond to new compounds? One is just beginning to see the power of modern genomic and other biotechnologies to reconstruct the past and predict the future. Recently genotyping of preserved specimens by PCR has in fact suggested that some resistance mutations did predate insecticide use (Newcomb et al., 2004). Likewise the tools of in vitro evolution, already used to predict resistance futures in the field of antibiotic resistant microorganisms, might also be useful in plotting possible futures for some key pesticide resistance enzymes.

Acknowledgments

The authors thank Thom Boyce, Bronwyn, Campbell, Chris Coppin, Erica Crone, Alan Devonshire, Peter East, Carol Hartley, Irene Horne, Kerrie-Anne Smyth, Tara Sutherland, and David Tattersall for sharing their insights into various topics covered in this review. They are also indebted to the tireless work of Narelle Dryden in compiling the literature. Undoubtedly, some significant relevant papers have been missed here, for which JGO takes full responsibility; the authors apologise to those authors whose work has been missed or inadvertently misrepresented.

References

Abdel-Aal, Y.A.I., Lampert, E.P., Roe, R.M., Semtner, P.J., 1992. Diagnostic esterases and insecticide resistance in the tobacco aphid, Myzus nicotianae Blackman (Homoptera: Aphididae). Pestic. Biochem. Physiol. 43, 123–133.

Abdel-Aal, Y.A.I., Soderlund, D.M., 1980. Pyrethroid hydrolysing esterases in southern armyworm larvae: tissue distribution, kinetic properties and selective inhibtion. Pestic. Biochem. Physiol. 14, 282–289.

Abdel-Aal, Y.A.I., Wolff, M.A., Roe, R.M., Lampert, E.P., 1990. Aphid carboxylesterases: biochemical aspects and importance in the diagnosis of insecticide resistance. Pestic. Biochem. Physiol. 38, 255–266.

Adams, M.D., Celniker, S.E., Holt, R.A., Evans, C.A., Gocayne, J.D., et al., 2000. The genome sequence of Drosophila melanogaster. Science 287, 2185–2195.

Aldridge, W.N., Reiner, E., 1972. Enzyme Inhibitors as Substrates: Interactions of Esterases with Esters of Organophosphorus and Carbamic Acids. North-Holland Publishing, Amsterdam.

Anazawa, Y., Tomita, T., Aiki, Y., Kozaki, T., Kono, Y., 2003. Sequence of a cDNA encoding acetylcholinesterase from susceptible and resistant two-spotted spider mite, Tetranychus urticae. Insect Biochem. Mol. Biol. 33, 509–514.

Anber, H.A.I., Oppenoorth, F.J., 1989. A mutant esterase degrading organophosphates in a resistant strain of the predacious mite Amblyseius potentillae (Garman). Pestic. Biochem. Physiol. 33, 283–297.

Anber, H.A.I., Overmeer, W.P.J., 1988. Resistance to organophosphates and carbamates in the predacious mite Amblyseius potentillae (Garman) due to insensitive acetylcholinesterase. Pestic. Biochem. Physiol. 31, 91–98.

Anderson, P.R., Oakeshott, J.G., 1984. Parallel geographic patterns of allozyme variation in two sibling Drosophila species. Nature 308, 729–731.

Anspaugh, D.D., Kennedy, G.G., Roe, R.M., 1995. Purification and characterization of a resistance-associated esterase from the Colorado potato beetle, Leptinotarsa decemlineata (Say). Pestic. Biochem. Physiol. 53, 84–96.

Arbeitman, M.N., Furlong, E.E.M., Imam, F., Johnson, E., Null, B.H., et al., 2002. Gene expression during the life cycle of Drosophila melanogaster. Science 297, 2270–2275.

Argentine, J.A., Lee, S.H., Sos, M.A., Barry, S.R., Clark, J.M., 1995. Permethrin resistance in a near isogenic stain of Colorado potato beetle. Pestic. Biochem. Physiol. 53, 97–115.

Armstrong, K.F., Suckling, D.M., 1988. Investigations into the biochemical basis of azinphosmethyl resistance in the light brown apple moth, Epiphyas postvittana (Lepidoptera, Tortricidae). Pestic. Biochem. Physiol. 32, 62–73.

Arnason, E., Chambers, G.K., 1987. Macromolecular interaction and the electrophoretic mobility of esterase-5 from Drosophila pseudoobscura. Biochem. Genet. 25, 287–307.

Arpigny, J.L., Jaeger, K.E., 1999. Bacterial lipolytic enzymes: classification and properties. Biochem. J. 343, 177–183.

Arrese, E.L., Canavoso, L.E., Jouni, Z.E., Pennington, J.E., Tsuchida, K., et al., 2001. Lipid storage and mobilization in insects: current status and future directions. Insect Biochem. Mol. Biol. 31, 7–17.

Auld, V.J., Fetter, R.D., Broadie, K., Goodman, C.S., 1995. Gliotactin, a novel transmembrane protein on peripheral glia, is required to form the blood–nerve barrier in Drosophila. Cell 81, 757–767.

Ayad, H., Georghiou, G.P., 1975. Resistance to organophosphates and carbamates in Anopheles albimanus based on reduced sensitivity of acetylcholinesterase. J. Econ. Entomol. 68, 295–297.

Bachurin, S.O., Shevtsova, E.P., Kireeva, E.G., Oxenkrug, G.F., Sablin, S.O., 2003. Mitochondria as a target for neurotoxins and neuroprotective agents. Ann. NY Acad. Sci. 993, 334–344.

Baker, F.C., Tsai, L.W., Reuter, C.C., Schooley, D.A., 1987. In vivo fluctuation of JH, JH acid, and ecdysteroid titer, and JH esterase activity, during development of 5th stadium *Manduca sexta*. *Insect Biochem. 17,* 989–996.

Baker, J.E., Fabrick, J.A., Zhu, K.Y., **1998a.** Characterization of esterases in malathion-resistant and susceptible strains of the pteromalid parasitoid *Anisopteromalus calandrae*. *Insect Biochem. Mol. Biol. 28,* 1039–1050.

Baker, J.E., Perez-Mendoza, J., Beeman, R.W., Throne, J.E., **1998b.** Fitness of a malathion-resistant strain of the parasitoid *Anisopteromalus calandrae* (Hymenoptera: Pteromalidae). *J. Econ. Entomol. 91,* 50–55.

Baker, J.E., Weaver, D.K., **1993.** Resistance in field strains of the parasitoid *Anisopteromalus calandrae* (Hymenoptera, Pteromalidae) and its host, *Sitophilus oryzae* (Coleoptera, Curculionidae), to malathion, chlorpyrifos-methyl, and pirimiphos-methyl. *Biol. Control 3,* 233–242.

Balakirev, E.S., Balakirev, E.I., Ayala, F.J., **2002.** Molecular evolution of the *Est-6* gene in *Drosophila melanogaster*: contrasting patterns of DNA variability in adjacent functional regions. *Gene 288,* 167–177.

Balakirev, E.S., Chechetkin, V.R., Lobzin, V.V., Ayala, F.J., **2003.** DNA polymorphism in the β-esterase gene cluster of *Drosophila melanogaster*. *Genetics 164,* 533–544.

Bannai, H., Tamada, Y., Maruyama, O., Nakai, K., Miyano, S., **2002.** Extensive feature detection of N-terminal protein sorting signals. *Bioinformatics 18,* 298–305.

Barber, M.D., Moores, G.D., Tatchell, G.M., Vice, W.E., Denholm, I., **1999.** Insecticide resistance in the currant-lettuce aphid, *Nasonovia ribisnigri* (Hemiptera: Aphididae) in the UK. *Bull. Entomol. Res. 89,* 17–23.

Baxter, G.D., Barker, S.C., **2002.** Analysis of the sequence and expression of a second putative acetylcholinesterase cDNA from organophosphate-susceptible and organophosphate-resistant cattle ticks. *Insect Biochem. Mol. Biol. 32,* 815–820.

Beeman, R.W., Schmidt, B.A., **1982.** Biochemical and genetic aspects of malathion-specific resistance in the Indian meal moth (Lepidoptera: Pyralidae). *J. Econ. Entomol. 75,* 945–949.

Behra, M., Cousin, X., Bertrand, C., Vonesch, J.L., Biellmann, D., *et al.,* **2002.** Acetylcholinesterase is required for neuronal and muscular development in the zebrafish embryo. *Nature Neurosci. 5,* 111–118.

Bellas, T.E., Bartell, R.J., Hill, A., **1983.** Identification of two components of the sex pheromone of the moth, *Epiphyas postvittana* (Lepidoptera, Tortricidae). *J. Chem. Ecol. 9,* 503–512.

Benedict, M.Q., Scott, J.A., Cockburn, A.F., **1994.** High-level expression of the bacterial *opd* gene in *Drosophila melanogaster*: improved inducible insecticide resistance. *Insect Mol. Biol. 3,* 247–252.

Berrada, S., Fournier, D., Cuany, A., Nguyen, T.X., **1994.** Identification of resistance mechanisms in a selected laboratory strain of *Cacopsylla pyri* (Homoptera, Psyllidae): altered acetylcholinesterase and detoxifying oxidases. *Pestic. Biochem. Physiol. 48,* 41–47.

Berticat, C., Boquien, G., Raymond, M., Chevillon, C., **2002.** Insecticide resistance genes induce a mating competition cost in *Culex pipiens* mosquitoes. *Genet. Res. 79,* 41–47.

Berticat, C., Marquine, M., Raymond, M., Chevillon, C., **2001.** Recombination between two amplified esterase alleles in *Culex pipiens*. *J. Hered. 92,* 349–351.

Beyssat-Arnaouty, V., Mouches, C., Georghiou, G.P., Pasteur, N., **1989.** Detection of organophosphate detoxifying esterases by dot-blot immunoassay in *Culex* mosquitoes. *J. Am. Mosquito Control Assoc. 5,* 196–200.

Bisset, J., Rodriguez, M., Soca, A., Pasteur, N., Raymond, M., **1997.** Cross-resistance to pyrethroid and organophosphorus insecticides in the southern house mosquito (Diptera: Culicidae) from Cuba. *J. Med. Entomol. 34,* 244–246.

Blackman, R.L., **1987.** Morphological discrimination of a tobacco-feeding form of *Myzus persicae* (Sulzer) (Hemiptera: Aphididae) and a key to New World *Myzus* (Nectarosiphon) species. *Bull. Entomol. Res. 77,* 713–730.

Blackman, R.L., Spence, J.M., Field, L.M., Devonshire, A.L., **1995.** Chromosomal location of the amplified esterase genes conferring resistance to insecticides in *Myzus persicae* (Homoptera, Aphididae). *Heredity 75,* 297–302.

Blackman, R.L., Spence, J.M., Field, L.M., Devonshire, A.L., **1999.** Variation in the chromosomal distribution of amplified esterase (FE4) genes in Greek field populations of *Myzus persicae* (Sulzer). *Heredity 82,* 180–186.

Bloch, G., Wool, D., **1994.** Methidathion resistance in the sweet potato whitefly (Aleyrodidae, Homoptera) in Israel: selection, heritability, and correlated changes of esterase activity. *J. Econ. Entomol. 87,* 1147–1156.

Bolliger, M.F., Frei, K., Winterhalter, K.H., Gloor, S.M., **2001.** Identification of a novel neuroligin in humans which binds to PsD-95 and has a widespread expression. *Biochem. J. 356,* 581–588.

Bonning, B.C., Hammock, B.D., **1996.** Development of recombinant baculoviruses for insect control. *Annu. Rev. Entomol. 41,* 191–210.

Borrás, T., Morozova, T.V., Heinsohn, S.L., Lyman, R.F., Mackay, T.F.C., *et al.,* **2003.** Transcription profiling in *Drosophila* eyes that overexpress the human glaucoma-associated trabecular meshwork-inducible glucocorticoid response protein/myocilin (TIGR/MYOC). *Genetics 163,* 637–645.

Botti, S., Felder, C., Sussman, J., Silman, I., **1998.** Electrotactins: a class of adhesion proteins with conserved electrostatic and structural motifs. *Protein Eng. 11,* 415–420.

Boublik, Y., Saint-Aguet, P., Lougarre, A., Arnaud, M., Villatte, F., *et al.,* **2002.** Acetylcholinesterase engineering for detection of insecticide residues. *Protein Eng. 15,* 43–50.

Bourguet, D., Fonseca, D., Vourch, G., Dubois, M.P., Chandre, F., *et al.,* **1998.** The acetylcholinesterase gene *Ace*: a diagnostic marker for the pipiens and quinquefasciatus forms of the *Culex pipiens* complex. *J. Am. Mosquito Control Assoc.* 14, 390–396.

Bourne, P.E., Shindyalov, I.N., 1998. A database of pairwise aligned 3-D structures for the acetylcholinesterases, lipases and other homologous proteins. In: Doctor, B.P., Gentry, M.K., Taylor, P., Quinn, D.M., Rotundo, R.L. (Eds.), Structure and Function of Cholinesterases and Related Proteins. Plenum, New York, pp. 457–462.

Brady, J.P., Richmond, R.C., 1990. Molecular analysis of evolutionary changes in the expression of *Drosophila* esterases. *Proc. Natl Acad. Sci. USA* 87, 8217–8221.

Brady, J.P., Richmond, R.C., 1992. An evolutionary model for the duplication and divergence of esterase genes in *Drosophila*. *J. Mol. Evol.* 34, 506–521.

Brady, J.P., Richmond, R.C., Oakeshott, J.G., 1990. Cloning of the esterase-5 locus from *Drosophila pseudoobscura* and comparison with its homologue in *D. melanogaster*. *Mol. Biol. Evol.* 7, 525–546.

Brogdon, W.G., Barber, A.M., 1990. Fenitrothion-deltamethrin cross-resistance conferred by esterases in Guatemalan *Anopheles albimanus*. *Pestic. Biochem. Physiol.* 37, 130–139.

Brown, T.M., Bryson, P.K., 1992. Selective inhibitors of methyl parathion-resistant acetylcholinesterase from *Heliothis virescens*. *Pestic. Biochem. Physiol.* 44, 155–164.

Bush, M.R., Abdel-Aal, Y.A.I., Rock, G.C., 1993. Parathion resistance and esterase activity in codling moth (Lepidoptera: Tortricidae) from North Carolina. *J. Econ. Entomol.* 86, 660–666.

Butenandt, A., Groschel, U., Karlson, P., Zillig, W., 1959. N-acetyl tyramine, its isolation from *Bombyx* cocoons and its chemical and biological properties. *Arch. Biochem. Biophys.* 83, 76–83.

Byrne, F.J., Devonshire, A.L., 1993. Insensitive acetylcholinesterase and esterase polymorphism in susceptible and resistant populations of the tobacco whitefly *Bemisia tabaci* (Genn.). *Pestic. Biochem. Physiol.* 45, 34–42.

Byrne, F.J., Devonshire, A.L., 1996. Biochemical evidence of haplodiploidy in the whitefly *Bemisia tabaci*. *Biochem. Genet.* 34, 93–107.

Byrne, F.J., Devonshire, A.L., 1997. Kinetics of insensitive acetylcholinesterases in organophosphate-resistant tobacco whitefly, *Bemisia tabaci* (Gennadius) (Homoptera: Aleyrodidae). *Pestic. Biochem. Physiol.* 58, 119–124.

Byrne, F.J., Gorman, K.J., Cahill, M., Denholm, I., Devonshire, A.L., 2000. The role of B-type esterases in conferring insecticide resistance in the tobacco whitefly, *Bemisia tabaci* (Genn.). *Pest Mgt Sci.* 56, 867–874.

Callaghan, A., Parker, P.J.A.N., Holloway, G.J., 1998. The use of variance in enzyme activity as an indicator of long-term exposure to toxicant-stressed environments in *Culex pipiens* mosquitoes. *Funct. Ecol.* 12, 436–441.

Campbell, B.E., 2001. The role of esterases in pyrethroid resistance in Australian populations of the cotton bollworm, *Helicoverpa armigera* (Hübner) (Lepidoptera: Noctuidae). Ph.D. thesis, Australian National University, Canberra.

Campbell, P.M., Harcourt, R.L., Crone, E.J., Claudianos, C., Hammock, B.D., et al., 2001. Identification of a juvenile hormone esterase gene by matching its peptide mass fingerprint with a sequence from the *Drosophila* genome project. *Insect Biochem. Mol. Biol.* 31, 513–520.

Campbell, P.M., Newcomb, R.D., Russell, R.J., Oakeshott, J.G., 1998a. Two different amino acid substitutions in the ali-esterase, E3, confer alternative types of organophosphorus insecticide resistance in the sheep blowfly, *Lucilia cuprina*. *Insect Biochem. Mol. Biol.* 28, 139–150.

Campbell, P.M., Oakeshott, J.G., Healy, M.J., 1998b. Purification and kinetic characterisation of juvenile hormone esterase from *Drosophila melanogaster*. *Insect Biochem. Mol. Biol.* 28, 501–515.

Campbell, P.M., Robin, G.C.D., Court, L.N., Dorrian, S.J., Russell, R.J., et al., 2003. Developmental expression and gene/enzyme identifications in the alpha-esterase gene cluster of *Drosophila melanogaster*. *Insect Mol. Biol.* 12, 459–471.

Campbell, P.M., Trott, J.F., Claudianos, C., Smyth, K.A., Russell, R.J., et al., 1997. Biochemistry of esterases associated with organophosphate resistance in *Lucilia cuprina* with comparisons to putative orthologues in other Diptera. *Biochem. Genet.* 35, 17–40.

Campbell, P.M., Yen, J.L., Masoumi, A., Russell, R.J., Batterham, P., et al., 1998c. Cross-resistance patterns among *Lucilia cuprina* (Diptera: Calliphoridae) resistant to organophosphorus insecticides. *J. Econ. Entomol.* 91, 367–375.

Casabé, N., Zerba, E.N., 1981. Esterases of *Triatoma infestans* and its relationship with the metabolism of organophosphorus insecticides. *Comp. Biochem. Physiol. C* 68, 255–258.

Casida, J.E., Gammon, D.W., Glickman, A.H., Lawrence, L.J., 1983. Mechanisms of selective action of pyrethroid insecticides. *Annu. Rev. Pharmacol.* 23, 413–438.

Chang, C.K., Whalon, M.E., 1986. Hydrolysis of permethrin by pyrethroid esterases from resistant and susceptible strains of *Amblyseius fallacis*. *Pestic. Biochem. Physiol.* 25, 446–452.

Chang, C.K., Whalon, M.E., 1987. Substrate specificities and multiple forms of esterases in the brown planthopper. *Pestic. Biochem. Physiol.* 27, 30–35.

Chapman, T., Bangham, J., Vinti, G., Seifried, B., Lung, O., et al., 2003. The sex peptide of *Drosophila melanogaster*: female post-mating responses analyzed by using RNA interference. *Proc. Natl Acad. Sci. USA* 100, 9923–9928.

Charpentier, A., Villatte, F., Fournier, D., 1998. Acetylcholinesterase increase in *Drosophila* as a mechanism of resistance to insecticide. In: Doctor, B.P., Gentry, M.K., Taylor, P., Quinn, D.M., Rotundo, R.L. (Eds.), Structure and Function of Cholinesterases and Related Proteins. Plenum, New York, pp. 503–507.

Chen, W.L., Sun, C.-N., 1994. Purification and characterization of carboxylesterases of a rice brown planthopper,

Nilaparvata lugens Stål. *Insect Biochem. Mol. Biol. 24,* 347–355.

Chen, Z.Z., Newcomb, R., Forbes, E., McKenzie, J., Batterham, P., **2001**. The acetylcholinesterase gene and organophosphorus resistance in the Australian sheep blowfly, *Lucilia cuprina. Insect Biochem. Mol. Biol. 31,* 805–816.

Chevillon, C., Bourguet, D., Rousset, F., Pasteur, N., Raymond, M., **1997**. Pleiotropy of adaptive changes in populations: comparisons among insecticide resistance genes in *Culex pipiens. Genet. Res. 70,* 195–203.

Chiang, S.W., Sun, C.N., **1996**. Purification and characterization of carboxylesterases of a rice green leafhopper *Nephotettix cincticeps* Uhler. *Pestic. Biochem. Physiol. 54,* 181–189.

Chou, K.C., **2001**. Using subsite coupling to predict signal peptides. *Protein Eng. 14,* 75–79.

Clarris, H.J., McKeown, S., Key, B., **2002**. Expression of neurexin ligands, the neuroligins and the neurexophilins, in the developing and adult rodent olfactory bulb. *Int. J. Devel. Biol. 46,* 649–652.

Claudianos, C., Crone, E., Coppin, C., Russell, R., Oakeshott, J., **2001**. A genomics perspective on mutant aliesterases and metabolic resistance to organophosphates. In: Marshall Clark, J., Yamaguchi, I. (Eds.), Agrochemical Resistance: Extent, Mechanism and Detection. American Chemical Society, Washington, DC, pp. 90–101.

Claudianos, C., Russell, R.J., Oakeshott, J.G., **1999**. The same amino acid substitution in orthologous esterases confers organophosphate resistance on the house fly and a blowfly. *Insect Biochem. Mol. Biol. 29,* 675–686.

Clements, K.M., Sorenson, C.E., Wiegmann, B.M., Neese, P.A., Roe, R.M., **2000**. Genetic, biochemical, and behavioral uniformity among populations of *Myzus nicotianae* and *Myzus persicae. Entomol. Exp. Applic. 95,* 269–281.

Coates, P.M., Mestriner, M.A., Hopkinson, D.A., **1975**. A preliminary genetic interpretation of the esterase isozymes of human tissues. *Ann. Hum. Genet. 39,* 1–20.

Collins, F.H., Sakai, R.K., Vernick, K.D., Paskewitz, S., Seeley, D.C., *et al.,* **1986**. Genetic selection of a plasmodium-refractory strain of the malaria vector *Anopheles gambiae. Science 234,* 607–610.

Collins, P.J., Rose, H.A., Wegecsanyi, M., **1992**. Enzyme activity in strains of the sawtoothed grain beetle (Coleoptera, Cucujidae) differentially resistant to fenitrothion, malathion, and chlorpyrifos-methyl. *J. Econ. Entomol. 85,* 1571–1575.

Combes, D., Fedon, Y., Grauso, M., Toutant, J.P., Arpagaus, M., **2000**. Four genes encode acetylcholinesterases in the nematodes *Caenorhabditis elegans* and *Caenorhabditis briggsae*: cDNA sequences, genomic structures, mutations and *in vivo* expression. *J. Mol. Biol. 300,* 727–742.

Combes, D., Fedon, Y., Toutant, J.P., Arpagaus, M., **2003**. Multiple *ace* genes encoding acetylcholinesterases of *Caenorhabditis elegans* have distinct tissue expression. *Eur. J. Neurosci. 18,* 497–512.

Conyers, C.M., Macnicoll, A.D., Price, N.R., **1998**. Purification and characterisation of an esterase involved in resistance to organophosphorus insecticides in the saw-toothed grain beetle, *Oryzaephilus surinamensis* (Coleoptera: Silvanidae). *Insect Biochem. Mol. Biol. 28,* 435–448.

Cooke, P.H., Oakeshott, J.G., **1989**. Amino acid polymorphisms for esterase-6 in *Drosophila melanogaster. Proc. Natl Acad. Sci. USA 86,* 1426–1430.

Costa, H.S., Brown, J.K., **1991**. Variation in biological characteristics and esterase pattern among populations of *Bemisia tabaci* and association of one population with silverleaf symptom induction. *Entomol. Exp. Applic. 61,* 211–219.

Crews-Oyen, A.E., Kumar, V., Collins, F.H., **1993**. Association of two esterase genes, a chromosomal inversion, and susceptibility to *Plasmodium cynomolgi* in the African malaria vector *Anopheles gambiae. Am. J. Trop. Med. Hyg. 49,* 341–347.

Cuany, A., Handani, J., Bergé, J., Fournier, D., Raymond, M., *et al.,* **1993**. Action of esterase B1 on chlorpyrifos in organophosphate-resistant *Culex* mosquitoes. *Pestic. Biochem. Physiol. 45,* 1–6.

Dang, H T., Gunning, R., **2002**. Resistance to *Bacillus thuringiensis* Delta-endotoxin CrylAc in Australian *Helicoverpa armigera* (Lepidoptera: Noctuidae). In: Akhurst, R.J., Beard, C.E., Hughes, P. (Eds.), Biotechnology of *Bacillus thuringiensis* and its Environmental Impact. CSIRO, Canberra, pp. 100–103.

Darboux, I., Barthalay, Y., Piovant, M., Hipeaujacquotte, R., **1996**. The structure–function relationships in *Drosophila* neurotactin show that cholinesterasic domains may have adhesive properties. *EMBO J. 15,* 4835–4843.

Davies, A.G., Game, A.Y., Chen, Z., Williams, T.J., Goodall, S., *et al.,* **1996**. Scalloped wings is the *Lucilia cuprina* Notch homologue and a candidate for the modifier of fitness and asymmetry of diazinon resistance. *Genetics 143,* 1321–1337.

Dawson, G.W., Griffiths, D.C., Pickett, J.A., Woodcock, C.M., **1983**. Decreased response to alarm pheromones by insecticide-resistant aphids. *Naturwissenschaften 70,* 254–255.

de Jersey, J., Nolan, J., Davey, P.A., Riddles, P.W., **1985**. Separation and characterization of the pyrethroid-hydrolyzing esterases of the cattle tick, *Boophilus microplus. Pestic. Biochem. Physiol. 23,* 349–357.

de Kort, C.A.D., Granger, N.A., **1981**. Regulation of the juvenile hormone titer. *Annu. Rev. Entomol. 26,* 1–28.

de Kort, C.A.D., Granger, N.A., **1996**. Regulation of JH titers: the relevance of degradative enzymes and binding proteins. *Arch. Insect Biochem. Physiol. 33,* 1–26.

De Malkenson, N.C., Wood, E.J., Zerba, E.N., **1984**. Isolation and characterization of an esterase of *Triatoma infestans* with a critical role in the degradation of organophosphorus esters. *Insect Biochem. 14,* 481–486.

De Silva, D., Hemingway, J., Ranson, H., Vaughan, A., 1997. Resistance to insecticides in insect vectors of disease: Estα3, a novel amplified esterase associated with amplified Estβ1 from insecticide resistant strains of the mosquito *Culex quinquesfasciatus*. *Exp. Parasitol.* 87, 253–259.

Dean, C., Scholl, F.G., Choih, J., Demaria, S., Berger, J., et al., 2003. Neurexin mediates the assembly of presynaptic terminals. *Nat. Neurosci.* 6, 708–716.

Delorme, R., Fournier, D., Chaufaux, J., Cuany, A., Bride, J.M., et al., 1988. Esterase metabolism and reduced penetration are causes of resistance to deltamethrin in *Spodoptera exigua* HUB (Noctuidae: Lepidoptera). *Pestic. Biochem. Physiol.* 32, 240–246.

Devonshire, A.L., 1977. The properties of a carboxylesterase from the peach-potato aphid, *Myzus persicae* (Sulz.), and its role in conferring insecticide resistance. *Biochem. J.* 167, 675–683.

Devonshire, A.L., 1989. The role of electrophoresis in the biochemical detection of insecticide resistance. In: Loxdale, H.D., Hollander, J.D. (Eds.), Electrophoretic Studies on Agricultural Pests. Clarendon Press, Oxford, pp. 363–377.

Devonshire, A.L., Field, L.M., Foster, S.P., Moores, G.D., Williamson, M.S., et al., 1998. The evolution of insecticide resistance in the peach-potato aphid, *Myzus persicae*. *Phil. Trans. Roy. Soc. London B* 353, 1677–1684.

Devonshire, A.L., Heidari, R., Bell, K.L., Campbell, P.M., Campbell, B.E., et al., 2003. Kinetic efficiency of mutant carboxylesterases implicated in organophosphate insecticide resistance. *Pestic. Biochem. Physiol.* 76, 1–13.

Devonshire, A.L., Moores, G.D., 1982. A carboxylesterase with broad substrate specificity causes organophosphorous, carbamate and pyrethroid resistance in peach potato aphids (*Myzus persicae*). *Pestic. Biochem. Physiol.* 18, 235–246.

Devonshire, A.L., Moores, G.D., 1989. Detoxication of insecticides by esterases from *Myzus persicae*: is hydrolysis important? In: Reiner, E., Aldridge, W.N., Hoskin, F.C.G. (Eds.), Esterases Hydrolysing Organophosphorus Compounds. Ellis Horwood, Chichester, pp. 180–192.

Devonshire, A.L., Sawicki, R.M., 1979. Insecticide resistant *Myzus persicae* as an example of evolution by gene duplication. *Nature* 280, 140–141.

Devonshire, A.L., Searle, L.M., Moores, G.D., 1986. Quantitative and qualitative variation in the messenger-RNA for carboxylesterases in insecticide-susceptible and resistant *Myzus persicae* (Sulz). *Insect Biochem.* 16, 659–665.

Dimopoulos, G., Richman, A., Muller, H.M., Kafatos, F.C., 1997. Molecular immune responses of the mosquito *Anopheles gambiae* to bacteria and malaria parasites. *Proc. Natl Acad. Sci. USA.* 94, 11508–11513.

Doctor, B.P., Taylor, P., Quinn, D.M., Rotundo, R.L., Gentry, M.K. (Eds.), 1998. Structure and Function of Cholinesterases and Related Proteins. Plenum, New York.

Dowd, P.F., Gagne, C.C., Sparks, T.C., 1987. Enhanced pyrethroid hydrolysis in pyrethroid-resistant larvae of the tobacco budworm, *Heliothis virescens* (F). *Pestic. Biochem. Physiol.* 28, 9–16.

Duchaud, E., Rusniok, C., Frangeul, L., Buchrieser, C., Givaudan, A., et al., 2003. The genome sequence of the entomopathogenic bacterium *Photorhabdus luminescens*. *Nat. Biotechnol.* 21, 1307–1313.

Dumancic, M.M., Oakeshott, J.G., Russell, R.J., Healy, M.J., 1997. Characterization of the *EstP* protein in *Drosophila melanogaster* and its conservation in Drosophilids. *Biochem. Genet.* 35, 251–271.

Dumas, D.P., Wild, J.R., Raushel, F.M., 1990. Expression of *Pseudomonas* phosphotriesterase activity in the fall armyworm confers resistance to insecticides. *Experientia* 46, 729–731.

Dyte, C.E., Rowlands, D.G., 1968. The metabolism and synergism of malathion in resistant and susceptible strains of *Tribolium castaneum* (Herbst) (Coleoptera, Tenebrionidae). *J. Stored Prod. Res.* 4, 157–173.

East, P., Graham, A., Whitington, G., 1990. Molecular isolation and preliminary characterisation of a duplicated esterase locus in *Drosophila buzzatii*. In: Barker, J.S.F., Starmer, W.T., MacIntyre, R.J. (Eds.), Ecological and Evolutionary Genetics of *Drosophila*. Plenum, New York, pp. 389–406.

El-Khatib, Z.I., Georghiou, G.P., 1985. Geographic variation of resistance to organophosphates, propoxur and DDT in the southern house mosquito, *Culex quinquefasciatus*, in California. *J. Am. Mosquito Control Assoc.* 1, 279–283.

El Saidy, M.F., Auda, M., Degheele, D., 1989. Detoxification mechanisms of diflubenzuron and teflubenzuron in the larvae of *Spodoptera littoralis* (Boisd). *Pestic. Biochem. Physiol.* 35, 211–222.

Elliott, M., 1977. Synthetic pyrethroids. In: Elliott, M. (Ed.), Synthetic Pyrethroids. American Chemical Society, Washington, DC, pp. 1–28.

Emanuelsson, O., Nielsen, H., Brunak, S., Von Heijne, G., 2000. Predicting subcellular localization of proteins based on their N-terminal amino acid sequence. *J. Mol. Biol.* 300, 1005–1016.

Enikolopov, G.N., Khechumyan, R.K., Kuzin, B.A., Korochkin, L.I., Georgiev, G.P., 1989. Molecular organisation of a family of genes specifying *Drosophila virilis* esterases: tissue specific expression of the gene for *Drosophila virilis* esterase S. *Soviet Genet.* 25, 265–273.

Estrada-Mondaca, S., Fournier, D., 1998. Stabilization of recombinant *Drosophila* acetylcholinesterase. *Protein Express. Purific.* 12, 166–172.

Eto, M., 1974. Organophosphorus Pesticides: Organic and Biological Chemistry. CRC Press, Cleveland, OH.

Ferrari, J.A., Morse, J.G., Georghiou, G.P., Sun, Y.Q., 1993. Elevated esterase-activity and acetylcholinesterase insensitivity in citrus thrips (Thysanoptera,

Thripidae) populations from the San Joaquin Valley of California. *J. Econ. Entomol. 86*, 1645–1650.

ffrench-Constant, R.H., Devonshire, A.L., **1988**. Monitoring frequencies of insecticide resistance in *Myzus persicae* (Sulzer) (Hemiptera, Aphididae) in England during 1985–86 by immunoassay. *Bull. Entomol. Res. 78*, 163–171.

ffrench-Constant, R.H., Pittendrigh, B., Vaughan, A., Anthony, N., **1998**. Why are there so few resistance-associated mutations in insecticide target genes? *Phil. Trans. Roy. Soc. London B 353*, 1685–1693.

Field, L.M., **2000**. Methylation and expression of amplified esterase genes in the aphid *Myzus persicae* (Sulzer). *Biochem. J. 349*, 863–868.

Field, L.M., Anderson, A.P., Denholm, I., Foster, S.P., Harling, Z.K., et al., **1997**. Use of biochemical and DNA diagnostics for characterising multiple mechanisms of insecticide resistance in the peach-potato aphid, *Myzus persicae* (Sulzer). *Pestic. Sci. 51*, 283–289.

Field, L.M., Blackman, R.L., **2003**. Insecticide resistance in the aphid *Myzus persicae* (Sulzer): chromosome location and epigenetic effects on esterase gene expression in clonal lineages. *Biol. J. Linnean Soc. 79*, 107–113.

Field, L.M., Blackman, R.L., Tyler-Smith, C., Devonshire, A.L., **1999**. Relationship between amount of esterase and gene copy number in insecticide-resistant *Myzus persicae* (Sulzer). *Biochem. J. 339*, 737–742.

Field, L.M., Devonshire, A.L., **1998**. Evidence that the E4 and FE4 esterase genes responsible for insecticide resistance in the aphid *Myzus persicae* (Sulzer) are part of a gene family. *Biochem. J. 330*, 169–173.

Field, L.M., Devonshire, A.L., ffrench-Constant, R.H., Forde, B.G., **1989**. Changes in DNA methylation are associated with loss of insecticide resistance in the peach-potato aphid *Myzus persicae* (Sulz.). *FEBS Lett. 243*, 323–327.

Field, L.M., Devonshire, A.L., Forde, B.G., **1988**. Molecular evidence that insecticide resistance in peach-potato aphids (*Myzus persicae* Sulz.) results from amplification of an esterase gene. *Biochem. J. 251*, 309–312.

Field, L.M., Javed, N., Stribley, M.F., Devonshire, A.L., **1994**. The peach potato aphid *Myzus persicae* and the tobacco aphid *Myzus nicotianae* have the same esterase-based mechanisms of insecticide resistance. *Insect Mol. Biol. 3*, 143–148.

Field, L.M., Williamson, M.S., Moores, G.D., Devonshire, A.L., **1993**. Cloning and analysis of the esterase genes conferring insecticide resistance in the peach-potato aphid, *Myzus persicae* (Sulzer). *Biochem. J. 294*, 569–574.

Fojan, P., Jonson, P.H., Petersen, M.T.N., Petersen, S.B., **2000**. What distinguishes an esterase from a lipase: a novel structural approach. *Biochimie 82*, 1033–1041.

Foster, S.P., Harrington, R., Devonshire, A.L., Denholm, I., Devine, G.J.K.M.G., et al., **1996**. Comparative survival of insecticide-susceptible and resistant peach-potato aphids *Myzus persicae* (Sulzer) (Hemiptera: Aphididae) in low temperature field trials. *Bull. Entomol. Res. 86*, 17–27.

Foster, S.P., Kift, N.B., Baverstock, J., Sime, S., Reynolds, K., et al., **2003**. Association of MACE-based insecticide resistance in *Myzus persicae* with reproductive rate, response to alarm pheromone and vulnerability to attack by *Aphidius colemani*. *Pest Mgt Sci. 59*, 1169–1178.

Fournier, D., Bergé, J.-B., Cardoso de Almeida, M.-L., Bordier, C., **1988**. Acetylcholinesterases from *Musca domestica* and *Drosophila melanogaster* brain are linked to membranes by a glycophospholipid anchor sensitive to an endogenous phospholipase. *J. Neurochem. 50*, 1158–1163.

Fournier, D., Bride, J.-M., Hoffmann, F., Karch, F., **1992**. Acetylcholinesterase: two types of modifications confer resistance to insecticide. *J. Biol. Chem. 267*, 14270–14274.

Fournier, D., Bride, J.-M., Mouchès, C., Raymond, M., Magnin, M., et al., **1987**. Biochemical characterization of the esterases A1 and B1 associated with organophosphate resistance in the *Culex pipiens* L. complex. *Pestic. Biochem. Physiol. 27*, 211–217.

Fournier, D., Mutero, A., **1994**. Modification of acetylcholinesterase as a mechanism of resistance to insecticides. *Comp. Biochem. Physiol. C 108*, 19–31.

Fremion, F., Darboux, I., Diano, M., Hipeau-Jacquotte, R., Seeger, M.A., et al., **2000**. Amalgam is a ligand for the transmembrane receptor neurotactin and is required for neurotactin-mediated cell adhesion and axon fasciculation in *Drosophila*. *EMBO J. 19*, 4463–4472.

Furk, C., Powell, D.F., Heyd, S., **1980**. Primicarb resistance in the melon and cotton aphid, *Aphis gossypii* Glover. *Plant Pathol. 29*, 191.

Gabay, J.E., Almeida, R.P., **1993**. Antibiotic peptides and serine protease homologs in human polymorphonuclear leukocytes: defensins and azurocidin. *Curr. Opin. Immunol. 5*, 97–102.

Game, A.Y., Oakeshott, J.G., **1990**. Associations between restriction site polymorphism and enzyme activity variation for esterase 6 in *Drosophila melanogaster*. *Genetics 126*, 1021–1031.

Gazave, L., Chevillon, C., Lenormand, T., Marquine, M., Raymond, M., **2001**. Dissecting the cost of insecticide resistance genes during the overwintering period of the mosquito *Culex pipiens*. *Heredity 87*, 441–448.

Genova, J.L., Fehon, R.G., **2003**. Neuroglian, gliotactin, and the Na^+/K^+ ATPase are essential for septate junction function in *Drosophila*. *J. Cell Biol. 161*, 979–989.

Georghiou, G.P., Pasteur, N., Hawley, M.K., **1980**. Linkage relationships between organophosphate resistance and a highly active esterase-B in *Culex pipiens quinquefaciatus* Say from California. *J. Econ. Entomol. 73*, 301–305.

Gibney, G., Camp, S., Dionne, M., Macpheequigley, K., Taylor, P., **1990**. Mutagenesis of essential functional residues in acetylcholinesterase. *Proc. Natl Acad. Sci. USA 87*, 7546–7550.

Gilbert, L.I., Granger, N.A., Roe, R.M., **2000**. The juvenile hormones: historical facts and speculations on future research directions. *Insect Biochem. Mol. Biol.* 30, 617–644.

Gilbert, M., Smith, J., Roskams, A.J., Auld, V.J., **2001**. Neuroligin 3 is a vertebrate gliotactin expressed in the olfactory ensheathing glia, a growth-promoting class of macroglia. *Glia 34*, 151–164.

Gnagey, A.L., Forte, M., Rosenberry, T.L., **1987**. Isolation and characterization of acetylcholinesterase from *Drosophila. J. Biol. Chem.* 262, 13290–13298.

Goh, D.K.S., Anspaugh, D.D., Motoyama, N., Rock, G.C., Roe, R.M., **1995**. Isolation and characterization of an insecticide-resistance-associated esterase in the tobacco budworm *Heliothis virescens* (F). *Pestic. Biochem. Physiol.* 51, 192–204.

Gomez, G.A., Hasson, E., **2003**. Transpecific polymorphisms in an inversion linked esterase locus in *Drosophila buzzatii. J. Mol. Evol.* 20, 410–423.

Grifman, M., Galyam, N., Seidman, S., Soreq, H., **1998**. Functional redundancy of acetylcholinesterase and neuroligin in mammalian neuritogenesis. *Proc. Natl Acad. Sci. USA 95*, 13935–13940.

Grisaru, D., Sternfeld, M., Eldor, A., Glick, D., Soreq, H., **1999**. Structural roles of acetylcholinesterase variants in biology and pathology. *Eur. J. Biochem.* 264, 672–686.

Guerrero, F.D., **2000**. Cloning of a horn fly cDNA, *HiαE7*, encoding an esterase whose transcript concentration is elevated in diazinon-resistant flies. *Insect Biochem. Mol. Biol.* 30, 1107–1115.

Guillemaud, T., Lenormand, T., Bourguet, D., Chevillon, C., Pasteur, N., *et al.*, **1998**. Evolution of resistance in *Culex pipiens*: allele replacement and changing environment. *Evolution 52*, 443–453.

Guillemaud, T., Makate, N., Raymond, M., Hirst, B., Callaghan, A., **1997**. Esterase gene amplification in *Culex pipiens. Insect Mol. Biol.* 6, 319–327.

Guillemaud, T., Rooker, S., Pasteur, N., Raymond, M., **1996**. Testing the unique amplification event and the worldwide migration hypothesis of insecticide resistance genes with sequence data. *Heredity 77*, 535–543.

Gunning, R.V., Moores, G.D., Devonshire, A.L., **1996a**. Insensitive acetylcholinesterase and resistance to thiodicarb in Australian *Helicoverpa armigera* Hübner (Lepidoptera: Noctuidae). *Pestic. Biochem. Physiol.* 55, 21–28.

Gunning, R.V., Moores, G.D., Devonshire, A.L., **1996b**. Esterases and esfenvalerate resistance in Australian *Helicoverpa armigera* (Hübner) (Lepidoptera: Noctuidae). *Pestic. Biochem. Physiol.* 54, 12–23.

Gunning, R.V., Moores, G.D., Devonshire, A.L., **1997**. Esterases and fenvalerate resistance in a field population of *Helicoverpa punctigera* (Lepidoptera: Noctuidae) in Australia. *Pestic. Biochem. Physiol.* 58, 155–162.

Gunning, R.V., Moores, G.D., Devonshire, A.L., **1998**. Inhibition of pyrethroid resistance related esterases by piperonyl butoxide in Australian *Helicoverpa armigera* (Lepidoptera: Noctuidae) and *Aphis gossypii* (Hemiptera: Aphididae). In: Jones, D.G. (Ed.), Piperonyl Butoxide: The Insecticide Synergist. Academic Press, San Diego, CA.

Gunning, R.V., Moores, G.D., Devonshire, A.L., **1999**. Esterase inhibitors synergise the toxicity of pyrethroids in Australian *Helicoverpa armigera* (Hübner) (Lepidoptera: Noctuidae). *Pestic. Biochem. Physiol.* 63, 50–62.

Gupta, A.P. (Ed.), **1990**. Morphogenetic Hormones of Arthropods: Recent Advances in Comparative Arthropod Morphology, Physiology, and Development. Part 1: Discoveries, Syntheses, Metabolism, Evolution, Modes of Action, and Techniques. Rutgers University Press, New Brunswick, NJ.

Hajos, J.P., Vermunt, A.M.W., Zuidema, D., Kuclsar, P., Varjas, L., *et al.*, **1999**. Dissecting insect development: baculovirus-mediated gene silencing in insects. *Insect Mol. Biol. 8*, 539–544.

Hammock, B.D., **1985**. Regulation of juvenile hormone titre: degradation. In: Kerkut, G.A., Gilbert, L.I. (Eds.), Comprehensive Insect Physiology, Biochemistry and Pharmacology, vol. 7. Pergamon, Oxford, pp. 431–472.

Hammock, B.D., Bonning, B.C., Possee, R.D., Hanzlik, T.N., Maeda, S., **1990**. Expression and effects of the juvenile-hormone esterase in a baculovirus vector. *Nature 344*, 458–461.

Hammock, B.D., McCutchen, B.F., Beetham, J., Choudary, P.V., Fowler, E., *et al.*, **1993**. Development of recombinant viral insecticides by expression of an insect-specific toxin and insect-specific enzyme in nuclear polyhedrosis viruses. *Arch. Insect Biochem. Physiol. 22*, 315–344.

Harel, M., Kryger, G., Rosenberry, T.L., Mallender, W.D., Lewis, T., *et al.*, **2000**. Three-dimensional structures of *Drosophila melanogaster* acetylcholinesterase and of its complexes with two potent inhibitors. *Protein Sci. 9*, 1063–1072.

Harold, J.A., Ottea, J.A., **1997**. Toxicological significance of enzyme activities in profenofos-resistant tobacco budworms, *Heliothis virescens* (F.). *Pestic. Biochem. Physiol. 58*, 23–33.

Harold, J.A., Ottea, J.A., **2000**. Characterization of esterases associated with profenofos resistance in the tobacco budworm, *Heliothis virescens* (F.). *Arch. Insect Biochem. Physiol. 45*, 47–59.

Hasui, H., Ozaki, K., **1984**. Electrophoretic esterase patterns in the brown planthopper, *Nilaparvata lugens* Stål (Hemiptera: Delphacidae) which developed resistance to insecticides. *Appl. Entomol. Zool. 19*, 52–58.

Hawkes, N.J., Hemingway, J., **2002**. Analysis of the promoters for the beta-esterase genes associated with insecticide resistance in the mosquito *Culex quinquefasciatus. Biochim. Biophys. Acta 1574*, 51–62.

Healy, M.J., Dumancic, M.M., Oakeshott, J.G., **1991**. Biochemical and physiological studies of soluble esterases from *Drosophila melanogaster. Biochem. Genet. 29*, 365–388.

Heidari, R., Devonshire, A.L., Campbell, B.E., Bell, K.L., Dorrian, S.J., *et al.*, **2004a**. Hydrolysis of organophosphorus insecticides by *in vitro* modified carboxylesterase

E3 from *Lucilia cuprina*. *Insect Biochem. Mol. Biol. 34,* 353–363.

Hemingway, J., **1982**. The biochemical nature of malathion resistance in *Anopheles stephensi* from Pakistan. *Pestic. Biochem. Physiol. 17,* 149–155.

Hemingway, J., **1983**. The genetics of malathion resistance in *Anopheles stephensi* from Pakistan. *Trans. Roy. Soc. Trop. Med. Hyg. 77,* 106–108.

Hemingway, J., **1985**. Malathion carboxylesterase enzymes in *Anopheles arabiensis* from Sudan. *Pestic. Biochem. Physiol. 23,* 309–313.

Hemingway, J., Callaghan, A., Amin, A.M., **1990**. Mechanisms of organophosphate and carbamate resistance in *Culex quinquefasciatus* from Saudi Arabia. *Med. Vet. Entomol. 4,* 275–282.

Hemingway, J., Callaghan, A., Kurtak, D.C., **1989**. Temephos resistance in *Simulium damnosum* Theobald (Diptera, Simuliidae): a comparative study between larvae and adults of the forest and savanna strains of this species complex. *Bull. Entomol. Res. 79,* 659–669.

Hemingway, J., Coleman, M., Paton, M., McCarroll, L., Vaughan, A., *et al.,* **2000**. Aldehyde oxidase is coamplified with the world's most common *Culex* mosquito insecticide resistance-associated esterases. *Insect Mol. Biol. 9,* 93–99.

Hemingway, J., Georghiou, G., **1983**. Studies on the acetylcholinesterase of *Anopheles albimanus* resistant and susceptible to organophosphate and carbamate insecticides. *Pestic. Biochem. Physiol. 19,* 167–171.

Hemingway, J., Hawkes, N., Prapanthadara, L., Jayawardenal, K.G., Ranson, H., **1998**. The role of gene splicing, gene amplification and regulation in mosquito insecticide resistance. *Phil. Trans. Roy. Soc. London B 353,* 1695–1699.

Hemingway, J., Karunaratne, S.H., **1998**. Mosquito carboxylesterases: a review of the molecular biology and biochemistry of a major insecticide resistance mechanism. *Med. Vet. Entomol. 12,* 1–12.

Hemingway, J., Malcolm, C.A., Kissoon, K.E., Boddington, R.G., Curtis, C.F., *et al.,* **1985**. The biochemistry of insecticide resistance in *Anopheles sacharovi*: comparative studies with a range of insecticide susceptible and resistant *Anopheles* and *Culex* species. *Pestic. Biochem. Physiol. 24,* 68–76.

Herath, P.R.J., Hemingway, J., Weerasinghe, I.S., Jayawardena, K.G.I., **1987**. The detection and characterization of malathion resistance in field populations of *Anopheles culicifacies* B in Sri Lanka. *Pestic. Biochem. Physiol. 29,* 157–162.

Hick, C.A., Field, L.M., Devonshire, A.L., **1996**. Changes in the methylation of amplified esterase DNA during loss and reselection of insecticide resistance in peach-potato aphids, *Myzus persicae*. *Insect Biochem. Mol. Biol. 26,* 41–47.

Hinton, A.C., Hammock, B.D., **2003**. Juvenile hormone esterase (JHE) from *Tenebrio molitor*: full-length cDNA sequence, *in vitro* expression, and characterization of the recombinant protein. *Insect Biochem. Mol. Biol. 33,* 477–487.

Hoffmann, F., Fournier, D., Spierer, P., **1992**. Minigene rescues acetylcholinesterase lethal mutations in *Drosophila melanogaster*. *J. Mol. Biol. 223,* 17–22.

Holloway, G.J., Crocker, H.J., Callaghan, A., **1997**. The effects of novel and stressful environments on trait distribution. *Funct. Ecol. 11,* 579–584.

Holmes, R.S., Masters, C.J., **1967**. The developmental multiplicity and isoenzyme status of cavian esterases. *Biochim. Biophys. Acta 132,* 379–399.

Horne, I., Sutherland, T.D., Harcourt, R.L., Russell, R.J., Oakeshott, J.G., **2002a**. Identification of an *opd* (organophosphate degradation) gene in an agrobacterium isolate. *Appl. Envir. Microbiol. 68,* 3371–3376.

Horne, I., Sutherland, T.D., Oakeshott, J.G., Russell, R.J., **2002b**. Cloning and expression of the phosphotriesterase gene *hocA* from *Pseudomonas monteilii* C11. *Microbiology 148,* 2687–2695.

Hortsch, M., Patel, N.H., Bieber, A.J., Traquina, Z.R., Goodman, C.S., **1990**. *Drosophila* neurotactin, a surface glycoprotein with homology to serine esterases, is dynamically expressed during embryogenesis. *Development 110,* 1327–1340.

Hou, X., Maser, R.L., Magenheimer, B.S., Calvet, J.P., **1996**. A mouse kidney- and liver-expressed cDNA having homology with a prokaryotic parathion hydrolase (phosphotriesterase)-encoding gene: abnormal expression in injured and polycystic kidneys. *Gene 168,* 157–163.

Huang, Q., Deveraux, Q.L., Maeda, S., Salvesen, G.S., Stennicke, H.R., *et al.,* **2000**. Evolutionary conservation of apoptosis mechanisms: lepidopteran and baculoviral inhibitor of apoptosis proteins are inhibitors of mammalian caspase-9. *Proc. Natl Acad. Sci. USA 97,* 1427–1432.

Hughes, P.B., Devonshire, A.L., **1982**. The biochemical basis of resistance to organophosphorus insecticides in the sheep blowfly, *Lucilia cuprina*. *Pestic. Biochem. Physiol. 18,* 289–297.

Hughes, P.B., Green, P.E., Reichmann, K.G., **1984**. Specific resistance to malathion in laboratory and field populations of the Australian sheep blowfly, *Lucilia cuprina* (Diptera: Calliphoridae). *J. Econ. Entomol. 77,* 1400–1404.

Hughes, P.B., Raftos, D.A., **1985**. Genetics of an esterase associated with resistance to organophosphorus insecticides in the sheep blowfly, *Lucilia cuprina* (Wiedemann) (Diptera: Calliphoridae). *Bull. Entomol. Res. 75,* 535–544.

Ichtchenko, K., Nguyen, T., Sudhof, T.C., **1996**. Structures, alternative splicing, and neurexin binding of multiple neuroligins. *J. Biol. Chem. 271,* 2676–2682.

Ishaaya, I., Casida, J.E., **1980**. Properties and toxicological significance of esterases hydrolyzing permethrin and cypermethrin in *Trichoplusia ni* larval gut and integument. *Pestic. Biochem. Physiol. 14,* 178–184.

Ishaaya, I., Casida, J.E., **1981**. Pyrethroid esterase(s) may contribute to natural pyrethroid tolerance of larvae of the common green lacewing. *Envir. Entomol. 10,* 681–684.

Ishaaya, I., Degheele, D., **1988**. Properties and toxicological significance of diflubenzuron hydrolase activity in *Spodoptera littoralis* larvae. *Pestic. Biochem. Physiol.* 32, 180–187.

Ishaaya, I., Klein, M., **1990**. Response of susceptible laboratory and resistant field strains of *Spodoptera littoralis* (Lepidoptera, Noctuidae) to teflubenzuron. *J. Econ. Entomol.* 83, 59–62.

Ishida, Y., Leal, W.S., **2002**. Cloning of putative odorant-degrading enzyme and integumental esterase cDNAs from the wild silkmoth, *Antheraea polyphemus*. *Insect Biochem. Mol. Biol.* 32, 1775–1780.

Ismail, S.M., Goin, C., Muthumani, K., Kim, M., Dahm, K.H., *et al.*, **2000**. Juvenile hormone acid and ecdysteroid together induce competence for metamorphosis of the Verson's gland in *Manduca sexta*. *J. Insect Physiol.* 46, 59–68.

Ismail, S.M., Satyanarayana, K., Bradfield, J.Y., Dahm, K.H., Bhaskaran, G., **1998**. Juvenile hormone acid: evidence for a hormonal function in induction of vitellogenin in larvae of *Manduca sexta*. *Arch. Insect Biochem. Physiol.* 37, 305–314.

Iwata, T., Hama, H., **1972**. Insensitivity of cholinesterase in *Nephotettix cincticeps* resistant to carbamate and organophosphorus insecticides. *J. Econ. Entomol.* 65, 643–644.

Jamroz, R.C., Guerrero, F.D., Pruett, J.H., Oehler, D.D., Miller, R.J., **2000**. Molecular and biochemical survey of acaricide resistance mechanisms in larvae from Mexican strains of the southern cattle tick, *Boophilus microplus*. *J. Insect Physiol.* 46, 685–695.

Jao, L.T., Casida, J.E., **1974**. Insect pyrethroid-hydrolyzing esterases. *Pestic. Biochem. Physiol.* 4, 465–472.

Jones, G., Venkataraman, V., Ridley, B., O'Mahony, P., Turner, H., **1994**. Structure, expression and gene sequence of a juvenile hormone esterase-related protein from metamorphosing larvae of *Trichoplusia ni*. *Biochem. J.* 302, 827–835.

Jung, S., Aliberti, J., Graemmel, P., Sunshine, M.J., Kreutzberg, G.W., *et al.*, **2000**. Analysis of fractalkine receptor CC(3)CR1 function by targeted deletion and green fluorescent protein reporter gene insertion. *Mol. Cell. Biol.* 20, 4106–4114.

Kadono-Okuda, K., Ridley, B., Jones, D., Jones, G., **2000**. Distinctive structural and kinetic properties of an unusual juvenile hormone-hydrolyzing esterase. *Biochem. Biophys. Res. Commun.* 272, 12–17.

Kambysellis, M.P., Johnson, F.M., Richardson, R.H., **1968**. Isozyme variability in species of the genus *Drosophila*. Distribution of the esterases in the body tissues of *Drosophila aldrichi* and *Drosophila mulleri* adults. *Biochem. Genet.* 1, 249–265.

Kamita, S.G., Hinton, A.C., Wheelock, C.E., Wogulis, M.D., Wilson, D.K., *et al.*, **2003**. Juvenile hormone (JH) esterase: why are you so JH specific? *Insect Biochem. Mol. Biol.* 33, 1261–1273.

Kang, D., Lundstrom, A., Liu, G., Steiner, H., **2002**. An azurocidin-like protein is induced in *Trichoplusia ni* larval gut cells after bacterial challenge. *Devel. Comp. Immunol.* 26, 495–503.

Kao, L.R., Motoyama, N., Dauterman, W.C., **1984**. Studies on hydrolases in various house fly strains and their role in malathion resistance. *Pestic. Biochem. Physiol.* 22, 86–92.

Kao, L.R., Motoyama, N., Dauterman, W.C., **1985**. The purification and characterisation of esterases from insecticide-resistant and susceptible house flies. *Pestic. Biochem. Physiol.* 23, 228–239.

Karoly, E.D., Rose, R.L., Thompson, D.M., Hodgson, E., Rock, G.C., *et al.*, **1996**. Monooxygenase, esterase, and gutathione transferase activity associated with azinphosmethyl resistance in the tufted apple bud moth, *Platynota idaeusalis*. *Pestic. Biochem. Physiol.* 55, 109–121.

Karunaratne, S.H.P.P., **1994**. Characterisation of multiple variants of carboxylesterases which are involved in insecticide resistance in the mosquito. *Culex quinquefasciatus*. Ph.D. thesis, University of London.

Karunaratne, S.H.P.P., Hemingway, J., **2000**. Insecticide resistance spectra and resistance mechanisms in populations of Japanese encephalitis vector mosquitoes, *Culex tritaeniorhynchus* and *Cx. gelidus* in Sri Lanka. *Med. Vet. Entomol.* 14, 430–436.

Karunaratne, S.H.P.P., Jayawardena, K.G., Hemingway, J., Ketterman, A.J., **1993**. Characterization of a beta-type esterase involved in insecticide resistance from the mosquito *Culex quinquefasciatus*. *Biochem. J.* 294, 575–579.

Karunaratne, S.H.P.P., Jayawardena, K.G., Hemingway, J., Ketterman, A.J., **1994**. Immunological cross-reactivity of a mosquito carboxylesterase-A2 antibody to other mosquito and vertebrate esterases and cholinesterase. *Biochem. Soc. Trans.* 22, 127S.

Karunaratne, S.H.P.P., Hemingway, J., **2001**. Malathion resistance and prevalence of the malathion carboxylesterase mechanism in populations of mosquito vectors of disease in Sri Lanka. *Bull. WHO* 79, 1060–1064.

Karunaratne, S.H.P.P., Hemingway, J., Jayawardena, K.G.I., Dassanayaka, V., Vaughan, A., **1995**. Kinetic and molecular differences in the amplified and non-amplified esterases from insecticide-resistant and susceptible *Culex quinquefasciatus* mosquitoes. *J. Biol. Chem.* 270, 31124–31128.

Karunaratne, S.H.P.P., Small, G.J., Hemingway, J., **1999**. Characterization of the elevated esterase-associated insecticide resistance mechanism in *Nilaparvata lugens* (Stål) and other planthopper species. *Int. J. Pest Mgt* 45, 225–230.

Karunaratne, S.H.P.P., Vaughan, A., Paton, M.G., Hemingway, J., **1998**. Amplification of a serine esterase gene is involved in insecticide resistance in Sri Lankan *Culex tritaeniorhynchus*. *Insect Mol. Biol.* 7, 307–315.

Kasai, Y., Konno, T., Dauterman, W.C., **1992**. Role of phosphorotriester hydrolases in the detoxication of organophosphorus insecticides. In: Chambers, J.E., Levi, P.E. (Eds.), Organophosphates: Chemistry, Fate and Effects. Academic Press, San Diego, CA, pp. 169–182.

Ketterman, A.J., Jayawardena, K.G., Hemingway, J., **1992**. Purification and characterization of a carboxylesterase involved in insecticide resistance from the mosquito *Culex quinquefasciatus*. *Biochem. J. 287*, 355–360.

Ketterman, A.J., Karunaratne, S.H.P.P., Jayawardena, K.G.I., Hemingway, J., **1993**. Qualitative differences between populations of *Culex quinquefasciatus* in both the esterases A-2 and B-2 which are involved in insecticide resistance. *Pestic. Biochem. Physiol. 47*, 142–148.

Khlebodarova, T.M., Gruntenko, N.E., Grenback, L.G., Sukhanova, M.Z., Mazurov, M.M., *et al.*, **1996**. A comparative analysis of juvenile hormone metabolizing enzymes in two species of *Drosophila* during development. *Insect Biochem. Mol. Biol. 26*, 829–835.

Klein, U., **1987**. Sensillum-lymph proteins from antennal olfactory hairs of the moth *Antheraea polyphemus* (Saturniidae). *Insect Biochem. 17*, 1193–1204.

Knight, A.L., Light, D.M., **2001**. Attractants from Bartlett pear for codling moth, *Cydia pomonella* (L.), larvae. *Naturwissenschaften 88*, 339–342.

Knudsen, J.T., Tollsten, L., Bergstrom, L.G., **1993**. Floral scents: a checklist of volatile compounds isolated by head-space techniques. *Phytochemistry 33*, 253–280.

Kochansky, J., Tette, J., Taschenberg, E.F., Cardé, R.T., Kaissling, K.-E., *et al.*, **1975**. Sex pheromone of the moth, *Antheraea polyphemus*. *J. Insect Physiol. 21*, 1977–1983.

Konno, T., Kasai, Y., Rose, R.L., Hodgson, E., Dauterman, W.C., **1990**. Purification and characterization of a phosphorotriester hydrolase from methyl parathion-resistant *Heliothis virescens*. *Pestic. Biochem. Physiol. 36*, 1–13.

Korochkin, L.I., Belyaeva, E.S., Matveeva, N.M., Kuzin, B.A., Serov, O.L., **1976**. Genetics of esterases in *Drosophila* Slow-migrating S-esterase in *Drosophila* of the *virilis* group. *Biochem. Genet. 14*, 161–182.

Korochkin, L.I., Ludwig, M., Tamarima, N., Uspenskyi, I., Yenikolopov, G., *et al.*, **1990**. Molecular genetic mechanisms of tissue-specific esterase isozymes and protein expression in *Drosophila*. In: Markert, C., Scandalios, J. (Eds.), Isozymes: Structure, Function, and Use in Biology and Medicine. Wiley-Liss, New York, pp. 399–440.

Korochkin, L.I., Ludwig, M.Z., Poliakova, E.V., Philinova, M.R., **1987**. Some molecular genetic aspects of cellular differentiation in *Drosophila*. *Soviet Sci. Rev. 1*, 411–466.

Korochkin, L.I., Panin, V.M., Pavlova, G.V., Kopantseva, M.R., Shostak, N.G., *et al.*, **1995**. A relatively small 5′ regulatory region of esterase-S gene of *Drosophila virilis* determines the specific expression as revealed in transgenic experiments. *Biochem. Biophys. Res. Commun. 213*, 302–310.

Kozaki, T., Shono, T., Tomita, T., Kono, Y., **2001**. Fenitroxon insensitive acetylcholinesterases of the housefly, *Musca domestica* associated with point mutations. *Insect Biochem. Mol. Biol. 31*, 991–997.

Kriek, N., Tilley, L., Horrocks, P., Pinches, R., Elford, B.C., *et al.*, **2003**. Characterization of the pathway for transport of the cytoadherence-mediating protein, PfEMP1, to the host cell surface in malaria parasite-infected erythrocytes. *Mol. Microbiol. 50*, 1215–1227.

Kuwahara, M., **1982**. Insensitivity of the acetylcholinesterase from the organophosphate-resistant Kanzawa spider mite, *Tetranychus kanzawai* Kishida (Acarina: Tetranychidae), to organophosphorus and carbamate insecticides. *Appl. Entomol. Zool. 17*, 486–493.

Kuwahara, M., Miyata, T., Saito, T., Eto, M., **1982**. Activity and substrate specificity of the esterase associated with organophosphorus insecticide resistance in the Kanzawa spider mite, *Tetranychus kanzawai* Kishida. *Appl. Entomol. Zool. 17*, 82–91.

Kwon, T.H., Kim, M.S., Choi, H.W., Joo, C.H., Cho, M.Y., *et al.*, **2000**. A masquerade-like serine proteinase homologue is necessary for phenoloxidase activity in the coleopteran insect, *Holotrichia diomphalia* larvae. *Eur. J. Biochem. 267*, 6188–6196.

La Du, B.N., Aviram, M., Billecke, S., Navab, M., Primo-Parmo, S., *et al.*, **1999**. On the physiological role(s) of the paraoxonases. *Chemico-Biol. Interact. 120*, 379–388.

Lai, E.C., **2003**. Lipid rafts make for slippery platforms. *J. Cell Biol. 162*, 365–370.

Lenormand, T., Bourguet, D., Guillemaud, T., Raymond, M., **1999**. Tracking the evolution of insecticide resistance in the mosquito *Culex pipiens*. *Nature 400*, 861–864.

Lenormand, T., Guillemaud, T., Bourguet, D., Raymond, M., **1998a**. Appearance and sweep of a gene duplication: adaptive response and potential for new functions in the mosquito *Culex pipiens*. *Evolution 52*, 1705–1712.

Lenormand, T., Guillemaud, T., Bourguet, D., Raymond, M., **1998b**. Evaluating gene flow using selected markers: a case study. *Genetics 149*, 1383–1392.

Lev-Lehman, E., Evron, T., Broide, R.S., Meshorer, E., Ariel, I., *et al.*, **2000**. Synaptogenesis and myopathy under acetylcholinesterase overexpression. *J. Mol. Neurosci. 14*, 93–105.

Lewis, G.A., Madge, D.S., **1984**. Esterase activity and associated insecticide resistance in the damson-hop aphid, *Phorodon humuli* (Schrank) (Hemiptera: Aphididae). *Bull. Entomol. Res. 74*, 227–238.

Li, F., Han, Z.J., **2002**. Two different genes encoding acetylcholinesterase existing in cotton aphid (*Aphis gossypii*). *Genome 45*, 1134–1141.

Liebl, E.C., Rowe, R.G., Forsthoefel, D.J., Stammler, A.L., Schmidt, E.R., *et al.*, **2003**. Interactions between the secreted protein Amalgam, its transmembrane receptor Neurotactin and the Abelson tyrosine kinase affect axon pathfinding. *Development 130*, 3217–3226.

Liu, H.F., Kubli, E., **2003**. Sex-peptide is the molecular basis of the sperm effect in *Drosophila melanogaster*. *Proc. Natl Acad. Sci. USA 100*, 9929–9933.

Liu, M.Y., Chen, J.S., Sun, C.N., **1984**. Synergism of pyrethroids by several compounds in larvae of the diamondback moth (Lepidoptera: Plutellidae). *J. Econ. Entomol. 77*, 851–856.

Lockridge, O., Blong, R.M., Masson, P., Froment, M.T., Millard, C.B., *et al.*, **1997**. A single amino acid substitution, Gly117His, confers phosphotriesterase (organophosphorus acid anhydrase hydrolase) activity on human butyrylcholinesterase. *Biochemistry 36*, 786–795.

Malcolm, C.A., Boddington, R.G., **1989**. Malathion resistance conferred by a carboxylesterase in *Anopheles culicifacies* Giles (Species-B) (Diptera, Culicidae). *Bull. Entomol. Res. 79*, 193–199.

Marullo, R., Lovei, G.L., Tallarico, A., Tremblay, E., **1988**. Quick detection of resistant phenotypes with high esterase activity in two species of aphids (Homoptera, Aphididae). *J. Appl. Entomol. 106*, 212–214.

Massoulie, J., Anselmet, A., Bon, S., Krejci, E., Legay, C., *et al.*, **1999**. The polymorphism of acetylcholinesterase: post-translational processing, quaternary associations and localization. *Chemico-Biol. Interact. 120*, 29–42.

Matsumura, F., Brown, A.W.A., **1963**. Studies on carboxylesterase in malathion-resistant *Culex tarsalis. J. Econ. Entomol. 56*, 381–388.

Matsumura, F., Hogendijk, C.J., **1964**. The enzymatic degradation of malathion in organophosphae resistant and susceptible strains of *Musca domestica. Entomol. Exp. Applic. 7*, 179–193.

Matsumura, F., Voss, G., **1964**. Mechanism of malathion and parathion resistance in the two spotted spider mite, *Tetranychus urticae. J. Econ. Entomol. 57*, 911–917.

Mazzarri, M.B., Georghiou, G.P., **1995**. Characterization of resistance to organophosphate, carbamate, and pyrethroid insecticides in field populations of *Aedes aegypti* from Venezuela. *J. Am. Mosquito Control Assoc. 11*, 315–322.

McCarroll, L., Paton, M.G., Karunaratne, S.H., Jayasuryia, H.T., Kalpage, K.S., *et al.*, **2000**. Insecticides and mosquito-borne disease. *Nature 407*, 961–962.

McClintock, T.S., Sammeta, N., **2003**. Trafficking prerogatives of olfactory receptors. *Neuroreport 14*, 1547–1552.

McKenzie, J.A., Parker, A.G., Yen, J.L., **1992**. Polygenic and single gene responses to selection for resistance to diazinon in *Lucilia cuprina. Genetics 130*, 613–620.

McKenzie, J.A., Whitten, M.J., Adena, M.A., **1982**. The effect of genetic background on the fitness of diazinon resistance genotypes of the Australian sheep blowfly, *L. cuprina. Heredity 49*, 1–9.

Meikle, D.B., Sheehan, K.B., Phillis, D.M., Richmond, R.C., **1990**. Localisation and longevity of seminal-fluid esterase-6 in mated female *Drosophila melanogaster. J. Insect Physiol. 36*, 93–101.

Merryweather, A.T., Crampton, J.M., Townson, H., **1990**. Purification and properties of an esterase from organophosphate-resistant strain of the mosquito *Culex quinquefasciatus. Biochem. J. 266*, 83–90.

Mikhailov, A.T., Torrado, M., **2000**. Carboxylesterases moonlight in the male reproductive tract: a functional shift pivotal for male fertility. *Frontiers Biosci. 5*, E53–E62.

Millard, C.B., Kryger, G., Ordentlich, A., Greenblatt, H.M., Harel, M., *et al.*, **1999**. Crystal structures of aged phosphonylated acetylcholinesterase: nerve agent reaction products at the atomic level. *Biochemistry 38*, 7032–7039.

Miota, F., Scharf, M.E., Ono, M., Marcon, P., Meinke, L.J., *et al.*, **1998**. Mechanisms of methyl and ethyl parathion resistance in the western corn rootworm (Coleoptera : Chrysomelidae). *Pestic. Biochem. Physiol. 61*, 39–52.

Miyata, T., Saito, T., **1976**. Mechanism of malathion resistance in the green rice leafhopper, *Nephotettix cincticeps* Uhler (Hemiptera: Deltocephalidae). *J. Pestic. Sci. 1*, 23–29.

Moores, G.D., Devine, G.J., Devonshire, A.L., **1994**. Insecticide-insensitive acetylcholinesterase can enhance esterase-based resistance in *Myzus persicae* and *Myzus nicotianae. Pestic. Biochem. Physiol. 49*, 114–120.

Moores, G.D., Gao, X., Denholm, I., Devonshire, A.L., **1996**. Characterisation of insensitive acetylcholinesterase in insecticide-resistant cotton aphids, *Aphis gossypii* Glover (Homoptera: Aphididae). *Pestic. Biochem. Physiol. 56*, 102–110.

Motoyama, N., Kao, L., Lin, P., Dauterman, W., **1984**. Dual role of esterases in insecticide resistance in the green rice leafhopper. *Pestic. Biochem. Physiol. 21*, 139–147.

Mouchès, C., Fournier, D., Raymond, M., Magnin, M., Bergé, J.-B., *et al.*, **1985**. Specific amplified DNA sequences associated with organophosphate insecticide resistance in mosquitoes of the *Culex pipiens* complex, with a note on similar amplification in the housefly, *Musca domestica* L. *C. R. Acad. Sci. Paris 301*, 695–700.

Mouchès, C., Magnin, M., Bergé, J.B., de Silvestri, M., Beyssat, V., *et al.*, **1987**. Overproduction of detoxifying esterases in organophosphate-resistant *Culex* mosquitoes and their presence in other insects. *Proc. Natl Acad. Sci. USA 84*, 2113–2116.

Mouchès, C., Pauplin, Y., Agarwal, M., Lemieux, L., Herzog, M., *et al.*, **1990**. Characterization of amplification core and esterase B1 gene responsible for insecticide resistance in *Culex. Proc. Natl Acad. Sci. USA 87*, 2574–2578.

Muehleisen, D.P., Benedict, J.H., Plapp, F.W., Carino, F.A., **1989**. Effects of cotton allelochemicals on toxicity of insecticides and induction of detoxifying enzymes in bollworm (Lepidoptera, Noctuidae). *J. Econ. Entomol. 82*, 1554–1558.

Murugasu-Oei, B., Rodrigues, V., Yang, X.H., Chia, W., **1995**. Masquerade: a novel secreted serine protease-like molecule is required for somatic muscle attachment in the *Drosophila* embryo. *Genes Devel. 9*, 139–154.

Mutero, A., Fournier, D., 1992. Post-translational modifications of *Drosophila* acetylcholinesterase. *J. Biol. Chem. 267*, 1695–1700.

Mutero, A., Pralavorio, M., Bride, J.M., Fournier, D., 1994. Resistance-associated point mutations in insecticide-insensitive acetylcholinesterase. *Proc. Natl Acad. Sci. USA 91*, 5922–5926.

Myers, M.A., Healy, M.J., Oakeshott, J.G., 1993. Effects of the residue adjacent to the reactive serine on the substrate interactions of *Drosophila* esterase 6. *Biochem. Genet. 31*, 259–278.

N'guessan, R., Darriet, F., Guillet, P., Carnevale, P., Traore-Lamizana, M., et al., 2003. Resistance to carbosulfan in *Anopheles gambiae* from Ivory Coast, based on reduced sensitivity of acetylcholinesterase. *Med. Vet. Entomol. 17*, 19–25.

Nabeshima, T., Kozaki, T., Tomita, T., Kono, Y., 2003. An amino acid substitution on the second acetylcholinesterase in the pirimicarb-resistant strains of the peach potato aphid, *Myzus persicae. Biochem. Biophys. Res. Commun. 307*, 15–22.

Nabeshima, T., Mori, A., Kozaki, T., Iwata, Y., Hidoh, O., et al., 2004. An amino acid substitution attributable to insecticide-insensitivity of acetylcholinesterase in a Japanese encephalitis vector mosquito, *Culex tritaeniorhynchus. Biochem. Biophys. Res. Commun. 313*, 794–801.

Nance, E., Heyse, D., Britton-Davidian, J., Pasteur, N., 1990. Chromosomal organization of the amplified esterase B1 gene in organophosphate-resistant *Culex pipiens quinquefasciatus* Say (Diptera, Culicidae). *Genome 33*, 148–152.

Narang, S., Seawright, J. A., 1982. Linkage relationships and genetic mapping in *Culex* and *Anopheles*. In: Steiner, W.W.M., Tabachnick, W.J., Rai, K.S., Narang, S. (Eds.), Recent Developments in the Genetics of Insect Disease Vectors. Stipes, Champaign, IL, pp. 231–280.

Nardini, M., Dijkstra, B.W., 1999. Alpha/beta hydrolase fold enzymes: the family keeps growing. *Curr. Opin. Struct. Biol. 9*, 732–737.

Needham, P.H., Sawicki, R.M., 1971. Diagnosis of resistance to organophosphorus insecticides in *Myzus persicae. Nature 230*, 125–126.

Newcomb, R.D., Campbell, P.M., Ollis, D.L., Cheah, E., Russell, R.J., et al., 1997. A single amino acid substitution converts a carboxylesterase to an organophosphorus hydrolase and confers insecticide resistance on a blowfly. *Proc. Natl Acad. Sci. USA 94*, 7464–7468.

Newcomb, R.D., East, P.D., Russell, R.J., Oakeshott, J.G., 1996. Isolation of alpha cluster esterase genes associated with organophosphate resistance in *Lucilia cuprina. Insect Mol. Biol. 5*, 211–216.

Newcomb, R.D., Gleeson, D.M., Yong, C.G., Russell, R.J., Oakeshott, J.G., 2004. Multiple mutations and gene duplications conferring organophosphorus insecticide resistance have been selected at the *Rop-1* locus of the sheep blowfly, *Lucilia cuprina. J. Mol. Evol.* (in press).

Nicolet, Y., Lockridge, O., Masson, P., Fontecilla-Camps, J.C., Nachon, F., 2003. Crystal structure of human butyrylcholinesterase and of its complexes with substrate and products. *J. Biol. Chem. 278*, 41141–41147.

O'Brien, P.J., Abdel-Aal, Y.A., Ottea, J.A., Graves, J.B., 1992. Relationship of insecticide resistance to carboxylesterases in *Aphis gossypii* (Homoptera, Aphididae) from midsouth cotton. *J. Econ. Entomol. 85*, 651–657.

Oakeshott, J.G., Boyce, T.M., Russell, R.J., Healy, M.J., 1995. Molecular insights into the evolution of an enzyme: esterase 6 in *Drosophila. Trends Ecol. Evol. 10*, 103–110.

Oakeshott, J.G., Claudianos, C., Russell, R.J., Robin, G.C., 1999. Carboxyl/cholinesterases: a case study of the evolution of a successful multigene family. *BioEssays 21*, 1031–1042.

Oakeshott, J.G., Healy, M.J., Game, A.Y., 1990. Regulatory evolution of the β-carboxyl esterases in *Drosophila*. In: Barker, J.S.F., Starmer, W.T., McIntyre, R.J. (Eds.), Ecological and Evolutionary Genetics of *Drosophila*. Plenum, New York, pp. 359–387.

Oakeshott, J.G., van Papenrecht, E.A., Boyce, T.M., Russell, R.J., Healy, M.J., 1993. Evolutionary genetics of *Drosophila* esterases. *Genetica 90*, 239–268.

Oakeshott, J.G., van Papenrecht, E.A., Claudianos, C., Morrish, B.C., Coppin, C., et al., 2001. An episode of accelerated amino acid change in *Drosophila* esterase-6 associated with a change in physiological function. *Genetica 110*, 231–244.

Odgers, W.A., Aquadro, C.F., Coppin, C.W., Healy, M.J., Oakeshott, J.G., 2002. Nucleotide polymorphism in the *Est6* promoter, which is widespread in derived populations of *Drosophila melanogaster*, changes the level of esterase 6 expressed in the male ejaculatory duct. *Genetics 162*, 785–797.

Odgers, W.A., Healy, M.J., Oakeshott, J.G., 1995. Nucleotide polymorphism in the 5′-promoter region of esterase-6 in *Drosophila melanogaster* and its relationship to enzyme activity variation. *Genetics 141*, 215–222.

Ollis, D.L., Cheah, E., Cygler, M., Dijkstra, B., Frolow, F., et al., 1992. The α/β hydrolase fold. *Protein Eng. 5*, 197–211.

Olson, P.F., Fessler, L.I., Nelson, R.E., Sterne, R.E., Campbell, A.G., et al., 1990. Glutactin, a novel *Drosophila* basement membrane-related gylcoprotein with sequence similarity to serine esterases. *EMBO J. 9*, 1219–1227.

Ono, M., Swanson, J.J., Field, L.M., Devonshire, A.L., Siegfried, B.D., 1999. Amplification and methylation of an esterase gene associated with insecticide-resistance in greenbugs, *Schizaphis graminum* (Rondani) (Homoptera: Aphididae). *Insect Biochem. Mol. Biol. 29*, 1065–1073.

Oppenoorth, F.J., 1990. Reevaluation of kinetic data for a paraoxon-hydrolyzing enzyme in a resistant strain of the predacious mite *Amblyseius potentillae. Pestic. Biochem. Physiol. 38*, 99–100.

Oppenoorth, F.J., van Asperen, K., **1960**. Allelic genes in the housefly producing modified enzymes that cause organophosphate resistance. *Science 132*, 298–299.

Ordentlich, A., Barak, D., Kronman, C., Ariel, N., Segal, Y., *et al.*, **1996**. The architecture of human acetylcholinesterase active centre probed by interactions with selected organophosphate inhibitors. *J. Biol. Chem. 271*, 11953–11962.

Ordentlich, A., Barak, D., Kronman, C., Ariel, N., Segal, Y., *et al.*, **1998**. Functional characteristics of the oxyanion hole in human acetylcholinesterase. *J. Biol. Chem. 273*, 19509–19517.

Ordentlich, A., Barak, D., Kronman, C., Flashner, Y., Leitner, M., *et al.*, **1993**. Dissection of the human acetylcholinesterase active centre determinants of substrate specificity. *J. Biol. Chem. 268*, 17083–17095.

Ottea, J.A., Ibrahim, S.A., Younis, A.M., Young, R.J., Leonard, B.R., *et al.*, **1995**. Biochemical and physiological mechanisms of pyrethroid resistance in *Heliothis virescens* (F). *Pestic. Biochem. Physiol. 51*, 117–128.

Parker, A.G., Campbell, P.M., Spackman, M.E., Russell, R.J., Oakeshott, J.G., **1996**. Comparison of an esterase associated with organophosphate resistance in *Lucilia cuprina* with an orthologue not associated with resistance in *Drosophila melanogaster*. *Pestic. Biochem. Physiol. 55*, 85–99.

Parker, A.G., Russell, R.J., Delves, A.C., Oakeshott, J.G., **1991**. Biochemistry and physiology of esterases in organophosphate-susceptible and organophosphate-resistant strains of the Australian sheep blowfly, *Lucilia cuprina*. *Pestic. Biochem. Physiol. 41*, 305–318.

Parker, C.G., Fessler, L.I., Nelson, R.E., Fessler, J.H., **1995**. Drosophila UDP-glucose:glycoprotein glucosyltransferase: sequence and characterization of an enzyme that distinguishes between denatured and native proteins. *EMBO J. 14*, 1294–1303.

Pasteur, N., Georghiou, G.P., **1981**. Filter paper test for rapid determination of phenotypes with high esterase activity in organophosphate resistant mosquitoes. *Mosquito News 41*, 181–183.

Pasteur, N., Georghiou, G.P., **1989**. Improved filter paper test for detecting and quantifying increased esterase activity in organophosphate-resistant mosquitoes (Diptera: Culicidae). *J. Econ. Entomol. 82*, 347–353.

Pasteur, N., Iseki, A., Georghiou, G.P., **1981a**. Genetic and biochemical studies of the highly active esterases A′ and B associated with organophosphate resistance in mosquitoes of the *Culex pipiens* complex. *Biochem. Genet. 19*, 909–919.

Pasteur, N., Marquine, M., Hoang, T.H., Nam, V.S., Failloux, A.B., **2001**. Overproduced esterases in *Culex pipiens quinquefasciatus* (Diptera: Culicidae) from Vietnam. *J. Med. Entomol. 38*, 740–745.

Pasteur, N., Sinegre, G., Gabinaud, A., **1981b**. Est-2 and Est-3 polymorphisms in *Culex pipiens* L. from Southern France in relation to organophosphate resistance. *Biochem. Genet. 19*, 499–508.

Pelham, H.R., **1996**. The dynamic organisation of the secretory pathway. *Cell Struct. Funct. 21*, 413–419.

Perrier, A.L., Massoulie, J., Krejci, E., **2002**. PRiMA: the membrane anchor of acetylcholinesterase in the brain. *Neuron 33*, 275–285.

Pezzementi, L., Johnson, K., Tsigelny, I., Cotney, J., Manning, E., *et al.*, **2003**. Amino acids defining the acyl pocket of an invertebrate cholinesterase. *Comp. Biochem. Physiol. B 136*, 813–832.

Phillips, J.P., Xin, J.H., Kirby, K., Milne, C.P., Krell, P., *et al.*, **1990**. Transfer and expression of an organophosphate insecticide-degrading gene from *Pseudomonas* in *Drosophila melanogaster*. *Proc. Natl Acad. Sci. USA 87*, 8155–8159.

Phythian, S.J., **1998**. Esterases. In: Kelly, D.R. (Ed.), Biotechnology, Vol. 8a: Biotransformations 1, 2nd edn. Wiley-VCH, Weinheim, pp. 194–241.

Picollo de Villar, M.I., Van Der Pas, L.J.T., Smissaert, H.R., Oppenoorth, F.J., **1983**. An unusual type of malathion-carboxylesterase in a Japanese strain of housefly. *Pestic. Biochem. Physiol. 19*, 60–65.

Pistillo, D., Manzi, A., Tino, A., Boyl, P.P., Graziani, F., *et al.*, **1998**. The *Drosophila melanogaster* lipase homologs: a gene family with tissue and developmental specific expression. *J. Mol. Biol. 276*, 877–885.

Plapp, F.W., Jr., Eddy, G.W., **1961**. Synergism of malathion against resistant insects. *Science 334*, 2043–2044.

Plepys, D., Ibarra, F., Francke, W., Lofstedt, C., **2002a**. Odour-mediated nectar foraging in the silver Y moth, *Autographa gamma* (Lepidoptera: Noctuidae): behavioural and electrophysiological responses to floral volatiles. *Oikos 99*, 75–82.

Plepys, D., Ibarra, F., Lofstedt, C., **2002b**. Volatiles from flowers of *Platanthera bifolia* (Orchidaceae) attractive to the silver Y moth, *Autographa gamma* (Lepidoptera: Noctuidae). *Oikos 99*, 69–74.

Poecke, R.M.P., van Posthumus, M.A., Dicke, M., **2001**. Herbivore-induced volatile production by *Arabidopsis thaliana* leads to attraction of the parasitiod *Cotesia rubecula*: chemical, behavioural, and gene-expression analysis. *J. Chem. Ecol. 27*, 1911–1928.

Pohlenz, H.-D., Boidol, W., Schuttke, I., Streber, W.R., **1992**. Purification and properties of an *Arthobacter oxydans* P52 carbamate hydrolase specific for the herbicide phenmedipham and nucleotide sequence of the corresponding gene. *J. Bacteriol. 174*, 6600–6607.

Poirié, M., Raymond, M., Pasteur, N., **1992**. Identification of two distinct amplifications of the esterase-B locus in *Culex pipiens* (L.) mosquitos from Mediterranean countries. *Biochem. Genet. 30*, 13–26.

Prabhakaran, S.K., Kamble, S.T., **1995**. Purification and characterization of an esterase isozyme from insecticide resistant and susceptible strains of German cockroach, *Blattella germanica* (L.). *Insect Biochem. Mol. Biol. 25*, 519–524.

Priester, T.M., Georghiou, G.P., Hawley, M.K., Pasternak, M.E., **1981**. Toxicity of pyrethroids to organophosphate-, carbamate-, and DDT-resistant mosquitoes. *Mosquito News 41*, 143–150.

Primo-Parmo, S.L., Sorenson, R.C., Teiber, J., Ladu, B.N., 1996. The human serum paraoxonase arylesterase gene (Pon1) is one member of a multigene family. *Genomics* 33, 498–507.

Qiao, C.L., Raymond, M., 1995. The same esterase B1 haplotype is amplified in insecticide-resistant mosquitoes of the *Culex pipiens* complex from the Americas and China. *Heredity* 74, 339–345.

Raftos, D.A., 1986. The biochemical basis of malathion resistance in the sheep blowfly, *Lucilia cuprina*. *Pestic. Biochem. Physiol.* 26, 302–309.

Raftos, D.A., Hughes, P.B., 1986. Genetic basis of a specific resistance to malathion in the Australian sheep blowfly, *Lucilia cuprina* (Diptera, Calliphoridae). *J. Econ. Entomol.* 79, 553–557.

Ranson, H., Claudianos, C., Ortelli, F., Abgrall, C., Hemingway, J., *et al.*, 2002. Evolution of supergene families associated with insecticide resistance. *Science* 298, 179–181.

Raymond, M., Beyssat-Arnaouty, V., Sivasubramanian, N., Mouchès, C., Georghiou, G.P., *et al.*, 1989. Amplification of various esterase Bs responsible for organophosphate resistance in *Culex* mosquitoes. *Biochem. Genet.* 27, 417–423.

Raymond, M., Callaghan, A., Fort, P., Pasteur, N., 1991. Worldwide migration of amplified insecticide resistance genes in mosquitoes. *Nature 350*, 151–153.

Raymond, M., Pasteur, N., Georghiou, G.P., Mellon, R.B.W.M.C., Hawley, M.K., 1987. Detoxification esterases new to California, USA, in organophosphate-resistant *Culex quinquefasciatus* (Diptera: Culicidae). *J. Med. Entomol.* 24, 24–27.

Raymond, M., Poulin, E., Boiroux, V., Dupont, E., Pasteur, N., 1993. Stability of insecticide resistance due to amplification of esterase genes in *Culex pipiens*. *Heredity 70*, 301–307.

Richmond, R.C., Nielsen, K.M., Brady, J.P., Snella, E.M., 1990. Physiology, biochemistry and molecular biology of the *Est-6* locus in *Drosophila melanogaster*. In: Barker, J.S.F., Starmer, W.T., MacIntyre, R.J. (Eds.), Ecological and Evolutionary Genetics of *Drosophila*. Plenum, New York, pp. 273–293.

Riddiford, L.M., 1996. Molecular aspects of juvenile hormone action in insect metamorphosis. In: Gilbert, L.I., Tata, J.R., Atkinson, B.G. (Eds.), Metamorphosis: Postembryonic Reprogramming of Gene Expression in Amphibian and Insect Cells. Academic Press, San Diego, CA, pp. 223–251.

Riskallah, M.R., 1983. Esterases and resistance to synthetic pyrethroids in the Egyptian cotton leafworm. *Pestic. Biochem. Physiol.* 19, 184–189.

Rivet, Y., Marquine, M., Raymond, M., 1993. French mosquito populations invaded by A2-B2 esterases causing insecticide resistance. *Biol. J. Linnean Soc. 49*, 249–255.

Robin, C., Russell, R.J., Medveczky, K.M., Oakeshott, J.G., 1996. Duplication and divergence of the genes of the alpha-esterase cluster of *Drosophila melanogaster*. *J. Mol. Evol. 43*, 241–252.

Robin, G.C., Claudianos, C., Russell, R.J., Oakeshott, J.G., 2000a. Reconstructing the diversification of alpha-esterases: comparing the gene clusters of *Drosophila buzzatii* and *D. melanogaster*. *J. Mol. Evol.* 51, 149–160.

Robin, G.C., Russell, R.J., Cutler, D.J., Oakeshott, J.G., 2000b. The evolution of an alpha-esterase pseudogene inactivated in the *Drosophila melanogaster* lineage. *Mol. Biol. Evol.* 17, 563–575.

Roe, R.M., Venkatesh, K., 1990. Metabolism of juvenile hormones: degradation and titre regulation. In: Gupta, A.P. (Ed.), Morphogenetic Hormones of Arthropods: Recent Advances in Comparative Arthropod Morphology, Physiology, and Development, Part 1. Rutgers University Press, New Brunswick, NJ, pp. 125–180.

Rooker, S., Guillemaud, T., Berge, J., Pasteur, N., Raymond, M., 1996. Coamplification of esterase A and B genes as a single unit in *Culex pipiens* mosquitoes. *Heredity* 77, 555–561.

Rosario-Cruz, R., Miranda-Miranda, E., Garcia-Vasquez, Z., Ortiz-Estrada, M., 1997. Detection of esterase activity in susceptible and organophosphate resistant strains of the cattle tick *Boophilus microplus* (Acari: Ixodidae). *Bull. Entomol. Res.* 87, 197–202.

Rossiter, L.C., Conyers, C.M., MacNicoll, A.D., Rose, H.A., 2001. Two qualitatively different B-esterases from two organophosphate-resistant strains of *Oryzaephilus surinamensis* (Coleoptera: Silvanidae) and their roles in fenitrothion and chlorpyrifos-methyl resistance. *Pestic. Biochem. Physiol.* 69, 118–130.

Roush, R.T., McKenzie, J.A., 1987. Ecological genetics of insecticide and acaricide resistance. *Annu. Rev. Entomol.* 32, 361–380.

Ruigt, G.É.S.F., 1985. Pyrethroids. In: Kerkut, G.A., Gilbert, L.I. (Eds.), Comprehensive Insect Physiology, Biochemistry and Pharmacology, vol. 12. Pergamon, Oxford, pp. 183–262.

Russell, R.J., Claudianos, C., Campbell, P.M., Horne, I.M., Sutherland, T.D., *et al.*, 2004. Two major classes of target site insensitivity mutations confer resistance to organophosphate and carbamate insecticides. *Pestic. Biochem. Physiol.* 79, 84–93.

Russell, R.J., Dumancic, M.M., Foster, G.G., Weller, G.L., Healy, M.J., *et al.*, 1990. Insecticide resistance as a model system for studying molecular evolution. In: Barker, J.S.F., Starmer, W.T., MacIntyre, R.J. (Eds.), Ecological and Evolutionary Genetics of *Drosophila*. Plenum, New York, pp. 293–314.

Russell, R.J., Robin, G.C., Kostakos, P., Newcomb, R.D., Boyce, T.M., *et al.*, 1996. Molecular cloning of an alpha-esterase gene cluster on chromosome 3R of *Drosophila melanogaster*. *Insect Biochem. Mol. Biol.* 26, 235–247.

Saad, M., Game, A.Y., Healy, M.J., Oakeshott, J.G., 1994. Associations of esterase-6 allozyme and activity variation with reproductive fitness in *Drosophila melanogaster*. *Genetica* 94, 43–56.

Sabourault, C., Guzov, V.M., Koener, J.F., Claudianos, C., Plapp, F.W., *et al.*, 2001. Overproduction of a P450 that metabolizes diazinon is linked to a loss-of-function in the chromosome 2 ali-esterase *(Md Alpha E7)* gene in resistant houseflies. *Insect Mol. Biol. 10*, 609–618.

Sakata, K., Miyata, T., **1994**. Biochemical characterization of carboxylesterase in the small brown planthopper *Laodelphax striatellus* (Fallén). *Pestic. Biochem. Physiol. 50*, 247–256.

Salama, M.S., Schouest, L.P., Miller, T.A., **1992**. Effect of diet on the esterase patterns in the hemolymph of the corn-earworm and the tobacco budworm (Lepidoptera, Noctuidae). *J. Econ. Entomol. 85*, 1079–1087.

Sawicki, R.M., Devonshire, A.D., Rice, A.D., Moores, G.D., Petzing, S.M., *et al.*, **1978**. The dectection and distribution of organophosphorus and carbamate insecticide-resistant *Myzus persicae* (Sulz.) in Britain in 1976. *Pestic. Sci. 9*, 189–201.

Scharf, M.E., Hemingway, J., Small, G.J., Bennett, G.W., **1997**. Examination of esterases from insecticide resistant and susceptible strains of the German cockroach, *Blattella germanica* (L.). *Insect Biochem. Mol. Biol. 27*, 489–497.

Scharf, M.E., Meinke, L.J., Siegfried, B.D., Wright, R.J., Chandler, L.D., **1999a**. Carbaryl susceptibility, diagnostic concentration determination, and synergism for US populations of western corn rootworm (Coleoptera: Chrysomelidae). *J. Econ. Entomol. 92*, 33–39.

Scharf, M.E., Meinke, L.J., Wright, R.J., Chandler, L.D., Siegfried, B.D., **1999b**. Metabolism of carbaryl by insecticide-resistant and-susceptible western corn rootworm populations (Coleoptera: Chrysomelidae). *Pestic. Biochem. Physiol. 63*, 85–96.

Scholl, F.G., Scheiffele, P., **2003**. Making connections: cholinesterase-domain proteins in the CNS. *Trends Neurosci. 26*, 618–624.

Schrag, J.D., Cygler, M., **1993**. 1.8 Å refined structure of the lipase from *Geotrichum candidum*. *J. Mol. Biol. 230*, 575–591.

Schulte, J., Tepass, U., Auld, V.J., **2003**. Gliotactin, a novel marker of tricellular junctions, is necessary for septate junction development in *Drosophila*. *J. Cell Biol. 161*, 991–1000.

Scott, J.G., Georghiou, G.P., **1986**. Malathion-specific resistance in *Anopheles stephensi* from Pakistan. *J. Am. Mosq. Control Assoc. 2*, 29–32.

Scott, J.G., Zhang, L., **2003**. The house fly aliesterase gene (MdalphaE7) is not associated with insecticide resistance or P450 expression in three strains of house fly. *Insect Biochem. Mol. Biol. 33*, 139–144.

Sergeev, P.V., Panin, V.M., Pavlova, G.V., Kopantseva, M.R., Shostak, N.G., *et al.*, **1995**. The expression of esterase-S gene of *Drosophila virilis* in *Drosophila melanogaster*. *FEBS Lett. 360*, 194–196.

Sergeev, P.V., Yenikolopov, G.N., Peunova, N.I., Kuzin, B.A., Khechumian, R.A., *et al.*, **1993**. Regulation of tissue-specific expression of the esterase S gene in *Drosophila virilis*. *Nucl. Acids Res. 21*, 3545–3551.

Shan, G., Ottea, J.A., **1998**. Contributions of monooxygenases and esterases to pyrethroid resistance in the tobacco budworm, *Heliothis virescens*. *Proc. Natl Cotton Counc. America 2*, 1148–1151.

Shan, G.M., Hammer, R.P., Ottea, J.A., **1997**. Biological activity of pyrethroid analogs in pyrethroid-susceptible and-resistant tobacco budworms, *Heliothis virescens* (F.). *J. Agric. Food Chem. 45*, 4466–4473.

Shanmugavelu, M., Baytan, A.R., Chesnut, J.D., Bonning, B.C., **2000**. A novel protein that binds juvenile hormone esterase in fat body tissue and pericardial cells of the tobacco hornworm *Manduca sexta* L. *J. Biol. Chem. 275*, 1802–1806.

Sharma, K.V., Koenigsberger, C., Brimijoin, S., Bigbee, J.W., **2001**. Direct evidence for an adhesive function in the noncholinergic role of acetylcholinesterase in neurite outgrowth. *J. Neurosci. Res. 63*, 165–175.

Shono, T., **1983**. Linkage group analysis of carboxylesterase in a malathion resistant strain of the housefly, *Musca domestica* L. (Diptera: Muscidae). *Appl. Entomol. Zool. 18*, 407–415.

Shufran, R.A., Wilde, G.E., Sloderbeck, P.E., **1996**. Description of three isozyme polymorphisms associated with insecticide resistance in greenbug (Homoptera: Aphididae) populations. *J. Econ. Entomol. 89*, 46–50.

Shulaev, V., Silverman, P., Raskin, I., **1997**. Airborne signalling by methyl salicylate in plant pathogen resistance. *Nature 385*, 718–721.

Siegfried, B.D., Ono, M., **1993a**. Mechanisms of parathion resistance in the greenbug, *Schizaphis graminum* (Rondani). *Pestic. Biochem. Physiol. 45*, 24–33.

Siegfried, B.D., Ono, M., **1993b**. Parathion toxicokinetics in resistant and susceptible strains of the greenbug (Homoptera, Aphididae). *J. Econ. Entomol. 86*, 1317–1323.

Siegfried, B.D., Swanson, J.J., Devonshire, A.L., **1997**. Immunological detection of greenbug (*Schizaphis graminum*) esterases associated with resistance to organophosphate insecticides. *Pestic. Biochem. Physiol. 57*, 165–170.

Siegfried, B.D., Zera, A.J., **1994**. Partial purification and characterization of a greenbug (Homoptera, Aphidiidae) esterase associated with resistance to parathion. *Pestic. Biochem. Physiol. 49*, 132–137.

Small, G.J., Hemingway, J., **2000a**. Differential glycosylation produces heterogeneity in elevated esterases associated with insecticide resistance in the brown planthopper *Nilaparvata lugens* Stål. *Insect Biochem. Mol. Biol. 30*, 443–453.

Small, G.J., Hemingway, J., **2000b**. Molecular characterization of the amplified carboxylesterase gene associated with organophosphorus insecticide resistance in the brown planthopper, *Nilaparvata lugens*. *Insect Mol. Biol. 9*, 647–653.

Smissaert, H.R., Abd El Hamid, F.M., Overmeer, W.P.J., **1975**. The minimum acetylcholinesterase (AChE) fraction compatible with life derived by aid of a simple model explaining the degree of dominance of resistance to inhibitors in AChE "mutants" *Biochem. Pharmacol. 24*, 1043–1047.

Smyth, K.A., Boyce, T.M., Russell, R.J., Oakeshott, J.G., 2000. MCE activities and malathion resistances in field populations of the Australian sheep blowfly (*Lucilia cuprina*). *Heredity 84*, 63–72.

Smyth, K.A., Russell, R.J., Oakeshott, J.G., 1994. A cluster of at least three esterase genes in *Lucilia cuprina* includes malathion carboxylesterase and two other esterases implicated in resistance to organophosphates. *Biochem. Genet. 32*, 437–453.

Smyth, K.A., Walker, V.K., Russell, R.J., Oakeshott, J.G., 1996. Biochemical and physiological differences in the malathion carboxylesterase activities of malathion-susceptible and resistant lines of the sheep blowfly, *Lucilia cuprina. Pestic. Biochem. Physiol. 54*, 48–55.

Soderlund, D. M., Casida, J. E., 1977a. Substrate specificity of mouse-liver microsomal enzymes in pyrethroid metabolism. In: Elliott, M. (Ed.), Synthetic Pyrethroids. American Chemical Society, Washington, DC, pp. 162–172.

Soderlund, D. M., Casida, J. E., 1977b. Stereospecificity of pyrethroid metabolism in mammals. In: Elliott, M. (Ed.), Synthetic Pyrethroids. American Chemical Society, Washington, DC, pp. 173–185.

Song, J.Y., Ichtchenko, K., Sudhof, T.C., Brose, N., 1999. Neuroligin 1 is a postsynaptic cell-adhesion molecule of excitatory synapses. *Proc. Natl Acad. Sci. USA 96*, 1100–1105.

Sorenson, R.C., Primo-Parmo, S.L., Camper, S.A., La Du, B.N., 1995. The genetic mapping and gene structure of mouse paraoxonase/arylesterase. *Genomics 30*, 431–438.

Sparagana, S.P., Bhaskaran, G., Barrera, P., 1985. Juvenile homone acid methyltransferase activity in imaginal discs of *Manduca sexta* prepupae. *Arch. Insect Biochem. Physiol. 2*, 191–202.

Speicher, S., Garcia-Alonso, L., Carmena, A., Martin-Bermudo, M.D., De La Escalera, S., et al., 1998. Neurotactin functions in concert with other identified CAMs in growth cone guidance in *Drosophila. Neuron 20*, 221–233.

Spencer, A.G., Price, N.R., Callaghan, A., 1998. Malathion-specific resistance in a strain of the rust red grain beetle *Cryptolestes ferrugineus* (Coleoptera: Cucujidae). *Bull. Entomol. Res. 88*, 199–206.

Sternfeld, M., Ming, G.L., Song, H.J., Sela, K., Timberg, R., et al., 1998. Acetylcholinesterase enhances neurite growth and synapse development through alternative contributions of its hydrolytic capacity, core protein, and variable C termini. *J. Neurosci. 18*, 1240–1249.

Swofford, D.L., 2000. PAUP: Phylogenetic Analysis Using Parsimony, v4. Sinauer Associates, Sunderlund, MA.

Tabuchi, K., Sudhof, T.C., 2002. Structure and evolution of neurexin genes: insight into the mechanism of alternative splicing. *Genomics 79*, 849–859.

Talley, T.T., Jianmongkol, S., Richardson, R., Radic, Z., Thompson, C.M., 1998. Isomalathion stereoisomers: insecticide impurities or new probe of cholinesterase? In: Doctor, B.P., Gentry, M.K., Taylor, P., Quinn, D.M., Rotundo, R.L. (Eds.), Structure and Function of Cholinesterases and Related Proteins. Plenum, New York, pp. 531–538.

Taylor, J.L., Radic, Z., 1994. The cholinesterases: from genes to proteins. *Annu. Rev. Pharmacol. Toxicol. 34*, 281–320.

Thomas, B.A., Church, W.B., Lane, T.R., Hammock, B.D., 1999. Homology model of juvenile hormone esterase from the crop pest, *Heliothis virescens. Proteins: Struct. Funct. Genet. 34*, 184–196.

Thompson, J.D., Higgins, D.G., Gibson, T.J., 1994. CLUSTAL W: improving the sensitivity of progressive multiple sequence alignment through sequence weighting, position-specific gap penalties and weight matrix choice. *Nucl. Acids Res. 22*, 4673–4680.

Tittiger, C., Walker, V.K., 1997. Isolation and characterization of an unamplified esterase B3 gene from malathion-resistant *Culex tarsalis. Biochem. Genet. 35*, 119–138.

Tomita, T., Hidoh, O., Kono, Y., 2000. Absence of protein polymorphism attributable to insecticide-insensitivity of acetylcholinesterase in the green rice leafhopper, *Nephotettix cincticeps. Insect Biochem. Mol. Biol. 30*, 325–333.

Toutant, J.P., 1989. Insect acetylcholinesterase: catalytic properties, tissue distribution and molecular forms. *Progr. Neurobiol. 32*, 423–446.

Townsend, M.G., Busvine, J.R., 1969. The mechanism of malathion resistance in the blowfly *Chrysomyia putoria. Entomol. Exp. Applic. 12*, 243–267.

Truman, J.W., Riddiford, L.M., 2002. Endocrine insights into the evolution of metamorphosis in insects. *Annu. Rev. Entomol. 47*, 467–500.

Tumlinson, J.H., Brennan, M.M., Doolittle, R.E., Mitchell, E.R., Brabham, A., et al., 1989. Identification of a pheromone blend attractive to *Manduca sexta* (L.) males in a wind-tunnel. *Arch. Insect Biochem. Physiol. 10*, 255–271.

van Meer, M.M.M., Bonning, B.C., Ward, V.K., Vlak, J.M., Hammock, B.D., 2000. Recombinant, catalytically inactive juvenile hormone esterase enhances efficacy of baculovirus insecticides. *Biol. Control 19*, 191–199.

Vaughan, A., Hawkes, N., Hemingway, J., 1997a. Co-amplification explains linkage disequilibrium of two mosquito esterase genes in insecticide-resistant *Culex quinquefasciatus. Biochem. J. 325*, 359–365.

Vaughan, A., Rocheleau, T., ffrench-Constant, R., 1997b. Site-directed mutagenesis of an acetylcholinesterase gene from the yellow fever mosquito *Aedes aegypti* confers insecticide insensitivity. *Exp. Parasitol. 87*, 237–244.

Vaughan, A., Rodriguez, M., Hemingway, J., 1995. The independent gene amplification of electrophoretically indistinguishable Beta-esterases from the insecticide-resistant mosquito *Culex quinquefasciatus. Biochem. J. 305*, 651–658.

VedBrat, S.S., Whitt, G.S., 1975. Isozyme ontogeny during development of the mosquito, *Anopheles albimanus*. In: Markert, C.L. (Ed.), Isozymes, vol. 3. Academic Press, New York, pp. 131–143.

Vermunt, A.M.W., Koopmanschap, A.B., Vlak, J.M., De Kort, C.A.D., 1998. Evidence for two juvenile hormone esterase-related genes in the Colorado potato beetle. *Insect Mol. Biol.* 7, 327–336.

Vernick, K.D., Collins, F.H., 1989. Association of a plasmodium-refractory phenotype with an esterase locus in *Anopheles gambiae*. *Am. J. Trop. Med. Hyg.* 40, 593–597.

Vernick, K.D., Collins, F.H., Seeley, D.C., Gwadz, R.W., Miller, L.H., 1988. The genetics and expression of an esterase locus in *Anopheles gambiae*. *Biochem. Genet.* 26, 367–379.

Villani, F., White, G.B., Curtis, C.F., Miles, S.J., 1983. Inheritance and activity of some esterases associated with organophosphate resistance in mosquitoes of the complex of *Culex pipiens* L. (Diptera: Culicidae). *Bull. Entomol. Res.* 23, 154–170.

Villatte, F., Bachmann, T.T., 2002. How many genes encode cholinesterase in arthropods? *Pestic. Biochem. Physiol.* 73, 122–129.

Villatte, F., Marcel, V., Estrada-Mondaca, S., Fournier, D., 1998. Engineering sensitive acetylcholinesterase for detection of organophosphate and carbamate insecticides. *Biosensors Bioelectron.* 13, 157–164.

Villatte, F., Ziliani, P., Estrada-Mondaca, S., Menozzi, P., Fournier, D., 2000a. Is acetyl/butyrylcholine specificity a marker for insecticide resistance mutations in insect acetylcholinesterase? *Pest Mgt Sci.* 56, 1023–1028.

Villatte, F., Ziliani, P., Marcel, V., Menozzi, P., Fournier, D., 2000b. A high number of mutations in insect acetylcholinesterase may provide insecticide resistance. *Pestic. Biochem. Physiol.* 67, 95–102.

Voet, D., Voet, J.G., 1990. Biochemistry. Wiley, New York.

Vogt, R.G., Riddiford, L.M., 1981. Pheromone binding and inactivation by moth antennae. *Nature* 293, 161–163.

Vogt, R.G., Riddiford, L.M., 1986. Scale esterase: a pheromone-degrading enzyme from scales of silk moth *Antheraea polyphemus*. *J. Chem. Ecol.* 12, 469–482.

Vogt, R.G., Riddiford, L.M., Prestwich, G.D., 1985. Kinetic properties of a sex pheromone-degrading enzyme: the sensillar esterase of *Antheraea polyphemus*. *Proc. Natl Acad. Sci. USA* 82, 8827–8831.

Vontas, J.G., Cosmidis, N., Loukas, M., Tsakas, S., Hejazi, M.J., et al., 2001. Altered acetylcholinesterase confers organophosphate resistance in the olive fruit fly *Bactrocera oleae*. *Pestic. Biochem. Physiol.* 71, 124–132.

Vontas, J.G., Hejazi, M.J., Hawkes, N.J., Cosmidis, N., Loukas, M., et al., 2002. Resistance-associated point mutations of organophosphate insensitive acetylcholinesterase, in the olive fruit fly *Bactrocera oleae*. *Insect Mol. Biol.* 11, 329–336.

Vontas, J.G., Small, G.J., Hemingway, J., 2000. Comparison of esterase gene amplification, gene expression and esterase activity in insecticide susceptible and resistant strains of the brown planthopper, *Nilaparvata lugens* (Stål). *Insect Mol. Biol.* 9, 655–660.

Wachendorff, U., Zoebelein, G., 1988. Diagnosis of insecticide resistance in *Phorodon humuli* (Homoptera, Aphididae). *Entomol. General.* 13, 145–155.

Walsh, S.B., Dolden, T.A., Moores, G.D., Kristensen, M., Lewis, T., et al., 2001. Identification and characterization of mutations in housefly (*Musca domestica*) acetylcholinesterase involved in insecticide resistance. *Biochem. J.* 359, 175–181.

Wang, K.Y., Liu, T.X., Yu, C.H., Jiang, X.Y., Yi, M.Q., 2002. Resistance of *Aphis gossypii* (Homoptera: Aphididae) to fenvalerate and imidacloprid and activities of detoxification enzymes on cotton and cucumber. *J. Econ. Entomol.* 95, 407–413.

Webb, E.C., 1992. Enzyme Nomenclature 1992: Recommendations of the Nomenclature Committee of the International Union of Biochemistry and Molecular Biology on the Nomenclature and Classification of Enzymes. International Union of Biochemistry and Molecular Biology. Academic Press, San Diego, CA.

Weill, M., Berticat, C., Raymond, N., Chevillon, C., 2000. Quantitative polymerase chain reaction to estimate the number of amplified esterase genes in insecticide-resistant mosquitoes. *Anal. Biochem.* 285, 267–270.

Weill, M., Fort, P., Berthomieu, A., Dubois, M.P., Pasteur, N., et al., 2002. A novel acetylcholinesterase gene in mosquitoes codes for the insecticide target and is non-homologous to the *ace* gene in *Drosophila*. *Proc. Roy. Soc. London B* 269, 2007–2016.

Weill, M., Lutfalla, G., Mogensen, K., Chandre, F., Berthomieu, A., et al., 2003. Insecticide resistance in mosquito vectors. *Nature* 423, 136–137.

Weill, M., Malcolm, C., Chandre, F., Mogensen, K., Berthomieu, A., et al., 2004. The unique mutation in *ace-1* giving high insecticide resistance is easily detectable in mosquito vectors. *Insect Mol. Biol.* 13, 1–7.

Welling, W., De Vries, A.W., Voerman, S., 1974. Oxidative cleavage of a carboxyester bond as a mechanism of resistance to malaoxon in houseflies. *Pestic. Biochem. Physiol.* 4, 31–43.

Wheelock, C.E., Wheelock, A.M., Zhang, R., Stok, J.E., Morisseau, C., et al., 2003. Evaluation of alpha-cyanoesters as fluorescent substrates for examining inter-individual variation in general and pyrethroid-selective esterases in human liver microsomes. *Anal. Biochem.* 315, 208–222.

White, M.M., Mane, S.D., Richmond, R.C., 1988. Studies of esterase 6 in *Drosophila melanogster*. 18. Biochemical differences between the slow and fast allozymes. *Mol. Biol. Evol.* 5, 41–62.

White, N.D.G., Bell, R.J., 1988. Inheritance of malathion resistance in a strain of *Tribolium castaneum* (Coleoptera, Tenebrionidae) and effects of resistance genotypes on fecundity and larval survival in malathion-treated wheat. *J. Econ. Entomol.* 81, 381–386.

Whyard, S., Russell, R.J., Walker, V.K., 1994. Insecticide resistance and malathion carboxylesterase in the sheep blowfly, *Lucilia cuprina*. *Biochem. Genet.* 32, 9–24.

Whyard, S., Walker, V.K., **1994**. Characterization of malathion carboxylesterase in the sheep blowfly *Lucilia cuprina*. *Pestic. Biochem. Physiol. 50*, 198–206.

Wierenga, J.M., Hollingworth, R.M., **1993**. Inhibition of altered acetylcholinesterases from insecticide-resistant Colorado potato beetles (Coleoptera: Chrysomelidae). *J. Econ. Entomol. 86*, 673–679.

Wirth, M.C., Georghiou, G.P., **1996**. Organophosphate resistance in *Culex pipiens* from Cyprus. *J. Am. Mosq. Control Assoc. 12*, 112–118.

Wirth, M.C., Marquine, M., Georghiou, G.P., Pasteur, N., **1990**. Esterases A2 and B2 in *Culex quinquefasciatus* (Diptera: Culicidae): role in organophosphate resistance and linkage studies. *J. Med. Entomol. 7*, 202–206.

Wolff, M.A., Abdel-Aal, Y.A.I., Goh, D.K.S., Lampert, E.P., Roe, R.M., **1994**. Organophosphate resistance in the tobacco aphid (Homoptera, Aphididae): purification and characterization of a resistance-associated esterase. *J. Econ. Entomol. 87*, 1157–1164.

Wood, E.J., de Villar, M.I.P., Zerba, E.N., **1985**. Role of microsomal carboxylesterase in reducing the action of malathion in eggs of *Triatoma infestans*. *Pestic. Biochem. Physiol. 23*, 24–32.

Xu, J., Qu, F., Weide, L., **1994**. Diversity of amplified esterase B genes responsible for organophosphate resistance in *Culex quinquefasciatus* from China. *J. Med. Colleges Peoples Liberation Army 9*, 20–23.

Yamamoto, K., Hamada, H., Shinkai, H., Kohno, Y., Koseki, H., *et al.*, **2003**. The KDEL receptor modulates the endoplasmic reticulum stress response through mitogen-activated protein kinase signaling cascades. *J. Biol. Chem. 278*, 34525–34532.

Yin, C.M., Zou, B.X., Jiang, M.G., Li, M.F., Qin, W.H., *et al.*, **1995**. Identification of juvenile-hormone-III bisepoxide (JHB(3)), juvenile-hormone-III and methyl farnesoate secreted by the corpus allatum of *Phormia regina* (Meigen) *in vitro* and function of JHB(3) either applied alone or as a part of a juvenoid blend. *J. Insect Physiol. 41*, 473–479.

Yu, S.J., **1991**. Insecticide resistance in the fall armyworm, *Spodoptera frugiperda* (Smith, J.E.). *Pestic. Biochem. Physiol. 39*, 84–91.

Zador, E., **1989**. Tissue specific expression of the acetylcholinesterase gene in *Drosophila melanogaster*. *Mol. Gen. Genet. 218*, 487–490.

Zera, A.J., Harshman, L.G., **2001**. The physiology of life history trade-offs in animals. *Annu. Rev. Ecol. Syst. 32*, 95–126.

Zera, A.J., Huang, Y., **1999**. Evolutionary endocrinology of juvenile hormone esterase: functional relationship with wing polymorphism in the cricket, *Gryllus firmus*. *Evolution 53*, 837–847.

Zera, A.J., Sanger, T., Hanes, J., Harshman, L., **2002**. Purification and characterization of hemolymph juvenile hormone esterase from the cricket, *Gryllus assimilis*. *Arch. Insect Biochem. Physiol. 49*, 41–55.

Zhao, G., Rose, R.L., Hodgson, E., Roe, R.M., **1996**. Biochemical mechanisms and diagnostic microassays for pyrethroid, carbamate, and organophosphate insecticide resistance/cross-resistance in the tobacco budworm, *Heliothis virescens*. *Pestic. Biochem. Physiol. 56*, 183–195.

Zhao, G.Y., Liu, W., Knowles, C.O., **1994**. Mechanisms associated with diazinon resistance in western flower thrips. *Pestic. Biochem. Physiol. 49*, 13–23.

Zhou, Z., Scharf, M.E., Parimi, S., Meinke, L.J., Wright, R.J., *et al.*, **2002**. Diagnostic assays based on esterase-mediated resistance mechanisms in western corn rootworms (Coleoptera: Chrysomelidae). *J. Econ. Entomol. 95*, 1261–1266.

Zhu, K.Y., Brindley, W.A., **1990a**. Acetylcholinesterase and its reduced sensitivity to inhibition by paraoxon in organophosphate-resistant *Lygus hesperus* Knight (Hemiptera, Miridae). *Pestic. Biochem. Physiol. 36*, 22–28.

Zhu, K.Y., Brindley, W.A., **1990b**. Properties of esterases from *Lygus hesperus* Knight (Hemiptera: Miridae) and the roles of the esterases in insecticide resistance. *J. Econ. Entomol. 83*, 725–732.

Zhu, K.Y., Brindley, W.A., **1992**. Enzymological and inhibitory properties of acetylcholinesterase purified from *Lygus hesperus* Knight (Hemiptera, Miridae). *Insect Biochem. Mol. Biol. 22*, 245–251.

Zhu, K.Y., Clark, J.M., **1995a**. Comparisons of kinetic properties of acetylcholinesterase purified from azinphosmethyl-susceptible and azinphosmethyl-resistant strains of Colorado potato beetle. *Pestic. Biochem. Physiol. 51*, 57–67.

Zhu, K.Y., Clark, J.M., **1995b**. Cloning and sequencing of a cDNA encoding acetylcholinesterase in Colorado potato beetle, *Leptinotarsa decemlineata* (Say). *Insect Biochem. Mol. Biol. 25*, 1129–1138.

Zhu, K.Y., Lee, S.H., Clark, J.M., **1996**. A point mutation of acetylcholinesterase associated with azinphosmethyl resistance and reduced fitness in Colorado potato beetle. *Pestic. Biochem. Physiol. 55*, 100–108.

Zhu, Y.C., Dowdy, A.K., Baker, J.E., **1999a**. Detection of single-base substitution in an esterase gene and its linkage to malathion resistance in the parasitoid *Anisopteromalus calandrae* (Hymenoptera: Pteromalidae). *Pestic. Sci. 55*, 398–404.

Zhu, Y.C., Dowdy, A.K., Baker, J.E., **1999b**. Differential mRNA expression levels and gene sequences of a putative carboxylesterase-like enzyme from two strains of the parasitoid *Anisopteromalus calandrae* (Hymenoptera: Pteromalidae). *Insect Biochem. Mol. Biol. 29*, 417–425.

Ziegler, R., Whyard, S., Downe, A.E.R., Wyatt, G.R., Walker, V.K., **1987**. General esterase, malathion carboxylesterase, and malathion resistance in *Culex tarsalis*. *Pestic. Biochem. Physiol. 28*, 279–285.

Zouros, E., van Delden, W., **1982**. Substrate-preference polymorphism at an esterase locus of *Drosophila mojavensis*. *Genetics 100*, 307–314.

Zouros, E., van Delden, W., Odense, R., van Dijk, H., **1982**. An esterase duplication in *Drosophila*: differences in expression of duplicate loci within and among species. *Biochem. Genet. 20*, 929–942.

Zschunke, F., Salmassi, A., Kreipe, H., Buck, F., Parwaresch, M.R., *et al.*, **1991**. cDNA cloning and characterization of human monocyte macrophage serine esterase-1. *Blood 78*, 506–512.

Relevant Websites

http://www.sanger.ac.uk – The Sanger Institute; includes Pfam nomenclature.

http://www.flygenome.yale.edu – full sequence of *Drosophila melanogaster* genome.

http://www.phenolist.slu.se – up-to-date listing of insect pheromones.

http://www.flybase.bio.indiana.edu – comparative data on insect genomics.

http://www.ncbi.nlm.nih.gov – National Center for Biotechnology Information.

http://www.ensembl.org – ENSEMBL database.

http://www.hgsc.bcm.tmc – Honeybee esterases sequences.

http://www.rcsb.org – Crystal structure of AChE.

http://www.openrasmol.org – Structural modeling software.

http://www.psc.edu – Multiple alignment software.

http://www.pherolist.slu.se – Components of sex pheromone blends.

http://www.hgsc.bcm.tmc.edu – *D. pseudoobscura* esterase sequences.

A7   Addendum: New Genomic Perspectives on Insect Esterases and Insecticide Resistance

J G Oakeshott and R J Russell, CSIRO
Entomology, Black Mountain, Canberra,
Australia
C Claudianos, Visual Neuroscience, Queensland
Brain Institute, University of Queensland, QLD,
Australia

© 2010, Elsevier BV. All Rights Reserved.

The 5 years since the first publication of our review have seen several advances in the molecular and biochemical analysis of insect carboxyl/cholinesterases (CCEs) and their role in insecticide resistance. Of particular significance has been the publication of several more insect genome sequences. The only genome sequences available in 2005 from which CCE gene complements could be abstracted were from *Drosophila melanogaster* and the malarial mosquito *Anopheles gambiae*. We now know the CCE gene complements of several more Diptera (11 further *Drosophila* species and the yellow fever mosquito *Aedes aegypti* (The Drosophila 12 Genomes Consortium, 2007; Strode *et al.*, 2008) plus the red flour beetle *Tribolium castaneum*, the honeybee *Apis mellifera*, the silkworm *Bombyx mori* and the endoparasitic jewel wasp *Nasonia vitripennis* (Claudianos *et al.*, 2006; Yu *et al.*, 2009; Teese *et al.*, 2010; Oakeshott *et al.*, 2010).

The basic topology of the insect CCE phylogeny that we presented in our review was not changed with the addition of these extra sequences (Oakeshott *et al.*, 2010). There remain three major groupings. One is an ancient group of six major clades implicated in neuro/developmental functions, the only catalytically competent members of which are the acetylcholinesterases (AChEs). The second is a group of four major clades of enzymes, many of which are known to be involved in hormone and semiochemical processing. The third group comprises three major clades of enzymes that contains most of the CCEs known to be involved in the detoxification of xenobiotics, including many insecticides. Very high levels of orthology are retained across the insect orders in the neurodevelopmental group but there is very little inter-order orthology in the other two.

Between 11 and 14 of the CCEs in every species sequenced belong to the highly conserved neurodevelopmental group but the number of CCEs in the other two groups ranges from as few as 13 up to over 50, honeybees having the fewest. Our understanding of this variation is limited by the paucity of empirical functional data for most insect CCEs. However, it has been suggested that the high level of eusociality and homeostasis of the nest environment of honeybees has reduced its need for as many genes involved in detoxification functions and some other environmental interactions (Claudianos *et al.*, 2006; Oakeshott *et al.*, 2010).

There has been little advance over the last 5 years in our understanding of the functions of the five clades of noncatalytic neurodevelopmental CCEs. All these CCEs are complex proteins comprising extracellular CCE domains plus trans-membrane and cytoplasmic domains and they are known collectively as cholinesterase-like adhesion molecules (CLAMs). Some, such as neurligins, neurotactins, and gliotactins, have known binding partners in the central nervous system (CNS) but essentially nothing is known about the cell biology of the others. Their behavioral phenotypes are generally unknown in insects although more is known about the functions of some vertebrate orthologs. Significantly, Biswas *et al.* (2008; 2010) recently associated neuroligins with learning and memory in the honeybees. Vertebrate neuroligins are also associated with learning and memory, with various mutations linked to autism, so there is now a prospect for using the honeybees as a model to understand the molecular biology underlying some crucial human phenotypes.

Many papers have been published on the molecular biology and/or biochemistry of insect AChEs over the last 5 years, the majority dealing with the molecular basis of target site resistance to organophosphate (OP) and carbamate (CB) insecticides. Russell *et al.* (2004) noted that at a biochemical and bioassay level, most resistance cases involving these two insecticides fall into two classes. Pattern 1 resistance is generally more effective for CBs than OPs, while Pattern 2 resistance is at least as effective for OPs as CBs and may even be specific to OPs in some cases. This classification has generally held true in later papers (see e.g., Kristensen *et al.*, 2006; Alout *et al.*, 2008; Alout and Weill, 2008). Russell *et al.* (2004) also noted that, where molecular data were also available, there was a trend for each class to be associated with particular amino acid substitutions in the AChE active site; often G151S for Pattern 1 or S371F and I161V/T, G265A/V, F330Y, G368A, F371W, or G436S for Pattern 2 (numbering according to the nomenclature for DmAChE; Harel *et al.*, 2000). These latter correlations are also apparent in some of the more recent studies (Kristensen *et al.*, 2006; Alout *et al.*, 2008; Alout and Weill, 2008), albeit not in others (e.g., Liming *et al.*, 2006; Alout *et al.*, 2007; Lee *et al.*, 2007). Some species-dependence of the molecular correlations might be expected because some specific aspects of sequence–structure–function relationships within the active site will inevitably vary to some degree across species, despite a very high level of active site conservation overall. Some additional AChE mutations have also been linked to resistance in the last 5 years. V182L and D344V in *Musca domestica* (Liming *et al.*, 2006), A239S in both *Plutella xylostella* (Lee *et al.*, 2007) and *Aphis gossypii* (Andrews *et al.*, 2004), and S276G in *Leptinotarsa decemlineata* (Kim *et al.*, 2007) have all been linked to Pattern 1 resistance, while F330V in *Cydia pomonella* (Cassanelli *et al.*, 2006) and *Culex pipiens* (Alout and Weill, 2008) has been associated with both Pattern 1 and 2 resistances.

Most insects have two paralogous AChE gene/enzyme systems, the exceptions being the higher Diptera where there is only one. Outside the higher Diptera, it is widely assumed that orthologous AChEs (generally denoted AChE-1s) carry out the essential function of recycling acetylcholine in the CNS and are the target for OP and CB insecticides in all species. Consistent with this, the great majority of target site resistance mutations continue to be found in AChE-1. However, isolated instances of AChE-2 associations are still being reported (e.g., Chen *et al.*, 2007). There is still very little empirical information otherwise on the function of AChE-2 and it remains unclear whether AChE-2 is directly, causally involved in synapse function or resistance.

Some quite spectacular evolutionary changes have recently been found in the hormone/semiochemical processing group of CCEs and one set of these involves a function that, *a priori*, would be expected to be quite highly conserved. The latter involves the juvenile hormone esterases (JHEs) that regulate the titre of juvenile hormone and in turn,

metamorphosis. It now appears that this function has evolved in different clades in the Diptera, Lepidoptera, and Hymenoptera (Crone *et al.*, 2007a, 2007b; Mackert *et al.*, 2008; Teese *et al.*, 2010; Tsubota *et al.*, 2010). Moreover there are three cases where there have been recent expansions of copy number in the specific lineage containing the JHE, up to 16 in the case of *Drosophila willistoni* (Bai *et al.*, 2007; Crone *et al.*, 2007a; Oakeshott *et al.*, 2010). The biology of juvenile hormone may thus be more complex than hitherto suspected, possibly including related but as yet uncharacterized signaling molecules. Investigation of the biochemistry underlying these evolutionary behaviors should be assisted by the recent resolution of the crystal structure of a lepidopteran JHE (Wogulis *et al.*, 2006).

Another unexpected feature of the hormone/semiochemical processing group of CCEs that recently emerged has been the relatively high frequency of noncatalytic esterases in the group. Catalytic incompetence can be reliably inferred from the primary sequence of a CCE by the loss of one or more members of the Ser–His–acid catalytic triad which is essential to the esterase reaction mechanism. One set of these CCEs involves dipteran glutactins, which comprise noncatalytic CCE domains fused to extracellular domains, as *per* the neurodevelopmental CLAMs seen earlier (Claudianos *et al.*, 2006; Strode *et al.*, 2008; Oakeshott *et al.*, 2010). The physiological function of the glutactins is still unknown but some Diptera have several glutactin, or glutactin-like, genes. Several Drosophila also have another set of noncatalytic CCEs, in this case without fused extracellular domains, which have evolved repeatedly and in many cases been preserved for too long to be functionless (Robin *et al.*, 2009). These occur in the β-esterase lineage, which includes known hormone-related functions. Robin *et al.* (2009) have shown that one lineage of noncatalytic β-esterases has survived for ~40 My, with clear evidence of sequence conservation through that time outside the catalytic triad. Several noncatalytic CCEs have also been found in the *B. mori* and *Helicoverpa armigera* genomes, mainly in the hormone/semiochemical processing clades, but nothing is yet known about their functional significance (Yu *et al.*, 2010; Teese *et al.*, 2010).

Another very recent paper reported the presence of a total of 19 paralogous CCEs in an Expressed Sequence Tag (EST) library from the antennae of the cotton leafworm *Spodoptera litteralis* (Durand *et al.*, 2010). Most of these fell in the hormone/semiochemical processing clades. In this moth at least it appears that numerous esterases could have roles in semiochemical processing.

Two basic mechanisms of esterase-mediated metabolic resistance to OPs had been elucidated at a molecular level by 2005, generally but not always involving CCEs in the detoxification group of clades. One mechanism involves overexpression of CCEs that bind but do not hydrolyse the pesticides, nevertheless effectively sequestering them away from the pesticide target site. There is now evidence for such a mechanism in other species (e.g., Gao *et al.*, 2006; Karunaratne *et al.*, 2007; Li *et al.*, 2007), possibly involving upregulation of the relevant genes in some cases rather than amplification of their copy number, which is the basis for the earlier, exemplar cases involving the aphid E4s and the Estαs and Estβs of culicine mosquitoes. The second mechanism involves structural mutations in carboxylesterases that convert them to OP hydrolases. By 2005, this had been shown in three higher Diptera and a wasp, in all cases due to specific mutations in one or other of two residues in the active site of the enzyme. An equivalent mutation in an orthologous CCE has now been reported to confer resistance on another fly species (de Carvalho *et al.*, 2006). Furthermore *in vitro* mutagenesis and heterologous expression has shown that one of the mutations in question can confer OP hydrolase activity on a CCE from another species, *C. pipiens*, so the mechanism may be more widespread (Cui *et al.*, 2007a, 2007b; and see also Magaña *et al.*, 2008). One of the mutations indeed exists as a polymorphism in insecticide-naïve populations of blowflies (Hartley *et al.*, 2006).

Evidence also continues to accumulate implicating esterases in metabolic resistance to synthetic ester pyrethroids (SPs) in a wide range of species (Gunning *et al.*, 2007; Huang and Han, 2007; Tan and McCaffery, 2007; van Leeuwen and Tirry, 2007; Zhang *et al.*, 2007; Achaleke *et al.*, 2009; Ahmad *et al.*, 2009), with higher intensities of specific zones of esterase isozymes associated with high levels of resistance to these SPs (Flores *et al.*, 2005). Young *et al.*, (2006) have also reported that CCEs can be inhibited by the presumptively diagnostic P450 inhibitor piperonyl butoxide, which could imply that esterase-based SP resistance is even more common than hitherto suspected. However, the evidence for this remains controversial (see e.g., Li *et al.*, 2007). There is still no case where an esterase-based SP resistance has been elucidated at a molecular level, although intensive *in vitro* analysis of an esterase from (a presumptively SP susceptible) blowfly has shown not only that the wild-type enzyme can metabolize some SPs with good catalytic efficiency but also that its efficiency is much reduced for the more insecticidal isomers of the more recent Type 2 SPs (Heidari *et al.*, 2005; Devonshire *et al.*, 2007). Significantly, *in vitro* mutagenesis revealed several single amino acid changes in the active site, which radically altered and in some cases reversed the isomer preferences, albeit all the mutations remained much less active with Type 2 than with Type 1 SPs.

There is also now more evidence implicating CCEs in metabolic resistance to CBs (Oakeshott *et al.*, 2005; Jackson *et al.*, 2009; Kwon *et al.*, 2009), but there remains very little evidence as to the molecular basis for this. Intriguingly, there are now reports associating greater intensity of certain esterase isozymes with resistance to the *Bacillus thuringiensis* cry toxins in insect-resistant transgenic crops, even though these toxins lack ester bonds (e.g., Gunning *et al.*, 2005). Some involvement of the esterase(s) in the binding of the toxins to caterpillar midguts has been suggested; molecular data are now needed to investigate this possibility further.

References

Achaleke, J., Martin, T., Ghogomu, R.T., Vaissayre, M., Brévault, T., 2009. Esterase-mediated resistance to pyrethroids in field populations of *Helicoverpa armigera* (Lepidoptera: Noctuidae) from Central Africa. *Pest Manag. Sci.* 65, 1147–1154.

Ahmad, M., Saleem, M.A., Sayyed, A.H., 2009. Efficacy of insecticide mixtures against pyrethroid- and organophosphate-resistant populations of *Spodoptera litura* (Lepidoptera: Noctuidae). *Pest Manag. Sci.* 65, 266–274.

Alout, H., Weill, M., 2008. Amino-acid substitutions in acetylcholinesterase 1 involved in insecticide resistance in mosquitoes. *Chem. Biol. Interact.* 175, 138–141.

Alout, H., Berthomieu, A., Cui, F., Tan, Y., Berticat, C., et al., 2007. Different amino-acid substitutions confer insecticide resistance through acetylcholinesterase 1 insensitivity in *Culex vishnui* and *Culex tritaeniorhynchus* (Diptera: Culicidae) from China. *J. Med. Entomol.* 44, 463–469.

Alout, H., Djogbénou, L., Berticat, C., Chandre, F., Weill, M., 2008. Comparison of *Anopheles gambiae* and *Culex pipiens* acetylcholinesterase 1 biochemical properties. *Comp. Biochem. Physiol. B. Biochem. Mol. Biol.* 150, 271–277.

Andrews, M.C., Callaghan, A., Field, L.M., Williamson, M.S., Moores, G.D., 2004. Identification of mutations conferring insecticide-insensitive AChE in the cotton-melon aphid, *Aphis gossypii glover*. *Insect Mol. Biol.* 13, 555–561.

Bai, H., Ramaseshadri, P., Palli, S.R., 2007. Identification and characterization of juvenile hormone esterase gene from the yellow fever mosquito, *Aedes aegypti*. *Insect Biochem. Mol. Biol.* 37, 829–837.

Biswas, S., Russell, R.J., Jackson, C.J., Vidovic, M., Ganeshina, O., et al., 2008. Bridging the synaptic gap: Neuroligins and Neurexin 1 in *Apis mellifera*. *PLoS ONE 3*, e3542. doi:10.1371/journal.pone.0003542.

Biswas, S., Reinhard, J., Oakeshott, J., Russell, R., Claudianos, M.V.S., 2010. Sensory regulation of *Neuroligins* and *Neurexin I* in the honeybee brain. *PLoS ONE 5*, e9133. doi:10.1371/journal.pone.0009133.

Cassanelli, S., Reyes, M., Rault, M., Carlo Manicardi, G., Sauphanor, B., 2006. Acetylcholinesterase mutation in an insecticide-resistant population of the codling moth *Cydia pomonella* (L.). *Insect Biochem. Mol. Biol. 36*, 642–653.

Chen, M., Han, Z., Qiao, X., Qu, M., 2007. Resistance mechanisms and associated mutations in acetylcholinesterase genes in *Sitobion avenae* (Fabricius). *Pestic. Biochem. Physiol. 87*, 189–195.

Claudianos, C., Ranson, H., Johnson, R.M., Biswas, S., Schuler, M.A., et al., 2006. A deficit of detoxification enzymes: pesticide sensitivity and environmental response in the honeybee. *Insect Mol. Biol. 15*, 615–636.

Crone, E.J., Sutherland, T.D., Campbell, P.M., Coppin, C.W., Russell, R.J., et al., 2007a. Only one esterase of *Drosophila melanogaster* is likely to degrade juvenile hormone *in vivo*. *Insect Biochem. Mol. Biol. 37*, 540–549.

Crone, E.J., Zera, A.J., Anand, A., Oakeshott, J.G., Sutherland, T.D., et al., 2007b. Jhe in *Gryllus assimilis*: cloning, sequence-activity associations and phylogeny. *Insect Biochem. Mol. Biol. 37*, 1359–1365.

Cui, F., Qu, H., Cong, J., Liu, X.-L., Qiao, C.-L., 2007a. Do mosquitoes acquire organophosphate resistance by functional changes in carboxylesterases? *FASEB J 21*, 3584–3591.

Cui, F., Weill, M., Berthomieu, A., Raymond, M., Qiao, C.-L., 2007b. Characterization of novel esterases in insecticide-resistant mosquitoes. *Insect Biochem. Mol. Biol. 37*, 1131–1137.

de Carvalho, R.A., Torres, T.T., de Azeredo-Espin, A.M.L., 2006. A survey of mutations in the *Cochliomyia hominivorax* (Diptera: Calliphoridae) esterase E3 gene associated with organophosphate resistance and the molecular identification of mutant alleles. *Vet. Parasitol. 140*, 344–351.

Devonshire, A.L., Heidari, R., Huang, H.Z., Hammock, B.D., Russell, R.J., et al., 2007. Hydrolysis of individual isomers of fluorogenic pyrethroid analogs by mutant carboxylesterases from *Lucilia cuprina*. *Insect Biochem. Mol. Biol. 37*, 891–902.

Durand, N., Carot-Sans, G., Chertemps, T., Montagne, N., Jacquin-Joly, E., et al., 2010. A diversity of putative carboxylesterases are expressed in the antennae of the noctuid moth *Spodoptera litteralis*. *Insect Mol. Biol. 19*, 87–97.

Flores, A.E., Albeldaño-Vázquez, W., Salas, I.F., Badii, M.H., Becerra, H.L., et al., 2005. Elevated [alpha]-esterase levels associated with permethrin tolerance in *Aedes aegypti* (L.) from Baja California, Mexico. *Pestic. Biochem. Physiol. 82*, 66–78.

Gao, J.-R., Yoon, K.S., Frisbie, R.K., Coles, G.C., Clark, J.M., 2006. Esterase-mediated malathion resistance in the human head louse, *Pediculus capitis* (Anoplura: Pediculidae). *Pestic. Biochem. Physiol. 85*, 28–37.

Gunning, R.V., Dang, H.T., Kemp, F.C., Nicholson, I.C., Moores, G.D., 2005. New resistance mechanism in *Helicoverpa armigera* threatens transgenic crops expressing *Bacillus thuringiensis* Cry1Ac toxin. *Appl. Environ. Microbiol. 71*, 2558–2563.

Gunning, R.V., Moores, G.D., Jewess, P., Boyes, A.L., Devonshire, A.L., et al., 2007. Temporal synergism by microencapsulation of piperonyl butoxide and γ-cypermethrin overcomes insecticide resistance in crop pests. *Pest Manag. Sci 63*, 276–281.

Harel, M., Kryger, G., Rosenberry, T.L., Mallender, W.D., Lewis, T., et al., 2000. Three-dimensional structures of *Drosophila melanogaster* acetylcholinesterase and of its complexes with two potent inhibitors. *Protein Sci 9*, 1063–1072.

Hartley, C.J., Newcomb, R.D., Russell, R.J., Yong, C.G., Stevens, J.R., et al., 2006. Amplification of DNA from preserved specimens shows blowflies were pre-adapted for the rapid evolution of insecticide resistance. *Proc. Natl. Acad. Sci. USA 103*, 8757–8762.

Heidari, R., Devonshire, A.L., Campbell, B.E., Dorrian, S.J., Oakeshott, J.G., et al., 2005. Hydrolysis of pyrethroids by carboxylesterases from *Lucilia cuprina* and *Drosophila melanogaster* with active sites modified by *in vitro* mutagenesis. *Insect Biochem. Mol. Biol. 35*, 597–609.

Huang, S., Han, Z., 2007. Mechanisms for multiple resistances in field populations of common cutworm, *Spodoptera litura* (Fabricius) in China. *Pestic. Biochem. Physiol. 87*, 14–22.

Jackson, C., Sanchez-Hernandez, J., Wheelock, C., Oakeshott, J., 2009. Carboxylesterases in the metabolism and toxicity of pesticides. In: Satoh, T., Gupta, R. (Eds.), Anticholinesterase Pesticides: Metabolism, Neurotoxicity and Epidemiology. Wiley, Hoboken.

Karunaratne, S.H.P.P., Damayanthi, B.T., Fareena, M.H.J., Imbuldeniya, V., Hemingway, J., 2007. Insecticide resistance in the tropical bedbug *Cimex hemipterus*. *Pestic. Biochem. Physiol. 88*, 102–107.

Kim, H.J., Yoon, K.S., Clark, J.M., 2007. Functional analysis of mutations in expressed acetylcholinesterase that result in azinphosmethyl and carbofuran resistance in Colorado potato beetle. *Pestic. Biochem. Physiol. 88*, 181–190.

Kristensen, M., Huang, J., Qiao, C.-L., Jespersen, J.B., 2006. Variation of *Musca domestica* L. acetylcholinesterase in Danish housefly populations. *Pest Manag. Sci. 62*, 738–745.

Kwon, D.H., Choi, B.R., Lee, S.W., Clark, J.M., Lee, S.H., 2009. Characterization of carboxylesterase-mediated pirimicarb resistance in *Myzus persicae*. *Pestic. Biochem. Physiol. 93*, 120–126.

Lee, D.-W., Choi, J.Y., Kim, W.T., Je, Y.H., Song, J.T., et al., 2007. Mutations of acetylcholinesterase1 contribute to prothiofos-resistance in *Plutella xylostella* (L.). *Biochem. Biophys. Res. Commun. 353*, 591–597.

Li, A.Y., Guerrero, F.D., Pruett, J.H., 2007. Involvement of esterases in diazinon resistance and biphasic effects of piperonyl butoxide on diazinon toxicity to *Haematobia irritans irritans* (Diptera: Muscidae). *Pestic. Biochem. Physiol. 87*, 147–155.

Liming, T., Mingan, S., Jiangzhong, Y., Peijun, Z., Chuanxi, Z., et al., 2006. Resistance pattern and point mutations of insensitive acetylcholinesterase in a carbamate-resistant strain of housefly (*Musca domestica*). *Pestic. Biochem. Physiol. 86*, 1–6.

Mackert, A., do Nascimento, A.M., Bitondi, M.M.G., Hartfelder, K., Simoes, Z.L.P., 2008. Identification of a juvenile hormone esterase-like gene in the honey bee *Apis mellifera* L. – expression analysis and functional assays. *Comp. Biochem. Physiol. B. Biochem. Mol. Biol. 150*, 33–44.

Magaña, C., Hernández-Crespo, P., Brun-Barale, A., Couso-Ferrer, F., Bride, J.-M., et al., 2008. Mechanisms of resistance to malathion in the medfly *Ceratitis capitata*. *Insect Biochem. Mol. Biol. 38*, 756–762.

Oakeshott, J.G., Devonshire, A.L., Claudianos, C., Sutherland, T.D., Horne, I., et al., 2005. Comparing the organophosphorus and carbamate insecticide resistance mutations in cholin- and carboxyl-esterases. *Chem. Biol. Interact. 157–158*, 269–275.

Oakeshott, J.G., Johnson, R.M., Berenbaum, M.R., Hanson, R., Cristino, R.S., et al., 2010. Metabolic enzymes associated with xenobiotic and chemosensory responses in *Nasonia vitripennis*. *Insect Mol. Biol. 19*, s1, 147–163.

Robin, G.C., Bardsley, L.M.J., Coppin, C., Oakeshott, J.G., 2009. Birth and death of genes in the β-esterase cluster of *Drosophila. J. Mol. Evol. 69*, 10–21.

Russell, R.J., Claudianos, C., Campbell, P.M., Horne, I., Sutherland, T.D., et al., 2004. Two major classes of target site insensitivity mutations confer resistance to organophosphate and carbamate insecticides. *Pestic. Biochem. Physiol. 79*, 84–93.

Strode, C., Wondji, C.S., David, J.-P., Hawkes, N.J., Lumjuan, N., et al., 2008. Genomic analysis of detoxification genes in the mosquito *Aedes aegypti*. *Insect Biochem. Mol. Biol. 38*, 113–123.

Tan, J., McCaffery, A.R., 2007. Efficacy of various pyrethroid structures against a highly metabolically resistant isogenic strain of *Helicoverpa armigera* (Lepidoptera: Noctuidae) from China. *Pest Manag. Sci. 63*, 960–968.

Teese, M.G., Campbell, P.M., Scott, C., Gordon, K.H.J., Southon, A. et al., 2010. Gene identification and proteomic analysis of the esterases of the cotton bollworm, *Helicoverpa armigera*. *Insect Biochem. Mol. Biol. 40*, 1–16.

The Drosophila 12 Genomes Consortium, 2007. Evolution of genes and genomes on the *Drosophila* phylogeny. *Nature 450*, 203–218.

Tsubota, T., Shimomura, M., Ogura, T., Seino, A., Nakakura, T. et al., 2010. Molecular characterisation and functional analysis of novel carboxyl/cholinesterases with CQSAG motif in the silkworm *Bombyx mori*. *Insect Biochem. Mol. Biol. 40*, 100–112.

van Leeuwen, T., Tirry, L., 2007. Esterase-mediated bifenthrin resistance in a multiresistant strain of the two-spotted spider mite, *Tetranychus urticae*. *Pest Manag. Sci. 63*, 150–156.

Wogulis, M., Wheelock, C.E., Kamita, S.G., Hinton, A.C., Whetstone, P.A., et al., 2006. Structural studies of a potent insect maturation inhibitor bound to the juvenile hormone esterase of *Manduca sexta*. *Biochemistry 45*, 4045–4057.

Young, S.J., Gunning, R.V., Moores, G.D., 2006. Effect of pretreatment with piperonyl butoxide on pyrethroid efficacy against insecticide-resistant *Helicoverpa armigera* (Lepidoptera: Noctuidae) and *Bemisia tabaci* (Sternorrhyncha: Aleyrodidae) 62, 114–119.

Yu, Q-Y., Lu, C., Li, W-L., Xiang, Z-H., Zhang, Z., 2009. Annotation and expression of carboxylesterases in the silkworm, *Bombyx mori*. *BMC Genomics 10*, 553.

Zhang, L., Gao, X., Liang, P., 2007. Beta-cypermethrin resistance associated with high carboxylesterase activities in a strain of house fly, *Musca domestica* (Diptera: Muscidae). *Pestic. Biochem. Physiol. 89*, 65–72.

8 Glutathione Transferases

H Ranson and J Hemingway, Liverpool School of
Tropical Medicine, Liverpool, UK

© 2005, Elsevier BV. All Rights Reserved.

8.1. Introduction

Glutathione transferases (GSTs) are found ubiquitously in aerobic organisms. They were first discovered in animals in 1961 where they were postulated to play a role in the detoxification of drugs (Booth *et al.*, 1961). Their central role in detoxification and drug resistance pathways in mammals has now been established but additional functions are continually being attributed to this complex enzyme family. For example, GSTs are important mediators in oxidative stress responses, are involved in the synthesis of prostaglandins, and facilitate intracellular transport of hydrophobic compounds. Mammalian GSTs are particularly well studied due to their role in cancer epidemiology and treatment. Several GSTs can metabolize environmentally derived carcinogens and polymorphisms in these genes are linked to cancer risk. In addition, the overexpression of GSTs in tumor cells can contribute to drug resistance (Landi, 2000).

The vast majority of GSTs are cytosolic dimeric proteins with typical molecular masses of around 50 kDa. A small subset of trimeric microsomal GSTs that are very distantly related to the cytosolic GSTs occur in mammals and insects. Most organisms contain multiple forms of GSTs with both specific and overlapping substrate preferences.

In 1985, GSTs had only been identified in a limited number of insect species including the grass grub, greater wax moth, housefly, and American cockroach (Mannervik, 1985). The importance of this enzyme family in many different types of insecticide resistance had been demonstrated and multiple forms were known to exist in individual insects. That GSTs have now been studied in over 30 different insect species is testimony to the importance of this enzyme family. This review describes the diversity of the GSTs family in insects and highlights some of the diverse functions of this complex enzyme family.

8.2. Methods Used to Study Insect GSTs

Although a wide range of biochemical techniques have been exhaustively applied to the isolation of

GSTs from numerous insect species, relatively few enzymes have been purified to homogeneity. Frequently, after several purification steps, multiple bands were still visible after sodium dodecyl sulfate (SDS) electrophoresis and, moreover, given our current knowledge of the extent of certain insect GST classes, a single band on an SDS gel may not always be indicative of a homogeneous enzyme preparation. The choice of purification techniques also led to a systematic bias in the classes of GSTs identified with some GST classes being more amenable to purification. The methods employed have now been shown to capture only a relatively small subset of the total GST population. Thus, it is no surprise that advances in gene cloning and recombinant protein expression techniques were seized upon by those studying the role of individual GST enzymes in insects. However, early molecular biology approaches to the isolation of insect GSTs were also biased towards the isolation of certain subsets of GSTs and it was not until the era of large-scale genome sequencing projects that the full extent of this enzyme family was appreciated (see Section 8.3). Some of the various approaches that have been used to isolate and characterize insect GSTs are outlined below.

8.2.1. Isolation of Insect GSTs

8.2.1.1. Biochemical purification   Early purification approaches used conventional column chromatographic methods such as ion exchangers, gel filtration, and hydroxyapatite chromatography to isolate GSTs. These approaches were very time consuming and often ineffective, and purification of many individual isoenzymes was really only feasible after the introduction of affinity chromatography (Yu, 1996). Original affinity chromatography columns used the glutathione conjugate of sulfobromophthalein immobilized to an agarose matrix as a ligand (e.g., Clark and Dauterman, 1982; Clark and Shamaan, 1984). Problems sometimes arose when recovering the bound enzyme from this matrix and most purification methods now use glutathione coupled via the sulfur atom or S-hexyl glutathione coupled via the α-amino group of the glutamyl residue as ligands (e.g., Grant et al., 1991; Prapanthadara et al., 1993). Enzymes that bind to the columns are usually eluted using glutathione and the GST activity of different fractions determined with the "general substrate" for all GSTs, i.e., 1-chloro-2,4-dinitrobenzene (CDNB). Once eluted, the multiple enzymes are sometimes further separated by nondenaturing polyacrylamide gel electrophoresis (PAGE), isoelectrofocusing, or chromatofocusing.

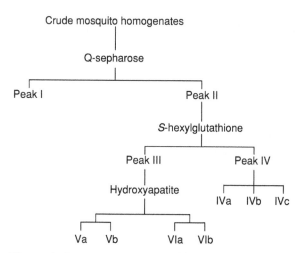

Figure 1   Partial purification of GSTs from the mosquito *A. gambiae* by column chromatography. Bound enzymes are shown on the right arm of the branch. (Adapted from Prapanthadara, L., Hemingway, J., Ketterman, A.J., **1995**. DDT-resistance in *Anopheles gambiae* (Diptera: culicidae) from Zanzibar, Tanzania, based on increased DDT-deyhdrochlorinase activity of glutathione S-transferases. *Bull. Ent. Res. 85*, 267–274.

There are two major limiting factors associated with the purification approaches outlined above. First, many GSTs do not bind to any of these affinity columns. It is generally reported that theta class GSTs are not retained by GSH-based affinity matrices (Meyer et al., 1991) but, even within a particular GST class, individual enzymes can sometimes be divided into bound and unbound fractions (Sawicki et al., 2003; Ortelli et al., 2004). Second, CDNB is not a good substrate for all GSTs. Several GSTs have now been characterized that have low or undetectable activity with this substrate (e.g., Singh et al., 2001; Sawicki et al., 2003).

Nevertheless, despite these limitations, the application of biochemical purification techniques provided the first indications of the diversity, in terms of both physical properties and substrate range, of this enzyme family in insects. For example, using the purification scheme outlined in **Figure 1**, eight different fractions with CDNB conjugation activity, each containing multiple isozymes, were obtained from the mosquito *Anopheles gambiae* (Prapanthadara et al., 1993).

A summary of the GSTs isolated biochemically from insects is given in **Table 1**. A review of the purification procedures used to isolate many of these enzymes is provided by Clark (1989).

8.2.1.2. Cloning of insect GSTs   The approaches that have led to the successful cloning of insect GSTs can be broadly divided into five categories: (1) use

Table 1 Examples of glutathione transferases isolated from insects by biochemical or molecular cloning techniques

Insect order	Species name	Common name	Number of enzymes isolated by biochemical techniques[a]	Number of genes cloned	Reference
Coleoptera	*Costelytra zealandica*	New Zealand grass grub	3		Clark *et al.* (1985)
Dictyoptera	*Periplaneta Americana*	American cockroach	5	1	Arruda *et al.* (1995)
Diptera	*Aedes aegypti*	Yellow fever mosquito	2, 3	1	Grant *et al.* (1991), Grant and Hammock (1992)
	Anopheles gambiae	African malaria mosquito	7	28	Prapanthadara *et al.* (1993), Reiss and James (1993), Ranson *et al.* (1997, 1998, 2001), Ding *et al.* (2003)
	Culicoides variipennis sonorensis	Biting midge		1	Abdallah *et al.* (2000)
	Drosophila melanogaster	Fruit fly	1, 2	13	Toung *et al.* (1990, 1993), Sawicki *et al.* (2003), Beall *et al.* (1992)
	Lucilia cuprina	Sheep blowfly	1	1	Board *et al.* (1994)
	Musca domestica	Housefly	4, 2, 9	7	Clark *et al.* (1986), Chien *et al.* (1995), Wei *et al.* (2001)
Hemiptera	*Nilaparvata lugens*	Rice brown planthopper	1	1	Vontas *et al.* (2002)
	Triatoma infestans	Kissing bug	1		Wood *et al.* (1986)
Lepidoptera	*Choristoneura fumiferana*	Spruce budworm		1	Feng *et al.* (1999)
	Diatraea saccharalis	Sugar borer	2		Tiwari *et al.* (1991)
	Eoreuma loftni	Mexican rice borer	3		Tiwari *et al.* (1991)
	Galleria mellonella	Greater wax moth	1, 4		Clark *et al.* (1977), Baker *et al.* (1994)
	Helicoverpa zea	Corn earworm	3, 1		Chien and Dauterman (1991)
	Manduca sexta	Tobacco hornworm	2	3	Snyder *et al.* (1995), Rogers *et al.* (1999)
	Orthosia gothica	Hebrew character moth	4		Egaas *et al.* (1995)
	Plutella xylostella	Diamondback moth	1, 4, 3	1	Chiang and Sun (1993), Huang *et al.* (1998)
	Spodoptera frugiperda	Fall armyworm	4		Kirby and Ottea (1995)

[a]In species with more than one report of the isolation of GST enzymes, where it can be difficult to determine the relationship between the enzymes described in each study, multiple entries are given. Note also that the figures given do not distinguish between homodimers and heterodimers.

of GST-specific antibodies to screen cDNA expression libraries; (2) PCR using degenerate primers; (3) screening of cDNA or gDNA libraries with GST nucleotide probes; (4) searching nucleotide sequence databases; and (5) fortuitous discovery. Of these, the latter approach accounted for a surprising proportion of the early insect GST gene sequences determined and is still a valuable source of new sequences for species without large volumes of sequence data. Since many GST classes are encoded by closely related multigene families (and these are often clustered in the genome; see Section 8.5.1), the identification of the first gene within a family has often had a cascade effect with multiple additional members of the gene family being identified in quick succession. The difficult step is isolating the first member of a particular gene class. In *Drosophila melanogaster* and *Musca domestica*, two of the first insect GSTs to be isolated were obtained by raising antibodies against GSTs purified by affinity chromatography from these species and using these to screen expression libraries (Toung et al., 1990; Fournier et al., 1992). A similar approach was used to isolate GSTs from the spruce budworm *Choristoneura fumiferana* (Feng et al., 1999) and the biting midge *Culicoides variipennis* (Abdallah et al., 2000), although in both of these cases, the antibodies were raised against unknown proteins rather than purified GSTs.

The degenerate PCR approach, using primers designed by homology to other known insect GSTs, was successfully employed to identify the first GST gene from the brown planthopper *Nilaparvata lugens* (Vontas et al., 2002) and the diamondback moth *Plutella xylostella* (Huang et al., 1998). *DmGST2* the first of a new class of GSTs to be identified in insects (now classified as the sigma class) was amplified as a secondary PCR product obtained during an unrelated project (Beall et al., 1992) and the orthologous gene from *A. gambiae* was also fortuitously isolated during a library screen using the coding region of a salivary gland-specific gene as a probe (Reiss and James, 1993). The recurrent isolation of GST clones using antibodies or DNA probes designed to bind to other protein or gene families may partly reflect the relatively high frequency with which some members of the GST family are expressed. For example, the accidentally amplified *DmGST2* described above has now been shown to account for about 2% of the total soluble protein in *Drosophila* (Singh et al., 2001).

Searching EST databases and partial or full genome sequence databases has led to the detection of fragments of GST-like sequence that have been used as a starting point for obtaining full-length gene sequences by PCR-based techniques (Snyder et al., 1995; Ranson et al., 2001; Sawicki et al., 2003).

Several insect GST genes have been expressed *in vitro* in order to characterize the recombinant proteins. The absence of posttranslational modification and their location in the cytosol make GST proteins amenable to expression in *Escherichia coli* (e.g., Ranson et al., 1997; Huang et al., 1998; Wei et al., 2001) although baculovirus expression systems have also been used (Feng et al., 1999). One of the potential limitations of using recombinant proteins to characterize GST activity is the absence of heterodimers. This problem has been elegantly addressed in plants by coexpressing two GST subunits and separating the recombinant proteins into homodimers and heterodimers by chromatofocusing or anion exchange chromatography. By constructing

tandem expression systems, the transcription and translation of both subunits are coordinated helping to ensure that they are expressed at the same level (Dixon et al., 1999).

8.2.2. Characterization of Substrate Specificity

A number of model substrates are widely used to characterize the substrate specificity of GSTs, the most common of which is the conjugation of reduced glutathione (GSH) to CDNB. This assay is simple to conduct and the rate of conjugation can be monitored by measuring the change in optical density at 340 nm. The conjugation of GSH to 1,2-dichloro-4-nitrobenzene (DCNB) can also be measured under similar assay conditions although the activity of most insect GST enzymes is higher with CDNB than DCNB. Other model substrates used to characterize GST activity include *p*-nitrophenol acetate (*p*-NPA), *p*-nitrophenyl bromide (*p*-NPB), and ethacrynic acid. The selenium independent peroxidase activity of GSTs is usually detected using cumene hydroperoxide or *t*-butyl hydroperoxide (*t*-BHP).

A vast array of GST inhibitors have been described (Mannervik and Danielson, 1988) and the effects of these on GST enzymes can be useful in distinguishing between different enzymes. Those commonly used to study insect GSTs include bromosulfophthalein, cibacron blue, and glutathione derivatives (Vontas et al., 2002). Quinones can also have strong inhibitory effects on GSTs and this can be problematic in purification of insect GSTs (Grant et al., 1991).

These model substrates and inhibitors have been proved very valuable for characterizing and comparing the various GST enzymes, and many early approaches to the classification of insect GSTs relied heavily on data generated from such experiments (see below). More recent studies on insect GSTs have focused mainly on their role in the detoxification of xenobiotics or in protection against oxidative stress, and assays have now been developed to assess the role of individual enzymes in these physiological processes. For example, spectrophotometric, chromatographic, and radiolabeling techniques can detect products generated by the GST catalyzed metabolism of many insecticides and products of lipid peroxidation reactions (Clark et al., 1986; Prapanthadara et al., 1995; Singh et al., 2001).

8.3. Classification and Nomenclature

Mammalian GST nomenclature was originally based on substrate specificities. Five GST enzyme types

were recognized and referred to as glutathione-
S-aryltransferase, glutathione-S-epoxidetransferase,
glutathione-S-alkyltransferase, glutathione-S-aralkyl-
transferase, and glutathione-S-alkenetransferase.
However, later studies on GSTs isolated from rat
livers found that each enzyme recognized a broad
and overlapping range of substrates and thus this
nomenclature was abandoned in favor of the use
of roman letters reflecting the order of elution
from a CM-cellulose ion exchanger (Habig *et al.*,
1974). Subsequently, the realization that GSTs
existed as both homo- and heterodimers led to a
further change in nomenclature to reflect that subunit composition and individual subunits of GSTs
were numbered in the order in which they were
discovered and classified. These GSTs were divided
into three distinct groups on the basis of their
isoelectric points: the basic (α-ε), near neutral (μ),
and acidic (π) groups (Mannervik, 1985). These
groups were later designated as classes and assigned
the Greek letters alpha, mu, and pi (Mannervik and
Danielson, 1988).

In early studies on insect GSTs, where multiple
enzymes were isolated from a species they were
generally assigned numbers according to their
order of elution from the various purification procedures employed or isoelectric points (Clark *et al.*,
1985; Prapanthadara *et al.*, 1993). Two immunologically distinct classes of GSTs were later recognized in houseflies and designated class I and class II
(Fournier *et al.*, 1992). The first insect GST to be
cloned was *DmGST1* from *D. melanogaster* (Toung
et al., 1990). The ortholog of this gene was later
cloned from *M. domestica* and proposed to be the
sole member of the class I GST gene family in this
species (Fournier *et al.*, 1992). However, soon after,
a cluster of eight GST genes (termed the *gstD* genes
and including the original class I *DmGST1* gene)
were identified in *D. melanogaster* (Toung *et al.*,
1993). Representatives of the class II family were
also cloned from *D. melanogaster* and *A. gambiae*
and in each species, the class II family was reported
to be encoded by a single gene (Beall *et al.*, 1992;
Reiss and James, 1993). As predicted by their immunological properties, these class II GSTs had very
low levels of sequence identity to the class I genes
and were proposed to have a different physiological
function. This was supported by immunohistology
studies on houseflies, which located the class I GSTs
to the hemolymph and the class II GSTs predominately to the indirect flight muscles in the thorax
(Franciosa and Berge, 1995).

The advent of large-scale EST and full genome
sequencing data has led to a marked increase in the
number of GST classes recognized in insects, plants,

Table 2 Classes of cytosolic glutathione transferases and
their representation in insects

Class	Lineage	Putative number in *A. gambiae*	Putative number in *D. melanogaster*
Alpha	Mammals		
Beta	Bacteria		
Delta	Insects	15	11
Epsilon	Insects	8	14
Kappa	Mammals		
Lambda	Plants		
Mu	Mammals		
Omega	Mammals, insects, nematodes	1	5
Pi	Mammals		
Phi	Plant		
Tau	Plant		
Theta	Mammals, insects, plants	2	4
Sigma	Molluscs, helminths, nematodes, insects, mammals	1	1
Zeta	Plants, nematodes, mammals, insects	1	2

and mammals. A total of nine different classes of
cytosolic GSTs have now been described (**Table 2**).
In insects, it became clear that the existing classification of insect GSTs into class I and class II was
inadequate and new guidelines were proposed for
the naming of insect GSTs along the lines of the
mammalian classification system (Chelvanayagam
et al., 2001).

8.3.1. Nomenclature Guidelines for Insect GSTs

As a general rule, GST sequences that share over
40% amino acid sequence identity are assigned
to the same class. A robust nomenclature requires
a degree of flexibility in class assignments, however, and ideally other properties such as phylogenetic relationships, immunological properties,
tertiary structure, and ability to form heterodimers
should be employed in classification decisions. As in
mammals, kinetic characteristics have proven a
poor criterion for classification with several GSTs
from different insect classes catalyzing similar reactions. At least six classes of insect GSTs have been
identified in both *A. gambiae* and *D. melanogaster*
(**Table 2**) (Ranson *et al.*, 2002). Despite the availability of full genome sequences for two insect species, the possibility that further insect GST classes

exist cannot be discounted. In particular, several of the assignments to the delta and epsilon classes are weakly supported by phylogenetic analysis and may be more correctly placed in a separate class (Ding *et al.*, 2003).

Individual GST subunits are assigned names indicating the species they were isolated from and the GST class. They are also given a number that may either reflect the order of discovery or the genomic organization. Thus, *AgGSTd12* is the twelfth member of the *A. gambiae* delta class of GSTs to be identified. The proteins, represented by capital letters, are nonitalicized and contain information on the composition of the dimer. So the designation AgGSTD12-12 represents a homodimer of two AgGSTD12 subunits, whereas AgGSTD11-12 would be a heterodimer made up of a subunit of AgGSTD11 and a subunit of AgGSTD12. In practice, except where ambiguity may arise, the species prefix is usually omitted.

8.3.2. Classes of Insect GSTs

8.3.2.1. Delta class   The delta class is one of the two largest classes of insect GSTs (**Table 2**). GSTs originally classified as belonging to insect class I have mostly been reclassified as delta GSTs. There are 15 delta GST genes in *A. gambiae* and 11 in *D. melanogaster* (Ranson *et al.*, 2002). GSTs from this class are found uniquely in insects and have been cloned from many species including *Lucilia cuprina* (Board *et al.*, 1994), *M. domestica* (Fournier *et al.*, 1992), *N. lugens* (Vontas *et al.*, 2002), and *C. variipennis* (Abdallah *et al.*, 2000). Multiple recombinant delta GST homodimers have been produced *in vitro* and biochemically characterized. Most, but not all, have activity with CDNB and the majority bind to glutathione or *S*-hexylglutathione columns. Delta class GSTs have been implicated in resistance to all the major classes of insecticide (Wang *et al.*, 1991; Tang and Tu, 1994; Vontas *et al.*, 2002).

8.3.2.2. Epsilon class   The epsilon class is also a large, insect-specific class. Members of this class have been cloned from *P. xylostella*, *M. domestica*, *Manduca sexta*, *A. gambiae*, and *D. melanogaster* (Snyder *et al.*, 1995; Huang *et al.*, 1998; Singh *et al.*, 2000; Ranson *et al.*, 2001; Wei *et al.*, 2001). The majority of epsilon GSTs that have been characterized have low levels of activity with CDNB and are often not retained by glutathione affinity matrices. Both these factors may explain the relatively recent discovery of this class. As with the delta class, epsilon GSTs play an important role in the detoxification

of insecticides (Huang *et al.*, 1998; Wei *et al.*, 2001; Ortelli *et al.*, 2004).

8.3.2.3. Omega class   The omega GST class was first identified in humans by bioinformatic techniques (Board *et al.*, 2000). Members of the omega class have also been found in the mouse, rat, *Caenorhabditis elegans*, *Schistosoma mansoni*, *D. melanogaster*, and *A. gambiae* (Board *et al.*, 2000; Ranson *et al.*, 2002). In most species, the omega GSTs appear to be encoded by a single gene but five putative omega GST genes have been identified in *D. melanogaster* (Ranson *et al.*, 2002). The human recombinant GSTO1-1 has low levels of activity with most typical GST substrates but high thiol transferase activity. Its physiological function is unclear but one suggestion is that it may play a housekeeping role in protecting against oxidative stress by removing *S*-thiol adducts from proteins (Board *et al.*, 2000).

8.3.2.4. Sigma class   The sigma class GSTs are related to the catalytically inactive cephalopod *S*-crystallin, a major component of eye lens (Ji *et al.*, 1995). A structural role has also been suggested for the insect sigma GSTs (formally known as the insect class II GSTs), as in both *D. melanogaster* and *M. domestica* these proteins are found predominately in the indirect flight muscles in close association with troponin-H. Both of these enzymes also possess a proline/alanine rich N-terminal extension, which may aid attachment to the flight muscle. Recently, however, insect sigma GSTs (with or without the N-terminal extension) have been shown to be catalytically active (Singh *et al.*, 2001). This class of proteins has low levels of activity with typical GST substrates but a high affinity for the lipid peroxidation product 4-hydroxynonenal (4-HNE). A means of reducing the levels of reactive oxygen species in highly metabolically active tissues such as flight muscles is essential and it is postulated that the high levels of GSTS1-1 found in this tissue may protect the flight muscle against by-products of oxidative stress (Singh *et al.*, 2001). Sigma GSTs are found in all major eukaryotic taxonomic groups except plants. In addition to the housefly and *Drosophila* sigma GSTs described above, members of this class lacking the N-terminal extension have been cloned or purified from *M. sexta*, *Blattella germanica*, *A. gambiae*, and *C. fumiferana* (Reiss and James, 1993; Snyder *et al.*, 1995; Arruda *et al.*, 1997; Feng *et al.*, 1999). In all cases to date, the sigma GSTs are encoded by a single gene, although in *A. gambiae*, two alternative transcripts from this gene have been detected (see Section 8.5.3.1).

8.3.2.5. Theta class Theta GST were first purified from mammalian livers. The first theta GSTs to be studied were unusual amongst the mammalian GSTs in that they had very low levels of activity with CDNB and were not retained by glutathione affinity matrices (Meyer *et al.*, 1991). Theta GSTs also occur in bacteria, plants, and insects and initially the theta GSTs were proposed to be the progenitor of all GST classes (Pemble and Taylor, 1992). However, as more GSTs were identified it became apparent that many GSTs, including the insect delta class, were inappropriately assigned to this class and proposals for the evolutionary pathway of GSTs were revised (see Section 8.8). Insects do possess several GST genes whose sequence identity and phylogeny warrant their inclusion in the theta class. These putative insect theta GSTs have not yet been biochemically characterized and their physiological role is unknown.

8.3.2.6. Zeta class The zeta class of GSTs was first identified by bioinformatic approaches. Zeta GSTs occur in many different species and their sequence is highly conserved, particularly at the N-terminus of the proteins where the SSCXWRVIAL motif is conserved in plants, insects, and mammals (Board *et al.*, 1997). The highly conserved structure of this protein suggests it plays an essential housekeeping role and in this regard, GSTZ1-1 catalyzes an important step in the tyrosine degradation pathway. Two putative zeta class GSTs have been identified in *D. melanogaster*, while a single zeta GST gene is found in *A. gambiae* (Ranson *et al.*, 2002).

8.3.2.7. Microsomal GSTs The microsomal GSTs have a different structure and genetic origin to the cytosolic GSTs but perform similar functions. The microsomal GST class is encoded by a single gene in *D. melanogaster*. Mutants that have this gene disrupted have reduced life spans but are still viable (Toba and Aigaki, 2000). Humans possess multiple microsomal GSTs some of which detoxify xenobiotic and lipid-derived products of reactive oxygen but other members of this class are involved in the synthesis of prostaglandins. Microsomal GSTs are sometimes referred to as membrane associated proteins in eicosanoid and glutathione metabolism (MAPEG) (Jakobsson *et al.*, 1999).

8.4. Protein Structure

There are two types of GST with very differing structures. The majority of insect GSTs are soluble dimeric proteins found in the cytosol. The second type, the microsomal class, are trimeric membrane associated proteins. At present, the three-dimensional structure has only been elucidated for four insect GSTs, three from the delta class and a single sigma class GST and much of our knowledge of the structure of insect GST proteins is extrapolated from knowledge of mammalian GSTs. Post-translational modification of GSTs appears to be a very rare phenomenon and has not been reported in insects.

8.4.1. Cytosolic GSTs

Each soluble GST is a dimer of subunits ranging from 24 to 28 kDa in size. Each subunit has two binding sites, the G site and the H site. The highly conserved G site binds the tripeptide glutathione and is largely composed of amino acid residues found in the N-terminal of the protein. The H site or substrate binding site is more variable in structure and is largely formed from residues in the C-terminal (Mannervik, 1985). An alignment of amino acid sequences of representative insect GSTs from different classes is shown in **Figure 2**. The active site residue at the N-termini of the protein is shown in bold. This residue interacts with and activates the sulfhydryl group of glutathione to generate the catalytically active thiolate anion (Armstrong, 1991). This nucleophilic thiolate anion is then capable of attacking substrates bound to the H site. In most mammalian GSTs and in all sigma GSTs, the active site residue is a tyrosine (Wilce and Parker, 1994; Agianian *et al.*, 2003). In the delta, epsilon, theta, and zeta classes this role is performed by a serine residue while in the omega class, cysteine is the active site residue (review: Sheehan *et al.*, 2001). Generally, mutation of the active site residue results in reductions in GST activity of up to 90% (Rushmore and Pickett, 1993; Caccuri *et al.*, 1997). However, it is noteworthy that GSTD3-3 in *D. melanogaster* lacks the 15 N-terminal amino acids and yet is still catalytically active (Sawicki *et al.*, 2003). The determination of the three-dimensional structures of GST subunits have identified additional residues that interact with glutathione (Wilce *et al.*, 1995; Agianian *et al.*, 2003) (**Figure 2**) and perhaps an alternative residue substitutes for the active site tyrosine in GSTD3-3. The *cis*-proline at residue 53 is conserved in all GSTs and appears to be critical for the correct formation of the GSH binding site.

The H site, which binds the hydrophobic substrate, is more variable in structure and contributes to the substrate specificity of the enzyme. The H site of the *Drosophila* sigma GST, GSTS1-1, is polar and unusually open in structure (Agianian *et al.*, 2003) whereas the delta class GSTs from *A. gambiae* have

                                                 β1

```
DeltaGST    ------------------------------------------------MDFYYLPGSAE
EpsilonGST  ------------------------------------------------MKLYKLDMSPP
SigmaGST    MADEAQAPPAEGAPPAEGEAPPPAEGAEGAVEGGEAAPPAEPAEPIKHSYTLFYFNVKAS
ThetaGST    ---------------------------------------------MSKNLKYYYDLMSQP
ZetaGST     --------------------------------------MANVDILPESQPILYSYWRSSC
OmegaGST    ---------------------------MSNGKHLAKGSSPPSLPDDGKLRLYSMRFCPY

                α1        β2                α1              β3        β4
DeltaGST    KRSVLMTAKNLGIELNKKLLNLQA--GEHLKPEFLKINPQHTIPTLVD---GDFALWESR
EpsilonGST  ARATMMVAEALGVKVDTVDVNLMK--GDHTTPEYLKKNPIHTVPLLED---GDLILHDSH
SigmaGST    AEPIRYLFAYGNQEYEDVRVTRDE--VPALKPT----MPMGQMPVLEV---DGKRVHQSL
ThetaGST    SRALWIFLEKTKLPYEKCLINLGK--GEHLTEEFKAINRFQKVPCITD---SQIKLAESV
ZetaGST     SWRVRIALNLKEIPYDIKPISLIKSGGEQHCNEYREVNPMEQVPALQI---DGHTLIESV
OmegaGST    AQRVHLMLDAKKIPYHAIYINLSE-----KPEWYLEKNPLGKVPALEIPGKEGVTLYESL
                                                              *
                α3                       α4
DeltaGST    AIMYYLVEKYGKNDSLFPKCPKKRAVINQRLYFDMG------------KSFADYYYPQIFAK
EpsilonGST  AIVTYLVDKY-GKSDALYPKDVKKRAQVDQKLYLDAT------ILFPRLRAVTF-LIFTE
SigmaGST    SMAREFAKTV-GLCGATEWEDLQIDLVDEINDFRLK--------LAVVSYEPEDE-----
ThetaGST    AIFRYLCREY-QVPDHWYPADSRRQALVDEYLEWQHHNTRATCAIYFQYVWLRPRMFGTK
ZetaGST     SIMYYLEETR-PQR-PLMPQDVLKRAKVREICEVVIA------SGVQPLQNLIVLIHVGE
OmegaGST    VLSDYIEEAYSAQQRKLYPADPFSKAQDRILIERFAG---------SVIGPYYRILFAAD
                          ←            →
                            α5                          α6
DeltaGST    APADE--PLYKKMEAARDELNTELEGH--QYVAGD-----SLTVADLALLASVSTFEVAG
EpsilonGST  GLKKPSDKMLKDIEEAYSILNSFLSTS--KYLAGD-----QLSLADISAVATVTSLVYVL
SigmaGST    IKEK--KLVTLNAEVIPFYLEKLEQT--VKENDG-----HLALG---KLTWADYVFAL
ThetaGST    VDPKQAEKYRGQMEGTLDFIEREYLGSGARFIAGD-----EITVADLLAACEIEQPRMAG
ZetaGST     EKKKE--WAQHWITRGFRAIEKLLSTSAGKFCVGD-----EITLADCCLVPQVFN-ARRF
OmegaGST    GIPPG---AITEFGAGLDIFEKELKARGTPYFGGDKPGMIDYMIWPWCERVDLLKFALGD
                                α7           α8
DeltaGST    -FDFSK--YANVAKWYANAKTVA-PGFDENWEGCLEFKKFFN-------------
EpsilonGST  PLDEAK--YPKVTAWLKTMKDLP-FVKSKNEPGVTQSGQWITSSLTSK-------
SigmaGST    TDMYN--YMVKRDLLEPYPALR-GWDAVNA-LEPIKAWIERRPVTEV------
ThetaGST    YDPCEG--RPNLTQWMARVRESTNPYYDQAHKLVNKFAQDTASKAKL-------
ZetaGST     HVDLRP--YPIILRIDRELEGHP-AFRAAHPSNQPDCPPEAAK-----------
OmegaGST    KYELDKERFGKLLQWRELMEKDD--AVKQSFISTEDHTKFLQSRKNGENNYDILA
```

Figure 2 Alignment of representative amino acid sequence from each insect GST class: delta GST = *L. cuprina* (Acc. No. P42860], epsilon GST = *P. xylostella* (Acc. No. U66342], sigma GST = *D. melanogaster* (Acc. No. M95198], theta GST = *A. gambiae* (Acc. No. AF15526], zeta GST = *A. gambiae* (Acc. No. AF15522], omega GST = *A. gambiae* (Acc. No. AY255856]. The active site residues are shown in bold and the conserved proline necessary for maintaining the GST fold is marked with an asterisk. The extent of the N and C domains is indicated by an arrow below the alignment. The two GSTs in this alignment for which three-dimensional structures have been resolved (Wilce *et al.*, 1995; Agianian *et al.*, 2003) are annotated as follows. Residues putatively involved in the binding of glutathione are underlined, those involved in substrate binding are shown in italics. Beta sheets are highlighted in blue, alpha helices in pink.

a mixture of polar and hydrophobic pockets (Oakley *et al.*, 2001). Again, elucidation of three-dimensional structures has predicted a number of residues involved in substrate binding (Wilce *et al.*, 1995; Oakley *et al.*, 2001; Agianian *et al.*, 2003). The putative H site residues in the *D. melanogaster* sigma GST and the delta class GST1 from *L. cuprina* are shown in **Figure 2**.

Representative crystal structures of vertebrate GSTs are available for all the cytosolic GST classes found in insects. The general structure is the same for all classes with the major differences occurring at the active site and at the intersubunit interface. The unique structural features of the different GST classes were recently comprehensively reviewed (Sheehan *et al.*, 2001) and will only be covered briefly here.

Each monomer consists of two distinct domains joined by a variable linker region. The N-terminal domain consists of four beta sheets and three flanking alpha helices and adopts a conformation similar to the thioredoxin domain found in many proteins that bind GSH or cysteine (Sheehan *et al.*, 2001). A conserved proline in the *cis* conformation (marked with an asterisk in **Figure 2**) is found in all GST sequences and appears to be important in maintaining the protein in a catalytically competent structure (Wilce *et al.*, 1995). The C-terminal domain is larger and consists of a variable number of alpha helices. The omega GST class is unusual in possessing an N-terminal extension that forms a unique structural unit with the C-terminal domain in human omega GSTs (Board *et al.*, 2000). Aligning the putative insect omega GSTs with the mammalian GST sequence suggests this may be a general characteristic of the omega GST structure. A 45 amino acid residue N-terminal extension is present in some sigma class GSTs (**Figure 2**) but, unlike the omega extension, this region of the protein is not essential for catalytic activity (Singh *et al.*, 2001; see Section 8.3.2.4).

The interface between the two subunits can be hydrophobic or hydrophilic and interactions between residues in both subunits are essential for dimer stability. Many GST classes have a V-shaped dimer interface with a "lock and key" motif linking the dimers. The "key," which can be aliphatic (zeta class and *D. melanogaster* sigma class) or aromatic (mammalian alpha, mu, and pi classes), from one subunit interacts with a hydrophobic pocket on the other subunit. The lock and key mechanism appears to be absent from the omega and theta classes of GSTs (Sheehan *et al.*, 2001).

Incompatibility in interfacial residues prevents heterodimers forming between two GST subunits from different classes. However, within a class, the formation of heterodimers can expand the range of functional proteins produced. The prevalence of GST heterodimers has not been widely studied in insects. However, an elegant series of experiments in plants, in which two GST subunits were consecutively expressed *in vitro*, found that even within a class not all GSTs will dimerize with each other to form heterodimers (Dixon *et al.*, 1999).

8.4.2. Microsomal GSTs

The microsomal GSTs are trimeric proteins with a molecular mass of approximately 50 kDa. The human MGST1 is a homotrimer found in the membranes of the endoplasmic reticulum and mitochondria. The three-dimensional structure of this protein has been resolved and is markedly different to the soluble GST classes. Each subunit consists of four transmembrane alpha helices. The homotrimer contains a single active site with the hydrophobic substrate binding site facing the cytosol (Schmidt-Krey *et al.*, 1999). The *D. melanogaster* microsomal GST class is encoded by a single gene and hence only homotrimeric proteins can be formed in this species (Toba and Aigaki, 2000). In contrast three distinct microsomal gene GSTs are transcribed in *A. gambiae* but their tertiary structure *in vivo* has not been resolved.

8.5. GST Gene Organization

The wide range of functions performed by GSTs is aided by the broad substrate specificities of individual enzymes and by the extensive nature of the GST supergene family in insects. Multiple gene duplications provide a degree of genetic security enabling individual GST genes to evolve to occupy new biochemical niches while still ensuring that essential

functions are covered. The insect-specific GST classes in particular are characterized by paralogous clusters of genes, indicating that expansion has occurred by local gene duplications (see below). The permissiveness of GSTs to alterations in their primary structure has also facilitated the diversification of this enzyme family. Very dramatic changes in substrate specificity can be achieved by a small number of amino acid substitutions (Ortelli *et al.*, 2004). In addition, several genetic mechanisms have been described in insects that contribute further levels of heterogeneity to the GST supergene family. In this section, we discuss current knowledge of the genomic organization of the GST supergene family in insects and describe some specific examples of alternative splicing, gene conversion, and allelic variation.

8.5.1. Clustering of Paralogous Genes

The physical location of GST genes in the genomes of *A. gambiae* and *D. melanogaster* is shown in **Figure 3**. In both species, the genes are found on all three chromosomes but several large clusters exist. In fact, of the total 69 putative GST genes present in these two species only 17 are present as singletons. In both species, the majority of epsilon class GSTs are found sequentially on both chromosome strands of a single polytene division and there is evidence of recent internal duplications within these gene clusters (Ding *et al.*, 2003; Sawicki *et al.*, 2003). The 11 *D. melanogaster* delta GSTs are sequentially arranged on chromosome 3R, division 87B. The orthologous class in *A. gambiae* is larger and individual members are scattered over all chromosomes, but two closely linked clusters each consisting of six genes are found on chromosome 2R. These two paralogous *A. gambiae* delta class GST clusters may be a result of a segmental duplication (Ding *et al.*, 2003).

8.5.2. Intron Size and Position

The frequency of introns in the insect GST supergene family varies widely between species, particularly within the insect-specific delta and epsilon classes. In the *D. melanogaster* delta and epsilon GSTs and the *M. domestica* delta GSTs, the majority of the genes are uninterrupted by introns within the coding sequence (Zhou and Syvanen, 1997; Sawicki *et al.*, 2003) although some have introns in their 5′UTRs (Lougarre *et al.*, 1999). In contrast, the coding sequence of all but one of the *A. gambiae* GSTs is interrupted by at least one intron. The intron size in *A. gambiae* ranges from 64 to 13 937 bp (Ding *et al.*, 2003). A common intron is found in the N-terminal

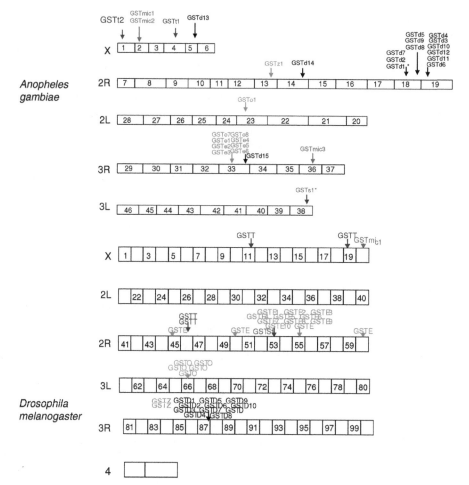

Figure 3   Chromosomal location of GST genes in two species of *Diptera*.

of 16 of the 28 cytosolic GST genes in this species (Ding *et al.*, 2003).

8.5.3. Mechanisms for Generating Additional Heterogeneity

8.5.3.1. Alternative splicing of GSTs   Alternative splicing of an RNA transcript can result in the production of different mRNAs, and thereby different proteins, from the same DNA segment. To date, alternative splicing of insect GST genes has only been reported in the *Anopheles* genus. A delta class GST *GSTd1* from *A. gambiae* (and its ortholog from *A. dirus*) is alternatively spliced to produce four distinct mature transcripts each encoding subunits with different biochemical properties (Ranson *et al.*, 1998; Jirajaroenrat *et al.*, 2001). In *A. gambiae*, the single sigma class GST gene is also alternatively spliced to produce two transcripts, both of which have been detected by RT-PCR (Ding *et al.*, 2003). In both of these cases, the mature transcripts share a common 5′ exon (constitutive exon) but differ in their use of 3′ exons (**Figure 4**). The constitutive

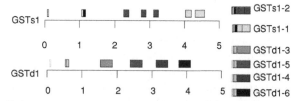

Figure 4   Alternative splicing of two *A. gambiae* GST genes. A schematic of the two genes is shown on the left and the mRNA transcripts on the right. The sigma class gene, *GSTs1*, produces two transcripts, *GSTs1-1* and *GSTs1-2*. The delta class *GSTd1* gene produces four distinct transcripts, *GSTd1-3* to *GSTd1-6*. Coding exons are shown by colored rectangles, 5′UTRs are shown by empty rectangles. The scale bar is in kilobases.

exon in each gene comprises the majority of the residues in the N-terminal domain whereas the variable exons encode the C-termini. As the carboxyl region of the GST subunit contains the majority of the residues involved in substrate binding (the H site), this form of alternative splicing is an efficient means of broadening the substrate recognition patterns of the GSTs with a minimal addition to the genome.

The factors governing choice of splice site in the *Anopheles* GSTs are unknown. For the delta GST at least, all four alternative transcripts are expressed throughout mosquito development and in both sexes. The expression pattern of this GST gene in different tissues has not yet been reported however, and it is possible that the choice of 3′ splice site is tissue dependent (Ranson *et al.*, 1998).

8.5.3.2. GST fusion genes

The delta class GSTs in *M. domestica* have an unusual genomic organization. Transcripts encoding five members of this class (*MdGST-1–MdGST-5*) have been detected in adult flies (Syvanen *et al.*, 1994; Zhou and Syvanen, 1997). Screening a genomic library for clones containing these GST genes identified a complex array of loci encoding MdGST-3 and MdGST-4. At least three genomic loci encoding MdGST-3 and two encoding MdGST-4 were found, although none of these were identical to the sequence of the corresponding cDNA clones. Furthermore, several loci that appear to have resulted from the fusion of the 5′ end of one of these genes with the 3′ of another were identified in two different strains of *M. domestica* (Zhou and Syvanen, 1997). The open reading frame (ORF) is preserved in all of these fusion genes but, surprisingly, the breakpoints are not conserved and there are no introns at fusion sites, or elsewhere in the coding sequence. The genetic mechanism that produces these fusion genes is unclear and it is not known whether or not these fusion genes encode functional GSTs. Nevertheless, it is possible that a special recombination mechanism associated with the delta class GST family in *M. domestica* may contribute additional potential for functional diversity (Zhou and Syvanen, 1997).

8.5.3.3. Allelic variation

Small numbers of amino acid substitutions can have dramatic effects on the substrate specificities of GST homodimers (Ketterman *et al.*, 2001) and thus the maintenance of allelic variation in a population can provide further functional diversification. In *A. gambiae*, two alleles of the epsilon class GST *GSTe1*, differing at 12 amino acid residues, are found in laboratory and field populations. One of these encodes a protein with very high levels of peroxidase activity and it has been proposed that this may provide a fitness advantage in the presence of insecticide exposure, thus explaining its frequency in *A. gambiae* populations (Ortelli *et al.*, 2003).

In addition to allelic variation, the formation of heterodimers can expand the range of functional proteins (see Section 8.4). The prevalence of GST heterodimers has not been widely studied in insects and it is often difficult to distinguish between genuine heterodimers and mixed populations of homodimers when characterizing GSTs isolated from insect tissues. An *in vitro* approach, similar to that adopted in plants (Dixon *et al.*, 1999), may help reveal the importance of heterodimers in generating diversity in the insect GST family.

8.5.4. GST Pseudogenes

Pseudogenes can be produced by deleterious mutations, which result in the silencing of a gene. This type of pseudogene, known as an unprocessed pseudogene, normally occurs in a duplicate gene and frequently pseudogenes are found proximal to the genes from which they are derived. Unprocessed pseudogenes are rarely transcribed although some are transcribed and not translated or, occasionally, translated into nonfunctional proteins. Two possible unprocessed pseudogenes have been described in the GST delta class of *A. gambiae*. One of these is untranscribed and the second is transcribed but predicted to encode a nonfunctional protein (Ding *et al.*, 2003). Both genes are located within the GST delta cluster on chromosome 2R. Originally, two of the *D. melanogaster* delta GSTs, *GSTD7* and *GSTD3*, were reported to be pseudogenes on the basis of abnormalities in their putative transcript lengths (Toung *et al.*, 1993). However, transcripts for both of these have since been detected by real-time polymerase chain reaction (RT-PCR) and recombinant proteins derived from these cDNAs encode catalytically active proteins (Sawicki *et al.*, 2003).

A different type of pseudogene, known as a processed pseudogene, is found in the GST supergene family in *D. melanogaster*. *Mgst1-psi* is a putative pseudogene of the single microsomal GST gene found in this species. This pseudogene contains a poly-A tail towards the 3′ end, has no introns, a premature stop codon, and a direct repeat at both ends. Presumably, it was derived from the reverse transcription and subsequent integration of *Mgst1* mRNA into the genome (Toba and Aigaki, 2000).

8.6. Functions of Insect GSTs

The primary function of GSTs is generally considered to be in the detoxification of both endogenous and xenobiotic compounds either by acting directly on these compounds or by catalyzing the secondary metabolism of a vast array of compounds that are oxidized by the cytochrome P450 family. Perhaps of equal importance is the role of GSTs in stress physiology. In addition, members of this enzyme family have been implicated in various biosynthetic pathways and may be involved in the intracellular transport of ligands. In insects, the majority of studies have focused on their role in the detoxification of foreign compounds, in particular insecticides and plant allelochemicals, and in many cases the endogenous substrates of insect GSTs are unknown. The application of functional genomic approaches to this enzyme family will hopefully redress this balance and enable the housekeeping functions of these enzymes to be fully elucidated.

8.6.1. Endogenous Substrates

The little that is known about the endogenous substrates of insect GSTs is generally inferred from functions that have been attributed to mammalian enzymes. An exception to this is the antennae-specific GST identified in *M. sexta*, which is proposed to play a vital role in olfactory function both by detoxifying compounds that can interfere with sex pheromone detection and, more specifically, by conjugating GSH to aldehyde pheromones thereby terminating signals from sex-pheromone odorants (Rogers *et al.*, 1999).

The wide range of substrates recognized by GSTs suggests they play a central role in cellular processes. One of these processes, GSH conjugation, is a key stage in the conversion of lipophilic compounds to water soluble metabolites that are more readily exported from the cell. GSTs are also important in the biosynthesis of prostaglandins (Meyer and Thomas, 1995). They can act as intracellular transporters of various nonsubstrate hydrophobic compounds in mammals and may be involved in similar processes in insects (Wilce and Parker, 1994). The zeta class of GST is one of the few enzymes with a clearly defined role in biosynthetic pathways in the cell. It catalyzes the isomerization of maleylacetoacetate to fumarylacetoacetate, an essential step in the catabolism of tyrosine (Blackburn *et al.*, 1998).

8.6.2. Protection against Oxidative Stress

GSTs play a vital role in the inactivation of toxic products of oxygen metabolism. Reactive oxygen species (ROS), including hydrogen peroxide, superoxide anions, and hydroxyl radicals, are generated during aerobic respiration. These ROS trigger a cascade of reactions that can be highly damaging to the cells. The generation of ROS initiates the conversion of polyunsaturated fatty acids to lipid hydroperoxides. These in turn can give rise to highly reactive α,β-unsaturated aldehydes, e.g., 4-HNE. 4-HNE is toxic at high concentrations but at low concentrations it has important signaling functions affecting cell proliferation, apoptosis,

and differentiation (Sawicki *et al.*, 2003). GSTs have a protective effect at several different stages in this pathway. Peroxidase activity has been detected in insect GSTs from the delta, epsilon, and sigma classes using the model substrate cumene hydroperoxide (Vontas *et al.*, 2001; Singh *et al.*, 2001; Ortelli *et al.*, 2004). Other delta and sigma insect GSTs can detoxify 4-HNE by conjugation with glutathione (Singh *et al.*, 2001; Sawicki *et al.*, 2003).

8.6.3. Detoxification of Xenobiotics

Insects, like all organisms, are continually being exposed to foreign chemical species. Some of these, such as insecticides, are in evolutionary terms relatively new threats, but other environmental pollutants, such as plant degradation products, have existed over a much longer timespan. GSTs can protect against these toxins by sequestration or, more commonly, by converting them to less toxic metabolites or conjugates that are tagged for export from the cell. Much attention has been focused on the role of insect GSTs in insecticide resistance because of the significant economic and public health problems associated with resistance. Insecticide resistance will be used to illustrate some of the different pathways of GST catalyzed detoxification.

8.6.3.1. Metabolism of insecticides   Elevated GST activity has been implicated in resistance to at least four classes of insecticides. Higher enzyme activity is usually due to an increase in the amount of one or more GST enzymes, either as a result of gene amplification or, more commonly, through increases in transcriptional rate, rather than qualitative changes in individual enzymes (review: Hemingway and Ranson, 2000). An overview of GST involvement in insecticide resistance is shown in **Figure 5**.

8.6.3.1.1. Organophosphates   This was the first insecticide class in which resistance due to increases in GST activity was detected. GSTs have been implicated in organophosphate resistance in many insect species (Hayes and Wolf, 1988). Recombinant GST enzymes from the diamondback moth and housefly have verified the role of these enzymes in the metabolism of organophosphate insecticides (Huang *et al.*, 1998; Wei *et al.*, 2001). Detoxification occurs via an O-dealkylation or O-dearylation reaction. In O-dealkylation, glutathione is conjugated with the alkyl portion of the insecticide, e.g., the demethylation of tetrachlorvinphos in resistant houseflies (Oppenoorth *et al.*, 1979), whereas the reaction of glutathione with the leaving group, important in the detoxification of parathion and methyl parathion in

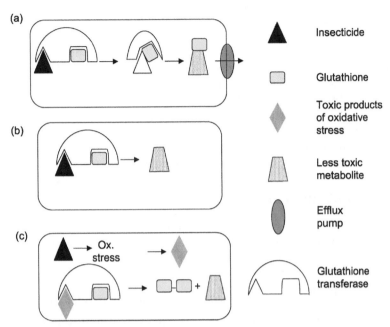

Figure 5 Overview of known GST involvement in insecticide resistance. (a) GSTs can detoxify insecticides via glutathione conjugation; the conjugates are then exported from the cell, e.g., demethylation of tetrachlorovinphos to demethyltetrachlorovinphos). (b) GSTs can detoxify the insecticide DDT via an elimination reaction. Glutathione acts as a proton source but no glutathione conjugate is produced (e.g., the dehydrochlorination of DDT to DDE). (c) GSTs with glutathione peroxidase activity can protect against insecticide induced oxidative stress.

the diamondback moth (Chiang and Sun, 1993), is an O-dearylation reaction.

GSTs can also catalyze the secondary metabolism of organophosphate insecticides. This insecticide class is usually applied in its noninsecticidal phosphorothionate form and activated to the insecticidal form by the action of cytochrome P450s in the insect. GSTs that can detoxify the active oxon analog have been described in mosquitoes and elevated activity of these enzymes is a cause of organophosphate resistance in *Anopheles subpictus* (Hemingway *et al.*, 1991).

8.6.3.1.2. Organochlorines DDT dehydrochlorination is a major route of detoxification in insects. The enzyme catalyzing this reaction was not initially recognized as a member of the GST family partly due to the inability to detect even a transient GSH conjugate and also because of the lack of DDT dehydrochlorinase activity in some purified GSTs. Evidence that DDT dehydrochlorinase is a GST was finally provided in the housefly. Three enzymes with CDNB conjugating activity were isolated from a DDT resistant strain of housefly and two of these were also active with DDT as a substrate (Clark and Shamaan, 1984). The DDT dehydrochlorinase reaction proceeds via a base abstraction of hydrogen, catalyzed by the thiolate anion generated in the

active site of the GST, leading to elimination of chlorine from DDT to generate DDE.

In addition to the housefly, an increased rate of DDT dehydrochlorination confers resistance to DDT in the mosquito *A. aegypti* (Grant *et al.*, 1991) and *A. gambiae* (Prapanthadara *et al.*, 1993).

8.6.3.2. Pyrethroids GSTs have not yet been shown to be involved in the direct metabolism of pyrethroid insecticides. Nevertheless, they play a very important role in conferring resistance to this insecticide class by detoxifying lipid peroxidation products induced by pyrethroids (Vontas *et al.*, 2001). A delta class GST present in elevated quantities in a pyrethroid resistant strain of *N. lugens* was expressed *in vitro*. The recombinant protein had high peroxidase activity and a role for this enzyme in the prevention or repair of oxidative damage was proposed (Vontas *et al.*, 2002). Elevated GSTs in other resistant insects may play a similar role. GSTs have also been suggested to protect against pyrethroid toxicity in insects by sequestering the insecticide (Kostaropoulos *et al.*, 2001).

8.6.3.3. Plant chemicals Plants produce a wide variety of allelochemicals to protect themselves from pathogens and herbivores. The detoxification of these compounds presents a major challenge for herbivorous insects and one consequence of this is that many insects feed on a very limited range of host plants. Many polyphagous species have highly evolved detoxification systems. Most studies on the metabolism of plant allelochemicals have focused on the role of cytochrome P450s (Fogleman *et al.*, 1998). However, GSTs can also metabolize many allelochemicals (review: Yu, 1996) and in some cases, the allelochemical acts as a regulator, inducing its own metabolism.

8.7. Regulation and Induction of GST Expression

In noninsect species, many GSTs are differentially regulated in response to various inducers or environmental signals or in a tissue developmental-specific manner. A similar complex pattern of expression is gradually being elucidated within the insect GST family. A detailed account of the effect of dietary compounds, insecticides, and laboratory inducers on GST expression is provided in two review articles (Clark, 1989; Yu, 1996). The main purpose of the following discussion is to outline some of the recent results from molecular studies on expression of individual GST genes and to highlight gaps in our knowledge.

8.7.1. Tissue and Life Stage Specificity of Expression

The majority of individual insect GSTs whose expression profiles have been determined are expressed constitutively in all life stages. Of course, the caveat here is that GSTs with a very restricted window of expression may have escaped detection in earlier studies. Nevertheless, in recent experiments looking at the global expression of the insect GSTs it was found that in *A. gambiae*, transcripts for all but one of the GST supergene family were detectable in 1-day-old adult mosquitoes by RT-PCR (Ding *et al.*, 2003). In these preliminary experiments no attempt was made to quantify the expression level in different developmental stages and it is apparent from several studies in other insects that the levels of individual enzymes can fluctuate widely during the life span of an insect. For example, using the less sensitive technique of Northern blotting, four delta class GSTs were undetectable in *D. melanogaster* adults (Toung *et al.*, 1993) and in the spruce budworm, *C. fumiferana*, a sigma GST was found at very low levels in feeding larvae but at high levels in diapausing larvae (Feng *et al.*, 1999).

Variations in the level of GST activity in different insect tissues have been reported in several species. In cases where the variation in activity is attributed

to individual enzymes, such studies can provide valuable insights into the functions of different GSTs. Thus, the finding that sigma GSTs from housefly and *Drosophila* were predominately located in the indirect flight muscles in association with troponin H suggested that the role of this GST class was structural rather than catalytic (subsequently, however, these GSTs have been found to play a very important role in the protection against oxidative stress; Franciosa and Berge, 1995; Clayton *et al.*, 1998). Very high levels of GST activity have been reported in the fat body and midguts of insects. Both these tissues are important sites for the detoxification of xenobiotics and further studies characterizing the individual enzymes present in these tissues are needed.

8.7.2. Induction of GSTs

In a number of studies, insects have been exposed to various xenobiotics to determine the effect of these chemicals on GST expression levels (Ottea and Plapp, 1984; Yu, 1984). In the majority of cases, a model substrate, such as CDNB, is used to assay GST activity from crude insect homogenates and hence the results are general measurements of the activity of a large subset of GSTs and do not reveal much about fluctuations in the level of individual enzymes. A more complete picture of factors governing GST expression is available for the plant GST family. By studying the expression profile of 10 different *Arabidopsis* GSTs in response to a variety of inducers and stress treatments different signaling mechanisms were demonstrated to regulate the expression of individual GSTs in response to different stimuli (Wagner *et al.*, 2002). Although not resolved at the level of individual enzymes, a study in the fall armyworm, *Spodoptera frugiperda*, showed that different enzyme fractions were induced to varying extents by treatment with different inducers (Kirby and Ottea, 1995). Injection of larvae with 8-methoxypsoralen enhanced the activity of GSTs active against DCNB, but not those active against CDNB, whereas injection with pentamethylbenzene had the opposite effect. Coadministration of the inducers prevented some of these induction effects leading the authors to conclude that multiple mechanisms are involved in the regulation of these enzymes.

Other chemicals that induce GSTs include insecticides, fumigants, barbiturates, and plant allelochemicals (review: Yu, 1996). Several of these, e.g., paraquat and pyrethroid insecticides, are also inducers of oxidative stress. Repressors of insect GST activity have also been identified in the diet of some insects. For example, GST activity in the tobacco budworm, *Heliothis virescens*, is decreased by a diet of wild tomato leaves (Clark, 1989).

8.7.3. Regulation of GST Expression

Most studies of the response of GSTs to environmental or artificial inducers suggest that regulation occurs at the transcriptional level. Several regulatory elements have been identified in the promoter regions of mammalian GSTs that may mediate their induction. These include xenobiotic response elements (XREs) and antioxidant or electrophile response elements (AREs/EpREs) (Rushmore and Pickett, 1993). Analysis of the promoters of the delta class GSTs from *D. melanogaster* identified several putative transcription factor binding sites including 12-O-tetradecanoylphorbol-13-acetate (TPA) responsive elements (Toung *et al.*, 1993). Other putative binding sites for transcription factors involved in developmental regulation were identified upstream of an epsilon GST in *A. dirus* (Udomsinprasert and Ketterman, 2002). In the absence of functional studies, the significance of these findings are unclear.

Genetic mechanisms responsible for the elevation in GST expression observed in many insecticide resistant strains have been the subject of many studies. In both the housefly and yellow fever mosquito (*Ae. aegypti*), modifications in *trans*-acting regulators have been implicated in increased GST expression. In *Ae. aegypti*, a mutation in a *trans*-acting repressor element is the proposed mechanism for the enhanced expression of a delta class GST in a DDT resistant strain (Grant and Hammock, 1992). A major regulatory gene on chromosome II of *M. domestica* has been proposed to regulate both GST and monooxygenase expression in insecticide resistant strains of houseflies (Plapp, 1984). Neither of these postulated regulators have been identified, although candidates have been proposed (Sabourault *et al.*, 2001). Genetic mapping of the major genes controlling GST-based DDT resistance in *A. gambiae* also provided tentative evidence for a *trans*-acting regulator (Ranson *et al.*, 2000) although in this species, mutations in promoter elements of the epsilon GST cluster are also associated with resistance (Ranson *et al.*, 2001).

The significance of alternative splicing in regulating GST expression has not been fully investigated. Four alternative transcripts of a delta class gene and two of a sigma GST have been detected in *A. gambiae* (see Section 8.5.3.1) and it is possible that different inducers or stress treatments may affect the choice of splice site.

Regulation of GST enzyme levels may also occur by alterations in the stability of mRNA transcripts. Most studies on gene regulation rely on the detection of steady state mRNA levels and the role of differential stability of transcripts is often overlooked. An exception to this was a comprehensive study in *D. melanogaster* in which factors responsible for the observed difference in the protein levels of two delta GSTs, GSTD1 and GSTD21, were investigated (Tang and Tu, 1995). Although the transcription rate of *GSTD2* (previously known as *GSTD21*) was higher than that of the neighboring gene *GSTD1*, the steady state level of *GSTD1* mRNA was over 20-fold higher than that of *GSTD2*, suggesting that the *GSTD2* mRNA is unstable. Treatment with pentobarbital (PB) enhanced the stability of *GSTD2* mRNA, but also may have prevented its efficient translation, as no corresponding increase in the levels of GSTD2 protein was observed after PB treatment.

In general, posttranslational modification has not been widely reported in the GST family and there have been no confirmed reports in the insect GSTs. A glycosylated form of a GST from the human filarial parasite *Onchocerca volvulus* has been detected. This GST is unusual as it is located in the parasite cuticle at the host–parasite interface and hence is speculated to be involved in protecting the parasite from reactive oxygen produced during the host's immune response (Sommer *et al.*, 2001). The detection of multiple forms of immunologically related GSTs during purification of housefly GSTs was initially interpreted as being suggestive of posttranslational modification (Fournier *et al.*, 1992). However, the extent of the insect GST family, and the presence of multiple members within both the delta and epsilon classes, suggests that these multiple forms actually represented unique GST subunits.

8.8. Evolution of GSTs

GSTs are thought to have evolved from a thioredoxin-like ancestor to protect cells from oxygen toxicity (Hayes and McLellan, 1999). Multiple rounds of gene duplication and subsequent diversification have led to the range of GST proteins found in eukaryotes. The thioredoxin-like fold found in the N-terminal domain of GSTs occurs in many other glutathione binding proteins, including glutathione reductase, chloride intracellular channels (CLICs), and glutaredoxin (Bushweller *et al.*, 1994; Xia *et al.*, 2001; Cromer *et al.*, 2002). The low levels of sequence identity between these proteins and GSTs led to the conclusion that the structural similarity was due to convergent rather than

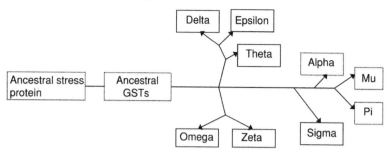

Figure 6 Suggested phylogenetic relationship between selected GST classes (those not present in insects are shown by dashed boxes). Schematic was modified from Sheehan *et al.* (2001) based on alignments of insect and mammalian GSTs in Ding *et al.* (2003).

divergent evolution (Wilce and Parker, 1994). However, the discovery of additional GST classes, in particular the omega class, warrants a reassessment of the relationship between these glutathione binding proteins and GSTs. For example, the CLIC1 sequence shows approximately 15% sequence identity with an omega class GST and structural similarity in both the N- and C-terminal domains suggesting that, despite the very different functions of these proteins, they have a common ancestor and belong in the same superfamily (Cromer *et al.*, 2002).

A schematic diagram showing the probable phylogenetic relationships between different GST classes is shown in **Figure 6**. The insect specific delta and epsilon GSTs have a common ancestor and are most closely related to the theta class to which they were originally assigned (Pemble and Taylor, 1992). The omega and zeta classes are also derived from a common ancestor and both of these classes are represented in a range of phylogenetic groups.

A comparative genomic analysis of the GSTs of the two Diptera, *A. gambiae* and *D. melanogaster*, found remarkably few clear orthologs between the two species (Ranson *et al.*, 2002). The orthologs that can be unambiguously identified were generally found within the noninsect-specific classes and these GSTs are presumably involved in essential steps in conserved physiological pathways, with subsequent constraints on their evolution. The two insect-specific classes, delta and epsilon, have independently radiated in *D. melanogaster* and *A. gambiae* suggesting that these GST classes play important roles in the adaptation of these insects to their specific biological niches rather than fulfilling general housekeeping functions. The majority of the delta and epsilon genes are tightly clustered in both species and the expanse of these GST classes is a result of local duplications of individual genes or blocks of genes (Holt *et al.*, 2002; Sawicki *et al.*, 2003).

8.9. Conclusions

Advances in genetics and biochemistry have unveiled a surprising complexity in the variety of the insect GST genes and enzymes in insects. The sequencing of insect genomes has been invaluable in unveiling the extent of the GST supergene family but clearly sequence analysis alone is not sufficient to uncover the *in vivo* function of individual enzymes. Furthermore, presently, full genome sequences are only publicly available for two Dipteran species. Extensive comparisons between the GST supergene family in these two species have enabled testable hypotheses about the broad roles of different GST classes to be formulated, but clearly the power of these comparative genomic approaches will be enhanced by expanding this analysis to other insect orders.

The role of GSTs in the detoxification of xenobiotics and, in particular insecticide detoxification, is rapidly being elucidated. However, much still remains to be learnt about the endogenous substrates of insect GSTs. Overexpression of individual GST genes via transient or stable transfection in insect cell lines or under expression via RNA interference or gene knockout studies will enable the role of GSTs in protecting cells against various external stimuli or chemical treatments to be studied. However, one of the biggest challenges for GST research is characterizing their roles in endogenous metabolic pathways. Deciphering the precise tissue and developmental expression profile of individual GST genes may provide some clues to their function and this type of data can now be more readily obtained for entire gene families via microarray-based experiments. Putative GST substrates are more readily assayed in an *in vitro* system, but frequently recombinant insect GSTs expressed in *E. coli* are sequestered in insoluble inclusion bodies. Overcoming this obstacle is a major priority for GST research.

References

Abdallah, M.A., Pollenz, R.S., Droog, F.N., Nunamaker, R.A., Tabachnick, W.J., *et al.*, **2000**. Isolation and characterization of a cDNA clone coding for a glutathione *S*-transferase class delta enzyme from the biting midge *Culicoides variipennis sonorensis* Wirth and Jones. *Biochem. Genet. 38*, 377–390.

Agianian, B., Tucker, P.A., Schouten, A., Leonard, K., Bullard, B., *et al.*, **2003**. Structure of a *Drosophila* sigma class glutathione *S*-transferase reveals a novel active site topography suited for lipid peroxidation products. *J. Mol. Biol. 326*, 151–165.

Armstrong, R.N., **1991**. Glutathione *S*-transferases: reaction mechanism, structure, and function. *Chem. Res. Toxicol. 4*, 131–140.

Arruda, L.K., Vailes, L.D., Benjamin, D.C., Chapman, M.D., **1995**. Molecular cloning of German cockroach (*Blattella germanica*) allergens. *Int. Arch. Allergy Immunol. 107*, 295–297.

Arruda, L.K., Vailes, L.D., Platts-Mills, T.A., Hayden, M.L., Chapman, M.D., **1997**. Induction of IgE antibody responses by glutathione *S*-transferase from the German cockroach (*Blattella germanica*). *J. Biol. Chem. 272*, 20907–20912.

Baker, W.L., Clark, A.G., Faulds, G., Nielsen, J.S., **1994**. Multiple glutathione *S*-transferases in *Galleria mellonella*. *Insect Biochem. Molec. Biol. 24*, 301–307.

Beall, C., Fyrberg, C., Song, S., Fyrberg, E., **1992**. Isolation of a *Drosophila* gene encoding glutathione *S*-transferase. *Biochem. Genet. 30*, 515–527.

Blackburn, A.C., Woollatt, E., Sutherland, G.R., Board, P.G., **1998**. Characterization and chromosome location of the gene GSTZ1 encoding the human Zeta class glutathione transferase and maleylacetoacetate isomerase. *Cytogenet. Cell Genet. 83*, 109–114.

Board, P., Russell, R.J., Marano, R.J., Oakeshott, J.G., **1994**. Purification, molecular cloning and heterologous expression of a glutathione *S*-transferase from the Australian sheep blowfly (*Lucilia cuprina*). *Biochem. J. 299*, 425–430.

Board, P.G., Baker, R.T., Chelvanayagam, G., Jermiin, L.S., **1997**. Zeta, a novel class of glutathione transferases in a range of species from plants to humans. *Biochem. J. 328*, 929–935.

Board, P.G., Coggan, M., Chelvanayagam, G., Easteal, S., Jermiin, L.S., *et al.*, **2000**. Identification, characterization, and crystal structure of the Omega class glutathione transferases. *J. Biol. Chem. 275*, 24798–24806.

Booth, J., Boyland, E., Sims, P., **1961**. An enzyme from rat liver catalyzing conjugation with glutathione. *Biochem. J. 79*, 516–524.

Bushweller, J.H., Billeter, M., Holmgren, A., Wuthrich, K., **1994**. The nuclear magnetic resonance solution structure of the mixed disulfide between *Escherichia coli* glutaredoxin (C14S) and glutathione. *J. Mol. Biol. 235*, 1585–1597.

Caccuri, A.M., Antonini, G., Nicotra, M., Battistoni, A., Bello, M.L., *et al.*, **1997**. Catalytic mechanism and role of hydroxyl residues in the active site of theta class glutathione *S*-transferases. Investigation of Ser-9 and Tyr-113 in a glutathione *S*-transferase from the Australian sheep blowfly *Lucilia cuprina*. *J. Biol. Chem. 272*, 29681–29686.

Chelvanayagam, G., Parker, M.W., Board, P.G., **2001**. Fly fishing for GSTs: a unified nomenclature for mammalian and insect glutathione transferases. *Chem. Bio. Interact. 133*, 256–260.

Chiang, F., Sun, C., **1993**. Glutathione transferase isozymes of diamondback moth larvae and their role in the degradation of some organophosphorous insecticides. *Pest. Biochem. Physiol. 45*, 7–14.

Chien, C., Dauterman, W.C., **1991**. Studies on glutathione *S*-transferase in *Helicoverpa* (= *Heliothis*) *zea*. *Insect Biochem. 21*, 857–864.

Chien, C., Motoyama, N., Dauterman, W.C., **1995**. Separation of multiple forms of acidic glutathione *S*-transferase isozymes in a susceptible and a resistant strain of housefly, *Musca domestica* (L.). *Arch. Insect Biochem. Physiol. 28*, 397–406.

Clark, A.G., **1989**. The comparative enzymology of the glutathione *S*-transferases from non-vertebrate organisms. *Comp. Biochem. Physiol. B 92*, 419–446.

Clark, A.G., Dauterman, W.C., **1982**. The characterization by affinity chromatography of a glutathione *S*-transferase from different strains of housefly. *Pestic. Biochem. Physiol. 17*, 307–314.

Clark, A.G., Dick, G.K., Martindale, S.M., Smith, J.N., **1985**. Glutatione *S*-transferases from the New Zealand grass grub, *Costelytra zealandica*. *Insect Biochem. 15*, 35–44.

Clark, A.G., Letoa, M., Ting, W.S., **1977**. The purification by affinity chromatography of a glutathione *S*-transferase from larvae of *Galleria mellonella*. *Life Sci. 20*, 141–148.

Clark, A.G., Shamaan, N.A., **1984**. Evidence that DDT dehydrochlorinase from the housefly is a glutathione *S*-transferase. *Pestic. Biochem. Physiol. 22*, 249–261.

Clark, A.G., Shamaan, N.A., Sinclair, M.D., Dauterman, W.C., **1986**. Insecticide metabolism by multiple glutathione *S*-transferases in two strains of the housefly, *Musca domestica* (L.). *Pestic. Biochem. Physiol. 25*, 169–175.

Clayton, J.D., Cripps, R.M., Sparrow, J.C., Bullard, B., **1998**. Interaction of troponin-H and glutathione *S*-transferase-2 in the indirect flight muscles of *Drosophila melanogaster*. *J. Muscle Res. Cell Motil. 19*, 117–127.

Cromer, B.A., Morton, C.J., Board, P.G., Parker, M.W., **2002**. From glutathione transferase to pore in a CLIC. *Eur. Biophys. J. 31*, 356–364.

Ding, Y., Ortelli, F., Rossiter, L., Hemingway, J., Ranson, H., **2003**. The *Anopheles gambiae* glutathione transferase family: annotation, phylogeny and gene expression profiles. *BMC Genom. 13*, 35.

Dixon, D.P., Cole, D.J., Edwards, R., **1999**. Dimerisation of maize glutathione transferases in recombinant bacteria. *Plant Mol. Biol. 40*, 997–1008.

Egaas, E., Sandvik, M., Svendsen, N.O., Skaare, JU., 1995. The separation and identification of glutathione S-transferase subunits from *Orthosia gothica*. *Insect Biochem. Mol. Biol. 25*, 783–788.

Feng, Q.L., Davey, K.G., Pang, A.S., Primavera, M., Ladd, T.R., et al., 1999. Glutathione S-transferase from the spruce budworm, *Choristoneura fumiferana*: identification, characterization, localization, cDNA cloning, and expression. *Insect Biochem. Mol. Biol. 29*, 779–793.

Fogleman, J.C., Danielson, P.B., Macintyre, R.J., 1998. The molecular basis of adaptation in *Drosophila*. In: Hecht, M.K. (Ed.), Evolutionary Biology. Plenum, New York, p. 15.

Fournier, D., Bride, J.M., Poirie, M., Berge, J.B., Plapp, F.W., Jr., 1992. Insect glutathione S-transferases. Biochemical characteristics of the major forms from houseflies susceptible and resistant to insecticides. *J. Biol. Chem. 267*, 1840–1845.

Franciosa, H., Berge, J.B., 1995. Glutathione S-transferases in housefly (*Musca domestica*): location of GST-1 and GST-2 families. *Insect Biochem. Mol. Biol. 25*, 311–317.

Grant, D.F., Dietze, E.C., Hammock, B.D., 1991. Glutathione S-transferase isozymes in *Aedes aegypti*: purification, characterization, and isozyme-specific regulation. *Insect Biochem. 21*, 421–433.

Grant, D.F., Hammock, B.D., 1992. Genetic and molecular evidence for a trans-acting regulatory locus controlling glutathione S-transferase-2 expression in *Aedes aegypti*. *Mol. Gen. Genet. 234*, 169–176.

Habig, W.H., Pabst, M.J., Jakoby, W.B., 1974. The glutathione S-transferases: the first enzymatic step in mercapturic acid formation. *J. Biol. Chem. 249*, 7130–7139.

Hayes, J.D., McLellan, L.I., 1999. Glutatione and glutathione-dependent enzymes represent a co-ordinately regulated defence against oxidative stress. *Free Radic. Res. 4*, 273–300.

Hayes, J.D., Wolf, C.R., 1988. Role of glutathione transferase in drug resistance. In: Sies, H., Ketterer, B. (Eds.), Glutathione Conjugation. Academic Press, London, p. 315.

Hemingway, J., Miyamoto, J., Herath, P.R.J., 1991. A possible novel link between organophosphorous and DDT insecticide resistance genes in *Anopheles*: supporting evidence from fenitrothion metabolism studies. *Pest. Biochem. Physiol. 39*, 49–56.

Hemingway, J., Ranson, H., 2000. Insecticide resistance in insect vectors of human disease. *Annu. Rev. Entomol. 45*, 371–391.

Holt, R.A., Subramanian, G.M., Halpern, A., Sutton, G.G., Charlab, R., et al., 2002. The genome sequence of the malaria mosquito *Anopheles gambiae*. *Science 298*, 129–149.

Huang, H.S., Hu, N.T., Yao, Y.E., Wu, C.Y., Chiang, S.W., et al., 1998. Molecular cloning and heterologous expression of a glutathione S-transferase involved in insecticide resistance from the diamondback moth, *Plutella xylostella*. *Insect Biochem. Mol. Biol. 28*, 651–658.

Jakobsson, P.J., Thoren, S., Morgenstern, R., Samuelsson, B., 1999. Identification of human prostaglandin E synthase: a microsomal, glutathione-dependent, inducible enzyme, constituting a potential novel drug target. *Proc. Natl Acad. Sci. 96*, 7220–7225.

Ji, X., von Rosenvinge, E.C., Johnson, W.W., Tomarev, S.I., Piatigorsky, J., et al., 1995. Three-dimensional structure, catalytic properties, and evolution of a sigma class glutathione transferase from squid, a progenitor of the lens S-crystallins of cephalopods. *Biochemistry 34*, 5317–5328.

Jirajaroenrat, K., Pongjaroenkit, S., Krittanai, C., Prapanthadara, L., Ketterman, A.J., 2001. Heterologous expression and characterization of alternatively spliced glutathione S-transferases from a single *Anopheles* gene. *Insect Biochem. Mol. Biol. 31*, 867–875.

Ketterman, A.J., Prommeenate, P., Boonchauy, C., Chanama, U., Leetachewa, S., et al., 2001. Single amino acid changes outside the active site significantly affect activity of glutathione S-transferases. *Insect Biochem. Mol. Biol. 31*, 65–74.

Kirby, M.L., Ottea, J.A., 1995. Multiple mechanisms for enhancement of glutathione S-transferase activities in *Spodoptera frugiperda* (Lepidoptera: Noctuidae). *Insect Biochem. Mol. Biol. 25*, 347–353.

Kostaropoulos, I., Papadopoulos, A.I., Metaxakis, A., Boukouvala, E., Papadopoulou-Mourkidou, E., 2001. Glutathione S-transferase in the defence against pyrethroids in insects. *Insect Biochem. Mol. Biol. 31*, 313–319.

Landi, S., 2000. Mammalian class theta GST and differential susceptibility to carcinogens: a review. *Mutat. Res. 463*, 247–283.

Lougarre, A., Bride, J.M., Fournier, D., 1999. Is the insect glutathione S-transferase I gene family intronless? *Insect Mol. Biol. 8*, 141–143.

Mannervik, B., 1985. The isoenzymes of glutathione transferase. *Adv. Enzymol. Relat. Areas Mol. Biol. 57*, 357–417.

Mannervik, B., Danielson, U.H., 1988. Glutathione transferases–structure and catalytic activity. *CRC Crit. Rev. Biochem. 23*, 283–337.

Meyer, D.J., Coles, B., Pemble, S.E., Gilmore, K.S., Fraser, G.M., et al., 1991. Theta, a new class of glutathione transferases purified from rat and man. *Biochem. J. 274*, 409–414.

Meyer, D.J., Thomas, M., 1995. Characterization of rat spleen prostaglandin H D-isomerase as a sigma-class GSH transferase. *Biochem. J. 311*, 739–742.

Oakley, A.J., Harnnoi, T., Udomsinprasert, R., Jirajaroenrat, K., Ketterman, A.J., et al., 2001. The crystal structures of glutathione S-transferases isozymes 1–3 and 1–4 from *Anopheles dirus* species B. *Protein Sci. 10*, 2176–2185.

Oppenoorth, F.J., Van der Pas, L.J.T., Houx, N.W.H., 1979. Glutathione S-transferases and hydrolytic activity in a tetrachlorovinphos-resistant strain of housefly and their influence on resistance *Pestic. Biochem. Physiol. 11*, 176–188.

Ortelli, F., Rossiter, L.C., Vontas, J., Ranson, H., Hemingway, J., 2003. Heterologous expression of four glutathione transferase genes genetically linked to a major insecticide resistance locus, from the malaria vector. *Anopheles gambiae Biochem. J. 373*, 957–963.

Ottea, J.A., Plapp, F.W., 1984. Glutathione *S*-transferases in the house fly: biochemical and genetic changes associated with induction and insecticide resistance. *Pest. Biochem. Physiol. 22*, 203–208.

Pemble, S.E., Taylor, J.B., 1992. An evolutionary perspective on glutathione transferases inferred from class-theta glutathione transferase cDNA sequences. *Biochem. J. 287*, 957–963.

Plapp, F.W., 1984. The genetic basis of insecticide resistance in the housefly: evidence that a single locus plays a major role in metabolic resistance to insecticides. *Pest. Biochem. Physiol. 22*, 194–201.

Prapanthadara, L., Hemingway, J., Ketterman, A.J., 1993. Partial purification and characterization of glutathione *S*-transferases involved in DDT resistance from the mosquito *Anopheles gambiae*. *Pest. Biochem. Physiol. 47*, 119–133.

Prapanthadara, L., Hemingway, J., Ketterman, A.J., 1995. DDT-resistance in *Anopheles gambiae* (Diptera: culicidae) from Zanzibar, Tanzania, based on increased DDT-deyhdrochlorinase activity of glutathione *S*-transferases. *Bull. Ent. Res. 85*, 267–274.

Ranson, H., Claudianos, C., Ortelli, F., Abgrall, C., Hemingway, J., *et al.*, 2002. Evolution of supergene families associated with insecticide resistance. *Science 298*, 179–181.

Ranson, H., Collins, F., Hemingway, J., 1998. The role of alternative mRNA splicing in generating heterogeneity within the *Anopheles gambiae* class I glutathione *S*-transferase family. *Proc. Natl Acad. Sci. USA 95*, 14284–14289.

Ranson, H., Jensen, B., Wang, X., Prapanthadara, L., Hemingway, J., *et al.*, 2000. Genetic mapping of two loci affecting DDT resistance in the malaria vector *Anopheles gambiae*. *Insect Mol. Biol. 9*, 499–507.

Ranson, H., Prapanthadara, L., Hemingway, J., 1997. Cloning and characterization of two glutathione *S*-transferases from a DDT-resistant strain of *Anopheles gambiae*. *Biochem. J. 324*, 97–102.

Ranson, H., Rossiter, L., Ortelli, F., Jensen, B., Wang, X., *et al.*, 2001. Identification of a novel class of insect glutathione *S*-transferases involved in resistance to DDT in the malaria vector *Anopheles gambiae*. *Biochem. J. 359*, 295–304.

Reiss, R.A., James, A.A., 1993. A glutathione *S*-transferase gene of the vector mosquito, *Anopheles gambiae*. *Insect Mol. Biol. 2*, 25–32.

Rogers, M.E., Jani, M.K., Vogt, R.G., 1999. An olfactory-specific glutathione-*S*-transferase in the sphinx moth *Manduca sexta*. *J. Exp. Biol. 202 (Pt 12)*, 1625–1637.

Rushmore, T.H., Pickett, C.B., 1993. Glutathione *S*-transferases, structure, regulation, and therapeutic implications. *J. Biol. Chem. 268*, 11475–11478.

Sabourault, C., Guzov, V.M., Koener, J.F., Claudianos, C., Plapp, F.W., Jr., *et al.*, 2001. Overproduction of a P450 that metabolizes diazinon is linked to a loss-of-function in the chromosome 2 ali-esterase (MdalphaE7) gene in resistant house flies. *Insect Mol. Biol. 10*, 609–618.

Sawicki, R., Singh, S.P., Mondal, A.K., Benes, H., Zimniak, P., 2003. Cloning, expression and biochemical characterization of one Epsilon-class (GST-3) and ten Delta-class (GST-1) glutathione *S*-transferases from *Drosophila melanogaster*, and identification of additional nine members of the Epsilon class. *Biochem. J. 370*, 661–669.

Schmidt-Krey, I., Murata, K., Hirai, T., Mitsuoka, K., Cheng, Y., *et al.*, 1999. The projection structure of the membrane protein microsomal glutathione transferase at 3 A resolution as determined from two-dimensional hexagonal crystals. *J. Mol. Biol. 288*, 243–253.

Sheehan, D., Meade, G., Foley, V.M., Dowd, C.A., 2001. Structure, function and evolution of glutathione transferases: implications for classification of non-mammalian members of an ancient enzyme superfamily. *Biochem. J. 360*, 1–16.

Singh, S.P., Coronella, J.A., Benes, H., Cochrane, B.J., Zimniak, P., 2001. Catalytic function of *Drosophila melanogaster* glutathione *S*-transferase DmGSTS1-1 (GST-2) in conjugation of lipid peroxidation end products. *Eur. J. Biochem. 268*, 2912–2923.

Singh, M., Silva, E., Schulze, S., Sinclair, D.A., Fitzpatrick, K.A., *et al.*, 2000. Cloning and characterization of a new theta-class glutathione-*S*-transferase (GST) gene, gst-3, from *Drosophila melanogaster*. *Gene 247*, 167–173.

Snyder, M.J., Walding, J.K., Feyereisen, R., 1995. Glutathione *S*-transferases from larval *Manduca sexta* midgut: sequence of two cDNAs and enzyme induction. *Insect Biochem. Mol. Biol. 25*, 455–465.

Sommer, A., Nimtz, M., Conradt, H.S., Brattig, N., Boettcher, K., *et al.*, 2001. Structural analysis and antibody response to the extracellular glutathione *S*-transferases from *Onchocerca volvulus*. *Infect. Immun. 69*, 7718–7728.

Syvanen, M., Zhou, Z.H., Wang, J.Y., 1994. Glutathione transferase gene family from the housefly *Musca domestica*. *Mol. Gen. Genet. 245*, 25–31.

Tang, A.H., Tu, C.P, 1994. Biochemical characterization of *Drosophila* glutathione *S*-transferases D1 and D21. *J. Biol. Chem. 269*, 27876–27884.

Tang, A.H., Tu, C.P., 1995. Pentobarbital-induced changes in *Drosophila* glutathione *S*-transferase D21 mRNA stability. *J. Biol. Chem. 270*, 13819–13825.

Tiwari, N.K., Singhai, S.S., Sharma, R., Meagher, R.L., Awasthi, Y.C., 1991. Purification and characterization of glutathione transferases of *Eorenma loftini* and *Diatraea sacohara* (Lepidoptera: Pyralidae). *J. Econ. Entomol. 84*, 1424–1432.

Toba, G., Aigaki, T., 2000. Disruption of the microsomal glutathione *S*-transferase-like gene reduces life span of *Drosophila melanogaster*. *Gene 253*, 179–187.

Toung, Y.P., Hsieh, T.S., Tu, C.P., 1990. *Drosophila* glutathione *S*-transferase 1-1 shares a region of sequence homology with the maize glutathione *S*-transferase III. *Proc. Natl Acad. Sci. USA 87*, 31–35.

Toung, Y.P., Hsieh, T.S., Tu, C.P., 1993. The glutathione *S*-transferase D genes. A divergently organized, intronless gene family in *Drosophila melanogaster. J. Biol. Chem. 268*, 9737–9746.

Udomsinprasert, R., Ketterman, A.J., 2002. Expression and characterization of a novel class of glutathione *S*-transferase from *Anopheles dirus. Insect Biochem. Mol. Biol. 32*, 425–433.

Vontas, J.G., Small, G.J., Hemingway, J., 2001. Glutathione *S*-transferases as antioxidant defence agents confer pyrethroid resistance in *Nilaparvata lugens. Biochem. J. 357*, 65–72.

Vontas, J.G., Small, G.J., Nikou, D.C., Ranson, H., Hemingway, J., 2002. Purification, molecular cloning and heterologous expression of a glutathione *S*-transferase involved in insecticide resistance from the rice brown planthopper, *Nilaparvata lugens. Biochem. J. 362*, 329–337.

Wagner, U., Edwards, R., Dixon, D.P., Mauch, F., 2002. Probing the diversity of the Arabidopsis glutathione *S*-transferase gene family. *Plant Mol. Biol. 49*, 515–532.

Wang, J.Y., McCommas, S., Syvanen, M., 1991. Molecular cloning of a glutathione *S*-transferase overproduced in an insecticide-resistant strain of the housefly (*Musca domestica*). *Mol. Gen. Genet. 227*, 260–266.

Wei, S.H., Clark, A.G., Syvanen, M., 2001. Identification and cloning of a key insecticide-metabolizing glutathione *S*-transferase (MdGST-6A) from a hyper insecticide-resistant strain of the housefly *Musca domestica. Insect Biochem. Mol. Biol. 31*, 1145–1153.

Wilce, M.C., Board, P.G., Feil, S.C., Parker, M.W., 1995. Crystal structure of a theta-class glutathione transferase. *EMBO J. 14*, 2133–2143.

Wilce, M.C., Parker, M.W., 1994. Structure and function of glutathione *S*-transferases. *Biochim. Biophys. Acta 1205*, 1–18.

Wood, E., Casabe, N., Melgar, F., Zebra, E., 1986. Distribution and properties of glutathione *S*-transferases from *T. infestans. Comp. Biochem. Physiol. 84*, 607–617.

Xia, B., Vlamis-Gardikas, A., Holmgren, A., Wright, P.E., Dyson, H.J., 2001. Solution structure of *Escherichia coli* glutaredoxin-2 shows similarity to mammalian glutathione-*S*-transferases. *J. Mol. Biol. 310*, 907–918.

Yu, S.J., 1984. Interactions of allelochemicals with detoxication enzymes of insecticide-susceptible and resistant fall army-worms. *Pestic. Biochem. Physiol. 22*, 60–68.

Yu, S.J., 1996. Insect glutathione *S*-transferases. *Zool. Stud. 35*, 9–19.

Zhou, Z.H., Syvanen, M., 1997. A complex glutathione transferase gene family in the housefly *Musca domestica. Mol. Gen. Genet. 256*, 187–194.

A8 Addendum: Glutathione Transferases

H Ranson, Liverpool School of Tropical Medicine, Liverpool, UK

© 2010, Elsevier BV. All Rights Reserved.

The past 5 years witnessed completion of many insect genome sequences that show the full extent of the glutathione transferase (GST) gene family in insects. There is extensive heterogeneity in both the total number and the representation of each of the different GST classes between insects from different orders (**Table A1**). The insect species with the largest number of cytosolic GSTs reported to date are the fruitfly, *Drosophila ananassae* (45 genes) (Low *et al.*, 2007) and the flour beetle, *Tribolium castaneum* (35 genes) (Richards *et al.*, 2008). In both cases, approximately half of the GSTs belong to the epsilon class. The honeybee, *Apis mellifera*, has a particularly small complement of GSTs with just eight putative cytosolic GSTs and two microsomal GSTs (Claudianos *et al.*, 2006). Furthermore, just one of the eight cytosolic GSTs belongs to the insect specific classes, delta and epsilon, which are typically the largest classes in the other insects. Given that the majority of GSTs implicated in insecticide detoxification are found within the delta and epsilon classes, a scarcity of these GSTs in the honeybee may, in part, explain its extreme susceptibility to insecticides. However, another species that is also very susceptible to insecticides is the silkworm, *Bombyx mori*, which has 12 GSTs belonging to either the delta or epsilon GST classes (Yu *et al.*, 2008). Interestingly, relatively few of silkworm delta and epsilon GSTs are expressed at high levels in the major detoxification tissues (the fat body and midgut), and many show quite specific tissue distribution suggesting that these enzymes may have specialized roles in the silkworm that remain to be elucidated.

The sigma GST class is represented by a single gene in three dipteran species but this class is much more extensive in the pea aphid (six sigma GSTs) (Ramsey, 2009) and Hymenoptera with over 40% of total cytosolic GSTs belonging to this class in both the honeybee and the parasitic wasp, *Nasonia vitripennis* (Claudianos *et al.*, 2006; Oakeshott, 2009). The Sigma GST class in the coleopteran *T. castaneum* has also undergone an expansion resulting in a total of seven Sigma GST genes (Richards *et al.*, 2008). Although there have been many publications describing Sigma GSTs from other insect ospecies in recent years (Li *et al.*, 2009; Huang *et al.*, 2009; Valles *et al.*, 2006), there has been little progress in identifying the endogenous role of this class of GSTs and hence the reasons for expansion in particular species remain unknown.

Even within a genus, there can be large variations in GST gene numbers. A comparative analysis of 12 *Drosophila* species identified between 30 and 45 GST genes in each species with most variation in copy number found in the delta and epsilon classes. Within these insect specific classes, a number of highly conserved genes were identified (*GSTd1*, *GSTd9*, *GSTe9*, and *GSTe4*) which are presumed to have unique roles either due to their substrate specificity or due to their spatial and/or temporal expression (Low *et al.*, 2007).

The majority of research on insect GSTs remains focused on their role in conferring insecticide resistance. However, despite the relative ease with which GST enzymes can be expressed *in vitro*, very few enzymes capable of metabolizing insecticides have been identified and many studies have implicated particular GSTs in conferring insecticide resistance on the basis of differential expression between resistant and susceptible strains alone. The role of GSTs in pyrethroid resistance remains unresolved. Several populations of pyrethroid resistant insects with highly elevated GST activity and/or increased expression of one or more GST genes have been identified but a role in primary or secondary metabolism of pyrethroids has not been definitively attributed to GST enzymes. The two original hypotheses (that GSTs may protect against oxidative stress induced by pyrethroid exposure (Vontas *et al.*, 2001), or may act as pyrethroid binding proteins thereby sequestering the insecticide (Kostaropoulos *et al.*, 2001)) are still viable propositions but this is clearly an area that warrants further investigation, given the increasing reliance of pest control strategies on pyrethroid insecticides. In contrast, GSTs have a clear role in DDT metabolism in insects. DDT dehydrochlorinase activity is a property of only a small subset of the insect GST family. For example, among the eight *Anopheles gambiae* and eight *Aedes aegypti* GSTs that have been biochemically

Table A1 Classification of cytosolic GST genes from insect species with whole genome sequences available

Order	Diptera				Hymenoptera		Coleoptera	Lepidoptera
Species	Drosophila melanogaster	Anopheles gambiae	Aedes aegypti	Acrythosiphon pisum	Apis mellifera	Nasonia vitripennis	Tribolium castaneum	Bombyx mori
	Fruitfly	Mosquito	Mosquito	Pea aphid	Honeybee	Paraistic wasp	Flour beetle	Silkworm
Delta	11	12	8	10	0	4		4
Epsilon	14	8	8	0	0	0	23	8
Omega	5	1	1	0	1	2	4	4
Sigma	1	1	1	6	4	8	6	2
Theta	4	2	4	2	1	3	1	1
Zeta	2	1	1	0	1	1	1	2
Others	0	3	3	0	0	0	0	2
Subtotal	37	28	27	18	8	18	35	23

characterized, only GSTE2–2 has high levels of DDTase activity and most have no detectable metabolism of this insecticide (Ortelli et al., 2003; Lumjuan et al., 2005). GSTE2–2 is over expressed in multiple DDT-resistant populations of mosquitoes and is a very strong candidate for conferring DDT resistance (David et al., 2005; Strode et al., 2008). The structure of this enzyme has been resolved and computational modeling has identified a putative DDT binding site of GSTE2–2, which is protected by a cap to shield the hydrophobic substrate from the aqueous environment (Wang et al., 2008).

There is still much to learn about the endogenous functions of insect GSTs. There have been very few novel functions attributed to individual insect GSTs in the past 5 years. One notable exception is the elucidation of the role of the *Drosophila* omega class GST (CG6781) as a pyrimidodiazepine synthase involved in eye pigment synthesis (Kim et al., 2006). Improvements in gene silencing and gain of function methodologies will hopefully accelerate our understanding of the *in vivo* roles of the insect GST family.

References

Claudianos, C., Ranson, H., Johnson, R.M., Biswas, S., Schuler, M.A., Berenbaum, M.R., Feyereisen, R., Oakeshott, J.G., 2006. A deficit of detoxification enzymes: pesticide sensitivity and environmental response in the honeybee. *Insect Mol. Biol.* 15, 615–636.

David, J.P., Strode, C., Vontas, J., Nikou, D., Vaughan, A., Pignatelli, P.M., Louis, C., Hemingway, J., Ranson, H., 2005. The *Anopheles gambiae* detoxification chip: a highly specific microarray to study metabolic-based insecticide resistance in malaria vectors. *Proc. Natl Acad. Sci. USA.* 102, 4080–4084.

Huang, Y., Krell, P.J., Ladd, T., Feng, Q., Zheng, S., 2009. Cloning, characterization and expression of two glutathione S-transferase cDNAs in the spruce budworm, *Choristoneura fumiferana. Arch. Insect Biochem. Physiol.* 70, 44–56.

Kim, J., Suh, H., Kim, S., Kim, K., Ahn, C., Yim, J., 2006. Identification and characteristics of the structural gene for the *Drosophila* eye colour mutant sepia, encoding PDA synthase, a member of the Omega class glutathione S-transferases. *Biochem. J.* 398, 451–460.

Kostaropoulos, I., Papadopoulos, A.I., Metaxakis, A., Boukouvala, E., Papadopoulou-Mourkidou, E., 2001. Glutathione S-transferase in the defence against pyrethroids in insects. *Insect Biochem. Mol. Biol.* 31, 313–319.

Li, X., Zhang, X., Zhang, J., Starkey, S.R., Zhu, K.Y., 2009. Identification and characterization of eleven glutathione S-transferase genes from the aquatic midge *Chironomus tentans* (Diptera: Chironomidae). *Insect Biochem. Mol. Biol.* 39, 745–754.

Low, W.Y., Ng, H.L., Morton, C.J., Parker, M.W., Batterham, P., Robin, C., 2007. Molecular evolution of glutathione S-transferases in the genus *Drosophila*. *Genetics 177*, 1363–1375.

Lumjuan, N., Mccarroll, L., Prapanthadara, L.A., Hemingway, J., Ranson, H., 2005. Elevated activity of an Epsilon class glutathione transferase confers DDT resistance in the dengue vector, *Aedes aegypti*. *Insect Biochem. Mol. Biol. 35*, 861–871.

Oakeshott, J.G., Johnson, R.M., Berenbaum, M.R., Ranson, H., Cristino, A.S., Claudianos, C., 2009. Metabolic enzymes associated with xenobiotic and chemosensory responses in *Nasonia vitripennis*. *Insect Mol. Biol.* (in press).

Oakeshott, J.G., Johnson, R.M., Berenbaum, M.R., Ranson, H., Cristino, A.S., Claudianos, C., 2010. Metabolic enzymes associated with xenobiotic and chemosensory responses in *Nasonia vitripennis* 19 (s1), 147–163. Published Online: Jan 15 2010 6:01AMDOI: 10.1111/j.1365-2583.2009.00961.

Ortelli, F., Rossiter, L.C., Vontas, J., Ranson, H., Hemingway, J., 2003. Heterologous expression of four glutathione transferase genes genetically linked to a major insecticide-resistance locus from the malaria vector *Anopheles gambiae*. *Biochem. J. 373*, 957–963.

Ramsey, J.S., Rider, D.S., Walsh, T.K., de Vos, M., Gordon, K.H.J., Ponnala, L., Macmil, S.L., Roe, B.A., Janer, G., 2009. Comparative analysis of detoxification enzymes in *Acyrthosiphon pisum* and *Myzus persicae*. *Insect Mol. Biol.* (in press).

Richards, S., Gibbs, R.A., Weinstock, G.M., Brown, S.J., Denell, R., Beeman, R.W., Gibbs, R., Bucher, G., Friedrich, M., Grimmelikhuijzen, C.J., Klingler, M., Lorenzen, M., Roth, S., Schroder, R., Tautz, D., Zdobnov, E.M., Muzny, D., Attaway, T., Bell, S., Buhay, C.J., Chandrabose, M.N., Chavez, D., Clerk-Blankenburg, K.P., Cree, A., Dao, M., Davis, C., Chacko, J., Dinh, H., Dugan-Rocha, S., Fowler, G., Garner, T.T., Garnes, J., Gnirke, A., Hawes, A., Hernandez, J., Hines, S., Holder, M., Hume, J., Jhangiani, S.N., Joshi, V., Khan, Z.M., Jackson, L., Kovar, C., Kowis, A., Lee, S., Lewis, L.R., Margolis, J., Morgan, M., Nazareth, L.V., Nguyen, N., Okwuonu, G., Parker, D., Ruiz, S.J., Santibanez, J., Savard, J., Scherer, S.E., Schneider, B., Sodergren, E., Vattahil, S., Villasana, D., White, C.S., Wright, R., Park, Y., Lord, J., Oppert, B., Brown, S., Wang, L., Weinstock, G., Liu, Y., Worley, K., Elsik, C.G., Reese, J.T., Elhaik, E., Landan, G., Graur, D., Arensburger, P., Atkinson, P., Beidler, J., Demuth, J.P., Drury, D.W., Du, Y.Z., Fujiwara, H., Maselli, V., Osanai, M., Robertson, H.M., Tu, Z., Wang, J.J., Wang, S., Song, H., Zhang, L., Werner, D., Stanke, M., Morgenstern, B., Solovyev, V., Kosarev, P., Brown, G., Chen, H.C., Ermolaeva, O., Hlavina, W., Kapustin, Y., et al., 2008. The genome of the model beetle and pest *Tribolium castaneum*. *Nature 452*, 949–955.

Strode, C., Wondji, C.S., David, J.P., Hawkes, N.J., Lumjuan, N., Nelson, D.R., Drane, D.R., Karunaratne, S.H., Hemingway, J., Black, W.C.T., Ranson, H., 2008. Genomic analysis of detoxification genes in the mosquito *Aedes aegypti*. *Insect Biochem. Mol. Biol. 38*, 113–123.

Valles, S.M., Perera, O.P., Strong, C.A., 2006. Gene structure and expression of the glutathione S-transferase, SiGSTS1, from the red imported fire ant, *Solenopsis invicta*. *Arch. Insect Biochem. Physiol. 61*, 239–245.

Vontas, J.G., Small, G.J., Hemingway, J., 2001. Glutathione S-transferases as antioxidant defence agents confer pyrethroid resistance in *Nilaparvata lugens*. *Biochem. J. 357*, 65–72.

Wang, Y., Qiu, L., Ranson, H., Lumjuan, N., Hemingway, J., Setzer, W.N., Meehan, E.J., Chen, L., 2008. Structure of an insect epsilon class glutathione S-transferase from the malaria vector *Anopheles gambiae* provides an explanation for the high DDT-detoxifying activity. *J. Struct. Biol. 164*, 228–235.

Yu, Q., Lu, C., Li, B., Fang, S., Zuo, W., Dai, F., Zhang, Z., Xiang, Z., 2008. Identification, genomic organization and expression pattern of glutathione S-transferase in the silkworm, *Bombyx mori*. *Insect Biochem. Mol. Biol. 38*, 1158–1164.

9 Insect G Protein-Coupled Receptors: Recent Discoveries and Implications

Y Park, Kansas State University, Manhattan,
KS, USA
M E Adams, University of California, Riverside,
CA, USA

© 2005, Elsevier BV. All Rights Reserved.

9.1. Introduction

G protein-coupled receptors (GPCRs) are integral membrane proteins that sense and transduce extracellular signals into cellular responses. The binding of small ligands to the extracellular domains of these proteins elicits a cascade of intracellular responses through activation of heterotrimeric G proteins. In insects, many signaling molecules and their corresponding biological functions have been identified, but knowledge of the signal transduction subsequent to receptor binding has been limited by the relative paucity of information about GPCR structure and function.

It is clear that remarkably diverse physiological processes are mediated by GPCRs. Some of these include sensing of environmental signals such as light, odors, and taste. Others sense extracellular signaling molecules such as calcium, nucleotides, neurotransmitters, biogenic amines, and neuroendocrine peptides. Evolution of GPCRs for specific functions in a vast array of organisms is one interesting area of study, for which insects in many ways are ideal for comparative studies. In addition to their value for basic science, GPCRs mediate vital physiological events in insects that may provide promising targets for development of pest control measures as well as therapeutic targets for promotion of human health.

The advent of the postgenomics era is providing excellent opportunities to explore insect GPCRs and their biological functions. Genomic surveys of GPCRs in the fruit fly *Drosophila melanogaster* and in the malaria mosquito *Anopheles gambiae* have identified at least ~200 and 276 GPCRs, respectively (Brody and Cravchik, 2000; Hill *et al.*, 2002a). Sequence analysis reveals an appreciable degree of groupwise homology relationships between insect and vertebrate GPCRs, suggesting many GPCRs have common ancestry dating more than 600 million years ago (Benton and Ayala, 2003). Data mining combined with functional studies of insect GPCRs has begun to reveal specific ligand–receptor interactions.

Deorphaning insect GPCRs by finding cognate ligands in *Drosophila* is now being followed by expansion of physiological studies in other species of insects that are more amenable to physiological experiments. The use of genomics as a springboard for understanding physiological mechanisms is complemented by advances in molecular probes, genetics, and imaging technologies, leading to a new era of "molecular physiology." This review will cover recent progress in the molecular and functional characterization of insect GPCRs.

9.2. Structure–Function Relationships for GPCRs

GPCRs are characterized by conserved core structures, consisting of an extracellular N-terminus, seven transmembrane (TM) α-helices each connected by alternating extracellular and intracellular loops (e1, e2, e3, i1, i2, and i3), and an intracellular C-terminus. The structure of the seven transmembrane protein bovine rhodopsin revealed by X-ray crystallography epitomizes the generalized bauplan predicted for GPCRs (Palczewski *et al.*, 2000; Teller *et al.*, 2001; review: Filipek *et al.*, 2003) (**Figure 1**). A number of important general structural features emerging from studies of vertebrate GPCRs are worth summarizing here as a prelude to discussions of insect GPCRs. Of course, caution is necessary in extrapolating knowledge from these works directly to other groups of phylogenetically diversified GPCRs.

The location of ligand binding sites varies considerably between different families of GPCRs. Indeed, GPCRs can be broadly categorized according to ligand size and location of the binding site. Such information is critical in the design of candidate drugs with either agonist or antagonist properties. For instance, GPCRs for small ligands such as retinoic acid, odorants, and biogenic amines appear to have a binding pocket surrounded by the TM bundle (Gotzes and Baumann, 1996; Vaidehi *et al.*, 2002). However, ligand binding to other groups of GPCRs involves extracellular epitopes of the protein, including the N-terminus and extracellular loops (Bockaert and Pin, 1999). In the case of the metabotropic glutamate and γ-aminobutyric acid B (GABA$_B$) receptors, the large N-terminus is composed of two lobes arranged for trapping ligands (Hermans and Challiss, 2001). In *Drosophila*, the N-terminal ectodomain of methuselah, an orphan GPCR associated with extended lifespan, consists of three β structure regions, which meet to form a shallow interdomain groove. This region characterized by an exposed tryptophan residue forms a putative ligand binding site by analogy with ectodomains of other GPCRs (West *et al.*, 2001).

The seven TM helices of GPCRs are arranged in an anticlockwise sequential manner when viewed from the extracellular aspect (**Figure 1**). Ligand binding induces changes in core structure, and in particular TM3 is thought to have a major role in receptor activation by altering the relative orientation of TM6 (Bockaert and Pin, 1999). Likewise, the conserved tripeptide signature of the TM3 cytoplasmic interface ([D,E]R[Y,W]) is considered to be important in receptor activation. The TM helices often are interrupted by distortions of the α-helix backbone, which may serve as hinge points

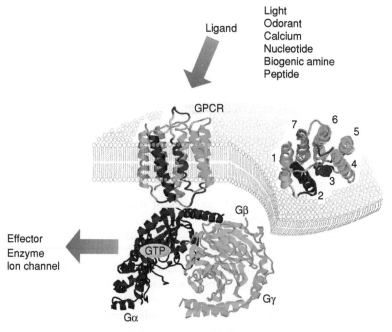

Figure 1 A diagram summarizing G protein-coupled receptor (GPCR) signaling. An array of seven transmembrane segments is shown according to the bovine rhodopsin structure. The trimeric G protein consisting of Gα, Gβ, and Gγ is coupled to the GPCR. Activation of the GPCR by ligands binding to the extracellular cell surface is amplified by intracellular effector proteins.

underlying conformational changes induced by ligand binding.

The cytoplasmic face of GPCRs interacts with heterotrimeric G-proteins, which dissociate upon receptor activation to trigger specific signal transduction pathways (**Figure 1**). Specificity for particular subtypes of heterotrimeric G-proteins is an important factor in determining downstream pathways. The *Drosophila* genome presents at least seven genes encoding Gα, six for Gβ, and two for Gγ subunits. The subunit Gα is in direct contact with the cytoplasmic interface of GPCRs (Bourne, 1997; Wess, 1998; Horn *et al.*, 2000; Moller *et al.*, 2001). Prediction of G protein subtype coupling specificity has been attempted in vertebrate GPCRs through identification of characteristic cytoplasm-facing sequence motifs (Bourne, 1997; Wess, 1998; Horn *et al.*, 2000; Moller *et al.*, 2001). This is a nontrivial exercise, considering reports showing GPCRs coupled with multiple G proteins as well as agonist-dependent changes in GPCR coupling with different G proteins (Robb *et al.*, 1994; Reale *et al.*, 1997; Cordeaux *et al.*, 2001; Cordeaux and Hill, 2002).

Proper functioning of GPCRs often may depend on oligomerization with other proteins (Gomes *et al.*, 2001; Angers *et al.*, 2002). Such interactions may serve to localize receptors for effective coupling with downstream pathways and/or to influence desensitization, which regulates receptor levels on the cell surface and temporal parameters of the biological response (Pierce and Lefkowitz, 2001; Brady and Limbird, 2002; Kohout and Lefkowitz, 2003).

9.3. Evolution of Insect GPCRs

In a pioneering study, Hewes and Taghert (2001) performed a comprehensive phylogenetic analysis of *Drosophila* peptide GPCRs covering the entire array of annotated genes in the *Drosophila* genome. The analysis revealed 42 putative peptide GPCRs arranged within 15 monophyletic subgroups together with vertebrate peptide GPCRs. The relationships established in this study also appear to be valid generally for the *Anopheles* mosquito (Hill *et al.*, 2002a). Evolutionary relationships between *Drosophila* and vertebrate peptide GPCRs have provided impetus for formulation of hypotheses to guide the process of deorphaning insect GPCRs. The success of this approach has been enhanced by many instances of previously established vertebrate ligand–receptor relationships. As this process advances, coevolutionary relationships between peptide GPCRs and their ligands have been brought into clearer focus.

Ligand–receptor coevolution has occurred as a consequence of reciprocal evolutionary influences between two interacting proteins – ligand and receptor – while the original definition of coevolution emphasized interactions between organisms (see Ridley, 1993). Assuming GPCRs share a common ancestral gene, diverse GPCRs must have arisen through multiple steps of gene duplication types of evolutionary processes (Chothia *et al.*, 2003). In phylogenetic analyses, closely related GPCRs have cognate ligands also related each other, indicating that an ancestral ligand–receptor partnership has effectively rooted new sets of ligand–receptors with a novel function. The ligand–receptor partners appear to be maintained during evolution without punctuated changes of their relationships; thus coevolution may be a common phenomenon.

In the coevolution of ligand–receptor partners, it has been proposed that divergence of the receptor precedes the evolution of its cognate novel neuropeptide ligand (Darlison and Richter, 1999). Such reasoning is based on the frequent occurrence of multiple receptors for a single peptide. This hypothesis, however, may be questioned because many peptide encoding genes have multiple repeats of isopeptides and processing sites. A good example is the tetrapeptide FMRFamide, for which eight homologous repeats appear in the *Drosophila* propeptide gene (Schneider and Taghert, 1988, 1990). There also appears to be a significant degree of flexibility in the system, evidenced first by the presence of multiple receptors for a given ligand that are coupled to different G proteins (Darlison and Richter, 1999; Park *et al.*, 2003) and second by differential selectivity of receptors to a number of isopeptides (Darlison and Richter, 1999). Further details regarding processes of ligand–receptor coevolution can be found in a number of recent publications (van Kesteren *et al.*, 1996; Darlison and Richter, 1999; Goh and Cohen, 2002).

While discerning evolutionary relationships among GPCRs is challenging, establishing such relationships for signaling peptides is a much more daunting task. The main problem lies in the fact that only a small number of functionally important, conserved amino acids within peptide sequences defines their evolutionary relationships. A valuable search tool in bioinformatics, the position-specific iterated BLAST (PSI-BLASAT: Altschul *et al.*, 1997), provides a powerful means of using small short, conserved sequence motifs for searching distantly related peptides. The method uses a score matrix generated from multiple alignments of the highest scoring hits in each round of the search. Conservation of consensus amino acids in the multiple sequence alignment implies functionally and evolutionarily important amino acid residues common for a group of signaling peptides.

A good example of ligand–receptor coevolution was found for a group of neuropeptides having a C-terminal PRXamide motif (Pro-Arg-Xxx with C-terminal amide). This motif is common to an array of peptides occurring in both insects and vertebrates, including neuromedin U (NMU), cardioacceleratory peptide 2b (CAP2b), pyrokinins, and ecdysis triggering hormone (ETH). This group of peptides activates a cluster of GPCRs related to vertebrate receptors for NMU and thyrotropin releasing hormone (Park *et al.*, 2002) (**Figure 3**). While phylogenetic relationships among receptors are established almost exclusively on the basis of amino acid sequence, the patterns of pharmacological cross-reactivity for these GPCRs to cognate PRXamide peptides provide a new and "functional" criterion for inferring evolutionary relationships for ligand–receptor pairs.

Despite the frequent correspondence of insect and vertebrate ligand–receptor relationships, a number of *Drosophila* GPCRs are found to be relatively unique (Hewes and Taghert, 2001), such as *Drosophila* GPCRs having bootstrap values of less than 300 in 1000 pseudosamples. For example, closely related GPCRs within the group CG13229, CG13803, and CG8985 are found to have no significant relationship with other vertebrate GPCRs. Recently, two of these receptors CG13803 and CG8985 were found to be activated by Dromyosuppressin (Johnson *et al.*, 2003a; Egerod *et al.*, 2003a) having the C-terminal motif FLRLamide, a member of FMRFamide-related peptides (FaRPs) containing a C-terminal RFamide. Such unique GPCRs could arise either through rapidly evolving branches of GPCRs, or through recent loss of the related GPCRs in the vertebrates.

9.4. Insect GPCRs

9.4.1. Classification of GPCRs

Classification of the GPCR superfamily by families and groups has been made based on a combination of natural ligand categories and receptor sequence homology (Kolakowski, 1994; Horn *et al.*, 2000; Hewes and Taghert, 2001). For the following discussions, these classifications are used to cover comprehensive categories of GPCRs in all organisms. Accordingly, insect GPCRs are classified into the following families: family A, rhodopsin and various peptide receptors; family B, secretin-like peptide receptors and methuselah (mth); family C, metabotropic glutamate receptors, atypical GPCRs such as

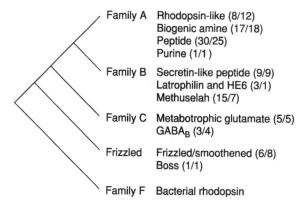

Figure 2 A conceptual phylogenetic relationship of the GPCR superfamily. The numbers of annotated genes for *Drosophila* (left) and *Anopheles* (right) are in parentheses. The conceptual tree is still not congruent and reconstructed based on Josefsson (1999).

frizzled/smoothened and bride of sevenless (boss), and odorant receptors (**Figure 2**).

Assuming all groups of GPCRs (or 7-TM proteins) are ancestrally related, a conceptual tree depicting evolutionary relationships among them can be drawn based on combined results from previous studies (**Figure 2**), while construction of a congruent tree is still controversial because of its high sequence divergency (Josefsson, 1999). The number of *Drosophila* genes encoding putative GPCRs as a fraction of the total in each group is shown in parenthesis for each category (**Figure 2**). In rooting the tree on family F, which includes bacterial rhodopsin, the group frizzled appears to be the most ancestral form. Family A with ~90 GPCRs is the largest group, and includes biogenic amines, peptides, rhodopsin, and ~30 orphan receptors (Hill *et al.*, 2002a). Given the excellent phylogenetic analysis for *Drosophila* peptide receptors in the previous study (Hewes and Taghert, 2001). We provide **Tables 1–3** summarizing the GPCRs that have been cloned so far. This table includes *Drosophila* data and GPCRs from other insects, but not the GPCRs of *A. gambiae* where experimental data is unavailable.

9.4.2. Family A: Biogenic Amine, Rhodopsin, and Neuropeptide Receptors

2.4.2.1. Biogenic amine receptors   Most small signaling molecules used as transmitters, modulators, and hormones are common to vertebrates and invertebrates, including insects. The most prominent are acetylcholine, glutamate, GABA, and the biogenic amines serotonin, dopamine, and histamine (**Table 1**). Norepinephrine and epinephrine, abundant in the vertebrates, are absent or present only in very low amounts in insects (Maxwell *et al.*,

Table 1   Insect G protein-coupled receptors for biogenic amines

Name	Synonym (GenBank accession number)	Species	Cellular response (expression system)	Expression pattern	Reference
Serotonin receptor					
5-HT2	CG1056, 5HT-dro1, 5HT2dro, Dm5HT2 (X81835)	*Drosophila melanogaster*		Coexpression with pair-rule gene *ftz* in embryo	Colas *et al.* (1999, 1995)
5HT7	CG12073, 5HT-dro1, 5HT7dro, Dm5HTdro1, 5HT-R1 (M55533)	*Drosophila melanogaster*	Increased cAMP (Sf9, NIH 3T3)	Adult head, larval ventral ganglion	Witz *et al.* (1990), Saudou *et al.* (1992), Obosi *et al.* (1996)
5-HT1A	CG16720, 5HT-dro2A (Z11489)	*Drosophila melanogaster*	Inhibition cAMP (COS-7)	Ventral unpaired median neurons	Saudou *et al.* (1992)
5-HT1B	CG15113, 5HT-dro2B (Z11490)	*Drosophila melanogaster*	Inhibition cAMP (COS-7)	Larval ventral ganglion	Saudou *et al.* (1992), Obosi *et al.* (1996)
CG8007	(AE003679, AAN13390)	*Drosophila melanogaster*			Brody and Cravchik (2000)
Aedes 5-HT7	(AF296125)	*Aedes aegypti*		Malpighian tubules, tracheolar cells	Pietrantonio *et al.* (2001)
Bm5HT	B150 *Bombyx* (X95604)	*Bombyx mori*			von Nickisch-Rosenegk *et al.* (1996)
Hv5HT	K15 *Heliothis* (X95605)	*Heliothis virescens*			von Nickisch-Rosenegk *et al.* (1996)
Dopamine receptor					
DopR1	CG9652, dDA1, DmDop1 (X77234)	*Drosophila melanogaster*	Increased cAMP (HEK-293)	Somata of optic lobe	Gotzes *et al.* (1994), Gotzes and Baumann (1996)
DopR2	CG7569, CG18741, DopR99B, DAMB (U34383)	*Drosophila melanogaster*	Ca^{2+} mobilization (*Xenopus* oocyte), Increased cAMP (*Xenopus* oocyte, HEK-292, S2)	Mushroom body	Feng *et al.* (1996), Han *et al.* (1996), Reale *et al.* (1997)
DD2R	CG17004 (AY150862)	*Drosophila melanogaster*	Increased cAMP (CRE-luciferase assay in HEK-293)	Head and body	Hearn *et al.* (2002)
AmDop1	(CAA73841)	*Apis mellifera*	Increased cAMP (HEK-293)		Blenau *et al.* (1998)
AmBAR6	D2 (AAM19330)	*Apis mellifera*		Mushroom body	Ebert *et al.* (1998), Blenau and Baumann (2001)
Muscarinic acetylcholine receptor					
AchR-60C	CG4356, mAChR-1, DM1 (M23412)	*Drosophila melanogaster*	Increased IP3 metabolism (mouse Y1)	Mushroom body, optic lobe, antennal lobe	Shapiro *et al.* (1989), Blake *et al.* (1993), Hannan and Hall (1996)
CG7918	(AE003677, AAF54188)	*Drosophila melanogaster*			Brody and Cravchik (2000)
Octopamine/tyramine receptor					
OAMB	CG3856, CG15698, DmOCT1B (AF065443)	*Drosophila melanogaster*	Increased cAMP (S2), increased cAMP (CHO-K1), Ca^{2+} mobilization (CHO-K1)	Mushroom body	Arakawa *et al.* (1990), Robb *et al.* (1994), Han *et al.* (1998)
BmOcr	B96 *Bombyx* (X95607)	*Bombyx mori*			von Nickisch-Rosenegk *et al.* (1996)

Continued

Table 1 Continued

Name	Synonym (GenBank accession number)	Species	Cellular response (expression system)	Expression pattern	Reference
HvOcr	K50 Heliothis (X95606)	Heliothis virescens	Inhibition cAMP (LLC-PK1)	Antennae	von Nickisch-Rosenegk et al. (1996)
TyrR	CG7485, Tyr-dro, honoka (X54794, AB073914)	Drosophila melanogaster	Inhibition cAMP, Ca^{2+} mobilization (NIH 3T3)	Antennae, CNS	Saudou et al. (1990), Kutsukake et al. (2000)
Amtyr1	(AJ245824)	Apis mellifera	Inhibition cAMP, Ca^{2+} mobilization (HEK-293)	Mushroom body, CNS	Blenau et al. (2000)
TyrLoc	Tyr-Loc (X69520)	Locusta migratoria	Inhibition cAMP, Ca^{2+} mobilization (murine erythroleukemia cell (MEL-C88L, S2)		Vanden Broeck et al. (1995), Poels et al. (2001)
MabOcR	(AAK14402)	Mamestra brassicae			

1978; Evans, 1980). Prominent amines in insects and crustaceans are octopamine and tyramine (Evans, 1980). Octopamine is thought to play a modulatory role in arousal, paralleling that of noradrenaline in the sympathetic system of vertebrates (Evans, 1993; Blenau and Baumann, 2001). Octopamine and tyramine, while only present as "trace amines" in the vertebrates, may nevertheless play functional roles (Borowsky et al., 2001). Among the biogenic amines known to be signaling molecules in insects, histamine is so far the only one for which GPCRs are unknown; electrophysiological data provide conclusive evidence for gating of ionotropic receptors by histamine. This contrasts with histamine receptors in the vertebrates, where both ionotropic and GPCR receptors are known (Roeder, 2003).

The evolutionary patterns of biogenic amine receptors among vertebrates and invertebrates are generally in accordance with monophyletic ancestries (Roeder, 2002), although recent gene duplication events have led to rapid expansions of certain vertebrate receptor groups. Furthermore, there are apparent shifts in ligand specificity in a number of receptor phylogenies, for which the ligands are structurally similar. This phenomenon is particularly apparent for Bombyx and Heliothis octopamine receptors, which occur in a monophyletic clade with tyramine receptors (Blenau et al., 1998; Ebert et al., 1998; Roeder, 2002).

Numerous functions of biogenic amines in insects are described at the behavioral and physiological levels, including modulation of sensory input and motor output, learning and memory, development,

and elevated states of arousal (Blenau and Baumann, 2001; Roeder, 2002). Modulation of olfactory and visual stimuli is indicated by expression patterns of amines in sensory lobes, and by physiological experiments. For example, sensitization of pheromone receptor cells by octopamine and serotonin is indicated in Bombyx mori and Manduca sexta (Pophof, 2000; Dolzer et al., 2001), and a possible role for serotonin in feedback from higher centers in the brain is indicated from morphological studies (Hill et al., 2002b). Octopamine plays a major hormonal role in energy metabolism during locust flight by acting directly on the fat body and indirectly through release of adipokinetic hormone (Orchard et al., 1993). Setting the day state in circadian cycling by serotonin is suggested for a cricket (Gryllus bimaculatus) (Saifullah and Tomioka, 2002), and dopamine receptors are thought to be targets of circadian modulation (Andretic and Hirsh, 2000). Egg diapause in B. mori appears to be controlled by dopaminergic pathways in the mother (Noguchi and Hayakawa, 2001).

In a growing number of cases, biogenic amine receptor-specific functions are reported. The honoka mutant in Drosophila carries a tyramine receptor defect and shows reduced responses to repellent odors. In addition, this mutant exhibited reduced tyramine modulation at the neuromuscular junction, while normal modulation by octopamine was observed (Kutsukake et al., 2000). A role for the serotonin receptor subtype 5-HT2 in embryonic development was shown by a Drosophila mutant (Colas et al., 1995, 1999). In these studies, serotonin signaling was found to be necessary for increased

Table 2 Insect G protein-coupled receptors for neuropeptides: family A group III and group V, and family B

Name	Synonym (GenBank accession no.)	Species	Active ligand[a]	Vertebrate homology group	Evidence/functional assay (expression system)[b]	Expression pattern	Reference
Family A group III							
CCKLR-17D3	CG6857 (AE003509, AAF48857)	*Drosophila melanogaster*		Cholecytokinin/gastrin R	Homology to CCKLR-17D1		McBride *et al.* (2001), Kubiak *et al.* (2002)
CCKLR-17D1	CG6881, CG6894, CG32540, dsk-r1 (AE003510, AAF48879)	*Drosophila melanogaster*	Sulfakinin	Cholecytokinin/gastrin R	Ca²⁺ mobilization (HEK-293, CHO-K1, SH-EP)		McBride *et al.* (2001), Kubiak *et al.* (2002)
NepYr	CG5811 (M81490)	*Drosophila melanogaster*		Neurokinin R	Inward current (*Xenopus* oocyte)		Li *et al.* (1992)
Leucokinin receptor	CG10626 (AE003566, NM139711, AY070926)	*Drosophila melanogaster*	Leucokinin	Neurokinin R	Ca²⁺ mobilization (S2)	Malpighian tubules, CNS	Radford *et al.* (2002)
Myokinin receptor	(AF228521)	*Boophilus microplus*		Neurokinin R	Homology cloning		Holmes *et al.* (2000)
Takr86C	CG6515, NKD (M77168)	*Drosophila melanogaster*	Tachykinin	Neurokinin R	Increased IP3 (mouse NIH-3T3), translocation of β-arrestin2	CNS	Monnier *et al.* (1992), Johnson *et al.* (2003b)
Takr99D	CG7887, DTKD (X62711)	*Drosophila melanogaster*	Tachykinin	Neurokinin R	Inward current (*Xenopus* oocyte), translocation of β-arrestin2	CNS	Li *et al.* (1991), Johnson *et al.* (2003b)
STKR	*Stomoxys* tachykinin receptor	*Stomoxys calcitrans*	Tachykinin-like	Neurokinin R	Ca²⁺ mobilization (S2)		Guerrero (1997), Torfs *et al.* (2001, 2002a)
LTKR	*Leucophaea* tachykinin receptor (AJ310390)	*Leucophaea maderae*		Neurokinin R	Homology cloning	Brain	Johard *et al.* (2001)
SNPF-R	CG7395, Drm-sNPF-R, NPFR76F, GPCR60 (NP_524176)	*Drosophila melanogaster*	Short neuropeptide F	Neurokinin R	Ca²⁺ mobilization (CHO-K1/Gα₁₆), inward current (*Xenopus* oocyte)		Mertens *et al.* (2002), Feng *et al.* (2003)
NPFR1	CG1147 (AF364400)	*Drosophila melanogaster*	Neuropeptide F	Neurokinin R	Ca²⁺ mobilization (CHO-K1/Gα₁₆), ¹²⁵I-NPF binding (CHO), translocation of β-arrestin2	CNS, midgut	Garczynski *et al.* (2002), Johnson *et al.* (2003b)

Continued

Table 2 Continued

Name	Synonym (GenBank accession no.)	Species	Active ligand[a]	Vertebrate homology group	Evidence/functional assay (expression system)[b]	Expression pattern	Reference
FMRFamide receptor	CG2114 (NM139501)	Drosophila melanogaster	FMRFamide	Neurokinin R	Ca^{2+} mobilization (CHO-K1/Gα$_{16}$), translocation of β-arrestin2	Trachea, gut, fat body, Malpighian tubules, ovary	Cazzamali and Grimmelikhuijzen (2002), Meeusen et al. (2002), Johnson et al. (2003b)
Allatostatin B receptor	CG14484, CG18192 (NP611241)	Drosophila melanogaster	Allatostatin B	Bombesin R	translocation of β-arrestin2		Johnson et al. (2003a)
PRXa receptor 1	CG9918 (AF522191)	Drosophila melanogaster	Cardioacceleratory peptide CAP2b	Neuromedin U R	Inward current (Xenopus oocyte)		Park et al. (2002)
PRXa receptor 2	CG8784 (AF522189)	Drosophila melanogaster	Hugin	Neuromedin U R	Inward current (Xenopus oocyte), Ca^{2+} mobilization (CHO-K1/Gα$_{16}$)		Park et al. (2002), Rosenkilde et al. (2003)
PRXa receptor 3	CG8795 (AF522190)	Drosophila melanogaster	Hugin	Neuromedin U R	Inward current (Xenopus oocyte), Ca^{2+} mobilization (CHO-K1/Gα$_{16}$)		Park et al. (2002), Rosenkilde et al. (2003)
HezPBAN R	AY319852	Helicoverpa zea	HezPBAN	Neuromedin U R	Ca^{2+} mobilization (Sf9)		Choi et al. (2003)
CAP2b receptor	CG14575 (AF522193)	Drosophila melanogaster	Cardioacceleratory peptide (CAP2b)	Neuromedin U R	Inward current (Xenopus oocyte), Ca^{2+} mobilization (CHO-K1/Gα$_{16}$)		Iversen et al. (2002), Park et al. (2002)
ETH receptor a	CG5911a (AY220741)	Drosophila melanogaster	Ecdysis triggering hormone (ETH)	Neuromedin U R/ thyrotropin releasing hormone R	Ca^{2+} mobilization (CHO-K1/Gα$_{16}$)		Iversen et al. (2002), Park et al. (2003)
ETH receptor b	CG5911b (AY220742)	Drosophila melanogaster	Ecdysis triggering hormone (ETH)	Neuromedin U/ thyrotropin releasing hormone R	Ca^{2+} mobilization (CHO-K1/Gα$_{16}$)		Iversen et al. (2002), Park et al. (2003)
Myosuppressin receptor 1	CG8985 (AF545042)	Drosophila melanogaster	Myosuppressin	Neuromedin U/ thyrotropin releasing hormone R	Ca^{2+} mobilization (CHO-K1/Gα$_{16}$), translocation of β-arrestin2, inhibition cAMP (HEK-292)		Egerod et al. (2003a), Johnson et al. (2003a)

Receptor	Gene (accession)	Species	Ligand	Assay / signaling	Expression	References
Myosuppressin receptor 2	CG13803 (AF544244)	*Drosophila melanogaster*	Myosuppressin	Ca^{2+} mobilization (CHO-K1/Gα_{16}), translocation of β-arrestin2, inhibition cAMP (HEK-292), GTPγS binding (HEK-292)		Egerod *et al.* (2003a), Johnson *et al.* (2003a)
Proctolin receptor	CG6986 (NM167020)	*Drosophila melanogaster*	Proctolin	Ca^{2+} mobilization (HEK-293), translocation of β-arrestin2	Brain, medulla, hindgut, pericardial cell, heart	Egerod *et al.* (2003b), Johnson *et al.* (2003b)
Family A group V Corazonin receptor	CG10698 (AF522192, AF373862)	*Drosophila melanogaster*	Corazonin	Inward current (*Xenopus* oocyte), Ca^{2+} mobilization (CHO-K1/Gα_{16}), translocation of β-arrestin2		Cazzamali *et al.* (2002), Park *et al.* (2002), Johnson *et al.* (2003b)
Bombyx AKH receptor	BAKHR (AF403542)	*Bombyx mori*	Adipokinetic hormone (AKH)	Ca^{2+} mobilization (CHO-K1/Gα_{16})		Staubli *et al.* (2002)
AKH receptor	CG11325, GRHR, DAKHR (AF522194, AAC61523)	*Drosophila melanogaster*	Adipokinetic hormone (AKH)	Inward current (*Xenopus* oocyte), Ca^{2+} mobilization (CHO-K1/Gα_{16})		Park *et al.* (2002), Staubli *et al.* (2002)
CCAP receptor	CG6111 (AF522188)	*Drosophila melanogaster*	Crustacean cardioactive peptide (CCAP)	Inward current (*Xenopus* oocyte), Ca^{2+} mobilization (CHO-K1/Gα_{16})		Park *et al.* (2002), Staubli *et al.* (2002)
Drostar1	CG7285 (AAG54080, NM140783)	*Drosophila melanogaster*	Somatostatin R/opioid R	GIRK current (*Xenopus* oocyte)	Corpora allata	Kreienkamp *et al.* (2002a), Johnson *et al.* (2003a)
Drostar 2	CG13702 (NM140782)	*Drosophila melanogaster*	Somatostatin R/opioid R	GIRK current (*Xenopus* oocyte)	Optic lobe, corpora allata	Kreienkamp *et al.* (2002a), Johnson *et al.* (2003a)
DAR-1	AlstR, EG:121E7.2, CG2872 (AF163775, NM166982)	*Drosophila melanogaster*	Galanin	GIRK current (*Xenopus* oocyte), Ca^{2+} mobilization (CHO-K1), GTPγS binding (CHO-K1)		Birgul *et al.* (1999), Larsen *et al.* (2001)
DAR-2	AR-2, CG10001	*Drosophila melanogaster*	Galanin	Ca^{2+} mobilization (CHO-K1), GTPγS binding (CHO-K1)	Gut	Lenz *et al.* (2000, 2001), Larsen *et al.* (2001)

Continued

Table 2 Continued

Name	Synonym (GenBank accession no.)	Species	Active ligand[a]	Vertebrate homology group	Evidence/functional assay (expression system)[b]	Expression pattern	Reference
Pea-AlstR	Periplaneta AlstR (AF336364)	Periplaneta americana	Allatostatin A	Galanin	GIRK current (Xenopus oocyte)		Auerswald et al. (2001)
BAR	Bombyx AlstR (AH011256)	Bombyx mori	Allatostatin A	Galanin	Ca^{2+} mobilization (CHO-K1/$G\alpha_{16}$)	Gut	Secher et al. (2001)
LGR1	CG7665, Fsh (U47005)	Drosophila melanogaster	nd	FSH/LH/TRH	FSH receptor homology cloning		Hauser et al. (1997), Nishi et al. (2000)
LGR2	CG8930, Rickets (AF142343)	Drosophila melanogaster	Bursicon	FSH/LH/TRH	Mutant insensitive to bursicon-like factor	Embryo and pupae	Eriksen et al. (2000), Nishi et al. (2000), Baker and Truman (2002)
Family B							
Mas-DH-R	Manduca DH receptor (U03489)	Manduca sexta	Mas-diuretic hormone	Corticotropin releasing hormone R	[^{3}H]Mas-DH binding and cAMP (COS-7), [^{125}I], Mas-DH binding and cAMP (Sf-9)		Reagan (1994, 1995)
Acd-DH-R	Acheta DH receptor (U15959)	Acheta domesticus	Acheta-diuretic hormone	Corticotropin releasing hormone R	[^{3}H]Acd-DH binding and cAMP (COS-7),		Reale et al. (1997)
Bm-DH-R	Bombyx DH receptor	Bombyx mori		Corticotropin releasing hormone R	Homology cloning		Ha et al. (2000)

[a]The most active or putative ligand suggested in the respective reference(s).
[b]Methods used for cloning and assay, or other evidence provided.

Table 3 Family C: Metabotropic glutamate receptors and GABA_B receptors

Name	Synonym	Evidence	Expression	Reference
mGluRA	CG11144, metabotropic glutamate receptor A, DmGluRA,	Inhibition of AC (HEK 293), GIRK current (*Xenopus* oocyte)	CNS	Raymond *et al.* (1999)
mGluRB	CG30361, CG18447, CG8692, metabotropic glutamate receptor B,	Homology		Brody and Cravchik (2000)
GABA-B-R1	CG15274, D-GABA_BR1	GIRK current (*Xenopus* oocyte)	CNS	Mezler *et al.* (2001)
GABA-B-R2	CG6706, D-GABA_BR2, GH07312	GIRK current (*Xenopus* oocyte)	CNS	Mezler *et al.* (2001)
GABA-B-R3	CG3022, GABA_BR3	Homology	CNS	Mezler *et al.* (2001)

cell adhesiveness required for cell–cell intercalation during gastrulation. Expression patterns of the serotonin 5-HT7-like receptor in *Aedes* Malpighian tubules and tracheolar cells are suggestive of possible functions in the control of diuresis and respiration (Pietrantonio *et al.*, 2001).

9.4.2.2. Rhodopsin Rhodopsins are a class of retinal-binding chromophores widespread in visual cells of animals. In each rhodopsin, the light-absorbing retinal binds to a pocket within the seven-transmembrane opsin protein. Incoming photons induce isomerization of retinal, which passes protons to its partner opsin, resulting in G protein activation. The eyes of insects are tuned to specific spectral ranges through diversity in genes encoding the opsin structure, whereby amino acid sequence variability occurs in the vicinity of the retinal binding pocket. The coupling of opsins to G-proteins leads to visual transduction through the opening or closing of cation channels in the visual cell. The evolution of genes encoding rhodopsins with different spectral properties have been described in various insect species (review: Briscoe and Chittka, 2001).

The insect compound eye is composed of cartridges called ommatidia, each organized as a set of eight photoreceptor retinula cells (R1–R8). In *Drosophila*, the compound eye expresses seven rhodopsin genes (*Rh1–Rh7*), which show unique expression patterns and spectral properties. A specialized portion of the plasma membrane in each retinula cell is convoluted via microvilli to form the rhodopsin-rich rhabdomere. The rhabdomere of each retinula cell faces the others within the cartridge to create the composite rhabdom, where light is absorbed. Rhodopsin Rh1, also called ninaE (neither inactivation nor after potential E), is expressed in retinula cells R1–R6 (O'Tousa *et al.*, 1985; Feiler *et al.*, 1988). Rh2 is expressed in ocelli (Cowman *et al.*, 1986), simple eyes located on the vertical side of the head, while Rh3 and Rh4 are expressed in R7 in nonoverlapping subsets of ommatidia (Fryxell and Meyerowitz, 1987). Rh5 and Rh6 are expressed

in R8 in nonoverlapping subsets of ommatidia (Chou *et al.*, 1996, 1999). Rh7 (CG5638) has been identified in the genome sequence, but has not been associated with a physiological function as yet (Brody and Cravchik, 2000).

Insect rhodopsins are classified generally into three categories according to their responsiveness to ultraviolet, blue, and green colors (Townson *et al.*, 1998). In some cases, rhodopsins sensitive to red color have been identified in the Lepidoptera and Hymenoptera (Briscoe and Chittka, 2001), where adaptive evolution appears to have taken place through replacement of amino acids adjacent to the retinal binding site and thus shifting the spectral sensitivity (Briscoe, 2002). Recent expansions of the rhodopsin gene in different species of insects are observed in the phylogenetic relationships among rhodopsins in the orders Hymenoptera, Diptera, and Lepidoptera (Montell, 1999; Salcedo *et al.*, 1999; Briscoe, 2001; Briscoe and Chittka, 2001; Hill *et al.*, 2002a).

The phototransduction cascade in invertebrates begins with light-absorbing rhodopsin coupled to G_q. This activates phosphoinositide turnover through activation of phospholipase C β (PLCβ), liberation of inositol trisphosphate (IP3), and the opening of transient receptor potential (TRP) channels, leading to inward cation (Na^+ and Ca^{2+}) flux and depolarizing potentials in retinula cells (Alvarez *et al.*, 1996; Montell, 1999; Bahner *et al.*, 2000; Hardie and Raghu, 2001). Invertebrate phototransduction contrasts with that of vertebrates, where at low light intensities a large "dark current" flows through cyclic guanidine monophosphate (cGMP) nucleotide gated sodium channels to maintain a relatively depolarized state (approximately $-40\,mV$). Under dark conditions, photoreceptor cells maintain high levels of cytosolic cGMP and hence a high number of open cyclic nucleotide gated sodium channels. Light absorbance by rhodopsin activates the G protein transducin (G_t), leading to elevation of phosphodiesterase, destruction of cGMP, and consequent closing of sodium channels (Shichida and Imai, 1998). Through this mechanism, vertebrate photoreceptors hyperpolarize in response to light stimuli.

9.4.2.3. Neuropeptide receptors in family A

Neuropeptides are ubiquitous signaling molecules acting mainly through GPCRs for activation of cellular responses. Numerous signaling peptides in insects have been characterized by their abilities to activate visceral activity, heartbeat, or glandular secretion (reviews: Gäde *et al.*, 1997; Nassel, 2002). Studies of neuropeptide signaling have been accomplished mainly in large insects where biochemical studies and physiological assays are relatively easy. Subsequent findings of neuropeptides in various species show that amino acid sequences of neuropeptides are generally well conserved. With the opening of the postgenomic era, genomic sequences in *Drosophila* and *Anopheles* revealed more than 23 and 35 genes, respectively, encoding neuropeptides (Hewes and Taghert, 2001; Riehle *et al.*, 2002). Peptidomics of the central nervous system (CNS) of *Drosophila* (Baggerman *et al.*, 2002) identified 28 neuropeptides including eight novel peptides using capillary liquid chromatography in conjunction with tandem mass spectrometry.

Studies of neuropeptide receptors in vertebrates have identified a large number of GPCRs grouped as family A with rhodopsins and biogenic amine receptors (Kolakowski, 1994). Phylogenetic analysis of *Drosophila* and *Anopheles* genome sequences for putative GPCRs revealed ~40 GPCRs in family A in each species (Hewes and Taghert, 2001; Hill *et al.*, 2002a). Analysis of phylogenetic relationships revealed that many insect neuropeptide receptors occur in monophyletic groups with vertebrate receptors, implying evolution from common ancestors in each group (Hewes and Taghert, 2001; Hill *et al.*, 2002a). Family A is further classified into several subgroups, and insect peptide GPCRs currently are placed into group III or group V (Kolakowski, 1994; Hewes and Taghert, 2001).

Numerous insect GPCRs in family A group III and group V have been cloned and expressed, and functional assays have been key to identification of their ligands (**Table 2**). The earliest efforts to clone neuropeptide GPCRs started with cDNA library screenings that targeted *Drosophila* homologs of the vertebrate neuropeptide Y and tachykinin receptors (Li *et al.*, 1991, 1992; Monnier *et al.*, 1992). Following identification of these receptors in *Drosophila*, subsequent investigations located their homologs in the stable fly (*Stomoxys Calcitrans*) and cockroach (*Leucophaea maderae*) (Guerrero, 1997; Torfs *et al.*, 2000, 2001, 2002a; Johard *et al.*, 2001).

Recent approaches in functional genomics have contributed to the deorphaning of a large number of insect GPCRs. Functional analyses have been designed for screening sets of GPCRs and ligands belonging to a phylogenetic clade. This approach assumes ligand–receptor coevolution (see Section 9.6) and utilizes an established ligand–receptor pair as a baseline for formulation of testable hypotheses.

Such an approach was taken to identify likely receptors for peptide ligands having C-terminal PRXamide motifs (Park *et al.*, 2002). The PRXa peptides in *Drosophila* and other insects fall into three classes: pyrokinins including the pheromone biosynthesis activating neuropeptide (PBAN) (–FXPRXa) (Raina *et al.*, 1989; Matsumoto *et al.*, 1990), cardioactive CAP2b-like peptides (–FPRXa) (Morris *et al.*, 1982; Huesmann *et al.*, 1995), and ETHs (–PRXa) (Zitnan *et al.*, 1996; Adams and Zitnan, 1997; Park *et al.*, 1999, 2002). In the vertebrates, PRXa motifs are found in the peptides NMU, vasopressin (AVP), and pancreatic polypeptide (PP), for which receptors already have been identified (Thibonnier *et al.*, 1994; Fujii *et al.*, 2000; Howard *et al.*, 2000). Functional analyses demonstrated that *Drosophila* GPCRs related to the monophyletic NMU receptor group are activated by all three classes of *Drosophila* PRXa peptides (**Figure 3**). PRXa receptors are now being investigated in other species of insects, such as *Manduca*, where the physiological functions of the neuropeptides are better characterized (Kim and Adams, unpublished data), while genuine functions of the PRXamide receptors in *Drosophila* also are being investigated.

Drosophila GPCRs related to the vasopressin receptor failed to respond to PRXa peptides, but instead was activated by crustacean cardioactive peptide (CCAP), corazonin, and adipokinetic hormone (AKH), none of which falls into the PRXa peptide group (Park *et al.*, 2002). *Drosophila* sulfakinin receptors also were identified through the logic of ligand–receptor coevolution upon finding homology in ligands between vertebrate cholecystokinin/gastrin and *Drosophila* sulfakinin (Kubiak *et al.*, 2002). A *Drosophila* neuropeptide F receptor was identified, and found to belong to one of four *Drosophila* GPCRs in the neuropeptide Y receptor subgroup (Garczynski *et al.*, 2002; Mertens *et al.*, 2002).

Another productive approach for deorphaning GPCRs is reverse physiology (see Section 9.5). Receptors for allatostatin type A (DAR-1) (Birgul *et al.*, 1999) and allatostatin type C (Drostar1, Drostar2) (Kreienkamp *et al.*, 2002a) were identified through isolation from tissue extracts of native

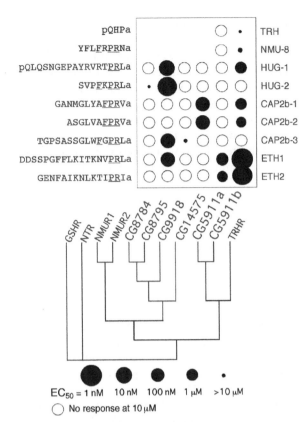

pQHPa — TRH
YFLFRPRNa — NMU-8
pQLQSNGEPAYRVRTPRLa — HUG-1
SVPFKPRLa — HUG-2
GANMGLYAFPRVa — CAP2b-1
ASGLVAFPRVa — CAP2b-2
TGPSASSGLWFGPRLa — CAP2b-3
DDSSPGFFLKITKNVPRLa — ETH1
GENFAIKNLKTIPRIa — ETH2

GSHR NTR NMUR1 NMUR2 CG8784 CG8795 CG9918 CG14575 CG5911a CG5911b TRHR

EC₅₀ = 1 nM   10 nM   100 nM   1 μM   >10 μM
○ No response at 10 μM

Figure 3 Ligand–receptor specificity for *Drosophila* PRXamide and related peptides. The amino acid sequence of each ligand is shown at the left, with the names of each peptide on the right. Circles indicate relative activity of ligands to each receptor. The larger the circle, the more potent is the peptide. Empty circles indicate "no response" at concentrations up to 10 μM, the highest concentration tested. The tree and names of neuropeptide receptors are: GSHR, GH secretagogue receptor; NTR, neurotensin receptor; NMUR1, neuromedin U receptor 1; NMUR2, neuromedin U receptor 2; TRHR, thyrotropin releasing hormone receptor; *Drosophila* PRXamide receptor GPCRs with CG numbers. Red letters are vertebrate peptides and GPCRs. Blue letters are *Drosophila* PRXamide peptides and GPCRs. (Data from Park, Y., Kim, Y.J., Adams, M.E., **2002**. Identification of G protein-coupled receptors for *Drosophila* PRXamide peptides, CCAP, corazonin, and AKH supports a theory of ligand–receptor coevolution. *Proc. Natl Acad. Sci. USA 99*, 11423–11428; and Park, Y., Kim, Y.J., Dupriez, V., Adams, M.E., **2003**. Two subtypes of ecdysis triggering hormone receptor in *Drosophila melanogaster*. *J. Biol. Chem. 278*, 17710–17715).

neuropeptide ligands that activate the receptor. Closely related genes in *Drosophila* and other insect species subsequently cloned and expressed, and showed their sensitivity to related allatostatin peptides (Lenz *et al.*, 2000, 2001; Auerswald *et al.*, 2001; Larsen *et al.*, 2001; Secher *et al.*, 2001). The reverse physiology approach also was successfully employed in identification of the *Drosophila* FMRFamide receptor (Meeusen *et al.*, 2002) and the leucokinin receptor in the snail *Lymnaea*

(Cox *et al.*, 1997). Subsequent searches for leucokinin receptor orthologous genes in insects led to identification of novel leucokinin receptors in the cattle tick, *Boophilus microplus* (Holmes *et al.*, 2000, 2003) and in *Drosophila* (Radford *et al.*, 2002).

A gene homologous to the adenosine-like receptor has been found in both *Drosophila* and *Anopheles* genomes (Brody and Cravchik, 2000; Hill *et al.*, 2002a), but biological data are unavailable as yet. The findings in *Drosophila* will facilitate GPCR studies in other arthropod species where biological functions of those neuropeptides have been well characterized (see Section 9.5).

9.4.3. Family B: Secretin-Like Receptors

Insect GPCRs belonging to family B consist of secretin-like receptors (five genes in *Drosophila*) (**Table 3**), the *methuselah* group (~11 genes), human epididymis-specific protein 6 (HE6; two genes), and latrophilin (one gene) (Brody and Cravchik, 2000; Hewes and Taghert, 2001; Hill *et al.*, 2002a). Two insect neuropeptides, diuretic hormone and amnesiac (Feany and Quinn, 1995; Zhong, 1995; Zhong and Pena, 1995), are similar to corticotropin releasing factor (CRF) and the pituitary adenylate cyclase activating peptide (PACAP), respectively, of vertebrate neuropeptides that are grouped with receptors in the family B. All five secretin-like receptors in the *Drosophila* family B (Brody and Cravchik, 2000; Hewes and Taghert, 2001) appear to have orthologs in the *Anopheles* genome (Hill *et al.*, 2002a).

The first receptor for an insect diuretic hormone (DH-R) was identified by screening a *Manduca* expression library with a tritiated ligand (Reagan, 1994). The sequence of the *Manduca* DH-R revealed that it is an ortholog of the vertebrate CRF receptor (Hewes and Taghert, 2001). Activation of the *Manduca* DH-R receptor expressed in COS-7 and Sf9 insect cells triggered cAMP elevation (Reagan, 1995). An orthologous gene from the house cricket, *Acheta domesticus*, was cloned and characterized in a functional expression system (Reagan, 1996). The orthologous *Drosophila* DH-R sequence reveals two closely related genes as candidates (CG32843 and CG4395), but no experimental evidence has appeared as yet.

The *methuselah* (*mth*) group of receptors, found by screening for *Drosophila* mutants with extended lifespan (Lin *et al.*, 1998), comprises a large group of GPCRs unique to insects. Phylogenetic analysis of the closely related species *A. gambiae* and *D. melanogaster* revealed recent expansions in the number of copies of the *mth* genes (Hill *et al.*, 2002a). Examination of *Drosophila mth* mutants suggested that this gene product controls synaptic efficacy at the glutamatergic neuromuscular junction by increasing the rate of neurotransmitter exocytosis from the presynaptic terminal (Song *et al.*, 2002). The crystal structure of the *mth* ectodomain revealed potential ligand binding features in a shallow interdomain groove, although the ligand(s) have yet to be identified (West *et al.*, 2001).

Two genes encoding the GPCRs grouped with the human epididymis-specific protein 6 (HE6) were identified in *Drosophila*, but not in *Anopheles* (Brody and Cravchik, 2000; Hill *et al.*, 2002a). This receptor in vertebrates has an unusually long extracellular region characteristic of cell adhesion proteins, indicating potential roles in cell adhesion and signaling (Stacey *et al.*, 2000; Kierszenbaum, 2003). The genomes of *Drosophila* and *Anopheles* each have genes orthologous to vertebrate latrophilin. Vertebrate latrophilin is a putative receptor or binding protein for α-latrotoxin, a major component of black widow spider (genus *Lactrodectus*) venom. It appears to be involved in triggering calcium independent exocytosis, thus is also known as CIRL (calcium-independent receptor for α-latrotoxin) (Henkel and Sankaranarayanan, 1999). Latrophilin is thought to be a component of a large NMDA-receptor-associated signaling complex (Kreienkamp *et al.*, 2002b).

9.4.4. Family C: Metabotropic Glutamate Receptor and GABA$_B$ Receptor

Along with its well-known role in excitatory neuromuscular transmission, the neurotransmitter glutamate mediates both excitation and inhibition within the insect CNS (Parmentier *et al.*, 1998; Washio, 2002). Likewise, GABAergic inhibitory transmission occurs both in the peripheral and central nervous systems. In the CNS, these transmitters mediate both fast and slow transmission via ionotropic and metabotropic receptors, respectively. While our understanding of glutamate and GABA transmission in invertebrates is derived mostly from work on ionotropic receptors, metabotropic receptors are receiving increased attention because of new information gathered from genome sequences.

Analysis of the *Drosophila* genome revealed two genes encoding putative metabotropic glutamate receptors (mGluR) and three genes for metabotropic GABA$_B$ receptors, along with two other related genes not yet annotated (Brody and Cravchik, 2000). The *Anopheles* genome shows orthologous relationships with corresponding *Drosophila* genes in this group (Hill *et al.*, 2002a).

Drosophila mGluR receptors appear to function in the CNS by coupling to G$_i$, inhibiting activation of

adenylyl cyclase (Parmentier *et al.*, 1996, 1998), although linkage to K$^+$ and Ca^{2+} channel modulation downstream could serve excitatory roles as well.

GABA is a major inhibitory neurotransmitter in the CNS of both vertebrates and invertebrates. Postsynaptic, ionotropic GABA$_A$ and GABA$_C$ receptors mediate fast inhibitory responses, while slow transmission occurs via metabotropic GABA$_B$ receptors. In vertebrates, GABA$_B$ receptors mediate reduction of transmitter release presynaptically (Slesinger *et al.*, 1997) and hyperpolarization postsynaptically (Isaacson and Vitten, 2003). Pharmacological studies of the GABA$_B$ receptor in the cockroach *Periplaneta americana* demonstrated hyperpolarizing, ionotropic responses of the fast coxal depressor motor neuron (Bai and Sattelle, 1995).

Heterodimerization of GABA$_B$R1 and R2 subunits is necessary for constitution of a functional receptor in vertebrates (Jones *et al.*, 1998; Kaupmann *et al.*, 1998; White *et al.*, 1998). Furthermore, it is proposed that ligand binding involves only the GABA$_B$R1 subunit (Kniazeff *et al.*, 2002). Orthologous relationships of the subtypes GABA$_B$R1 and R2 are extended to *Drosophila* and the nematode *Caenorhabditis elegans* (Kniazeff *et al.*, 2002). As in the vertebrates, functional *Drosophila* GABA$_B$ receptors require heterodimerization of GABA$_B$R1 and R2 subunits (Mezler *et al.*, 2001). A third *Drosophila* subunit, GABA$_B$R3, appears to be an insect-specific subtype expressed in the CNS.

9.4.5. Other GPCRs: Odorant Receptor, Gustatory Receptor, Atypical 7TM Proteins

Analysis of genomic sequences in *Drosophila* and *Anopheles* indicates that the largest groups of GPCRs in insects are odorant and gustatory receptors (~70 genes for each) (Clyne *et al.*, 1999, 2000; Scott *et al.*, 2001; Hill *et al.*, 2002a). These receptors appear to have evolved quickly and comprise largely divergent subgroups having little if any conserved amino acid sequences in most cases. Evidence for functionality as odorant receptors has been provided by heterologous expression and bioassay (Stortkuhl and Kettler, 2001; Wetzel *et al.*, 2001). A gustatory receptor sensing trehalose was identified from a *Drosophila* mutant (Ishimoto *et al.*, 2000; Dahanukar *et al.*, 2001). Homology based cDNA library screening led to cloning of a number of putative odorant receptors in *Heliothis virescens*, and to demonstration of tissue-specific expression in antennae. This work thus has expanded olfactory receptor studies from *Drosophila* to other insect species (Krieger *et al.*, 2002).

Another group of putative receptors having seven transmembrane topology is classified separately as "atypical 7TM proteins." This class includes frizzled (*fz*), smoothened, and bride of sevenless (bos), most of which were initially identified as *Drosophila* mutants. The function of *fz*, four homologous copies of which are found in *Drosophila* and *Anopheles*, appears to be in establishing morphogenic symmetry and cell polarity during development (Bhanot *et al.*, 1999; Strutt, 2001).

9.5. Intracellular Signaling Pathways Triggered by GPCRs

Activation of GPCRs by extracellular ligands at the cell surface leads to a rich diversity of intracellular signal transduction pathways, beginning with activation of heterotrimeric guanine nucleotide-binding proteins (G-proteins), comprised of α, β, and γ subunits (**Figure 1**). The α subunit binds guanosine nucleotides, GDP in the resting state and GTP following activation of the GPCR with which the G-protein is associated. Upon GPCR activation, GTP binding to the α subunit of the G-protein leads to dissociation of the α and $\beta\gamma$ subunits. The $\beta\gamma$ dimer does not dissociate further under physiological conditions. Both α and $\beta\gamma$ subunits can bind independently to target proteins to initiate cellular responses.

The type of intracellular signaling that ensues GPCR activation depends on the associated G protein, most importantly the $G\alpha$ subunit. Coupling with $G\alpha_s$ activates adenylyl cyclase (AC), whereas coupling with $G\alpha_i$ and $G\alpha_o$ suppresses AC activity. Coupling with $G\alpha_q$ activates phospholipase C, initiating phosphoinositide turnover and intracellular Ca^{2+} mobilization. Prediction of coupling specificity between GPCR and G protein subtypes has been attempted using precedent data from vertebrate GPCRs (Bourne, 1997; Wess, 1998; Horn *et al.*, 2000; Moller *et al.*, 2001). To what extent such methods are predictive for insect GPCR signaling remains an open question.

Numerous instances of $G\beta\gamma$ signaling also have been documented in the context of $G\alpha_{i/o}$ signaling. The target proteins for $G\beta\gamma$ include AC, PLCβ, potassium channels, and phosphatidylinositol 3-kinase. Perhaps the best known of these is K channel modulation following parasympathetic activation of heart muscarinic acetylcholine receptors to slow heartbeat frequency (Kofuji *et al.*, 1995). This response results from dissociation of $G\alpha_i$ and $G\beta\gamma$, with the latter acting directly as a ligand to open G protein coupled inwardly rectifying potassium channels (GIRK) and hyperpolarize myocardial cells (Reuveny *et al.*, 1994; Wickman *et al.*, 1994). This signaling pathway provides a particularly clear example of how direct G protein modulation of ion channels leads to an important physiological response.

Analysis of the *Drosophila* genome sequence reveals seven, six, and two genes that are homologous to vertebrate Gα, Gβ, and Gγ subunits, respectively (**Table 4**). The seven Gα homologous genes are classified as $G\alpha_q$, $G\alpha_s$, and $G\alpha_{i/o}$, the latter including $G\alpha_t$, the well-known transducin involved in phototransduction (Simon *et al.*, 1991; Neves *et al.*, 2002) (**Figure 4**). Genes homologous to the Gβ subunit appear to have larger diversity than their vertebrate counterparts (**Figure 4**). Only two copies of the Gγ subunit are found in *Drosophila* (**Figure 4**). Alternatively spliced forms of these genes appear to provide additional flexibility to signaling systems (de Sousa *et al.*, 1989; Ray and Ganguly, 1994; Talluri *et al.*, 1995).

Immunolocalization of G protein α subunits in *Drosophila* demonstrated that expression of $G\alpha_q$, $G\alpha_s$, and $G\alpha_{i/o}$ is highly restricted in the CNS (Wolfgang *et al.*, 1990). Expression in photoreceptor synaptic terminals and antennal lobes suggests functional roles in transmission of primary sensory information. Distinct expression patterns in embryonic stages implies involvement of G protein-mediated sensory transmission in embryonic development (Wolfgang *et al.*, 1991). Expression profiles of Gα subtypes in *M. sexta* also were examined by immunohistochemistry, where they were shown to be expressed in subsets of CNS cells (Copenhaver *et al.*, 1995).

A salient characteristic of GPCR-induced signaling is the opportunity for amplification along multiple steps of the pathway. For example, a single GPCR binding event leads to dissociation of α and $\beta\gamma$ subunits, each of which can trigger downstream signaling pathways. In instances where the G protein subunit triggers second messengers through activation of enzymes such as AC or PLC, a second level of amplification occurs. Downstream activation of additional enzymes such as kinases or phosphatases provides a third level of amplification. Thus activation of a very small number of GPCRs can elicit highly significant biochemical responses in target cells. The diversity of downstream signal transduction pathways is summarized in several reviews, and generalized current concepts based vertebrate studies may well be applicable to insects (Simon *et al.*, 1991; Morris and Malbon, 1999; Neves *et al.*, 2002).

9.6. Assignment of GPCR Functions

GPCRs regulate a plethora of cellular functions by modulating the activities of diverse intracellular

Table 4 *Drosophila* trimeric G protein subunits Gα, Gβ, and Gγ

Group	Name	Synonym	Expression pattern	Reference
Gα subunit				
Gq	Gα49B	CG17759, dgq, DGqα, DGαq	Chemosensory cells, photoreceptor cells	Lee *et al.* (1990, 1994), Talluri *et al.* (1995), Bahner *et al.* (2000), Ratnaparkhi *et al.* (2002)
	CG17760			
Gi/o	Gαo47A	CG2204, DGαo, dgo, Goα	Brain, ovary, cardioblast, CNS	de Sousa *et al.* (1989), Schmidt *et al.* (1989), Wolfgang *et al.* (1990, 1991), Fremion *et al.* (1999)
	Gαi65A	CG10060, Gαi , Giα, DGα1,	CNS	Wolfgang *et al.* (1990, 1991)
Gs	Gα73B	CG12232, Gfα	Embryonic stage, midgut	Quan *et al.* (1993)
	Gαs60A	CG2835, Gαs, Gsα, Dgsα, dgs, Gs	Neuropil in CNS	Wolfgang *et al.* (1990, 1991, 2001)
	Concertina	CG17678, cta	Embryonic mesoderm	Parks and Wieschaus (1991)
Gβ subunit				
	CG2812			
	CG3004			
	EG:86E4.3	CG17766		
	Gβ13F	CG10545, dgβ, Gβ, G-protein β 13F	Eye	Yarfitz *et al.* (1988, 1991)
	Gβ5			
	Gβ76C	CG8770		
Gγ subunit				
	Gγ1	CG8261, D-Gγ1, dg1α	CNS	Ray and Ganguly (1994)
	Gγ30A	CG18511, CG3694, Gγe	Eye	Schulz *et al.* (1999)

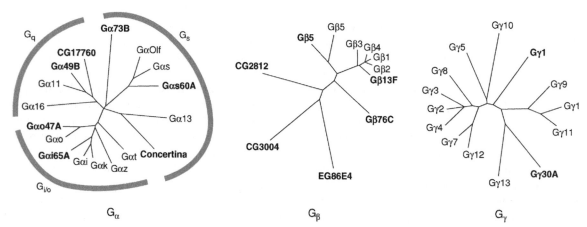

Figure 4 Phylogram for *Drosophila* G proteins with vertebrate G proteins. *Drosophila* G proteins are in bold letters.

target proteins, including ion channels, metabolic enzymes, and transporters. These actions lead to cellular responses important for embryonic, postembryonic, and reproductive development, homeostatic processes, responses to environmental stimuli, and induction of various context-driven behavioral states.

Strategies for functional identification of GPCRs have gone through several phases since the initial breakthroughs with β-adrenergic and muscarinic acetylcholine receptors (Dixon *et al.*, 1986; Kubo *et al.*, 1986; Yarden *et al.*, 1986). These first discoveries began with direct protein purification and partial amino acid sequencing of receptors with known ligands. This allowed for construction of oligonucleotide probes, cDNA library screening, and eventual cloning and sequencing of entire receptors. The next wave of GPCR identification occurred through expression cloning of, among others, neurokinin2 and serotonin receptors using early stage oocytes of the African clawed toad, *Xenopus laevis* (Masu *et al.*, 1987; Julius *et al.*, 1988).

Taking advantage of sequence information from earlier work, homology cloning based on nucleotide sequence identity provided relatively easy and productive approaches to further receptor identification. Such homology searches led to the identification of the first insect GPCR, the muscarinic acetylcholine receptor (Shapiro *et al.*, 1989). Finally, availability of entire genome sequences from an increasing number of organisms has led to an explosive phase of *in silico* discoveries of putative GPCRs (Kim and Carlson, 2002), while homology search using cDNA library screening and degenerate polymerase chain reaction (PCR) is still highly effective for GPCR identification in species for which whole genome sequences are not yet available.

Whereas early efforts focused on receptors with known function, the recent era of genome discovery has generated numerous "orphan receptors" with unknown ligands and uncertain functions. For example, large numbers of putative GPCR in insects have been identified and analyzed from the genome sequences of *Drosophila* and *Anopheles* (Hewes and Taghert, 2001; Hill *et al.*, 2002a). Homology analysis of putative *Drosophila* GPCRs and particularly for neuropeptide receptors (Hewes and Taghert, 2001) has provided a foundation for GPCR deorphaning based on comparisons with mammalian genomes. This analysis, combined with improvements in heterologous expression systems and reporter constructs, has increased the rate of GPCR deorphaning during the past several years.

Deorphaning putative GPCRs through identification of endogenous ligands depends primarily on heterologous expression within a suitable bioassay system, and characterization of interactions between candidate ligands through biochemical or functional methods. The choice of expression and assay systems for a given GPCR gene entail practical considerations based on knowledge of previous studies. Since the assay primarily focuses on ligand–receptor interaction rather than endogenous downstream pathways, it often uses an exogenously introduced reporter system. The most frequently used expression systems have been the *Xenopus* oocyte expression system and various cell lines. The expression system needs to be chosen according to certain criteria: paucity of receptor endogenous to the expression system, which might cross-react with ligands under investigation, robust expression of the heterologous receptor, localization of the receptor on the cell membrane, appropriate posttranslational processing, and efficient coupling to downstream signaling pathways for activation of reporters. **Tables 1–3** summarize assay systems currently used for studies of insect GPCRs.

9.6.1. Functional GPCR Assays: Coupling with Reporters

Given the diversity of functions regulated by GPCRs, it is not surprising that they are targeted by a high percentage of pharmaceutical drugs currently on the market. For this reason, considerable efforts have been devoted to development of methods for high-throughput screening of novel GPCR agonists and antagonists. It seems likely that similar approaches may lead to future insecticidal agents.

The most straightforward functional assays for GPCRs in oocytes or cell lines register Ca^{2+} mobilization as an increased fluorescence or luminescence. This cellular response is most commonly a result of $G\alpha_q$ activation and phosphoinositol metabolism, leading to IP3 mediated Ca^{2+} release from intracellular stores (Neves *et al.*, 2002). Commonly used Ca^{2+} sensors generate fluorescence or luminescence signals (see Section 2.5.3). Many orphan GPCRs are coupled to other $G\alpha$ subunits that do not mobilize calcium. Consequently, efforts have been made to utilize more promiscuous G proteins coupled to the PLCβ pathway. One very successful method has been the use of a $G\alpha_q$ variant known as $G\alpha_{16}$, which couples widely divergent GPCRs to the phosphoinositide turnover pathway via PLC, leading to Ca^{2+} mobilization (review: Stables *et al.*, 1997; Kostenis, 2001) (**Figures 5** and **6**). For example, expression of a *Drosophila* odorant receptor with exogenous $G\alpha_{16}$ in *Xenopus* oocytes was the system used to identify its cognate odorant ligand (Wetzel *et al.*, 2001). Other approaches have involved engineering of chimeric G proteins (e.g., $G\alpha_{16/z}$) which couple a diversity of GPCRs to PLCβ (Liu *et al.*, 2003).

For fluorescence assays, cells expressing GPCRs are pretreated with membrane permeant calcium indicators such as Fura2-AM, fast green AM, or fluo-3 AM (Meeusen *et al.*, 2002; Choi *et al.*, 2003; Johnson *et al.*, 2003a). For luminescence responses in cells expressing aequorin, pretreatment with the cofactor coelenterazine is necessary. Fluorescence or luminescence microplate readers having injection ports can be used for introduction of the cells into microplate wells that contain various concentrations of ligand. For high-throughput assay, simultaneous injection and reading of a plate for 96 or 384 wells also has been developed (Le Poul *et al.*, 2002).

Coupling to G_s normally leads to increases in AC activity. Experiments with the *Xenopus* oocyte expression system have demonstrated that $G\alpha_s$ activation can gate exogenous inwardly rectifying potassium channels GIRK1 and GIRK2 through the direct action of $G\beta\gamma$ (Kofuji *et al.*, 1995; Lim *et al.*,

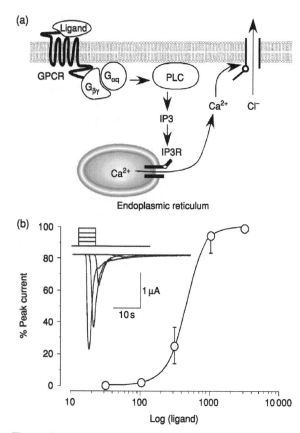

Figure 5 (a) A presumed signaling pathway in the *Xenopus* oocyte assay system for orphan GPCRs. GPCR expressed from injected cRNA transduce signal through phospholipase C (PLC), inositol triphosphate (IP3), and IP3 receptor triggering Ca^{2+} mobilization from intracellular Ca^{2+} storage. Ca^+ activates chloride channels, resulting in inward currents. (b) Progressively higher concentrations of ligands generate larger currents. Mean values and corresponding standard deviations from multiple experiments are plotted as a function of ligand concentration.

1995; Peleg *et al.*, 2002). Expression of these exogenous GIRK subtypes in the proper stoichiometery can lead to robust inward K^+ currents, and hence provide a convenient assay for activation of $G\alpha_s$.

Other approaches for detecting $G\alpha_s$ activation have included measurements of increased cAMP levels (i.e., serotonin receptor 5HTR7) (Witz *et al.*, 1990; Saudou *et al.*, 1992; Obosi *et al.*, 1996). For $G\alpha_i$ activation, inhibition of AC activity can be detected through a reduction in forskolin-inducible cAMP increase (i.e., tyramine receptors) (Kutsukake *et al.*, 2000; Poels *et al.*, 2001).

9.6.2. Expression of GPCRs in *Xenopus* Oocytes

Advantages of using unfertilized *Xenopus* oocytes are many, not the least of which is the absence of endogenous GPCRs (Lee and Durieux, 1998). Injection of complementary RNA prepared with 5′ and 3′

untranslated regions of *Xenopus* β-globin gene results in sufficient expression in stage 5 or 6 oocytes for robust bioassays (Jespersen *et al.*, 2002). The oocyte contains sufficient machinery to link activation of exogenously introduced GPCRs to cellular responses.

The most routine use of oocytes is coupling of heterologously expressed GPCRs with endogenous G_q, leading to mobilization of intracellular Ca^{2+} through PLCβ activation and phosphoinositide turnover. This results in opening of Ca^{2+} dependent Cl^- channels on the oocyte plasma membrane and large inward currents that can be quantified by the two-electrode voltage clamp technique (**Figure 5**). This assay can be made considerably more flexible through the introduction of $G\alpha_{16}$, chimeric G proteins, or coupling to GIRK K^+ channels.

A precise perfusion system is required for generation of large sets of data using oocytes. Park *et al.* (2002) used a liquid chromatography pump and associated Rheodyne injection valve, supplying a continuous flow of buffer at a flow rate of $1.2\,ml\,min^{-1}$. Various concentrations of ligands were applied in standard 100 µl volumes of ligand through the injection valve, providing a 5 s exposure of ligand solution to the oocyte (Park *et al.*, 2002). This system allows precise application of a given concentration of ligand in a continuous flow of buffer, while caution must be exercised in the use of lipophilic ligands that may persist in the walls of the perfusion system even after extensive washes.

A challenge using *Xenopus* oocytes for GPCR assay is rapid desensitization following receptor activation. The degree of desensitization is highly variable and dependent on the specific GPCR expressed (Park *et al.*, 2002). In the case of *Drosophila* ETH receptors, the assay in oocytes was confounded by desensitization (Park, unpublished data). Even very low concentrations that did not elicit measurable currents were sufficient to cause desensitization. In such a case, alternative expression and assay systems using cell lines are preferable, where measurement of β-arrestin activity has proved to be a very effective method (see Section 9.6.4).

9.6.3. Assay of GPCR Activation in Cell Lines

Expression of GPCRs in cell lines currently is the method of choice both for deorphaning receptors and for large-scale, rapid-throughput drug screening for agonists and antagonists. Considerations in the choice of cell lines include high-level expression and proper processing of the exogenous GPCR gene, knowledge of the endogenous signaling systems operating, and the efficacy of the reporter system.

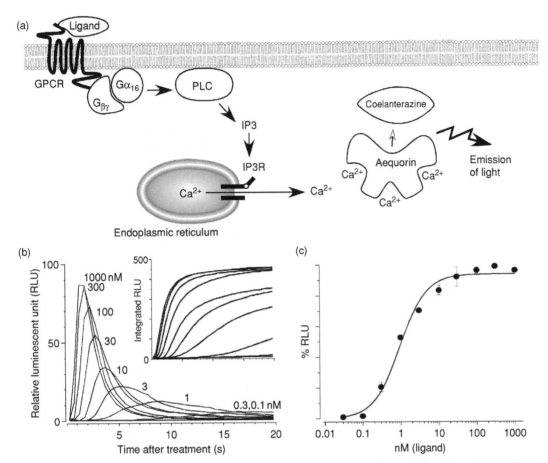

Figure 6 (a) A presumed signaling pathway operating in the aequorin assay for orphan GPCRs. The expressed heterologous GPCR transduces ligand binding through $G\alpha_{16}$, phospholipase C (PLC), inositol triphosphate (IP_3), and IP_3 receptor, resulting in Ca^{2+} mobilization from intracellular stores. Binding of Ca^{2+} to aequorin with the cofactor coelanterazine results in emission of light. (b) Progressively higher concentrations of ligand generate a more intense light response. (c) Integrated luminescence during a given time is used to generate the concentration–response curve.

Cell lines generally used for expression of insect GPCRs are Chinese hamster ovary (CHO-K1) cells, human embryonic kidney (HEK) cells, and Schneider's *Drosophila* (S2) cells (see **Tables 1** and **2**). All of these cells are known for their efficient coupling of GPCR activation to endogenous downstream signaling pathways. For instance, GPCRs expressed in CHO-K1 cells can be coupled to the PLC–calcium mobilization pathway, presumably via activation of G_q. A summary of downstream coupling mechanisms and signaling pathways for GPCRs in each cell line can be found at http://www.tumor-gene.org. Characteristics of *Drosophila* S2 cells are found in two recent publications (Van Poyer *et al.*, 2001; Torfs *et al.*, 2002b).

Both ligand binding assays and functional assays can be performed using cell line expression systems, depending on the need and situation. Affinity constants, receptor density, and kinetics of binding of heterologously expressed receptors can be investigated by using radioligands and compared with data

obtained from native receptors (e.g., Johnson *et al.*, 2003b; Meeusen *et al.*, 2002). An advantage of the binding assay is that the biochemical properties of the expressed GPCR can be determined independent of coupling efficiency to downstream signaling and reporter activation.

9.6.4. Other GPCR Reporter Assays

While functional assays for GPCRs have been improved and refined through introduction of various reporter molecules into the expression system, rapid desensitization can severely attenuate responses. An effective strategy for dealing with this problem is to detect the interactions of β-arrestin with the GPCR. In this instance, desensitization of the GPCR can be monitored using bioluminescence resonance energy transfer (BRET). GPCR fused to *Renilla* luciferase is used for the cell line stably expressing arrestin fused to green fluorescent protein (GFP) (Barak *et al.*, 1997; Bertrand *et al.*, 2002). Activation of the GPCR leads to interaction between the arrestin::GFP and GPCR::luciferase resulting in the BRET from the donor luciferase to the acceptor GFP. GPCR activation can be also measured by downstream gene regulation. Activation of GPCR coupled to AC and elevation of cAMP is monitored by expression of reporter gene controlled by cAMP responsive elements (CREs) (Chen *et al.*, 1995; Durocher *et al.*, 2000). It is thought that the CRE is activated though both $G\alpha_s$ and $G\alpha_q$ pathways (Durocher *et al.*, 2000). A simpler approach is to simply follow migration of a β-arrestin–GFP fusion protein to the cell membrane following GPCR activation (Barak *et al.*, 1997; Johnson *et al.*, 2003b).

The use of heterologous expression systems to obtain pharmacological profiles of GPCRs should be interpreted with appropriate caution, since it is often likely that critical transduction components unique to the natural physiological system will be absent. Thus the ligand affinities of certain GPCRs may be under the strong influence of native G protein subtypes with which they are associated (Kostenis, 2001). Even more complexity can be introduced in instances where relative stoichiometries of receptors to other cellular signaling components affect the assay system (Kenakin, 1997, 2003). Finally, the availability of specific G proteins appears to affect ligand binding affinity, as in the case of dopamine receptors (Reale *et al.*, 1997; Cordeaux and Hill, 2002; Gazi *et al.*, 2003).

9.6.5. Identifying Biological Functions of GPCRs

Understanding the biological function of a particular signaling pathway has been advanced through examination of ligand actions in large insects amenable to physiological experimentation. Important tools for endocrinological studies have been classical physiological experiments involving ligation, surgical extirpation and implantation of organs and tissues, parabiosis, and transplantations followed by bioassay-guided chemical isolation of the ligand on a particular system. Study of receptors mediating these processes provides further in-depth understanding of the mode of action of chemical signals at the molecular and cellular levels.

Biochemical and pharmacological data obtained through heterologous expression of a GPCR provides the first step in understanding molecular mechanisms of a signaling pathway and association of a putative authentic receptor with its ligand and physiological actions. Experimental evidence supporting an endogenous ligand–receptor interaction could include: highest affinity functional or biochemical interaction between the ligand and the receptor (usually at nanomolar to picomolar

concentrations of ligand), spatial and temporal expression patterns of the GPCR that concurs with biological data, and mutant phenotypes that are in accordance with biological function of the signaling pathway.

Chemical isolation of unknown ligands for heterologously expressed orphan GPCRs is an increasingly useful method for functional analysis, especially with recently developed tools in proteomics. This approach is referred as reverse physiology (Birgul *et al.*, 1999) or reverse pharmacology (Civelli *et al.*, 2001) where the natural ligand for a receptor is sought. This approach contrasts with general approaches in which receptors are sought for a given ligand. Reverse physiology was successful in associating the peptide allatostatin A with its physiological receptor by isolating the peptide activating a *Drosophila* GPCR that is homologous to the somatostatin receptor of vertebrates (Birgul *et al.*, 1999). Three steps of peptide purification by cation-exchange column and reverse phase chromatography succeeded in isolation of a fraction active on the GPCR of interest. The final active fraction purified was analyzed by a matrix-assisted laser desorption ionization post source decay (MALDI-PSD) time-of-flight spectrum for obtaining the sequence. A similar approach was successful for deorphaning *Drosophila* CG2114 as an FMRFamide receptor (Meeusen *et al.*, 2002).

Multidisciplinary approaches have been successful in a number of cases. Dow and his group studied signaling systems *Drosophila* Malpighian tubules mediating diuretic functions (review: Dow and Davies, 2003). Three neuroendocrine peptides and their signaling pathways implicated in the activation of fluid secretion were examined: cardioacceleratory peptide, diuretic hormone, and leucokinin. A combination of physiology, pharmacology, and *Drosophila* genetics elucidated downstream signaling pathways at the molecular and cellular levels. Specific GPCRs activated by leucokinin (CG10626) (Radford *et al.*, 2002) and cardioacceleratory peptide (CG14575) (Park *et al.*, 2002) were identified in functional assays. In the case of the leucokinin receptor, immunohistochemistry has been used to study its expression in stellate cells, a subtype of Malpighian tubule epithelial cell previously shown to respond to leucokinin by Ca^{2+} mobilization activating chloride current and fluid secretion (Rosay *et al.*, 1997; Radford *et al.*, 2002).

9.7. Conclusions

We find ourselves in an exciting era of exponential growth of knowledge regarding GPCR structure and function. Knowledge of these receptors obtained in basic science could well to be applicable to the development of innovative pest control measures. Disruption of neuropeptide signaling systems through use of peptide hormone mimetics targeting the GPCRs also shows increasing promise (Nachman *et al.*, 2001, 2002a, 2002b). Chemical properties of pyrokinin analogs have been improved for increased cuticle penetrability and for resistance to peptidases (Nachman *et al.*, 2002a, 2000b). A novel method for delivering compounds to target GPCRs was attempted through use of a fusion protein combining lectin with allatostatin (Fitches *et al.*, 2002). By this means, allatostatin is delivered to insect hemolymph following oral ingestion and inhibits feeding and growth of the insect tomato moth (*Lacanobia oleracea*). With developing new biotechnology, GPCRs, particularly those mediating peptide signaling, may be excellent targets for pest control strategies providing an optimal range of specificity.

References

Adams, M.E., Zitnan, D., **1997**. Identification of ecdysis-triggering hormone in the silkworm *Bombyx mori*. *Biochem. Biophys. Res. Commun. 230*, 188–191.

Altschul, S.F., Madden, T.L., Schaffer, A.A., Zhang, J., Zhang, Z., *et al.*, **1997**. Gapped BLAST and PSI-BLAST: a new generation of protein database search programs. *Nucl. Acids Res. 25*, 3389–3402.

Alvarez, C.E., Robison, K., Gilbert, W., **1996**. Novel Gq alpha isoform is a candidate transducer of rhodopsin signaling in a *Drosophila* testes-autonomous pacemaker. *Proc. Natl Acad. Sci. USA 93*, 12278–12282.

Andretic, R., Hirsh, J., **2000**. Circadian modulation of dopamine receptor responsiveness in *Drosophila melanogaster*. *Proc. Natl Acad. Sci. USA 97*, 1873–1878.

Angers, S., Salahpour, A., Bouvier, M., **2002**. Dimerization: an emerging concept for G protein-coupled receptor ontogeny and function. *Annu. Rev. Pharmacol. Toxicol. 42*, 409–435.

Arakawa, S., Gocayne, J.D., McCombie, W.R., Urquhart, D.A., Hall, L.M., *et al.*, **1990**. Cloning, localization, and permanent expression of a *Drosophila* octopamine receptor. *Neuron 4*, 343–354.

Auerswald, L., Birgul, N., Gäde, G., Kreienkamp, H.J., Richter, D., **2001**. Structural, functional, and evolutionary characterization of novel members of the allostatin receptor family from insects. *Biochem. Biophys. Res. Commun. 282*, 904–909.

Baggerman, G., Cerstiaens, A., De Loof, A., Schoofs, L., **2002**. Peptidomics of the larval *Drosophila melanogaster* central nervous system. *J. Biol. Chem. 277*, 40368–40374.

Bahner, M., Sander, P., Paulsen, R., Huber, A., **2000**. The visual G protein of fly photoreceptors interacts with the PDZ domain assembled INAD signaling complex via direct binding of activated Galpha(q) to phospholipase cbeta. *J. Biol. Chem. 275*, 2901–2904.

Bai, D., Sattelle, D., **1995**. A GABA$_B$ receptor on an identified insect motor neurone. *J. Exp. Biol. 198*, 889–894.

Baker, J.D., Truman, J.W., **2002**. Mutations in the *Drosophila* glycoprotein hormone receptor, rickets, eliminate neuropeptide-induced tanning and selectively block a stereotyped behavioral program. *J. Exp. Biol. 205*, 2555–2565.

Barak, L.S., Ferguson, S.S., Zhang, J., Caron, M.G., **1997**. A beta-arrestin/green fluorescent protein biosensor for detecting G protein-coupled receptor activation. *J. Biol. Chem. 272*, 27497–27500.

Benton, M.J., Ayala, F.J., **2003**. Dating the tree of life. *Science 300*, 1698–1700.

Bertrand, L., Parent, S., Caron, M., Legault, M., Joly, E., *et al.*, **2002**. The BRET2/arrestin assay in stable recombinant cells: a platform to screen for compounds that interact with G protein-coupled receptors (GPCRS). *J. Recept. Signal Transduct. Res. 22*, 533–541.

Bhanot, P., Fish, M., Jemison, J.A., Nusse, R., Nathans, J., *et al.*, **1999**. Frizzled and Dfrizzled-2 function as redundant receptors for Wingless during *Drosophila* embryonic development. *Development 126*, 4175–4186.

Birgul, N., Weise, C., Kreienkamp, H.J., Richter, D., **1999**. Reverse physiology in *Drosophila*: identification of a novel allatostatin-like neuropeptide and its cognate receptor structurally related to the mammalian somatostatin/galanin/opioid receptor family. *EMBO J. 18*, 5892–5900.

Blake, A.D., Anthony, N.M., Chen, H.H., Harrison, J.B., Nathanson, N.M., *et al.*, **1993**. *Drosophila* nervous system muscarinic acetylcholine receptor: transient functional expression and localization by immunocytochemistry. *Mol. Pharmacol. 44*, 716–724.

Blenau, W., Balfanz, S., Baumann, A., **2000**. Amtyr1: characterization of a gene from honeybee (*Apis mellifera*) brain encoding a functional tyramine receptor. *J. Neurochem. 74*, 900–908.

Blenau, W., Baumann, A., **2001**. Molecular and pharmacological properties of insect biogenic amine receptors: lessons from *Drosophila melanogaster* and *Apis mellifera*. *Arch. Insect Biochem. Physiol. 48*, 13–38.

Blenau, W., Erber, J., Baumann, A., **1998**. Characterization of a dopamine D1 receptor from *Apis mellifera*: cloning, functional expression, pharmacology, and mRNA localization in the brain. *J. Neurochem. 70*, 15–23.

Bockaert, J., Pin, J.P., **1999**. Molecular tinkering of G protein-coupled receptors: an evolutionary success. *EMBO J. 18*, 1723–1729.

Borowsky, B., Adham, N., Jones, K.A., Raddatz, R., Artymyshyn, R., *et al.*, **2001**. Trace amines: identification of a family of mammalian G protein-coupled receptors. *Proc. Natl Acad. Sci. USA 98*, 8966–8971.

Bourne, H.R., **1997**. How receptors talk to trimeric G proteins. *Curr. Opin. Cell Biol. 9*, 134–142.

Brady, A.E., Limbird, L.E., **2002**. G protein-coupled receptor interacting proteins: emerging roles in localization and signal transduction. *Cell. Signal. 14*, 297–309.

Briscoe, A.D., 2001. Functional diversification of lepidopteran opsins following gene duplication. *Mol. Biol. Evol.* 18, 2270–2279.

Briscoe, A.D., 2002. Homology modeling suggests a functional role for parallel amino acid substitutions between bee and butterfly red- and green-sensitive opsins. *Mol. Biol. Evol.* 19, 983–986.

Briscoe, A.D., Chittka, L., 2001. The evolution of color vision in insects. *Annu. Rev. Entomol.* 46, 471–510.

Brody, T., Cravchik, A., 2000. *Drosophila melanogaster* G protein-coupled receptors. *J. Cell Biol. 150,* F83–F88.

Cazzamali, G., Grimmelikhuijzen, C.J., 2002. Molecular cloning and functional expression of the first insect FMRFamide receptor. *Proc. Natl Acad. Sci. USA* 99, 12073–12078.

Cazzamali, G., Saxild, N., Grimmelikhuijzen, C., 2002. Molecular cloning and functional expression of a *Drosophila* corazonin receptor. *Biochem. Biophys. Res. Commun.* 298, 31–36.

Chen, W., Shields, T.S., Stork, P.J., Cone, R.D., 1995. A colorimetric assay for measuring activation of Gs- and Gq-coupled signaling pathways. *Analyt. Biochem.* 226, 349–354.

Choi, M.Y., Fuerst, E.J., Rafaeli, A., Jurenka, R., 2003. Identification of a G protein-coupled receptor for pheromone biosynthesis activating neuropeptide from pheromone glands of the moth *Helicoverpa zea*. *Proc. Natl Acad. Sci. USA* 100, 9721–9726.

Chothia, C., Gough, J., Vogel, C., Teichmann, S.A., 2003. Evolution of the protein repertoire. *Science 300,* 1701–1703.

Chou, W.H., Hall, K.J., Wilson, D.B., Wideman, C.L., Townson, S.M., et al., 1996. Identification of a novel *Drosophila* opsin reveals specific patterning of the R7 and R8 photoreceptor cells. *Neuron 17,* 1101–1115.

Chou, W.H., Huber, A., Bentrop, J., Schulz, S., Schwab, K., et al., 1999. Patterning of the R7 and R8 photoreceptor cells of *Drosophila*: evidence for induced and default cell-fate specification. *Development 126,* 607–616.

Civelli, O., Nothacker, H.P., Saito, Y., Wang, Z., Lin, S.H., et al., 2001. Novel neurotransmitters as natural ligands of orphan G-protein-coupled receptors. *Trends Neurosci.* 24, 230–237.

Clyne, P.J., Warr, C.G., Carlson, J.R., 2000. Candidate taste receptors in *Drosophila*. *Science 287,* 1830–1834.

Clyne, P.J., Warr, C.G., Freeman, M.R., Lessing, D., Kim, J., et al., 1999. A novel family of divergent seventransmembrane proteins: candidate odorant receptors in *Drosophila*. *Neuron 22,* 327–338.

Colas, J.F., Launay, J.M., Kellermann, O., Rosay, P., Maroteaux, L., 1995. *Drosophila* 5-HT2 serotonin receptor: coexpression with fushi-tarazu during segmentation. *Proc. Natl Acad. Sci. USA 92,* 5441–5445.

Colas, J.F., Launay, J.M., Vonesch, J. L., Hickel, P., Maroteaux, L., 1999. Serotonin synchronises convergent extension of ectoderm with morphogenetic gastrulation movements in *Drosophila*. *Mech. Devel.* 87, 77–91.

Copenhaver, P.F., Horgan, A.M., Nichols, D.C., Rasmussen, M.A., 1995. Developmental expression of heterotrimeric G proteins in the nervous system of *Manduca sexta*. *J. Neurobiol.* 26, 461–484.

Cordeaux, Y., Hill, S.J., 2002. Mechanisms of cross-talk between G-protein-coupled receptors. *Neurosignals 11,* 45–57.

Cordeaux, Y., Nickolls, S.A., Flood, L.A., Graber, S.G., Strange, P.G., 2001. Agonist regulation of D(2) dopamine receptor/G protein interaction: evidence for agonist selection of G protein subtype. *J. Biol. Chem.* 276, 28667–28675.

Cowman, A.F., Zuker, C.S., Rubin, G.M., 1986. An opsin gene expressed in only one photoreceptor cell type of the *Drosophila* eye. *Cell 44,* 705–710.

Cox, K.J., Tensen, C.P., Van der Schors, R.C., Li, K.W., van Heerikhuizen, H., et al., 1997. Cloning, characterization, and expression of a G-protein-coupled receptor from *Lymnaea stagnalis* and identification of a leucokinin-like peptide, PSFHSWSamide, as its endogenous ligand. *J. Neurosci.* 17, 1197–1205.

Dahanukar, A., Foster, K., van der Goes van Naters, W.M., Carlson, J.R., 2001. A Gr receptor is required for response to the sugar trehalose in taste neurons of *Drosophila*. *Nat. Neurosci.* 4, 1182–1186.

Darlison, M.G., Richter, D., 1999. Multiple genes for neuropeptides and their receptors: co-evolution and physiology. *Trends Neurosci.* 22, 81–88.

de Sousa, S.M., Hoveland, L.L., Yarfitz, S., Hurley, J.B., 1989. The *Drosophila* Go alpha-like G protein gene produces multiple transcripts and is expressed in the nervous system and in ovaries. *J. Biol. Chem.* 264, 18544–18551.

Dixon, R.A., Kobilka, B.K., Strader, D.J., Benovic, J.L., Dohlman, H.G., et al., 1986. Cloning of the gene and cDNA for mammalian beta-adrenergic receptor and homology with rhodopsin. *Nature 321,* 75–79.

Dolzer, J., Krannich, S., Fischer, K., Stengl, M., 2001. Oscillations of the transepithelial potential of moth olfactory sensilla are influenced by octopamine and serotonin. *J. Exp. Biol.* 204, 2781–2794.

Dow, J.A.T., Davies, S.A., 2003. Integrative physiology and functional genomics of epithelial function in a genetic model organism. *Physiol. Rev. 83,* 687–729.

Durocher, Y., Perret, S., Thibaudeau, E., Gaumond, M.H., Kamen, A., et al., 2000. A reporter gene assay for high-throughput screening of G-protein-coupled receptors stably or transiently expressed in HEK293 EBNA cells grown in suspension culture. *Analyt. Biochem.* 284, 316–326.

Ebert, P.R., Rowland, J.E., Toma, D.P., 1998. Isolation of seven unique biogenic amine receptor clones from the honey bee by library scanning. *Insect Mol. Biol.* 7, 151–162.

Egerod, K., Reynisson, E., Hauser, F., Cazzamali, G., Williamson, M., et al., 2003a. Molecular cloning and functional expression of the first two specific insect myosuppressin receptors. *Proc. Natl Acad. Sci. USA* 100, 9808–9813.

Egerod, K., Reynisson, E., Hauser, F., Williamson, M., Cazzamali, G., et al., 2003b. Molecular identification of the first insect proctolin receptor. *Biochem. Biophys. Res. Commun. 306*, 437–442.

Eriksen, K.K., Hauser, F., Schiott, M., Pedersen, K.M., Sondergaard, L., et al., 2000. Molecular cloning, genomic organization, developmental regulation, and a knock-out mutant of a novel leu-rich repeats-containing G protein-coupled receptor (DLGR-2) from *Drosophila melanogaster. Genome Res. 10*, 924–938.

Evans, P.D., 1980. Biogenic amines in the insect nervous system. *Adv. Insect Physiol. 15*, 317–473.

Evans, P.D., 1993. Molecular tudies on insect octopamine receptors. In: Pichon, Y. (Ed.), Comparative Molecular Neurobiology. Birkhauser, Basel, pp. 286–296.

Feany, M.B., Quinn, W.G., 1995. A neuropeptide gene defined by the *Drosophila* memory mutant amnesiac. *Science 268*, 869–873.

Feiler, R., Harris, W.A., Kirschfeld, K., Wehrhahn, C., Zuker, C.S., 1988. Targeted misexpression of a *Drosophila* opsin gene leads to altered visual function. *Nature 333*, 737–741.

Feng, G., Hannan, F., Reale, V., Hon, Y.Y., Kousky, C.T., et al., 1996. Cloning and functional characterization of a novel dopamine receptor from *Drosophila melanogaster. J. Neurosci. 16*, 3925–3933.

Feng, G., Reale, V., Chatwin, H., Kennedy, K., Venard, R., et al., 2003. Functional characterization of a neuropeptide F-like receptor from *Drosophila melanogaster. Eur. J. Neurosci. 18*, 227–238.

Filipek, S., Teller, D.C., Palczewski, K., Stenkamp, R., 2003. The crystallographic model of rhodopsin and its use in studies of other G protein-coupled receptors. *Annu. Rev. Biophys. Biomol. Struct. 32*, 375–397.

Fitches, E., Audsley, N., Gatehouse, J.A., Edwards, J.P., 2002. Fusion proteins containing neuropeptides as novel insect contol agents: snowdrop lectin delivers fused allatostatin to insect haemolymph following oral ingestion. *Insect Biochem. Mol. Biol. 32*, 1653–1661.

Fremion, F., Astier, M., Zaffran, S., Guillen, A., Homburger, V., et al., 1999. The heterotrimeric protein Go is required for the formation of heart epithelium in *Drosophila. J. Cell Biol. 145*, 1063–1076.

Fryxell, K.J., Meyerowitz, E.M., 1987. An opsin gene that is expressed only in the R7 photoreceptor cell of *Drosophila. EMBO J. 6*, 443–451.

Fujii, R., Hosoya, M., Fukusumi, S., Kawamata, Y., Habata, Y., et al., 2000. Identification of neuromedin U as the cognate ligand of the orphan G protein-coupled receptor FM-3. *J. Biol. Chem. 275*, 21068–21074.

Gäde, G., Hoffmann, K.H., Spring, J.H., 1997. Hormonal regulation in insects: facts, gaps, and future directions. *Physiol. Rev. 77*, 963–1032.

Garczynski, S.F., Brown, M.R., Shen, P., Murray, T.F., Crim, J.W., 2002. Characterization of a functional neuropeptide F receptor from *Drosophila melanogaster. Peptides 23*, 773–780.

Gazi, L., Nickolls, S.A., Strange, P.G., 2003. Functional coupling of the human dopamine D(2) receptor with Galpha(i1), Galpha(i2), Galpha(i3) and Galpha(o) G proteins: evidence for agonist regulation of G protein selectivity. *Br. J. Pharmacol. 138*, 775–786.

Goh, C.S., Cohen, F.E., 2002. Co-evolutionary analysis reveals insights into protein–protein interactions. *J. Mol. Biol. 324*, 177–192.

Gomes, I., Jordan, B.A., Gupta, A., Rios, C., Trapaidze, N., et al., 2001. G protein coupled receptor dimerization: implications in modulating receptor function. *J. Mol. Med. 79*, 226–242.

Gotzes, F., Balfanz, S., Baumann, A., 1994. Primary structure and functional characterization of a *Drosophila* dopamine receptor with high homology to human D1/5 receptors. *Recept. Channels 2*, 131–141.

Gotzes, F., Baumann, A., 1996. Functional properties of *Drosophila* dopamine D1-receptors are not altered by the size of the N-terminus. *Biochem. Biophys. Res. Commun. 222*, 121–126.

Guerrero, F.D., 1997. Transcriptional expression of a putative tachykinin-like peptide receptor gene from stable fly. *Peptides 18*, 1–5.

Ha, S.D., Kataoka, H., Suzuki, A., Kim, B.J., Kim, H.J., et al., 2000. Cloning and sequence analysis of cDNA for diuretic hormone receptor from *Bombyx mori. Mol. Cells 10*, 13–17.

Han, K.A., Millar, N.S., Davis, R.L., 1998. A novel octopamine receptor with preferential expression in *Drosophila* mushroom bodies. *J. Neurosci. 18*, 3650–3658.

Han, K.A., Millar, N.S., Grotewiel, M.S., Davis, R.L., 1996. DAMB, a novel dopamine receptor expressed specifically in *Drosophila* mushroom bodies. *Neuron 16*, 1127–1135.

Hannan, F., Hall, L.M., 1996. Temporal and spatial expression patterns of two G-protein coupled receptors in *Drosophila melanogaster. Invertebr. Neurosci. 2*, 71–83.

Hardie, R.C., Raghu, P., 2001. Visual transduction in *Drosophila. Nature 413*, 186–193.

Hauser, F., Nothacker, H.P., Grimmelikhuijzen, C.J., 1997. Molecular cloning, genomic organization, and developmental regulation of a novel receptor from *Drosophila melanogaster* structurally related to members of the thyroid-stimulating hormone, follicle-stimulating hormone, luteinizing hormone/choriogonadotropin receptor family from mammals. *J. Biol. Chem. 272*, 1002–1010.

Hearn, M.G., Ren, Y., McBride, E.W., Reveillaud, I., Beinborn, M., et al., 2002. A *Drosophila* dopamine 2-like receptor: molecular characterization and identification of multiple alternatively spliced variants. *Proc. Natl Acad. Sci. USA 99*, 14554–14559.

Henkel, A.W., Sankaranarayanan, S., 1999. Mechanisms of alpha-latrotoxin action. *Cell Tissue Res. 296*, 229–233.

Hermans, E., Challiss, R.A., 2001. Structural, signalling and regulatory properties of the group I metabotropic glutamate receptors: prototypic family C G-protein-coupled receptors. *Biochem. J. 359*, 465–484.

Hewes, R.S., Taghert, P.H., 2001. Neuropeptides and neuropeptide receptors in the *Drosophila melanogaster* genome. *Genome Res. 11*, 1126–1142.

Hill, C.A., Fox, A.N., Pitts, R.J., Kent, L.B., Tan, P.L., *et al.*, 2002a. G protein-coupled receptors in *Anopheles gambiae*. *Science* 298, 176–178.

Hill, E.S., Iwano, M., Gatellier, L., Kanzaki, R., 2002b. Morphology and physiology of the serotonin-immuno-reactive putative antennal lobe feedback neuron in the male silkmoth *Bombyx mori*. *Chem. Senses* 27, 475–483.

Holmes, S.P., Barhoumi, R., Nachman, R.J., Pietrantonio, P.V., 2003. Functional analysis of a G protein-coupled receptor from the Southern cattle tick *Boophilus microplus* (Acari: Ixodidae) identifies it as the first arthropod myokinin receptor. *Insect Mol. Biol.* 12, 27–38.

Holmes, S.P., He, H., Chen, A.C., Ivie, G.W., Pietrantonio, P.V., 2000. Cloning and transcriptional expression of a leucokinin-like peptide receptor from the southern cattle tick, *Boophilus microplus* (Acari: Ixodidae). *Insect Mol. Biol.* 9, 457–465.

Horn, F., van der Wenden, E.M., Oliveira, L., IJzerman, A.P., Vriend, G., 2000. Receptors coupling to G proteins: is there a signal behind the sequence? *Proteins* 41, 448–459.

Howard, A.D., Wang, R., Pong, S.-S., Mellin, T.N., Strack, A., *et al.*, 2000. Identification of receptors for neuromedin U and its role in feeding. *Nature* 406, 70–74.

Huesmann, G.R., Cheung, C.C., Loi, P.K., Lee, T.D., Swiderek, K.M., *et al.*, 1995. Amino acid sequence of CAP-2b, an insect cardioacceleratory peptide from the tobacco hawkmoth *Manduca sexta*. *FEBS Lett.* 371, 311–314.

Isaacson, J.S., Vitten, H., 2003. GABA$_B$ receptors inhibit dendrodendritic transmission in the rat olfactory bulb. *J. Neurosci.* 23, 2032–2039.

Ishimoto, H., Matsumoto, A., Tanimura, T., 2000. Molecular identification of a taste receptor gene for trehalose in *Drosophila*. *Science* 289, 116–119.

Iversen, A., Cazzamali, G., Williamson, M., Hauser, F., Grimmelikhuijzen, C.J., 2002. Molecular identification of the first insect ecdysis triggering hormone receptors. *Biochem. Biophys. Res. Commun.* 299, 924–931.

Jespersen, T., Grunnet, M., Angelo, K., Klaerke, D.A., Olesen, S.P., 2002. Dual-function vector for protein expression in both mammalian cells and *Xenopus laevis* oocytes. *Biotechniques* 32, 536–538, 540.

Johard, H.A., Muren, J.E., Nichols, R., Larhammar, D.S., Nassel, D.R., 2001. A putative tachykinin receptor in the cockroach brain: molecular cloning and analysis of expression by means of antisera to portions of the receptor protein. *Brain Res.* 919, 94–105.

Johnson, E.C., Bohn, L.M., Barak, L.S., Birse, R.T., Nassel, D.R., *et al.*, 2003a. Identification of *Drosophila* neuropeptide receptors by G protein-coupled receptors-beta-arrestin2 interactions. *J. Biol. Chem.* 278, 52172–52178.

Johnson, E.C., Garczynski, S.F., Park, D., Crim, J.W., Nassel, D.R., *et al.*, 2003b. Identification and characterization of a G protein-coupled receptor for the neuropeptide proctolin in *Drosophila melanogaster*. *Proc. Natl Acad. Sci. USA* 100, 6198–6203.

Jones, K.A., Borowsky, B., Tamm, J.A., Craig, D.A., Durkin, M.M., *et al.*, 1998. GABA(B) receptors function as a heteromeric assembly of the subunits GABA(B)R1 and GABA(B)R2. *Nature* 396, 674–679.

Josefsson, L.G., 1999. Evidence for kinship between diverse G-protein coupled receptors. *Gene* 239, 333–340.

Julius, D., MacDermott, A.B., Axel, R., Jessell, T.M., 1988. Molecular characterization of a functional cDNA encoding the serotonin 1c receptor. *Science* 241, 558–564.

Kaupmann, K., Malitschek, B., Schuler, V., Heid, J., Froestl, W., *et al.*, 1998. GABA(B)-receptor subtypes assemble into functional heteromeric complexes. *Nature* 396, 683–687.

Kenakin, T., 1997. Differences between natural and recombinant G protein-coupled receptor systems with varying receptor/G protein stoichiometry. *Trends Pharmacol. Sci.* 18, 456–464.

Kenakin, T., 2003. A guide to drug discovery: predicting therapeutic value in the lead optimization phase of drug discovery. *Nature Rev. Drug Discov.* 2, 429–438.

Kierszenbaum, A.L., 2003. Epididymal G protein-coupled receptor (GPCR): two hats and a two-piece suit tailored at the GPS motif. *Mol. Reprod. Devel.* 64, 1–3.

Kim, J., Carlson, J.R., 2002. Gene discovery by e-genetics: *Drosophila* odor and taste receptors. *J. Cell Sci.* 115, 1107–1112.

Kniazeff, J., Galvez, T., Labesse, G., Pin, J.P., 2002. No ligand binding in the GB2 subunit of the GABA(B) receptor is required for activation and allosteric interaction between the subunits. *J. Neurosci.* 22, 7352–7361.

Kofuji, P., Davidson, N., Lester, H.A., 1995. Evidence that neuronal G-protein-gated inwardly rectifying K+ channels are activated by G beta gamma subunits and function as heteromultimers. *Proc. Natl Acad. Sci. USA* 92, 6542–6546.

Kohout, T.A., Lefkowitz, R.J., 2003. Regulation of G protein-coupled receptor kinases and arrestins during receptor desensitization. *Mol. Pharmacol.* 63, 9–18.

Kolakowski, L.F., Jr., 1994. GCRDb: a G-protein-coupled receptor database. *Recept. Channels* 2, 1–7.

Kostenis, E., 2001. Is Galpha16 the optimal tool for fishing ligands of orphan G-protein-coupled receptors? *Trends Pharmacol. Sci.* 22, 560–564.

Kreienkamp, H.J., Larusson, H.J., Witte, I., Roeder, T., Birgul, N., *et al.*, 2002a. Functional annotation of two orphan G-protein-coupled receptors, Drostar1 and -2, from *Drosophila melanogaster* and their ligands by reverse pharmacology. *J. Biol. Chem.* 277, 39937–39943.

Kreienkamp, H.J., Soltau, M., Richter, D., Bockers, T., 2002b. Interaction of G-protein-coupled receptors with synaptic scaffolding proteins. *Biochem. Soc. Trans.* 30, 464–468.

Krieger, J., Raming, K., Dewer, Y.M., Bette, S., Conzelmann, S., *et al.*, 2002. A divergent gene family encoding candidate olfactory receptors of the moth *Heliothis virescens*. *Eur. J. Neurosci.* 16, 619–628.

Kubiak, T.M., Larsen, M.J., Burton, K. J., Bannow, C.A., Martin, R.A., et al., 2002. Cloning and functional expression of the first Drosophila melanogaster sulfakinin receptor DSK-R1. Biochem. Biophys. Res. Commun. 291, 313–320.

Kubo, T., Fukuda, K., Mikami, A., Maeda, A., Takahashi, H., et al., 1986. Cloning, sequencing and expression of complementary DNA encoding the muscarinic acetylcholine receptor. Nature 323, 411–416.

Kutsukake, M., Komatsu, A., Yamamoto, D., Ishiwa-Chigusa, S., 2000. A tyramine receptor gene mutation causes a defective olfactory behavior in Drosophila melanogaster. Gene 245, 31–42.

Larsen, M.J., Burton, K.J., Zantello, M.R., Smith, V.G., Lowery, D.L., et al., 2001. Type A allatostatins from Drosophila melanogaster and Diplotera punctata activate two Drosophila allatostatin receptors, DAR-1 and DAR-2, expressed in CHO cells. Biochem. Biophys. Res. Commun. 286, 895–901.

Le Poul, E., Hisada, S., Mizuguchi, Y., Dupriez, V.J., Burgeon, E., et al., 2002. Adaptation of aequorin functional assay to high throughput screening. J. Biomol. Screen. 7, 57–65.

Lee, A., Durieux, M.E., 1998. The use of Xenopus laevis oocytes for the study of G protein-coupled receptors. In: Lynch, K.R. (Ed.), Identification and Expression of G Protein-Coupled Receptors. Wiley-Liss, New York, pp. 73–96.

Lee, Y.J., Dobbs, M.B., Verardi, M.L., Hyde, D.R., 1990. dgq: a Drosophila gene encoding a visual system-specific G alpha molecule. Neuron 5, 889–898.

Lee, Y.J., Shah, S., Suzuki, E., Zars, T., O'Day, P.M., et al., 1994. The Drosophila dgq gene encodes a G alpha protein that mediates phototransduction. Neuron 13, 1143–1157.

Lenz, C., Williamson, M., Grimmelikhuijzen, C.J., 2000. Molecular cloning and genomic organization of a second probable allatostatin receptor from Drosophila melanogaster. Biochem. Biophys. Res. Commun. 273, 571–577.

Lenz, C., Williamson, M., Hansen, G.N., Grimmelikhuijzen, C.J., 2001. Identification of four Drosophila allatostatins as the cognate ligands for the Drosophila orphan receptor DAR-2. Biochem. Biophys. Res. Commun. 286, 1117–1122.

Li, X.J., Wolfgang, W., Wu, Y.N., North, R.A., Forte, M., 1991. Cloning, heterologous expression and developmental regulation of a Drosophila receptor for tachykinin-like peptides. EMBO J. 10, 3221–3229.

Li, X.J., Wu, Y.N., North, R.A., Forte, M., 1992. Cloning, functional expression, and developmental regulation of a neuropeptide Y receptor from Drosophila melanogaster. J. Biol. Chem. 267, 9–12.

Lim, N.F., Dascal, N., Labarca, C., Davidson, N., Lester, H.A., 1995. A G protein-gated K channel is activated via beta 2-adrenergic receptors and G beta gamma subunits in Xenopus oocytes. J. Gen. Physiol. 105, 421–439.

Lin, Y.J., Seroude, L., Benzer, S., 1998. Extended lifespan and stress resistance in the Drosophila mutant methuselah. Science 282, 943–946.

Liu, A.M., Ho, M.K., Wong, C.S., Chan, J.H., Pau, A.H., et al., 2003. Galpha(16/z) chimeras efficiently link a wide range of G protein-coupled receptors to calcium mobilization. J. Biomol. Screen. 8, 39–49.

Masu, Y., Nakayama, K., Tamaki, H., Harada, Y., Kuno, M., et al., 1987. cDNA cloning of bovine substance-K receptor through oocyte expression system. Nature 329, 836–838.

Matsumoto, S., Kitamura, A., Nagasawa, H., Kataoka, H., Orikasa, C., et al., 1990. Functional diversity of a neurohormone produced by the subesophageal ganglion: molecular identity of melanization and reddish coloration hormone and pheromone biosynthesis activating neuropeptide. J. Insect Physiol. 36, 427–432.

Maxwell, G.D., Tait, J.F., Hildebrand, J.G., 1978. Regional synthesis of neurotransmitter candidates in the CNS of the moth Manduca sexta. Comp. Biochem. Physiol. C 61, 109–119.

McBride, E.W., Reveillaud, I., Ren, Y., Kopin, A.S., 2001. Molecular cloning, genomic organization and expression of a Drosophila cholecystokinin-like receptor. Annu. Drosophila Res. Conf. 42, 432C.

Meeusen, T., Mertens, I., Clynen, E., Baggerman, G., Nichols, R., et al., 2002. Identification in Drosophila melanogaster of the invertebrate G protein-coupled FMRFamide receptor. Proc. Natl Acad. Sci. USA 99, 15363–15368.

Mertens, I., Meeusen, T., Huybrechts, R., De Loof, A., Schoofs, L., 2002. Characterization of the short neuropeptide F receptor from Drosophila melanogaster. Biochem. Biophys. Res. Commun. 297, 1140–1148.

Mezler, M., Muller, T., Raming, K., 2001. Cloning and functional expression of GABA(B) receptors from Drosophila. Eur. J. Neurosci. 13, 477–486.

Moller, S., Vilo, J., Croning, M.D., 2001. Prediction of the coupling specificity of G protein coupled receptors to their G proteins. Bioinformatics 17, S174–S181.

Monnier, D., Colas, J.F., Rosay, P., Hen, R., Borrelli, E., et al., 1992. NKD, a developmentally regulated tachykinin receptor in Drosophila. J. Biol. Chem. 267, 1298–1302.

Montell, C., 1999. Visual transduction in Drosophila. Annu. Rev. Cell Devel. Biol. 15, 231–268.

Morris, A.J., Malbon, C.C., 1999. Physiological regulation of G protein-linked signaling. Physiol. Rev. 79, 1373–1430.

Morris, H.R., Panico, M., Karplus, A., Lloyd, P.E., Riniker, B., 1982. Elucidation by FAB-MS of the structure of a new cardioactive peptide in Aplysia. Nature 300, 643–645.

Nachman, R.J., Strey, A., Isaac, E., Pryor, N., Lopez, J.D., et al., 2002a. Enhanced in vivo activity of peptidase-resistant analogs of the insect kinin neuropeptide family. Peptides 23, 735–745.

Nachman, R.J., Teal, P.E., Strey, A., 2002b. Enhanced oral availability/pheromonotropic activity of peptidase-resistant topical amphiphilic analogs of pyrokinin/PBAN insect neuropeptides. *Peptides 23*, 2035–2043.

Nachman, R.J., Teal, P.E., Ujvary, I., **2001**. Comparative topical pheromonotropic activity of insect pyrokinin/PBAN amphiphilic analogs incorporating different fatty and/or cholic acid components. *Peptides 22*, 279–285.

Nassel, D.R., **2002**. Neuropeptides in the nervous system of *Drosophila* and other insects: multiple roles as neuromodulators and neurohormones. *Prog. Neurobiol. 68*, 1–84.

Neves, S.R., Ram, P.T., Iyengar, R., **2002**. G protein pathways. *Science 296*, 1636–1639.

Nishi, S., Hsu, S.Y., Zell, K., Hsueh, A.J., **2000**. Characterization of two fly LGR (leucine-rich repeat-containing, G protein-coupled receptor) proteins homologous to vertebrate glycoprotein hormone receptors: constitutive activation of wild-type fly LGR1 but not LGR2 in transfected mammalian cells. *Endocrinology 141*, 4081–4090.

Noguchi, H., Hayakawa, Y., **2001**. Dopamine is a key factor for the induction of egg diapause of the silkworm, *Bombyx mori. Eur. J. Biochem. 268*, 774–780.

O'Tousa, J.E., Baehr, W., Martin, R.L., Hirsh, J., Pak, W.L., *et al.*, **1985**. The *Drosophila* ninaE gene encodes an opsin. *Cell 40*, 839–850.

Obosi, L.A., Schuette, D.G., Europe-Finner, G.N., Beadle, D.J., Hen, R., *et al.*, **1996**. Functional characterisation of the *Drosophila* 5-HTdro1 and 5-HTdro2B serotonin receptors in insect cells: activation of a G(alpha)s-like protein by 5-HTdro1 but lack of coupling to inhibitory G-proteins by 5-HTdro2B. *FEBS Lett. 381*, 233–236.

Orchard, I., Jan-Marino, R., ALange, A.B., **1993**. A multifunctional role for octopamine in locust flight. *Annu. Rev. Entomol. 38*, 227–249.

Palczewski, K., Kumasaka, T., Hori, T., Behnke, C.A., Motoshima, H., *et al.*, **2000**. Crystal structure of rhodopsin: a G protein-coupled receptor. *Science 289*, 739–745.

Park, Y., Kim, Y.J., Adams, M.E., **2002**. Identification of G protein-coupled receptors for *Drosophila* PRXamide peptides, CCAP, corazonin, and AKH supports a theory of ligand–receptor coevolution. *Proc. Natl Acad. Sci. USA 99*, 11423–11428.

Park, Y., Kim, Y.J., Dupriez, V., Adams, M.E., **2003**. Two subtypes of ecdysis triggering hormone receptor in *Drosophila melanogaster. J. Biol. Chem. 278*, 17710–17715.

Park, Y., Zitnan, D., Gill, S.S., Adams, M.E., **1999**. Molecular cloning and biological activity of ecdysis-triggering hormones in *Drosophila melanogaster. FEBS Lett. 463*, 133–138.

Parks, S., Wieschaus, E., **1991**. The *Drosophila* gastrulation gene concertina encodes a G alpha-like protein. *Cell 64*, 447–458.

Parmentier, M.L., Joly, C., Restituito, S., Bockaert, J., Grau, Y., *et al.*, **1998**. The G protein-coupling profile of metabotropic glutamate receptors, as determined with exogenous G proteins, is independent of their ligand recognition domain. *Mol. Pharmacol. 53*, 778–786.

Parmentier, M.L., Pin, J.P., Bockaert, J., Grau, Y., **1996**. Cloning and functional expression of a *Drosophila* metabotropic glutamate receptor expressed in the embryonic CNS. *J. Neurosci. 16*, 6687–6694.

Peleg, S., Varon, D., Ivanina, T., Dessauer, C.W., Dascal, N., **2002**. G(alpha)(i) controls the gating of the G protein-activated K(+) channel, GIRK. *Neuron 33*, 87–99.

Pierce, K.L., Lefkowitz, R.J., **2001**. Classical and new roles of beta-arrestins in the regulation of G-protein-coupled receptors. *Nat. Rev. Neurosci. 2*, 727–733.

Pietrantonio, P.V., Jagge, C., McDowell, C., **2001**. Cloning and expression analysis of a 5HT7-like serotonin receptor cDNA from mosquito *Aedes aegypti* female excretory and respiratory systems. *Insect Mol. Biol. 10*, 357–369.

Poels, J., Suner, M.M., Needham, M., Torfs, H., De Rijck, J., *et al.*, **2001**. Functional expression of a locust tyramine receptor in murine erythroleukaemia cells. *Insect Mol. Biol. 10*, 541–548.

Pophof, B., **2000**. Octopamine modulates the sensitivity of silkmoth pheromone receptor neurons. *J. Comp. Physiol. A 186*, 307–313.

Quan, F., Wolfgang, W.J., Forte, M., **1993**. A *Drosophila* G-protein alpha subunit, Gf alpha, expressed in a spatially and temporally restricted pattern during *Drosophila* development. *Proc. Natl Acad. Sci. USA 90*, 4236–4240.

Radford, J.C., Davies, S.A., Dow, J.A., **2002**. Systematic G-protein-coupled receptor analysis in *Drosophila melanogaster* identifies a leucokinin receptor with novel roles. *J. Biol. Chem. 277*, 38810–38817.

Raina, A.K., Jaffe, H., Kempe, T.G., Keim, P., Blacher, R.W., *et al.*, **1989**. Identification of a neuropeptide hormone that regulates sex pheromone production in female moths. *Science 244*, 796–798.

Ratnaparkhi, A., Banerjee, S., Hasan, G., **2002**. Altered levels of Gq activity modulate axonal pathfinding in *Drosophila. J. Neurosci. 22*, 4499–4508.

Ray, K., Ganguly, R., **1994**. Organization and expression of the *Drosophila melanogaster* D-G gamma 1 gene encoding the G-protein gamma subunit. *Gene 148*, 315–319.

Raymond, V., Hamon, A., Grau, Y., Lapied, B., **1999**. DmGluRA, a *Drosophila* metabotropic glutamate receptor, activates G-protein inwardly rectifying potassium channels in *Xenopus* oocytes. *Neurosci. Lett. 269*, 1–4.

Reagan, J.D., **1994**. Expression cloning of an insect diuretic hormone receptor, a member of the calcitonin/secretin receptor family. *J. Biol. Chem. 269*, 9–12.

Reagan, J.D., **1995**. Functional expression of a diuretic hormone receptor in baculovirus-infected insect cells:

evidence suggesting that the N-terminal region of diuretic hormone is associated with receptor activation. *Insect Biochem. Mol. Biol. 25*, 535–539.

Reagan, J.D., **1996**. Molecular cloning and function expression of a diuretic hormone receptor from the house cricket, *Acheta domesticus*. *Insect Biochem. Mol. Biol. 26*, 1–6.

Reale, V., Hannan, F., Hall, L.M., Evans, P.D., **1997**. Agonist-specific coupling of a cloned *Drosophila melanogaster* D1-like dopamine receptor to multiple second messenger pathways by synthetic agonists. *J. Neurosci. 17*, 6545–6553.

Reuveny, E., Slesinger, P.A., Inglese, J., Morales, J.M., Iniguez-Lluhi, J.A., *et al.*, **1994**. Activation of the cloned muscarinic potassium channel by G protein beta gamma subunits. *Nature 370*, 143–146.

Ridley, M., **1993**. Evolution. Blackwell Scientific Publications, Oxford.

Riehle, M.A., Garczynski, S.F., Crim, J.W., Hill, C.A., Brown, M.R., **2002**. Neuropeptides and peptide hormones in *Anopheles gambiae*. *Science 298*, 172–175.

Robb, S., Cheek, T.R., Hannan, F.L., Hall, L.M., Midgley, J.M., *et al.*, **1994**. Agonist-specific coupling of a cloned *Drosophila* octopamine/tyramine receptor to multiple second messenger systems. *EMBO J. 13*, 1325–1330.

Roeder, T., **2002**. Biochemistry and molecular biology of receptors for biogenic amines in locusts. *Microsc. Res. Tech. 56*, 237–247.

Roeder, T., **2003**. Metabotropic histamine receptors: nothing for invertebrates? *Eur. J. Pharmacol. 466*, 85–90.

Rosay, P., Davies, S.A., Yu, Y., Sozen, A., Kaiser, K., *et al.*, **1997**. Cell-type specific calcium signalling in a *Drosophila* epithelium. *J. Cell Sci. 110*, 1683–1692.

Rosenkilde, C., Cazzamali, G., Williamson, M., Hauser, F., Sondergaard, L., *et al.*, **2003**. Molecular cloning, functional expression, and gene silencing of two *Drosophila* receptors for the *Drosophila* neuropeptide pyrokinin-2. *Biochem. Biophys. Res. Commun. 309*, 485–494.

Saifullah, A.S., Tomioka, K., **2002**. Serotonin sets the day state in the neurons that control coupling between the optic lobe circadian pacemakers in the cricket *Gryllus bimaculatus*. *J. Exp. Biol. 205*, 1305–1314.

Salcedo, E., Huber, A., Henrich, S., Chadwell, L.V., Chou, W.H., *et al.*, **1999**. Blue- and green-absorbing visual pigments of *Drosophila*: ectopic expression and physiological characterization of the R8 photoreceptor cell-specific Rh5 and Rh6 rhodopsins. *J. Neurosci. 19*, 10716–10726.

Saudou, F., Amlaiky, N., Plassat, J.L., Borrelli, E., Hen, R., **1990**. Cloning and characterization of a *Drosophila* tyramine receptor. *EMBO J. 9*, 3611–3617.

Saudou, F., Boschert, U., Amlaiky, N., Plassat, J.L., Hen, R., **1992**. A family of *Drosophila* serotonin receptors with distinct intracellular signalling properties and expression patterns. *EMBO J. 11*, 7–17.

Schmidt, C.J., Garen-Fazio, S., Chow, Y.K., Neer, E.J., **1989**. Neuronal expression of a newly identified *Drosophila melanogaster* G protein alpha 0 subunit. *Cell Reg. 1*, 125–134.

Schneider, L.E., Taghert, P.H., **1988**. Isolation and characterization of a *Drosophila* gene that encodes multiple neuropeptides related to Phe-Met-Arg-Phe-NH2 (FMRFamide). *Proc. Natl Acad. Sci. USA 85*, 1993–1997.

Schneider, L.E., Taghert, P.H., **1990**. Organization and expression of the *Drosophila* Phe-Met-Arg-Phe-NH2 neuropeptide gene. *J. Biol. Chem. 265*, 6890–6895.

Schulz, S., Huber, A., Schwab, K., Paulsen, R., **1999**. A novel Ggamma isolated from *Drosophila* constitutes a visual G protein gamma subunit of the fly compound eye. *J. Biol. Chem. 274*, 37605–37610.

Scott, K., Brady, R., Jr., Cravchik, A., Morozov, P., Rzhetsky, A., *et al.*, **2001**. A chemosensory gene family encoding candidate gustatory and olfactory receptors in *Drosophila*. *Cell 104*, 661–673.

Secher, T., Lenz, C., Cazzamali, G., Sorensen, G., Williamson, M., *et al.*, **2001**. Molecular cloning of a functional allatostatin gut/brain receptor and an allatostatin preprohormone from the silkworm *Bombyx mori*. *J. Biol. Chem. 276*, 47052–47060.

Shapiro, R.A., Wakimoto, B.T., Subers, E.M., Nathanson, N.M., **1989**. Characterization and functional expression in mammalian cells of genomic and cDNA clones encoding a *Drosophila* muscarinic acetylcholine receptor. *Proc. Natl Acad. Sci. USA 86*, 9039–9043.

Shichida, Y., Imai, H., **1998**. Visual pigment: G-protein-coupled receptor for light signals. *Cell. Mol. Life Sci. 54*, 1299–1315.

Simon, M.I., Strathmann, M.P., Gautam, N., **1991**. Diversity of G proteins in signal transduction. *Science 252*, 802–808.

Slesinger, P.A., Stoffel, M., Jan, Y.N., Jan, L.Y., **1997**. Defective gamma-aminobutyric acid type B receptor-activated inwardly rectifying K^+ currents in cerebellar granule cells isolated from weaver and Girk2 null mutant mice. *Proc. Natl Acad. Sci. USA 94*, 12210–12217.

Song, W., Ranjan, R., Dawson-Scully, K., Bronk, P., Marin, L., *et al.*, **2002**. Presynaptic regulation of neurotransmission in *Drosophila* by the g protein-coupled receptor methuselah. *Neuron 36*, 105–119.

Stables, J., Green, A., Marshall, F., Fraser, N., Knight, E., *et al.*, **1997**. A bioluminescent assay for agonist activity at potentially any G-protein-coupled receptor. *Analyt. Biochem. 252*, 115–126.

Stacey, M., Lin, H.H., Gordon, S., McKnight, A.J., **2000**. LNB-TM7, a group of seven-transmembrane proteins related to family-B G-protein-coupled receptors. *Trends Biochem. Sci. 25*, 284–289.

Staubli, F., Jorgensen, T.J., Cazzamali, G., Williamson, M., Lenz, C., *et al.*, **2002**. Molecular identification of the insect adipokinetic hormone receptors. *Proc. Natl Acad. Sci. USA 99*, 3446–3451.

Stortkuhl, K.F., Kettler, R., **2001**. Functional analysis of an olfactory receptor in *Drosophila melanogaster*. *Proc. Natl Acad. Sci. USA 98*, 9381–9385.

Strutt, D.I., **2001**. Asymmetric localization of frizzled and the establishment of cell polarity in the *Drosophila* wing. *Mol. Cell 7*, 367–375.

Talluri, S., Bhatt, A., Smith, D.P., **1995**. Identification of a *Drosophila* G protein alpha subunit (dGq alpha-3) expressed in chemosensory cells and central neurons. *Proc. Natl Acad. Sci. USA 92*, 11475–11479.

Teller, D.C., Okada, T., Behnke, C.A., Palczewski, K., Stenkamp, R.E., **2001**. Advances in determination of a high-resolution three-dimensional structure of rhodopsin, a model of G-protein-coupled receptors (GPCRs). *Biochemistry 40*, 7761–7772.

Thibonnier, M., Auzan, C., Madhun, Z., Wilkins, P., Berti-Mattera, L., *et al.*, **1994**. Molecular cloning, sequencing, and functional expression of a cDNA encoding the human V-1a vasopressin receptor. *J. Biol. Chem. 269*, 3304–3310.

Torfs, H., Detheux, M., Oonk, H.B., Akerman, K.E., Poels, J., *et al.*, **2002a**. Analysis of C-terminally substituted tachykinin-like peptide agonists by means of aequorin-based luminescent assays for human and insect neurokinin receptors. *Biochem. Pharmacol. 63*, 1675–1682.

Torfs, H., Oonk, H.B., Broeck, J.V., Poels, J., Van Poyer, W., *et al.*, **2001**. Pharmacological characterization of STKR, an insect G protein-coupled receptor for tachykinin-like peptides. *Arch. Insect Biochem. Physiol. 48*, 39–49.

Torfs, H., Poels, J., Detheux, M., Dupriez, V., Van Loy, T., *et al.*, **2002b**. Recombinant aequorin as a reporter for receptor-mediated changes of intracellular Ca2+-levels in *Drosophila* S2 cells. *Invertebr. Neurosci. 4*, 119–124.

Torfs, H., Shariatmadari, R., Guerrero, F., Parmentier, M., Poels, J., *et al.*, **2000**. Characterization of a receptor for insect tachykinin-like peptide agonists by functional expression in a stable *Drosophila* Schneider 2 cell line. *J. Neurochem. 74*, 2182–2189.

Townson, S.M., Chang, B.S., Salcedo, E., Chadwell, L.V., Pierce, N.E., *et al.*, **1998**. Honeybee blue- and ultraviolet-sensitive opsins: cloning, heterologous expression in *Drosophila*, and physiological characterization. *J. Neurosci. 18*, 2412–2422.

Vaidehi, N., Floriano, W.B., Trabanino, R., Hall, S.E., Freddolino, P., *et al.*, **2002**. Prediction of structure and function of G protein-coupled receptors. *Proc. Natl Acad. Sci. USA 99*, 12622–12627.

van Kesteren, R.E., Tensen, C.P., Smit, A.B., van Minnen, J., Kolakowski, L.F., *et al.*, **1996**. Co-evolution of ligand–receptor pairs in the vasopressin/oxytocin superfamily of bioactive peptides. *J. Biol. Chem. 271*, 3619–3626.

Van Poyer, W., Torfs, H., Poels, J., Swinnen, E., De Loof, A., *et al.*, **2001**. Phenolamine-dependent adenylyl cyclase activation in *Drosophila* Schneider 2 cells. *Insect Biochem. Mol. Biol. 31*, 333–338.

Vanden Broeck, J., Vulsteke, V., Huybrechts, R., De Loof, A., **1995**. Characterization of a cloned locust tyramine receptor cDNA by functional expression in permanently transformed *Drosophila* S2 cells. *J. Neurochem. 64*, 2387–2395.

von Nickisch-Rosenegk, E., Krieger, J., Kubick, S., Laage, R., Strobel, J., *et al.*, **1996**. Cloning of biogenic amine receptors from moths (*Bombyx mori* and *Heliothis virescens*). *Insect Biochem. Mol. Biol. 26*, 817–827.

Washio, H., **2002**. Glutamate receptors on the somata of dorsal unpaired median neurons in cockroach, *Periplaneta americana*, thoracic ganglia. *Zool. Sci. 19*, 153–162.

Wess, J., **1998**. Molecular basis of receptor/G-protein-coupling selectivity. *Pharmacol. Therapeut. 80*, 231–264.

West, A.P., Jr., Llamas, L.L., Snow, P.M., Benzer, S., Bjorkman, P.J., **2001**. Crystal structure of the ectodomain of Methuselah, a *Drosophila* G protein-coupled receptor associated with extended lifespan. *Proc. Natl Acad. Sci. USA 98*, 3744–3749.

Wetzel, C.H., Behrendt, H.J., Gisselmann, G., Stortkuhl, K.F., Hovemann, B., *et al.*, **2001**. Functional expression and characterization of a *Drosophila* odorant receptor in a heterologous cell system. *Proc. Natl Acad. Sci. USA 98*, 9377–9380.

White, J.H., Wise, A., Main, M.J., Green, A., Fraser, N.J., *et al.*, **1998**. Heterodimerization is required for the formation of a functional GABA(B) receptor. *Nature 396*, 679–682.

Wickman, K.D., Iniguez-Lluhl, J.A., Davenport, P.A., Taussig, R., Krapivinsky, G.B., *et al.*, **1994**. Recombinant G-protein beta gamma-subunits activate the muscarinic-gated atrial potassium channel. *Nature 368*, 255–257.

Witz, P., Amlaiky, N., Plassat, J.L., Maroteaux, L., Borrelli, E., *et al.*, **1990**. Cloning and characterization of a *Drosophila* serotonin receptor that activates adenylate cyclase. *Proc. Natl Acad. Sci. USA 87*, 8940–8944.

Wolfgang, W.J., Hoskote, A., Roberts, I.J., Jackson, S., Forte, M., **2001**. Genetic analysis of the *Drosophila* Gs(alpha) gene. *Genetics 158*, 1189–1201.

Wolfgang, W.J., Quan, F., Goldsmith, P., Unson, C., Spiegel, A., *et al.*, **1990**. Immunolocalization of G protein alpha-subunits in the *Drosophila* CNS. *J. Neurosci. 10*, 1014–1024.

Wolfgang, W.J., Quan, F., Thambi, N., Forte, M., **1991**. Restricted spatial and temporal expression of G-protein alpha subunits during *Drosophila* embryogenesis. *Development 113*, 527–538.

Yarden, Y., Rodriguez, H., Wong, S.K., Brandt, D.R., May, D.C., *et al.*, **1986**. The avian beta-adrenergic receptor: primary structure and membrane topology. *Proc. Natl Acad. Sci. USA 83*, 6795–6799.

Yarfitz, S., Niemi, G.A., McConnell, J.L., Fitch, C.L., Hurley, J.B., **1991**. A G beta protein in the *Drosophila* compound eye is different from that in the brain. *Neuron 7*, 429–438.

Yarfitz, S., Provost, N.M., Hurley, J.B., **1988**. Cloning of a *Drosophila melanogaster* guanine nucleotide regulatory protein beta-subunit gene and characterization of its expression during development. *Proc. Natl Acad. Sci. USA 85*, 7134–7138.

Zhong, Y., **1995**. Mediation of PACAP-like neuropeptide transmission by coactivation of Ras/Raf and cAMP signal transduction pathways in *Drosophila*. *Nature 375*, 588–592.

Zhong, Y., Pena, L.A., **1995**. A novel synaptic transmission mediated by a PACAP-like neuropeptide in *Drosophila*. *Neuron 14*, 527–536.

Zitnan, D., Kingan, T.G., Hermesman, J., Adams, M. E., **1996**. Identification of ecdysis-triggering hormone from an epitracheal endocrine system. *Science 271*, 88–91.

A9 Addendum: Insect G Protein-Coupled Receptors

Y Park, Kansas State University, Manhattan, KS, USA

M E Adams, University of California, Riverside, CA, USA

© 2010, Elsevier BV. All Rights Reserved.

During the past 5 years, our understanding of insect G protein-coupled receptor (GPCR) structure and function has advanced in two main areas. One is through extended surveys of the genes encoding GPCRs in diverse insect species with known genome sequences. Another has come through analysis of signaling pathways involving various GPCRs using advanced molecular tools. We summarize these two areas in the midst of rapidly accumulating new knowledge gained through interdisciplinary approaches.

A9.1. GPCR Genes in Comprehensive Genome Surveys

Following completion of the landmark *Drosophila melanogaster* and *Anopheles gambiae* genome projects in early 2000, results of many additional genome sequencing efforts have been reported for important insect and related arthropod species. While current information from NCBI indicates genome sequences for 48 insect species are now available and or in progress, a large number of species are in initial stages of genome analysis, but are not yet entered into the NCBI database. Of added benefit are manual annotations of GPCRs and their peptide ligands for several species following completion of their genome sequences (Hauser *et al.*, 2006, 2008).

One of the most striking outcomes of genome surveys has been the discovery of frequent deletions and duplications of genes encoding GPCRs in different taxa. Such phenomena are most apparent in comparisons of GPCRs and their ligands in *D. melanogaster* and *Tribolium castaneum*, which are relatively well annotated. For example, the *T. castaneum* genome contains an ancestrally conserved arginine–vasopressin (AVP)-like peptide and its cognate GPCR, whereas this pair is missing in *D. melanogaster*. Expanded surveys of AVP-like ligands and receptors revealed their appearance in the parasitic wasp *Nasonia* and the locust, where these signaling molecules were originally described, but lacking in other holometabolous orders and the aphid. Similarly, genes encoding GPCRs for corazonin, kinin, and allatostatin-A and their putative ligands present in *Drosophila* are missing in the *T. castaneum* genome (Li *et al.*, 2008). Loss of a gene and its corresponding function may be a common phenomenon in the evolutionary fine tuning of multiple signaling pathways that often function redundantly.

On the other hand, duplication of a GPCR gene may lead to acquisition of a new function or to loss of a copy over evolutionary time. For example, phylogenetic analysis of genes encoding the crustacean cardioactive peptide (CCAP) receptor indicates that at least three independent gene duplications occurred in holometabolous insects (Li and Park, unpublished data), followed by deletion of extra copies in *Drosophila* and hymenopteran species. Similar phenomena can be found in other instances, such as GPCRs for AKH, cholecystokinin, and pyrokinin.

In conclusion, surveys of numerous insect genomes have revealed that evolutionary constraints and functional requirements of GPCRs appear to be more flexible than previously thought, as indicated by frequent gene duplications and deletions of GPCR genes. More studies are necessary to determine whether such phenomena are confined to certain evolutionary lineages or functional categories or are more generalized.

A9.2. New Functions for GPCRs Revealed in Recent Studies

In the past, assignment of biological functions for GPCRs was based largely on *in vitro* assays focused on functions of cognate ligands. Examples include use of the prothoracic gland, corpus allatum, or hindgut for assignment of "-tropic", "-static", "kinin", or "-suppressin" activities for various peptides. While useful in revealing the potential functions of peptides, there are potential pitfalls in the logic associated with such approaches, including the possibility or even likelihood of multiple functions for "orphan" peptides. Conversely, certain receptors may function pleiotropically under the influence of multiple ligands. Such is the case for the peptide FMRFamide acting through the myosuppressin (FLRFamide) receptor in the prothoracic gland of *Bombyx mori* (Yamanaka *et al.*, 2006). Therefore, a comprehensive understanding of endogenous functions for GPCRs requires multiple approaches. The following sections provide a number of examples whereby adoptions of new multidisciplinary methods are improving our understanding of GPCR functions.

Investigation of mutant phenotypes carrying random transposable element insertions has been a powerful approach to determining the functional significance of peptidergic signaling pathways. Another useful and simpler genetic approach involves cell-knockout through targeted expression of apoptosis gene(s) using the two component GAL4-UAS system derived from yeast. This involves use of a transgene carrying a cell-specific promoter to drive expression of GAL4, which in turn activates UAS in conjunction with one or more apoptosis genes (e.g., UAS::hid or UAS::rpr). While this technique has yielded much information, cautions in need of consideration are (1) Does the GAL4 driver contain sufficient and necessary promoter components such that the cell-knockout is confined to specific target cells? (2) Is the time course of cell-knockout completed at the stage of development being observed? It must be kept in mind that different cell-types may not respond to such insults in the same manner or along the same time course. For example, in a study of CCAP function, a *ccap*-cell knockout strategy targeted to peripheral neurons appeared to be considerably less effective than on central neurons (Dulcis *et al.*, 2005). (3) The phenotype caused by a cell-knockout phenotype could be generated by removal of unknown co-expressed genes in the cell. A good example is found again in *ccap*-cell knockouts showing partial eclosion deficiencies. *ccap* cells co-express a number of neuropeptides, including myoinhibitory peptides, burs (burs-α), partner of burs (burs-β), and diuretic hormone (CT-like). On the basis of the fact that the *ccap*-null mutant showed no observable phenotype (Ewer *et al.*, 2004), the eclosion deficiency in the *ccap*-cell knockout is likely caused by loss of gene products other than the CCAP itself.

The GAL4-UAS system has been further developed for use of UAS::RNAi (RNA hairpin) constructs to suppress transcript levels of specific target genes. Use of this strategy has been facilitated by availability of public stocks that offer large collections of UAS::RNAi transgenic fly lines (VDRC and NIG-Fly). While effective in many instances, this approach also has potential pitfalls related to tissue-specific efficacy and to off-target effects. Numerous other *Drosophila* tools are being developed and applied for expanding our understanding of GPCR functions and signaling pathways. Flies expressing the biogenic calcium sensor GCaMP, a cAMP sensor, temperature- or light-activated channels, and conditional expression of GAL4 are providing new opportunities to ascertain GPCR functions by manipulating cells producing either cognate ligand or receptors. Through use of GAL4 to drive expression of GCaMP, an ensemble of peptideric downstream actions of ecdysis triggering hormone (ETH) has been revealed by examining cells expressing two isoforms of ETH receptors (Kim *et al.*, 2006). A second strategy involves use of GAL4 to drive expression of a cAMP sensor (Epac-camp) to investigate physiological functions of cells producing the pigment dispersing factor PDF (Shafer *et al.*, 2008).

Finally, RNAi has become a readily available technique for investigations of many non-*Drosophila* insects. This is particularly true of postembryonic and systemic RNAi, in which injection of double-stranded RNA at any stage in the body cavity suppresses target gene expression. In this regard, the red flour beetle, *T. castaneum*, has become a powerful tool for assigning functions to numerous GPCRs and neuropeptides (Aikins *et al.*, 2008).

References

Aikins, M.J., Schooley, D.A., Begum, K., Detheux, M., Beeman, R.W., Park, Y., 2008. Vasopressin-like peptide and its receptor function in an indirect diuretic signaling pathway in the red flour beetle. *Insect Biochem. Mol. Biol.* 38(7), 740–748.

Dulcis, D., Levine, R.B., Ewer, J., 2005. Role of the neuropeptide CCAP in Drosophila cardiac function. *J. Neurobiol.* 64(3), 259–274.

Ewer, J., del Campo, M.L., Clark, A.C., 2004. Control of larval ecdysis behavior: complex regulation by partially redundant neuropeptides. *A Dros. Res. Conf.* 45, 109.

Hauser, F., Cazzamali, G., Williamson, M., Blenau, W., Grimmelikhuijzen, C.J.P., 2006. A review of neurohormone GPCRs present in the fruitfly *Drosophila melanogaster* and the honey bee *Apis mellifera*. *Prog. Neurobiol.* 80(1), 1–19.

Hauser, F., Cazzamali, G., Williamson, M., Park, Y., Li, B., Tanaka, Y., Predel, R., Neupert, S., Schachtner, J., Verleyen, P., Grimmelikhuijzen, C.J.P., 2008. A genome-wide inventory of neurohormone GPCRs in the red flour beetle *Tribolium castaneum*. *Front. Neuroendocrinol.* 29(1), 142–165.

Kim, Y.J., Zitnan, D., Cho, K.H., Schooley, D.A., Mizoguchi, A., Adams, M.E., 2006. Central peptidergic ensembles associated with organization of an innate behavior. *Proc. Natl. Acad. Sci. USA 103*(38), 14211–14216.

Li, B., Predel, R., Neupert, S., Hauser, F., Tanaka, Y., Cazzamali, G., Williamson, M., Arakane, Y., Verleyen, P., Schoofs, L., Schachtner, J., Grimmelikhuijzen, C.J.P., Park, Y., 2008. Genomics, transcriptomics, and peptidomics of neuropeptides and protein hormones in the red flour beetle *Tribolium castaneum. Genome Res.* 18(1), 113–122.

Shafer, O.T., Kim, D.J., Dunbar-Yaffe, R., Nikolaev, V.O., Lohse, M.J., Taghert, P.H., 2008. Widespread receptivity to neuropeptide PDF throughout the neuronal circadian clock network of Drosophila revealed by real-time cyclic AMP imaging. *Neuron 58*(2), 223–237.

Yamanaka, N., Zitnan, D., Kim, Y.J., Adams, M.E., Hua, Y.J., Suzuki, Y., Suzuki, A., Satake, H., Mizoguchi, A., Asaoka, K., Tanaka, Y., Kataoka, H., 2006. Regulation of insect steroid hormone biosynthesis by innervating peptidergic neurons. *Proc. Natl. Acad. Sci. USA 103*(23), 8622–8627.

Subject Index

Cross-reference terms in italics are general cross-references, or refer to subentry terms within the main entry (the main entry is not repeated to save space). Readers are also advised to refer to the end of each article for additional cross-references – not all of these cross-references have been included in the index cross-references.

The index is arranged in set-out style with a maximum of four levels of heading. Major discussion of a subject is indicated by bold page numbers. Page numbers suffixed by *T* and *F* refer to Tables and Figures respectively. *vs.* indicates a comparison.

To save space in the index the following abbreviations have been used:

    ETH – ecdysis triggering hormone
    GPCRs – G protein-coupled receptors
    PBAN – pheromone biosynthesis activating neuropeptide
    PDV – polydnaviruses
    PTTH – prothoracicotropic hormone
    QSAR – qualitative structure-activity relation
    RDL – resistance to dieldrin

This index is in letter-by-letter order, whereby hyphens and spaces within index headings are ignored in the alphabetization. Prefixes and terms in parentheses are excluded from the initial alphabetization.

Printed and bound by CPI Group (UK) Ltd, Croydon, CR0 4YY

03/10/2024

01040321-0019